HIGH-SPEED SIGNAL PROPAGATION

ADVANCED BLACK MAGIC

ISBN 0-13-084408-X

Prentice Hall Modern Semiconductor Design Series

James R. Armstrong and F. Gail Gray
    VHDL Design Representation and Synthesis

Mark Gordon Arnold
    Verilog Digital Computer Design: Algorithms into Hardware

Jayaram Bhasker
    A VHDL Primer, Third Edition

Eric Bogatin
    Signal Integrity – Simplified

Kanad Chakraborty and Pinaki Mazumder
    Fault-Tolerance and Reliability Techniques for High-Density Random-Access Memories

Ken Coffman
    Real World FPGA Design with Verilog

Alfred Crouch
    Design-for-Test for Digital IC's and Embedded Core Systems

Daniel P. Foty
    MOSFET Modeling with SPICE: Principles and Practice

Nigel Horspool and Peter Gorman
    The ASIC Handbook

Howard Johnson and Martin Graham
    High-Speed Digital Design: A Handbook of Black Magic

Howard Johnson and Martin Graham
    High-Speed Signal Propagation: Advanced Black Magic

Pinaki Mazumder and Elizabeth Rudnick
  Genetic Algorithms for VLSI Design, Layout, and Test Automation

Farzad Nekoogar
  From ASICs to SOCs

Farzad Nekoogar
  Timing Verification of Application-Specific Integrated Circuits (ASICs)

David Pellerin and Douglas Taylor
  VHDL Made Easy!

Samir S. Rofail and Kiat-Seng Yeo
  Low-Voltage Low-Power Digital BiCMOS Circuits: Circuit Design, Comparative Study, and Sensitivity Analysis

Frank Scarpino
  VHDL and AHDL Digital System Implementation

Wayne Wolf
  Modern VLSI Design: System-on-Chip Design, Third Edition

Kiat-Seng Yeo, Samir S. Rofail, and Wang-Ling Goh
  CMOS/BiCMOS ULSI: Low Voltage, Low Power

Brian Young
  Digital Signal Integrity: Modeling and Simulation with Interconnects and Packages

Bob Zeidman
  Verilog Designer's Library

HIGH-SPEED SIGNAL PROPAGATION

ADVANCED BLACK MAGIC

Howard Johnson
Martin Graham

**PRENTICE HALL
PROFESSIONAL TECHNICAL REFERENCE
UPPER SADDLE RIVER, NJ 07458
WWW.PHPTR.COM**

Library of Congress Cataloging-in-Publication Data

Johnson, Howard W.
    High-speed signal propagation: advanced black magic / Howard Johnson, Martin Graham.
      p. cm
    Includes bibliographical references and index.
    ISBN 0-13-084408-X
    1. Telecommunication cables. 2. Telecommunication--Traffic. 3. Signal processing. I.
Graham, Martin. II. Title.
TK5103.15 .J64 2002
621.382'2--dc21

                                                                                                             2002041041

Editorial/Production Supervision: *Nick Radhuber*
Acquisitions Editor: *Bernard Goodwin*
Editorial Assistant: *Michelle Vincente*
Marketing Manager: *Dan DePasquale*
Manufacturing Buyer: *Alexis Heydt-Long*
Cover Design: *Talar Boorujy*
Cover Design Director: *Jerry Votta*

© 2003 by Pearson Education, Inc.
Publishing as Prentice Hall Professional Technical Reference
One Lake Street
Upper Saddle River, NJ 07458

Prentice Hall books are widely used by corporations and government agencies for training, marketing, and resale.

The publisher offers discounts on this book when ordered in bulk quantities. For more information, contact Corporate Sales Department, phone: 800-382-3419; fax: 201-236-7141; email: corpsales@prenhall.com Or write: Corporate Sales Department, Prentice Hall PTR, One Lake Street, Upper Saddle River, NJ 07458.

Product and company names mentioned herein are the trademarks or registered trademarks of their respective owners.

All rights reserved. No part of this book may be reproduced, in any form or by any means, without permission in writing from the publisher.

Printed in the United States of America

10 9 8 7 6 5 4 3 2

ISBN 0-13-084408-X

Pearson Education LTD.
Pearson Education Australia PTY, Limited
Pearson Education Singapore, Pte. Ltd
Pearson Education North Asia Ltd
Pearson Education Canada, Ltd.
Pearson Educación de Mexico, S.A. de C.V.
Pearson Education—Japan
Pearson Education Malaysia, Pte. Ltd

Even if you're on the right track, you'll get run over if you just sit there.

—Will Rogers (1879–1935)

Photo courtesy of Will Rogers Memorial Museums, Claremore, Oklahoma

TABLE OF CONTENTS

Preface xxi
Glossary of Symbols xxvii

1   Fundamentals 1
   1.1   Impedance of Linear, Time-Invariant, Lumped-Element Circuits 1
   1.2   Power Ratios 2
   1.3   Rules of Scaling 5
       1.3.1   Scaling of Physical Size 6
           1.3.1.1   Scaling Inductors 8
           1.3.1.2   Scaling Transmission-Line Dimensions 8
       1.3.2   Power Scaling 9
       1.3.3   Time Scaling 10
       1.3.4   Impedance Scaling with Constant Voltage 12
       1.3.5   Dielectric-Constant Scaling 14
           1.3.5.1   Partially Embedded Transmission Lines 15
       1.3.6   Magnetic Permeability Scaling 15
   1.4   The Concept of Resonance 16
   1.5   Extra for Experts: Maximal Linear System Response to a Digital Input 22

2   Transmission Line Parameters 29
   2.1   Telegrapher's Equations 31
       2.1.1   *So Good It Works on Barbed Wire* 34
       2.1.2   The No-Storage Principle and Its Implications for
                Returning Signal Current 35
   2.2   Derivation of Telegrapher's Equations 38
       2.2.1   Definition of Characteristic Impedance Z_C 39
       2.2.2   Changes in Characteristic Impedance 40
       2.2.3   Calculation of Impedance Z_C From Parameters R, L, G, And C 41
       2.2.4   Definition of Propagation Coefficient γ 44
       2.2.5   Calculation of Propagation Coefficient γ from Parameters R, L, G, and C ... 46
   2.3   Ideal Transmission Line 48
   2.4   DC Resistance 55
   2.5   DC Conductance 57

2.6 Skin Effect 58
   2.6.1 What Causes the Skin Effect, and What Does It Have to Do With Skin? 58
   2.6.2 Eddy Currents within a Conductor 61
   2.6.3 High and Low-Frequency Approximations for Series Resistance 63
2.7 Skin-Effect Inductance 66
2.8 Modeling Internal Impedance 67
   2.8.1 Practical Modeling of Internal Impedance 70
   2.8.2 Special Issues Concerning Rectangular Conductors 73
2.9 Concentric-Ring Skin-Effect Model 75
   2.9.1 *Modeling Skin Effect* 76
   2.9.2 *Regarding Modeling Skin Effect* 79
2.10 Proximity Effect 79
   2.10.1 Proximity Factor 81
   2.10.2 Proximity Effect for Coaxial Cables 84
   2.10.3 Proximity Effect for Microstrip and Stripline Circuits 85
   2.10.4 Last Words on Proximity Effect 85
      2.10.4.1 *Proximity Effect II* 85
      2.10.4.2 *2-D Quasistatic Field Solvers* 87
2.11 Surface Roughness 90
   2.11.1 Severity of Surface Roughness 90
   2.11.2 Onset of Roughness Effect 91
   2.11.3 Roughness of Pcb Materials 91
   2.11.4 Controlling Roughness 92
2.12 Dielectric Effects 94
   2.12.1 Dielectric Loss Tangent 98
   2.12.2 Rule of Mixtures 99
   2.12.3 Calculating the Loss Tangent for a Uniform Dielectric Mixture 101
   2.12.4 Calculating the Loss Tangent When You Don't Know q 103
   2.12.5 Causality and the Network Function Relations 105
   2.12.6 Finding $|e_r|$ to Match a Measured Loss Tangent 110
   2.12.7 Kramers-Kronig Relations 114
   2.12.8 Complex Magnetic Permeability 115
2.13 Impedance in Series with the Return Path 115
2.14 Slow-Wave Mode On-Chip 117

3 Performance Regions 121

3.1 Signal Propagation Model 121
   3.1.1 Extracting Parameters for RLGC Simulators 127
3.2 Hierarchy of Regions 128
   3.2.1 *A Transmission Line Is Always a Transmission Line* 130
3.3 Necessary Mathematics: Input Impedance and Transfer Function 132
3.4 Lumped-Element Region 135
   3.4.1 Boundary of Lumped-Element Region 136
   3.4.2 Pi Model 137

- 3.4.3 Taylor-Series Approximation of H (Lumped-Element Region) 139
- 3.4.4 Input impedance (Lumped-Element Region) 140
- 3.4.5 Transfer Function (Lumped-Element Region) 143
- 3.4.6 Step Response (Lumped-Element Region) 145
- 3.5 RC Region 148
  - 3.5.1 Boundary of RC Region 149
  - 3.5.2 Input Impedance (RC Region) 151
  - 3.5.3 Characteristic Impedance (RC Region) 152
  - 3.5.4 General Behavior within RC Region 153
  - 3.5.5 Propagation Coefficient (RC Region) 155
  - 3.5.6 Transfer Function (RC Region) 155
    - 3.5.6.1 Propagation Function of RC Line with Open-Circuited Load 155
    - 3.5.6.2 Propagation Function of RC Line with Matched End Termination 156
    - 3.5.6.3 Propagation Function of RC Line with Matched Source Termination 156
    - 3.5.6.4 Propagation Function of RC Line with Resistive End Termination 157
  - 3.5.7 Normalized Step Response (RC Region) 157
  - 3.5.8 Tradeoffs Between Distance and Speed (RC Region) 159
  - 3.5.9 Closed-Form Solution for Step Response (RC Region) 159
  - 3.5.10 Elmore Delay Estimation (RC Region) 160
- 3.6 LC Region (Constant-Loss Region) 166
  - 3.6.1 Boundary of LC Region 166
  - 3.6.2 Characteristic Impedance (LC Region) 167
  - 3.6.3 Influence of Series Resistance on TDR Measurements 169
  - 3.6.4 Propagation Coefficient (LC Region) 173
  - 3.6.5 Possibility of Severe Resonance within the LC Region 176
    - 3.6.5.1 Alternate Interpretation of Equation [3.17] 178
    - 3.6.5.2 Practical Effect of Resonance 179
  - 3.6.6 Terminating an LC Transmission Line 179
    - 3.6.6.1 End Termination 180
    - 3.6.6.2 Source Termination 181
    - 3.6.6.3 Both-Ends Termination 181
    - 3.6.6.4 Subtle Differences Between Termination Styles 181
    - 3.6.6.5 Application of Termination Equations to Other Regions 183
  - 3.6.7 Tradeoffs Between Distance And Speed (LC Region) 183
  - 3.6.8 Mixed-Mode Operation (LC and RC Regions) 184
- 3.7 Skin-Effect Region 185
  - 3.7.1 Boundary of Skin-Effect Region 185
  - 3.7.2 Characteristic Impedance (Skin-Effect Region) 186
  - 3.7.3 Influence of Skin-Effect on TDR Measurement 188
  - 3.7.4 Propagation Coefficient (Skin-Effect Region) 189
  - 3.7.5 Possibility of Severe Resonance within Skin-Effect Region 193
    - 3.7.5.1 Subtle Differences Between Termination Styles 194
    - 3.7.5.2 Application of Termination Equations to Other Regions 194
  - 3.7.6 Step Response (Skin-Effect Region) 195

    3.7.7   Tradeoffs Between Distance and Speed (Skin-Effect Region) 199
  3.8   Dielectric Loss Region.. 200
    3.8.1   Boundary of Dielectric-Loss-Limited Region............................... 200
    3.8.2   Characteristic Impedance (Dielectric-Loss-Limited Region)...................... 202
    3.8.3   Influence of Dielectric Loss on TDR Measurement.................... 205
    3.8.4   Propagation Coefficient (Dielectric-Loss-Limited Region)......... 206
    3.8.5   Possibility of Severe Resonance within Dielectric-Loss Limited Region... 210
        3.8.5.1   Subtle Differences Between Termination Styles....................... 211
        3.8.5.2   Application of Termination Equations to Other Regions..................... 211
    3.8.6   Step Response (Dielectric-Loss-Limited Region)......................... 212
    3.8.7   Tradeoffs Between Distance and Speed (Dielectric-Loss Region) 216
  3.9   Waveguide Dispersion Region.. 216
    3.9.1   Boundary of Waveguide-Dispersion Region 217
  3.10   Summary of Breakpoints Between Regions... 218
  3.11   Equivalence Principle for Transmission Media 221
  3.12   Scaling Copper Transmission Media 224
  3.13   Scaling Multimode Fiber-Optic Cables 229
  3.14   Linear Equalization: Long Backplane Trace Example............................... 230
  3.15   Adaptive Equalization: Accelerant Networks Transceiver 234

4   Frequency-Domain Modeling 237

  4.1   *Going Nonlinear* 237
  4.2   Approximations to the Fourier Transform 239
  4.3   Discrete Time Mapping 241
  4.4   Other Limitations of the FFT 243
  4.5   Normalizing the Output of an FFT Routine 243
    4.5.1   Deriving the DFT Normalization Factors 244
  4.6   Useful Fourier Transform-Pairs 245
  4.7   Effect of Inadequate Sampling Rate 247
  4.8   Implementation of Frequency-Domain Simulation................................... 249
  4.9   Embellishments... 251
    4.9.1   What if a Large Bulk-Transport Delay Causes the Waveform to
        Slide Off the end of the Time-Domain Window? 251
    4.9.2   How Do I Transform an Arbitrary Data Sequence? 251
    4.9.3   How Do I Shift the Time-Domain Waveforms? 252
    4.9.4   What If I Want to Model a More Complicated System?............................ 252
    4.9.5   What About Differential Modeling? 252
  4.10   Checking the Output of Your FFT Routine 253

5   Pcb (printed-circuit board) Traces.. 255

  5.1   Pcb Signal Propagation 257
    5.1.1   Characteristic Impedance and Delay... 257
    5.1.2   Resistive Effects... 258
        5.1.2.1   DC Resistance of Pcb Trace... 258

- 5.1.2.2 AC Resistance of Pcb Trace...258
- 5.1.2.3 Calculation of Perimeter of Pcb Trace 261
- 5.1.2.4 Very Low Impedance Pcb Trace..262
- 5.1.2.5 Calculation of Skin-Effect Loss Coefficient for Pcb trace 262
- 5.1.2.6 *Popsicle-Stick Analysis* 262
- 5.1.2.7 *Nickel-Plated Traces* 266
- 5.1.3 Dielectric Effects..268
  - 5.1.3.1 Estimating the Effective Dielectric Constant for a Microstrip 269
  - 5.1.3.2 Propagation Velocity..270
  - 5.1.3.3 Calculating the Effective Loss Tangent for a Microstrip 270
  - 5.1.3.4 Dielectric Properties of Laminate Materials (core and prepreg) 271
  - 5.1.3.5 *Variations in Dielectric Properties with Temperature*.............275
  - 5.1.3.6 Passivation and Soldermask 277
  - 5.1.3.7 Dielectric Properties of Soldermask Materials........................280
  - 5.1.3.8 Calculation of Dielectric Loss Coefficient for Pcb Trace 280
- 5.1.4 Mixtures of Skin Effect and Dielectric Loss 281
- 5.1.5 Non-TEM Modes 282
  - 5.1.5.1 *Strange Microstrip Modes*..282
  - 5.1.5.2 Simulation of Non-TEM Behavior..286
- 5.2 Limits to Attainable Distance 288
  - 5.2.1 *SONET Data Coding*..291
- 5.3 Pcb Noise and Interference 294
  - 5.3.1 Pcb: Reflections 294
    - 5.3.1.1 *Both Ends Termination* 295
    - 5.3.1.2 Pcb: Lumped-Element Reflections..297
    - 5.3.1.3 *Potholes*..300
    - 5.3.1.4 Inductive Potholes..303
    - 5.3.1.5 *Who's Afraid of the Big, Bad Bend?* 304
    - 5.3.1.6 *Stubs and Vias* 305
    - 5.3.1.7 *Parasitic Pads* 306
    - 5.3.1.8 *How Close Is Close Enough?*..309
    - 5.3.1.9 *Placement of End Termination*...312
    - 5.3.1.10 *Making an Accurate Series Termination*...............................314
    - 5.3.1.11 *Matching Pads* 315
  - 5.3.2 Pcb Crosstalk 318
    - 5.3.2.1 Purpose of Solid Plane Layers 318
    - 5.3.2.2 Variations with Trace Geometry..318
    - 5.3.2.3 Directionality 319
    - 5.3.2.4 NEXT: Near-End or Reverse Crosstalk 320
    - 5.3.2.5 FEXT: Far-End or Forward Crosstalk.....................................321
    - 5.3.2.6 Special Considerations..322
    - 5.3.2.7 *Directionality of Crosstalk*..323
- 5.4 Pcb Connectors 326
  - 5.4.1 *Mutual Understanding* 326

5.4.2	*Through-Hole Clearances*	328
5.4.3	*Measuring Connectors*	330
5.4.4	*Tapered Transitions*	332
5.4.5	*Straddle-Mount Connectors*	335
5.4.6	*Cable Shield Grounding*	336
5.5	Modeling Vias	338
5.5.1	Incremental Parameters of a Via	338
5.5.2	Three Models for a Via	341
5.5.3	Dangling Vias	343
5.5.4	Capacitance Data	345
5.5.4.1	Three-Layer Via Capacitance	345
5.5.4.2	Effect of Back-Drilling	346
5.5.4.3	Effect of Multiple Planes	347
5.5.5	Inductance Data	351
5.5.5.1	Through-Hole Via Inductance	351
5.5.5.2	Via Crosstalk	354
5.6	*The Future of On-Chip Interconnections*	359

6 Differential Signaling ... 363

6.1	Single-Ended Circuits	363
6.2	Two-Wire Circuits	368
6.3	Differential Signaling	370
6.4	Differential and Common-Mode Voltages and Currents	374
6.5	Differential and Common-Mode velocity	376
6.6	Common-Mode Balance	377
6.7	Common-Mode Range	378
6.8	Differential to Common-Mode Conversion	378
6.9	Differential Impedance	380
6.9.1	Relation Between Odd-Mode and Uncoupled Impedance	383
6.9.2	Why the Odd-Mode Impedance Is Always Less Than the Uncoupled Impedance	383
6.9.3	*Differential Reflections*	384
6.10	Pcb Configurations	385
6.10.1	*Differential (Microstrip) Trace Impedance*	386
6.10.2	Edge-Coupled Stripline	389
6.10.3	*Breaking Up a Pair*	397
6.10.4	Broadside-Coupled Stripline	399
6.11	Pcb Applications	404
6.11.1	Matching to an External, Balanced Differential Transmission Medium	404
6.11.2	Defeating ground bounce	405
6.11.3	Reducing EMI with Differential Signaling	405
6.11.4	Punching Through a Noisy Connector	407
6.11.4.1	*Differential Signaling (Through Connectors)*	408
6.11.5	Reducing Clock Skew	409

		6.11.6	Reducing Local Crosstalk 411
		6.11.7	A Good Reference about Transmission Lines............................... 413
		6.11.8	*Differential Clocks* 413
		6.11.9	*Differential Termination* 414
		6.11.10	*Differential U-Turn* 417
		6.11.11	*Your Layout Is Skewed* 419
		6.11.12	*Buying Time* 420
	6.12	Intercabinet Applications 422
		6.12.1	Ribbon-Style Twisted-Pair Cables 423
		6.12.2	Immunity to Large Ground Shifts 424
		6.12.3	Rejection of External Radio-Frequency Interference (RFI) 426
		6.12.4	Differential Receivers Have Superior Tolerance to Skin Effect and Other High-Frequency Losses 427
	6.13	LVDS Signaling 429
		6.13.1	Output Levels 429
		6.13.2	Common-Mode Output 430
		6.13.3	Common-Mode Noise Tolerance 430
		6.13.4	Differential-Mode Noise Tolerance 431
		6.13.5	Hysteresis 431
		6.13.6	Impedance Control 432
		6.13.7	Trace Radiation 435
		6.13.8	Risetime 435
		6.13.9	Input Capacitance 435
		6.13.10	Skew 435
		6.13.11	Fail-Safe 436

7 Generic Building-Cabling Standards 439

7.1	Generic Cabling Architecture 442
7.2	SNR Budgeting 446
7.3	Glossary of Cabling Terms 446
7.4	Preferred Cable Combinations 449
7.5	FAQ: Building-Cabling Practices 449
7.6	Crossover Wiring 451
7.7	Plenum-Rated Cables 452
7.8	Laying cables in an Uncooled Attic Space 453
7.9	FAQ: Older Cable Types 453

8 100-Ohm Balanced Twisted-Pair Cabling 457

8.1	UTP Signal Propagation 459
	8.1.1	UTP Modeling 460
	8.1.2	Adapting the Metallic-Transmission Model 462
8.2	UTP Transmission Example: 10BASE-T 465
8.3	UTP Noise and Interference 471
	8.3.1	UTP: Far-End Reflections 471

	8.3.2	UTP: Near-End Reflections	475
	8.3.2.1	UTP: (Structural) Return Loss	477
	8.3.2.2	Modeling Structural Return Loss	480
	8.3.3	UTP: Hybrid Circuits	481
	8.3.4	UTP: Near-End Crosstalk	487
	8.3.5	UTP: Alien crosstalk	490
	8.3.6	UTP: Far-End Crosstalk	490
	8.3.7	Power sum NEXT and ELFEXT	493
	8.3.8	UTP: Radio-Frequency Interference	493
	8.3.9	UTP: Radiation	496
8.4		UTP Connectors	497
8.5		Issues with Screening	501
8.6		Category-3 UTP at Elevated Temperature	502

9 150-Ohm STP-A Cabling ... 505

9.1	150-Ω STP-A Signal Propagation	506
9.2	150-Ω STP-A Noise and Interference	506
9.3	150-Ω STP-A: Skew	507
9.4	150-Ω STP-A: Radiation and Safety	508
9.5	150-Ω STP-A: Comparison with UTP	509
9.6	150-Ω STP-A Connectors	509

10 Coaxial Cabling ... 513

10.1		Coaxial Signal Propagation	515
	10.1.1	Stranded Center-Conductors	522
	10.1.2	*Why 50 Ohms?*	523
	10.1.3	*50-Ohm Mailbag*	526
10.2		Coaxial Cable Noise and Interference	528
	10.2.1	Coax: Far-End Reflected Noise	528
	10.2.2	Coax: Radio Frequency Interference	529
	10.2.3	Coax: Radiation	529
	10.2.4	Coaxial Cable: Safety Issues	530
10.3		Coaxial Cable Connectors	532

11 Fiber-Optic Cabling ... 537

11.1		Making Glass Fiber	538
11.2		Finished Core Specifications	539
11.3		Cabling the Fiber	541
11.4		Wavelengths of Operation	543
11.5		Multimode Glass Fiber-Optic Cabling	544
	11.5.1	Multimode Signal Propagation	546
	11.5.2	Why Is Graded-Index Fiber Better than Step-Index?	551
	11.5.3	Standards for Multimode Fiber	552
	11.5.4	What Considerations Govern the Use of 50-micron Fiber?	554

11.5.5	Multimode Optical Performance Budget	555
11.5.5.1	Multimode Dispersion Budget	555
11.5.5.2	Multimode Attenuation Budget	566
11.5.6	Jitter	568
11.5.7	Multimode Fiber-Optic Noise and Interference	570
11.5.8	Multimode Fiber Safety	571
11.5.9	Multimode Fiber with Laser Source	571
11.5.10	VCSEL Diodes	573
11.5.11	Multimode Fiber-Optic Connectors	575
11.6	Single-Mode Fiber-Optic Cabling	576
11.6.1	Single-Mode Signal Propagation	577
11.6.2	Single-Mode Fiber-Optic Noise and Interference	578
11.6.3	Single-Mode Fiber Safety	578
11.6.4	Single-Mode Fiber-Optic Connectors	578

12 Clock Distribution 579

12.1	*Extra Fries, Please*	582
12.2	Arithmetic of Clock Skew	584
12.3	Clock Repeaters	589
12.3.1	Active Skew Correction	593
12.3.2	Zero-Delay Clock Repeaters	594
12.3.3	Compensating for Line Length	595
12.4	Stripline vs. Microstrip Delay	596
12.5	Importance of Terminating Clock Lines	599
12.6	Effect of Clock Receiver Thresholds	601
12.7	Effect of Split Termination	602
12.8	Intentional Delay Adjustments	605
12.8.1	Fixed Delay	605
12.8.2	Adjustable Delays	607
12.8.3	Automatically Programmable Delays	609
12.8.4	*Serpentine Delays*	610
12.8.5	Switchback Coupling	612
12.9	Driving Multiple Loads with Source Termination	616
12.9.1	*To Tee or Not To Tee*	619
12.9.2	*Driving Two Loads*	625
12.10	Daisy-Chain Clock Distribution	627
12.10.1	Case Study of Daisy-Chained Clock	629
12.11	*The Jitters*	634
12.11.1	When Clock Jitter Matters	636
12.11.1.1	Clock Jitter Rarely Matters within the Boundaries of a Synchronous State Machine	636
12.11.1.2	Clock Jitter Propagation	636
12.11.1.3	Variance of the Tracking Error	640
12.11.1.4	Clock Jitter in FIFO-Based Architectures	643

		12.11.1.5	What Causes Jitter	644
		12.11.1.6	Random and Deterministic Jitter	645
	12.11.2		Measuring Clock Jitter	648
		12.11.2.1	*Jitter Measurement*	651
		12.11.2.2	*Jitter and Phase Noise*	654
12.12			Power Supply Filtering for Clock Sources, Repeaters, and PLL Circuits	656
	12.12.1		*Healthy Power*	659
	12.12.2		*Clean Power*	661
12.13			*Intentional Clock Modulation*	663
	12.13.1		*Signal Integrity Mailbag*	665
	12.13.2		*Jitter-Free Clocks*	667
12.14			Reduced-Voltage Signaling	668
12.15			Controlling Crosstalk on Clock Lines	669
12.16			*Reducing Emissions*	670

13 Time-Domain Simulation Tools and Methods ... 673

13.1	*Ringing in a New Era*	673
13.2	Signal Integrity Simulation Process	674
	13.2.1 How Much Modeling Do You Need?	676
	13.2.2 What Happens After Parameter Extraction?	676
	13.2.3 A Word of Caution	677
13.3	The Underlying Simulation Engine	678
	13.3.1 Evolving Forward	680
	13.3.2 Pitfalls of SPICE-Like Algorithms	680
	13.3.3 Transmission Lines	682
	13.3.4 Interpreting Your Results	684
	13.3.5 Using SPICE Intelligently	685
13.4	*IBIS (I/O Buffer Information Specification)*	685
	13.4.1 What Is IBIS?	686
	13.4.2 Who Created IBIS?	686
	13.4.3 What Is Good About IBIS?	687
	13.4.4 What's Wrong with IBIS?	687
	13.4.5 What You Can Do to Help	688
13.5	*IBIS: History and Future Direction*	689
	13.5.1 IBIS Historical Overview	689
	13.5.2 Comparison to SPICE	690
	13.5.3 Future Directions	690
13.6	IBIS: Issues with interpolation	691
13.7	IBIS: Issues with SSO Noise	695
13.8	Nature of EMC Work	697
	13.8.1 *EMC Simulation*	698
13.9	*Power and Ground Resonance*	699

Collected References ... 703

Points to Remember 710

Appendix A - *Building a Signal Integrity Department* 731

Appendix B - Calculation of Loss Slope 733

Appendix C - Two-Port Analysis 735
    Simple Cases Involving Transmission Lines 737
    Fully Configured Transmission Line 739
    Complicated Configurations 741

Appendix D - Accuracy of Pi Model 743
    Pi-Model Operated in the LC Region 745

Appendix E - erf( ) 747

Index 749

PREFACE

NOTE FROM THE AUTHOR

Welcome, and thank you for your interest in *High-Speed Signal Propagation: Advanced Black Magic*. This is an advanced-level reference text for experienced digital designers who want to press their designs to the upper limits of speed and distance.

If you need to transmit faster and further than ever before, this book is here to help. You'll find it packed with practical advice.

The material in this book has been honed during my many years of work as chief technical editor of standards for both Fast Ethernet and Gigabit Ethernet—projects which, I hope, have touched your life in a favorable way. During those and many other projects, the models and concepts described here have been of invaluable service to me. Now I'd like to pass them on to you.

When you are done reading, share your knowledge with those around you as my technical mentor, Martin Graham, has done with me. Educate your coworkers. Educate your management. Above all, continue to educate yourself. If this book inspires you to advance your understanding with even one laboratory measurement, then I will know you are on the right track.

I would also like to say it has been a great pleasure teaching and working with many of you through my classes and lectures. Above all, I appreciate those who take the time to share with me their thoughts, their concerns, their dreams, and their problems. It always interests me to hear about real experiences from real engineers.

I wish you the best of luck on your next design.

See you on the Internet,
Dr. Howard Johnson
www.sigcon.com

Topics Covered

Printed circuit traces	Limits to attainable speed and distance
	RC and LC mode propagation
	Skin effect and dielectric loss design charts and equations
	Proximity effect
	Surface roughness
	Non-TEM mode of propagation
	Step response
	Effect of vias
Differential signaling	Edge-coupled and broadside-coupled differential pairs
	Effect of bends
	Intrapair skew
	Differential trace geometry impedance
	Crosstalk
	Radiation
Inter-cabinet connections	Coaxial cables
	Twisted-pair cables
	Fiber optics
	Equalizers
	General building wiring for LAN applications
Clock distribution	Special requirements for clocks
	Clock repeaters
	Multidrop clock distribution
	Clock jitter
	Power filtering for clock sources
Simulation	Frequency-domain simulation method
	Applicability of Spice and IBIS

How This Book Is Organized

Each chapter in this book treats a specialized topic having to do with high-speed signal propagation. They may be studied in any order.

Chapters 1 and 2 present the underlying physical theory of various transmission-line parameters, including the skin effect, proximity effect, dielectric loss, and surface roughness.

Chapter 3 develops a generalized frequency-response model common to all conductive media.

Chapter 4 outlines the calculation of time-domain waveforms from frequency-domain transfer functions.

Chapters 5 through 11 discuss specific transmission media, including single-ended pcb traces, differential media, general building wiring standards, unshielded twisted-pair wiring, 150-Ω shielded twisted-pair wiring, coaxial cables, and fiber.

Chapter 12 addresses miscellaneous issues concerning clock distribution.

Chapter 13 explores the limitations of Spice and IBIS simulation methods.

PREREQUISITES

A basic understanding of the frequency domain representation of linear systems is assumed. Readers without the benefit of formal training in analog circuit theory can use and apply the formulas and examples in this book. Readers who have completed a first-year class in introductory linear circuit theory will comprehend the material at a deeper level.

RELATION TO PRIOR BOOKS

This book is a companion to the original book by Johnson and Graham, *High-Speed Digital Design: A Handbook of Black Magic*, Prentice-Hall, 1993. The two books may be used separately or together. They cover different material.

The original book deals with a broad spectrum of high-speed phenomena. It builds a solid understanding of ringing, crosstalk, ground bounce, and power supply noise as they exist on printed circuit boards. It emphasizes basic circuit configurations where these effects may be easily understood and learned. It treats supplementary subjects including chip packages, oscilloscope probe, and power systems for high-speed digital products.

This *High-Speed Signal Propagation* book is more highly specialized, delving into issues relevant to transmission at the upper limits of speed and distance. If you need to transmit faster and further than ever before, this book shows you how.

High-Speed Digital Design and *High-Speed Signal Propagation* together comprise a good reference set for persons working with high-speed digital technology.

Those of you familiar with my other books will recognize similarities in style. Notably, I've tried to impart, as best I can, the same sense of realism born of long experience.

ACKNOWLEDGEMENTS

Literally thousands of people have taken the time to communicate with me about high-speed issues, either through email or in person at my seminars. These conversations have inspired me to investigate and collect together the material in this book. To all of you, I owe a debt of gratitude.

The following people contributed specific comments or questions that are discussed in the text (in alphabetical order): Sal Aguinaga, James C. Bach, Eric V. Berger, Raymond Bullington, Doug Butler, Tim Canales, Bruce Carsten, Code Cubitt, Dave Cuthbert, Bill Daskalakis, Martin Graham, Paul Greene, Gary Griffin, Bob Haller, John Lehew, John Lin, Raymond P. Meixner, Craig Miller, Mitch Morey, Dan Nitzan, Bhavesh Patel, Dipak Patel, Jim Rautio, Ravi, Boris Shusterman, Kevin Slattery, Bob Stroupe, Bill Stutz (twice), and Fabrizio Zanella. Thanks to all of you for many hours of good correspondence.

I especially thank those who volunteered for the difficult task of reviewing the text. This group of intrepid individuals spotted numerous errors and suggested many new topics for exploration. They deserve a large measure of credit for helping make this a more useful text (in alphabetical order): Jacob Ben Ary at Aquanet, Greg Dermer at Easystreet, Steve Ems at Lecroy, Alexandre Guterman at Nortel, Valery Kugel at Juniper, Professor Will Moore at Oxford University, Jose Moreira at Agilent, Gopa Parameswaran at Cisco, Bob Ross at Mentor Graphics, Bert Simonovich at Nortel, Palani Subbiah at Cypress, and Geoff Thompson at Nortel.

My editors at Prentice-Hall, Bernard Goodwin, Nicholas Radhuber, and Carol J. Lallier, have contributed their professional expertise (and patience) during the long process required to complete this project.

Without my dutiful and highly accurate assistant Jennifer Epps this book would not have been possible.

All the articles adapted for publication in this book are reprinted with permission from EDN magazine, a publication of Reed Business Information, Electronic Design Magazine, a division of Penton Media, Inc., or PC Design Magazine, a publication of UP Media Group, Inc., as noted in the header of each article, respectively.

Bob Ross, Mentor Graphics Corp., past chair of the EIA IBIS Open Forum, wrote a fine discussion about the future of IBIS modeling for Chapter 13. Bruce Archambeault contributed the article in Chapter 12 about reducing emissions, which I only edited. Brad Cole and Matt Hudale of Ansoft simulated the capacitance of many via configurations for Chapter 5. To Gopa Parameswaran at Cisco, thanks for your simulations of via capacitance, although your data did not appear in the final version of the book. Steve Ems and Robert Talambiras of Lecroy piqued my interest in non-TEM modes of propagation during a visit to my ranch in October of 2000. Roger Billings of Wideband Corporation deserves mention as the world record-holder for fastest data conveyed across barbed-wire cabling (Chapter 2).

Jeff Sonntag at Accelerant Networks was the first to focus my attention on how backplane performance changes with temperature (Chapter 5). Thanks also to Jim Tavacoli at Accelerant for sending the cool pictures showing adaptive equalization at work in Chapter 3.

The discussions I have been privileged to hold with Michael King, Ed Sayre, and Doug Smith have been of enormous value to me. Thank you for your friendship.

My technical mentor Dr. Martin Graham of U.C. Berkeley has contributed his enduring support, encouragement, and technical assistance over the past twenty years, as well as having been the first to direct my attention toward the general features of transmission line attenuation and how it varies with frequency. Thank you, Martin.

To my wonderful and understanding wife Liz, thank you for taking care of all the details of my life so that I could have time to write a book like this.

Regardless of the assistance of others, any remaining errors are entirely mine.

> Information contained in this work has been obtained by Prentice Hall from sources believed to be reliable. However, neither Prentice Hall nor its author guarantees the accuracy or completeness of any information published herein and neither Prentice Hall nor its authors shall be responsible for any errors, omission, or damages arising out of this information.

NOTATIONAL CONVENTIONS

This book uses metric units, except for some common printed-circuit board dimensions, which are denoted in English units. Variables and general function names appear in *italics*. Constants, enumerators, and specific well-known functions appear in ordinary type (e.g., $f(x) = 1 + \sin x$). Matrix and vector-field quantities occur in **boldface** type.

CONTACTING THE AUTHOR

Should you spot something out of place in the text, or merely wish to discuss the finer points of transmission-line theory, I may be reached at my ranch high in the mountains near the town of Twisp, Washington: howiej@sigcon.com.

A great place to keep up with the latest developments in high-speed signaling is my web site, *www.sigcon.com*. At that site I maintain a growing collection of articles about high-speed digital phenomena, and information about my schedule of public seminars. If you would like to read even more about signal integrity issues, sign up to receive my newsletter. An errata page for this book is located on the site.

> Howard Johnson, PhD, is the author of *High-Speed Digital Design: A Handbook of Black Magic,* Prentice-Hall, 1993, *Fast Ethernet: Dawn of a New Network*, Prentice-Hall, 1996, and the *Signal Integrity* columnist for *EDN* magazine. He frequently conducts technical workshops for digital engineers at Oxford University and other sites worldwide.

GLOSSARY OF SYMBOLS

Units and Special Constants

Ω Unit of resistance; one amp flowing through an impedance of one ohm creates a voltage of one volt.

S   International unit of conductance, the Sieman. Equivalent to $1/\Omega$. In older texts this unit is sometimes called a mho, written with an upside-down ohm symbol.

c Velocity of light in a vacuum, $2.998 \cdot 10^8$ m/s.

ϵ_0 Electric permittivity of free space, $8.854 \cdot 10^{-12}$ F/m.

μ_0 Magnetic permeability of free space, $4\pi \cdot 10^{-7}$ H/m.

j Square root of -1.

Time-Dependent and Frequency-Dependent Variables

ω Frequency (rad/s).

ω_{LE} Upper boundary of lumped-element mode (rad/s).

ω_{LC} Onset frequency for LC mode (rad/s).

ω_δ Onset frequency for skin effect (rad/s).

ω_{rough} Onset frequency for roughness effect (rad/s).

ω_λ Onset frequency for non-TEM waveguide modes (rad/s).

δ Skin depth (always a function of frequency) (m).

$i(t)$ Current that varies with time (A).

$I(\omega)$ Fourier or Laplace transform of current waveform $i(t)$.

$v(t)$ Voltage that varies with time (V).

$V(\omega)$ Fourier or Laplace transform of voltage waveform $v(t)$.

Transmission-Line Parameters

R Series resistance per meter (Ω/m).

L Series inductance per meter (H/m).

G Shunt conductance per meter (S/m).

C Shunt capacitance per meter (F/m).

	Glossary of Symbols
z	Series impedance per meter, composed of z_i and z_e (complex-valued) (Ω/m).
z_i	Internal series impedance per meter representing magnetic flux trapped within the conductors, and power dissipated within the conductors (Ω/m).
z_e	External series impedance per meter representing magnetic flux occupying the space surrounding the conductors, and power dissipated in the region surrounding the conductors (Ω/m). The real part of z_e represents, in a vague sense, the radiation resistance of the structure.
L_i	Internal inductance per meter ($j\omega L_i = j\cdot\text{Im } z_i$) representing magnetic flux trapped within the conductors (H/m).
L_e	External inductance per meter ($j\omega L_e = j\cdot\text{Im } z_e$) representing magnetic flux occupying the space surrounding the conductors (H/m).
R_i	Internal resistance per meter ($R_i = \text{Re } z_i$) representing power dissipated within the conductors (Ω/m).
y	Shunt conductance per meter (complex-valued) (S/m).
θ	Angle formed by the imaginary and real parts of complex electric permittivity (rad). Equals the arctangent of $-\epsilon''/\epsilon'$. For small angles, $\theta \approx \tan\theta = -\epsilon''/\epsilon'$.
$\tan\theta$	Often called the *loss tangent* for a dielectric material. Ratio of the imaginary and real parts of complex electric permittivity ($-\epsilon''/\epsilon'$).
Z_C	Characteristic impedance of a transmission line, a function of frequency (Ω). The variable Z_C is reserved as an expression for the characteristic impedance when it is important to emphasize the variations with frequency, usually but not always shown as $Z_C(\omega)$. The variable Z_0 is interpreted as a single-valued constant showing the value of characteristic impedance at the particular frequency ω_0 (as in the expression $Z_0 = 50 \, \Omega$).
v	Velocity of propagation (m/s), also voltage (V).
R_{DC}	DC resistance of conductors (Ω/m).
w	Trace width (m).
h	Trace height above the nearest reference plane (m).
t	Trace thickness (m).
b	(For striplines) total height between the reference planes (m).
s	Separation between facing surfaces of two traces (m).
r	Radius of round wire (m).
p	Length around perimeter of conductor (m).
a	Cross-sectional area of conductor (m^2).

AC-Transmission-Line Parameters Specified at Frequency ω_0

ω_0 Frequency at which AC line parameters are specified (rad/sec). Presumably this value lies above the LC and skin-effect mode onset frequencies, but below the onset of multiple waveguide modes of operation.

R_0 AC resistance at frequency ω_0 (Ω/m).

L_0 Series inductance at frequency ω_0 (H/m).

C_0 Shunt capacitance at frequency ω_0 (F/m).

θ_0 Angle formed by the real and imaginary parts of complex electric permittivity (rad) at frequency ω_0. Equals the arctangent of $-\epsilon''/\epsilon'$ at frequency ω_0.
For small angles, $\theta_0 \approx \tan \theta_0 = -\epsilon''/\epsilon'$.

Z_0 Characteristic impedance at the particular frequency ω_0 (Ω). The variable Z_0 is interpreted as a single-valued constant (as in the expression $Z_0 = 50\ \Omega$). The variable Z_C is reserved as an expression for the characteristic impedance when it is important to emphasize the variations with frequency, usually but not always shown as $Z_C(\omega)$.

v_0 Velocity of propagation (inverse of group delay) at frequency ω_0 (m/s).

Material Parameters

ϵ Electric permittivity (in general, a complex value) (F/m).

ϵ' Real part of $\epsilon \triangleq \epsilon' - j\epsilon''$ (F/m).

ϵ'' Negative of the imaginary part of $\epsilon \triangleq \epsilon' - j\epsilon''$ (F/m).

ϵ_r Relative electric permittivity defined as ϵ/ϵ_0 (in general a complex value).

χ_e Electric susceptibility, $\chi_e \triangleq \epsilon_r - 1$ (in general a complex value).

μ Magnetic permeability (in general a complex value) (H/m).

σ Bulk conductivity (S/m).
For annealed copper, as prepared for ordinary wires: $\sigma = 5.80 \cdot 10^7$ S/m.
For electro-deposited (pure) copper as used on pcb-traces $\sigma = 5.98 \cdot 10^7$ S/m.
For aluminum as prepared for ordinary wires: $\sigma = 3.54 \cdot 10^7$
For electro-deposited (pure) aluminum: $\sigma = 3.77 \cdot 10^7$
For silver (pure): $\sigma = 6.29 \cdot 10^7$.
For steel, SAE1045: $\sigma = 5.80 \cdot 10^6$.
For steel, stainless: $\sigma = 1.16 \cdot 10^6$ (varies widely).

	For nickel (pure): $\sigma = 1.16 \cdot 10^7$.
	NOTE: Owing to wide differences in the composition of steel alloys, the relative magnetic permeability of "ordinary" steel ranges from 100 to 10,000 or more. This variation in permeability induces dramatic variations in the skin depth, and thus the surface resistivity, of the material.
ρ	Bulk resistivity (Ω-m) (inverse of σ).
η	Intrinsic impedance: ratio of electric to magnetic field intensity in a plane wave propagating within the material (Ω).

General Notation

l	Length of structure (m).
$\mathcal{L}$	Length of rising or falling edge $\mathcal{L} = t_{10-90} v_0$ (m).
t_{10-90}	Rise or fall time, 10 to 90 percent (s).
V_{CC}	Power supply voltage for digital logic (V).

CHAPTER 1

FUNDAMENTALS

1.1 IMPEDANCE OF LINEAR, TIME-INVARIANT, LUMPED-ELEMENT CIRCUITS

The following relations are well worth committing to memory, as I will use them many times over during the course of this text. If they are not already familiar to you, see [1], or any other general electrical engineering reference.

The impedance magnitude of a linear, time-invariant lumped-element inductor, as measured with a sinusoidal input at frequency f, is

$$X_L \triangleq 2\pi f L \qquad [1.1]$$

where X_L is the impedance magnitude (ohms),
f is the measurement frequency (Hz),
L is the inductance (Henries).

The impedance magnitude of a linear, time-invariant, lumped-element capacitor, as measured with a sinusoidal input at frequency f, is

$$X_C \triangleq \frac{1}{2\pi f C} \qquad [1.2]$$

where X_C is the impedance magnitude (ohms),
f is the measurement frequency (Hz),

C is the capacitance (Farads).

When dealing with step waveforms (like digital signals), an appropriate frequency of analysis for estimating the effective impedance of a parasitic element, over a time scale corresponding to a rising (or falling) edge, is

$$f_{knee} \triangleq \frac{0.5}{t_{10\text{-}90\%}} \qquad [1.3]$$

where f_{knee} is the assumed frequency for sinusoidal analysis (Hz),

$t_{10\text{-}90\%}$ is the 10% to 90% rise (or fall) time of the circuit (seconds).

The *knee frequency*, f_{knee}, is a crude estimate of the highest frequency content within a particular digital signal (Figure 1.1). Signaling channels whose parasitic impedances are not significant at all frequencies up to and including the knee frequency tend to pass digital signals undistorted.

At the knee frequency the average spectral power density of a random digital signal with Gaussian rising and falling edges lies 6.8 dB below the straight slope shown in Figure 1.1. Obviously, there is no crisply defined knee at this point. Some authors define the bandwidth of a digital signal as $0.35/t_{10\text{-}90\%}$, which happens to lie at the −3 dB point in Figure 1.1, closer to the meat of the spectrum associated with a rising edge. For some calculations this may produce a more accurate result, however, I still prefer to use [1.3], thinking of it as a conservative *over*-estimate of bandwidth. If the parasitic elements in my circuit remain insignificant up to [1.3] then I know they can safely be ignored. If, on the other hand, the parasitic elements begin to give difficulty at frequency [1.3] I immediately turn to more accurate techniques (such as Spice, Fourier-Transforms, or Laplace Transforms) to ferret out the exact circuit response.

POINTS TO REMEMBER

➢ The *knee frequency*, $f_{knee} \triangleq 0.5/t_{10\text{-}90\%}$ Hz, is a crude estimate of the highest frequency content within a particular digital signal.

➢ The frequency $0.35/t_{10\text{-}90\%}$ Hz may better approximate the meat of the spectrum associated with a rising edge.

1.2 POWER RATIOS

The decibel equivalent for any power ratio is

$$\text{Power ratio in dB} = 10 \log_{10}\left(\frac{p_1}{p_2}\right) \text{ dB} \qquad [1.4]$$

where p_1/p_2 is a power ratio.

1.2 • Power Ratios

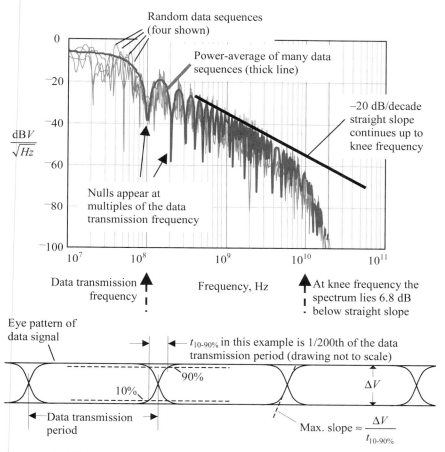

Figure 1.1—The power spectral density of random data signal has nulls at multiples of the data rates and a maximum effective bandwidth related to its rise and fall time.

In two circuits of equal impedance, the ratio of their powers may be determined as the square of the ratios of the voltages. This definition of a power ratio, in units of decibels, *assumes equal impedances in the two circuits*:

$$\text{Power ratio in dB} = 20\log_{10}\left(\frac{v_1}{v_2}\right) \text{dB} \qquad [1.5]$$

where v_1/v_2 is the ratio of RMS voltages across two circuit elements of equal resistance, or in the case of identical signal shapes in the two circuits, such as sine waves, the ratio of voltage amplitudes across two circuits having equal complex-valued impedances, and the same formula and conditions apply to the determination of a power ratio from a ratio of currents.

If the impedances in the two circuits differ, then the power ratio must be corrected for that difference.

$$\text{Power ratio in dB} = 20\log_{10}\left(\frac{v_1}{v_2}\right) - 10\log_{10}\left(\frac{|z_1|}{|z_2|}\right) \text{dB}$$

$$\text{Power ratio in dB} = 20\log_{10}\left(\frac{i_1}{i_2}\right) + 10\log_{10}\left(\frac{|z_1|}{|z_2|}\right) \text{dB}$$

[1.6]

where v_1/v_2 is the ratio of RMS voltages across two circuit elements having resistances z_1 and z_2 respectively, or in the case of identical signal shapes in the two circuits, such as sine waves, the ratio of voltage amplitudes across two circuits having proportional complex-valued impedances.

i_1/i_2 is the ratio of RMS currents flowing through two circuit elements having resistances z_1 and z_2 respectively, or in the case of identical signal shapes in the two circuits, such as sine waves, the ratio of current amplitudes across two circuits having proportional complex-valued impedances.

In some cases it is convenient to use decibel notation to express a ratio of the voltages (or currents) within two circuits without consideration of the level of power.

$$\text{Voltage ratio in dB} = 20\log_{10}\left(\frac{v_1}{v_2}\right) \text{dB}$$

$$\text{Current ratio in dB} = 20\log_{10}\left(\frac{i_1}{i_2}\right) \text{dB}$$

[1.7]

where v_1/v_2 is the ratio of voltage amplitudes within the two circuits. Presumably, the two circuits carry signals with identical shapes, such as sine waves.

i_1/i_2 is the ratio of current amplitudes within the two circuits. Presumably, the two circuits carry signals with identical shapes, such as sine waves.

The magnitude of a voltage, current, or power relative to a standard reference level may be expressed in absolute terms using decibel notation. For example, the notation dBmV refers to the following quantity:

$$\text{Voltage amplitude relative to 1 mV} = 20\log_{10}\left(\frac{v_1}{1\,\text{mV}}\right) \text{dBmV}$$

[1.8]

where v_1 represents the voltage amplitude of a signal generally assumed to be sinusoidal.

Units of nepers occasionally appear in transmission-line problems. One neper equals 8.685889638065 dB. Definitions of power, voltage, and current ratios in units of nepers use the natural (base-e) logarithm $ln()$.

$$\text{Power ratio in nepers} = \frac{1}{2} \ln\left(\frac{p_1}{p_2}\right) \text{ neper}$$

$$\text{Voltage ratio in nepers} = \ln\left(\frac{v_1}{v_2}\right) \text{ neper} \qquad [1.9]$$

$$\text{Current ratio in nepers} = \ln\left(\frac{i_1}{i_2}\right) \text{ neper}$$

where p_1/p_2 is a power ratio,

v_1/v_2 is the ratio of voltage amplitudes within the two circuits. Presumably, the two circuits carry signals with identical shapes, such as sine waves.

i_1/i_2 is the ratio of current amplitudes within the two circuits. Presumably, the two circuits carry signals with identical shapes, such as sine waves.

$\ln(x)$ is the natural logarithm function (log base e of x).

Historical note

The term *neper* was named after John Neper, 1550–1617, "Scottish mathematician and inventor of logarithms" (*Webster's Encyclopedic Unabridged Dictionary of the English Language*, 1989). His name has been somewhat confused in history with that of Sir Charles James Napier, 1782–1853, a British general and the inventor of a "diagram for showing the deviation of a magnetic compass from magnetic north at any heading" (ibid). The result today is that you may occasionally see the unit *neper* spelled *napier*. According to the American National Standards Institute, the correct spelling is *neper*. Further confusing the issue is that the correct spelling for the term that means "natural logarithm" is *Napierian logarithm*.

POINT TO REMEMBER

➢ One neper equals 8.685889638065 dB.

1.3 RULES OF SCALING

Digital designers do not restrict their attention to any particular range of power, impedance, physical size, or frequency. Unlike designers in other application areas, the field of digital design cuts across many orders of magnitude of the frequency, power, and size domains. For example, a designer of battleship rudder control systems might build a whole career on an understanding of large electromechanical and hydraulic machinery in the frequency range of from 0.01 to 1 Hz. An audio engineer, on the other hand, might specialize in the 10 to 100,000 Hz band, working with desktop-size amplifier boxes. Digital designers know no such bounds. Every few years their horizon expands, taking them to ever-higher realms of frequency and to ever-smaller physical scales of operation. Digital systems today span a vast array of physical scales, from integrated circuit dimensions smaller than one micron to intercontinental distances, and a vast array of frequencies, from deep-sea submarine communications at 0.1 Hz to fiber-optic links running at 100 GHz and beyond.

If you wish to exist in such a world, you must make your knowledge portable so that you may more easily move from one domain to another. The theory that makes this possible is the mathematical theory of scaling.

The rules of scaling (listed below) show how to transform one system into an equivalent system operating at a different scale. This scaling may distort the physical sizes or the component values of the circuit elements involved, but the principles of circuit operation will remain the same: Low-impedance series elements in the first system will translate into (relatively) low-impedance elements in the new system, and so forth. A system that transfers power efficiently (or blocks the transmission of power) at one scale may be transformed into a new system that performs the same function on a different scale.

You can use these rules to establish analogies between systems that are known to work in one range of power, impedance, physical size, and frequency and others that you may wish to construct on a different scale.

1.3.1 Scaling of Physical Size

When you change the physical dimensions of a distributed circuit, modifying all geometrical dimensions x, y, and z by a common factor k, without changing either the electric permittivity or the magnetic permeability, you find that all inductances change by a factor of k and all capacitances change by the same factor.

If the circuit is *passive* and *lossless* (that is, composed only of inductive and capacitive effects, with no resistances) it will have been scaled in time, whereby the new circuit should behave the same as the old circuit, only its step response stretched (or compressed) in time by the factor k. A network-analyzer plot of the new system will show the same frequency response as the old, only shifted in frequency by a factor of $1/k$. A resonance at frequency f_1 in the original circuit appears at frequency f_1/k in the new circuit. As a consequence, physically enlarging a system of physical conductors lowers its resonant frequencies, while shrinking the system physically raises them.

This principle of physical scaling explains why high-frequency microwave connectors must be made so darn small—the tiny dimensions are required to push the parasitic resonances up to frequencies above the bandwidth of the signal.

The presence of significant resistance in a circuit complicates the situation. When a circuit is physically scaled, a new resistor R_n' replaces each resistor R_n in the old circuit. It would be nice if each new resistance R_n' operating at the new frequency f/k presented the same impedance as the old resistance R_n operating at the old frequency f:

1.3.1 • Scaling of Physical Size

$$R_n'(f/k) = R_n(f) \qquad [1.10]$$

If [1.10] held true, the new circuit would behave identically to the old, except for the factor-of-k dilation (or compression) of time. If you are dealing merely with lumped-element components, then resistive scaling is trivial—just don't change the values of resistances in the circuit.

Unfortunately, if you are trying to model damping effects due to the skin-effect resistance of physical conductors, the resistive scaling becomes more complicated. Here's how it works.

At frequencies *below* the onset of skin effect the bulk resistance of a conducting object varies with its *length* and inversely with its *cross-sectional area*. When you scale all physical dimensions by k, the length goes as k, while the inverse of cross-sectional area goes as $1/k^2$, with the result that bulk resistance changes as $1/k$. This effect changes the losses within the circuit in such as way that the new circuit will not necessarily perform like the old.

At frequencies *above* the skin-effect threshold the AC resistance of a long, skinny object like a wire varies with three terms: *length*, *circumference*, and *frequency*. AC resistance varies proportional to length, inversely with circumference (not cross-sectional area), and because of changes in the skin depth, proportional to the square root of frequency. When you change all physical dimensions by k and simultaneously change the operating frequency by a factor of $1/k$, the length goes as k, the inverse of circumference goes as $1/k$, and the skin depth term goes as $\sqrt{1/k}$. The net result is that skin-effect resistance appears modified in the new circuit by the ratio $\sqrt{1/k}$.

The difficulties associated with resistive scaling of conductors can be fixed (theoretically) by building the new conductors from a new material, where the conductivity of the new material is reduced from the old by a factor of k. This substitution causes the resistance to remain unchanged (which is what you want) at any scale factor. While this theory is intellectually appealing, it does not appear to have much practical significance except for the idea that one could make 1:4 blowups of small copper structures using brass (which has 1/4 the conductivity of copper), and they would have exactly the same electrical properties as the original, including the percentage penetration of skin depth, and so on.

Fortunately, in many problems where this rule of scaling applies, the resistance hardly matters. For example, whenever you are working with low-loss elements (like the metal lead-frame on a semiconductor package), the resistance is vanishingly small in the first place, with the result that physical scaling changes mostly just the inductive and capacitive parameters. Thus, it is practical to manufacture a 1:100 enlargement of a portion of a chip package and directly measure the inductance and capacitance of the enlarged structure. The values of L and C corresponding to the IC structure will be precisely 100 times less.

The rule of physical scaling applies well to low-loss conducting structures like metal plates, conducting wires, connector pins, and semiconductor packages. This rule is valid over any range of physical scales for which useful conducting objects may be constructed. It breaks down for certain structures near the atomic level, for which the conducting surfaces cannot be scaled due to the inherent quantization of atomic matter. As far physicists know, this law applies to structures of galactic dimensions, although such structures have not been tested to verify conformance with the rule.

Examples of physical scaling

An SOIC package, being generally smaller than a DIP package, exhibits less lead inductance.

A big via has more parasitic capacitance than a small via.

1.3.1.1 Scaling Inductors

Given two different sized inductors built from the same material and having the same geometry, the larger of the two will have more inductance. At any fixed frequency f below the onset of the skin effect the resistance of the larger inductor will be less. At frequencies above the onset of the skin effect the resistances of the two inductors, determined by the ratio of the length to the circumference of their conductors, will be the same. In either case the quality factor (Q), being related to the ratio $j\omega L/R_{SERIES}$, will be markedly better in the larger inductor.

Unfortunately, the larger inductor will also have more parasitic capacitance, reducing its parallel-resonant frequency inversely with the scale factor.

1.3.1.2 Scaling Transmission-Line Dimensions

If we apply the rule of physical scaling to the problem of transmission lines, an interesting result appears. Imagine a printed-circuit board (pcb) trace having a certain length, width, thickness, and height above ground. Now scale all physical dimensions by the common factor k. According to the rule of physical scaling, the resulting structure will possess a total inductance (measured at low frequencies with the far end shorted) and capacitance (measured at low frequencies with the far end open-circuited) of k times the original values. Also note that the total length will now be k times the original length. Looking at the ratio of total inductance to length, note that their ratio is unaffected, because the total inductance value and the length value scale similarly. The same thing happens to the ratio of total capacitance to length: it is unaffected by physical scaling.

To make the next leap in this argument you need to recall that the impedance and per-unit-length delay parameters are functions of the per-unit-length values of inductance and capacitance for the line. If, as a result of scaling, you have failed to affect the per-unit-length values of inductance and capacitance, then you will also have failed to affect the impedance and per-unit-length delay.

This is the fascinating result: scaling the width, height, and thickness of a transmission line by a common factor k has no effect on the impedance or per-unit-length delay. You may have affected the resistive losses in a peculiar fashion, but not the impedance or delay.

This result applies to transmission lines operated in the LC mode, where the series inductance greatly exceed the series resistance, and the shunt capacitance greatly outweighs the effect of shunt conductance, conditions generally true in most pcb-trace problems.

POINTS TO REMEMBER

> ➤ Large objects have more inductance and capacitance than small ones.
> ➤ High-frequency connectors must be very small to push the parasitic resonances up to frequencies above the bandwidth of the signal.

> Simultaneously enlarging the height and width of a transmission line has no effect on the characteristic impedance or per-unit-length delay.

1.3.2 Power Scaling

In a linear circuit, one may amplify (or attenuate) the voltage and current levels at all points within a circuit simply by scaling all independent voltage and current sources (including fixed DC sources) by a common factor k. Assuming all impedances remain fixed, this adjustment multiplies all power levels within the circuit by the factor k^2. Dependent sources need not be adjusted, as they will automatically change their output levels in response to a change in stimulus.

In a nonlinear circuit, to obtain the same scaling effect, one must also scale all I-V curves for nonlinear elements according to the following rule: Each point in the (i,v) relation for the original element is mapped to a new point (ki, kv) representing the new element. Such scaling of the nonlinear properties may in general not be physically possible. The control laws for nonlinear controlled sources must also be similarly scaled. The resulting circuit will behave identically to the first, remembering that all voltages and currents will be modified by the factor k and power will be modified by a factor k^2.

This rule applies to all circuits. The possible range of scaling is limited on the low-power end by the quantization of electrical charge. That limit is reached in some advanced semiconductor memory storage circuits. On the high-power end power scaling is limited by the maximum current density that may be supported by the good conductors in the circuit and also by the maximum power density that may be dissipated by any resistive or semiconducting elements within the circuit. For example, within an integrated circuit at gate dimensions of 0.35 μm or less, the problem of aluminum atom migration commonly limits the total current that may be carried by a thin aluminum trace.[1] The aluminum migration problem, while severe at the physical scale of an integrated circuit, causes no difficulties at the scale normally used for pcb production.

For another example, the power density present at the active transmitting facet of a high-performance 1 mW data communications laser diode is greater than that of an electric oven-heating element on full broil. Attempts to scale up the power beyond this limit result mostly in melted laser diodes.

Example of power scaling

All other specifications being equal, 3.3V logic dissipates less than half the power of 5V logic.

POINT TO REMEMBER

> Lower-voltage logic is remarkably power-efficient.

---

[1] The passage of electrons at incredibly high current density though a tiny aluminum wire can actually knock some of the aluminum atoms out of place, leading to eventual failure of the conductor.

1.3.3 Time Scaling

This is one of the more interesting rules of scaling, as it indicates how a circuit must be modified to obtain high-speed operation.

In a *linear, time-invariant, passive* circuit, changing all inductances and capacitances by a fixed factor k, leaving the resistances fixed, scales the system step response in time. The new step response is the same as the old in every respect except for a scaling of the independent variable, time, by a factor of k. If the inductances and capacitances are increased, the new response is slower than the old. If the inductances and capacitances are made smaller, the new response is faster than the old.

Time scaling does not change the power, voltage, or current levels within a circuit. It merely stretches the scale of time for all voltage and current waveforms in the original circuit. This principle is the key to speeding up digital systems. Before we delve into a detailed example showing the time-scaling principle at work, let's look at some pesky details having to do with nonlinear and time-varying circuits.

If a passive circuit is nonlinear, and the nonlinear elements are assumed to be instantaneous in their actions (or a composite of instantaneous nonlinear effects combined with other linear, time-invariant circuits), the same result holds. No modification is needed to the I-V characteristics of the nonlinear components, although a scaling of the reactive elements internal to each nonlinear model is required. For example, if you wish to scale the basic SPICE level-1 model of an individual FET, the basic equations governing the conductance of the gate region need not be modified, but all parasitic inductances and capacitances must be scaled. Within a physical semiconductor structure such as a FET, other effects, such as the transit time for electrons drifting across the gate region, also require adjustment.

If the system includes time-varying voltage or current sources, a time dilation (or compression) of all the independent source voltages (or currents), in conjunction with time scaling of all other reactive elements in the circuit, will produce an exact time scaling of every other voltage and current in the system. Except for the time-scale factor, the new system will perform exactly k times more (or less) rapidly than the old.

Often a circuit includes time-varying impedances, like the totem-pole switching elements within a digital driver circuit. In this case a time dilation (or compression) of the independent *driving functions* for all the impedance variations, in conjunction with time scaling of all the other reactive elements in the circuit, will produce the same general time-scaling result. This is the form in which the time-scaling principle is most important within the realm of digital systems work.

To scale a transmission line in time you must modify the total transmission-line inductance and capacitance by a factor of k—meaning that the length of every transmission structure must be scaled by the factor k. Transmission-line impedance, being the square root of the ratio of inductance to capacitance, remains unchanged. Combining this requirement with the other ideas in this section, we arrive at a final statement of time scaling appropriate for use in digital systems work.

Beginning with a circuit composed of one or more digital drivers, some passive L, C, and R elements, and one or more transmission lines, you may scale the speed of operation by doing the following:

1. Scale lumped L and C components by the factor k,

1.3.3 • Time Scaling

2. Lumped resistances remain fixed,
3. Scale the delay of all distributed transmission elements by the factor k,
4. Transmission-line impedances remain fixed,
5. Scale all internal logic delays and timing specifications by the factor k,
6. Scale the rise/fall times of all logic drivers by the factor k,
7. Scale the timing for all external inputs by the factor k, and
8. Scale all clock intervals by the factor k.

The new circuit will behave identically to the first, but at a different scale of time (time t in the old circuit maps to time kt in the new).

The preceding rules apply to all circuits with deterministic inputs. There exist certain infinite-bandwidth random signals that cannot be so easily time-scaled. One example is white Gaussian noise. To see the difficulty, imagine white Gaussian noise applied to a low-pass filter. White noise goes in, and a random low-pass filtered signal comes out. In this circuit the output signal power is a function of the *equivalent noise power bandwidth* of the filter. If you scale the inductance and capacitance values within the filter by a constant k, the new bandwidth changes to $1/k$ times the old bandwidth, and the output power changes to $1/k$ times the old output power. That change of power violates the spirit of pure time scaling.

To rectify this situation, we are forced to conclude that the *spectral power density* of a white Gaussian noise signal should be scaled proportional to k. This problem is peculiar only to random signals with infinite bandwidth (and therefore infinite power). As soon as you specify a *particular* cutoff bandwidth for any random signal, you obtain a signal with finite total power. Consistent application of the scaling principle then requires that you scale the cutoff frequency of the noise signal proportional to $1/k$. Once the cutoff frequency is so scaled, you have in effect spread the same amount of power evenly over a bandwidth $1/k$ times the original, so you should see k times the power density concentrated within each spectral interval (i.e., the spectral power density works out to the correct value). For an infinite-bandwidth white noise signal, you achieve this same result by scaling the *spectral power density* of the source.

The principle of time scaling for noise signals points out that without modification of the spectral power density of the white noise at the input to a circuit, a slower circuit always exhibits less noise in its output than a faster one. This is an immutable principle of high-frequency circuit design. It's the reason that FDDI optical transceivers can operate at 125 Mb/s with a greater receive sensitivity than Gigabit Ethernet optical transceivers working at 1250 Mb/s.

Example of time scaling

The shorter the risetime, the shorter the permissible length of unterminated transmission line stubs.

As promised earlier, let's now look in detail at how the time-scaling principle is used. I shall pick a simple example, involving a bus, four transceivers, and some connectors. There is one common clock used on the bus, driven from the center. Suppose you wish to increase the operating speed by a factor of four, and you want to maintain precisely the same signal fidelity as in the original bus. The pure time-scaling approach requires that you should first obtain faster drivers, with faster rise/fall times and shorter setup-hold requirements, all scaled to 1/4 their original values. That much seems obvious.

You shall also have to shrink the bus delay by a factor of four. Unfortunately, there are only two means of accomplishing this goal: Either shorten the bus by a factor of four or implement the bus with a better dielectric material having a lower value of ϵ_r. If the bus is presently implemented in FR-4 material ($\epsilon_r = 4.5$), the alternate option alone is not sufficient to achieve your goal of a 4:1 speedup. Some shortening of the bus structure will be required.

The parasitic effects of the connectors will have to be scaled as well. According to the physical scaling principle, this could be accomplished by shrinking the connector geometry. In practice, the geometry may already be too small to permit further shrinkage, so one is forced to procure more expensive connectors with various features built in to ameliorate the parasitic effects.

If you are successful at simultaneously shrinking all the bus parameters together, the new bus will perform precisely as the old, only four times faster. Should you fail to fully reduce one of the parameters, you will have to make up the shortfall by overscaling one of the other factors.

Terminating the bus relieves you of the problem of having to scale absolutely everything. A properly terminated bus can be sped up without requiring a shortening of the structure, hence the popularity of terminations. In this case the signal waveforms will not be precisely the same as in the old bus, but as long as the bus lines settle adequately, the system still works.

POINTS TO REMEMBER

➢ Shrinking *every parameter* of an unterminated structure speeds its settling time in direct proportion to the scale factor.

➢ Terminated structures circumvent the link between physical size and signal quality.

1.3.4   *Impedance Scaling with Constant Voltage*

From time to time, you will want to modify the impedance of a circuit. This happens when you change the widths of pcb traces or when you wish to reimplement a long-distance communication system on a different style of copper cabling.

It is possible to change the impedance of a circuit without disturbing the shape of its voltage or current waveforms. This is done by first changing the impedance of every element in the circuit, and then changing the voltage and current excitations to the system to stimulate it in a manner analogous to the original circuit. I say *analogous* rather than *identical* because in this situation it is not possible to completely preserve both current and voltage waveforms. Either the voltage waveforms or the current waveforms (or some combination of both) must be modified as part of this procedure. I describe next the most common form of impedance scaling—impedance scaling with constant voltage.

In the constant-voltage method, all voltages in the new system will remain as they were in the old system. All currents in the new system will be scaled by the factor $1/k$, as compared to their values in the old system.

1.3.4 • Impedance Scaling

This is accomplished by first scaling all the impedances in the system. For linear impedances, we simply multiply the impedance value by a factor of k. This means that each resistance must take on a value that is k times as great as in the original circuit. Each inductance must do the same. Each capacitance in the new circuit must be scaled to $1/k$ times its original.

The procedure for constructing the new circuit is as follows:

1. Scale all resistances and inductances by the factor k,
2. Scale all capacitances by the factor $1/k$, in order that their impedances, $1/(2\pi f C)$, should be k times as great,
3. Scale the impedance of all distributed transmission elements by the factor k (this is a consequence of rules 1 and 2),
4. Leave all independent voltage sources the same, and
5. Scale all independent current sources by the factor $1/k$.

The new circuit will behave identically to the first, but with all currents scaled by $1/k$. This change in current implies that every instantaneous power level in the new circuit will be $1/k$ times the equivalent power in the old circuit. Changing the impedance can have a major impact on power consumption.

It is also possible to scale impedance while holding constant the currents. In this case the voltages scale with k, and power also scales with k. Since most digital logic operates with voltage-mode sources (that is, sources whose output impedance is much lower than the impedance of the receivers), the constant-voltage scaling law is the one most commonly used.

This rule as stated applies to all *linear* lumped-element circuits, whether they be time invariant or time varying. In a linear circuit the scaling of impedance does not affect the *ratios* of impedances—so no changes occur in circuit loss or signal fidelity throughout the system. For nonlinear circuits the situation is different. For a nonlinear system it is not easy to talk sensibly about an impedance ratio, because the dynamic impedance of a nonlinear element changes as a function of its bias current. For a nonlinear circuit to scale properly you would have to modify the I-V relation for every nonlinear device according to the law that every point i in the old relation is mapped to a new point i/k. In the case of many digital driver/receiver problems this condition is easily met simply by specifying a different size FET output cell, and the conclusions derived from the application of this principle are applicable; that is, increasing all circuit impedances while weakening the drivers will directly reduce power consumption without affecting signal fidelity.

If you set k less than unity, you find that a lower-impedance source coupled to a lower-impedance transmission line can drive larger capacitive loads.

This rule applies to *lumped* or *distributed* circuits for which the impedances are under your direct control and may be scaled. This rule does not apply in any practical way to problems involving radiation into free space, like antennas, for which it is not physically possible to directly modify the impedances related to unchanging physical factors like the permittivity of free space ϵ_0 or the magnetic permeability of free space μ_0.

Example of impedance scaling

The lower the impedance of a transmission line, the more loads it can bear.

POINT TO REMEMBER

> ➢ A lower-impedance source coupled to a lower-impedance transmission line can drive larger capacitive loads.

1.3.5 Dielectric-Constant Scaling

Increasing the dielectric constant surrounding a circuit slows its operation and decreases its impedance. For circuits that are entirely embedded within a uniform, homogeneous, dielectric material, increasing the dielectric constant by a fixed factor k increases all circuit capacitances by precisely k. This has the effect of increasing all transmission-line delays by a factor of $k^{1/2}$ and reducing all transmission-line impedances by a factor of $1/k^{1/2}$. Circuits partially surrounded by a high-dielectric-constant medium enjoy the same benefits, but to a lesser degree.

Certain circuits benefit from the slow-speed operation made possible by high-dielectric-constant circuit board materials. One prime example is a microwave resonator. An effective resonator may be formed from a section of pcb trace cut to a length of 1/4 wavelength. These sorts of resonant elements are commonly used for microwave oscillators, filters, couplers, and mixers. For resonator applications, the higher the dielectric constant of the material surrounding the resonator trace, the shorter that trace may be made. High-dielectric-constant materials therefore serve a useful miniaturization function in microwave circuits. In general, high-dielectric-constant substrate materials accentuate the distributed nature of pcb traces, causing them to exhibit transmission-line behavior at lower than normal frequencies.

Certain other circuits benefit from the lower impedance made possible by high-dielectric-constant insulating materials. The premier example is a bypass capacitor. The purpose of a bypass capacitor is to provide a low-impedance connection between power and ground. The higher the dielectric constant of the material used to separate its plates, the higher the capacitance and the lower its impedance (at all frequencies sufficiently far below the first series resonant frequency of the capacitor). High-dielectric-constant materials are desirable between the plates of lumped-element bypass capacitors.

Finally, many circuits decidedly do *not* benefit from the use of high-dielectric-constant materials. Pcbs used for high-speed digital computers belong to this class of circuits. In the digital application, delay is your enemy, not your friend. Even between the solid power and ground planes of a digital design, high-dielectric-constant materials rarely help. Some mixed-signal designs and multichip module (MCM) designs force the use of a high-dielectric-constant substrate material, such as alumina ($\epsilon_r = 10$). In those cases one usually implements narrower trace widths than would normally be the case with an FR-4 substrate ($\epsilon_r = 4.3$) in order to counteract the impedance-lowering tendencies of the high dielectric constant.

The rule of dielectric scaling applies to any circuit for which all conducting surfaces are in intimate contact with, and totally embedded in, a uniform, homogeneous dielectric material. The term *embedded* means that all electric fields associated with the circuit fall entirely within the surrounding dielectric medium. This rule is particularly applicable to stripline traces, and it is a useful approximation for microstrip behavior. In the interior of a capacitor whose plate geometry is small enough to act as a lumped-element circuit the scaling rule applies. In the interior of a capacitor whose plate geometry is too large to act as a lumped-element circuit the rule still applies; however, the inductive effects associated with the distributed nature of the circuit may interfere with your ability to directly measure an increase in total capacitance proportional to k.

> **Example of dielectric scaling**
>
> All other things being equal, changing from an FR-4 substrate ($\epsilon_r = 4.3$) to an Alumina substrate ($\epsilon_r = 10$) increases the raw, unloaded pcb trace delay by 52% and lowers the trace impedance by 34%.
>
> Changing to a lower dielectric constant will *increase* the characteristic impedance and *decrease* the delay. If you then broaden the line width to obtain the same characteristic impedance as in the original circuit you will find the skin effect losses are reduced.

1.3.5.1 Partially Embedded Transmission Lines

Many practical transmission-line structures lie only partially embedded within a dielectric material. For these lines the per-unit-length capacitance does not scale precisely with k. For example, imagine a stripline positioned very near the edge of a pcb. This stripline enjoys the benefits of a solid plane above and a solid plane below, confining most of the electric field energy surrounding the stripline to the dielectric medium between the planes. There are, however, some pathological paths for the electric fields that may be drawn, starting with the stripline, penetrating sideways through the dielectric medium to the outside air surrounding the board, and returning through the air to the planes on the top (or bottom) surface of the board. Obviously, these paths are not completely embedded within the dielectric medium. Some electric field energy will be stored along these paths, with the result that the effective electric permittivity experienced by the stripline will fall somewhat between the electric permittivity of the pcb dielectric medium and the electric permittivity of air. Air being the fastest dielectric medium known, you may conclude that a stripline positioned near the edge of a pcb will exhibit slightly higher impedance and slightly less delay than its more deeply embedded cousins. Its impedance and delay will also be somewhat less sensitive to variations in the dielectric material. Microstrip traces, because they lie on the outside surface of a pcb exposed on one side to air, exhibit this effect to an even greater degree. Striplines positioned well back into the interplane region, at least $20h$ away from the edge of the planes (where h is the trace height above the nearest plane), perform for all practical (digital) purposes as completely embedded striplines.

POINT TO REMEMBER

> ➢ Reducing the dielectric constant of your transmission-line substrate increases the characteristic impedance and decreases the delay.

1.3.6   *Magnetic Permeability Scaling*

Increasing the magnetic permeability surrounding a circuit slows its operation and increases its impedance. For circuits that are entirely embedded within a uniform, homogeneous, permeable material, increasing the permeability by a fixed factor k increases all circuit inductance by precisely k. This has the effect of increasing all transmission-line delays by a factor of $k^{1/2}$ and increasing all transmission-line impedances by a factor of $k^{1/2}$. Circuits partially surrounded by a highly permeable medium enjoy the same benefits but to a lesser degree.

This trick is used in the design of delay lines. One notable application appears in all analog color television receivers. Within the television receiver the *luminance* signal (the black-and-white information) has a bandwidth of approximately 6 MHz. Buried within that same band is a *chroma* signal. The chroma signal controls the relative intensities of the RGB electron guns. As received, the luminance and chroma signals are aligned in time. As these signals are processed inside the receiver, the chroma signal is extracted and decoded from the main signal by a set of filters having a bandwidth of approximately 500 KHz. These extraction filters delay the chroma information by roughly 500 ns relative to the luminance. A matching delay is then applied to the luminance signal to realign it with the chroma information prior to driving the RGB guns. The delay in older television sets was nothing more than a big, slow transmission line. It was typically implemented in the form of a wire helically wrapped around a long stick of material with high magnetic permeability. The highly permeable material slows the propagation time of the transmission line, thereby shrinking its physical length to a practical and manufacturable size. By whatever factor the delay is increased, so also goes the transmission-line impedance. This helps raise the impedance up into a range easily driven by analog tube circuits. (Use of a high-dielectric-permittivity material in this case would have lowered the impedance to an unusable extent).

Without the help of a highly permeable material, a comparable air-dielectric transmission line would have to be some 500 feet in length.

POINT TO REMEMBER

➢ Adjustments to magnetic permeability are rarely made in digital circuits.

1.4 THE CONCEPT OF RESONANCE

Atoms resonate. Oceans resonate. Your vocal chords resonate. The Tacoma Narrows bridge (Figure 1.2) used to resonate in high winds. Vibrating, periodic phenomena abound in nature, especially within high-speed digital electronics.

The technical requirements for resonance are simple. First, there must be two or more reservoirs for the storage of some amount of energy E along with a relatively loss-free mechanism for exchanging the energy between them. Such a system almost always harbors one or more modes of oscillatory behavior whereby the total stored system energy can slosh back and forth among the various storage modes.

Second, there must be a source of external power coupled into (at least) one of the reservoirs. The coupling mechanism must be capable of delivering power P into the system without unduly disturbing the resonant behavior.

Third, the external stimulation must excite the periodic resonant behavior you wish to excite. Otherwise, the system won't start resonating. In technical terms, this means that for linear systems the spectral power density of the source must overlap with the natural

1.4 • Resonance

Figure 1.2—The Tacoma Narrows bridge in Washington State, known locally as "Galloping Gertie," heaving under the influence of heavy winds.

resonant frequency (or frequencies) of the system. For nonlinear phenomena, this spectral condition is usually helpful but not always required (see [2]).

Resonant systems may be characterized by their quality factor, Q. The Q measures the reluctance of a system to give up its naturally stored energy. A system with high Q tends to resonate, or ring, for many cycles. A system with low Q damps quickly.

The value of Q is technically defined as the ratio of energy stored to the average energy dissipated per radian of oscillation. The range of Q values typically encountered in digital electronics ranges from less than unity (no resonance) to more than 10^6 (a highly resonant crystal oscillator).

Let's do a specific example. Consider the simple RLC resonator depicted in Figure 1.3. It is representative of a short, unterminated pcb connection with a heavy capacitive load at the receiver. The resistor R_{SERIES} represents the output impedance of the driving gate. In

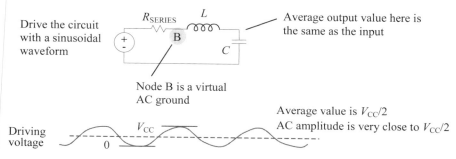

Figure 1.3—A short, unterminated pcb trace with a heavy capacitive load resembles this simple LC series-resonant circuit.

such a circuit, for a single resistance in series with the resonant elements, the Q equals

$$Q_{\text{SERIES}} = \frac{\sqrt{L/C}}{R_{\text{SERIES}}} \qquad [1.11]$$

We'll assume in this example the Q is greater than one. Drive this circuit with a periodic waveform, and let the sine wave frequency match the natural frequency of oscillation. Assume the driver rise/fall times are just fast enough to perform this function, so the driving

> ### A Child's Playground Swing Set
>
> A child's swing set possesses at least two main reservoirs for energy storage. The first reservoir is the gravitational potential energy of the child in proportion to the child's height above the earth and also in proportion to the child's mass. At the peak of each swing cycle, when the child is lifted maximally above the ground and the swing momentarily stops moving, most of the energy in the system is stored as the gravitational potential energy of the child's mass.
>
> The other reservoir is the kinetic energy of the child's motion in proportion to the square of the child's velocity and also in proportion to the child's mass. At the nadir of each cycle, when the child is nearest the ground and moving the fastest, most of the energy in the system is stored as the kinetic energy of the child's motion.
>
> The chains and supporting structure of the swing provide a low-loss mechanism for energy to flow back and forth between these two modes as the child swings up, down, and back again. A well-designed swing will exhibit a natural resonant frequency in the neighborhood of 0.25 Hz.
>
> To operate the system in a continuous fashion, there must be a source of power. Presumably, this is your arm (or maybe your foot). Look carefully at this interaction. Once you get the child in motion, *you need to touch the child for only a small fraction of each cycle*. This is of prime importance to the operation of the system. It prevents your presence from unduly disturbing the resonant behavior. Were you to grab hold of the swing continuously, running back and forth along with the child, the swing would merely follow the motion of your arms. If you had arms like Arnold Schwarzenegger, it would be difficult if not impossible to detect any resonant behavior in the swing itself. Technically speaking, in that circumstance your arms would have overpowered the natural forces within the system, defeating the resonance. Highly resonant systems are usually weakly coupled to their driving elements.
>
> The last point worth noting in this example is that (for linear systems) the frequency of excitation must reasonably match the natural frequency of oscillation of the swing. When my daughter was 3 years old, she was strong enough to push a swing but did not comprehend that it mattered at what rate the swing was pushed. Her early attempts to entertain her playmates were met with the disappointment of occasionally getting bashed by the swing. Today, at the age of 10, she intuitively grasps the idea that to maximize the power delivered to the swing, the rate of excitation must match the rate of natural oscillation. If she grows up to learn calculus, I shall be able to explain to her why this must be so.

1.4 • Resonance

signal looks pretty sinusoidal. From these conditions we may derive the output amplitude of the voltage across the capacitive load. It may surprise you to find that the output amplitude can grow quite large.

To begin the analysis, break down the driving signal into two components: a steady-state DC offset and a sinusoid at frequency f. The amplitude of the DC offset is $V_{CC}/2$ (that's the average value of the output). The amplitude of the fundamental frequency is also $V_{CC}/2$. When added to the DC offset, this brings the peak excursions just up to V_{CC} and down to zero. We will do the DC and AC analyses separately.

The DC analysis for this circuit is easy. Since input and output are connected with an inductor, the average value of the output must equal the average value of the input. Therefore, the DC component of the output waveform is precisely $V_{CC}/2$.

The AC analysis for this circuit is not too hard either. It may be derived using an energy-balance equation. This equation will balance the energy delivered to the circuit during one radian of oscillation with the energy dissipated within the circuit over the same period.

Once the circuit has reached steady-state operation, the power delivered to the circuit per radian of oscillation is $P/2\pi f$. The amount of energy dissipated per radian of oscillation (from the definition of Q) is E/Q. These two quantities must balance:

$$\frac{P}{2\pi f} = \frac{E}{Q} \qquad [1.12]$$

where P is the power delivered to the circuit,

E is the total energy stored within the circuit, and

Q is the quality factor for the resonator, equal to $\sqrt{L/C}/R_{SERIES}$.

You are driving the circuit at its resonant frequency, and you know that the input impedance of a perfect series LC resonator operated at its resonant frequency is zero. Therefore, the AC voltage at point (B) in the circuit must be zero. That implies that the full AC driving voltage of $\pm V_{CC}/2$ must appear across resistor R_{SERIES}. The power P delivered to the circuit must therefore equal $\frac{1}{2}(V_{CC}/2)^2/R_{SERIES}$.

We also know that the total energy E sloshes back and forth between the inductor and the capacitor. At the peak of the output voltage, all the energy is retained within the capacitor, and none in the inductor (the current through L at this moment is zero). If the AC component of the output swings from $+V_{OUT}$ to $-V_{OUT}$, the energy E stored within the capacitor at the peaks must equal $\frac{1}{2}CV_{OUT}^2$.

Substituting all into equation [1.12], we can now write an energy-balance equation in terms of the AC input amplitude $V_{CC}/2$ and the AC output amplitude V_{OUT}.

$$\frac{\frac{1}{2}(V_{CC}/2)^2/R_{SERIES}}{2\pi f} = \frac{\frac{1}{2}CV_{OUT}^2}{Q} \qquad [1.13]$$

Rearranging terms,

$$\frac{V_{OUT}^2}{(V_{CC}/2)^2} = \frac{Q}{2\pi f R_{SERIES} C} \qquad [1.14]$$

Recognizing that $2\pi f = 1/\sqrt{LC}$,

$$\frac{V_{OUT}^2}{(V_{CC}/2)^2} = \frac{Q\sqrt{LC}}{R_{SERIES} C} \qquad [1.15]$$

Combining the C terms on the right-hand side,

$$\frac{V_{OUT}^2}{(V_{CC}/2)^2} = \frac{Q\sqrt{L/C}}{R_{SERIES}} \qquad [1.16]$$

Finally, recognizing for a series-resonant circuit the definition of $Q = \frac{\sqrt{L/C}}{R_{SERIES}}$,

$$\frac{V_{OUT}^2}{(V_{CC}/2)^2} = Q^2 \qquad [1.17]$$

And taking a square root of both sides,

$$\frac{V_{OUT}}{(V_{CC}/2)} = Q \qquad [1.18]$$

We see that the amplification factor for this circuit, that is, the ratio of the AC output amplitude to the amplitude of the AC fundamental in the driving signal, equals the circuit Q. Therein lies the importance of Q. It defines (roughly) the amount of amplification that may occur within a resonant circuit. Note that the complete output waveform is a superposition of the DC response $V_{CC}/2$ and the amplified AC signal. The preceding analysis applies only when $Q \gg 1$.

Now let's move on to a different problem. Figure 1.4 provides some useful approximations for the maximum degree of overshoot expected when a resonant circuit is driven by a *step input*.

As illustrated in Figure 1.4, the higher the Q, the greater the overshoot. For Q of 1/2 or below the circuit is over-damped and no visible overshoot occurs. Values of Q greater than 1/2 create overshoot. These approximations apply to any resonant circuit that has unit response at DC and is possessed of only one main resonance, at a frequency well below the knee frequency of the driving step source, with a Q greater than one.

An excessive value of R_{SERIES} pushes the Q below 1/2, giving the circuit a sluggish, overdamped response. The overdamped response in such a circuit is dominated by the action of the resistor R_{SERIES} charging the capacitor. Given particular values of L and C, the best

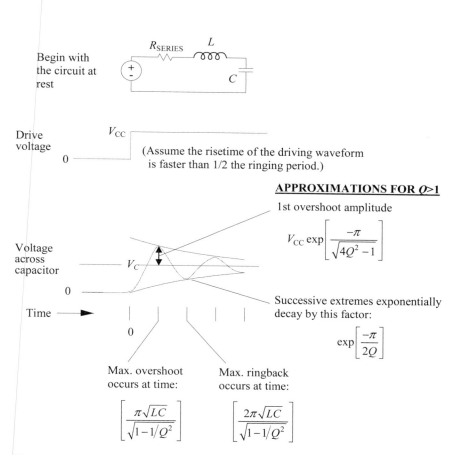

Figure 1.4—The higher the Q, the greater the overshoot.

choice of R_{SERIES} for digital applications is usually whatever value forces Q closest to unity. That gives the best risetime without undue amounts of overshoot. In other applications, such as the feedback filter for a control system, one might strive for a Q somewhat below 1/2.

Occasionally you will encounter an LC tank circuit driven by a current source and loaded by some parallel resistance. For a single resistance in parallel with the resonant elements, the Q equals

$$Q_{PARALLEL} = \frac{R_{PARALLEL}}{\sqrt{L/C}} \qquad [1.19]$$

For any LC resonator that includes series and parallel elements, the inverses of the Q's add

$$Q_{TOTAL} = \frac{1}{(1/Q_{SERIES}) + (1/Q_{PARALLEL})} \qquad [1.20]$$

Distributed transmission lines also harbor resonant modes. As in all resonance problems, they are caused by multiple reservoirs of energy coupled together with a low-loss mechanism.

To understand the energy reservoir formations along a transmission line, imagine a line driven with a specific risetime t_r. Every point p_n along the line is surrounded by a quantity of dielectric medium that serves as a storage repository for electric field energy. It is also surrounded by a magnetically permeable medium that serves as a storage repository for magnetic field energy. In circuit terms, the voltage and current at point p_n form two independent circuit variables, each of which can serve as an energy storage reservoir. Therefore, every transmission line has at least *two* energy storage reservoirs.

Furthermore, if two points p_n and p_m along the line are separated by a delay greater than t_r, then over a scale of time commensurate with one rising edge, they each represent distinct, uncoupled, independent energy storage reservoirs. The longer the line, the more independent reservoirs exist and the more possibilities develop for resonant behavior.

Common strategies for combating resonance in transmission lines are threefold. Combinations of the three techniques are often employed.

1. Shrink the line until it is very short, with a delay much shorter than a rising edge. This couples together all the voltages and all the currents along the line so that they can no longer independently vary without influencing each other. Any reasonable real-valued impedance at one end of the line will then fix the global relationship between the voltage and current, nailing them inextricably together. If done properly, the line will be reduced to a system that for all practical purposes possesses only one energy reservoir and therefore cannot resonate.

2. Terminate the transmission line at one end, the other, or both, using a resistance equal to the characteristic impedance of the line. The termination absorbs part of the power within line, effectively reducing the circuit Q to a very low value. Low-Q circuits cannot resonate.

3. Add sufficient distributed loss along the transmission line to control the Q. This can be done by selecting the conductor geometry and materials such that the skin-effect resistance is sufficient to do the job or by manipulating the dielectric losses of the substrate.

POINT TO REMEMBER

> Low-Q, dissipative circuits can't resonate. This is a desirable feature for a digital transmission path.

1.5 EXTRA FOR EXPERTS: MAXIMAL LINEAR SYSTEM RESPONSE TO A DIGITAL INPUT

The remainder of this chapter I shall fill with a theorem from linear system theory concerning the worst-case response of linear time-invariant systems when driven by digital

1.5 • Maximal Linear System Response

signals. This theorem is useful for evaluating, among other things, worst-case crosstalk voltages and worst-case power supply noise voltages. The theorem begins with a rather abstract situation that reduces in the end to a more familiar digital situation.

The mathematical model underlying the theorem appears in Figure 1.5. This figure shows an ideal source, producing a waveform $\xi(t)$ with an output limited to the range of $\pm A$. No other requirements are placed upon $\xi(t)$. It may have infinitely fast rise and fall times, it can transition at any time (not limited only to clock intervals), and it may attain intermediate voltages within the range of $\pm A$. The signal $\xi(t)$ is a *superset* of the class of all digital signals.

The ideal source output is then presented to a linear, time-invariant system with transfer function $H(f)$, corresponding impulse response $h(t)$, and step response $\tilde{h}(t) = \int_{-\infty}^{t} h(v)dv$. The output of the linear, time-invariant system is signal $y(t)$.

The theorem states a bound Y on the amplitude of $|y(t)|$ that holds for all possible signals $\xi(t)$. Furthermore, this is a constructive theorem in that it also develops at least one particular function $\xi(t)$ that is guaranteed to excite $|y(t)|$ to its maximum value. The bound Y on the output amplitude is

$$Y = A \int_{-\infty}^{+\infty} |h(t)| dt \qquad [1.21]$$

One signal $\xi(t)$ that will cause the bound to be reached is

$$\xi(t) = A \cdot \mathrm{sgn}(h(-t)) \qquad [1.22]$$

where the function sgn() is the sign function, $\mathrm{sgn}(x) \triangleq \begin{cases} 1 & \text{if } x \geq 0 \\ -1 & \text{otherwise} \end{cases}$

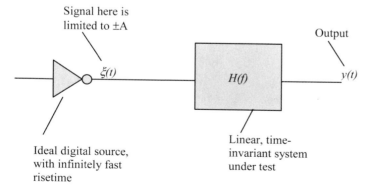

Figure 1.5—The maximum output from system H is bounded by equation [1.21]

The proof of this solution follows from an examination of the convolution integral for linear, time-invariant systems. The output $y(T)$ at time T in response to any excitation is thus defined:

$$y(T) = \int_{-\infty}^{T} h(T-t)\xi(t)\,dt \qquad [1.23]$$

If we wish to maximize the output at any particular time T, this integral grants us enormous flexibility to do so. We are free to assign whatever values to $\xi(t)$ we choose, at every point t in time, provided that no value may fall outside the range $\pm A$. Note that the various values of $\xi(t)$ each are used but once inside the integral, with no complicating constraints. They may all be independently assigned. Therefore, the way to maximize the integral is simply to maximize the integrand for each value of t.

For example, the best choice for $\xi(0)$ would be whatever value maximizes its elemental contribution to the integral, $h(T-0)\xi(0)$. If the algebraic sign of $h(T)$ is positive, choose $\xi(0) = +A$ (the largest possible positive value). If the algebraic sign of $h(T)$ is negative, choose $\xi(0) = -A$. That's all there is to it.

In general, the values for $\xi(t)$ that maximize y at time T are $\xi(t) = A\,\mathrm{sgn}(h(T-t))$. In the special case where you wish to maximize the output at time $T = 0$, equation [1.22] results. Equation [1.21] results from inserting solution [1.22] into equation [1.23], evaluating the result at $T = 0$, and recognizing that $h(t)\,\mathrm{sgn}(h(t)) = |h(t)|$.

Similar reasoning holds in the reverse algebraic direction. To minimize y, use $\xi(t) = -A\,\mathrm{sgn}(h(T-t))$. For a linear system with symmetrical input (no DC offset), the worst-case positive and negative amplitudes are equal.

Let's extend this theorem to treat some special forms of impulse response. For example, what if the impulse response is everywhere positive? In that case the particular signal that maximizes the response is simply $\xi(t) = A$, a full-valued, steady-state input. This says that systems with purely positive impulse responses are special. In response to a bounded input, their outputs can never exceed the steady-state maximum output. In other words, they have *no overshoot*.

The condition $h(t) \geq 0$ may also be expressed in terms of the step response. The equivalent condition imposed on the step response is that the step response be monotonic. These three properties, therefore, always go together:

> Nonnegative impulse response
> Monotonic step response
> No overshoot in response to bounded input

What if the impulse response goes negative, but just for a short interval? That would place a dip in the step response, a short interval of nonmonotonic behavior. What happens to the maximum output signal in this case? I'd like to show you this answer and express it in terms directly observable on the step response waveform. To do so requires a little more calculus.

The answer may be found directly from equation [1.23], substituting our known solution for $\xi(t)$, choosing to maximize the response at point $T = 0$, and substituting a new time variable $u = -t$.

1.5 • Maximal Linear System Response

$$Y = y(0) = \int_0^{+\infty} h(u) A \operatorname{sgn}(h(u)) du \qquad [1.24]$$

Suppose now that you are given a list of the intervals of time $[pstart_n, pend_n]$ over which $h(u)$ were positive, and another list of intervals $[qstart_m, qend_m]$ over which $h(u)$ were known to be negative. You could then rewrite the integral as a sum of component integrals:

$$Y = \sum_n \int_{pstart_n}^{pend_n} h(u) A \operatorname{sgn}(h(u)) du + \sum_m \int_{qstart_m}^{qend_m} h(u) A \operatorname{sgn}(h(u)) du \qquad [1.25]$$

By the construction of the *p*-intervals, the value of $A \operatorname{sgn}(h(u))$ within them is always A, so that factor may be pulled out from under the integral. Similarly, over all the *q*-intervals the value of $A \operatorname{sgn}(h(u))$ is $-A$, so that factor may be pulled out as well. The resulting expression looks like this:

$$Y = \sum_n A \int_{pstart_n}^{pend_n} h(u) du + \sum_m (-A) \int_{qstart_m}^{qend_m} h(u) du \qquad [1.26]$$

Now you have a series of ordinary integrals of the impulse function of a linear system. Express these integrals in terms of differences between various points on the step response $\tilde{h}$:

$$Y = \sum_n A \left[\tilde{h}(pend_n) - \tilde{h}(pstart_n) \right] + \sum_m (-A) \left[\tilde{h}(qend_m) - \tilde{h}(qstart_m) \right] \qquad [1.27]$$

If you have followed me so far, you will note that each of the terms within the left-hand summation is positive. Each of the terms within the rightmost summation is positive as well, by virtue of having been multiplied by $-A$. I can therefore eliminate all references to the sign of A by using the absolute value symbol:

$$Y = A \left\{ \sum_n \left| \tilde{h}(pend_n) - \tilde{h}(pstart_n) \right| + \sum_m \left| \tilde{h}(qend_m) - \tilde{h}(qstart_m) \right| \right\} \qquad [1.28]$$

What this equation tells you to do is to first mark out all the maxima and minima on the step response. These are the points where the slope (the impulse response) changes sign. The take the *absolute value* of the successive differences between each of these extrema and add them up ($|\Delta_1|+|\Delta_2|+|\Delta_3|\ldots$). Multiply by A and you have your answer. That's the biggest output you will ever see in response to a bounded input (Figure 1.6).

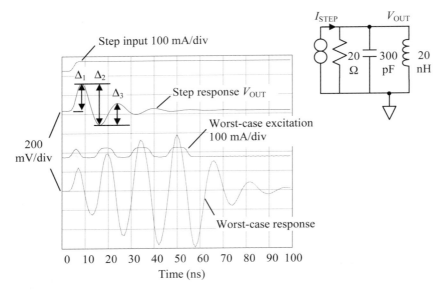

Figure 1.6—The response to a bounded excitation can grow larger than step response.

Applying this theory to a real digital system takes a few more manipulations. First, we need to treat the issue of DC offset, because the voltages 0 and V_{CC}, not $\pm A$, generally bound a digital signal. Finally, we need to treat the issue of finite risetime.

The DC offset issue easily succumbs to a superposition analysis. To generate a solution, break the input signal into two components, a DC offset ($V_{CC}/2$) and an AC signal (with amplitude $A = V_{CC}/2$). Use your intuition to derive the DC steady-state output (assuming the driver constantly put out $V_{CC}/2$). Use the maximal value theorem [1.22] to tell you how much the output may range above or below the DC steady-state output.[2]

The mathematical model used to solve the finite-risetime part of the problem appears in Figure 1.7.

Here the signal $\xi(t)$ is put through a low-pass filter function $G(f)$ whose purpose is to round off the (possibly) square edges of $\xi(t)$ such that $x(t)$ will never display a transition faster than that allowed by the step response of the filter $G(f)$. For a suitable choice of $G(f)$, the space of all signals available at $x(t)$ includes pretty nearly all the things that could come out of a digital driver with rise and fall time t_r. You will want to set up the filter $G(f)$ to have a monotonic step response, and unity DC gain, in order that the excursion limits at x match those at ξ.

As a practical matter, what I do is make a step response measurement from the system under test *using the system's own driver as the step source*. The system driver therefore implements the function $G(f)$ for me. I then mark off the extrema of the step response, measure the magnitudes of the inter-extreme excursions, sum their values, and divide by two (I measured the step response with a driver with range [0, V_{CC}] for which $A = V_{CC}/2$; ergo, I

---

[2] This use of superposition works only for *linear* systems.

1.5 • Maximal Linear System Response

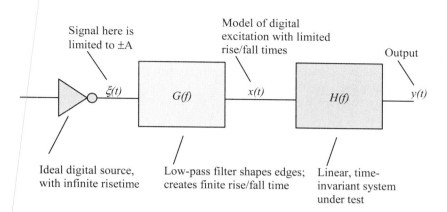

Figure 1.7—The magnitude of the largest possible response at *y(t)* is a function of the step response of the combined filter *G(f)H(f)*.

scale the measured output by 1/2). The result tells me how much the output may range above or below the DC steady-state output.

This theorem applies to linear, time-invariant systems. It does not apply to limited, clamped, or nonlinear systems, or to systems that change significantly during the time within which the output is created.

It does, however, apply reasonably well to systems with nonlinear totem-pole drivers provided that the driver after each rising edge rapidly attains an output near V_{CC} before the first reflections return from the end of the line. Under this condition the behavior of the driver *during the time the signal reflects off the driver* is indistinguishable from a simple, linear source connected to V_{CC}, so the theorem applies. The same reasoning applies to drivers that rapidly pull low, attaining a near-zero state prior to the arrival of the first reflections from the far end of the line.

The space of candidate functions ξ used in the construction of the maximal solution includes *all* functions that slam back and forth between the given limits, without regard to the *times* at which they may do so. This space is a *superset* of the collection of all synchronous waveforms clocked at some fixed rate. Therefore, the upper bound *Y* is just that, an *upper bound* on the maximal excursion. It is possible in some situations that the maximal solution function ξ and its corresponding filtered version *x* cannot be produced by the target system and are therefore of little or no concern in worst-case analysis. This can happen in systems that have risetimes much shorter than the baud interval. In these cases it's a good idea to examine the waveform ξ so you can decide whether or not to worry about it.

In the design of high-speed LAN communications systems I have found this theorem to be of great value. In LAN systems the risetime and the baud interval are closely related, and the difference between the calculated *Y* and the actual, real-world upper bound is negligible.

This discussion highlights an important point about crosstalk in cable systems—the long, lingering, diddling crosstalk signals that show up in response to a step input can, if excited with a worst-case input, easily be made to build up to significant voltages. Equation [1.28] predicts the worst-case crosstalk response from a single measurement of the step response of the system.

Equation [1.28] is also of great benefit in predicting the worst-case response of a power system to random surges of current. From a single measurement of the step response of a digital system (swinging from minimum current to maximum current in one quick step) equation [1.28] predicts the worst-case power-supply excursion. This result will always meet or exceed the excursions computed based upon simulations of the power system response using sinusoidal or square-wave excitations.

POINT TO REMEMBER

> Resonance affects all physical structures, including bridges (Figure 1.8).

Figure 1.8—The Tacoma Narrows bridge collapsed on July 1, 1940.

For further study see: www.sigcon.com

CHAPTER 2

TRANSMISSION LINE PARAMETERS

Metallic interconnections reign supreme for the conveyance of both data and power within digital systems. This chapter addresses the parameters of all metallic interconnections. Although this chapter is oriented towards the use of copper media, the same theory with slight modifications to account for the difference in conductivity also applies to interconnections built from aluminum, steel, nickel, and other metals. It may also be applied to hydrocarbon-based conductors and polysilicon, although in those cases, due to the high resistance of the conductors, generally the RC mode of propagation will predominate.

The following chapters concentrate on four types of copper-based transmission media used for high-speed data transmission (Table 2.1). As explained in the chapters that follow, the performance of each cabling type varies as a function of its length, quality, and other factors. Nevertheless, certain key features noted in the table stand out in a gross comparison of their capabilities.

Unshielded 100-Ω twisted-pair (UTP) cabling is the best choice for high-volume, cost-sensitive LAN applications. UTP delivers an aggregate bandwidth (all four pairs) of 1 Gb/s at 100 meters, while meeting FCC and EN emissions requirements. The connector and cross-connect technologies for this style of cable are mature and available at low cost, in large measure because the ISO 11801 generic building wiring guidelines (see Chapter 7, "Generic Building-Cabling Standards") officially sanction UTP cabling. In high-volume applications, UTP is ideal.

Versions of UTP cabling are available with exterior shields (called screens). These configurations are popular in Europe.

In low-volume applications, manufacturers sometimes choose not to undertake the development of UTP transceivers because of the complexity and risk associated with the mixed-signal technology required to fully exploit the benefits of UTP. Low-volume applications have historically selected coaxial or 150-Ω shielded twisted-pair (150-Ω STP-

Table 2.1—Popular Transmission Media for Digital Applications

Cable type	Signals per cable	Best features	Worst features
UTP	4	Useful up to 250 Mb/s per pair; connectors are inexpensive; new types constantly being developed.	Significant mixed-signal technology is required for high-speed use.
150-Ω STP-A (IBM Type I)	2	Useful up to 1 Gb/s per pair; interfaces directly with differential transceivers.	Bulky and extremely difficult to install.
Coax	1	Useful beyond 1 Gb/s; interfaces directly with high-speed digital logic.	Not standardized for building wiring; difficult to install.
Pcb traces	1	Useful to 10 Gb/s and beyond.	High-frequency losses severely limit attainable distances.

A) cabling because it is easier to develop transceivers for those cables, even though the ultimate per-unit cost of the cabling and connectors are higher.

One hundred-fifty-Ω STP-A is rated for single-pair operation up to 1 GHz. Designing a 150-Ω STP-A transceiver is a snap. The 150-Ω STP-A cable is well enough balanced and well enough shielded that it can directly accept signals from a high-speed differential driver and still pass FCC and EN emissions requirements. The primary disadvantages of 150-Ω STP-A are its extreme bulkiness and the difficulty of installation. Like UTP, 150-Ω STP-A was originally sanctioned by the ISO 11801 generic building wiring guidelines, but support has been withdrawn in favor of newer category 5E, 6, and 7 twisted-pair wiring standards. Additional information about the use of UTP and 150-Ω STP-A cable styles appears in Chapter 7.

Coaxial cabling is the simplest means of interconnecting two systems and potentially delivers the highest bandwidth. Coaxial cabling is also generally the poorest performing in terms of radiated emissions. Above 100 MHz, data scrambling has been used to guarantee that ordinary cables will not radiate in excess of FCC or EN limits.[3] The primary disadvantages of coaxial cabling are the difficulty of installation, the lack of LAN industry standardization for the manufacture of coaxial cables, and the lack of standard tests for installation compliance. In the LAN market, coax is dead. In the television and audio/visual markets, however, coaxial cable is still very much in use.

Printed-circuit board (pcb) traces, of course, are used for extremely high-frequency connections within a single pcb or on back-planes connecting one or more pcbs.

All metallic transmission media share the performance model described in Chapter 3.

---

[3] Unscrambled transmission systems radiate horribly because simple repetitive structures within the data stream, like the idle pattern, tend to concentrate all their radiated power at harmonics of the basic pattern repetition rate. These concentrated harmonics then leak from the coaxial cable, where they are easily detected by FCC or EN test antennas. In contrast, scrambled transmission systems spread their radiated power across a wide frequency range, limiting the peak radiation in any one radio-frequency band.

2.1 TELEGRAPHER'S EQUATIONS

Figure 2.1 illustrates a variety of possible transmission structures. Popular configurations for long-distance digital communications include various types of coaxial cables and twisted-pair cables with and without shielding. Pcbs today most often use either microstrip or stripline traces. Transmission-line interconnections built on-chip typically use structures that resemble the microstrip, with either polysilicon or metal layers used for the conductors and a doped semiconductor substrate used as the underlying solid return path.

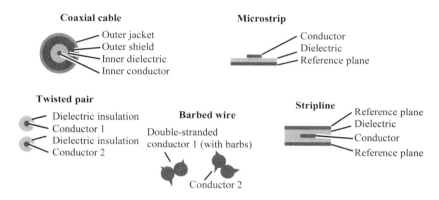

Figure 2.1—Useful transmission-line structures are available with a variety of cross sections.

Surprisingly, you can model all these media with the same equations. Provided there exists a well-defined uniform path for the flow of both signal and return current, and further provided that the conductors are closely spaced in comparison to the wavelength of the signals conveyed, the *telegrapher's equations* accurately model the propagation of electrical currents and voltages along the structure. The telegrapher's equations were developed in the late 19th century to explain the behavior of high-speed signals on long telegraph lines *and they still apply today*.[4]

A typical signal-transmission application comprises one or more long, uniform transmission lines hooked together with certain source and load impedances representing connectors and packages. To properly model such a system, you use the telegrapher's equations to create a two-port circuit model of each independent, long transmission structure, and then you combine the two-port models to determine the behavior of the entire system. The use of two-port analysis to predict overall system behavior is discussed extensively in Appendix C, "Two-Port Analysis." This chapter concentrates exclusively on the performance of one individual uniform transmission structure.

In general the telegrapher's equations may be successfully applied to any transmission media having the following characteristics:

> At least two conductors, insulated from each other,

> Having a uniform cross section along the entire length of the structure,

---

[4] In the 19th century "high-speed" Morse-code transmissions were accomplished in a batch processing mode at roughly 100 bits per second using automatic paper-tape transmission and recording devices.

> With a cross-sectional geometry small compared to the wavelength of the signals conveyed, and
> A length long compared to the spacing between the conductors.

Any such structure, if properly terminated, may be used to convey high-speed data. One example of an unusual transmission structure is a railroad track. This works only in dry areas of the country where the tracks are not shorted together by ground moisture. The tracks have to be designed to maintain electrical continuity along their entire length.

A fascinating example of such a track is the Bay Area Rapid Transit (BART) system built in the early 1980's in the San Francisco region. This system uses light-gauge railroad tracks held off the ground on concrete pylons to avoid ground moisture. In the early versions of the system, digital control data was transmitted through the rails in a differential mode from the master control center to each train. The trains picked up and responded to the control signals through their metal wheels. The axles on each car were of course insulated from side to side to prevent shorting out the tracks. This system worked reasonably well until, after a few months of operation, the rust buildup on the tracks began to interfere with the rail-to-wheel electrical contact, producing intermittent behavior that foiled the system.

Another interesting transmission structure is a barbed-wire fence. The wires have to be supported on naturally insulating wooden fence poles, or metal poles with plastic insulators, to avoid shorting all the wires together. Such a structure might incorporate multiple conductors running perfectly parallel for miles. It might surprise you to know that practical communications systems working on barbed wire have been demonstrated at speeds as high as 1000 Mb/s (see "So Good It Works on Barbed Wire" in this chapter).

The reason the telegrapher's equations are so powerful and so widely used lies in the

---

TEM Wave Configuration

The assumption of independence between the elements turns out to be quite important. If the signal leapfrogs, or jumps, directly from one element to another element without passing directly through the continuum of elements in their natural order, then the cascading equations used to build the telegrapher's model do not work. The telegrapher's equations only work for systems that don't have leapfrogging.

The assumption of independence between sections is equivalent to an assumption that the electromagnetic fields from one section do not interfere with subsequent sections. In the limit as the length of each segment is reduced, this principle of non-interference suggests that the lines of electric and magnetic flux from each segment must stick out sideways in a flat plane perpendicular to the direction of signal flow. They are not permitted to have any forward-and-back (longitudinal) components. This field configuration is called a transverse-electric-and-magnetic (TEM) pattern.

The telegrapher's equations in their basic form work only for transmission structures that support a simple TEM field pattern. Such structures include coaxial cables, pcb striplines, and all other configurations of two or more isolated conductors having a uniform cross section, a homogeneous dielectric, and a conductor spacing much smaller than the wavelength of the signals conveyed.

2.1.1 • Barbed Wire

simple but flexible assumptions upon which they are built. In essence, the telegrapher's equations assume any transmission line may be modeled as a succession of small, independent elements, each with a transverse-electric-and-magnetic (TEM) wave configuration (see "TEM Wave Configuration" box). Each element represents a very short length of the line. Because each element is short, its performance is simple to describe. The telegrapher's equations then mathematically model the complete line as an infinite cascade of these short elements, taken in the limit as the length of each element approaches zero and the number of elements approaches infinity.

The telegrapher's equations model each short element of the transmission structure as a combination of two quantities (Figure 2.2):

➢ An impedance z in series with the signal-and-return current, and
➢ An admittance y shunting the signal conductor to the return conductor.

Once you know z and y, the telegrapher's equations give you the input impedance and frequency response of the complete transmission structure. From those values you may compute the system response to any digital input.

For a typical transmission structure the series impedance z comprises the series resistance of the signal conductor, the series resistance of the return path, and the inductance represented by the combined loop of outgoing and returning current. The shunt admittance y comprises the parasitic capacitance between the signal and return conductors, and any DC leakage through the dielectric insulation separating the two. The series impedance and shunt admittance are defined in units of ohms-per-unit-length and Sieman-per-unit-length[5] respectively. Both quantities vary with frequency.

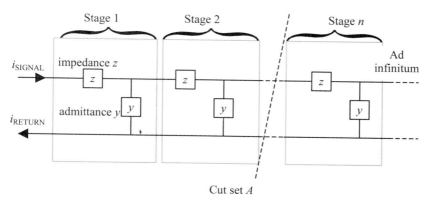

Figure 2.2—The telegrapher's equations are based on this infinitely cascaded circuit model.

---

[5] One Sieman, the international standard unit of conductivity, is the inverse of an ohm, the unit of resistivity. An element of resistance r ohms has a conductivity of $1/r$ Siemans.

2.1.1 So Good It Works on Barbed Wire

Article first published in *EDN Magazine*, July 5, 2001

In 1995, I had the privilege of serving as the chief technical editor of the Fast Ethernet specification.[6] In that capacity, I got to know many of the design teams working on chips to support the standard. The group from Broadcom was working particularly hard on an all-digital implementation of a subset of Fast Ethernet called 100BASE-T4 (different from the 100BASE-TX version that ultimately prevailed in the market).

Broadcom's T4 chip contained one of the first all-digital adaptive equalizers built for use on a fast serial link. Adaptive equalization was necessary in this case because the severely limited bandwidth of 100m of fairly low-grade category-3 data wiring filters out all the fast edges in a signal, turning a crisp transmitted signal into slush at the far end of the cable. A good adaptive equalizer reverses the filtering effect of the cable, restoring the received data to its normal appearance.

By 1995, the use of digital adaptive equalization at lower speeds was well-established, as in telephone and satellite modems. These products used programmable-DSP cores operating at speeds of approximately 100 Kbps. In contrast to the programmable approach, Broadcom used dedicated digital state machines to increase the speed of those same algorithms by a factor of 1000.

> **The "big four" transmission-line properties are impedance, delay, high-frequency loss, and crosstalk.**

The value of the Broadcom T4 design was its ability to work at high speeds on horrible cables. In many cases, customers could use it on the cables they already had without upgrading to better cables. To demonstrate the power of what they felt was the world's best chip, the engineers at Broadcom decided to demonstrate the operation of their transceiver using the world's worst cable.

At Interop that year, Broadcom set up a 2×4-ft glass case containing eight parallel strands of barbed wire. The wires were configured as four differential pairs, each running straight from side to side, suspended in air. The wires were ugly and rusty, and had nasty little barbs all over them. A transmitter and a reel of category-3 data cabling were on one side of the case. The data cabling led to the glass case where it coupled onto the four barbed-wire pairs. The other side of the case coupled through more category-3 cabling to a receiver.

During the show, lo and behold, Broadcom's demonstration flawlessly conveyed 100 Mbps of data through the barbed wire. "Buy our parts" was the message the Broadcom marketing folks wanted to impress on their audience.

I'd like you to receive a different message: Only four properties really affect the performance of most digital transmission structures. The "big four" transmission-line properties are impedance, delay, high-frequency loss, and crosstalk.

---

[6] Thanks to Paul Sherer at 3Com for sponsoring this work.

2.1.2 • The No-Storage Principle

1. Crosstalk in a barbed-wire configuration is controlled by enforcing a large spacing between the pairs, as compared with the much smaller spacing between the individual wires of each pair. The glass case in Broadcom's demonstration was easily large enough to accommodate such spacing, so crosstalk wasn't a problem.

2. What about the high-frequency loss? It wasn't a problem either. The T4 system divides its data among the four pairs so that each pair operates at only 25 Mbps. At that low frequency the skin-effect resistance of 4 ft of barbed wire is insignificant, and the overall high-frequency loss in the glass case at 25 Mbps was practically nil.

Roger Billings extended the barbed-wire demo concept to even higher speeds, successfully conveying 1 Gb/s with his transceiver. Photo courtesy of Wideband Corporation, manufacturer of transceivers for Gigabit Ethernet and other standards.

3. The signal delay is *less* on barbed wire than on an equivalent length of PVC-insulated category-3 wiring due to the use of an air dielectric between the barbed strands. This difference in delay was insignificant, however, because of the serial nature of the communications architecture.

4. Finally, if you consider the characteristic impedance, you find that this quantity is just a fixed number, such as 75 or 150 Ω. For most cables, it varies little with frequency between 1 and 100 MHz, but it varies significantly with the spacing between the wires. You can intentionally set the spacing to create almost any impedance you want. Inside the glass case, the spacing between barbed strands was set to create an impedance of 100 Ω—the *same impedance* as in the category-3 UTP cabling on either side of the glass case. Thus, the case introduced no impedance discontinuity.

In summary, the barbed wire had zero impact on signal quality. The signals went through perfectly undistorted. The only thing the barbed wire did was impress the heck out of Broadcom's customers.

Next time you look at a transmission line, I hope you'll focus on the big four properties: characteristic impedance, high-frequency loss, delay, and crosstalk. These properties determine how well a transmission structure functions, regardless of the physical appearance or configuration of its conductors.

2.1.2 The No-Storage Principle and Its Implications for Returning Signal Current

The theory of lumped-element electrical circuits requires that no individual device store current, meaning that the currents into and out of every device must sum to zero. Expanding the no-storage principle to an entire network of devices, you may conclude that no network of lumped-element devices can store current. The current into the network equals the current out. Always.

In Figure 2.2 cut set A divides one stage from the next. In circuit theory, such a line completely partitioning one circuit network from another is called a cut set. Provided there are no other hidden paths for current between the two networks, the no-storage principle demands that the sum of currents crossing A must at all times precisely equal zero. In Figure 2.2 only two lines cross A, which makes the situation very simple—the currents on the upper and lower wires crossing A must at all times be equal in magnitude but opposite in polarity.

Gustav Robert Kirchoff, the founder of modern circuit analysis, recognized the no-storage principle as central to his analysis of lumped-element circuits. By drawing tiny circles (cut sets) around each node in a network, he concluded that the sum of currents into and out of each circle must be zero. His conclusion is codified as Kirchoff's current law (KCL):

The sum of currents into any network node is zero.

You may draw a number of conclusions from the no-storage principle.

> For every signal current there must be an equal and opposite return current.
> Signal currents emanating from a digital chip generate equal but opposite return currents on the power and ground pins.
> The return current is equally as important as the signal current for noise, crosstalk, and EMC analysis.
> Current always makes a loop—if it goes out, it must return.
> The sum of currents into any network node is zero (KCL).

Let's examine the assumptions behind these important conclusions to see when and under what conditions they apply. First I shall address a classic paradox involving the applicability of the no-storage principle: the vertical monopole 1/4-wave antenna. In that situation there exists a signal conductor, but apparently no return path, at least none in the sense of Kirchoff's lumped-element circuit analysis. Current flows on the signal conductor (the antenna wire) but nowhere else. How can this be?

Maxwell's equations provide a neat solution to this paradox. Maxwell determined that the concept of current comprises two distinct possibilities. The first possibility represents the physical movement of charged particles, as in ordinary current flowing on a wire. The second possibility is called displacement current. It represents the net effect of changes in the electric flux[7] entering or leaving a conducting body. The reason we must include displacement current in our electromagnetic field calculations is that the changing lines of electric flux, while they do not directly convey charged particles, push on the mobile charge carriers within other nearby conducting bodies, causing current to flow in them. In that sense, a changing electric flux acts just like ordinary current in that it pushes and pulls on the charges in other conductors, and so must be included in any accounting of total current flow. For near-field radiation problems, you may consider the displacement current flowing from one conducting body to another to be equivalent to the current that would flow through an appropriately sized parasitic capacitor connected directly between those two bodies.

---

[7] Electric flux is the integral of electric field intensity normal to a surface.

2.1.2 • The No-Storage Principle

The reason Kirchoff gets away with ignoring displacement current is that he stipulates a priori that there shall be no stray electromagnetic fields flowing among the elements of any of his systems. This is what it means to be a *lumped-element system*. In that case the *only* possibility for the flow of current *between elements* is the direct transport of charged particles (on wires). Of course, Kirchoff understood that within the elements themselves (capacitors in particular), analysis of displacement current must be included, but in order to derive his simplified (and extremely useful) laws of circuit operation he set aside the possibility of external fields, thereby eliminating displacement current from his analysis.

Once you include the displacement current in the antenna analysis, you see that the current flowing into the antenna (the signal current) is matched by an equal and opposite current flowing out of the antenna (the displacement current). If you wish, you may track the electric flux through all of space to it various destinations, sum the totality of all currents induced in grounded objects that receive said electric flux, and account for the flow of physical current back through the Earth to the grounding rods of the transmitter through which the returning current flows back into the transmitting circuitry. Current (including displacement current) always makes a loop. If it's going out, it must be coming back. The only trick is, some of the returning current flows not through wires but through the action of parasitic capacitance.

Let us proceed one step further in the examination of the applicability of Kirchoff's laws by looking at the importance of the TEM assumption. The TEM assumption implies that the lines of electric and magnetic flux stick out sideways in a plane perpendicular to the direction of signal flow. This perpendicular field structure implies that the flux from one segment does not materially affect the next segment. In other words, TEM propagation precludes direct electromagnetic coupling (leapfrogging) between sections. Therefore, the telegrapher's equations need not contemplate any displacement current between sections, and Figure 2.2 need not show any connections between sections other than the upper (signal) and lower (return) wires.

Structures that do not adhere to the TEM mode of operation are not properly represented by Figure 2.2 and do not respond to the simple form of telegrapher's analysis presented here. Examples of non-TEM structures include microwave waveguide tubes with no central conductor, free-space radio propagation channels with multipath interference, and multimode fiber-optic cables.

As a final note, the conclusion about equal and opposite return currents is certainly affected by the degree to which Figure 2.2 properly represents the whole circuit. In systems where there exist a multitude of return-current pathways, different amounts of the returning signal current will flow along each path. To the extent that you provide one well-placed, intentional, continuous return-current path near the outgoing signal conductor, the loop inductance of the intentional return path in combination with your signal path will be much less than the effective loop inductance of more remote return pathways. Because high-frequency current flows most heavily on the path of least inductance, *most* of the return current will naturally flow on your intentional return pathway, so the magnitude of current in the intentional return path will be *approximately* the same as in the signal path. Systems in which there is no continuous return path or in which the return path is interrupted do not adhere to the TEM mode of operation. Such systems are not properly represented by Figure 2.2 and do not adhere to the telegrapher's analysis presented here.

POINTS TO REMEMBER

> ➤ Two long conductors insulated from each other with a uniform cross section make a good transmission line.
> ➤ The telegrapher's equations represent only the TEM mode of signal propagation.
> ➤ You can model almost any transmission line with the telegrapher's equations.

2.2 DERIVATION OF TELEGRAPHER'S EQUATIONS

The telegrapher's discrete equivalent circuit model for a continuous transmission line appears in Figure 2.3. This model breaks the transmission line into a cascade of small segments or blocks of a standard length. Each model comprises a series impedance z and a shunt admittance y.

The series impedance z consists of a resistor R in combination with an inductor L. The shunt admittance y consists of a conductance G in parallel with a capacitor C. Each of the values R, L, G, and C represent the cumulative amount of resistance, inductance, capacitance, or conductance measured *per unit length* in the transmission line, where the standard of measurement conforms to the size of the blocks in Figure 2.3. The standard length is traditionally set in units related to the size of the transmission structure. For example, in pcb problems the standard length might be inches or mm, whereas in a power-transmission problem the standard length would more likely be miles or km. Provided that you maintain the same units consistently throughout your calculations, the standard unit of length becomes irrelevant in the final formulas for impedance and transmission loss. For the sake of concreteness this book employs a unit length of one meter.

The telegrapher's equations may be derived from the cascaded lumped-element equivalent circuit model. I like deriving the equations this way because ladder circuits are fairly easy to analyze and don't require the use of partial differential equations. The two transmission-line quantities of most importance that may be gleaned from the telegrapher's equations are the input impedance Z_C and transfer coefficient γ (gamma).

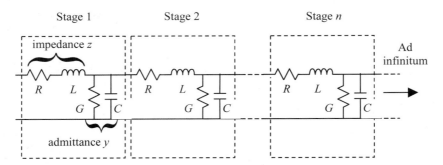

Figure 2.3—Series impedance z consists of resistor R in combination with inductor L; shunt admittance y consists of conductance G in parallel with capacitor C.

2.2.1 Definition of Characteristic Impedance Z_C

Let the symbol Z_C represent the *characteristic impedance* of a transmission line. Characteristic impedance is defined mathematically as the ratio of voltage to current experienced by a signal traveling in one direction along a transmission line. At every point within such a traveling waveform the incremental amounts of voltage v and current i *caused by that waveform* bear the same relationship to one another, $v/i = Z_C$. The quantity Z_C is a function of frequency.

In this book the variable Z_0 is interpreted as a single-valued constant showing the value of characteristic impedance at some particular frequency ω_0 (as in the expression $Z_0 = 50\ \Omega$). The variable Z_C is reserved as an expression for the characteristic impedance as a function of frequency, usually but not always shown as $Z_C(\omega)$.

The characteristic impedance Z_C of a transmission line does not in general equal its input impedance Z_{in}. The difficulty with relating Z_C to Z_{in} is that transmission lines support (at least) two modes of signal flow. These modes include a signal moving to the right and a signal moving to the left. Both modes co-exist, superimposed on top of each other. Within the traveling components of each individual mode the voltage and current always bear the proper relationship, but where the modes cross the ratio of total voltage to total current can take on any value.

What you need to remember about Z_{in} and Z_C are three facts:

1. Lacking any intentional (or reflected) signals emanating from the far end of a transmission line (i.e., when there is only one mode of signal flow), Z_{in} equals Z_C.
2. If reflections are present, the correlation between Z_{in} and Z_C evaporates.
3. Only under special circumstances where there are no reflections can you infer Z_C from measurements of Z_{in}.

The typical procedure for measuring characteristic impedance uses a *time-domain reflectometer* (TDR). The TDR setup injects a rising or falling edge of known open-circuit amplitude v and known source impedance Z_S into the transmission line under test while observing the signal coupled into the input of the transmission line (Figure 2.4). Provided that the rise (or fall) time of the measuring apparatus completes well before one round-trip delay of the transmission line, the observed signal will achieve some steady-state value a prior to the arrival of the first reflections from the far end of the line. The ratio of a to v indicates the input impedance of the structure, which under the circumstances stated should equal the characteristic impedance of the transmission line.

$$Z_C = Z_S \frac{a/v}{1 - a/v} \quad [2.1]$$

where a is the steady-state step amplitude of the measured signal (V),
v is the open-circuit voltage of the step source (V),
Z_S is the output impedance of the step source (Ω), and
Z_C is the characteristic impedance of the transmission line (Ω).

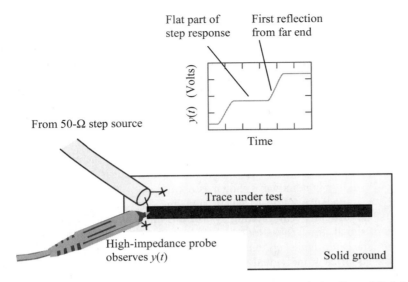

Figure 2.4—A simple TDR experiment observes the transmission line while injecting a step of known amplitude and source impedance.

Professional-quality TDR instruments conduct their observations from a point near the source end of the 50-Ω cable, but the principle remains the same.

When injecting a step into a line of finite length, the front end of the line initially experiences only one mode of signal transmission (excepting any static initial conditions), so the line temporarily displays an input impedance equal to Z_C. After one round-trip delay, reflections may arrive from the far end. The reflections and the resulting measured input impedance will then depend on the configuration of the load at the far end of the line—hence your measurement of Z_C must be completed prior to the arrival of the first reflection.

POINT TO REMEMBER

> ➢ The telegrapher's equations are derived from a cascaded lumped-element equivalent circuit model.

2.2.2 Changes in Characteristic Impedance

This brief discussion of characteristic impedance has so far glossed over an important point, namely, that the characteristic impedance Z_C may change as a function of frequency. In that case the measured step-response waveform will not display a perfectly flat top, and you must get into the business of deciding where along a sloping waveform to pick the one true point from which to calculate the characteristic impedance. You should know that there is no good way to determine such a point. Your only recourse is to loosely relate the characteristic impedance at frequency ω (rad/s) to the step-response amplitude averaged over an interval of time equal to $1/\omega$.

2.2.5 • Calculation of Propagation Coefficient

delay. Restated in terms of the propagation coefficient, the real part of γ controls the magnitude of *H*, and the imaginary part controls the phase (see [2.16]).

The variable α (real part of γ) is expressed in units of nepers per unit length. An attenuation of one neper per unit length ($\alpha = 1$) equals −8.6858896 dB of gain per unit length. This amount of attenuation scales a signal by $1/e$ as it passes through each unit length of transmission line.

The variable β (imaginary part of γ) is expressed in units of radians per unit length. A phase delay of one radian per unit length ($\beta = 1$) equals −57.295779 degrees of phase shift per unit length. After 2π lengths of line, a signal with this amount of phase delay is rotated back to its original phase.

Parameters α, β, and γ all vary with frequency. When these variables appear without their frequency arguments, you are expected to remember that a variation with frequency is still implied.

Sometimes it is convenient to work with α or β individually.

$$|H(\omega)| = e^{-\alpha}$$
$$\angle H(\omega) = -\beta \quad [2.16]$$

Combining [2.14] and [2.15] you may now express the complete one-way transfer function $H(\omega,l)$ of a line of length l as a function of its complex logarithm.

$$H(\omega,l) = e^{-l \cdot \gamma(\omega)} = e^{-l(\alpha + j\beta)} \quad [2.17]$$

POINTS TO REMEMBER

- Signals propagating on a transmission line decay exponentially with distance.
- The per-unit-length attenuation factor $H(\omega)$ is called the *propagation function* of a transmission line.
- The propagation coefficient $\gamma(\omega)$ is defined as the (negative of the) natural logarithm of $H(\omega)$.
- The propagation coefficient $\gamma(\omega)$ may be broken down into its real and imaginary parts (α and β).
- The real part of $\gamma(\omega)$ defines the attenuation per unit length of a transmission structure in nepers per unit length.
- The imaginary part $\gamma(\omega)$ defines the phase shift per unit length of a transmission structure in radians per unit length.

2.2.5 Calculation of Propagation Coefficient γ from Parameters R, L, G, and C

Figure 2.5 adds one unit-sized discrete transmission block to the head of a long, continuous transmission line with input impedance Z_C. Defining z' as the impedance looking to the right of line A, the resistor-divider theorem computes the transmission coefficient $\tilde{H}$.

$$\tilde{H} = \frac{z'}{z+z'} \qquad [2.18]$$

The term z' equals the parallel combination of admittance y and impedance Z_C, where the impedance Z_C is defined in the limit of small block size by [2.9]. Making this substitution into [2.18],

$$\tilde{H} = \frac{\dfrac{1}{y+\sqrt{y/z}}}{z+\dfrac{1}{y+\sqrt{y/z}}} \qquad [2.19]$$

Multiply top and bottom by $\left(y+\sqrt{y/z}\right)$.

$$\tilde{H} = \frac{1}{zy+\sqrt{zy}+1} \qquad [2.20]$$

Equation [2.20] expresses the transfer function of one discrete block of unit size. As before, an individual discrete block only approximates the behavior of a continuous transmission line, so I will again take a limit as I split the single unit-sized block into a succession of n blocks, each of length $1/n$. This change modifies z and y in [2.20] to produce new values z/n and y/n. The combined response of the cascade of n blocks equals the response of an individual block of size $1/n$ raised to the nth power.

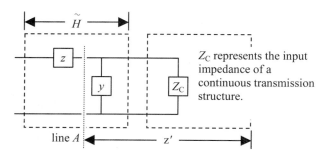

Figure 2.5—Define z' as the impedance of everything to the right of line A.

2.2.5 • Calculation of Propagation Coefficient

$$H = \lim_{n \to \infty} \left[\frac{1}{(z/n)(y/n) + \sqrt{(z/n)(y/n)} + 1} \right]^n \quad [2.21]$$

Simplify this expression by inverting the fraction, negating the exponent, and dividing out the common term n from part of the denominator.

$$H = \lim_{n \to \infty} \left(\frac{(zy/n) + \sqrt{zy}}{n} + 1 \right)^{-n} \quad [2.22]$$

Next you need a well-known mathematical fact [25].

$$\lim_{n \to \infty} \left[(a/n) + 1 \right]^{-n} = e^{-a} \quad [2.23]$$

In equation [2.22] the expression $(zy/n) + \sqrt{zy}$ plays the role of a in [2.23]. The final form of the result takes the limit as n approaches infinity, forcing the first part of the expression zy/n towards zero, leaving only $\sqrt{zy}$ to go into the exponent.

$$H = e^{-\sqrt{zy}} \quad [2.24]$$

From [2.24] you may extract the propagation coefficient γ.

$$\gamma = \sqrt{zy} \quad [2.25]$$

This expression may become more recognizable if you substitute [2.2] and [2.3] for z and y:

$$\gamma(\omega) = \sqrt{(j\omega L + R)(j\omega C + G)} \quad [2.26]$$

The factor $\gamma(\omega) \triangleq \sqrt{(j\omega L + R)(j\omega C + G)}$ is called the *propagation coefficient* of the transmission line. The propagation coefficient may be subdivided into its real and imaginary parts, $\gamma \triangleq \alpha + j\beta$, where α denotes the attenuation, in nepers per unit length, and β denotes the phase delay in radians per unit length. Both α and β vary with frequency.

The final form of the telegrapher's equation predicts, given *R, L, G,* and *C,* the amplitude and phase response for a single mode of propagation on any transmission line.

$$H(\omega, l) = e^{-l\sqrt{(j\omega L + R)(j\omega C + G)}} \quad [2.27]$$

The trick to successfully modeling transmission structures lies not merely in understanding the telegrapher's equation, but in understanding what compromises and simplifications you can make in deriving useful approximations for the basic line parameters R, L, G, and C. The remainder of this chapter discusses the calculation of basic line parameters.

> **Signals Flow Both Ways**
>
> In a TEM system with two conductors there are exactly two wave modes. These modes correspond to a changing signal flowing linearly down the structure from left to right or from right to left (figure wave). Each half of Figure 2.6 shows a succession of snapshots of the signal current along the line, taken at successive times. In the top half, a rising-edge wave front advances from left to right. In both cases, at all times and all positions, the current on the return conductor mimics the current on the signal conductor, but with opposite polarity.
>
> The transmission response and input impedance of the line are the same in both directions.

POINT TO REMEMBER

➢ The *propagation coefficient* $\gamma(\omega) \equiv \sqrt{(j\omega L + R)(j\omega C + G)}$.

2.3 IDEAL TRANSMISSION LINE

An ideal transmission line has these properties:

➢ No distortion
➢ No attenuation

Sufficient conditions for building an ideal transmission line are that you have two perfect conductors with zero resistance, uniform cross section, separation much smaller than the wavelength of the signals conveyed, and a perfect (lossless) dielectric. Voltages impressed upon one end of such an ideal transmission line will propagate forever, at constant velocity, without distortion or attenuation.

The *propagation velocity*, or *transmission velocity*, of a line is rated in units of m/s. The symbol for propagation velocity is *v*. This quantity indicates how far your signals will travel in every unit of time. For the case of perfect, zero-resistance conductors surrounded by a perfect vacuum, the propagation velocity equals *c*, the velocity of light in a vacuum, approximately $2.998 \cdot 10^8$ m/s. You may like to remember this constant in units convenient to your work, and with a precision commensurate with your needs, for instance 0.2998 m/ns (.3 m/ns), or 0.2998 mm/ps (.3 mm/ps), or perhaps 0.983 foot/ns (1 foot/ns).

2.3 • Transmission Line

The presence of any magnetic or dielectric materials near the conductors slows the propagation of electrical signals. If the conductors are embedded in a homogeneous dielectric material the velocity of propagation reduces to:

$$v = \frac{c}{\sqrt{\epsilon_r \mu_r}} \qquad [2.28]$$

where [2.28] applies only to ideal, distortionless, lossless lines,

v is the velocity of propagation (m/s),

c is the velocity of light in vacuum (m/s),

for lossless dielectric materials ϵ_r is a purely real quantity equal to the relative electric permittivity (a.k.a. dielectric constant) of the material surrounding the conductors, and

μ_r, a purely real quantity, is the relative magnetic permeability of the lossless material surrounding the conductors.

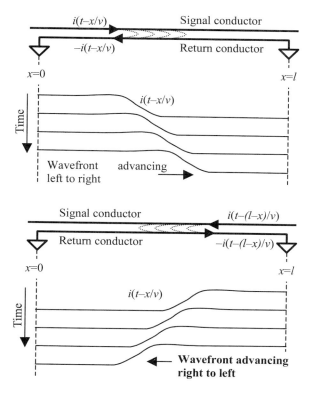

Figure 2.6—Every TEM structure supports at least two modes.

Most digital applications involve the use of nonmagnetic dielectric insulating materials for which $\mu_r = 1$; in that case expression [2.28] becomes

$$v = \frac{c}{\sqrt{\epsilon_r}} \qquad [2.29]$$

> where [2.29] applies only to ideal, distortionless, lossless lines *surrounded by a homogeneous nonmagnetic insulating material.*

If for some reason the insulating material surrounding the conductor is inhomogeneous (for example, a microstrip having dielectric material on one side and air on the other), then the dielectric material still has a slowing effect but not as great as if the material had completely surrounded the conductors.

Propagation delay per unit length is the inverse of velocity. This quantity indicates how much time elapses as your signal propagates some standard distance. If you are working in length-units of meters, the propagation delay of light in a vacuum is $3.336 \cdot 10^{-9}$ s/m.

Expressions [2.28] and [2.29] relate the velocity of propagation to the properties of the dielectric medium surrounding the conductors. The propagation velocity is also related to the quantities *R, L, G,* and *C*.

Using the mathematical language developed in the previous section, a lossless line requires $R = G = 0$. For this special case the formulas for impedance and propagation coefficient reduce to

$$Z_C = \sqrt{\frac{j\omega L}{j\omega C}} = \sqrt{\frac{L}{C}} \qquad [2.30]$$

$$\gamma(\omega) = \sqrt{(j\omega L)(j\omega C)} = j\omega\sqrt{LC} \qquad [2.31]$$

> where the characteristic impedance Z_C is a real quantity (Ω) independent of frequency,
>
> *L* (H/m) and *C* (F/m) are the per-unit-length values of inductance and capacitance respectively, and
>
> ω is the frequency of operation (rad/s).

The real and imaginary parts of γ tell you the attenuation in units of nepers/m and phase delay in units of rad/m respectively. The transfer function for a unit length of ideal, distortionless, lossless transmission line is a simple linear-phase delay:

$$H(j\omega) = e^{-j\omega\sqrt{LC}} \qquad [2.32]$$

The real part of the propagation coefficient in [2.31] equals zero at all frequencies, indicating zero loss. The imaginary part of the propagation coefficient equals $\omega\sqrt{LC}$, a simple linear-phase delay. The prescribed delay per unit length equals $\sqrt{LC}$, and the velocity is the inverse of delay.

$$v = \frac{1}{\sqrt{LC}} \qquad [2.33]$$

Take careful note of the units in these expressions. If L and C are rated in units of H/u and F/u respectively, where u is an arbitrary measure of length, then the propagation delay $\sqrt{LC}$ appears in units of s/u, the velocity in units of u/s, and the propagation coefficient in units of (complex) nepers per u.

Equating the two derivations of propagation velocity [2.28] and [2.33] reveals an important relationship between the quantities L, C, and the electrical properties of the insulating material.

$$\frac{c}{\sqrt{\epsilon_r \mu_r}} = \frac{1}{\sqrt{LC}} \qquad [2.34]$$

Equation [2.34] prescribes certain limits on your ability to independently modify L or C. No matter what modifications you make to one or the other of these variables, the product remains constant.[11] Equation [2.34] explains a peculiar property of stripline traces: A trace made wider to increase C enjoys a corresponding decrease in L, leaving the propagation velocity unchanged.

Microstrip traces behave differently. Due to their inhomogeneous dielectric, modifications to trace width affect (slightly) the relative proportions of the electric field in the air versus in the dielectric material underneath the trace, and thus have a modest impact on trace velocity.

Datasheets for cables often rate transmission velocity in percent of c. A relative velocity of 66% would indicate a velocity v and delay $1/v$ of

$$v = \left(2.998 \cdot 10^8 \; m/s\right) \cdot 0.66 = 1.978 \cdot 10^8 \; m/s$$
$$\frac{1}{v} = \frac{3.33 \cdot 10^{-9} \; s/m}{0.66} = 5.04 \; ns/m \qquad [2.35]$$

Example calculation of characteristic impedance and delay:

Measure the capacitance and inductance of RG-58/U coaxial cable, as shown in Figure 2.7, and then compute the delay. First cut off a 10-in. section of RG-58/U coaxial cable. Using a good-quality impedance meter, measure its capacitance. The correct value is 26 pF, which works out to 2.6 pF/in.

---

[11] Provided that the signal conductors remain uniform in cross section and embedded in a homogeneous dielectric.

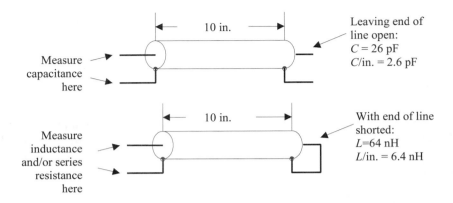

Figure 2.7—Measure capacitance with the far end of the line open-circuited and inductance with the far end of the line short-circuited.

Next short out one end of the same 10-in. section and measure (from the other end) its inductance.[12] Here the correct value is 64 nH, which works out to 6.4 nH/in.

Using a very sensitive four-terminal ohmmeter, you may notice that the center conductor of this cable also has a series resistance of 0.009 Ω, or 0.9 mΩ/in. Ideal transmission lines have zero resistance, but for our purposes right now the 10-in. segment of RG-58/U is close enough to ideal.

The per-unit-delay is $\sqrt{LC} = \sqrt{(2.6\,pF/in.)(6.4\,nH/in.)} = 129\ ps/in.$, precisely the value for RG-58/U listed in the Belden Wire and Cable Master Catalog 885.

Our test sample of RG-58/U has a characteristic impedance of

$$Z_C = \sqrt{\frac{6.4\,\text{nH/in.}}{2.6\,\text{pF/in.}}} = 50\,\Omega \qquad [2.36]$$

I am often asked why so much current is required to drive a long transmission line. To what good purpose is the current put? The function of the input current is simply this: to charge the parasitic capacitance of the line.

Figure 2.8 diagrams the patterns of voltage and current existing on a transmission line at two distinct snapshots separated in time by t_d. Between snapshots, the rising wave front advances by physical distance vt_d, where v is the propagation velocity of the transmission line.

Each snapshot shows current flowing along the signal conductor from the driver to the leading edge of the wave front. Near the leading edge, current pours through the local parasitic capacitance of the line, changing its state, and then moves backwards along the return connection towards the driver. In each snapshot the current is being used to charge

---

[12] One assumes the measurement of inductance is made at a frequency high enough that the skin effect has confined the flow of current to the surface of the conductors. Otherwise, a slightly higher value of inductance results, which slightly modifies the computed value of delay. If the conductors had zero resistivity, this change in inductance would never be observed, as the current would in no case penetrate the conductors.

2.3 • Transmission Line

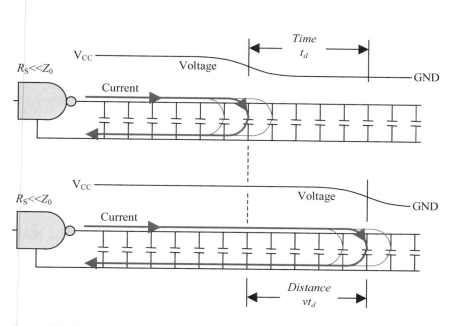

Figure 2.8—Return current builds up simultaneously with the signal current as a rising edge propagates through a transmission structure.

the parasitic capacitance of some chunk of line coinciding with the position of the rising edge at that moment.

Note that current flows only through the parasitic capacitance near the rising edge of the travelling waveform *because that's where the voltage is changing*. Current flows through capacitors only in response to changing voltages.

From the first moment when the driver begins to change state, a pattern immediately develops with current going into the line, charging up the first section of parasitic capacitance, and returning to the driver. As time progresses, the pattern stretches further down the line, but always with the coincidence of outgoing and return current. An understanding of this principle is fundamental to grasping the action of a transmission line. Current always makes a loop. *It does not* go down the signal conductor, reach the end, and then begin to make its way back. Rather, the outbound and returning paths build up together from the beginning. By the time the signal wave front reaches the end, the line has already established a complete pattern of outgoing and return current all along the structure.

Next let's look at how *much* current is required to support the movement of the advancing wave front. The figure shows that during interval t_d the rising waveform must provide sufficient charge to all the capacitors along a span of length vt_d. The change in voltage along this span goes from 0 to V_{CC}. The total amount of capacitance contained within this span is Cvt_d. The total amount of charge required is therefore $q = V_{CC}Cvt_d$. To provide this amount of charge within time t_d requires a current of $I = q/t_d = V_{CC}Cv$.

Substituting $Cv = \dfrac{C}{\sqrt{LC}} = \dfrac{1}{\sqrt{L/C}}$, the expression for I may be rewritten as $I = \dfrac{V_{CC}}{\sqrt{L/C}}$, which is precisely the amount of current you would expect during the first round trip, assuming a transmission-line characteristic impedance of $Z_C = \sqrt{L/C}$.

The above discussion shows that the characteristic impedance $Z_C = \sqrt{L/C}$ of an ideal transmission line is entirely accounted for by the current required to change the state of the line at the required propagation velocity v. The characteristic impedance of an *ideal transmission line* remains constant at all frequencies. It has no imaginary part and is not a function of frequency. It is a function only of the physical geometry of the transmission line and the dielectric constant of the insulation.

Practical values for characteristic impedance commonly range from a few ohms (between the inner and outer shields of triaxial cable) to 300 ohms (in a balanced configuration used for TV antenna connections).

For your amusement (and because I get this question a lot), Figure 2.9 illustrates the patterns of current flow existing at various snapshots in time as a traveling wave pulse flows down a long transmission structure.

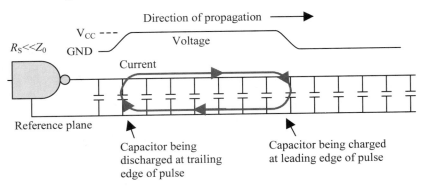

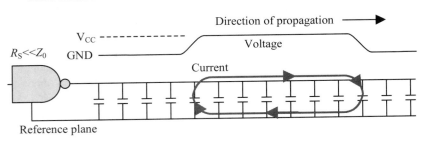

Figure 2.9—The current associated with a short-duration pulse creates a pattern reminiscent of a rolling tank tread.

POINTS TO REMEMBER

> A lossless transmission line requires $R = G = 0$.
> For the special case of a lossless line $Z_c = \sqrt{L/C}$ and $\gamma(\omega) = j\omega\sqrt{LC}$.

2.4 DC RESISTANCE

Practical transmission lines[13] incorporate series resistance. The nonzero resistance of practical transmission lines dissipates a portion of the signal power, causing both attenuation (loss) and distortion in propagating signals. This section shows how to estimate the DC resistance (per meter) of a transmission line.

$$R_{DC} = \frac{k_a \rho}{a} \quad \Omega/m$$

or, [2.37]

$$R_{DC} = \frac{k_a}{\sigma a} \quad \Omega/m$$

where a is the cross-sectional area of the conductor (m²),

k_a is a correction factor that accounts for the additional DC resistance of the return path,

ρ is the resistivity of the conductor (Ω–m), or

σ is the conductivity of the conductor (S/m).

For annealed copper at room temperature $\rho = 1.724 \cdot 10^{-8}$ Ω–m and $\sigma = 5.800 \cdot 10^7$ S/m.

Note that the *effective DC resistance* of a transmission-line structure is the sum of the DC resistance of the signal conductor plus the DC resistance of the return path. In [2.37] the factor $1/\sigma a$ accounts only for the resistance of the signal conductor. The factor k_a accounts for the additional resistance of the return path. The factor k_a equals a quotient, the numerator of which is the sum of the DC resistances of the signal and return paths, and the denominator of which is the DC resistance of the signal path alone.

For example, a pcb trace with a wide, flat return path has $k_a = 1$, because the resistance of the wide, flat return path is so low it adds almost nothing to the overall DC resistance of the structure.

A twisted-pair configuration, because the outbound (signal) and return (~signal) wires are the same, has $k_a = 2$.

A coaxial cable has a DC resistance equal to the sum of the resistances of the center conductor and shield. In many cases the center conductor may be composed of plated or

---

[13] Thin-film superconducting transmission lines have essentially no series resistance provided that you keep them cold. For high-speed digital applications the requirement for cold-temperature operation renders them, in this author's opinion, not practical.

clad materials. The shield often incorporates a braided construction with or without an inner foil wrap. When working with coaxial cables it's usually best to just accept the worst-case specification provided by the manufacturer for R_{DC} and not try to compute it.

The series resistance for long cables is often specified in ohms per 1,000 feet or ohms per 100 meters. When dealing with twisted-pair cables, the specified resistance usually includes the series resistance of both the outgoing and return wires. For coaxial cables the manufacturer often separately specifies the series resistances of the center conductor and outer shield. In any calculations of signal loss you must always add the resistance of both outgoing and return pathways, because current flows equally in both, and both dissipate power.

Here are eight rules of thumb for estimating the DC resistance of round copper wires:

1. The American Wire Gauge (AWG) system is a logarithmic measure of the diameter of round wires. The larger the gauge, the smaller the wire.
2. Six AWG points halves the diameter.
3. Area being proportional to diameter-squared, three AWG points halves the cross-sectional area.
4. Three AWG points doubles the wire resistance.
5. A round conductor of size #24 American Wire Gauge (24-AWG) has a nominal diameter of 0.507 mm (0.02 in.) and a resistance at room temperature of 0.085 Ω/m (26 Ω/1000 ft).
6. Twisted-pair 24-AWG cable has a total series resistance (adding both wires in series) of 0.170 Ω/m (52 Ω/1000 ft) at room temperature.
7. RG-58/U coaxial cable using a stranded core of AWG 20 has a resistance at room temperature of 0.034 Ω/m (10.3 Ω/1000 ft).
8. The resistance of copper increases 0.39% with every 1 °C increase in temperature. Over a 70 °C temperature range, that amounts to a variation of 31%.

Here are some handy equations for working with AWG sizes:

$$\text{Diameter in inches} = 10^{-(AWG+10)/20}$$
$$\text{Diameter in cm} = 2.54 \cdot 10^{-(AWG+10)/20}$$

[2.38]

$$R_{DC} \text{ per } 100\,\text{m} = \frac{0.0220}{(\text{Diameter in cm})^2} \quad (\Omega\,@\,25°C)$$

$$R_{DC} \text{ per } 1000\,\text{ft} = \frac{0.0104}{(\text{Diameter in inches})^2} \quad (\Omega\,@\,25°C)$$

[2.39]

$$R_{DC} \text{ per } 100\,\text{m} = 0.341 \cdot 10^{(AWG-10)/10} \quad (\Omega\,@\,25°C)$$
$$R_{DC} \text{ per } 1000\,\text{ft} = 1.04 \cdot 10^{(AWG-10)/10} \quad (\Omega\,@\,25°C)$$
[2.40]

The resistance (per unit length) of a printed-circuit trace varies with copper thickness and trace width. The trace thickness is rated in plating weight, typically reported in ounces. A 1-oz plating corresponds to a thickness of 34.8 μm. The thickness scales in proportion to plating weight.[14] The resistance of an electrodeposited pure copper pcb trace may be calculated from its thickness and width.

$$R_{DC} = \frac{1.669 \cdot 10^{-8}}{W \cdot T} \; \Omega/m$$

or,

$$R_{DC} = \frac{4.798 \cdot 10^{-4}}{W \cdot T_{OZ}} \; \Omega/m$$

[2.41]

where R_{DC} s the series resistance of the line (Ω/m),

W is the width of line (m),

T is the thickness of line (m), which equals $3.48 \cdot 10^{-5}$ meters (1.37 mil) for one ounce of plating weight, or

T_{OZ} is the plating weight of the line in ounces (oz/ft^2).

A 1/2-oz copper pcb trace with 100-μm (4-mil) width has a DC resistance of 9.6 Ω/m.

POINT TO REMEMBER

> The nonzero resistance of practical transmission lines dissipates a portion of the signal power, causing both attenuation (loss) and distortion in propagating signals.

2.5 DC CONDUCTANCE

Leakage due to wet or imperfect insulation between the signal conductors in long cables creates a per-unit-length amount of shunt conductance G. If present, G dissipates power, attenuating and distorting signals in a manner similar to series resistance. Because the shunt conductance G is practically zero at DC for the types of insulators used in most modern digital transmission applications, the shunt conductance term G is rarely used.

For low-voltage digital applications involving pcbs or long UTP, STP-A, or coaxial cables of all types you may safely assume the DC value of shunt conductance G is zero.

---

[14] Plating weight in units of ounces refers to the number of ounces of material deposited on a flat surface one foot square.

Dielectric loss models for high-frequency applications generally incorporate AC dielectric losses into the definition of complex permittivity. This technique imbues the capacitance term C with both real and imaginary parts. When the imaginary part of C is multiplied by $j\omega$ in the equations, it turns real, generating a high-frequency conductance term $-\omega\,\mathrm{Im}(C)$. Such a term has the same practical effect as the use of a G term that varies in proportion to frequency (see Section 2.12, "Dielectric Effects").

POINTS TO REMEMBER

> - The shunt conductance G is practically zero at DC for the types of insulators used in most modern digital transmission applications.
> - Dielectric loss models for high-frequency applications incorporate AC dielectric losses into the definition of complex permittivity, creating a capacitance term C with both real and imaginary parts.

2.6 SKIN EFFECT

High-frequency current in a practical conductor does not flow uniformly throughout the cross-sectional area of the conductor. Magnetic fields within the conductor adjust the distribution of current, forcing it to flow only in a shallow band just underneath the surface of the conductor. The redistribution of current increases the apparent resistance of the conductor. In transmission lines we call this increase in resistance the *skin effect*. The thickness of the conduction band δ is called the *skin depth* of the conductor.

The skin effect is an inductive mechanism related to the rate of change of magnetic fields within the conductor, so it becomes increasingly intense at higher frequencies. Below a certain cutoff frequency ω_δ the skin effect is still present, but not at a noticeable level. Above ω_δ, the skin effect squeezes the current into progressively more shallow regions at the periphery of the conductor, increasing the apparent AC resistance of the conductor without limit in proportion to the square root of frequency.

The skin effect governs the behavior of all conductors. As an example, Figure 2.10 depicts the series resistance of RG-58/U coaxial cable plotted as a function of frequency. The plot uses log-log axes. Figure 2.10 also shows, on the same axes, the series inductive reactance $2\pi f L$ of the cable, assuming $L = 253$ nH/m.

At frequencies below the skin-effect cutoff frequency ω_δ the series resistance maintains a constant value equal to the total DC resistance of the conductors (both the inner signal conductor and shield are added in series). Beyond ω_δ the effective series resistance rises with the square root of frequency.

2.6.1 What Causes the Skin Effect, and What Does It Have to Do With Skin?

To understand the skin effect, you must first understand how eddy currents operate within a solid conductor. Figure 2.11 depicts a sheet of conducting material, illuminated from the left

2.6.1 • What Causes the Skin Effect

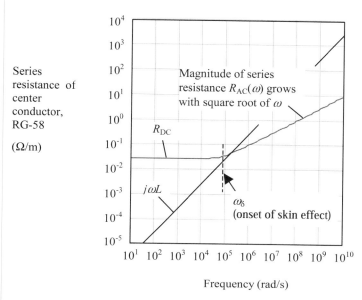

Series resistance of center conductor, RG-58 (Ω/m)

Figure 2.10—Above the onset of the skin effect the series resistance of the center conductor markedly increases.

side with a changing magnetic field. In a good conductor the presence of such a changing magnetic field sets up circulating eddy currents within the material, shown as dotted lines. The eddy currents circulate around the incoming lines of magnetic force, whose lines of flux point to the right. Although I've drawn the eddy currents as individual loops, in reality the loops merge continuously together across the sheet, forming an overall counterclockwise circulation of current around the perimeter of the sheet.

The eddy currents generate their own magnetic fields. The lines of magnetic flux associated with these secondary fields point to the left, always bucking the direction of the

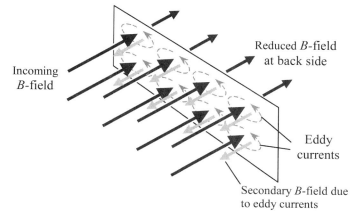

Figure 2.11—A changing magnetic flux creates eddy currents in a conducting plane.

incoming field.[15] When the resistivity of the material is high, or the material is thin, the eddy currents remain small and the intensity of the secondary field comprises no more than a small perturbation to the incoming field.

If the resistivity is lowered, or the sheet thickened, the eddy currents increase and the secondary magnetic fields grow to a more noticeable magnitude.[16] An observer located behind the sheet begins to see partial cancellation of the incoming and secondary fields. In other words, a good conducting sheet acts as a magnetic shield that attenuates the passage of magnetic flux.

Note that the above discussion relates only to magnetic fields perpendicular to the sheet. Magnetic lines of force running parallel to the sheet do not penetrate it and so do not create eddy currents.

In summary, magnetic lines of force intercepting a conducting object cause eddy currents, and the eddy currents reduce the intensity of the perpendicular magnetic field at the surface of the conductor. The eddy currents reduce the intensity of magnetic fields observed on the other (back) side of the sheet as well. The ratio of the field amplitude at the back of the sheet to the incoming field amplitude on the front is called the shielding effectiveness of the sheet.

If you add a second sheet in parallel with the first, the second sheet attenuates the field amplitude a second time. If you cascade a set of n parallel sheets, the overall structure should exponentially improve the shielding by n times. If you simply use a sheet n-times thicker, you get a similar result—the field on the back side decays exponentially with the thickness of the sheet.

I have left many details out of this general discussion in order to communicate one, simple main result: *The magnetic fields within a conductor decay exponentially as you move from the surface of any conductor towards the interior.* The whole story is considerably more complicated, as it involves consideration of both the electric and magnetic fields, the wave impedance of the impinging wave, the characteristic impedance of the conducting medium, and reflections back and forth across the thickness of the sheet. A good, practical summary of the issues appears in [3]. Detailed, mathematically complete descriptions may be found in [4] and [5].

The thickness of material required to reduce the internal magnetic field intensity within the conductor by a factor of $1/e$ is called the skin depth of the material. For a good conductor the skin depth δ depends on the frequency of operation, the conductivity, and the magnetic permeability of the material. In practice you will find that materials many skin depths thick make excellent magnetic shields, while materials less thick than one skin depth provide only partial attenuation of magnetic fields.

$$\delta = \sqrt{\frac{2}{\omega \mu \sigma}} \qquad [2.42]$$

where $\quad \delta$ is the skin depth over which magnetic fields are attenuated by a factor of $1/e$ (m),

---

[15] To do otherwise would lead to the possibility of perpetual motion.
[16] But never larger than the amplitude of the incoming field.

2.6.2 • Eddy Currents

$\omega = 2\pi f$ is the frequency of operation (rad/s),

μ is the magnetic permeability of the conducting material (H/m), and

σ is the conductivity of the conducting material (S/m).

For annealed copper at room temperature $\sigma = 5.80 \cdot 10^7$ S/m.

For nonmagnetic materials $\mu = 4\pi \cdot 10^{-7}$ H/m.

Equation [2.42] prescribes changes in the skin depth in inverse proportion to the square root of frequency. Table 2.2 lists the skin depths of several materials at various frequencies.

As you can see from the table, the thickness of a 1/2-oz copper layer (0.017 mm) spans many skin depths at frequencies above 100 MHz (the primary region of interest to high-speed digital designers). You may conclude that pcb traces and planes appear many skin depths thick at high frequencies and that these objects act, at high frequencies, as excellent magnetic shields.

2.6.2 Eddy Currents within a Conductor

Figure 2.12 shows a cutaway view of the of eddy currents present within the body of a circular conductor. This figure assumes the frequency of operation $\omega \ll \omega_\delta$, so the total current $I = \cos(\omega t)$ remains distributed fairly evenly across the entire cross-sectional area of the conductor.

Table 2.2—Skin Depth of Conductive Media

	Carbon steel SAE1045	Stainless steel[1]	Nickel	Copper, annealed	Aluminum	Soil (Earth)[2]
Relative magnetic permeability	1000	500	100	1	1	1
Conductivity (S/m)	$5.80 \cdot 10^6$	$1.16 \cdot 10^6$	$1.16 \cdot 10^7$	$5.80 \cdot 10^7$	$3.55 \cdot 10^7$	0.1
Frequency (Hz)			Skin depth (mm)			
60	0.85	2.7	1.9	8.5	11.	205.
10^3	0.21	0.66	0.47	2.1	2.7	50.
10^4	0.066	0.21	0.15	0.66	0.85	16.
10^5	0.021	0.066	0.047	0.21	0.27	5.0
10^6	0.0066	0.021	0.015	0.066	0.085	1.6
10^7	0.0021	0.0066	0.0047	0.021	0.027	0.52
10^8	0.00066	0.0021	0.0015	0.0066	0.0085	0.20
10^9	0.00021	0.00066	0.00047	0.0021	0.0027	0.16
10^{10}	0.000066	0.00021	0.00015	0.00066	0.00085	0.16

NOTE (1)—Values for steel alloys vary tremendously. Some stainless steels have almost no magnetic properties. See [3], p. 163, and [4], p. 244.

NOTE (2)—Above 100 MHz the Earth is no longer a good conductor, so the formula for skin depth changes.

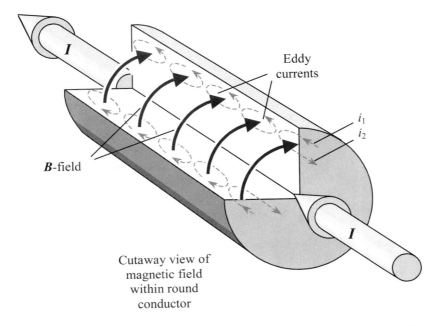

Figure 2.12—A changing magnetic field within the body of a conductor creates internal eddy currents that affect the distribution of current within the conductor.

The changing current I induces a changing magnetic field (curved arrows). These magnetic lines of force cut through the **body of the conductor, creating circulating eddy currents internal** to the conductor (dotted lines). As ω slowly increases, the magnetic fields grow in proportion to ω, and so do the eddy currents.

The end of the conductor is cut away in the drawing to show eddy current i_1 flowing near the periphery of the conductor in a direction aligned with the major direction of current flow and eddy current i_2 flowing interior to the conductor in the opposite direction. Even though these currents are depicted in the cutaway view in only two radial positions, they occur at all radial positions around the conductor. The eddy currents circulate end to end everywhere throughout the conductor. The eddy currents increase the effective current density around the periphery of the conductor while decreasing it in the middle.

If you raise ω well above ω_δ the eddy current effect becomes extremely pronounced, finally expunging all current (and all magnetic flux) from the center of the conductor. In the limit at very high frequencies the skin effect restricts the flow of current to a shallow band just under the surface of the conductor. The effective depth of the conduction band (skin depth) is given by [2.42]. If at frequency ω the skin depth of the material is much less than half the conductor thickness, then current hardly penetrates at all to the interior of the conductor.

Figure 2.13 plots the skin depth for copper versus frequency, marking out the particular frequencies at which the skin depth takes hold for various types of conductors.

2.6.3 • Approximations for Series Resistance

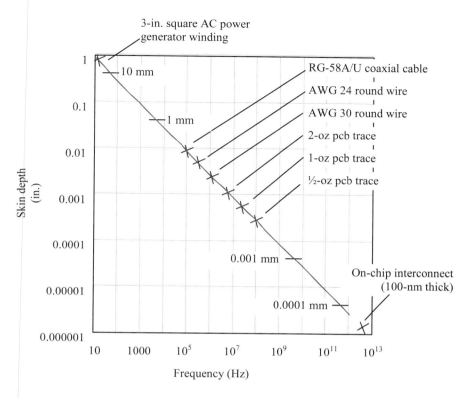

Figure 2.13—The X positions mark the skin-effect onset frequency for each conductor type.

2.6.3 High and Low-Frequency Approximations for Series Resistance

Once a conductor falls under the grip of the skin effect, current flows only in a shallow band of effective depth δ around the periphery of the conductor. In mathematical language, the skin effect restricts the flow of current to an annular ring of thickness δ and perimeter p, having cross-sectional area equal to $p\delta$. The effective resistance of a conductor in the skin-effect region is

$$\mathrm{Re}[R_{AC}] = \frac{k_p k_r}{p \delta \sigma} \qquad [2.43]$$

where $\mathrm{Re}[R_{AC}]$ is the real part of the skin-effect impedance (Ω/m), assuming you are operating at a frequency high enough that wire radius greatly exceeds the skin depth (i.e., $\omega \gg \omega_\delta$),

p is the perimeter of a cross-section of the signal conductor (m),[17]

δ is the skin depth of the signal conductor at the frequency of operation (m),

σ is the conductivity of the signal conductor (S/m),

k_p is a correction factor determined by the proximity effect (see Section 2.10, "Proximity Effect"), and

k_r is a correction factor determined by the roughness effect (see Section 2.11, "Surface Roughness").

For round conductors well separated from the path of returning signal current, the factor k_p equals unity. For conductors that are smooth on a scale comparable to the skin depth the factor k_r also equals unity.

Substituting [2.42] for δ yields an expression that shows how R_{AC} changes with frequency. Since the conductor's apparent resistance is inversely proportional to the depth of current flow (the skin depth), and equation [2.43] says that the skin depth varies inversely proportional to the square root of frequency, you may conclude that the skin-effect resistance grows proportional to the square root of ω.

$$\mathrm{Re}[R_{AC}] = \frac{k_p k_r}{p\sqrt{\dfrac{2}{\omega\mu\sigma}} \cdot \sigma} = \frac{k_p k_r \sqrt{\omega\mu}}{p\sqrt{2\sigma}} \quad [2.44]$$

where $\mathrm{Re}[R_{AC}]$ is the real part of the skin-effect impedance (Ω/m), assuming $\omega \gg \omega_\delta$,

k_p is a correction factor determined by the proximity effect (see Section 2.10, "Proximity Effect"), and

k_r is a correction factor determined by the roughness effect (see Section 2.11, "Surface Roughness").

p is the perimeter of a cross-section of the signal conductor (m),

ω is the frequency of operation (rad/s),

μ is the magnetic permeability of the signal conductor (H/m), and

σ is the conductivity of the signal conductor (S/m).

For annealed copper at room temperature $\sigma = 5.800 \cdot 10^7$ S/m.

For non-magnetic materials $\mu = 4\pi \cdot 10^{-07}$ H/m.

Equation [2.44] is a high-frequency approximation for the series resistance of a wire. It applies to conductors of any reasonable convex cross-sectional shape, well separated from their return path, at ω sufficiently far above ω_δ that the skin depth remains far less than half the conductor thickness.

---

[17] For example, the perimeter of a round conductor with diameter d is πd. The perimeter of a rectangular conductor having width w and thickness t is $2(w+t)$.

2.6.3 · Approximations for Series Resistance

The low-frequency approximation for series resistance is [2.37]. It applies when ω is sufficiently small that the conductor remains much thinner than the skin depth.

The crossover frequency ω_δ is that frequency where the low-frequency and high-frequency models of series resistance coincide. Since in this frequency range the roughness factor is not usually significantly different from unity, it has been omitted from the equation.

$$\omega_\delta = \frac{2}{\mu\sigma}\left(\frac{k_a p}{k_p a}\right)^2 \qquad [2.45]$$

where ω_δ is the frequency (rad/s) where the skin-effect resistance predicted by [2.44] equals the DC resistance,

k_p is a correction factor determined by the proximity effect (see Section 2.10, "Proximity Effect"), and

k_a is the DC resistance correction factor (see Section 2.4, "DC Resistance"),

μ is the magnetic permeability of the signal conductor (H/m),

σ is the conductivity of the signal conductor (S/m),

p is the perimeter of a cross-section of the signal conductor (m), and

a is the cross-sectional area of the signal conductor (m²).

At the crossover frequency the actual resistance of a round wire appears 2.09 dB larger than either the DC or skin-effect models would predict.

Table 2.3 lists the crossover frequencies and the skin depth at which each crossover occurs for variously shaped conductors. The calculation of series resistance at mid-frequencies is a matter of some delicacy addressed in later sections.

Table 2.3—Skin-Effect Onset Frequency

Conductor geometry	Skin-effect onset ω_δ (rad/s)	Skin depth $\delta(\omega_\delta)$ (m)
Round, radius r	$\dfrac{8}{\mu\sigma r^2}\dfrac{k_a^2}{k_p^2}$	$\dfrac{r}{2}\dfrac{k_p}{k_a}$
Rectangular, $w{\times}t$	$\dfrac{8}{\mu\sigma}\left(\dfrac{w+t}{wt}\right)^2\dfrac{k_a^2}{k_p^2}$	$\dfrac{wt}{2(w+t)}\dfrac{k_p}{k_a}$
Square, $w{\times}w$	$\dfrac{32}{\mu\sigma w^2}\dfrac{k_a^2}{k_p^2}$	$\dfrac{w}{4}\dfrac{k_p}{k_a}$
Thin, $w{\times}t$, $t{\ll}w$	$\dfrac{8}{\mu\sigma t^2}\dfrac{k_a^2}{k_p^2}$	$\dfrac{t}{2}\dfrac{k_p}{k_a}$

POINTS TO REMEMBER

➤ Magnetic fields within a conductor adjust the distribution of high frequency current, forcing it to flow only in a shallow band just underneath the surface of the conductor.
➤ The effective depth of penetration of current is called the skin depth.
➤ The increase in the apparent resistance of the conductor caused by this redistribution of current is called the skin effect.
➤ At frequencies above the skin-effect onset frequency ω_δ the effective series resistance of a conductor rises with the square root of frequency.

2.7 SKIN-EFFECT INDUCTANCE

Whenever you alter the path of current, you alter the inductance. Because the skin effect modifies the distribution of current within the conductor, *it must also change the inductance* of that conductor. You can observe this in very careful measurements of transmission-line inductance at high and low frequencies.

At frequencies well above the skin-effect onset frequency ω_δ the current in a transmission line distributes itself in whatever way minimizes the overall inductance of the circuit (current follows the path of least inductance). The least-inductive distribution for a transmission line concentrates current around the periphery of the conductors, with little or no flux interior to the conductors. The value of inductance so obtained is called the *external inductance* of the transmission line L_e.

The external inductance is defined at a frequency sufficiently high that the skin depth shrinks to much less than the wire thickness, but also sufficiently low that the wire still operates in a TEM mode (or for microstrips, a quasi-TEM mode—see "Non-TEM Modes" in Section 5.1.5. The name *external inductance* applies because the definition assumes the skin effect has expunged all magnetic flux from the interior of each conductor; thus the circuit responds only to magnetic flux appearing *external* to the conductors themselves. The external inductance is the value of series inductance, in Henries per meter, computed by a 2-D field solver under the assumption that current rides on the surface of each conductor without penetrating deeply into the body of any of the conductors.

At frequencies well below ω_δ, current in a transmission line redistributes itself to minimize the resistance of the circuit (current follows the path of least resistance). Because this distribution is not the same as the least-inductive distribution, the value of low-frequency inductance must *by definition* be higher than the *minimum* inductance L_e. The difference in inductance between low and high frequency values is called the *internal inductance* of the transmission line, L_i. It carries this name to remind you that the shift in inductance has to do with the penetration of flux internal to the conducting elements of the line.

The redistribution of current within the conductors at low frequencies affects their inductance, but not their capacitance. The creation of an analogous "internal capacitance" would require that electric fields penetrate the body of the conductors, something that does not happen for good metallic conductors at any reasonable operating frequency. When working with metallic conductors you may assume that charge remains bound to the surfaces at all frequencies of interest to digital designers. Around the skin-effect onset frequency there is no change in capacitance.

If a material of very low conductivity violates the assumption $\sigma \gg \omega\varepsilon$, then electric fields will penetrate the conductor. Above a certain frequency $\omega_\chi = \sigma/\varepsilon$, there will develop an electric-field skin effect accompanied by changes in capacitance somewhat analogous to the changes in internal inductance caused by the magnetic skin effect. The partial penetration of electric fields into a lightly doped silicon substrate is the root cause of the slow-wave effect (see Section 2.14, "Slow-Wave Mode On-Chip").

Models for predicting the series resistance and inductance of conductors are presented in the next section.

POINTS TO REMEMBER

- The distribution of current at high frequencies minimizes inductance.
- At DC, the path of least DC resistance creates a slightly higher inductance.
- Good models for skin effect take into account changes in both resistance and inductance with frequency.

2.8 MODELING INTERNAL IMPEDANCE

The total per-unit-length series impedance z of any transmission line may be partitioned into external and internal series impedances: $z = z_e + z_i$.

The external series impedance z_e represents mostly the inductance L_e formed by magnetic flux occupying the spaces between the signal conductors (i.e., in dielectric insulating medium).

$$z_e = j\omega L_e \qquad [2.46]$$

where L_e is the external series inductance (H/m).

The external series impedance z_e also incorporates a small real part representing resistive losses encountered in the space surrounding the conductors (like radiation resistance). The resistive component of z_e, I shall ignore under the assumption that conductors normally used for digital applications are designed not to significantly radiate.

Practical conductors posses two additional contributions to their series impedance: L_i and R_i. The internal inductance L_i represents the magnetic flux that penetrates the walls of the conductor (see Figure 2.14). The term R_i represents the *resistance* of the conductor. The internal *impedance* z_i captures both resistive and inductive terms:

$$z_i = R_i + j\omega L_i \qquad [2.47]$$

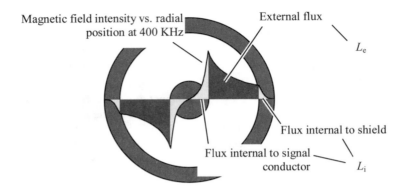

Figure 2.14—In this coaxial configuration a noticeable proportion of the low-frequency magnetic flux lies within the signal conductor and shield.

where R_i is the internal series resistance (Ω/m), and
L_i is the internal series inductance (H/m).

As you move to higher and higher frequencies the magnetic fields (and associated currents) penetrate the walls of the conductors less and less significantly, thus shrinking the internal inductance and raising the internal resistance. Figure 2.15 captures the changes to internal inductance and resistance that happen within an RG-58/U coaxial cable.

Below ω_δ (the skin-effect onset frequency), both resistance and inductance hold constant. The internal resistance is determined by the conductivity and cross-sectional area of the conductors, while the internal inductance depends on the conductor geometry. For round conductors, the low frequency value of internal inductance is $\mu/8\pi$. For other conductor shapes (like square or rectangular conductors), the internal inductance is undoubtedly less; however, the exact value of internal inductance for an arbitrary shape is not easily calculated.

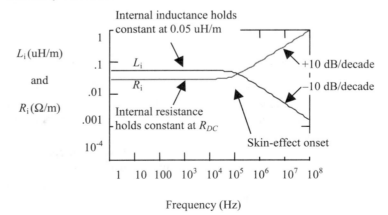

Figure 2.15—Below the onset of the skin effect both L_i and R_i hold constant.

2.8 • Modeling Internal Impedance

Above ω_δ, the internal resistance R_i grows in proportion to the square root of frequency, while the internal inductance L_i shrinks at the same rate. This result holds for all *good conductors,* meaning materials for which ($\sigma \gg \omega\varepsilon$), a property shared by most metals up to extremely high (i.e., optical) frequencies. In all but the lowest-impedance transmission configurations L_e exceeds L_i by a substantial factor even at DC, with the difference becoming even more pronounced about ω_δ. Not far above ω_δ, the internal inductance shrinks to insignificance.

Table 2.4 summarizes the asymptotic values of R_i and L_i for a *round conductor,* assuming it is located sufficiently far from its return path so $k_p = 1$.

A complete derivation of the distribution of current within a round conductor appears in [19] and [15]. Unfortunately, the derivation may be obtained only at the high mental cost of invoking Bessel functions. These derivations apply to the case of one round conductor exposed to a uniform electric field oriented along the axis of the conductor. The conductor must be made of a "good conductor" material for which σ greatly exceeds $\omega\varepsilon$ at all frequencies of interest (a good assumption for almost all metals). Under these assumptions, the value of L_i due to redistribution of current within the center conductor equals precisely $\mu/8\pi$. The complete form of the internal impedance z_i for a round conductor is

$$z_i(\omega) = \frac{\eta(\omega)}{2\pi r} \frac{I_0\left(r\sqrt{j}\sqrt{\omega\mu\sigma}\right)}{I_1\left(r\sqrt{j}\sqrt{\omega\mu\sigma}\right)} \text{ ohm/m} \quad [2.48]$$

where η is the intrinsic impedance of a good conducting material,

I_0 and I_1 are modified Bessel functions of order zero and one respectively,

ω is the frequency of operation (rad/s),

μ is the magnetic permeability of the conducting material (H/m),

σ is the conductivity of the conducting material (S/m), and

r is the radius of the wire (m).

The value of μ for nonmagnetic materials is $4\pi \cdot 10^{-7}$ (H/m).

For annealed copper at room temperature, $\sigma = 5.800 \cdot 10^7$ (S/m).

In general, the intrinsic impedance of a material is given by $\eta(\omega) = \sqrt{j\omega\mu/(\sigma + j\omega\varepsilon)}$. In a good conductor, however, the term σ greatly exceeds $j\omega\epsilon$ at all frequencies, so the intrinsic impedance reduces to $\eta(\omega) = \sqrt{j}\sqrt{\omega\mu/\sigma}$. The intrinsic impedance of a good conductor therefore takes on the phase angle of $\sqrt{j}$, which precisely equals $\pi/4$.

When evaluating [2.48], you need to know that many software implementations of the modified Bessel functions I_0 and I_1 exist for real-valued arguments, but few accept complex arguments. That problem may be solved by resorting to two specially tabulated Bessel functions, $\text{ber}_n(v)$ and $\text{bei}_n(v)$, defined in the following way:

Table 2.4—Round-Wire Values for R_i and L_i, Assuming $k_p = 1$ and $k_a = 1$

	$\omega \ll \omega_\delta$	$\omega \gg \omega_\delta$
Internal resistance R_i (Ω/m)	$\dfrac{1}{\sigma a}$	$\dfrac{\sqrt{\omega \mu}}{p\sqrt{2\sigma}}$
Internal inductance L_i (H/m)	$\dfrac{\mu}{8\pi}$	$\dfrac{\sqrt{\mu}}{p\sqrt{2\omega \sigma}}$
Internal impedance $z_i = R_i + j\omega L_i$ (Ω/m)	$\dfrac{1}{\sigma a} + j\omega \dfrac{\mu}{8\pi}$	$\dfrac{\sqrt{\omega \mu}}{p\sqrt{\sigma}} e^{j\pi/4}$

NOTE—Where a is the cross-sectional area (m^2), p is the perimeter of a cross-section of the signal conductor (m), σ is the conductivity (S/m), and μ is the permeability (H/m) of the conductor. The operating frequency is ω (rad/s). The conductor is assumed smooth and well separated from a large, low-resistance return path.

$$I_0\left(\sqrt{j}\, v\right) = J_0\left(e^{j\frac{3}{4}\pi} v\right) = \text{ber}_0(v) + j\,\text{bei}_0(v)$$
$$I_1\left(\sqrt{j}\, v\right) = (-j) J_1\left(e^{j\frac{3}{4}\pi} v\right) = \text{bei}_1(v) - j\,\text{ber}_1(v)$$
[2.49]

The functions $\text{ber}_n(v)$ and $\text{bei}_n(v)$ are available in MathCad, where they go by the name of the Bessel Kelvin functions. These same functions are tabulated elsewhere [6], [7].

Figure 2.16 plots the internal impedance predicted by [2.48]. Please keep in mind that equation [2.48] works only for round wires. Figure 2.16 depicts the real and imaginary parts of the complex internal impedance, R_i and $j\omega L_i$, slightly different from the parameters R_i and L_i (without the $j\omega$) shown in Figure 2.15. One fascinating aspect of this figure is how, at high frequencies, the real and imaginary parts of complex internal impedance converge to a common value. This convergence reveals a deep connection between the real and imaginary parts of any network function. Any causal, minimum-phase impedance function growing at a flat rate of +10 dB/decade must have a phase angle of exactly $\pi/4$. The phase angle of $\pi/4$ bestows upon the function equal real and imaginary parts.

2.8.1 Practical Modeling of Internal Impedance

Although the closed-form model [2.48] is mathematically quite beautiful, it isn't widely used for two reasons:

➢ It is too computationally intensive.
➢ It works only for round wires, not rectangular.

2.8.1 • Practical Modeling of Internal Impedance

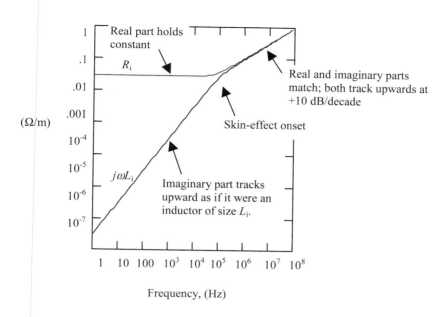

Figure 2.16—Equation [2.48] for round wires predicts both real and imaginary parts of the internal impedance.

For most signal-integrity simulations, an approximation will suffice. I'll present two possibilities here. The basic approximation produces the correct asymptotic behavior at DC and at high frequencies, but lacks accuracy in the transition region. The better approximation provides improved accuracy in the transition region. Both approximations use the same parameters:

- ω Frequency of operation (rad/s).
- R_{DC} The total DC series resistance per meter of the transmission line (Ω/m),
- ω_0 A particular frequency (rad/s), chosen well above the onset of the skin effect but below the onset of surface roughness and non-TEM modes, and
- R_0 The real part of the skin-effect impedance at the particular frequency ω_0. (Ω/m). R_0 should be set to take into account the proximity effect, if present, although in the following discussion for round-wire problems not in proximity to other conductors, you may assume $k_p \approx 1$. This value may be computed using [2.43] or [2.44].

Both approximations use the same expression for the high-frequency (AC) series impedance of the wire. Note that this expression comprises equal real and imaginary parts, meaning a phase angle of 45 degrees, which is correct for a causal, minimum-phase network function with a slope of +10 dB/decade.

$$R_{AC} = (1+j) R_0 \sqrt{\frac{\omega}{\omega_0}} \quad \Omega/m \qquad [2.50]$$

In both cases the real part of z_i will be identified as the internal resistance, while the imaginary part of z_i will be interpreted as the internal inductance: $L_i = \text{Im}[z_i]/\omega$.

Here are the two approximations for z_i.

Simple approximation:

$$z_{\text{SIMPLE}}(\omega) = R_{DC} + R_{AC} \quad \Omega/\text{m} \qquad [2.51]$$

Better approximation:

$$z_{\text{BETTER}}(\omega) = \sqrt{(R_{DC})^2 + (R_{AC})^2} \quad \Omega/\text{m} \qquad [2.52]$$

Figure 2.17 compares both approximations to the Bessel-function solution for a round wire. The asymptotic behavior of both functions is good. Both functions produce the correct impedance at very high frequencies, with the correct upward tilt of +10 dB/decade and a phase angle of +45 degrees. They also both produce the correct DC resistance at low frequencies. The difference between them lies in their behavior near the skin-effect onset frequency and in the predicted value of internal inductance.

In particular, at ω_δ the real part of z_{SIMPLE} looms far too large, exceeding R_{DC} by 6.02 dB. At that same frequency the real part of z_{BETTER} is only 2.09 dB larger than R_{DC}, much closer to the correct value (according to the Bessel-function solution) of 2.04 dB.

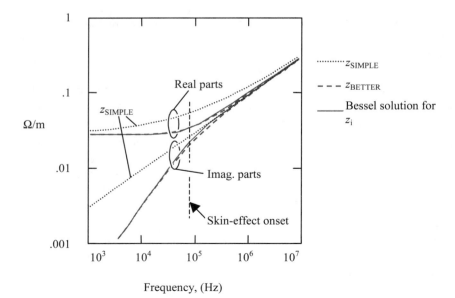

Figure 2.17—Two decades to either side of the skin-effect onset frequency, the real part of z_{SIMPLE} still errs by 10%.

The second problem with z_{SIMPLE} concerns its predicted value of L_i at low frequencies. The imaginary part of z_{SIMPLE} should have a positive slope of +20 dB/decade at low frequencies. Instead, it has a slope of only +10 dB/decade, an artifact that effectively predicts infinite inductance at DC.

These shortcomings combine to produce fractional errors in z_{SIMPLE} on the order of 50% at ω_δ, falling off to approximately 10% at frequencies two decades on either side of ω_δ. Don't use z_{SIMPLE} near the transition region. If your simulation operates exclusively at frequencies above ω_δ, then approximation z_{SIMPLE} may be somewhat improved by assuming $R_{DC} = 0$.

If you require a more accurate approximation, try z_{BETTER}. In the simulation of RG-58/U coaxial cable the peak error in z_i induced by the use of z_{SIMPLE} is 56.7%, where the error is expressed as a percentage of $|z_i|$. The error in z_{BETTER} under the same conditions is only 6.7%. Because the internal impedance z_i represents only a part of the overall series impedance z_s, calculating the same errors as a fraction of the overall series impedance z_s results in values of 25.4% and 1.4% for z_{SIMPLE} and z_{BETTER}, respectively. If you need even better accuracy in the transition region than provided by z_{BETTER}, use a two-dimensional full-wave electromagnetic field solver.

2.8.2 Special Issues Concerning Rectangular Conductors

For rectangular conductors there exist no closed-form solutions for internal inductance. Nor can the internal inductance be evaluated using ordinary quasi-static 2-D calculations, because the phase of the current deep inside the conductor differs substantially from the phase on the surface. This difference in phase introduces new complications into the mathematics of the problem—complications not handled by commonly available 2-D field simulation software.

Fortunately, the exact value of L_i matters little in the computation of many signal-integrity results, because the external inductance L_e always exceeds the internal inductance L_i for all but the very lowest-impedance configurations. As the frequency rises above ω_δ, the difference becomes even more pronounced so that any error in L_i is quickly swamped. The largest simulation errors lie near the skin-effect onset frequency, falling off on either side of the onset. For signal-integrity simulations of transmission lines at reasonable impedances (i.e., $L_e \gg L_i$) or at frequencies well above ω_δ, this author uses approximation z_{BETTER} for rectangular conductors.

Below the skin-effect onset, the magnitude of the internal inductance L_i plays a role in determining the phase delay of the transmission line. For nonround conductors, approximation z_{BETTER} makes some particular assumptions about the low-frequency value of L_i. The values of L_i implied by the use of approximation z_{BETTER} are found by evaluating $\text{Im}(z_{BETTER})/\omega$ in the limit as $\omega \to 0$.

$$L_i(0) = \text{Lim}_{\omega \to 0} \frac{\text{Im}\left[\sqrt{(R_{DC})^2 + \left((1+j)R_0\sqrt{\frac{\omega}{\omega_0}}\right)^2}\right]}{\omega} \quad [2.53]$$

The evaluation proceeds by first factoring R_{DC} from under the radical.

$$L_i(0) = \text{Lim}_{\omega \to 0} \frac{\text{Im}\left[R_{DC}\sqrt{1 + 2j\left(\frac{R_0}{R_{DC}}\right)^2 \frac{\omega}{\omega_0}}\right]}{\omega} \quad [2.54]$$

Take the first two terms in the Taylor's series approximation for $\sqrt{1+x} = 1 + (1/2)x + ...$

$$L_i(0) = \text{Lim}_{\omega \to 0} \frac{\text{Im}\left[R_{DC}\left(1 + j\left(\frac{R_0}{R_{DC}}\right)^2 \frac{\omega}{\omega_0}\right)\right]}{\omega} \quad [2.55]$$

Take the imaginary part and simplify the expression.

$$L_i(0) = \frac{R_0^2}{R_{DC}\,\omega_0} \quad [2.56]$$

Substitute [2.37] for R_{DC}. Substitute [2.43] for real part of the skin-effect impedance, R_0, evaluated at frequency ω_0. Assume $k_a = k_r = 1$.

$$L_i(0) = \left(\frac{k_p}{p\delta\sigma}\right)^2 (a\sigma)\frac{1}{\omega_0} \quad [2.57]$$

Plug in definition of the skin depth.

$$L_i(0) = \left(\frac{k_p}{p}\frac{\sqrt{\omega_0\mu\sigma}}{\sqrt{2}}\frac{1}{\sigma}\right)^2 (a\sigma)\frac{1}{\omega_0} \quad [2.58]$$

The final result is

$$L_i(0) = \frac{\mu a}{2}\left[\frac{k_p}{p}\right]^2 \qquad [2.59]$$

Table 2.5 lists the low-frequency internal inductance values asserted by approximation z_{BETTER} for various conductor geometries. Please don't make the mistake of assuming these are authoritative values of $L_i(0)$ for the listed geometries (except for the round wire, which is exact). These are merely *reasonable assumptions* made by the approximation z_{BETTER}.

If you need more accurate models in the vicinity of the skin-effect onset, you should use a 3-D full-wave electromagnetic field solver.

POINTS TO REMEMBER

- Merely adding the DC and AC models of resistance produces substantial errors at frequencies near the onset of the skin effect, and predicts the wrong value of internal inductance.
- The second-order approximation [2.52] better matches both the real and imaginary parts of the skin effect at frequencies near the transition region.

2.9 CONCENTRIC-RING SKIN-EFFECT MODEL

The skin effect does not depend on any mysterious underlying forces. It may be completely predicted based on one simple, lumped-element equivalent model called the concentric-ring model.

Table 2.5—Assumptions Made by z_{BETTER}

Conductor Geometry	$L_i(0)$
Round wire of any radius not in close proximity to a return path ($k_p = 1$)	$\dfrac{\mu}{8\pi}$ H/m
Wire of any convex cross section with perimeter p and area a	$\dfrac{\mu a}{2}\left[\dfrac{k_p}{p}\right]^2$ H/m
Rectangular wire of size $w \times t$	$\dfrac{\mu}{8}\dfrac{wtk_p^2}{(w+t)^2}$ H/m
Square wire of size $w \times w$	$\dfrac{\mu k_p^2}{32}$ H/m
Very thin rectangular wire, $t \ll w$	$\approx \dfrac{\mu k_p^2}{8}\dfrac{t}{w}$ H/m

Imagine a round conductor divided lengthwise into concentric tubes, like the growth rings on a tree trunk. In this model the current proceeds absolutely parallel to the wire's central axis. Because current naturally passes straight down the trunk, parallel to the dividing boundaries between the rings, you may insulate the rings from each other without affecting the circuit. You now have a collection of n distinct, isolated conductors, shorted together only at each end of the trunk.

You may now separately consider the inductance of each ring. The inner rings, like long, skinny pipes, have more inductance than the outer rings, which are fatter. You know that at high frequencies, current follows the path of least inductance. Therefore, at high frequencies you should expect more current in the outer tree rings than in the inner. That is exactly what happens. At high frequencies the current crowds into the outermost rings. At low frequencies current partitions itself according to the resistance of each ring, while at high frequencies it partitions itself according of the inductance of each ring.

This simple concentric-ring idea motivates the idea of a redistribution of current at higher frequencies. What it doesn't do, however, is properly indicate the magnitude of the effect. The skin-effect mechanism is far more powerful than just the ratio of individual concentric-ring inductances. Mutual inductance between the rings actually bunches the current much more tightly onto the outer rings than you might at first imagine. The general setup for constructing a tree-ring model, including the mutual inductance, is discussed in the article "Modeling Skin Effect." The concentric-ring circuit model, if taken to an extreme (hundreds of thousands of rings), properly predicts both skin-effect resistance and skin-effect inductance, at low and high frequencies, for a circular conductor.

2.9.1 Modeling Skin Effect

Article first published in *EDN Magazine*, April 12, 2001

Why does high-frequency current flow only on the outer surface of a printed-circuit trace?
—Dipak Patel

Magnetic fields cause the behavior you describe. The technical name for this property is the *skin effect*. It happens in all conductors. If you really like mathematics, the following section will help you to better understand *why* the skin effect happens. If not, this might be a good time to step out for a cup of tea.

> **At high frequencies magnetic interactions between conductors become significant.**

I'll start our discussion with a perfect coaxial cable. Figure 2.18 divides the center conductor of this cable into a series of three concentric rings with radii r_0, r_1, and r_2 (meters). A lumped-element model of this simple circuit demonstrates that high-frequency signal current flows only on ring number 2.

At DC, the longitudinal voltage drop per meter across each conductor n equals the current i_n times its resistance per meter. You can express this relation in matrix terms by defining a square matrix **R** relating the circuit voltages **V** to the currents **I**:

2.9.1 • Modeling Skin Effect

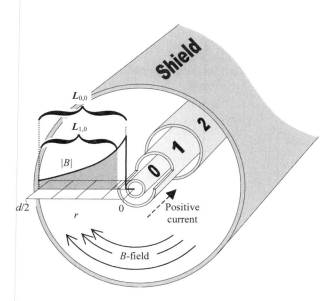

Figure 2.18—The total magnetic flux within the shaded region equals $L_{1,0}$.

$$\mathbf{R}_{n,m} = \begin{cases} \dfrac{1}{\pi\left(r_n^2 - r_{n-1}^2\right)\sigma} & \text{on the diagonal where } n = m \\ 0 & \text{otherwise} \end{cases} \quad [2.60]$$

$$\mathbf{V} = \mathbf{RI}$$

where vector $\mathbf{V}$ represents the longitudinal voltage across the ends of each conductor (V/m),
vector $\mathbf{I}$ represents the current in each conductor (A),
$\mathbf{R}$ is a matrix of resistance values (Ω/m),
σ is the conductivity of the center conductor (S/m), and
r_n represents the radius of each concentric ring (m), with $r_0 = 0$.

At high frequencies magnetic interactions between the conductors become significant. Figure 2.18 illustrates the pattern of magnetic fields between the center conductor and shield. The magnetic lines of force (*B*-field) form concentric circles around the conductive rings. The drawing plots the field intensity, $|B|$, versus radial position, r, assuming a positive signal current of 1A flowing in ring 0 with the return current flowing in the shield. The field strength is zero within the interior of ring 0 and zero outside the shield, and varies with $1/r$ (Ampere's law) in between. The exact field intensity for a current of 1A on conductor m is $B_m(r) = \dfrac{\mu}{2\pi}\dfrac{1}{r}$ for $r_m < r < d/2$, where μ is the magnetic permeability of the dielectric material (usually $4\pi 10^{-7}$ H/m).

You calculate the mutual inductance per meter between conductors n and m (for $n \geq m$) using Faraday's law as the integral of the magnetic-field strength, B_m, taken over the range from conductor n (at radial position r_n) to the shield (at radial position $d/2$). Integrating $1/r$ yields $\ln(r)$ and the following matrix equations for mutual inductance (values for $n < m$ are found using symmetry: $L_{n,m} = L_{m,n}$):

$$L_{n,m} = \begin{cases} \dfrac{\mu}{2\pi} \ln\left(\dfrac{d}{2r_n}\right) & \text{for } n \geq m \\ \dfrac{\mu}{2\pi} \ln\left(\dfrac{d}{2r_m}\right) & \text{for } n < m \end{cases} \quad [2.61]$$

The system equation for the whole coaxial circuit sums both resistive and inductive terms as $\mathbf{V} = (\mathbf{R} + j\omega\mathbf{L})\mathbf{I}$. As an example, the following is the inductance matrix for a three-ring model of an RG-58/U coaxial cable:

$$\mathbf{L} = \begin{bmatrix} 476 & 339 & 256 \\ 339 & 339 & 256 \\ 256 & 256 & 256 \end{bmatrix} \text{ nH/m} \quad [2.62]$$

Now comes the main point of this article: The terms in the right-hand column of $\mathbf{L}$ are all the same. Why? Because ring 2 concentrates *all its flux* into the space between ring 2 and the shield. Therefore, all other rings couple 100% to this flux.

The constancy of the right-hand column greatly simplifies the solution to the system equation. To solve this equation, you must find a pattern of currents $\mathbf{I}$ such that $(\mathbf{R} + j\omega\mathbf{L})\mathbf{I}$ generates the same longitudinal voltage across every ring. You need the same voltage across every ring because the rings are all connected together at their ends. If you operate at a frequency so high that the $\mathbf{R}$ term becomes insignificant compared to $j\omega\mathbf{L}$, the solution is simple. Just fill in the last element of $\mathbf{I}$, leaving all others zero. This solution peels off only the right-hand column of $\mathbf{L}$, properly generating the same voltage for every ring. This is one of the few matrix problems you can solve by inspection.

The simple solution says that at high frequencies the signal current flows only on the outer ring as governed by matrix $\mathbf{L}$. At DC, the current distributes itself more evenly, according to matrix $\mathbf{R}$. At middle frequencies, you get a mixture of both effects. That's the nature of the skin effect.

Real-world conductors behave in a similar manner, as if they were made from a continuum of infinitely thin concentric rings. At higher and higher frequencies, the current squeezes more and more tightly against the surface of the conductor, progressively decreasing the useful current-carrying cross section of the conductor and raising its effective resistance.

2.9.2 Regarding Modeling Skin Effect

Email correspondence received July 17, 2001
SE-HO YOU writes

You have shown an inductance matrix and indicated that since all the entries in the right column of the matrix are the same, ring number 2 concentrates all its flux into the space between ring number 2 and the shield.

Would you please explain *why* flux concentrates around ring number 2? The behavior of the flux comes first, I think. Then, the inductance matrix should be just a mathematical expression of how nature works. Could you explain more physically?

Reply

Eddy currents flowing in the outer ring create a magnetic shield through which flux cannot penetrate. The shielding effect of the outermost ring therefore prevents flux from reaching the inner rings. Receiving no electromagnetic impulsion from changing flux, no current flows on the inner rings.

At low frequencies the eddy currents in the outer ring are impeded by the resistance of the copper, so the shielding effect is imperfect. With an imperfect shield, *some* magnetic fields do reach the interior and *some* current does indeed flow on the inner rings.

As you go to higher and higher frequencies, however, the shielding effect becomes more pronounced. The improved shielding effect successively robs the inner rings of more and more flux (and current).

2.10 PROXIMITY EFFECT

High-frequency current in a round wire flows mostly on the surface of the wire, but not in a uniform distribution around the perimeter. The magnetic fields from the wire and its associated return current distribute the current around the perimeter in a slightly nonuniform way (Figure 2.19), which in turn increases the apparent resistance of the conductors above and beyond what you would expect from the action of the skin effect alone. In parallel conductors we call this additional increase in resistance the *proximity effect*.

The proximity effect stands apart from the skin effect, which is what holds high-frequency current in a shallow band of depth δ around the perimeter of a conductor (see previous section). The proximity effect also stands apart from Ampere's discovery that adjacent wires carrying opposing DC currents repel. While Ampere's forces push the atomic lattice structure of the two wires apart, and the skin effect binds current tightly to the surface, the proximity effect merely redistributes the AC current density around the perimeters of the two wires. The proximity effect exerts no net mechanical force on the wires.

The proximity effect is an inductive mechanism caused by changing magnetic fields. It perturbs the flow of high-frequency currents. It ignores steady currents that generate static magnetic fields. Above that frequency where the proximity effect takes hold, the distribution

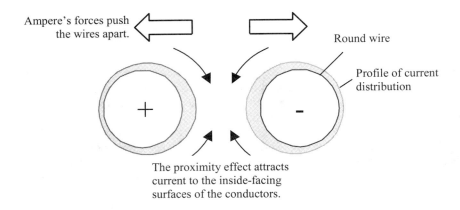

Figure 2.19—The proximity effect redistributes the high-frequency current around the perimeter of a conductor.

of current around the perimeter of the conductor attains a minimum-inductance configuration and does not vary further with frequency (until you reach the onset of non-TEM modes of propagation).

Figure 2.12 illustrates the magnetic fields within a single conductor, showing how self-induced eddy currents restrict the flow of current to a shallow band just underneath the surface of the conductor. The proximity effect operates by a similar mechanism. In the case of the proximity effect magnetic fields from a first wire induce eddy currents on a second wire, redistributing the current on the surface of a second conductor (Figure 2.20). The current on the second conductor is still bound tightly to a shallow band just underneath the surface by its own internal skin effect, but the proximity effect redistributes the current around the perimeter of the second conductor.

The diagram shows eddy currents circulating about the magnetic field, penetrating the top of the second conductor, and more eddy currents circulating on the bottom where the magnetic lines of force exit. The net effect of these eddy currents is to concentrate more current on the inside-facing surfaces of the conductors and less on the outward-facing sides. For a good conductor, at frequencies in excess of ω_δ, the concentration builds until the magnetic field no longer penetrates the second conductor.

The skin effect and the proximity effect are two manifestations of the same principle: that magnetic lines of flux cannot penetrate a good conductor. The difference between the effects is that the skin effect is the reaction to magnetic fields generated by current flowing *within* the affected conductor, while the proximity effect is the reaction to magnetic fields generated by current flowing in *other conductors*. Both effects operate by the same guiding principle, namely, that no high-frequency magnetic fields shall penetrate the conductor. The frequency at which the proximity effect takes hold is therefore the same as that frequency ω_δ where the skin effect takes hold. Figure 2.21 shows the pattern of magnetic lines of force surrounding two round conductors at a frequency well above ω_δ. As you can see, the lines of magnetic force lie tangent to the surfaces of the conductors.

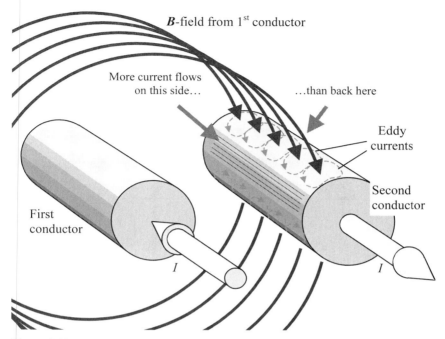

Figure 2.20—A changing magnetic field from a first conductor induces eddy currents on the surface of the second conductor, changing its distribution of current.

In any 2-D magnetic field picture the density of the magnetic lines of force indicates the strength of the magnetic field. Around the perimeter of the conductors, the density of current at a particular point on the surface of the conductor profile is proportional to the magnetic field at that point. In Figure 2.21 the magnetic lines of force clearly lie closer together on the inside-facing sides of the two conductors, indicating a preponderance of field intensity, and therefore current, on the inside-facing sides of the conductors.

2.10.1 Proximity Factor

The proximity factor represents the increase in the apparent AC resistance of conductors above and beyond what you would expect from the action of the skin effect alone. The proximity factor appears in the equation for AC resistance (see Section 2.6, "Skin Effect"):

$$\operatorname{Re}[R_{AC}] = \frac{k_p \sqrt{\omega \mu}}{p \sqrt{2\sigma}} \qquad [2.63]$$

where k_p is a correction factor determined by the proximity effect,

the correction factor for the roughness effect is ignored, (see Section 2.11, "Surface Roughness"),

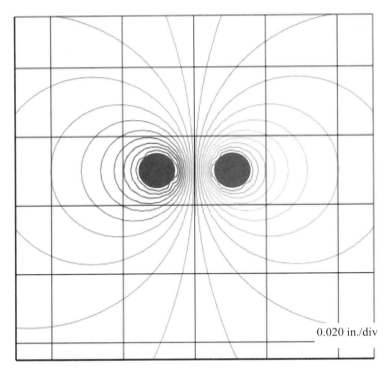

Figure 2.21—This cross-sectional view of the magnetic field in the vicinity of two round conductors shows the magnetic lines of force lie tangent to the surfaces of the conductors. Perturbations in the smoothness of the magnetic lines of force are artifacts of the spatial quantization used by a field solver to generate this diagram.

ω is the frequency of operation (rad/s),

$\text{Re}[R_{AC}]$ is the series resistance of the conductor taking into account both skin effect and proximity effect (Ω/m), and assuming $\omega \gg \omega_\delta$,

p is the perimeter of a cross-section of the signal conductor (m),

μ is the magnetic permeability of the conductor (H/m), and

σ is the resistivity of the wire (S/m).

For annealed copper $\sigma = 5.800 \cdot 10^7$ S/m.

For non-magnetic materials $\mu = 4\pi \cdot 10^{-7}$ H/m.

The factor k_p is technically defined as the ratio of (1) the actual AC resistance to (2) the resistance calculated assuming a uniform distribution of current around the perimeter of the signal conductor and ignoring the resistance of the return conductor.

The general behavior of k_p is as follows.

> Any conductor well separated from a low-resistance return path has $k_p \approx 1$.

2.10.1 · Proximity Factor

> Differential configurations have $k_p \approx 2$ (and also $k_a \approx 2$).

> As the conductor and its return path are brought more closely together, k_p increases.

> Whatever the value of k_p, at frequencies below ω_δ it has no effect on the transmission response; only the DC resistance matters at frequencies below ω_δ.

For a pair of round wires, the magnitude of the proximity factor k_p is determined by the ratio s/d, where s is the wire separation between centers and d is the wire diameter, as shown in Figure 2.22. At large ratios of s/d the proximity factor asymptotically approaches 2, representing the fact that in a twisted-pair cable the series resistance is always twice the resistance of a single wire alone. At a ratio of $s/d = 2.0$, corresponding roughly to the typical configuration of a 100-ohm twisted-pair cable, the proximity factor for round wires is 2.30. The proximity factor soars when the two wires almost touch.

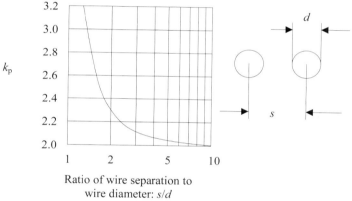

Figure 2.22—The proximity factor for round wires increases as they are brought together. (Adapted from Frederick Terman, *Radio Engineer's Handbook*, McGraw-Hill, New York, 1943, p. 43).

Figure 2.13 depicts for various conductors the frequency at which the skin and proximity effects take hold. For a conductor operating below its onset frequency, the magnetic forces due to changing currents in the conductor are too small, compared to the resistive forces, to influence the pattern of current flow. Low-frequency current therefore follows the *path of least resistance*. The path of least resistance fills the volume of every conductor, flowing uniformly throughout the cross section of the conductor.

Above the onset frequency, the magnetic forces, which grow in proportion to frequency, exceed the resistive forces, so it is the magnetic (inductive) effect that determines the path of current. Above the onset frequency, current flows in the *path of least inductance*.

The general principles of high-frequency current flow follow:

> Current in a conductor at high frequencies distributes itself to neutralize all the internal magnetic forces.

- At high frequencies magnetic lines of force will not penetrate a conducting surface.
- At high frequencies the lines of magnetic force flow tangent to every conducting surface.
- In more technical terms, the component of the magnetic field normal to a (good) conducting surface is (nearly) zero.
- Current at high frequencies distributes itself in that pattern which minimizes the total potential energy stored in the magnetic field surrounding the conductors.
- Current at high frequencies distributes itself in that pattern which minimizes the total inductance of the loop formed by the outgoing and returning current.
- Given a choice, returning signal current prefers to flow near the outbound signal path.
- In a high-speed pcb with a solid reference plane the return current for each signal flows on the underlying reference plane following closely underneath the signal trace.

All of the above viewpoints are correct, and they are all equivalent. The effective series resistance of a conductor in the zone far above ω_δ is usually found using a tool called a 2-D electromagnetic field solver. Such tools calculate the proximity effect for conductors of arbitrary shape. Some hints about the operation of this class of tools is provided in Section 2.10.4.2, "2-D Quasi-Static Field Solvers."

Near the onset region, the exact interactions between the skin effect and proximity effect are not easily modeled. Most 2-D modeling software works well below the onset frequency where only the DC resistance matters. It also works well above the onset frequency where only the perimeter, skin depth, and proximity factor k_p matter. Near the onset region, the software uses a generic mixing function to transition gradually from one mode to the other. The model is inexact near the transition region. In the modeling of typical pcb configurations this inaccuracy makes little difference, because the interesting problems usually occur well above the onset frequency. The differences between mixing functions are discussed in Section 2.8.1, "Practical Modeling of Internal Impedance."

2.10.2 Proximity Effect for Coaxial Cables

Coaxial cables, owing to the concentricity of their conductors, exhibit a very simple distribution of high-frequency current. Around the center conductor the current is evenly distributed. Around the shield it is also evenly distributed albeit at a lower current density related to the ratio between the diameter of the center conductor and the diameter of the shield.

The calculation of k_p takes into account the sum of series resistances of the signal conductor and the shield:

$$k_p = \left(1 + \frac{d_{CENTER}}{d_{SHIELD}}\right) \qquad [2.64]$$

where d_{CENTER} is the outside diameter of the center conductor (m), and

d_{SHIELD} is the inside diameter of the shield conductor (m).

2.10.3 Proximity Effect for Microstrip and Stripline Circuits

Microstrip and stripline circuits fall subject to the proximity effect in the same way as circular conductors. The proximity effect draws signal current towards the reference plane-facing side of the (microstrip or stripline) trace, and simultaneously concentrates the returning signal current on the reference plane, forming it into a narrow band that flows on the plane, mostly staying underneath the signal trace.

Chapters 5 and 6 include tables of proximity factors for both single-ended and differential configurations.

2.10.4 Last Words on Proximity Effect

I shall close this section with two brief articles about the proximity effect. The first article outlines one algorithm for computing the proximity effect. This algorithm can be implemented in any general-purpose math spreadsheet. The second article concerns general limitations that apply to all 2-D simulators.

2.10.4.1 Proximity Effect II

High-Speed Digital Design Online Newsletter, **Vol. 4, Issue 3**
Bill Stutz writes

I am reading your excellent book and would like to ask a couple of questions related to the return current issue presented in Vol.3, #11 of your newsletter.

Do you have any references dealing with the proximity effect in more detail (other than Terman) and more specifically with the current density distribution in a ground plane under a high-frequency signal trace?

I am interested in both a numerical calculation of the distribution as well as the use of partial differential equations to directly derive the form of the solution.

Thank you in advance for your reply.

Reply

Thanks for your interest in High-Speed Digital Design.

To answer both questions, I will have to point you in the direction of the literature on electromagnetic field simulation. Both effects are consequences of Maxwell's equations. Both can be observed by simulating various field patterns. There is no simple, correct explanation for either effect except merely to present, in a behavioral sense, what the currents tend to do.

Good references for E&M field behavior include [24] and [5]. WARNING: both are highly mathematical.

Here I'll outline one method of simulating the fields for the proximity-effect problem, in the hope that you may glean from this method some insights into how currents flow in solid conductors.

My method is to first model the surface of each conductor involved in the problem. Since the skin effect causes high-frequency currents to flow only near the surface of each conductor, a surface-only model should be adequate for most transmission-line problems. Furthermore, if we assume that the traces proceed in the z direction, and that the current distribution around the circumference of each conductor is constant with linear position z, we need only model the x-y cross-sectional view of the surface of each conductor. So, I begin by drawing a line around the cross section of each conductor (traces and planes) and breaking that line into a succession of little line segments. I assume the current density over each element is constant. [ed. note—A significant enhancement of the above technique linearly interpolates the current density between points spaced around the perimeter of the conductor. In either case the effective current density is specified by only one value per segment.]

Now I can represent the problem as one of finding the current for each little element that satisfies a number of constraints:

1. The sum of all currents in the signal conductor equals I_0,
2. The sum of all currents in the return conductor(s) equals $-I_0$,
3. On the surface of the signal conductor, the total magnetic flux penetrating each segment is zero (true if the voltage potential everywhere along the surface of the conductor remains constant, which it pretty much does for a good conductor), and
4. On the surface of the return conductor(s), the total magnetic flux penetrating each segment is zero.

Another way to state constraints 3 and 4 is to say that the magnetic field is parallel to the conducting surfaces at all points (or that the component of the magnetic field perpendicular to the surfaces must be zero).

In terms of a solution algorithm, if you have N conductor segments, it would be convenient to work with exactly N constraints. We know that 1 and 2 together comprise precisely two constraints, so for constraints 3 and 4, one would normally pick a total of $N-2$ points at which to evaluate these constraints. Now you have N variables and N constraints, and there exists one unique solution.

One simple solution procedure calculates the perpendicular magnetic field at each constraint point as a linear function of the currents on all the other segments, constructs a big matrix representing all the constraints, and then inverts it to find a final solution. This is neither the most efficient nor the most accurate solution method, but it's the easiest to visualize. If you know how to write an expression for the magnetic field surrounding one element (a long, straight wire) you can program this in MathCad.

Another way to find the solution is to guess some basic distribution of currents and then calculate the magnetic fields. If this current pattern were to exist in nature, then wherever the perpendicular magnetic field between two adjacent current elements is nonzero, it would cause an increase in the current in one element and a decrease in the other. Make appropriate adjustments. Then recompute the field patterns, and adjust again. Iterate until you arrive at the final solution. This is how nature solves the problem.

2.10.4 • Last Words on Proximity Effect

Either approach leads to the conclusion that for the case of a small, thin trace located near a solid reference plane, the correct distribution of currents on the solid plane is given by

$$\frac{(I_0/\pi h)}{1+(d/h)^2} \qquad [2.65]$$

where I_0 = the total signal current, in amps.

h = height of trace about the nearest reference plane, and

d = horizontal distance away from the centerline of the trace.

This relation is the basis for my simple crosstalk estimates (and especially the conclusion that, given d much bigger than h, the crosstalk falls off with $1/d^2$). A closed-form derivation of [2.65] appears in my newsletter vol.4, #8, Proximity Effect III.

Modeling a realistic-sized trace above a solid plane, you will find that the current density is slightly greater on the reference-plane side of the trace than on the reverse side. This phenomenon is sometimes called the proximity effect. The same thing happens for two skinny wires placed in close proximity—the current tends to concentrate on the two facing surfaces. The proximity effect is a simple manifestation of the general rule that, given a choice, high-speed current tends to concentrate near its return path.

I have presented just an outline of the simulation procedure. If you want more details, another good place to look is the EMI/EMC Computational Modeling Handbook [23]. I suggest you start your research there.

2.10.4.2 2-D Quasistatic Field Solvers

Article first published in *EDN Magazine*, September 27, 2001

I love signal-integrity simulators. Unfortunately, they don't always produce the right answers. For example, most signal-integrity software packages calculate the impedance and loss of transmission lines using a 2-D, quasi-static, discrete field solver. The field solver depends on six crucial assumptions. If your system violates any of these assumptions, the simulator produces wrong answers.

The Fringing-Field Assumption

Two-dimensional field solvers do not calculate fringing fields at the ends of conductors. This omission seems reasonable as long as the main effects in the middle of the line vastly exceed the fringing-field effects at the ends. To satisfy this requirement, the length of a transmission line must vastly exceed the separation between its conductors.

> **Simulators don't always produce the right answers.**

For typical pcb-trace dimensions, the fringing-field assumption holds. For example, a 2.5-cm (1-in.) transmission line placed 125 μm (0.005 in.) above the reference plane has a length-to-separation ratio of 200 to 1. Under these conditions, the end effects probably have a less than 1% overall effect on the behavior of the line.

The Assumption of Uniformity

If, in addition to being long, the transmission line possesses a uniform cross section, you may assume by symmetry that in every 2-D cross-sectional slice of the line the per-unit-length values of R, L, G, and C are the same. The software can then perform its impedance and crosstalk-coupling analysis only once for a single 2-D cross-sectional slice.

The assumption of uniformity reduces the complexity of the simulation problem from a full 3-D simulation problem to a problem involving only one cross section (a 2-D problem).

Any percentage imperfections or wobbles in the width and height of the line directly impact the model's accuracy. For example, if your trace width is specified as 3-mil (±1), the modeling error could be as great as ±33%.

The Quasi-Static Assumption

When your simulator calculates the distribution of current (or charge) across the face of one particular 2-D slice of a transmission line, it ignores the phase. Quasi-static analysis assumes the phase of current is uniform everywhere across the slice. The quasi-static assumption works only when transmission-line waves propagate in a TEM (transverse electromagnetic) mode, which requires that the signal wavelength greatly exceed the conductor separation.

Typical pcb traces at frequencies less than 10 GHz comply with the quasi-static assumption. For example, an FR-4 stripline placed 125 μm (0.005 in.) above the nearest reference plane has a wavelength-to-separation ratio at 10 GHz of better than 100 to 1. Under these conditions, any quasi-static behavior probably has a less than 1% overall effect on the behavior of the line.

The Small Skin-Depth Assumption

The inductance of a transmission line changes slightly at frequencies near the onset of the skin effect. To avoid having to contemplate frequency-varying values for inductance, most programs assume that your design operates at a frequency far above the onset of the skin effect so that changes in inductance become insignificant.

At such a high frequency, the skin depth is small, so current flows only in a shallow band just beneath the surface of each conductor—not in the middle of the conductor. The small-skin-depth assumption allows the software to calculate values of the current distribution only around the (1-D) perimeter of each conductor in a particular 2-D slice instead of computing the exact distribution throughout the entire (2-D) body of the 2-D slice. This assumption reduces the complexity of the simulation from a 2-D problem to a 1-D problem.

For rectangular traces at pcb dimensions of $w = 0.008$ in. and $t = 0.00065$ in. (½-oz copper), the skin-effect onset happens at the following frequency:

$$f_\delta = \frac{1}{2\pi} \frac{8\rho}{\mu} \left(\frac{w+t}{wt} \right)^2 = \frac{1}{2\pi} \frac{8 \cdot 6.787 \cdot 10^{-7}\, \Omega\text{-in}}{3.192 \cdot 10^{-8}\, \text{Wb/A-in}} \left(\frac{(.0080 + .00065)\,\text{in}}{(.0080 \cdot .00065)\,\text{in}^2} \right)^2 = 75\,\text{MHz} \quad [2.66]$$

At frequencies well above the skin-depth onset, the (1-D) perimeter calculations yield the correct answer. Furthermore, standard assumptions about how the skin

2.10.4 • Last Words on Proximity Effect

effect works reasonably extrapolate the changes in inductance at lower frequencies. If, however, you are simulating conductors at frequencies near the skin-effect-onset frequency, a more comprehensive 2-D simulation of the complete current distribution may be necessary.

The Discrete Assumption

A 2-D quasi-static field solver represents the perimeter of each conductor within a 2-D cross-sectional slice as a collection of short line segments. It represents the current density around the perimeter as a 1-D vector, with each element of the vector specifying the current density in one segment. Simulators make different assumptions about the interpolation of current values as you move within a segment and between segments around the perimeter of a conductor.

Obviously, this discrete approach to the problem works only when the size of the discrete segments is small compared with the curvature of the conductors. Commercial simulators rarely describe in a forthcoming manner the degree of imperfection that their discrete approximations introduce.

The Round-Corner Assumption

Field simulators generate slightly erroneous results at corners. Most generate better-looking results (with less spurious peaking at the corners) if you round off the corners of your conductors before doing the computations, because doing so reduces the curvature of the simulated structure. However, if your corners aren't rounded in the real world, you may wonder what effect the artificial rounding has on the accuracy of the results. I do.

For further reading, see [23], [24], and [5].

POINTS TO REMEMBER

➤ The proximity effect distributes AC current unevenly around the perimeter of a conductor.

➤ The proximity factor increases the apparent AC resistance of a conductor above and beyond what you would expect from the action of the skin effect alone.

➤ Above that frequency where the proximity effect takes hold, the distribution of current around the perimeter of the conductor attains a minimum-inductance configuration and does not vary further with frequency.

➤ The skin effect and the proximity effect are two manifestations of the same principle: that magnetic lines of flux cannot penetrate a good conductor.

➤ Field simulators base their calculations on many assumptions, and don't always produce the right answers.

2.11 SURFACE ROUGHNESS

At a microscopic scale no surface appears perfectly smooth. All materials exhibit surface irregularities and bumps. One measure of surface roughness is the root-mean-square (RMS) height h_{RMS} of the surface bumps.

At low frequencies the depth of penetration of current (the skin depth) exceeds h_{RMS}. Low-frequency currents therefore submarine below the surface bumps, unaffected by surface roughness.

High-frequency currents, on the other hand, remain tightly bound to the surface at all times. At frequencies so high that the skin depth δ shrinks to less than h_{RMS}, the current follows the contours of the surface, over hill and down dale, as it flows along the conductor. At these high frequencies the apparent resistance of the material increases to a value representative of the additional distance over which the current must flow to traverse the contours of the surface (Figure 2.23).

2.11.1 Severity of Surface Roughness

If the average incline of the conducting surface is 60 degrees (as if patterned with an infinite array of equilateral ridges perpendicular to the direction of current flow), the ultimate increase in surface resistance would be 100%. That's a pretty severe effect (Figure 2.24).

The exact dependence of skin-effect resistance on surface roughness defies analysis, as it depends not only on the height of the bumps, but also on their horizontal extent, spacing, and exact shape. Long, sinuous undulations don't cause much of a problem, whereas short, choppy, steep-faced bumps significantly increase the path length of the current. An infinite array of equilateral 3-D pyramids induces less of an effect than a series of equilateral ridges because, as anyone who hikes in the wilderness can tell you, you don't have to traverse every peak in order to make your way through the mountains.

Talk with your pcb vendor to explore the variety of surface treatments available that are compatible with your dielectric material.

At low frequencies the skin-depth exceeds the RMS roughness of the conductor, allowing current to submarine below the rough spots.

At high frequencies the skin-depth shrinks to less than the RMS roughness of the conductor, forcing current to traverse every hill and dale on the surface.

Figure 2.23—Surface roughness increases the apparent high-frequency resistance of a conductor.

2.11.2 Onset of Roughness Effect

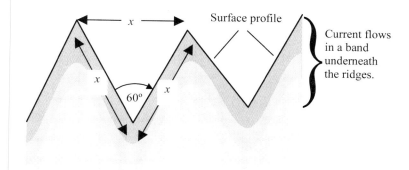

Figure 2.24—Current tightly bound to a series of equilateral ridges traverses distance 2*x* for every distance *x* made good along the surface.

The onset frequency for the RMS surface roughness effect depends on the square of the RMS surface roughness:

$$\omega_{\text{rough}} = \frac{2}{\mu \sigma h_{\text{RMS}}^2} \qquad [2.67]$$

where μ is the magnetic permeability (H/m),

σ is the conductivity (S/m) of the substrate,

h_{RMS} is the RMS bump height (m), and

ω_{rough} is the frequency at which the roughness effect has assumed 60% of its ultimate effect (rad/s).

For electro-deposited copper at room temperature $\mu = 4\pi \cdot 10^{-7}$ H/m and $\sigma = 5.98 \cdot 10^7$ S/m.

At the onset frequency the effective resistance of the conductor has progressed 60% of the way from the low-frequency asymptote (no effect) to the high-frequency asymptote (full effect). Figure 2.25 indicates the variation in roughness effect induced by an infinite array of equilateral ridges perpendicular to the direction of current flow. The equilateral ridges induce an ultimate effect of 100%, doubling the effective resistance of the material. Less severe geometries induce smaller ultimate effects, although probably with similar variations versus frequency.

2.11.3 Roughness of Pcb Materials

Surface roughness plays an especially important role when working with pcb traces. This happens to a greater degree than with extruded conductors because pcb traces are often either chemically etched, leaving a naturally rough surface, or pressed onto rough substrate

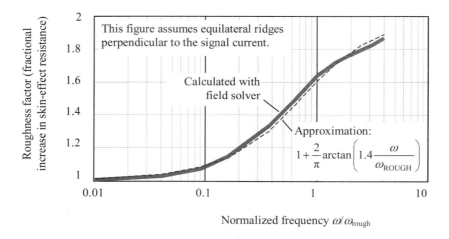

Figure 2.25—At the onset frequency ω_{rough} the surface roughness has attained 60% of its ultimate effect (data taken from Harper, *Electronic Packaging and Interconnection Handbook*, 3rd ed., McGraw-Hill, 2000).

materials, leaving a mechanically imprinted pattern of roughness. Increased resistance in some cases as high as 10% to 50% above and beyond ordinary skin-effect considerations has been attributed to surface roughness.

The roughness of the copper layers used in pcb materials is often rated in terms of the average peak-to-valley height. The relation between peak-to-valley height and RMS roughness is not clear for practical materials. Taking the worst-case equilateral-ridge geometry as a guide, you might expect the RMS roughness to be on the order of 0.29 times the average peak-to-valley roughness.

Typical roughness treatments suitable for use with FR-4 dielectrics have been reported ranging from 6 to 18 μm (0.24 to 0.71 mil). Choosing the low end of this roughness range, and assuming an average peak-to-valley height of 6 μm corresponds to an RMS height of 1.7 μm (0.07 mil), the onset frequency works out to just over 1 GHz. Below 1 GHz you wouldn't notice the roughness effect; above it you would. Moving to a different material at the high end of the roughness range drops the onset frequency by a factor of nine, lowering it into the vicinity of a couple-of-hundred MHz.

2.11.4 Controlling Roughness

Vendors of pcb materials refer to the *toothing profile* of their cores when speaking about surface roughness. Toothing profiles are purposefully etched into the copper to facilitate adhesion between layers. Smoother materials exhibit less of a roughness effect [8], [9].

You may not have much choice about the surface treatment used on the inside surfaces of a core layer—the core manufacturers make them pretty rough. What you can control are the surfaces on the outside of the core that are processed by your pcb fabricator. Stack your board so these are the heavy current-carrying surfaces on the bottom side (reference-plane-facing side) of your highest-frequency traces. That's where the current density is the highest and where surface roughness matters the most.

2.11.4 • Controlling Roughness

Pcb fabricators have access to many different surface treatments. The reverse-treat foil (RTF) process makes a surface that looks like the Himalayas. Sulfuric peroxide treatments add a dense forest of bushy trees to the Himalayas. The various oxide treatments (black, brown, and red oxide) produce cubic-looking crystalline shapes.

One of the most aggressive treatments, from the standpoint of good surface adhesion, is the double-treat process. It grows long, dendritic fingers that stick straight out from the copper surface, but leaves the underlying surface fairly smooth (Figure 2.26). Of all the choices, I like this best. My theory is that current on a double-treated surface will remain mostly bound to the smooth underlying surface, flowing like a river around the dendritic columns.

Table 2.6 lists values for the surface roughness of copper as normally apply to various substrate materials [20], [10]. Microwave designers often object to FR-4 materials because of their horrible loss tangent and the relatively rough surface treatments normally used to make the copper layers stick reliably to the dielectric material. For digital designs, however, surface roughness and dielectric losses don't render the material unusable—they merely restrict the distances at which it can be used. In the table, the values given for roughness are the average peak-to-valley height, a number not easily translated into the RMS deviation needed for estimating the onset frequency. Taking the worst-case equilateral-ridge geometry as a guide, you might expect the RMS roughness to be on the order of 0.29 times the average peak-to-valley roughness.

POINTS TO REMEMBER

> At a microscopic level, all materials exhibit surface irregularities and bumps.
> Toothing profiles are purposefully etched into the copper to facilitate adhesion between layers.
> Roughness on a scale comparable to the skin depth increases the mean length of the path of current, increasing the resistance.

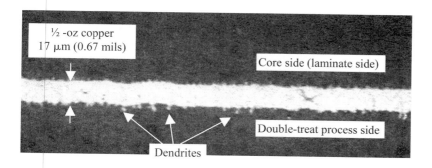

Figure 2.26—Dendritic fingers produced by the double-treat process stick up like dead trees (edge-view photomicrograph of 1/2-oz copper layer courtesy of Teradyne).

Table 2.6—Properties of pcb materials

Substrate	Dielectric constant (1 MHz)	Loss tan.	Surface roughness (avg. peak-to-valley, μm)
Air	1.0005	0	
Typical FR-4 applications	4.5	0.01–.025	6–18
RT-duroid® 5880 rolled Cu electro-deposited	2.16–2.24	0.0005–0.0015	0.75–1.0 4.25-8.75
RT-duroid® 6010 rolled Cu electro-deposited	10.2–10.7	0.001–0.006	0.75–1.0 4.25-8.75
Alumina 99.5% 96% 85%	10.1 9.6 8.5	0.0001–2 0.0006 0.0015	0.05–.25 5–20 30–50
Si (high resistivity)	11.9	0.001–0.01	0.025
GaAs	12.85	0.0006	0.025
Single-crystal sapphire	9.4, 11.6[1]	0.00004–0.00007	0.005–0.025
Fused quartz	3.8	0.0001	0.006–0.025

NOTE 1—Sapphire is an anisotropic dielectric material with different dielectric constants in two directions.
NOTE 2—Some of the material in this chart has been adapted from [20].

2.12 DIELECTRIC EFFECTS

Put a piece of bare FR-4 circuit board material (with no copper on either side) into a microwave oven. Bake it on full power for 1 minute. The microwaves noticeably warm the board. Next try a ceramic baking dish. It heats up too (but probably not as much).

In fact, just about any insulating material heats up in a microwave oven. The amount of incident electromagnetic power converted by a dielectric material into heat is called *dielectric loss*. When an insulating material is used as part of a transmission line, dielectric loss translates into signal attenuation. The higher the dielectric loss, the more attenuation your signals will suffer.

This section describes the dielectric properties of matter, leading to a mathematical definition of the term *dielectric loss tangent*. In this section I shall restrict the discussion to the types of materials commonly used in high-speed signal transmission applications.

The discussion of dielectric loss begins with a detailed examination of the phase of magnitude of currents flowing in solid materials. Figure 2.27 depicts the measurement of the

2.12 • Dielectric Effects

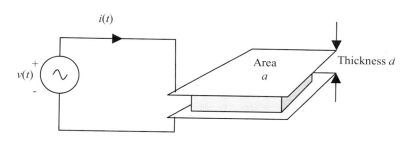

Figure 2.27—A small block of material generates currents both in phase and in quadrature to an externally applied electric field.

current through a solid block of material under the influence of an externally applied electric field.[18] Equation [2.68] expresses the magnitude and phase of the current in phasor notation.

$$I = V \frac{a}{d}(\sigma + j\omega \epsilon) \qquad [2.68]$$

where $V(\omega)$ is the source voltage as a function of frequency,

$I(\omega)$ is the current as a function of frequency,

ω is in units of rad/s,

a is the surface area of the block held between two thick copper plates (m^2),

d is the thickness of the block under test (m),

σ is the conductivity of the material (S/m), and

ϵ is the permittivity of the material, in units of Farads per meter (F/m).

In model [2.68] the amplitude of the *in-phase* current is controlled by parameter σ, while the amplitude of the *quadrature* current is controlled by parameter $\omega \epsilon$. The amount of in-phase current is called the *conduction current*. It represents the degree to which the test sample behaves like a resistor. The amount of in-quadrature current is called the *displacement current*. It represents the degree to which the test sample behaves like a capacitor.

A material is classified as a good conductor if $\sigma \gg \omega \epsilon$, meaning that the conduction current (resistive behavior) is much more significant that the displacement current (capacitive behavior). In a good conductor the current flows mostly in-phase with V. For good conductors, both σ and ϵ stay fairly constant over a broad range of frequencies, which is what makes [2.68] such a good model for conductors.

---

[18] If the permittivity of the block is *linear* then you may express the current in phasor notation. Most insulating materials used in digital applications are extremely linear over the range of typically applied voltages.

> **Critical Frequency for Conducting Versus Insulating Mode**
>
> The boundary ω_c between conducting and insulating modes varies tremendously. Good metallic conductors (like copper) remain conductive up to optical frequencies. Modern dielectric materials of which pcbs and cables are manufactured have extremely low conductivity, meaning that at frequencies as low as a few Hertz they still display predominantly a capacitive, or dielectric, behavior. The critical frequency for the soil of the Earth lies somewhere around 100 MHz. That's why in the 50-meter (20 MHz) Ham radio band you can depend on the conductivity of the Earth to serve as a good ground plane for a vertical monopole transmitting antenna, but at much higher frequencies you can't. In the higher bands you must construct a metallic ground lattice for your antenna or use a dipole or loop antenna that doesn't require a ground.

No matter how terrific a conductor you may have, there is always a critical frequency $\omega_c = \sigma/\epsilon$ above which the inexorably growth of ω causes $\omega\epsilon$ to vastly exceed σ. Beyond the critical frequency the material loses its conductive properties, the displacement current (capacitive behavior) rapidly becomes much more significant than the conduction current (resistive behavior), and the material behaves mostly like a capacitor instead of like a resistor (see box about critical frequency). The material may remain highly conductive above ω_c, but the capacitive behavior becomes even stronger. Any material operated at a frequency well above ω_c is classified as a *good insulator*.

In many insulating materials at frequencies well above ω_c the conduction current, although it remains much smaller than the displacement current, tracks upwards almost in direct proportion to frequency. It tracks in such a way that the ratio $\sigma/\omega\epsilon$ remains almost constant. For these materials it is convenient to express the current in terms of a quadrature term $j\omega\epsilon'$, where ϵ' stays fairly flat with frequency, and an in-phase term $\sigma = \omega\epsilon''$, where ϵ'' also stays fairly flat with frequency.

$$I = V\frac{a}{d}(\omega\epsilon'' + j\omega\epsilon') \qquad [2.69]$$

You can rearrange the above equation to emphasize the insulating qualities of a material by lumping both conduction and displacement terms under the umbrella of the $j\omega$ operator, like this:

$$I = V\frac{a}{d}\left(j\omega(\epsilon' - j\epsilon'')\right) \qquad [2.70]$$

In the form of [2.70] the term $\epsilon = \epsilon' - j\epsilon''$ is called the *complex electric permittivity* of a material. The real part ϵ' defines the displacement current, while the imaginary part $-\epsilon''$ takes on the role of defining the conduction current. For a good insulator, the imaginary part should be much smaller than the real part.

2.12 • Dielectric Effects

For many good insulators the ratio ϵ''/ϵ' remains so stable across such a wide range of frequencies that it becomes convenient to specify the material properties in terms of the real part ϵ' and the ratio ϵ''/ϵ'. In this case the ratio ϵ''/ϵ' is given the name *dielectric loss tangent*, or sometimes just the *loss tangent*. The loss tangent is used to determine the attenuation of a physical transmission line due to dielectric losses.

Except for some very unusual quasi-stable situations involved in laser physics, the imaginary part of complex permittivity is always negative, meaning that the value of ϵ'' is always positive. The loss tangent is always reported as a positive value.

When dealing with the low-permittivity materials commonly used for pcbs and cable insulation, it is convenient to express the permittivity in terms relative to the permittivity of free space.

The permittivity of free space (a perfect vacuum) is $\epsilon_0 = 8.854 \cdot 10^{-12}$. The value of ϵ_0 is entirely real, as free space has no dielectric loss. The permittivity of air differs from ϵ_0 by less than one part in 2,000, so for purposes of signal integrity analysis, air and free space are practically the same thing. All other materials have permittivities greater than ϵ_0 and involve some degree of loss.

The *complex relative permittivity* ϵ_r of any material may be expressed as the ratio of the complex permittivity ϵ to the permittivity of free space ϵ_0.

$$\epsilon_r \triangleq \frac{\epsilon'}{\epsilon_0} - j \frac{\epsilon''}{\epsilon_0} \quad [2.71]$$

Complex relative permittivity carries with it both real and imaginary parts $\epsilon_r \triangleq \epsilon_r' - j\epsilon_r''$. If you want to refer exclusively to the real part of complex relative permittivity, it is called (in the pcb industry) the *dielectric constant*. The dielectric constant may be measured for any insulating material as a ratio of capacitances. Specifically, if you construct a capacitor using an insulating material between the plates, and then make a second capacitor with the same physical dimensions but using only a perfect vacuum between the plates, the ratio of the first capacitance to the second equals the dielectric constant of the insulating material. In other words, the dielectric constant is that ratio by which an insulating material enhances, or increases, the effective capacitance of a structure.

The dielectric constant of a perfect vacuum is unity, and dry air at standard (sea-level) temperature and pressure is 1.0005. The dielectric constant of all other solid materials is greater than one.

Even if not explicitly stated, the dielectric constant is always defined at one particular frequency. For pcb materials, the usual specification frequencies are 1 KHz, 1 MHz, or 1 GHz. The values at these three frequencies differ, and you must ensure when working with a particular value of dielectric constant that you know the frequency at which your value is specified.

Symbols commonly used for the dielectric constant are k, D_k, ϵ_r' or just plain ϵ_r. The symbol ϵ_r thus appears somewhat confused as to whether it is a purely real quantity (dielectric constant) or a complex quantity (complex relative permittivity). The rationale behind allowing such confusion is that the imaginary part of relative permittivity is small compared to the real part. When you see ϵ_r used as a dielectric constant, you will just have to remember that ϵ_r carries with it a small, but sometimes important, imaginary component that is being temporarily ignored.

The term *dielectric constant* as used in other industries is sometimes defined as the real part of ordinary (not relative) permittivity and having units of F/m.

Some authors define the term *electric susceptibility* χ_e, which is nothing other than the complex relative electric permittivity minus one.

$$\chi_e = \epsilon_r - 1 \qquad [2.72]$$

Electric susceptibility is used extensively in the analysis of nonlinear materials. In general it is a function of both frequency and amplitude. Materials used for insulators in transmission-line applications are almost always extremely linear. As a consequence, signal integrity analysts have little use for χ_e.

POINTS TO REMEMBER

➢ All insulators exhibit some degree of dielectric loss.
➢ Make sure you know the frequency at which a value of dielectric constant is specified.

2.12.1 Dielectric Loss Tangent

Dielectric losses in a transmission line scale in proportion to both frequency and length. For any particular construction there exists a certain speed-length product beyond which the material will absorb too much of your signal, leading to system malfunction. The relationship between the dielectric loss specification and the limiting speed-length product is described in the next chapter. This section serves only to explain the way dielectric loss is specified.

The dielectric loss for materials used to construct pcbs and cables is commonly rated in terms of a dielectric loss tangent, which is the absolute value of the tangent of the phase angle formed by the components of complex permittivity (Figure 2.28):

$$\tan \theta \triangleq \frac{-\operatorname{Im}(\epsilon)}{\operatorname{Re}(\epsilon)} = \frac{\epsilon''}{\epsilon'} \qquad [2.73]$$

where $\tan \theta$ is the loss tangent (dimensionless) of a material with complex permittivity ϵ (F/m), and

ϵ' and ϵ'' are the real part and the negative of the imaginary part, respectively, of the complex permittivity ϵ.

The loss tangent precisely equals the *dissipation factor*, which is a term used in the capacitor industry to specify the quality of capacitors used in power-supply applications. The related term *power factor* is defined a little differently. Power factor is the ratio of active power to

2.12.2 • Rule of Mixtures

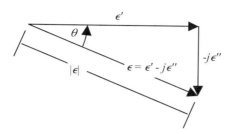

Figure 2.28—The components of complex permittivity are defined such that ϵ'' is always positive, and the angle θ spanned from $|\epsilon|$ to ϵ' is also always positive.

apparent power. In phasor notation it is the ratio $VI^*/|V||I|$, where the asterisk denotes complex conjugation. If the path for current in a circuit is a dielectric material with complex permittivity ϵ, then the power factor equals the ratio $\epsilon''/|\epsilon|$, which is the sine of the phase angle formed by the components of complex permittivity. For loss tangents less than 10%, the relative difference between dissipation factor and power factor is less than one part in 200.

There is some uncertainty in the literature about whether to call the phase angle of complex permittivity δ or θ. As a result, the loss tangent has been variously called the tangent of δ, tangent of θ, or simply tan-δ or tan-θ. To avoid confusion with the notation for skin depth (δ), I shall use the term tan-θ.

POINTS TO REMEMBER

- ➤ Dielectric losses in a transmission line scale in proportion to both frequency and length.
- ➤ The dielectric loss tangent is the tangent of the phase angle formed by the real and imaginary components of complex permittivity.
- ➤ For small loss tangents, l is approximately the same as the ratio of the imaginary part to the real part of complex permittivity.

2.12.2 Rule of Mixtures

A new dielectric material made by uniform mechanical mixing of two or more constituent materials carries a permittivity ϵ equal to the weighted average, on a volumetric basis, of the permittivities of the constituent materials:

$$\epsilon_e = \sum_n v_n \epsilon_n \qquad [2.74]$$

where v_n are the volumetric filling factors of the materials involved (assumed to all sum to unity),

ϵ_n are the complex permittivities of the various materials, and

ϵ_e is the effective complex permittivity of the resulting mixture.

The form of equation [2.74] applies equally well to relative permittivities (and also to dielectric constants, the dielectric constant being just the real part of relative permittivity).

$$\epsilon_{re} = \sum_n v_n \epsilon_{r,n} \quad [2.75]$$

where v_n are the volumetric filling factors of the materials involved (assumed to all sum to unity),

$\epsilon_{r,n}$ are the complex relative permittivities (dimensionless) of the various materials, and

ϵ_{re} is the effective complex relative permittivity (dimensionless) of the resulting mixture.

Taking only the real parts of [2.75], the same equation works when combining dielectric constants.

Note that definition of [2.74] excludes chemical mixtures for which the electrical properties of the newly created compound may differ vastly from the electrical properties of the constituents. Equation [2.74] only contemplates uniform mechanical mixtures in which the constituents remain chemically intact, but uniformly mixed *on a scale smaller than the wavelengths of the signals involved.*

As an example, consider the popular FR-4 pcb material. It is made from a mixture of epoxy and glass fibers. The (relative) dielectric constants of epoxy and glass measured at 1 MHz are approximately 3.45 and 5.8 respectively.[19] The glass fibers are on the order of a couple of mils in diameter, which qualifies as quite small compared to signals up to 100 GHz. Equation [2.75] applied to typical epoxy-glass mixtures says the effective dielectric constant ϵ'_{re} of the mixture should lie somewhere between 3.45 and 5.8.

$$\epsilon'_{re} = a(3.45) + (1-a)(5.8) \quad [2.76]$$

where a is the relative proportion of epoxy resin, by volume, in the mixture, and

ϵ'_{re} represents the effective dielectric constant of the resulting mixture.

Typical values for the resin content by volumetric percentage range from approximately 0.33 to 0.75, resulting in dielectric constants at 1 MHz ranging from a high of 5.0 to a low of 4.0. Beware that resin content is sometimes listed by percentage of weight, which is not the same as percentage of volume.

Equation [2.76] must sometimes be modified to account for the possible presence of air remaining trapped inside the board after lamination. Any residual air must be

---

[19] TC Edwards [20], p. 73, reports values of 3.8 and 6.3, but doesn't say at what frequency. Brzozowski [21] reports values of 5.8 for e-glass at 1 MHz. Clyde F. Coombs [22] quotes values of 3.45 and 6.2 (p. 31.29), measured presumably at 1 KHz.

incorporated into [2.74] as a third constituent with a dielectric constant of unity, occupying part of the total volume. Because air has the lowest possible dielectric constant, the net effect of air in the mixture *decreases* the dielectric constant of the resulting mixture. Water has an even worse effect because of its large dielectric constant and large loss tangent. Modern vacuum-bag lamination processes in well-controlled manufacturing environments virtually eliminate the possibility of trapped air, though, so air is no longer much of an issue for striplines in high-quality boards.

Air remains a major issue for microstrips, however, because microstrip traces are exposed to (or in close proximity to) air on one side. Although the electric fields between the microstrip and its nearest reference plane remain totally embedded in the dielectric substrate, the electric lines of force spewing up above the trace reside mostly in the air. The effective dielectric constant for the structure must then be a mixture of the dielectric constant of FR-4 and air. The same reasoning applies to twisted-pair cables.

For complicated geometries like microstrips and twisted-pair cables, the constituent dielectrics occupy distinct regions of space, in violation of the uniform-mixing requirement for [2.74]. The spatial distribution of dielectrics is the same at every point along the transmission line (a crucial property), but the dielectrics are not finely mixed. In such a case a new form of the dielectric mixing rule applies with the various dielectric constants weighted not by volume but according to the proportion of the total stored electric field energy contained within each region. These calculations are performed by electromagnetic field simulation software. The net result of field calculations is an effective dielectric constant ϵ'_{re} that takes into account the dielectric constants $\epsilon'_{r,n}$ and geometries associated with each of the constituent regions.

POINTS TO REMEMBER

> A mixed dielectric carries a permittivity ϵ equal to the weighted average, on a volumetric basis, of the permittivities of the constituent materials.
> Any air or water present in a dielectric mixture will change the dielectric constant of the resulting mixture.

2.12.3 Calculating the Loss Tangent for a Uniform Dielectric Mixture

Suppose you are given a collection of dielectric materials with complex relative permittivities $\epsilon_{r,1}, \epsilon_{r,2}...\epsilon_{r,N}$, and that the materials are mixed in relative volumetric proportions of $a_1, a_2...a_N$. The loss tangent of the mixture is defined as the ratio of the imaginary part to the real part of the effective complex relative permittivity of the mixture. The effective complex relative permittivity of the mixture may be found using [2.75]:

$$\epsilon_{re} = \sum_n a_n \epsilon_{r,n} \qquad [2.77]$$

where $a_1, a_2...a_N$ are the relative volumetric proportions of the various materials (all summing to one),

$\epsilon_{r,1}, \epsilon_{r,2}...\epsilon_{r,N}$ are the complex relative permittivities of the constituent materials, and

ϵ_{re} is the effective complex relative permittivity of the resulting mixture.

Separating equation [2.77] into its real and imaginary parts,

$$\epsilon'_{re} = \sum_n a_n \epsilon'_{r,n}$$
$$\epsilon''_{re} = \sum_n a_n \epsilon''_{r,n} \qquad [2.78]$$

And forming the effective loss tangent ratio,

$$\tan \theta_e = \frac{\epsilon''_{re}}{\epsilon'_{re}} = \frac{\sum_n a_n \epsilon''_{r,n}}{\epsilon'_{re}} \qquad [2.79]$$

where $\tan \theta_e$ represents the effective loss tangent of the resulting mixture (dimensionless).

You can simplify equation [2.79] somewhat by substituting $\epsilon''_{r,n} \approx \epsilon'_{r,n} \tan \theta_n$ for the imaginary term associated with each constituent.

$$\tan \theta_e \approx \frac{\sum_n a_n \epsilon'_{r,n} \tan \theta_n}{\epsilon'_{re}} \qquad [2.80]$$

where a_n are the relative volumetric proportions of the constituent materials,

$\epsilon'_{r,n}$ are the dielectric constants of the constituent materials, (dimensionless),

$\tan \theta_n$ are the loss tangents of the constituent materials (dimensionless),

ϵ'_{re} is the effective dielectric constant of the resulting mixture (dimensionless), and

$\tan \theta_e$ represents effective the loss tangent of the resulting mixture (dimensionless).

The above expression may be used to determine the loss tangent of any uniform mechanical mixture, given that *you know the weighting factors* a_n, and also given that *you have identified all the constituents*.

As pointed out by Alina Deutsch [11], "...dielectric loss [in pcbs] is very much dependent on any solvents that get trapped inside the multilayer stack [as well as the] characteristics of the epoxy and fiberglass reinforcing material." Even small percentages of lossy contaminates can easily double the measured loss in a finished product as compared to the calculated loss based on assumed bulk-material parameters. When in doubt, directly measure the dielectric loss on a test board.

In mixtures involving only two constituents, when the second constituent is air, equation [2.80] may be simplified by noting that for air, tan θ equals zero. The effective volumetric mixing fraction for the non-air constituent is then called the *filling factor q*.

$$\tan \theta_e = \frac{\epsilon'_{r,1}}{\epsilon'_{re}} q \tan \theta_1 \qquad [2.81]$$

where q is the filling factor (relative volumetric proportion) of the first material,

$\epsilon'_{r,1}$ is the dielectric constants of the non-air constituent material, (dimensionless),

$\tan \theta_1$ is the loss tangent of the non-air constituent material (dimensionless),

ϵ'_{re} is the effective dielectric constant of the resulting mixture (dimensionless), and

$\tan \theta_e$ is the effective loss tangent of the resulting mixture (dimensionless).

Equation [2.81] may be used to determine the loss tangent for any mixture of one constituent and air, *given that you know the filling factor q*. If you don't know q, read the next section.

When working microstrip and embedded microstrip problems associated with FR-4 materials, parameters $\epsilon'_{r,1}$ and $\tan \theta_1$ represent the composite properties of a uniform mixture of epoxy resin and glass fibers.

POINT TO REMEMBER

> The loss tangent for a mixed dielectric can be calculated from the loss tangents and filling factors of the constituent materials.

2.12.4 Calculating the Loss Tangent When You Don't Know q

This section describes a specialized method of determining the effective loss tangent for microstrip and twisted-pair transmission lines.

The design process for these media usually begins with two types of information: first, the dielectric constant $\epsilon_{r,1}$ of the insulating material, and second, the geometric configuration of the conductors, insulating regions, and air-filled regions in the vicinity of the conductors.

Given these input parameters, any of a number of software-driven field-solver algorithms can calculate the impedance and propagation velocity of the transmission structure. From the propagation velocity v you can easily determine the effective dielectric constant of the structure:

$$\epsilon'_{re} \triangleq (c/v)^2 \qquad [2.82]$$

where c is the velocity of light in vacuum, equal to $2.998 \cdot 10^8$ m/s,

v is the propagation velocity of the structure given by your field simulator (m/s),

NOTE: v equals the inverse of line delay (for delay measured in units of s/m), and

ϵ'_{re} is the effective dielectric constant of the transmission structure.

Provided that you know the loss tangent of the insulting material, you now have in hand three of the four parameters needed to apply [2.81]. The missing piece of information is the filling factor q. Fortunately, you can work backwards from $\epsilon'_{r,1}$ and ϵ'_{re} to determine q. Begin by noting that the value of q is defined so that the mixing function [2.75] produces the correct dielectric constant ϵ_{re}:

$$\epsilon'_{re} = q\epsilon'_{r,1} + (1-q) \cdot \epsilon'_{r,air} \qquad [2.83]$$

where ϵ'_{re} is the effective dielectric constant of the transmission line,

$\epsilon'_{r,1}$ is the dielectric constant of the insulating material used to build the line,

q is the filling factor you wish to find, and

$\epsilon'_{r,air} \cong 1$ represents the dielectric constant of air.

Solving [2.83] to find q,[20]

$$q = \frac{\epsilon'_{re} - 1}{\epsilon'_{r,1} - 1} \qquad [2.84]$$

Plugging this value of q into [2.81] yields a very useful expression for $\tan \theta_e$.

$$\tan \theta_e = \frac{\epsilon'_{r,1}}{\epsilon'_{re}} \frac{\epsilon'_{re} - 1}{\epsilon'_{r,1} - 1} \tan \theta_1 \qquad [2.85]$$

where $\epsilon'_{r,1}$ represents the dielectric constant of the material used to build the structure,

---

[20] Equation [2.84] works provided that k_1 exceed unity. In the rare but theoretically conceivable instance that the insulating material has unit dielectric constant but nonzero loss tangent, the result of equation [2.84] would be undefined.

tan θ_1 is the loss tangent of the dielectric material used to build the structure,

ϵ'_{re} is the effective dielectric constant of the resulting structure, and

tan θ_e is the effective loss tangent of the resulting structure.

POINT TO REMEMBER

> ➢ The filling factor for a dielectric-air mixture may be inferred from the velocity of propagation.

2.12.5 *Causality and the Network Function Relations*

The real and imaginary portions of any realizable network function bear certain subtle yet incontrovertible relations to each other. The original and best description of these relations appears in *Network Analysis and Feedback Amplifier Design* [12]. Another good reference is *The Fourier Integral and Its Applications* [13] page 206. A mere shadow of the power and originality of these early descriptions may also be found in more modern texts on the subject of analog filter design.

The rules laid down by Bode declare that you cannot change the magnitude of a network function without also making a corresponding change in the phase. The separate parts, magnitude and phase, are inseparably linked.

Every permittivity function ϵ must abide by the network-function relations (see box *Network Function Relations*). These relations become especially important when supplying loss tangent and dielectric constant data to signal integrity simulation programs. Here I am thinking of the H-SPICE W-element model for a lossy transmission line. In some versions the model allows the arbitrary prescription of loss tangent and dielectric constant data as a function of frequency. If the real and imaginary parts of the permittivity so defined do not bear the proper relationships to each other, the implied time-domain waveforms associated with the specified permittivity may be noncausal or nonreal or nonminimum-phase. Any of these conditions will throw SPICE into fits, producing a bogus simulation.

The difficulty with crafting a response both causal and real is the extremely subtle interplay of the real and imaginary components of $H(p)$. Fortunately, [12] provides considerable guidance about the relationship between the magnitude and phase of any realizable network function. Of special interest are his conclusions regarding the behavior of the magnitude and phase of any real, causal, and minimum-phase network function when drawn on a log-log scale, as a function of the imaginary-axis frequencies $j\omega$. Bode proves that in places where the slope on a log-log plot of a network function $A(j\omega)$ is fairly constant over a wide band, the phase of $A(j\omega)$ must equal $\pi/2$ times the slope.

In accordance with this familiar result, good capacitors with impedance slopes of -20 dB/decade (that's a slope of -1 on a log-log plot) must always posses a phase angle of precisely $-\pi/2$ radians. Good inductors, on the other hand, with impedance slopes of $+20$ dB/decade (that's a slope of $+1$ on a log-log plot), always have a phase angle of $+\pi/2$ radians.

Network Function Relations

Let $h(t)$ represent the impulse response, and $H(p)$ the Laplace transform, of a *linear*, *time-invariant*, *real-valued*, and *causal* network function.

Linearity says the material responds in a proportional way to scaled inputs.

Time-invariant means the material reacts the same way, every time.

Real-valued forces the real part of $H(j\omega)$ to be an even function of ω and the imaginary part to be odd.

Causal implies that $h(t)$ is zero for all times prior to time zero. In other words, the material doesn't react until you do something to it.

Minimum phase precludes zeroes in the interior of the right-hand plane of the function $H(p)$.

Together, all the foregoing indicate that $H(p)$ is analytic (i.e., has a defined derivative at all points) and has no singularities or zeros in the right-half plane save those located at points p_i on the imaginary axis for which $Lim_{p \to p_i}(p - p_i)H(p) = 0$.

The magnitude α (in nepers) and phase θ (in radians) are defined as a function of frequency ω according to the real and imaginary parts, respectively, of the natural logarithm of the network function such that

$$\alpha(\omega) + j\beta(\omega) = -\ln H(j\omega)$$

The magnitude and phase of every such network function satisfy the Hilbert transform relations:

$$\beta(\omega) = \frac{\omega}{\pi} \int_{-\infty}^{\infty} \frac{\alpha(u)}{u^2 - \omega^2} du$$

$$\alpha(\omega) = \alpha(0) - \frac{\omega^2}{\pi} \int_{-\infty}^{\infty} \frac{\beta(u)}{u(u^2 - \omega_0^2)} du$$

The general theory of phase/magnitude relations also applies to the case of a dielectric material with a loss tangent that remains constant over a wide range of frequencies. Bode's results say that if the phase of the complex permittivity[21] is $-\theta$ over a wide region, then the log-log slope of the magnitude of the permittivity over that region must be very close to $-(2/\pi)\theta$. From this simple observation you may conclude that the only way to make a complex relative permittivity function with a constant loss tangent at all frequencies is like this:

---

[21] According to Figure 2.28 the phase of the permittivity function is negative ($-\theta$).

2.12.5 • Network Function Relations

$$\epsilon_r(\omega) = a(j\omega)^{-\frac{2}{\pi}\theta}$$

$$= a(j)^{-\frac{2}{\pi}\theta} \omega^{-\frac{2}{\pi}\theta}$$

$$= ae^{j\frac{\pi}{2}\left(-\frac{2}{\pi}\theta\right)} \omega^{-\frac{2}{\pi}\theta}$$ [2.86]

$$= ae^{-j\theta} \omega^{-\frac{2}{\pi}\theta}$$

where ϵ_r is the complex relative permittivity of a hypothetical material with constant loss tangent at all frequencies,

θ is the phase angle formed by the components of complex permittivity (rad), and

a is an arbitrary real constant.

Examining the form of [2.86], several objections come to mind. First, the relative permittivity goes to zero at high frequencies. This behavior is not physically permitted, as nothing can have a dielectric constant less than one. Second, the relative permittivity goes to infinity at zero, which implies infinite amounts of capacitance near DC. Neither trait seems physically desirable.

It is important to realize that these objections are not shortcomings of the formulation [2.86]; they are shortcomings of the concept of constant loss tangent. The lesson learned from examining [2.86] is that complex permittivity cannot have a constant loss tangent at all frequencies. The frequency range over which the loss tangent is zero may be very large, but cannot be infinite.

Whether or not the shortcomings of [2.86] cause difficulty depends on the simulation technology you use. If you choose frequency-domain simulation, then [2.86] works fine, because a frequency-domain simulator evaluates ϵ_r only over a limited range of frequencies stipulated by the needs of the Fast-Fourier Transform (FFT). As long as ϵ_r remains well behaved over the required range, peculiar behavior at infinity or DC doesn't matter.

For example, in an FFT with one million points the ratio between the highest and lowest frequencies sampled (excluding DC) is 500,000 to 1. That may sound like a huge ratio, but over that range the slope in magnitude induced by [2.84] is so gentle that for a loss tangent of 0.02 (typical for FR-4) less than a 20% variation in the dielectric constant results. No big deal.

A frequency-domain simulator handles DC as a special case (just like SPICE does a separate DC analysis before starting its AC analysis). At DC the limiting value of the admittance $j\omega\epsilon_r$ stipulated by [2.84] is zero, so it works just like any other capacitor—it's an open circuit.

$$Lim_{\omega \to 0} j\omega a(j\omega)^{-\frac{2}{\pi}\theta}$$

$$= Lim_{\omega \to 0} a(j\omega)^{1-\frac{2}{\pi}\theta}$$ [2.87]

$$= 0$$

If all you know about a material is its worst-case loss tangent over the frequency range of interest and the dielectric constant at one particular frequency, then for frequency-domain simulation purposes, equation [2.84] functions beautifully. It generates the worst amount of dispersion possible for the transmission line, and it guarantees a real, causal, minimum-phase response.

When using [2.84], set θ such that $\tan(\theta)$ equals the specified loss tangent, and set $a = k_0 \omega_0^{\frac{2}{\pi}\theta} / \cos\theta$ to provide the appropriate value of dielectric constant at reference frequency ω_0. For loss tangents less than 0.05, you may assume $\theta = \tan\theta$ to within better than one part in 1,000.

Given ω_0, k_0, and θ, here are two modified forms of [2.86].

$$\epsilon_r(\omega) = \frac{k_0}{\cos\theta}\left(\frac{j\omega}{\omega_0}\right)^{-\frac{2}{\pi}\theta}$$

$$= k_0\left(\frac{\omega}{\omega_0}\right)^{-\frac{2}{\pi}\theta}(1+j\theta) \qquad [2.88]$$

where ϵ_r is the complex relative permittivity of a hypothetical material with constant loss tangent at all frequencies,

θ is the phase angle formed by the components of complex permittivity, and

k_0 is the dielectric constant (dimensionless) at frequency ω_0 (rad/s).

If you are doing SPICE simulation (using the W-element model or any other model that accepts a table of frequency-varying values for dielectric constant and loss tangent), then you may not have precise control over the frequencies at which SPICE attempts to perform its analysis. You may in this case want to produce a table of complex permittivity values that has constant loss-tangent in a given frequency band but does not tend towards infinity as $\omega \to 0$.

A reasonable solution to the problem of infinities is to feather the phase linearly to zero below some critical frequency ω_1. Provided that ω_1 falls well below the lowest frequency of interest in your simulations, the feathering will have little or no effect on the actual time-domain results, but will eliminate the infinite amplitude values at DC.

Here is a magnificent permittivity function with constant phase θ_1 above frequency ω_1, but with the phase feathered linearly to zero below ω_1. The magnitude and phase are matched to produce a real, causal, minimum-phase result (Figure 2.29). The function α is developed through meticulous application of the integral equations associated with the network relations in [12] and [13].

2.12.5 • Network Function Relations

$$\alpha(\omega) = \frac{\theta_1}{\pi}\left\{\left(1+\frac{\omega}{\omega_1}\right)\ln\left(1+\frac{\omega}{\omega_1}\right)+\left(1-\frac{\omega}{\omega_1}\right)\ln\left|1-\frac{\omega}{\omega_1}\right|\right\}$$

$$\theta(\omega) = \begin{cases} \theta_1 \dfrac{\omega}{\omega_1} & \text{if } \omega < \omega_1 \\ \theta_1 & \text{otherwise} \end{cases} \qquad [2.89]$$

$$\epsilon_r(\omega) = e^{-\alpha(w)-j\theta(\omega)}$$

where ϵ_r is the complex relative permittivity (dimensionless),

α is the log attenuation (*negative* of the log-magnitude) of ϵ_r (nepers),

θ is the angle formed by the components of ϵ_r (i.e., the *negative* of the phase of ϵ_r) (rad),

ω is the frequency of operation (rad/s), and

θ_1 is the limiting value of phase above ω_1 (rad/s).

Many other models for complex permittivity can be used. As long as the dielectric constant and loss tangent from the model match your material properties over the frequency range of interest, and the model scales in a causal, minimum-phase fashion across all frequencies, it should work. Of particular interest is one model proposed by Svensson and Dermer [14] involving a continuous array of poles located on the real axis. Other authors have used a finite array of poles located on the real axis, with the positions and weights adjusted by optimizing procedure to best match measured parameters.

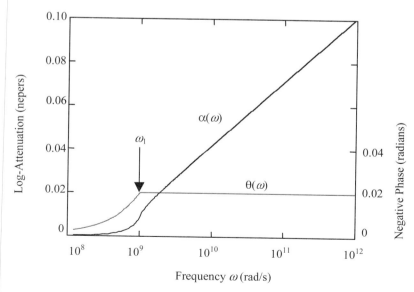

Figure 2.29—Feathering the phase to zero below ω_1 prevents the log attenuation from going to infinity at DC, but leaves the remainder of the curve unchanged. ($\theta_1 = 0.02$, $\omega_1 = 10^9$ rad/s).

POINT TO REMEMBER

> ➤ The real and imaginary portions of any realizable network function bear certain subtle yet incontrovertible relations to each other. Specifying one without the other leads to non-realizeable circuit results.

2.12.6 Finding $|\epsilon_r|$ to Match a Measured Loss Tangent

Although measured values for loss tangent can be quite accurate, measured values for the dielectric constant are notoriously *inaccurate*. The accuracy of the loss tangent measurement flows from its definition as the ratio of two currents in quadrature, both defined at the same frequency. Various precautions may be taken during measurement to ensure the accuracy of the measured results.

Dielectric constant measurements, on the other hand, are taken over vastly *differing* frequencies, leading to certain natural inaccuracies in the calibration of the test setup.

Suppose, then, that you have access to loss tangent data for which no dependable dielectric constant data are available. In this case you must synthesize a magnitude function to match the specified phase information, matching in the sense of being real, causal, and minimum-phase. Attempts to do so using the network relations in their integral-equation form may prove unsatisfactory due to the obvious divide-by-zero problems inherent in those equations. In the world of abstract mathematics, skilled mathematicians can overcome these divide-by-zero errors with suitable limiting arguments, but in the world of practical, everyday engineering calculations your typical math-processing spreadsheet programs just can't do those sorts of integrals.

To get you past the integral-equation difficulties, I shall describe an FFT technique for computing Hilbert-transform pairs. This technique generates a matching magnitude function for any arbitrary phase function. The disadvantage of this technique is that in order to ensure accuracy of the finished result, it requires a *lot* of points in the FFT.

Begin by deciding the range of frequencies over which the magnitude must be synthesized. This range spans from some low value f_1, below which the circuit reacts essentially at DC, to some high value f_2 corresponding to the bandwidth of the digital signals involved. A digital transmission line with propagation delay t_p mandates f_1 no greater than $1/(2\pi t_p)$. Signal rise/fall times of t_r require f_2 no less than $1/(2t_r)$. For the method to work successfully, the FFT must sample at a frequency at least 20 times higher than f_2 and contain a number of points N at least equal to $400 f_2/f_1$.

The *oversampling ratio* in the FFT sampling frequency is the factor by which the FFT sampling frequency f_s exceeds the bandwidth f_2 of the digital signals in your simulation. In this application the oversampling ratio controls the degree of distortion in the upper band edge of the finished permittivity function. This distortion is introduced by the Hanning window, which is included to suppress Gibb's phenomena in the time-domain response. An oversampling ratio of 20:1 produces distortion at f_2 of roughly 1% of the log magnitude of the permittivity. The upper band-edge distortion changes approximately in inverse proportion to the oversampling ratio.

The *number of points* in the FFT controls the spacing of the frequency-sampling grid used to represent the permittivity function. The frequency-sampling spacing is f_s/N. In the

2.12.6 • Match a Measured Loss Tangent

low-frequency zone near f_1 you should use a spacing no greater than $f_1/20$, which implies a lower bound on N.

$$\frac{f_s}{N} < \frac{f_1}{20} \rightarrow N > 20\frac{f_s}{f_1} \quad [2.90]$$

Substitute $20f_2$ for f_s (assuming you followed my advice on f_s),

$$N > 400\frac{f_2}{f_1} \quad [2.91]$$

Now substitute the expressions for f_1 and f_2 to get the final form:

$$N > 400\frac{1/2t_r}{1/2\pi t_p} = 400\pi\frac{t_p}{t_r} \quad [2.92]$$

According to [2.92], the number of points N used in your FFT operations must exceed 400π times the number of rise times stored in the transmission line at any given instant. The value of N is usually selected to be the next highest power of two above the bound set by [2.92].

Once you have selected an appropriate N, define indexes for the FFT operations:

$$\text{time index: } n = 0, 1 \ldots N-1$$
$$\text{frequency index: } k = 0, 1 \ldots \frac{N}{2} \quad [2.93]$$

Evaluate the loss tangent on a dense grid of frequencies, and fill in values for the phase of the permittivity.

$$f_k = f_s\frac{K}{N}$$
$$H_k = j\arctan\left(\tan\theta(f_k)\right) \quad [2.94]$$

Notice here that I have defined H only for positive frequencies in accordance with the way real-valued FFT routines usually operate. The frequency-domain parameters are specified as a vector of complex values using the frequency-domain index k, while the time-domain values are specified as a vector of real values using the time-domain index n.

Next you must window the function H to bring the imaginary part of the response down to zero at frequency sample point $N/2$. If you don't apply the window, then Gibb's phenomenon will induce horrible-looking wiggles in the implied time-domain response. I've chosen the Hanning window in this case as a reasonable compromise between the accuracy of the frequency-domain results below f_2 and the suppression of Gibb's phenomenon in the time-domain response.

$$H_k = H_k \left(\frac{1}{2} + \frac{1}{2} \cos\left(2\pi \frac{k}{N} \right) \right) \qquad [2.95]$$

Apply the inverse FFT to vector H, producing a real-valued time-domain response vector h of length N.

$$h = \text{InverseFFT}(H) \qquad [2.96]$$

At this point I must diverge to discuss briefly some of the properties of the Fourier transform. Any time-domain function may be broken into its odd and even constituents $h = h_e + h_o$, where h_o is a strictly odd function[22] and h_e is strictly even. The odd part h_o controls the imaginary part of the frequency response, while the even part h_e controls the real part of the frequency response. In the case at hand, h_o controls the phase of the permittivity, while h_e controls the log magnitude.

As computed, h is *already strictly odd*, as befits its origin from a purely imaginary frequency specification. To the function h you can therefore add *any strictly even function h_e* without distorting the phase of the permittivity.

Your task is to find a purely even function h_e such that the time-domain response $h + h_e$ becomes zero for all negative times. In mathematical terms,

$$h(-t) + h_e(-t) = 0 \text{ for all } t > 0 \qquad [2.97]$$

Substituting the definitions of odd and even functions for negative times produces a new equation:

$$-h(t) + h_e(t) = 0 \qquad [2.98]$$

Adding $2h$ to both sides,

$$h(t) + h_e(t) = 2h(t) \qquad [2.99]$$

Equation [2.99] demonstrates that when the function h_e is chosen to null the time-domain response for all negative times, then for all positive times the function $h + h_e$ simply equals $2h$. I know this sounds too easy, but that's how you find the time-domain response associated with the logarithm (log magnitude and phase) of the frequency-response of the permittivity. So, the next step is to create a new vector g which zeroes h for all negative times and doubles it for all positive times.

---

[22] In continuous time, an odd function satisfies $h(t) = -h(-t)$. In discrete time, the relation is $h_k = -h_{N-k}$.

2.12.6 • Match a Measured Loss Tangent

$$g_k = 2h_k$$
$$g_{N-k} = 0$$
$$g_0 = -\sum_k 2h_k$$
[2.100]

Next transform g from the time to the frequency domain, producing a complex-valued vector of length $1 + (N/2)$

$$G = ForwardFFT(g)$$
[2.101]

Vector G represents the magnitude (nepers) and phase (radians) of a real-valued, causal, and minimum-phase permittivity response. Values for dielectric constant and loss tangent may be extracted from G.

$$\epsilon_r(f_k) = \mathrm{Re}(e^{G_k})$$
$$\tan\theta(f_k) = \tan(\mathrm{Im}(\mathbf{G}_k))$$
[2.102]

Figure 2.30 illustrates the results obtainable with the FFT method. In this example the input phase is taken directly from [2.89] with parameters $\tan\theta = 0.02$ and $\omega_1 = 10^9$ rad/s. The desired phase curve is marked on the figure as "design-goal $\theta(\omega)$." For this particular phase specification, [2.89] stipulates the correct log-attenuation curve, marked in the figure as "design-goal $\alpha(\omega)$."

The problem parameters are to synthesize a good phase-and-magnitude pair for all frequencies up to 10^{11} rad/s (15.9 GHz), using only the input phase curve and the FFT method. For this purpose, the FFT sample frequency is chosen as $2 \cdot 10^{12}$ rad/s (318 GHz). The FFT has 65,536 points, which establishes a sampling-frequency grid spacing of $30.5 \cdot 10^6$ rad/s = 4.80 MHz. Such parameters are appropriate for a 6-inch trace carrying 10 Gb/s with rise/fall times of 30 ps.

The FFT procedure windows the phase curve as shown, and then produces a matching log-attenuation curve marked as the "FFT-method log-atten." In the band from DC to 10^{11} rad/s the FFT-method log-attenuation curve matches the design goal to within 1%. The FFT-method log-attenuation curve and the FFT-method windowed phase curve must be used together; do not mix and match the design goal components with the FFT-method components.

In [2.100] I have taken the liberty of adjusting g_0 to guarantee that g has zero DC content. This sets the real part of G_0 to zero, which in turn ensures that the DC value of $|\epsilon_r|$ equals unity. After computing the FFT, you may then scale the dielectric constant without changing the loss tangent. Choose a constant A such that $A|\epsilon_r(f_0)|$ gives you the particular dielectric constant you need at one particular frequency f_0 and thereafter use $A|\epsilon_r(f_k)|$ and $\tan\theta(f_k)$.

To evaluate $|\epsilon_r|$ and $\tan\theta$ at frequencies not on the dense grid established by the FFT, use linear interpolation.

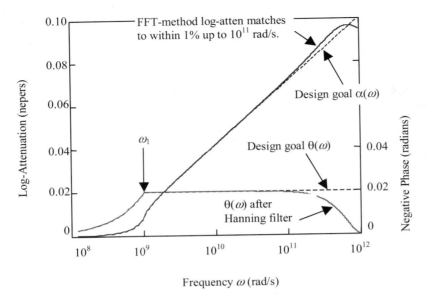

Figure 2.30—The FFT method synthesizes a log-attenuation curve to closely match any given phase curve ($\theta_1 = 0.02$, $\omega_1 = 10^9$ rad/s).

In the event you do not have accurate loss tangent data extending down to DC, use what data you have, filling in all unknown values down to frequency f_1 with a constant loss tangent, and below that feathering the loss tangent linearly to zero at DC.

POINT TO REMEMBER

> You can calculate a variation in dielectric constant to match any specified loss tangent.

2.12.7 Kramers-Kronig Relations

The Kramers-Kronig relations (see [5] page 83, and [15]) are of the same ilk as the network function relations in the previous section. They are a specialized form of those relations, crafted specifically to express the relation between the real and imaginary parts of complex permittivity (ϵ'_r and ϵ''_r respectively).

$$\epsilon''_r(\omega) = \frac{2\omega}{\pi} \int_0^\infty \frac{1-\epsilon'_r(u)}{u^2-\omega^2} du$$

$$\epsilon'_r(\omega) = 1 + \frac{2}{\pi} \int_0^\infty \frac{u\,\epsilon''_r(u)}{u^2-\omega^2} du$$

[2.103]

If you are familiar with the standard relations for the real and imaginary parts of a causal network function (see [12] and [13] page 200), you may recognize some differences here from the normal presentation. In particular, the integrand of the first equation here operates on the quantity $1-\epsilon'_r$ instead of ϵ'_r, a trivial modification that does not mathematically affect the outcome of the integral, but does improve the computational stability of the integration, the point being that at extremely high frequencies the term $1-\epsilon'_r$ tends toward zero. Another difference is that here ϵ''_r by definition equals the *negative* of the imaginary part of complex permittivity. A third difference is that here both integrations have been folded around zero.

POINT TO REMEMBER

> ➤ The Kramers-Kronig relations constrain the behavior of the real and imaginary parts of complex permittivity.

2.12.8 Complex Magnetic Permeability

In the science of magnetic materials there exist concepts analogous to complex permittivity; however, since digital applications work for the most part with nonmagnetic materials, we shall not explore that territory. Most conductors (except iron, steel, and nickel) and most insulating materials used in digital applications have a relative magnetic permeability of $1.000 + j0.000$.

2.13 IMPEDANCE IN SERIES WITH THE RETURN PATH

A single-ended transmission structure (as opposed to a differential transmission structure) comprises two conductors—one for the signal current and one for the returning signal current. These conductors are called the signal conductor and the return conductor respectively. High-speed digital pcbs use a solid reference plane for the return conductor. Although many digital engineers focus their attention on the signal conductor, both conductors play equally important roles in the transmission of high-speed signals.

Consider a transmission line with series impedance and shunt conductance z and y respectively per unit length (Figure 2.31), modeled as shown by per-unit-length parameters R, L, G, and C. To the return conductor of that structure add an impedance z_g per unit length. The values of input impedance Z_C and transmission coefficient γ are modified to become

$$Z_C = \sqrt{\frac{z+z_g}{y}} \qquad [2.104]$$

$$\gamma = \sqrt{(z+z_g)(y)} \qquad [2.105]$$

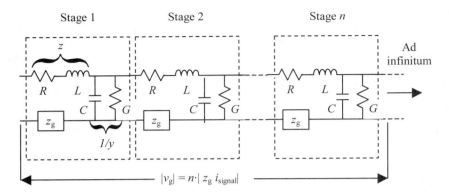

Figure 2.31—Any impedance placed in series with the return path increases Z_0, increases γ, and induces a ground shift voltage v_g.

As indicated by equations [2.104] and [2.105], the impedance z_g effectively adds to the total per-unit-length series impedance of the structure. Impedance z_g increases the apparent input impedance of the structure and also increases the high-frequency loss.

A lumped-element series-connected discontinuity inserted into a transmission structure causes equal amounts of disruption whether it is inserted in series with the signal path or the return path.

In high-speed pcb problems the return-path impedance comprises both the resistance and inductance of the solid reference planes. The resistive component of the return-path impedance would be negligible if the returning signal current were allowed to spread perfectly across the entire reference plane; however, that doesn't happen. At high frequencies the returning signal current bunches together, flowing only in a narrow band directly underneath the signal trace on the nearest reference plane. In a typical 50-ohm microstrip or stripline at height h above the nearest solid reference plane, 80% of the returning signal current in the reference plane flows within about $3h$ on either side of the signal trace. The reference-plane resistance adds to the total power dissipation of the structure, increasing the signal attenuation (see Section 2.10.3, "Proximity Effect for Microstrips and Striplines"). The skin-effect loss calculations presented in Chapter 5 take into account the location and profile of current on the reference plane. Such calculations are a normal part of any 2-D field solver that reports skin-effect loss.

What is not properly taken into account in most 2-D field solvers is the inductance of the return pathway. Most field solvers assume the reference plane is either infinite in extent (using the image-plane method) or a closed surface wrapped around the trace at a safe distance (using finite-element simulation of the wrapped reference plane). Either approach produces near-correct values for the overall per-unit-length inductance of the structure. What these approaches do not do is apportion that inductance into that part which appears in series with the signal conductor and that part in series with the ground. While the apportionment of inductance does not affect the line impedance or the transmission coefficient, it does affect the ground voltages measured from end to end across the structure, which in turn has a major effect on electromagnetic radiation and susceptibility.

When working with differential pairs, this discussion should remind you to add the series resistance of *both* wires when calculating the system loss. In working with coaxial cables, you must sum the resistance of the inner conductor and the resistance of the shield to find the total effective series resistance.

POINT TO REMEMBER

> ➢ An impedance in series with the return path affects the signal just as much as an impedance in series with the signal conductor.

2.14 SLOW-WAVE MODE ON-CHIP

The term *slow-wave mode* applies exclusively to on-chip interconnections implemented in a metal-insulator-semiconductor (MIS) configuration. On such interconnections the substrate resistance adds substantially to the signal loss and can sometimes have the peculiar effect of greatly slowing signal propagation. The resulting *slow-wave mode* occurs when the substrate conductivity is adjusted so that electromagnetic fields only *partially* penetrate the substrate. The wave velocity then becomes a function of the substrate, not just the good dielectric insulation between the trace and the top layer of the substrate.

Figure 2.32 illustrates a classic on-chip MIS transmission line, comprising a metal trace, a 1-μm silicon dioxide insulating layer, and a 200-μm semiconducting substrate. The solid metal layer on the back of the substrate is called *backside metallization*. In this example I'll assume a worst-case value for the conductivity of the silicon substrate layer, about 50 S/m. At a frequency of 1 GHz, the intrinsic impedance η and skin depth δ of the substrate are

$$|\eta| = \left| \sqrt{\frac{j\omega\mu}{\sigma + j\omega\epsilon_0\epsilon_{r,substrate}}} \right| = \left| \sqrt{\frac{j(2\pi \cdot 10^9)(4\pi \cdot 10^{-7})}{50 + j(2\pi \cdot 10^9)(8.854 \cdot 10^{-12})(12)}} \right| = 12.6 \ \Omega \qquad [2.106]$$

$$\delta = \sqrt{\frac{2}{\omega\mu\sigma}} = \sqrt{\frac{2}{(2\pi \cdot 10^9)(4\pi \cdot 10^{-7})(50)}} = 2.25 \ \text{mm} \qquad [2.107]$$

where $\epsilon_0\epsilon_{r,substrate}$ is the complex permittivity of the substrate (F/m),

μ is the magnetic permeability (H/m), and

σ is the conductivity (S/m) of the substrate.

For nonmagnetic substrate materials, $\mu = 4\pi \cdot 10^{-7}$ H/m.

At 1 GHz the low intrinsic impedance of the substrate (12.6 ohms) prevents electric fields from penetrating. The transmission line therefore inherits a large amount of capacitance in accordance with the small distance h_1 between the trace and the *top* of the substrate.

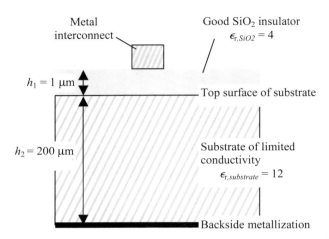

Figure 2.32—Metal-insulator-semiconductor (MIS) transmission lines suffer from the limited conductivity of the substrate.

Magnetic fields behave differently. At 1 GHz the penetration of magnetic fields (skin depth) greatly exceeds the thickness h_2 of the substrate, so the magnetic fields and their associated returning signal currents permeate the entire semiconducting substrate layer. The complete penetration of magnetic fields creates a large amount of inductance in accordance with the relatively large distance h_2 between the trace and the *bottom* of the substrate.

The difficulty with this circuit is that the electric and magnetic fields have become separated. In a perfect, homogeneous dielectric material, where the electric and magnetic fields both penetrate to the same depth, the velocity of propagation always equals $1/\sqrt{\epsilon\mu}$.

In this slow-wave example, the electric fields penetrate to a depth of h_1, while the magnetic fields penetrate all the way down to h_2, disconnecting the homogeneous assumption. The resulting combination of large capacitance and large inductance creates an absurdly slow velocity of signal propagation, much slower than would be indicated by the permittivity of either the insulator or the substrate acting alone.

In Figure 2.32 the velocity of propagation for a 1-GHz sine wave approaches 1/5 the speed of light in air. Furthermore, the complicated frequency dependencies associated with the slow-wave effect create significant phase distortion in the received waveform.

The slow-wave effect has been reported by numerous authors [16], [17], [18]. The consensus view about how to fix the problem is quite clear. You have three choices:

1. Raise the substrate conductivity by doping until it acts like a good, low-impedance return path. This approach shrinks the skin depth, forcing currents to flow mostly near the top surface of the semiconducting substrate. The line delay then depends only on the permittivity of the insulator (about 4.0 for silicon dioxide).

2. Decrease the substrate conductivity (by doping) until it acts like a good, high-impedance insulator. The electric and magnetic fields then completely penetrate the substrate layer together. The line delay in this case then depends

2.14 • Slow-Wave Mode

mostly on the permittivity of the substrate (about 12.0 for lightly doped silicon). A high-speed chip requires backside metallization for this approach to work.

3. Add intentional metallic return paths near the signal traces.

In the world of pcb design, the dielectric materials have such a low conductivity that one almost always adopts the configuration of solution 2. This approach implies the existence of a solid-metal reference plane somewhere in the layer stack. The reference plane serves the same purpose in a pcb as the backside metallization in a chip—it defines a good return-current path for all signals.

Occasionally, a board designer will implement solution 3. For example, in a 100BASE-TX interface, for the connection between the isolation transformer and the RJ-45 plug, you might use a co-planar differential pair with no underlying reference plane. The differential pair comprises a signal trace and an associated return-current conductor, so it meets the definition of solution 3.

In no case do pcb designers worry about the slow-wave mode or the implications of solution 1, because pcbs never use crummy, partly conducting substrates. That's one of the nice benefits of working at the printed-circuit design level.

The only way to separate the electric and magnetic fields on a pcb is to implement a nonuniform trace. For example, attaching hundreds of little cross-bars (like cilia) to a pcb trace adds a substantial amount of capacitance without changing the inductance, creating an absurdly low impedance and high delay.

POINT TO REMEMBER

> In an on-chip MIS configuration, if the electric fields penetrate to a depth of h_1, while the magnetic fields penetrate to a futher depth h_2, the resulting combination of large capacitance and large inductance creates an absurdly slow velocity of signal propagation.

For further study see: www.sigcon.com

CHAPTER 3

PERFORMANCE REGIONS

Figure 3.1 displays the propagation function of six distinct types of coaxial cabling, plus one typical pcb trace. The horizontal axis shows the operating frequency in units of Hz. The vertical axis shows cable attenuation in units of dB. Both axes use logarithmic scales.[23]

Each curve may be divided into distinct regions, with a characteristic shape to the loss function in each region. The hierarchy of regions, in order of increasing frequency, proceeds generally in the same order for all copper media:

- RC region
- LC region
- Skin-effect region
- Dielectric loss region
- Waveguide dispersion region

Within each region the requirements for termination differ, as do the tradeoffs between length and speed. Remarkably, a common signal propagation model accurately describes almost any type of metallic transmission media across all four regions.

3.1 SIGNAL PROPAGATION MODEL

This model computes the transfer function and impedance of any forms of metallic cabling. It is appropriate for use with cables made in multiwire, ribbon, UTP, STP, or coaxial format.

---

[23] Because the decibel is already a logarithmic unit, the vertical axis is actually a double-log (logarithm of a logarithm) of the cable propagation function.

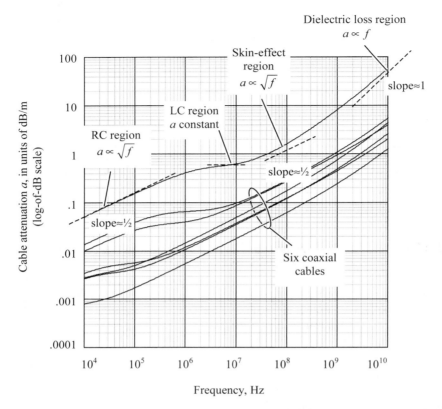

Figure 3.1—The attenuation curve for all copper transmission media is divided into distinct regions, with a characteristic relation in each region between the attenuation a and frequency.

It also works for pcb traces, both striplines and microstrips, up to a frequency of approximately 10 GHz.

Table 3.1 describes the six parameters to the model, and further explanation is provided in the notes that follow.

Table 3.1—Parameters for Metallic Transmission Model

Name	Meaning	Units	Ref
R_{DC}	DC resistance of conductors	Ω/m	[2.37]
ω_0	Frequency at which AC line parameters are specified	rad/s	see notes
R_0	Real part of AC resistance at frequency ω_0 (real part of z_i)	Ω/m	[2.43]
θ_0	Angle formed by the real and imaginary parts of complex electric permittivity (arctg $-\varepsilon''/\varepsilon'$) at frequency ω_0, rad; for small angles, $\theta_0 \approx \tan \theta_0$	rad	[2.73], [2.85]
Z_0	Characteristic impedance at frequency ω_0	Ω	[2.11]
v_0	Velocity of propagation (inverse of group delay) at frequency ω_0	m/s	[2.29]

3.1 • Signal Propagation Model

This model was developed over the course of many years, combining the best ideas from many versions of the international cabling standards with my own research.

This model is available in computer-executable form. It has been coded in MathCad 2000 syntax. A working version of the model along with a set of example spreadsheets and cable parameters is available through *http://www.sigcon.com*.

NOTES:

> ➢ Frequency ω_0 is assumed sufficiently high that the skin effect has substantially eliminated the internal inductance of the signal conductors, and the proximity effect has caused the distribution of current on the surface of the conductors to assume its high-frequency, magnetically dominated form, but not so high that you must concern yourself with non-TEM waveguide modes. In some geometrical arrangements such a setting is impossible. For example, a gigantic 3mm 50-ohm microstrip as typically used in microwave circuits might exhibit significant non-TEM behavior at frequencies as low as 10 GHz. Typical pcb geometries for digital circuits, being much smaller, enjoy the benefits of a much higher onset frequency for non-TEM behavior.

> ➢ For simulations of data transmission systems operating above the onset of the skin effect, ω_0 would ideally be located at or near one-half the transmission symbol frequency.

> ➢ For simulations of data transmission systems operating entirely within the RC dispersion region a value for R_{DC} and line capacitance C are required, but the other parameters may be ignored or set to default values:

$$\omega_0 = 10^6,$$

$$R_0 = 0, \text{ and}$$

$$\theta_0 = 0. \qquad [3.1]$$

$$Z_0 = \frac{1}{2.998 \cdot 10^8 \cdot C}$$

$$v_0 = 2.998 \cdot 10^8 \text{ m/s}$$

These substitutions are equivalent to accepting C as a constant value independent of frequency and then setting the minimum possible value of L consistent with the speed of light.

> ➢ If you have set ω_0 commensurate with the rise/fall bandwidth of your digital signals, then parameter Z_0 represents the average impedance over a section of transmission line with a delay comparable to the duration of one rising (or falling) edge. Parameter Z_0 is assumed to be purely real—the model supplies the imaginary parts of the overall characteristic impedance $Z_C(\omega)$.

- Parameter R_{DC} includes the total resistance of both outgoing and return conductors. For ribbon cables the total DC resistance sums the resistance of the outbound wire plus $1/N$ times the resistance of the N return wires. For twisted-pair cabling the total DC resistance is twice the resistance of either wire alone. For coaxial cables the total DC resistance is the sum of center conductor and shield resistances.
- Parameter R_0 similarly includes the total resistance of both outgoing and return conductors, with the path of current flow determined by the proximity effect.
- Sometimes R_{DC} is specified directly on the data sheet, in which case you should use that value instead of calculating it yourself from the wire sizes. Datasheet values take into account the resistivity of various alloys and coatings used in the construction of the conductors. Always use datasheet values when modeling copper-coated steel wire or tinned or silver-coated conductors.
- Parameter θ_0 models the effective dielectric loss for the mixture of insulating materials and air surrounding the conductors.
- The model assumes uniform values for all parameters along the length of the transmission line.
- The model does not take into account temperature variations. All parameters must be specified at their worst-case values.
- The transmission line attenuation is maximized when you combine low Z_0, low v_0, large R_{DC}, large R_0, and large θ_0.

Here are the model equations:

First use the specification of the real part of the skin-effect resistance, R_0, which is specified only at a single frequency, to produce a resistance model covering a range of frequencies.

$$R_{AC}(\omega) = R_0 \sqrt{\frac{2j\omega}{\omega_0}} \quad \Omega/\text{m} \quad [3.2]$$

where R_0 is the AC resistance of the wire at frequency ω_0, in ohms, and

R_{AC} is the complex-valued impedance due to the combination of skin effect and proximity effect.

The factor $\sqrt{\dfrac{2j\omega}{\omega_0}}$ indicates that the skin-effect resistance grows proportional to the square root of frequency.

Another way to write the same equation is to substitute $\sqrt{2j} = (1+j)$, which leads to

3.1 • Signal Propagation Model

$$R_{AC}(\omega) = R_0(1+j)\sqrt{\frac{\omega}{\omega_0}} \quad \Omega/m \quad [3.3]$$

The factor $(1+j)$ indicates that for all positive frequencies the real and imaginary parts of the complex skin-effect impedance are equal. You may also check that the real part of the complex skin-effect impedance equals R_0 when $\omega = \omega_0$.

Next we must model the crossover from DC resistance to AC resistance. I use a square-root-of-sum-of-squares type mixing function. This mixing function matches the measured skin-effect data presented in [28] and also produces the correct low-frequency value for the internal inductance of a round wire, 50 nH/m, given in [26] and [27].

$$R(\omega) = \sqrt{(R_{DC})^2 + (R_{AC}(\omega))^2} \quad \Omega/m \quad [3.4]$$

The transmission line is characterized by a nominal external inductance L_0 per meter, which I define as a constant, calculated from the characteristic impedance and velocity. This is the value of inductance measured at frequency ω_0. Variations in inductance with frequency (which occur as part of the internal inductance) are incorporated into the imaginary part of R.

External inductance: $\quad L_0 \triangleq \dfrac{Z_0}{v_0} \quad$ H/m $\quad [3.5]$

where Z_0 is the nominal characteristic impedance at frequency ω_0, and

v_0 is the velocity of propagation at frequency ω_0.

The transmission line is characterized by a nominal value of shunt capacitance C_0 per meter, which is calculated from the characteristic impedance and velocity.

$$C_0 \triangleq \frac{1}{Z_0 v_0} \quad F/m \quad [3.6]$$

Formulas [3.5] and [3.6] are mathematical inverses of the telegrapher's relations $Z_0 = \sqrt{L_0/C_0}$ [2.30] and $v_0 = 1/\sqrt{L_0 C_0}$ [2.33].

Value [3.6] is next expanded into a frequency-varying model of the complex capacitance $C(\omega)$ of the transmission line. The imaginary part of $j\omega C(\omega)$ represents the capacitive reactance. The real part of $j\omega C(\omega)$ represents dielectric losses within the transmission line. The ratio of the real part of $j\omega C(\omega)$ to the imaginary part of $j\omega C(\omega)$ equals the dielectric loss tangent. A nonzero loss tangent induces a slow degradation in effective capacitance with increasing frequency.

$$C(\omega) = C_0 \left(\frac{j\omega}{\omega_0}\right)^{-\frac{2\theta_0}{\pi}} \quad \text{F/m} \quad [3.7]$$

The transmission-line propagation coefficient, per meter, is defined in numerous texts [29]:

$$\gamma(\omega) = \left[(j\omega L_0 + R(\omega))(j\omega C(\omega))\right]^{1/2} \quad \text{neper/m} \quad [3.8]$$

The transfer function for a transmission line of length l meters is computed from the propagation coefficient,

$$H(\omega, l) = e^{-l \cdot \gamma(\omega)} \quad \text{(dimensionless)} \quad [3.9]$$

The attenuation in dB is defined as the negative of the transfer gain in dB.

$$Attenuation(\omega) = -20\log(|H(\omega,l)|) = \frac{20 \cdot l}{\ln(10)} \text{Re}(\gamma(\omega)) \quad \text{dB} \quad [3.10]$$

The characteristic impedance modeled as a function of frequency is

$$Z_C(\omega) = \left[(j\omega L_0 + R(\omega))/(j\omega C(\omega))\right]^{1/2} \quad \Omega \quad [3.11]$$

These equations correctly model the transfer function and impedance at low frequencies (in the dispersion-limited mode), mid-frequencies (in the skin-effect-limited mode), and at extremely high frequencies (in the dielectric-loss-limited mode).

In the mid-to-high frequency range the parameter R_{DC} provides an amount of loss that is flat with frequency. The parameter R_0 provides an amount of loss that grows (in dB) in proportion to the square root of frequency. The parameter θ_0 provides an amount of loss that grows (in dB) in direct proportion to frequency. At all frequencies the magnitude and phase responses match to produce a causal, minimum-phase response.

POINTS TO REMEMBER

> - The signal propagation model computes the transfer function and impedance of cables made in multiwire, ribbon, UTP, STP, or coaxial format. It also works for pcb traces, both striplines and microstrips, up to a frequency of approximately 10 GHz.
> - The parameter R_{DC} provides an amount of loss that is flat with frequency.
> - The parameter R_0 provides an amount of loss that grows (in dB) in proportion to the square root of frequency.

- The parameter θ_0 provides an amount of loss that grows (in dB) in direct proportion to frequency.
- At all frequencies the magnitude and phase responses match to produce a causal, minimum-phase response.

3.1.1 Extracting Parameters for RLGC Simulators

Transmission-line parameters R, L, G, and C may be extracted from the expressions in the previous section.

Some simulators require a discrete table of parameters. Each row of the table represents the four values R, L, G, and C sampled at one particular frequency. The number of rows (frequency sample points) may be unlimited, with the frequencies spaced either linearly or exponentially on some kind of dense grid. The simulator interpolates between the listed points to do its work. Assuming you have the list of sample frequencies ω_n in units of rad/s, Table 3.2 shows how to prepare the vectors **R**, **L**, **G**, and **C** from which you may prepare your table. The expressions properly account for the influence of the inductive component of skin effect (internal inductance) on the total inductance **L** and the influence of the imaginary part of $C(\omega)$ on **G**.

Other simulators expect you to provide values of only six parameters from which they extrapolate the full range of R, L, G, and C at all frequencies. I'm a little suspicious of these types of simulators because there is no consistent, standard way to perform the extrapolation. I have heard reports of some simulators generating noncausal waveforms. You should check carefully the step response of your simulator (especially under a condition with lots of dielectric loss but very little resistive loss) to make sure it gives you the correct causal dielectric response, as shown in Chapter 4, "Frequency-Domain Modeling." Some don't. One popular combination of parameters from which other values may be extrapolated appears in Table 3.3.

There is no consistent standard as to whether the value L_0 includes the internal inductance or not. You'll have to check the documentation of your simulator. The equations above assume it is *not* included in L_0; therefore L_0 in Table 3.3 just equals L_0, which in my system of definitions is the external inductance. If you wish to include the internal inductance, then you should evaluate the total line inductance at a frequency well below the skin-effect onset (e.g., 1 rad/s) as shown in [3.12].

Table 3.2—R, L, G, and C Vectors Sampled at Frequencies ω_n

Parameter	Value	Units
Series resistance	$\mathbf{R}_n \triangleq \mathrm{Re}[j\omega_n L_0 + R(\omega_n)]$	Ω/m
Series inductance	$\mathbf{L}_n \triangleq \dfrac{\mathrm{Im}[j\omega_n L_0 + R(\omega_n)]}{\omega_n}$	H/m
Shunt conductance	$\mathbf{G}_n \triangleq \mathrm{Re}[j\omega_n C(\omega_n)]$	S/m
Shunt capacitance	$\mathbf{C}_n \triangleq \dfrac{\mathrm{Im}[j\omega_n C(\omega_n)]}{\omega_n}$	F/m

Table 3.3—Six *R, L, G,* and *C* Parameters from Which Other Values May Be Extrapolated

Parameter	Value	Units
Series resistance	$\mathbf{R}_0 \triangleq R_{DC}$	Ω/m
Series inductance	$\mathbf{L}_0 \triangleq L_0$	H/m
Shunt conductance	$\mathbf{G}_0 \triangleq 0$	S/m
Shunt capacitance	$\mathbf{C}_0 \triangleq C_0$	F/m
Series skin-effect resistance	$\mathbf{R}_s \triangleq \dfrac{R_0}{\sqrt{\omega_1/2\pi}}$	$\Omega/(m\text{-}Hz^{1/2})$
Shunt conductance (dielectric loss)	$\mathbf{G}_d \triangleq 2\pi C_0 \theta_0$ for $\tan\theta_0 < 0$	$S/(m\text{-}Hz)$

$$L_{ext} + L_{int} \triangleq \lim_{\omega \to 0} \frac{\text{Im}[j\omega L_0 + R(\omega)]}{\omega} \qquad [3.12]$$

3.2 HIERARCHY OF REGIONS

The transmission loss associated with any conductive transmission media increases monotonically with frequency. Sweeping from low frequencies to high, the slope of the loss curve changes in a predictable way as you pass the onset of various regions of operation. The progression of regions, and the transmission performance within each region, is the subject of this chapter.

Alternate forms of transmission structures exist, such as fiber-optic waveguides and various forms of RF waveguides, that cannot convey DC signals. In these alternate structures the loss function must be necessarily be nonmonotonic, leading to a different hierarchy of performance regions. The discussion of regions presented here applies only to conductive transmission structures as normally used in digital applications.

Figure 3.2 illustrates the general arrangement of performance regions pertaining to copper media. The particular data shown in this diagram represents a 150-μm (6-mil), 50-Ω FR-4 pcb stripline. The waveguide dispersion region for this trace begins at frequencies higher than shown on the chart.

The distinguishing features of each region may be determined by analysis of the transmission-line propagation coefficient [3.13], propagation function [3.14], and characteristic impedance [3.15].

$$\gamma(\omega) = \left[(j\omega L_0 + R(\omega))(j\omega C(\omega))\right]^{1/2} \quad \text{neper/m} \qquad [3.13]$$

$$H(\omega, l) = e^{-l \cdot \gamma(\omega)} \quad \text{(dimensionless)} \qquad [3.14]$$

3.2 • Hierarchy of Regions

$$Z_C(\omega) = \left[\left(j\omega L_0 + R(\omega) \right) / \left(j\omega C(\omega) \right) \right]^{1/2} \quad \Omega \qquad [3.15]$$

where $R(\omega)$, L_0, and $C(\omega)$ represent the per-meter parameters of resistance, inductance, and capacitance respectively,

the line conductance G is assumed zero, and

the propagation function H at frequency ω (rad/s) varies exponentially with the product of the length l and the propagation coefficient γ.

Near DC the magnitude of the inductive reactance, ωL dwindles to insignificance in comparison to the DC resistance. All that matters below this point is the relation between the DC resistance of the line and its capacitance. Lines at such low frequencies are said to operate in the RC region.

At higher frequencies the inductive reactance grows, eventually exceeding the magnitude of the DC resistance, forcing the line into the LC region.

Beyond the LC transition the internal inductance of the conductors (a mere fraction of the total inductance) becomes significant compared to the DC resistance. This development forces a redistribution of current within the bodies of the conductors. The redistribution of current heralds the arrival of the skin-effect region.

Dielectric losses are present at all frequencies, growing progressively more severe at higher frequencies. These losses become noticeable only when they rise to a level comparable with the resistive losses, a point after which the line is said to operate in the dielectric-loss-limited region.

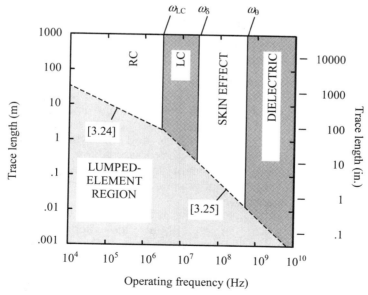

Figure 3.2—Performance regions for a 150-μm (6-mil), 50-Ω, FR-4 stripline.

At frequencies so high that the wavelength of the signals conveyed shrinks to a size comparable with the cross-sectional dimensions of the transmission line, other non-TEM modes of propagation appear. These modes do not by themselves portend a loss of signal power, but they can create objectionable phase distortion (i.e., dispersion of the rising and falling edges) that limits the maximum speed of operation. The region in which non-TEM modes must be taken into consideration is called the waveguide region.

At any frequency, regardless of the mode of operation, a transmission line can always be shortened to a length $l_{LE}(\omega)$ below which the line operates not in a distributed fashion, but in a mode reminiscent of a simple lumped-element circuit. The lumped-element region appears as a broad band underlying all the other regions in Figure 3.2, bounded by two dotted-line segments describing the function $l_{LE}(\omega)$.

As the length of a transmission line continues to shrink, at a point several orders of magnitude below $l_{LE}(\omega)$ it acts as a perfect wire.

POINTS TO REMEMBER

> Sweeping from low frequencies to high, the loss curve for a transmission line changes in a predictable way as you pass the onset of various regions of operation.

> The distinguishing features of each region are determined by the propagation coefficient, propagation function, and characteristic impedance.

> The regions usually appear in this order: lumped-element, RC, LC, skin-effect, dielectric, and waveguide.

3.2.1 A Transmission Line Is Always a Transmission Line

Article first published in *EDN Magazine*, April 4, 2002

Toss one end of a stout rope to a circus strongman. Then back up, pulling the rope taut as you go. When you are standing about 50 ft apart, flick the rope with a rapid up-and-down motion. If the man at the other end holds the line taut, you will observe a familiar pattern of wave propagation. Your up-and-down stroke first passes quickly from you to the strongman. At his end, the waveform reflects, sending an inverted copy of the original pulse back towards you. One round-trip delay after the initial flick, you feel the echo of your (attenuated and inverted) original excitation. Then, the residual signal bounces back and forth many times with an exponentially decaying amplitude.

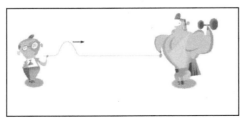

This simple physical analogy reveals much about the behavior of pcb transmission lines. It shows propagation of the input signal, reflection at the far end, and residual ringing.

It also reveals a temporal disconnection between the ends of a long transmission line. In the example, your strongman stands so far away that the propagation delay across the taut rope easily exceeds the rise-and-fall time of your input signal. Under these conditions, when you first flick the rope, you feel only the mass and tautness of the rope, not the strongman.

Your interaction with the strongman proceeds in three stages. First, you interact with the rope. Then, the rope conveys your inputs of force and velocity to the load. Finally, your signal (now delayed and possibly attenuated) interacts with the load. This sequence corresponds precisely to the behavior of an electrical source, a transmission line, and its load—provided that the delay of the line exceeds the rise (or fall) time of the source.

To expose the temporal disconnection in more interesting terms, suppose I drape a black curtain halfway between you and the strongman. With the curtain in place, as long as I don't change the tautness of the rope, you can't tell whether the rope is anchored to a man, a block of wood, or another long section of rope. Obviously, you can infer from the size and timing of the echo the characteristics of the far-end load, but before the echo returns, in the first moment after you create an outgoing waveform, you feel only the mass and tautness of the rope, not the anchor.

> **When you first flick a rope, you feel only the mass and tautness of the rope, not the anchor at the far end.**

Electrical transmission lines exhibit precisely the same effect. The input impedance of a long transmission line, in the brief interval of time before the echo returns, depends only on the characteristics of the line itself, not on the load.

"But what," asks a student, "about a short transmission line? In that case, doesn't the driver see the load directly? Does the input impedance thus behave one way on a long transmission line but differently when the load is adjacent to the driver? How does it know what to do?"

To answer this question, I want you to walk over to your strongman and clench the rope right next to his hands. Pull hard. What you feel now is the strength of his grip, not the rope. At a short distance, no matter what kind of rope you use, thick or thin, the same result applies: You feel the strongman, not the rope.

Keep in mind that in both cases, the rope remains a rope. It doesn't suddenly change character. It still conveys your forces to the strongman, only it does so with such speed that the returning signal influences you instantaneously. Before you even begin to create part of a rising edge, the returning (and opposing) force holds the rope back down. The instantaneously returning forces, in contrast to the temporally disconnected reflections of the previous case, are responsible for the change in behavior.

Similarly, in the world of high-speed digital design, a pcb trace of any length always remains a transmission line. It supports two modes of propagation, going out to the load and back. When the line is short, these two modes of propagation still exist, only their temporal superposition creates the illusion of a direct connection between source and load.

POINTS TO REMEMBER

> ➢ A pcb trace of any length always remains a transmission line, supportig two modes of propagation (out and back).
>
> ➢ When a transmission line is short, two modes of propagation still exist, only their temporal superposition creates the illusion of a direct connection between source and load.

3.3 NECESSARY MATHEMATICS: INPUT IMPEDANCE AND TRANSFER FUNCTION

The performance of the linear, time-invariant transmission circuit shown in Figure 3.3 depends on four crucial factors: the characteristic impedance of the line Z_C [3.15], the raw one-way propagation function of the transmission line H [3.14], the source impedance Z_S, and the load impedance Z_L. All four of these complex phasor quantities vary with frequency. Although the figure is drawn representing a single-ended coaxial configuration the same considerations apply to any form of conductive transmission circuit.

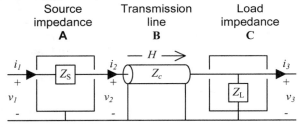

Figure 3.3—A transmission line complete with source and load impedances may be modeled as a cascade of three two-port circuits.

The input impedance v_2/i_2 of a loaded transmission line is derived in Appendix C, "Two-Port Analysis."

$$Z_{\text{in,loaded}} = Z_C \left\{ \frac{\left(\dfrac{H^{-1}+H}{2}\right) + \dfrac{Z_C}{Z_L}\left(\dfrac{H^{-1}-H}{2}\right)}{\left(\dfrac{H^{-1}-H}{2}\right) + \dfrac{Z_C}{Z_L}\left(\dfrac{H^{-1}+H}{2}\right)} \right\} \quad [3.16]$$

The gain G of the circuit of Figure 3.3, taking into account all the relevant loading effects and end-to-end reflections, is (see Appendix C)

$$G = \frac{v_3}{v_1} = \frac{1}{\left(\dfrac{H^{-1}+H}{2}\right)\left(1+\dfrac{Z_S}{Z_L}\right)+\left(\dfrac{H^{-1}-H}{2}\right)\left(\dfrac{Z_S}{Z_C}+\dfrac{Z_C}{Z_L}\right)} \quad [3.17]$$

where G is the overall system gain from the open-circuited output of the driver v_1 to the load v_3,

Z_S is the source impedance of the driver (Ω),

Z_L is the load impedance at the far end of the transmission structure (Ω),

Z_C is the characteristic impedance of the transmission structure (Ω), and

H is the one-way propagation function of the raw, unloaded transmission structure [3.14].

Formulation [3.17] is most useful for analyzing the behavior of lumped-element and RC-mode transmission structures. An alternate form applies best for the analysis of the *low-loss* structures (LC, skin-effect, and dielectric-loss regions).

$$G = \frac{1}{2}\frac{(1-\Gamma_1)(1+\Gamma_2)}{1-H^2\Gamma_1\Gamma_2} \quad [3.18]$$

where G is the overall system gain from the open-circuited output of the driver v_1 to the load v_3,

$\Gamma_1 = \dfrac{Z_S - Z_C}{Z_S + Z_C}$ is the reflection coefficient at the source end of the transmission structure,

$\Gamma_2 = \dfrac{Z_L - Z_C}{Z_L + Z_C}$ is the reflection coefficient at the load end of the transmission structure, and

H is the one-way propagation function of the raw, unloaded transmission structure [3.14].

If your objective is the undistorted conveyance of a signal from source to load, then you must ensure that the propagation function [3.17 or 3.18] remains flat over the band of frequencies covering the bulk of the spectral content of your data signal. The requirement for "undistorted conveyance" is equivalent to asking that each transition arrive intact, at full size, without significant dispersion of the rising or falling edge, and without any lingering aftereffects, like reflections or ringing. Such a waveform is useful for *first-incident-wave switching*, meaning that the first edge arrives with sufficient fidelity to be immediately and reliably used.

On the other hand, if you are willing to wait a few round-trip times for your signal to settle, then you do not need your signals to arrive with first-incident-wave quality. You can in this case tolerate significant imperfections in the frequency response of your channel, all of which are cured by waiting for the steady-state condition to emerge, after which the signal may be reliably used. Waiting solves almost every signal integrity problem. If you can wait long enough, any transmission line will settle to a steady state. Of course, I assume the reason you are reading this book is that you cannot afford to wait! In that case, you have a direct interest in the gain flatness of the propagation function.

The bulk of the useful spectral content of a random data sequence spans a range from DC (zero frequency) up to a maximum upper bandwidth of

$$f_{knee} \approx \frac{0.5}{t_r} \text{ Hz}$$
$$\omega_{knee} \approx 2\pi \frac{0.5}{t_r} \text{ rad/s}$$

[3.19]

The *knee frequency*, f_{knee}, is a crude estimate of the highest frequency content within a particular digital signal. Presuming the propagation function G [3.17] remains flat to within x percent over the range $0 < f < f_{knee}$, the expected distortion in the received waveform will be on the order of x percent.

The best assumption for the midpoint of the spectral content associated with the rising and falling edges of a digital signal is a little less than the maximum bandwidth:

$$f_{edge} \approx \frac{.35}{t_r} \text{ Hz}$$
$$\omega_{edge} \approx 2\pi \frac{.35}{t_r} \text{ rad/s}$$

[3.20]

A time-domain reflectometry (TDR) setup measures the gain from v_1 to v_2 (see Appendix C):

$$TDR = \frac{v_2}{v_1} = \frac{\left(\frac{H^{-1}+H}{2}\right) + \left(\frac{H^{-1}-H}{2}\right)\frac{Z_C}{Z_L}}{\left(\frac{H^{-1}+H}{2}\right)\left(1+\frac{Z_S}{Z_L}\right) + \left(\frac{H^{-1}-H}{2}\right)\left(\frac{Z_S}{Z_C}+\frac{Z_C}{Z_L}\right)}$$

[3.21]

where *TDR* is the overall system gain from the open-circuited output of the driver v_1 to the time-domain reflectometry observation point v_2,

Z_S is the source impedance of the driver (Ω),

Z_L is the load impedance at the far end of the transmission structure (Ω),

Z_C is the characteristic impedance of the transmission structure (Ω), and

H is the one-way propagation function of the raw, unloaded transmission structure [3.14].

The following sections detail the performance characteristics of each region and the means necessary to maintain acceptable flatness in [3.17] over the intended range of operation.

The step-response approximations for each region work best for transmission media with wide, well-defined regions. For example, the skin-effect approximation presented below represents the performance of Belden 8237 beautifully over a range of four orders of magnitude, from 10^5 to 10^9 Hz. As you approach the edge of each region, however, the step response begins to mutate into a new shape characteristic of the next region.

The modeling of complete systems, including arbitrary source and load impedances and combinations of operation in all regions, is considered at the end of this chapter.

POINT TO REMEMBER

> The undistorted conveyance of a signal from source to load requires a propagation function that remains flat over a band of frequencies covering the bulk of the spectral content of the data signal.

3.4 LUMPED-ELEMENT REGION

At any frequency, regardless of the mode of operation, a transmission line can always be shortened to a length below which the line operates not in a distributed fashion, but in a mode reminiscent of a simple lumped-element circuit (Section 3.2).

The mathematical extent of the lumped-element region includes all combinations of ω and l for which the magnitude of the propagation coefficient $l\gamma(\omega)$ remains less than Δ.

$$|l\gamma(\omega)| < \Delta \qquad [3.22]$$

where Δ is an arbitrary constant typically set to about 1/4.

l is the length of the transmission line, m, and

$\gamma(\omega)$ is the propagation coefficient of the transmission line (complex neper/m) at frequency ω (rad/s).

For typical digital transmission applications the quantity $|\gamma(\omega)|$ increases monotonically from DC so that inequality [3.22] need be checked only at the maximum length and maximum anticipated frequency of operation. If the transmission line satisfies [3.22] at that maximum point, it will similarly satisfy [3.22] at all shorter lengths and lower frequencies.

Beware the fallacy that a short transmission line, even one short enough to fall into the lumped-element region, never requires termination. Such is not the case. Even a perfect zero-length transmission line may resonate horribly if used to interconnect a ferociously reactive combination of source and load (see Section 3.4.6, "Step Response (Lumped-Element Region)").

The classification of a transmission line in the lumped-element region does not determine how the line is going to act. It determines merely how the line may be analyzed.

POINT TO REMEMBER

> The classification of a transmission line in the lumped-element region does not determine how the line is going to act. It determines merely how the line may be analyzed.

3.4.1 Boundary of Lumped-Element Region

A *exact* physical interpretation of equation [3.22] is fairly difficult to comprehend; however, by making a few reasonable assumptions you may approximate the boundary of the lumped-element region in the following way. First assume that the propagation coefficient for a transmission line is given by [3.13], and that parameters R, L, and C are constants that do not much vary with frequency. Substituting [3.13] for the propagation coefficient in [3.22], you may derive the following equation relating ω and l_{LE}.

$$l_{LE} = \left| \frac{1/4}{\sqrt{(j\omega L + R) \cdot j\omega C}} \right| \quad [3.23]$$

Since the boundaries of the lumped-element region is by definition a rather fuzzy concept in the first place, you needn't bother with precise calculation of [3.23], substituting instead two asymptotic approximations that handle the cases where $j\omega L$ is either much smaller than, or much larger than, R. These are the boundaries drawn in Figure 3.2.

$$l_{LE} \approx \frac{\Delta}{\sqrt{\omega R_{DC} C}} \quad \text{for } \omega < R_{DC}/L \quad [3.24]$$

$$l_{LE} \approx \frac{\Delta}{\omega\sqrt{LC}} \quad \text{for } \omega > R_{DC}/L \quad [3.25]$$

where ω is frequency of operation, rad/s,

R_{DC} is the series DC resistance of the transmission line (signal and return resistances added together), Ω/m,

L is the transmission-line series inductance per meter, H/m,

C is the transmission-line shunt capacitance per meter, F/m,

l_{LE} is the upper boundary of the lumped-element region, m, and

Δ is an arbitrary constant customarily set to 1/4.

NOTE: For differential configurations, define R_{DC} as the sum of the resistances of the outbound and returning conductors, and L and C as the inductance and capacitance respectively of the differential transmission line thus formed, $L = Z_{DIFFERENTIAL}/v_{DIFFERENTIAL}$ and $C = 1/(Z_{DIFFERENTIAL} \cdot v_{DIFFERENTIAL})$.

3.4.2 • Pi Model

If the length l is known and you wish to determine the maximum frequency of lumped-element operation, the constraints in [3.24] and [3.25] may be inverted to produce

$$\omega_{LE} < \left(\frac{\Delta}{l}\right)^2 \frac{1}{R_{DC}C} \quad \text{for} \quad l > \frac{\Delta}{R_{DC}}\sqrt{\frac{L}{C}} \qquad [3.26]$$

$$\omega_{LE} < \left(\frac{\Delta}{l}\right)\frac{1}{\sqrt{LC}} \quad \text{for} \quad l < \frac{\Delta}{R_{DC}}\sqrt{\frac{L}{C}} \qquad [3.27]$$

Here you can see the value of the approximations used in [3.24] and [3.25], as the direct inversion of any equation utilizing the full, frequency-varying form of $\gamma(\omega)$ would indeed be a formidable undertaking.

In physical terms, constraint [3.24] asks that the RC time constant $l^2 R_{DC} C$ formed by the structure's DC resistance and capacitance remain far smaller than the time constant $1/\omega$ associated with the highest frequency of operation. Constraint [3.25] asks that the LC delay of the transmission structure $l\sqrt{LC}$ remain much shorter than $1/\omega$. Working together, these two constraints ensure that the delay internal to the transmission structure itself remains far smaller than the signal rise or fall time. Under these conditions the transmission line enjoys a peculiarly tight coupling between the source and load impedances. When connecting obnoxiously reactive components, the system can still exhibit vigorous amounts of simple harmonic resonance, but what it cannot do is create the sort of lingering, unexpectedly late reflections that happen on a truly distributed circuit.

Transmission lines short enough to operate in the lumped-element region rarely require termination except in unusual situations involving very low-impedance drivers coupled either through exorbitant amounts of packaging inductance or connected through transmission lines to heavily reactive loads.

POINTS TO REMEMBER

➢ A transmission line can always be shortened to a length below which it operates in the lumped-element region.

➢ Transmission lines short enough to operate in the lumped-element region rarely require termination.

3.4.2 Pi Model

The *pi-model* circuit approximates the behavior of a short transmission line (Figure 3.4). This circuit is equivalent to a second-order Taylor series solution of the transmission equations (see Appendix D, "Accuracy of Pi Model").

The name *pi model* derives from the schematic configuration of the three circuit elements in the model whose positions resemble the three strokes of the Greek letter π.

Over the domain prescribed by [3.22], the fractional accuracy E of predictions made by the pi model is approximately bounded by

$$|E| < \left| \frac{\Delta^3}{6} \left(\frac{Z_S}{Z_C} + \frac{Z_C}{Z_L} \right) \right| \qquad [3.28]$$

where Z_S, Z_C, and Z_L represent the source impedance of the driver, the characteristic impedance of the transmission line, and the impedance of the load respectively.

In typical cases where the ratios Z_S/Z_C and Z_C/Z_L each remain less than two and $\Delta = 1/4$ the accuracy is better than 1% (see Appendix D).

Those experienced in analog design may be concerned that the inductive and capacitive components of the pi model could under some circumstances resonate severely. If you have made this discovery I should like to congratulate you on your astute observation and also address your discomfort by pointing out that condition [3.27] precludes the realization of such resonance by restricting the range of applicability of the model to only those frequencies well below the resonant frequency.

POINT TO REMEMBER

> The pi model applies to any transmission line electrically short compared to the signal wavelength, and where the time constant $l^2 R_{DC} C$ remains small compared to the signal period.

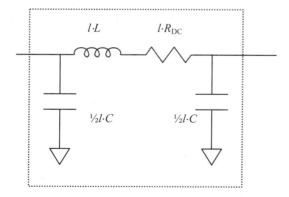

Figure 3.4—The pi model applies to any transmission line electrically short compared to the signal wavelength, and where the time constant $l^2 R_{DC} C$ remains small compared to the signal period.

3.4.3 Taylor-Series Approximation of H (Lumped-Element Region)

Operation within the lumped-element region assumes a line delay short compared to the rise and fall time of the externally applied signals. Both source and load therefore exert an almost instantaneous influence over the system behavior, creating the illusion of a direct connection between the two. Keep in mind, though, that even on a short line two fully independent modes of propagation still exist (out and back). Equations [3.13] through [3.17] still apply (see Section 3.2.1, "A Transmission Line Is Always a Transmission Line").

Within the lumped-element region the limited magnitude of the propagation coefficient $l\gamma$ renders possible the use of Taylor-series expansions for H and H^{-1} in this region. These expansions greatly simplify the analysis.

$$H \approx 1 - (l\gamma) + \frac{(l\gamma)^2}{2} - \frac{(l\gamma)^3}{6} + \ldots \quad [3.29]$$

$$H^{-1} \approx 1 + (l\gamma) + \frac{(l\gamma)^2}{2} + \frac{(l\gamma)^3}{6} + \ldots \quad [3.30]$$

where H is the one-way propagation function of the transmission line (a complex, dimensionless quantity),

l is the transmission-line length (m), and

γ is the per-unit-length propagation coefficient (complex neper/m).

Calculations of circuit behavior made using the Taylor-series approximation depend only upon ordinary differential equations as opposed to the partial differential equations required to construct a fully distributed model. Therein lies one advantage of operation in the lumped-element region—it's relatively easy to determine in this region how a transmission line will respond.

This text uses only the first four terms of the Taylor-series expansion to generate the following approximations used in calculation of lumped-element input impedance and system gain.

$$\frac{H^{-1} + H}{2} \approx 1 + \frac{(l\gamma)^2}{2} \quad [3.31]$$

$$\frac{H^{-1} - H}{2} \approx (l\gamma) + \frac{(l\gamma)^3}{6} \quad [3.32]$$

where H is the one-way propagation function of the transmission line (a complex, dimensionless quantity),

l is the transmission-line length (m), and

γ is the per-unit-length propagation coefficient (complex neper/m).

POINT TO REMEMBER

> ➢ Within the lumped-element region you may use Taylor-series expansions for H and H^{-1}.

3.4.4 Input Impedance (Lumped-Element Region)

On a lumped-element line the load exerts an almost instantaneous influence over the input impedance of the structure (see Section 3.2.1, "A Transmission Line Is Always a Transmission Line").

This happens because, by definition, the transit delay of a lumped-element structure is limited to a short fraction of the signal risetime. In the time the signal progresses only partway through a transition, information about the changing input propagates to the far end of the line, interacts with the load, and reflects back to the source. The source therefore receives almost instantaneous feedback about the conditions at the load.

On very long lines the same general scenario applies with the exception that the longer transit delay effectively disconnects the source and load in a temporal sense. Information about the load reflects back to the source too late to affect the progress of an individual rising or falling edge. The reflection in such a case, rather than being considered part of the "input impedance" of the line, may be considered separately as a distinct "reflection." From a mathematical viewpoint, whether you consider the reflection as a perturbation in the input impedance of the structure or as a structurally distinct reflection from a distant load makes no difference, provided that you properly account for the reflected-wave effect somehow in your calculations.

Let's next examine the input impedance of a lumped-element structure under various conditions of loading. To begin, apply the lumped-element Taylor-series approximations [3.31] and [3.32] to the general equation for input impedance [3.16], neglecting all but the constant and linear terms.

$$Z_{\text{in,loaded}} \approx Z_C \left\{ \frac{1 + \frac{Z_C}{Z_L}(l\gamma)}{(l\gamma) + \frac{Z_C}{Z_L}} \right\} \qquad [3.33]$$

Assuming the line to be very lightly loaded $(Z_L \gg Z_C)$ causes the right-hand terms in the numerator and denominator to vanish, leaving you with a very simple expression for the input impedance.

$$Z_{\text{in,open-circuited}} \approx Z_C \left\{ \frac{1}{l\gamma} \right\} \qquad [3.34]$$

Plugging in the definitions of γ [3.13] and Z_C [3.15] reveals a classic result—that the input impedance of a short, unloaded transmission line looks entirely capacitive.

3.4.4 • Input impedance

$$Z_{\text{in,open-circuited}} \approx \frac{1}{l \cdot j\omega C} \qquad [3.35]$$

The amount of capacitance equals the total distributed capacitance (*lC*) of the line. The *approximately equal* sign in [3.35] exists to remind you that this equation applies only under very special circumstances:

> ➢ The line delay is short compared to the signal rise/fall time (perhaps 1/6 or at most no more than 1/3 of the rise/fall time).
> ➢ The line is lightly loaded at its endpoint.

Violation of either constraint invalidates [3.35]. I mention this because one of the most widely held misconceptions about transmission lines is the belief that the input impedance of a transmission line looks capacitive. It holds this appearance only when short enough to qualify as a lumped-element structure and when lightly loaded. Under other conditions the input impedance varies considerably.

For example, let's see what happens if the line is short-circuited to ground at the far end. You may be asking yourself, What good is a transmission line shorted at the far end? The application for such a line is the grounding of an integrated circuit (IC) pin on a PC layout that does not permit room for a ground via contiguous with the IC pin pad. In congested situations it is common for (inexperienced) layout persons to connect the IC ground pin with a short trace to a ground via (see Figure 3.5). The following discussion computes the effective input impedance of the trace leading to ground, as viewed from the perspective of the IC ground ball.

Assuming the line to be shorted to ground at the ground via implies that $\left(Z_{\text{L}} \ll Z_{\text{C}}\right)$.

This assumption inflates the right-hand terms in both numerator and denominator of [3.33], causing them to dominate the equation and leaving you with this simple expression for the input impedance.

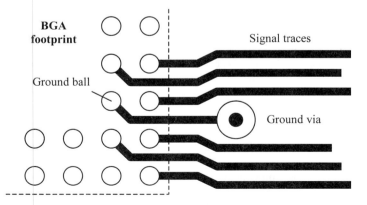

Figure 3.5—This ground via lies too far from the ball-grid array (BGA) ground ball.

$$Z_{\text{in,short-circuited}} \approx Z_{\text{C}}\{l\gamma\} \qquad [3.36]$$

Again plugging in definitions [3.13] and [3.15] reveals that the input impedance of a short, grounded transmission line looks either inductive or resistive, depending on the ratio of $j\omega L$ to R.

$$Z_{\text{in,short-circuited}} \approx l \cdot (j\omega L + R) \qquad [3.37]$$

Equation [3.37] tells you that the input impedance of a short line shorted to ground at the far end looks like nothing more or less than the total series impedance, inductive plus resistive, of the structure. Formula [3.37] provides the basis for making calculations of the *ground-bounce* (also called *simultaneous switching noise*) generated when a current $i(t)$ passes through a short trace leading to ground.

At the frequencies associated with most high-speed digital grounding problems, the inductance of a pcb trace is almost always much more significant than the resistance, leading to this simplified conclusion:

The input impedance of a short, grounded pcb trace looks entirely inductive.

The amount of inductance equals the total distributed inductance (lL) of the transmission line formed by the pcb trace and its associated return path. The approximately equal sign in [3.37] exists to remind you that this equation applies only under very special circumstances:

➢ The line delay is short compared to the signal rise/fall time (perhaps 1/6 or at most no more than 1/3 of the rise/fall time).
➢ The line is grounded at its endpoint.

You should check [3.33] to verify that when the line is properly terminated $(Z_{\text{L}} = Z_{\text{C}})$ the numerator and denominator return the same result so that the effective input impedance of the structure equals precisely Z_{C}. This is the situation on an end-terminated transmission line as typically implemented on a pcb.

I am compelled to point out that for lines operated at frequencies below the onset of the LC region, the input impedance $Z_{\text{C}}(\omega)$ is not at all constant but a strongly frequency-varying quantity with a phase angle of 45 degrees. Accurately matching the impedance Z_{C} in this region is not a trivial exercise. Fortunately, lines short enough to qualify for lumped-element analysis hardly require termination.

POINTS TO REMEMBER

➢ The input impedance of a short, unloaded transmission line looks entirely capacitive.
➢ The input impedance of a short, grounded pcb-trace looks entirely inductive.

3.4.5 Transfer Function (Lumped-Element Region)

Substitute expressions [3.31] and [3.32] into [3.17] to compute the overall circuit gain G.

$$G = \left[\left(1 + \frac{Z_S}{Z_L}\right) + l\gamma\left(\frac{Z_S}{Z_C} + \frac{Z_C}{Z_L}\right) + \frac{(l\gamma)^2}{2}\left(1 + \frac{Z_S}{Z_L}\right) + \frac{(l\gamma)^3}{6}\left(\frac{Z_S}{Z_C} + \frac{Z_C}{Z_L}\right)\right]^{-1} \quad [3.38]$$

From [3.38] you can (given sufficient mental effort) deduce the conditions necessary to achieve gain flatness. To begin your investigation of this equation, observe that as the term $l\gamma$ approaches zero it renders negligible all the terms associated with various powers of $l\gamma$. The propagation function G in that case asymptotically approaches $Z_L/(Z_S + Z_L)$, precisely the transfer gain you would expect if the source and load were directly connected. This simple mathematical deduction indicates that as the length l of the line is foreshortened, eventually you come to a length so short that the transmission line exerts no observable influence over the outcome—the response at that point depends only on the configuration of source and load. In other words,

> *Any transmission line can be shortened to the point where it acts as a perfect connection.*

Another conclusion you might draw from the analysis so far is that the gain $Z_L/(Z_S + Z_L)$ is in some sense the *best you can do* given the source and load. No directly attached configuration of transmission lines will be able to improve on this response. A corollary to this principle is simply stated:

> *If the source can't drive the load in the first place, then hooking the source and load together with a transmission line isn't likely to make things better.*

Within the lumped element region you may assume the extent x of the line is set to a value such that the magnitude of the coefficient $l\gamma$ remains less than $\Delta = 1/4$ at all frequencies of interest. Under such an assumption you might reasonably expect to ignore the second- and third-order terms in [3.38], but you could hardly ignore the first-order term, particularly if either of the ratios Z_S/Z_C or Z_C/Z_L exceeds unity. This observation may be cast into a statement of the conditions under which the wiring exerts negligible influence over the connection of source and load.

$$\left| l\gamma \frac{Z_S}{Z_C} \right| \ll 1 \quad [3.39]$$

$$\left| l\gamma \frac{Z_C}{Z_L} \right| \ll 1 \quad [3.40]$$

$$|l\gamma| < 0.25 \quad [3.41]$$

Conditions [3.39] through [3.41] may be simplified and perhaps made more recognizable by inserting the definitions of γ [3.13] and Z_C [3.15].

$$|Z_S| \ll \left|\frac{1}{l \cdot j\omega C}\right| \quad . \quad [3.42]$$

$$|l \cdot (j\omega L + R)| \ll |Z_L| \quad [3.43]$$

Formulas [3.42] and [3.43] stipulate the two conditions that are required above and beyond [3.41] such that the line not exert any deleterious influence over signal quality, namely,

1. The source impedance of the driver must be far less than the impedance represented by the total shunt capacitance $1/(l \cdot j\omega C)$ of the line [3.42], and

2. The total series impedance of the line $l \cdot (j\omega L + R)$ must remain far less than the impedance of the load [3.43].

If either condition is not satisfied, then the presence of the transmission line may substantially affect the signal amplitude and/or quality.

In the event the line is driven by an ideal voltage source $(Z_S = 0)$ and coupled to an ideal load at the endpoint $(Z_L = \infty)$, equation [3.38] reduces to this form:

$$G_{Z_S=0, Z_L=\infty} = \left[1 + \frac{(l\gamma)^2}{2}\right]^{-1} \quad [3.44]$$

Equation [3.44] provides a direct relationship between the definition of the lumped-element boundary $|l\gamma| < \Delta$ and the circuit performance expected under ideal conditions. Given $\Delta = 0.25$, the unloaded gain of the transmission line at the lumped-element boundary can differ from unity by no more than $\Delta^2/2 = .032$. Beware that the appearance of good performance with ideal loading does *not* imply that the transmission line would achieve satisfactory performance under other conditions of loading.

It is usually sufficient to check the conditions [3.39] through [3.41] at the highest anticipated frequency of operation. For digital circuits, that frequency corresponds to ω_{knee}. If the circuit performs satisfactorily at ω_{knee}, then conditions will generally improve at lower frequencies, provided that the following conditions are met:

1. The propagation coefficient γ increases monotonically with frequency [3.41].

2. Z_S may act in a capacitive manner (increasing at lower frequencies) as long as it does so no faster than $1/\omega$. This condition implies that Z_S contain no parallel resonance at frequencies less than ω_{knee}. In ordinary digital circuits Z_S is inductive and tending rapidly towards zero at DC, so that if [3.42] is met at ω_{knee}, then the inequality improves quadratically as ω tends to zero.

3. Z_L may act in an inductive manner (decreasing at lower frequencies) as long as it does so no faster than $j\omega L + R$. This condition places restrictions on the type and configuration of inductive loads that may be satisfactorily driven by a lumped-element transmission line. In ordinary digital circuits Z_L is capacitive and tending to infinity at DC, so that if [3.43] is met at any particular frequency ω_1, the inequality improves as $1/\omega^2$ when ω is reduced from ω_1 to ω_{LC} and then inversely with ω for ω below ω_{LC}.

Taken together, the preceding three conditions suggest that any ordinary transmission line that is satisfactorily terminated for operation at frequencies at the lumped-element boundary will also work satisfactorily at frequencies within the lumped-element region. This principle should relieve you of certain worries about the peculiar way in which the transmission-line impedance spikes to infinity near DC. Even when Z_C goes to infinity, as long as conditions [3.19] through [3.41] are met, the circuit will still function.

POINTS TO REMEMBER

➤ Any transmission line can be shortened to the point where it acts as a perfect connection.

➤ If the source can't drive the load in the first place, then hooking the source and load together with a transmission line isn't likely to make things better.

➤ Conditions necessary such that a short, lumped-element transmission line not affect signal quality are given by [3.39] through [3.41].

3.4.6 Step Response (Lumped-Element Region)

The best way to illustrate the general effect of a lumped-element transmission line is by example (Figure 3.6).

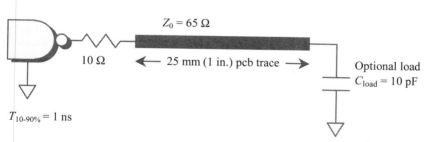

Figure 3.6—This short trace meets the conditions for a lumped-element circuit, yet when loaded with 10 pF it rings horribly.

Driver

➤ Risetime: 1 ns

➤ Output impedance: 10 ohms (shown explicitly as a 10-Ω resistor)

> Operating frequency (corresponds to center of spectral lobe associated with each rising or falling edge): $\omega_{edge} = 2\pi(.35)/t_r = 2.2 \cdot 10^9$ rad/s

Transmission line

- Characteristic impedance: $Z_0 = 65$ ohm (microstrip)
- Length: 25 mm (1 in.)
- Effective dielectric constant: 3.8
- High-frequency propagation velocity: $v_0 = c/\sqrt{3.8} = 1.54 \cdot 10^8$ m/s
- DC resistance: 3 Ω/m

Load

- When present, a 10-pF capacitive load

From the high-frequency values of Z_0 and v_0 you can determine the transmission line parameters L and C:

$$L = Z_0/v_0 = 422 \text{ nH/m}$$
$$C = 1/(Z_0 v_0) = 100 \text{ pF/m}$$

[3.45]

Presuming that the inductive effect of the line far outweighs the resistance (in this case it does), you may approximate the magnitude of the propagation coefficient using only the inductive term.

$$|l\gamma| = l\omega\sqrt{LC}$$
$$= .025 \cdot 2.2 \cdot 10^9 \cdot \sqrt{(422 \cdot 10^{-9})(100 \cdot 10^{-12})}$$
$$= .357$$

[3.46]

This value of the propagation coefficient puts you just outside the official boundary of the lumped-element region at a point where you should expect the pi-model approximations (and indeed equation [3.38]) to be accurate only to about 3%.

Next check conditions [3.42] and [3.43] to see whether the line significantly influences signal quality. Condition [3.42] appears to be satisfied.

$$\left|\frac{1}{l \cdot j\omega C}\right| = \frac{1}{.025(2.2 \cdot 10^9)(100 \cdot 10^{-12})} = 182 \ \Omega$$
$$|Z_S| = 10 \ \Omega$$

[3.47]

3.4.6 • Step Response

Checking [3.43] for the case of no load capacitance (infinite Z_L), you can see that it is clearly true as well—leading to the important conclusion that for rise and fall times no faster than 1 ns, this microstrip in the *absence of a load* induces practically no distortion in the transmitted waveform.

Re-checking [3.43] for the case of a 10-pF load capacitance leads to an opposite conclusion. In this case $|Z_L|$ exceeds the series impedance of the line by but a small amount (a factor of only 2:1) implying that the transmission line will have a noticeable effect on the shape of the received signal.

$$|Z_L| = \frac{1}{(2.2 \cdot 10^9)(10 \cdot 10^{-12})} = 45.5 \, \Omega$$

$$|l(j\omega L)| = .025(2.2 \cdot 10^9)(422 \cdot 10^{-9}) = 23.2 \, \Omega \quad [3.48]$$

Indeed, when the received waveform is plotted under the two conditions, showing the outcome with no load and with a 10-pF load, the unloaded condition shows an undistorted outcome, while the 10-pF load displays prodigious amounts of ringing (Figure 3.7).

Those readers steeped in the art of analog design may find some satisfaction in comparing this result with the simple pi-model approximation (Figure 3.4). The values for inductance and capacitance in that figure are

$$l \cdot L = .025 \cdot 422 \, \text{nH/m} = 10.6 \, \text{nH} \quad [3.49]$$

$$\frac{1}{2} l \cdot C = .025 \cdot 100 \, \text{pF/m} = 2.5 \, \text{pF} \quad [3.50]$$

When driven by a low-impedance source, the capacitor on the left in Figure 3.4 plays only a small role in determining the result. The main effect is an R-L-C series-resonant circuit formed by the output resistance of the driver, the 10.6 nH series inductance, and the 2.5 pF capacitor on the right, which connects *in parallel with the load*.

In an unloaded state the resonant frequency ω_{res} associated with the inductive and capacitive components lies far above the bandwidth of the driver.

$$\omega_{res} = \frac{1}{\sqrt{(l \cdot L)\left(\frac{1}{2} l \cdot C\right)}} = \frac{1}{\sqrt{10.6 \cdot 10^{-9} \cdot 2.5 \cdot 10^{-12}}} = 6.1 \cdot 10^9 \, \text{rad/s} \quad [3.51]$$

An R-L-C resonance certainly exists at 6.1 Grad/s, but since this is well above the spectral center of gravity of the rising and falling edges coming out of the driver (2.2 Grad/s), you never see it. Condition [3.41] ensures that under lightly loaded conditions the line is always sufficiently short to guarantee the resonance falls above the frequency of operation.

With a 10-pF load applied, the situation changes. The new load capacitance swells the total amount of capacitance on the right side of the pi model, reducing the resonant

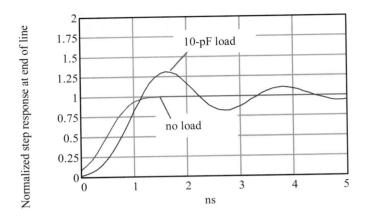

$T_{10\text{-}90\%} = 1$ ns, $R_S = 10$ ohms, $Z_0 = 65$ ohm, Line length = 25 mm (1 in.)

Figure 3.7—Even a short transmission line displays prodigious amounts of ringing when combined with a low-impedance driver and a heavy capacitive load.

frequency to within a stone's throw of the bandwidth of the driver. In this example the new resonant frequency is a function of the series inductance of the line (10.6 nH) and the parallel combination of 2.5 pF and 10 pF.

$$\omega_{res} = \frac{1}{\sqrt{(l \cdot L)\left(\frac{1}{2}l \cdot C + C_{load}\right)}} = \frac{1}{\sqrt{10.6 \cdot 10^9 \cdot 12.5 \cdot 10^{-12}}} = 2.7 \cdot 10^9 \text{ rad/s} \quad [3.52]$$

Such a resonance should have a period of $2\pi/(2.7 \cdot 10^9) = 2.3$ ns, corresponding nicely to the ringing period displayed in Figure 3.7.

POINT TO REMEMBER

> Even a short transmission line may resonate horribly if used to interconnect a ferociously reactive combination of source and load.

3.5 RC REGION

Near DC the magnitude of the inductive reactance of a transmission line ωL dwindles to insignificance in comparison to its DC resistance. All that matters below this point is the relation between the DC resistance of the line and its capacitance (Section 3.2).

In more technical terms, the RC, or *dispersive,* region includes all combinations of ω and l for which the line behaves in a distributed manner but for which the frequency remains sufficiently low that the magnitude of the per-unit-length series inductance (ωL) fails to rise to any significance compared to the per-unit-length DC resistance (R_{DC}). In this lowest-frequency distributed region both attenuation and phase shift depend solely on the transmission line's DC resistance and capacitance.

A transmission line operated in this region is termed an *RC transmission line*, or sometimes a *dispersive transmission line*. The partial differential equations describing the behavior of such a line are called *diffusion equations*, and the line is sometimes called a *diffusion line*. The term *diffusion line* is popular with physicists and mechanical engineers.

POINT TO REMEMBER

> The terms RC transmission line, dispersive transmission line, and diffusion line all mean the same thing.

3.5.1 Boundary of RC Region

As illustrated in Figure 3.2, the RC region extends from DC up to freqeuncy ω_{LC}. At this frequency the reactive component of the propagation coefficient (ωL) equals in magnitude the resistive component (R_{DC}).

$$\omega_{LC} = \frac{R_{DC}}{L} \qquad [3.53]$$

where ω_{LC} is the upper limit of the RC region, rad/s,

R_{DC} is the series DC resistance of the transmission line (signal and return resistances added together), Ω/m,

L is the series inductance, H/m.

NOTE: For differential configurations define R_{DC} as the sum of the resistances of the outbound and returning conductors, and L as the inductance of the differential transmission line thus formed, $L = Z_{\text{DIFFERENTIAL}}/v_{\text{DIFFERENTIAL}}$.

If at the maximum length and maximum anticipated operating frequency your transmission line satisfies conditions [3.24] and [3.25], then you may bypass any consideration of the peculiar properties of the RC region and use instead a lumped-element approximation (i.e., the pi model) to determine how your system will respond.

Let's look for a moment at the confluence of [3.24] and [3.53] as illustrated in Figure 3.2. The exact length at the intersection of these two constraints is found by substituting the LC mode cutoff frequency [3.53] into constraint [3.24]. This calculation reveals a critical length l_{RC} below which you need *never* concern yourself with the distributed RC mode of operation.

$$l_{\text{RC}} = \frac{\Delta}{\sqrt{\omega_{\text{LC}} R_{\text{DC}} C}} = \frac{\Delta}{\sqrt{\frac{R_{\text{DC}}}{L} R_{\text{DC}} C}} = \frac{\Delta}{R_{\text{DC}}} \sqrt{\frac{L}{C}} \qquad [3.54]$$

where l_{RC} represents the critical length below which distributed RC behavior never appears, m,

Δ is an arbitrary constant customarily set to 1/4,

R_{DC} is the series DC resistance of the transmission line (signal and return resistances added together), Ω/m,

L is the series inductance, H/m, and

C is the shunt capacitance, F/m.

Equation [3.54] tells you that the distributed RC mode of operation matters only at lengths sufficiently great that the total DC resistance of the line lR_{DC} becomes noticeable compared to the high-frequency value of impedance $\sqrt{L/C}$. For example, in Figure 3.2 a horizontal line drawn at the 1-m level passes from the lumped-element mode directly into the LC mode, skipping the distributed-RC mode altogether. This happens because a 1/2-oz, 50-Ω pcb trace with a width of 150 microns (6 mils) and a length of 1 m has a DC resistance of only 6.3 Ω, much less than the line impedance of 50 Ω. Such a line will never exhibit any observable RC-mode phenomenon. Designers of pcbs therefore needn't concern themselves with the RC mode of behavior.

Conversely, a much longer pcb trace of the same type, perhaps 100 m in length, would exhibit the signs of distributed RC operation over the band from a few kilohertz to several megahertz.

The critical length defined in [3.54] varies dramatically with the size of the conductors. If a given cross section is scaled by a factor of k in all dimensions, the L and C parameters remain unchanged, but the resistance changes inversely with the square of k. This quadratic change in resistance precipitates a change in l_{RC} by a factor of k^2.

Larger conductors, such as 24-gauge telephone wire, will not exhibit observable RC behavior until you reach lengths on the order of 100 meters. Smaller conductors, such as found within integrated circuits, may exhibit observable RC behavior at lengths much shorter than the dimensions of the chip package. The RC effect within integrated circuits is exacerbated by the low conductivity of some materials (like polysilicon) used to construct the transmission structures.

POINTS TO REMEMBER

> ➤ The RC region extends in frequency from DC up to that point ω_{LC} where the magnitude of the line inductance ($\omega_{\text{LC}} L$) equals the DC resistance (R_{DC}).
> ➤ For any transmission line there exists a critical length below which you need *never* concern yourself with the distributed RC mode of operation.

3.5.2 Input Impedance (RC Region)

In general the input impedance [3.16] of a transmission structure varies strongly with the length of the line and the type of load connected. This can at times be a distinct disadvantage, particularly when you wish to equalize the propagation function of a transmission line by using a reactive source or load impedance.

For example, if the input impedance of a transmission structure Z_{IN} were known, you could design a suitable reactive source-impedance network Z_S such that the propagation function $Z_{IN}/(Z_{IN} + Z_S)$ achieved some arbitrary equalization goal. Unfortunately, variations in Z_{IN} with line length and loading complicate the design of a good equalizer (see [3.16]).

There are, however, some special cases under which the problem can be simplified. These cases share a common principle of controlling (or ignoring) reflections.

The first case has already been mentioned in Section 2.2.1, "Definition of Characteristic Impedance Z_C." It happens during that brief interval of time after your driver has generated its first rising edge but before the first reflection arrives from the far end. During this brief interval the input impedance of the line equals precisely Z_C independent of the line length. A source-end equalizer designed to work into an impedance of Z_C would for a brief time operate properly until the moment the first reflection arrived, after which the performance of the equalizer would change.

The second case happens when you eliminate reflections by providing an end-termination load Z_L. The end-termination impedance must be adjusted such that $Z_L = Z_C$. Once the termination is in place, the input impedance of the structure equals precisely Z_C independent of line length. You should verify this fact using [3.16]. A source-equalizing network Z_S may then be designed such that the propagation function $Z_C/(Z_C + Z_S)$ achieves your equalization goal.

In the third case, a source-termination network with an impedance Z_S equal to Z_C may be used. This network provides a driving-point impedance at the load that does not vary with line length. In this case the end-termination impedance Z_L must be calculated so that $Z_L/(Z_L + Z_C)$ achieves your equalization goal.

The fourth method assumes a very long transmission line, sufficiently long that the transfer gain H falls to significantly less than one. Under this assumption, the inverse-gain H^{-1} vastly exceeds H, so much so that the term H may be neglected in equation [3.16]. Suitable rearrangement of the terms in [3.16] shows that under this assumption Z_{IN} equals Z_C.

POINTS TO REMEMBER

> The input impedance of a line *without reflections* is predictable and independent of line length.

> An RC transmission line may be equalized using a suitable reactive source or load impedance network.

3.5.3 Characteristic Impedance (RC Region)

The full expression for characteristic impedance (neglecting the conductance G) includes contributions from both the inductance L and resistance R of the line.

$$Z_C = \sqrt{\frac{j\omega L + R}{j\omega C}} \qquad [3.55]$$

Taking advantage of the defining assumption of the RC mode, namely that the term ωL remains small compared to R, the characteristic impedance in the heart of the RC territory may be approximated.

(RC assumption) $\qquad Z_C \approx \sqrt{\dfrac{R}{j\omega C}} \qquad [3.56]$

Expression [3.56] is a complex function of frequency with a phase angle of –45° and a magnitude slope of -10 dB per decade.

Depending on the frequency regime in which you choose to operate a line, it behaves either as an RC transmission line (at low frequencies) or a low-loss transmission line (at high frequencies). These cases are distinguished by the relative magnitudes of the inductive and resistive impedances in [3.55]: At frequencies below ω_{LC} the characteristic impedance varies as given by [3.56]. Above ω_{LC} the characteristic impedance may be derived from [3.55] under the assumption that ωL vastly exceeds R, leading to the conclusion that at high frequencies the characteristic impedance asymptotically approaches a constant value of $Z_0 = \sqrt{L/C}$.

Example of an RC Transmission Line

You probably have *RC* transmission lines in your house. The two wires running from the nearest central telephone switching office to your phone are usually AWG-24 wire. These wires are twisted in a configuration yielding the following line parameters:

$$\begin{aligned} R &= 0.165 \quad \text{ohm/in} \\ L &= 400 \quad \text{nH/in} \\ C &= 40 \quad \text{pF/in} \end{aligned} \qquad [3.57]$$

At 1600 Hz (the center of the voice band), the characteristic impedance of this configuration is

$$\begin{aligned} Z_C &= \sqrt{(j\omega L + R)/j\omega C} \\ &= \sqrt{(j2\pi \cdot 1600 \cdot 400 \cdot 10^{-9} + .165)/j2\pi \cdot 1600 \cdot 40 \cdot 10^{-12}} \\ &= 640.658 \angle -44.3° \end{aligned} \qquad [3.58]$$

This calculation explains the myth that telephone wires have a characteristic impedance of roughly 600 ohms. That's the correct impedance in the voice band, but at high frequencies the characteristic impedance Z_0 is 100 ohms.

Note that the characteristic impedance in the RC region varies markedly with frequency. Designing a good-quality termination for operation in the RC region is not a trivial project. For best operation, the termination must match Z_C over the entire frequency range spanned from ω_{LE} and ω_{LC}. The difficulty of designing a network to match Z_C over this range (i.e., the number of poles required in the network) is proportional to the number of octaves of frequency spanned.

Fortunately, good terminations for RC structures become necessary only if you are constructing a hybrid circuit for two-way communication (see Section 8.3.3, "UTP: Hybrid Circuits") or an equalizer. For general-purpose one-way communication, an RC structure may be left open-circuited at the far end, relying on the natural damping effect of the series resistance to quell reflections.

POINT TO REMEMBER

> Within the RC region, characteristic impedance is a complex function of frequency with a phase angle of –45° and a magnitude slope of -10 dB per decade.

3.5.4   General Behavior within RC Region

Figure 3.8 illustrates the propagation function of a unit-sized RC transmission line ($R = 1\Omega/\text{m}$, $C = 1\text{F/m}$, and $l = 1\text{m}$). The response is given under three conditions of loading. In all three cases the line is driven with a low-impedance voltage source. The solid curve shows the propagation function with the end open-circuited. This curve displays the least overall loss at high frequencies. It represents the most often-used configuration for general digital connections made on-chip using transmission structures operated in the RC mode.

The dashed curve shows what happens to the transfer gain of your transmission line (dashed line) when you apply a perfectly matched end-termination impedance equal at frequencies to $Z_C(\omega)$. The matched end termination degrades the line response in two ways: It reduces the available signal at the far end of the line, and it introduces a disagreeable tilt to the propagation function. Of the two problems, the tilt in the propagation function is the more serious, because it will introduce significant amounts of intersymbol interference. Sufficiently large intersymbol interference causes bit errors.

Binary signaling tolerates a tilt of no more than about 3 dB (and certainly never more than 6 dB) in the channel attenuation over the band occupied by the coded data. Any tilt greater than that amount must be flattened out with an equalizer (see Section 3.14, "Linear Equalization: Long Backplane Trace Example").

In summary, the only good reasons for using a matched end termination with an RC transmission line are

> As part of a hybrid circuit (see Section 8.3.3, "UTP: Hybrid Circuits").

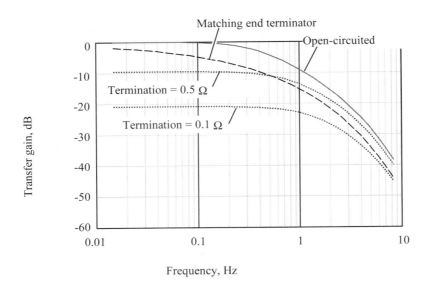

Figure 3.8—A matching end terminator on an RC transmission line produces a sloping transfer function.

> To stabilize the input impedance of the channel as part of an equalization circuit.

The dotted curves in Figure 3.8 correspond to two different values of resistive loads, equal to 1/2 or 1/10 respectively of the aggregate series resistance of the transmission line. In both cases the purely resistive load reduces the level of the output signal, but while *flattening* the overall attenuation curve instead of making it worse, as with the matched termination. The flattened attenuation curve, whose flat zone now reaches to a higher frequency than previously, makes possible simple binary signaling at a higher bandwidth than would have been possible with an open-circuited (or end-terminated) RC transmission line.

The resistive load establishes a classic gain-bandwidth tradeoff whereby you can improve the bandwidth of your channel at the expense of reducing the signal amplitude. The upper limit to which the bandwidth may be pushed is established by the onset of the LC mode of operation, which takes precedence at that point where the inductance of the line becomes significant in comparison to the series resistance. Because the proper value for a resistive end termination in the LC mode equals the high-frequency value of characteristic impedance, $Z_0 = \sqrt{L/C}$, you will see numerous examples of lines designed for operation in the crossover zone between RC and LC modes using a purely resistive end termination equal to Z_0. This value has the property of eliminating reflections at very high frequencies within the LC band while also providing a relatively flat propagation function within the RC band. This is the circuit most often used in digital telephone circuits operating at a few tens or hundreds of kilobits/s over twisted-pair cabling at distances up to approximately 1 km.

Points to Remember

- A perfectly-matched end termination applied to an RC transmission line renders the input impedance of the structure indepedant of line length. This advantage comes at the expense of a terrible degradation of the transfer response.
- A fixed resistance at the end of an RC transmission line flattens the gain curve, providing more usable bandwidth at the expense of a reduction in the received signal amplitude, and a greater variation with line length in the input impedance of the structure.
- A purely resistive end termination equal to $Z_0 = \sqrt{L/C}$ eliminates reflections within the LC band while also providing a relatively flat propagation function within the RC band.

3.5.5 Propagation Coefficient (RC Region)

The transfer equations for the RC mode are derived beginning with [3.13] and [3.14], neglecting the contribution of the line inductance parameter L, and also neglecting variations in R and C that generally occur only at higher frequencies.

RC mode
$$\gamma(\omega) = \sqrt{R(j\omega C)} \quad \text{neper/m} \qquad [3.59]$$

RC mode
$$H(\omega, l) = e^{-l\sqrt{R(j\omega C)}} \quad \text{(dimensionless)} \qquad [3.60]$$

3.5.6 Transfer Function (RC Region)

Substitute expression [3.60] into [3.17] along with appropriate assumptions about the source and load impedances to calculate the overall propagation function.

3.5.6.1 Propagation Function of RC Line with Open-Circuited Load

From [3.17], assuming $Z_S = 0$ and $Z_L = \infty$,

$$G = \left[\left(\frac{H^{-1}+H}{2}\right)\left(1+\frac{0}{\infty}\right)+\left(\frac{H^{-1}-H}{2}\right)\left(\frac{0}{Z_C}+\frac{Z_C}{\infty}\right)\right]^{-1} = \frac{2}{H^{-1}+H} \qquad [3.61]$$

Substituting [3.14] for H, the formula for G may be reduced using a hyperbolic cosine:

$$G(\omega) = \frac{1}{\cosh(l \cdot \gamma(\omega))} \qquad [3.62]$$

where G is the propagation function of the transmission-line circuit composed of a voltage source, RC transmission line, and open-circuited load,

the hyperbolic cosine function $\cosh(x) = (e^x + e^{-x})/2$,

l is the line length, m,

the transmission line propagation coefficient $\gamma(\omega) = \sqrt{(R)(j\omega C)}$, and

R and C are the values of resistance and capacitance, in units of Ω/m and F/m, respectively.

3.5.6.2 Propagation Function of RC Line with Matched End Termination

From [3.17], assuming ZS = 0 and ZL = ZC,

$$G = \left[\left(\frac{H^{-1}+H}{2}\right)\left(1+\frac{0}{Z_C}\right) + \left(\frac{H^{-1}-H}{2}\right)\left(\frac{0}{Z_C}+\frac{Z_C}{Z_C}\right)\right]^{-1} = H \qquad [3.63]$$

Substituting [3.14] for H, the formula for G may be evaluated:

$$G(\omega) = e^{-l\cdot\gamma(\omega)} \qquad [3.64]$$

where G is the propagation function of the transmission-line circuit composed of a voltage source, RC transmission line, and perfectly matched load,

l is the line length, m,

The transmission line propagation coefficient $\gamma(\omega) = \sqrt{(R)(j\omega C)}$ (see [3.59]), and

R and C are the values of resistance and capacitance, in units of Ω/m and F/m, respectively.

3.5.6.3 Propagation Function of RC Line with Matched Source Termination

From [3.17], assuming ZS = ZC and ZL = ∞,

$$G = \left[\left(\frac{H^{-1}+H}{2}\right)\left(1+\frac{Z_C}{\infty}\right) + \left(\frac{H^{-1}-H}{2}\right)\left(\frac{Z_C}{Z_C}+\frac{Z_C}{\infty}\right)\right]^{-1} = H \qquad [3.65]$$

This arrangement gives the same result as [3.63].

3.5.6.4 Propagation Function of RC Line with Resistive End Termination

From [3.17], assuming ZS = 0 and ZL = Z0, the high-frequency value of characteristic impedance.

$$G = \left[\left(\frac{H^{-1}+H}{2}\right)\left(1+\frac{0}{Z_0}\right)+\left(\frac{H^{-1}-H}{2}\right)\left(\frac{0}{Z_C}+\frac{Z_C}{Z_0}\right)\right]^{-1} \quad [3.66]$$

$$G = \left[\left(\frac{H^{-1}+H}{2}\right)+\left(\frac{H^{-1}-H}{2}\right)\left(\frac{Z_C}{Z_0}\right)\right]^{-1} \quad [3.67]$$

Equation [3.67] admits little in the way of further simplification. If you are considering a resistive termination of value Z0 for a line of length l that crosses between the RC and LC modes, you can calculate from DC analysis a value for $G(0) = Z_0/(Z_0 + lR_{DC})$, while the LC mode gain is given approximately by equation [3.95] $G(\omega > \omega_{LC}) \approx e^{-l\frac{R_{DC}}{2Z_0}}$. The difference between these two values defines a minimum perturbation in the propagation function of your circuit between the DC and LC modes.

If the transfer response is not suitably flat using a resistive end termination, try applying resistive terminations to both ends of the line. Although this technique does not produce a perfect response, it cuts by about half the percentage variation in gain between DC and LC modes at the expense of a reduction by by half in the received signal amplitude.

POINT TO REMEMBER

> A resistive termination at both ends of an RC-LC mixed-mode transmission line provides flatter gain than termination at only one end or the other.

3.5.7 Normalized Step Response (RC Region)

Figure 3.9 illustrates the step response of the same unit-sized RC transmission line used in the previous section ($R = 1\Omega/m$, $C = 1F/m$, and $l = 1m$).

The solid curve shows the propagation function with the end open-circuited. It represents the most often-used configuration for general digital connections made on-chip using transmission structures operated in the RC mode.

The dashed curve shows the degraded risetime that occurs when you apply a perfectly matched end-termination impedance equal to $Z_C(\omega)$.

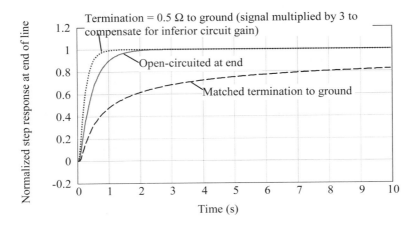

Figure 3.9—On an RC transmission line a resistive termination produces the fastest but smallest step edge.

The dotted curve illustrates the superior risetime achievable with a resistive termination at the cost of a reduced signal amplitude. In this case the DC gain of the circuit is 1/3, so the signal in the graph has been multiplied by three so you can easily compare the 10% to 90% risetime of the result against the other two signals.

Table 3.4 presents the normalized step response of an RC transmission line under selected conditions of loading. You may scale the values in this table to fit your transmission structure by multiplying all the time values listed in the table by the factor RCl^2, where R and C represent the resistance and capacitance per unit length of your transmission structure in units of Ω/m and F/m respectively, and l is its length in meters.

Table 3.4—Normalized Step Response of RC Transmission Line

Percentile (of step height)	End-termination condition			
	Open	Matched	$Z_L = (lR)/2$ (gain = 1/3)	$Z_L = (lR)/10$ (gain = 1/11)
10%	0.105	0.185	0.069	0.053
20%	0.160	0.304	0.101	0.073
50%	0.354	1.099	0.198	0.139
80%	0.726	7.790	0.376	0.252
90%	1.006	31.664	0.507	0.338
NOTE—All times are expressed in seconds.				

POINT TO REMEMBER

➢ A fixed resistance at the end of an RC transmission line improves the settling time at the expense of a reduction in the received signal amplitude.

3.5.8 Tradeoffs Between Distance and Speed (RC Region)

The scaling equation associated with Table 3.4 indicates that the risetime scales with the square of length. Thus, the speed of operation achievable within the RC region scales *inversely* with the square of transmission-line length. For example, given a fixed receiver architecture and assuming the bit error rate (BER) of the receiver is dominated by errors due to intersymbol interference, a 10% increase in length increases the line attenuation by 10%. To regain a normal signal amplitude at the end of a transmission line operated in the RC mode, the 10% increase in attenuation must be offset by a 20% reduction in the system operating speed.

POINT TO REMEMBER

> The speed of operation achievable within the RC region scales *inversely* with the square of transmission-line length.

3.5.9 Closed-Form Solution for Step Response (RC Region)

An analytic solution for the step response of an RC transmission line is known for only one configuration of loading. Unfortunately, this solution corresponds to the least popular configuration: a line with a matched end termination. The step response for other configurations is best derived using the full system-modeling equations given in Chapter 4.

The derivation begins with equation [3.64], which presents the propagation function (impulse response) of an RC line with a perfectly matched end termination. Multiplying that result by the Fourier-transform operator $1/j\omega$ produces the step response G:

$$G(\omega) = \frac{1}{j\omega} e^{-l\sqrt{j\omega RC}} \qquad [3.68]$$

Define a constant $k = l\sqrt{RC}$:

$$G(\omega) = \frac{1}{j\omega} e^{-k\sqrt{j\omega}} \qquad [3.69]$$

The inverse Fourier transform of this entity is known:

$$\mathcal{F}^{-1}\left[\frac{1}{j\omega} e^{-k\sqrt{j\omega}}\right] = \left[u(t)\mathrm{erfc}(\tfrac{1}{2} k t^{-1/2})\right] \qquad [3.70]$$

Reinserting the value of k produces the step response g^* as a function of time.

$$g(t) = \left[u(t) \, erfc\left(\tfrac{1}{2} l \sqrt{\frac{RC}{t}} \right) \right]$$
[3.71]

3.5.10 Elmore Delay Estimation (RC Region)

W. C. Elmore in 1948 published a marvelous way to estimate the delay of RC circuits [30]. Adaptations of his original method are used today as the primary means of validating on-chip timing.

The Elmore delay-estimation method applies to well-damped circuits composed of any number of series resistances and shunt capacitances. *It does not apply* (in its original form) to circuits involving inductance, resonance, overshoot, or any form of poorly damped or nonmonotonic behavior.

The simple one-stage RC low-pass filter in Figure 3.10 illustrates the general principle involved, which is that there exists some effective width τ for the shaded region such that the area A of the shaded region equals the step voltage ΔV multiplied times τ. The variable τ is defined

$$\tau = A/\Delta V$$
[3.72]

Elmore asserts that under a wide range of conditions the 50% input-to-output delay of the circuit approximately equals τ. For example, if the output waveform had a linear slope, the relation $t_{50\%} = \tau$ would hold exactly.

The Elmore delay-estimation procedure first determines the area A, and then computes τ using [3.72]. The determination of A hinges on the observation that whenever node $x(t)$ steps through a voltage range ΔV the total charge on capacitor C changes by an amount ΔQ equal to $C\Delta V$. Furthermore, this exact amount of charge ΔQ *must have passed through resistor R*.

The passage of charge ΔQ through resistor R causes a momentary voltage difference

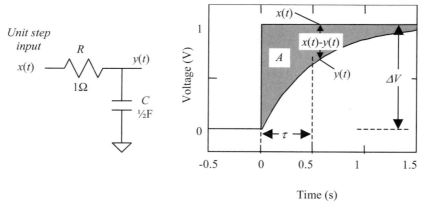

Figure 3.10—The area of the shaded region equals the step height ΔV multiplied by the effective width τ.

3.5.10 • Elmore Delay Estimation

between the ends of the resistor:

$$x(t)-y(t)=i(t)R \qquad [3.73]$$

where $i(t)$ is the current waveform necessary to charge the capacitor.

By definition the total integrated current $i(t)$ equals ΔQ, the step change in charge:

$$\Delta Q = \int_0^\infty i(t)dt \qquad [3.74]$$

Substituting equation [3.73] for $i(t)$ in [3.74],

$$\Delta Q = \int_0^\infty \frac{x(t)-y(t)}{R}dt \qquad [3.75]$$

The integral term equals the area A of the shaded region divided by R:

$$\Delta Q = \frac{A}{R} \qquad [3.76]$$

Rearranging the terms and recognizing $\Delta Q = C\Delta V$,

$$A = RC\Delta V \qquad [3.77]$$

Equation [3.77] suggests that the total integrated area A between curves $x(t)$ and $y(t)$ is a function only of the step change in voltage ΔV, the resistance R, and the amount of capacitance C downstream of R. It is not a function of the exact shape of the rising edge, the speed of the driver, or any other factors.

Substituting $A = \tau \Delta V$ from [3.72] into the left side of [3.77] produces the Elmore delay equation, which relates the effective width of the shaded region τ to both R and C.

$$\tau \Delta V = RC \Delta V \qquad [3.78]$$

$$\tau = RC \qquad [3.79]$$

The final step in the process assumes the 50% input-to-output delay of the circuit approximately equals the effective width of the shaded region, τ.

The Elmore estimate, crude as it may be, is surprisingly effective. It is exact in cases where the output waveform is monotonic and a strictly delayed version of the input waveform. To prove this exactness I must prove that for $y(t) = x(t - \tau)$ the area A between x and y equals $\tau \Delta V$. Such a case is illustrated in Figure 3.11, where the area A is broken into a set of rectangles of equal height and width whose total area clearly equals $\tau \Delta V$.

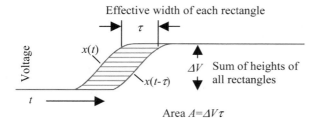

Figure 3.11—The total area contained between waveforms $x(t)$ and $x(t - \tau)$ scales linearly with τ.

When the input risetime significantly exceeds the response time of the circuit (that is, τ is much less than $t_{10-90\%}$), the output waveform tends to look like a strict delay of the input waveform and the Elmore delay becomes exact.

When the input risetime equals or falls short of the response time of the circuit, the output waveform becomes distorted and the Elmore delay serves only as an estimate of delay. For well-damped RC circuits with monotonic inputs, the Elmore delay τ always *over*estimates the actual circuit delay.

Figure 3.12 illustrates the step response of an RC low-pass filter showing the Elmore delay estimate and the actual 50% delay. The Elmore estimate of delay is $\tau = RC = 1$. At that point in time the actual step response has ascended to 63.2% of its final value. The 50%-point delay for the step response of a one-stage RC low-pass filter occurs at 0.69τ.

The Elmore delay τ by definition exists at that point in time that equalizes the areas A_1 and A_2 marked on Figure 3.12. Step-response waveforms having symmetrical precursor and postcursor waveforms tend to attain 50% amplitude precisely as indicated by the Elmore delay estimate. Step-response waveforms that have a mostly concave-down shape tend to attain 50% amplitude sooner than indicated by the Elmore delay estimate.

When composing networks of multiple RC sections, the Elmore principle still applies, but its application becomes slightly more complicated. Figure 3.13 depicts a multi-element configuration. The resistor R_1 in this figure charges all N capacitors downstream of its own

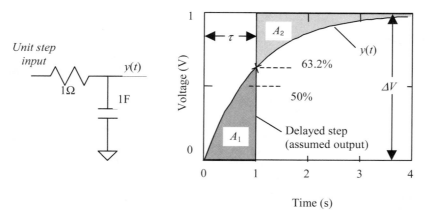

Figure 3.12—The effective delay of an RC circuit roughly equals the product RC.

3.5.10 • Elmore Delay Estimation

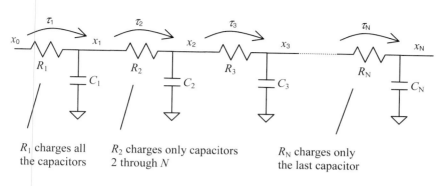

Figure 3.13—Each resistor charges successively less capacitance.

position. The Elmore estimated delay τ_1 from point x_0 to x_1 is therefore $\tau_1 = R_1 \sum_{m=1}^{N} C_m$. Resistor R_2 charges only capacitors numbered 2 through N, so the estimated delay from point x_1 to x_2 is $\tau_2 = R_2 \sum_{m=2}^{N} C_m$. Working down row, the total delay for the whole circuit is estimated:

$$\tau = \sum_{n=1}^{N} R_n \sum_{m=n}^{N} C_m \qquad [3.80]$$

The special case of a very large number of stages representing a continuous transmission line operated in the RC dispersion region lends itself to some simplifications. In this case we will assume the line has a total series resistance of R and a total parasitic capacitance to the reference plane of C. The line is driven by a low-impedance source on the left and unterminated at the far end. An N-element circuit model for this configuration would look like Figure 3.13 with all the resistances set to R/N and all the capacitances set to C/N.

Observe that in the N-element model the first resistor charges the full line capacitance C, while the last resistor in the chain charges almost nothing. In general the *average* capacitance charged by any resistor along the structure equals $C/2$. The *average* estimated delay at each stage is therefore $(R/N)(C/2)$, and the total delay for N such stages must equal $\frac{1}{2}RC$. This is the Elmore delay estimate for an unterminated transmission line operated in the RC dispersion region when driven by a low-impedance source.

Figure 3.14 illustrates the step response of a continuous dispersive RC transmission line, computed using the FFT method (see Chapter 4). The symbol for a dispersive transmission line is a resistor with a line under it representing the distributed capacitance to the nearest reference plane. The transmission line in Figure 3.14 has a total series resistance of 1Ω and a total shunt capacitance of 1F. It is driven by a low impedance at one end and is open-circuited at the other. The Elmore estimate of delay is $\tau = \frac{1}{2}RC = \frac{1}{2}$ sec. At that point in time the actual step response has ascended to 65.2% of its final value. The 50%-point delay for the step response of an unloaded RC transmission line occurs at 0.71τ. Chip

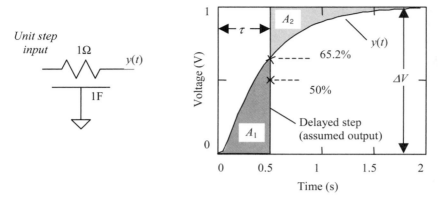

Figure 3.14—The Elmore delay estimate for a distributed RC transmission line equals ½RC.

designers often assume the delay of an RC transmission line is $0.35RC$, which comes from the formula $0.71(1/2)RC$.

Elmore delay calculations for a continuous RC dispersion-limited transmission line coincide with delay calculations for the C-R-C "pi" network indicated in Figure 3.14. The pi network is a very reasonable lumped-element approximation for a short RC dispersion-limited transmission line, where short means the Elmore delay is a small fraction of the signal risetime.

To apply the Elmore delay concept to a tree-structured circuit, you must identify all the resistors R_i that lie along a path from the input to the output under analysis. Then, for each series resistor R_i you must identify the total downstream capacitance C_i charged by that resistance and sum all the R_iC_i time constants. For the schematic in Figure 3.15 the path from input to x_4 traverses resistors R_1, R_2, R_3, and R_4. The Elmore delay estimate for this path is

$$\tau_{0-4} = R_1\left(C_1 + C_2 + C_3 + C_4 + C_5 + C_6\right)$$
$$+ R_2\left(C_2 + C_3 + C_4\right)$$
$$+ R_3\left(\frac{1}{2}C_3 + C_4\right)$$
$$+ R_4\left(C_4\right)$$

[3.81]

Resistors R_5 and R_6 are not involved in this calculation, but capacitors C_5 and C_6 are involved.

In the calculation of τ_{0-4} note that resistor R_3 effectively feeds only half the distributed capacitance of the RC transmission line, whereas resistors R_2 and R_1, being upstream of the distributed element must feed the entire distributed capacitance C_3.

The Elmore estimate from input to x_6 is

3.5.10 • Elmore Delay Estimation

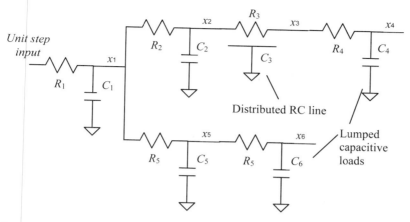

Figure 3.15—The Elmore algorithm quickly computes a reasonable upper bound for the delay of complicated tree and bus structures.

$$\tau_{0-6} = R_1\left(C_1 + C_2 + C_3 + C_4 + C_5 + C_6\right)$$
$$+ R_5\left(C_5 + C_6\right) \qquad [3.82]$$
$$+ R_6\left(C_6\right)$$

The appeal of the Elmore delay-estimating algorithm is its ability to quickly compute a reasonable upper bound on the delay of complicated tree and bus structures used on-chip. SPICE could of course be used to find a more accurate estimate, but the computational complexity of SPICE precludes its use for day-to-day on-chip floor-planning purposes.

In practice, the Elmore delay estimate may be successfully applied to well-damped RC networks, but not for transmission lines having appreciable amounts of inductance or for nonmonotonic inputs. Ismail [31] has worked out a promising approach that extends the Elmore delay-estimation algorithm to account for transmission line inductance.

POINTS TO REMEMBER

> ➤ The Elmore delay approximation takes into account only the resistance and capacitance of a transmission configuration. It applies to well-damped circuits composed of any number of series resistances, shunt capacitances, and distributed RC transmission lines.

> ➤ The Elmore delay approximation does not apply circuits involving inductance, resonance, overshoot, or any form of poorly damped or nonmonotonic behavior.

> ➤ The Elmore delay for a lumped resistance R feeding a total downstream capacitance of C is RC.

➤ The Elmore delay for a distributed RC transmission line having total resistance R and distributed capacitance C is (1/2)RC.

3.6 LC REGION (CONSTANT-LOSS REGION)

At higher frequencies the inductive reactance grows, eventually exceeding the magnitude of the DC resistance, forcing the line into the LC region (Section 3.2).

3.6.1 Boundary of LC Region

As illustrated in Figure 3.2, the LC region extends to all frequencies above frequency ω_{LC}. At that frequency the reactive component of the propagation coefficient (ωL) equals in magnitude the resistive component (R).

$$\omega_{LC} = \frac{R_{DC}}{L} \qquad [3.83]$$

where ω_{LC} is the lower limit of the LC region, rad/s,

R_{DC} is the series DC resistance of the transmission line (signal and return resistances added together), Ω/m, and

L is the series inductance, H/m

NOTE: For differential configurations, define R_{DC} as the sum of the resistances of the outbound and returning conductors, and L as the inductance of the differential transmission line thus formed, $L = Z_{DIFFERENTIAL}/v_{DIFFERENTIAL}$.

At all frequencies above ω_{LC} (including the skin effect, dielectric, and waveguide regions) the line inductance plays a significant role in its operation and must be taken into account in system modeling.

If your maximum operating frequency exceeds ω_{LC}, but if at that frequency your transmission line length falls short of the boundary described by [3.25], then you may bypass any consideration of the peculiar properties of the LC region and use instead a lumped-element approximation (i.e., the pi model) to determine how your system will respond.

The distinguishing feature of the LC mode is that within this region the line attenuation does not much vary with frequency. Some transmission lines possess a broad, well-defined LC region in which the attenuation does not change over a wide band. Unfortunately, in the practical transmission lines used in most digital designs the LC region is relatively narrow (or sometimes nonexistent) so that the LC region shows up as nothing more than a subtle "flat spot" in the curve of attenuation vs. frequency, pinched between the RC region, where attenuation again grows with the square root of frequency, and the skin-effect region, where the attenuation grows with the square root of frequency (see Figure 3.1).

The narrowness of the LC region comes about because of the relationship between the external inductance and the internal inductance of a typical transmission-line configuration (see Section 2.8, "Modeling Internal Impedance"). The onset of LC mode happens at a frequency where the transmission line's total inductance (internal plus external) becomes significant compared to its resistance. The onset of the skin-effect mode, on the other hand, depends on the relation of only the internal inductance to the resistance. Because in most practical transmission lines used for digital applications the total cross-sectional area of the main TEM wave-carrying region exceeds the cross-sectional area of the signal conductor(s) by only a modest factor (on the order of 10 or less), the external inductance exceeds the internal inductance by a similar modest factor. The onset frequency of the skin-effect region therefore exceeds the onset of the LC mode by the same modest factor. Nevertheless, the study of pure LC-mode propagation serves as a useful gateway to the understanding of other regions.

To achieve a different result, you would have to use very thin conductors. For example, some twisted-pair and coaxial conductors are constructed from a high-permeability steel core surrounded by a thin layer of copper plating. The high permeability (and thus very shallow skin depth) of the steel core rejects all currents at even moderate frequencies, forcing all the current to flow in the thin copper-plated region. As long as skin depth of the copper region remains larger than the plating thickness, the line enjoys the benefits of a constant resistance over a wide band of frequencies. Such a transmission line has a generally increased, but very flat, loss in the LC region. This trick is used in some 1000BASE-CX Gigabit Ethernet shielded twisted-pair cables to flatten the loss characteristics of the cable, simplifying the equalization circuitry at each end.

POINTS TO REMEMBER

> ➤ The LC region begins where the magnitude of the line inductance (ωL) exceeds the DC resistance (R_{DC}).
> ➤ Within the LC region the line attenuation does not much vary with frequency.
> ➤ Unfortunately, in the practical transmission lines used in most digital designs the LC region is relatively narrow (or sometimes nonexistent).

3.6.2 Characteristic Impedance (LC Region)

The full expression for characteristic impedance (neglecting the conductance G) includes contributions from both the inductance L and resistance R_{DC} of the line.

$$Z_C = \sqrt{\frac{j\omega L + R_{DC}}{j\omega C}} \qquad [3.84]$$

In the limit as you proceed to frequencies far above ω_{LC}, the contribution of the term R becomes negligible, leading to this approximation for characteristic impedance in the heart of the LC territory.

(LC assumption) $Z_0 \triangleq \mathrm{Lim}_{\omega \to \infty} \sqrt{\dfrac{j\omega L + R_{DC}}{j\omega C}} = \sqrt{\dfrac{L}{C}}$ [3.85]

I name this parameter Z_0 to distinguish it from the actual value of characteristic impedance Z_C, which of course varies with frequency. At a frequency three times greater than ω_{LC} the difference in magnitude between Z_0 and Z_C is on the order of 5%. At a frequency 10 times higher than ω_{LC} the difference is on the order of 0.5%. Near ω_{LC} the value of [3.84] takes on a substantial phase angle and varies noticeably.

The general relationship between the characteristic impedance and input impedance of a transmission structure is given by [3.33] and discussed in Section 3.5.2, "Input Impedance (RC Region)," and in Section 2.1 "A Transmission Line Is Always a Transmission Line."

Figure 3.16 illustrates the transition in impedance between RC and LC modes for a hypothetical pcb trace. This figure takes into account only the DC resistance, capacitance, and inductance of the trace (i.e., it ignores both skin effect and dielectric losses). The trace is a 50-ohm stripline constructed of 1/2-oz copper (17.4 micron thickness) on FR-4 with a width of 150 microns (6 mils, $v_0 = 1.446 \cdot 10^8$ m/s, $L_0 = 346$ nH/m, $C_0 = 138$ pF/m).

At frequencies below ω_{LC} (3 MHz in this example) both real and imaginary parts of the characteristic impedance vary with the inverse square root of frequency. Above ω_{LC} the imaginary part goes to zero and the overall impedance flattens out to approach a high-frequency asymptote of $Z_0 \triangleq \sqrt{L/C}$, which in this case equals 50 Ω.

Figure 3.17 magnifies the region near the LC-mode transition, showing on a linear scale both real and imaginary parts of the characteristic impedance. A decade above ω_{LC} the impedance magnitude has moved to within a tiny fraction of 50 ohms; however, the characteristic impedance at this point (30 MHz) still carries an imaginary quantity of about -2 ohms. In absolute terms this imperfection represents a 4% variation from 50 ohms, which will lead to a 2% reflection coefficient if the line is sourced (or terminated) with a pure 50-ohm resistance.

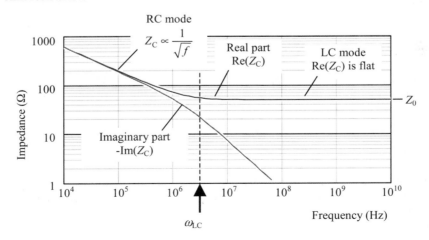

Figure 3.16—The characteristic impedance of an LC line asymptotically approaches Z_0 at frequencies above ω_{LC}.

3.6.3 • Influence of Series Resistance

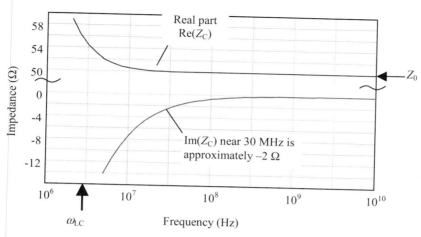

Figure 3.17—One decade above ω_{LC} the imaginary part of Z_C carries a value of approximately $-2\ \Omega$.

POINTS TO REMEMBER

- At frequencies far above ω_{LC} the characteristic impedance of a pure LC-mode transmission line asymptotically approaches $Z_0 \triangleq \sqrt{L/C}$.
- At a frequency ten times greater than ω_{LC} the complex value of characterisitic impedance differs from Z_0 by only about 4%.

3.6.3 Influence of Series Resistance on TDR Measurements

Figure 3.18 displays the TDR response of the same hypothetical pcb trace used in previous figures. The response begins with a half-sized step, persisting for one round-trip delay, after which the response rises to a full level. The trace length of 0.5 meters creates a round-trip delay of approximately 7 ns.

The series resistance present within the trace tilts the TDR steps. The values of DC resistance represented in Figure 3.18 are 0, 6.4, and 12.8 Ω/m (0, 0.163 and 0.325 Ω/in.), corresponding to stripline widths in 1/2-ounce copper of infinity, 150 µm, and 75 µm, (∞, 6, and 3 mils) respectively. The greater the series resistance of the trace, the greater the tilt it induces in the TDR response. You can predict the degree of tilt from an analysis of the characteristic impedance.

At frequencies just above ω_{LC} the characteristic impedance of the trace in response to a step edge can be modeled as a resistance Z_0 that trends gradually upward over time. This conclusion may be extracted from [3.85] by first factoring out from under the radical the asymptotic impedance Z_0, which by definition equals $\sqrt{L/C}$.

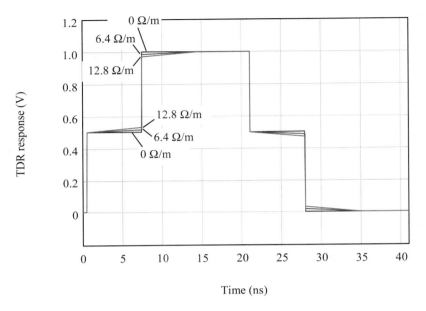

Figure 3.18—Series resistance tilts the TDR steps.

$$Z_C = \sqrt{\frac{L}{C}} \sqrt{1 + \frac{R_{DC}}{j\omega L}} = Z_0 \sqrt{1 + \frac{R_{DC}}{j\omega L}} \qquad [3.86]$$

Next use a first-order Taylor-series approximation for the remaining square root. This approximation is valid only at frequencies well above ω_{LC}.

$$Z_C = Z_0 \left(1 + \frac{1}{2} \frac{R_{DC}}{j\omega L} \right) = Z_0 \left(1 + \frac{1}{j\omega \tau} \right) \qquad [3.87]$$

where the time constant $\tau \triangleq 2L/R_{DC}$.

Converting [3.87] to an expression that is a function of time shows that in response to a step input of current, the characteristic impedance of the line can be approximated in its first moments ($t \ll 2L/R_{DC}$) as an initial value equal Z_0 followed by a gradual up-trend in impedance:

$$Z_C(t) = Z_0 \left(1 + \frac{t}{\tau} \right) \qquad [3.88]$$

3.6.3 • Influence of Series Resistance

Example Calculation of Impedance Slope

Let $Z_0 = 50 \ \Omega$, $L = 346$ nH/m, and $R_{DC} = 12.8 \ \Omega/\text{m}$.

In response to a step input of current, the initial apparent impedance is $Z_0 = 50 \ \Omega$. At a time 5 ns after the initial step edge, the apparent impedance has increased to

$$Z_C\big|_{t=5\,\text{ns}} \approx Z_0\left(1 + \frac{t}{(2L/R_{DC})}\right) = 50\left(1 + \frac{5 \cdot 10^{-9}}{2} \frac{12.8}{346 \cdot 10^{-9}}\right) = 54.62 \ \Omega$$

The trend in impedance differs slightly from the trend in the TDR voltage, because the TDR setup delivers an input that comes not from a perfect current source, but from a source with finite impedance R_S.

Given model [3.88] for the effective input impedance of an LC-mode transmission line in response to a step of current, you can relate the slope dx/dt of the slanted sections in a TDR waveform $x(t)$ to the series resistance R_{DC} of the transmission structure.

$$\frac{dx}{dt}\left\{\Delta V \frac{Z_C}{Z_C + R_S}\right\}\bigg|_{\text{at } t=0} \approx \Delta V \frac{R_S Z_0}{(R_S + Z_0)^2} \frac{1}{\tau} \qquad [3.89]$$

where dx/dt is the slope of the slanted section of the TDR response (V/s),

ΔV is the open-circuit step-size of the TDR output (V),

R_S is the source impedance of the TDR instrument (usually 50 ohms),

Z_0 is the high-frequency asymptote for characteristic impedance (ohms) equal to $\sqrt{L/C}$,

τ is a time constant defined equal to $2L/R_{DC}$, and

R_{DC} is the series DC resistance of the line (not taking into account skin effect).

Be forewarned that [3.89] applies only when the resistance R_{DC} remains fixed and does not vary with frequency. In cases where the series resistance of the line varies significantly over the time interval of interest, as it would for example in consideration of the skin effect, then the AC value of skin-effect resistance must be used in the estimation of τ and in the calculation of [3.89].

Example Calculation of TDR Slope

Let $Z_0 = 50 \ \Omega$, $L = 346$ nH/m, and $R_{DC} = 12.8 \ \Omega/\text{m}$.

With a 50-Ω TDR apparatus having a 1-V open-circuit output voltage, the apparent slope in the TDR waveform is

$$\frac{dx}{dt}\left\{\Delta V \frac{Z_C}{Z_C + R_S}\right\}\bigg|_{\text{at } t=0} \approx \Delta V \frac{50 \cdot 50}{(50 + 50)^2} \frac{1}{(2 \cdot 346 \cdot 10^{-9}/12.8)} = 4.62 \cdot 10^9 \ \text{V/s}$$

In a period of 5 ns you would expect the TDR waveform to creep upwards by

$$(4.62 \cdot 10^9 \text{ V/s})(4 \text{ ns}) = 0.018 \text{ V}$$

This amount agrees with the plot in Figure 3.19.

By similar reasoning, a fixed conductance G creates a *negative* slope to the TDR waveform, with a time-constant $\tau \triangleq 2C/G$.

The last two figures in this section illustrate the effect of limited TDR risetime and the effect of a 50-ohm termination placed at the far end of the same line used in previous figures. The DC resistance in both cases equals 12.8 Ω/m. Figure 3.19 assumes risetimes of 35, 100, 350, and 1000 ps. Because of the sloping effect in the steps, measurements of impedance made with the smallest risetime reveal the smallest value of Z_C. Slower measurements, to the extent that the step has time to drift upwards before a plateau in the driving waveform has been reached, tend to indicate slightly larger values. This figure highlights one of the difficulties of measuring characteristic impedance using the TDR method. Severe degrees of tilt make it very difficult to define one correct measurement procedure that is best for all applications.

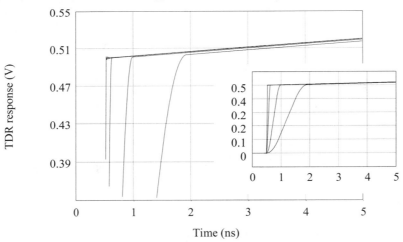

Figure 3.19—The inexorable upward creep of the TDR response causes measurements taken with slower risetimes to indicate slightly larger values of Z_C (R_{DC} = 12.8 Ω/m).

Figure 3.20 depicts the effect of a perfect 50-ohm load at the far end of the line. The end termination arrests the evolution of the sloped part of the TDR response, creating a perfectly flat top for all times beyond one round-trip delay.

POINT TO REMEMBER

> ➤ A fixed series resistance induces an upward tilt in the TDR measurement, indicating a gradually increasing impedance at lower frequencies.

3.6.4 • Propagation Coefficient

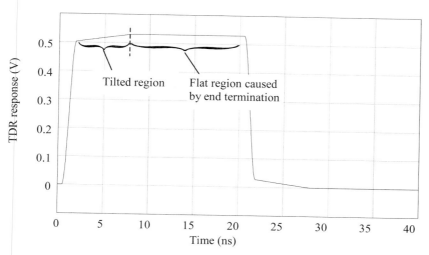

Figure 3.20—An end termination arrests the evolution of the tilted portion of the TDR response (R_{DC} = 12.8 Ω/m).

3.6.4 *Propagation Coefficient (LC Region)*

Figure 3.21 shows the propagation coefficient for a hypothetical pcb trace with fixed DC resistance, but no skin effect or dielectric loss. This figure plots the real and imaginary parts of the propagation coefficient versus frequency. The plot is drawn with log-log axes to highlight the polynomial relationships between portions of the curves.

In the RC region below ω_{LC} both the real part of the propagation coefficient (log of attenuation) and the imaginary part (phase in radians) rise together in proportion to the square root of frequency. Within the RC region the attenuation and phase are coupled together in such a way that any line with a significant phase shift also naturally inherits a significant attenuation. It is therefore impossible to construct a passive high-Q resonator from an RC-mode transmission line without adding reactive components to the ends of the line.

In the LC region above ω_{LC} the situation changes. There is no longer any coupling between attenuation and phase within the LC region. In this region the imaginary part (phase) grows linearly with increasing frequency, while the real part (attenuation) stays fixed. It is quite possible within this region to construct a line with an enormous phase delay and yet very low attenuation. Such a line makes a terrific high-Q resonator. The existence of such resonant structures is the fundamental reason why terminations are so often needed on pcb traces used in high-speed applications.

At frequencies far above the LC mode boundary ω_{LC} the imaginary part of γ approaches a linear-phase ramp equal to $\omega\sqrt{LC}$, while the real part of γ stabilizes at a constant value equal to $R_{DC}/(2Z_0)$. This conclusion is derived from [3.13] by first factoring out from under the radical a factor of $j\omega$, and also the inverse of propagation velocity v_0, defined as $1/v_0 \triangleq \sqrt{LC}$.

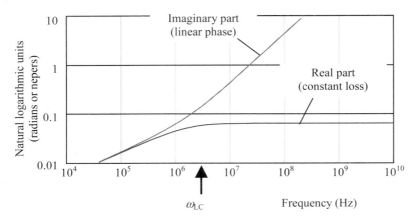

Figure 3.21—Above ω_{LC} the real and imaginary parts of the propagation coefficient separate, decoupling the attenuation (real part) from the phase (imaginary part).

$$\gamma(\omega) = j\omega\sqrt{LC}\sqrt{1+\frac{R_{DC}}{j\omega L}} = \frac{j\omega}{v_0}\sqrt{1+\frac{R_{DC}}{j\omega L}} \qquad [3.90]$$

Next approximate the remaining square root. This approximation is valid only at frequencies well above ω_{LC}.

$$\gamma(\omega) \approx j\omega\sqrt{LC}\left(1+\frac{1}{2}\frac{R_{DC}}{j\omega L}\right) = j\omega\sqrt{LC} + \frac{R_{DC}}{2\sqrt{L/C}} \qquad [3.91]$$

Substituting $\sqrt{L/C} \to Z_0$ leads to the aforementioned conclusion.

The linear-phase term implies that the one-way propagation function H of an LC-mode transmission line, for frequencies above ω_{LC}, acts mostly as a large time-delay element. The bulk propagation delay in units of seconds per meter is given by

$$t_p \triangleq 1/v_0 = \sqrt{LC} \text{ s/m} \qquad [3.92]$$

Equation [3.92] may also be stated in terms of the effective relative permittivity ϵ_{re} associated with the transmission line configuration $t_p = \sqrt{\epsilon_{re}}/c$ s/m, assuming the speed of light $c = 2.998 \cdot 10^8$ m/s and further assuming the relative magnetic permeability μ_r equals one (see [2.28]).

The linear-phase property, in conjunction with the constant-loss property, implies that the one-way propagation function H of an LC-mode transmission line, for frequencies above ω_{LC}, acts as nothing more than a simple time-delay element with a fixed attenuation. The

3.6.4 • Propagation Coefficient

delay varies in proportion to the length of the transmission line. Doubling the length doubles the delay.

The transfer loss in nepers per meter is given by the real part of the propagation function. This value is called the *resistive loss coefficient* (note that this definition takes into account only the DC resistance—consideration of the skin-effect resistance will increase actual line losses):

$$\alpha_{r,DC} \triangleq \text{Re}[\gamma(\omega)] = \frac{1}{2}\frac{R_{DC}}{Z_0} \text{ neper/m} \qquad [3.93]$$

$$\alpha_{r,DC} \triangleq \text{Re}[\gamma(\omega)] = 4.34\frac{R_{DC}}{Z_0} \text{ dB/m} \qquad [3.94]$$

The low-pass filtering action of the skin effect implies that the step response will be dispersed, slurring crisp rising edges into edges of limited rise and fall time, according to the properties of the filter.

The magnitude of H is determined from the real part of the propagation function (see [2.16]):

$$|H(\omega,l)| = e^{-l \cdot \text{Re}(\gamma(\omega))} = e^{-l\frac{1}{2}\frac{R_{DC}}{Z_0}} \qquad [3.95]$$

where $|H(\omega,l)|$ is the magnitude of the line attenuation implied by [3.91] (this term does not vary with frequency),

l is the length of the transmission line in meters,

R_{DC} is the DC resistance of the line, and

Z_0 is the high-frequency asymptotic characteristic impedance of the line.

The signal loss represented by H, measured in nepers (or decibels), varies in proportional to the length of the transmission line. Double the distance yields twice the loss in neper or decibel units (one neper equals 8.6858896 dB).

The properties in Table 3.5 hold only for LC-mode transmission lines operated at frequencies well above ω_{LC}. Also note that R_{DC} represents only the DC resistance of the line. Subsequent sections discuss the implications of the skin effect, which increases the total line resistance at high frequencies, and dielectric loss, which can further increases line attenuation.

POINTS TO REMEMBER

> ➤ The propagation function for an LC-mode transmission line is a simple delay with a fixed attenuation.
> ➤ Doubling the length of an LC-mode transmission line doubles the delay, and doubles the attenuation (in dB).

Table 3.5—Summary of LC Transmission Line Properties at Frequencies Well Above ω_{LC}

Property	Formula	Ref. Equation
Asymptotic value of characteristic impedance	$Z_0 = \sqrt{\dfrac{L}{C}}$	[3.85]
Bulk transport delay per meter	$t_p \triangleq 1/v_0 = \sqrt{LC} = \sqrt{\epsilon_{re}}/c$ s/m	[3.92]
Attenuation in nepers per meter (1 neper = 8.6858896 decibels)	$\dfrac{1}{2}\dfrac{R_{DC}}{Z_0}$ neper/m	[3.93]
Attenuation in decibels per meter	$4.34\dfrac{R_{DC}}{Z_0}$ dB/m	[3.94]
Transfer gain at length l	$\|H(\omega,l)\| = e^{-l\frac{R_{DC}}{2Z_0}}$	[3.95]
Inductance per meter	$L = Z_0/v_0$	[3.5]
Capacitance per meter	$C = \dfrac{1}{Z_0 v_0}$	[3.6]

3.6.5 Possibility of Severe Resonance within the LC Region

In the LC region it is possible for a signal traveling on a long transmission line to accumulate a substantial amount of phase delay without suffering much attenuation. This property indicates that a transmission line may in certain circumstances act as an extremely high-Q resonant circuit.

Figure 3.22 illustrates the resonance associated with a hypothetical pcb trace having a fixed DC resistance, but no skin effect or dielectric loss. The trace is driven at the near end by a driver having a source impedance of 0, 10, or 20 ohms and left open-circuited at the far end. All LC transmission lines exhibit similar resonant peaks when driven by a source impedance less than Z_0. The peaks are located at frequencies (in Hertz) equal to odd multiples of $1/(4v_0)$.

The heights of the peaks are limited partly by the series resistance of the line and partly by the damping effect of the source impedance at the driver. The greater the source impedance of the driver, the greater its damping effect on the circuit. Because in this example only the DC resistance of the line is taken into account (i.e., ignoring skin effect and dielectric effects), all the peaks occur at the same height.

The peaks are predicted by Equation [3.17]. In the present case involving an open-circuited (i.e., non-loaded) transmission line this equation may be simplified by eliminating from the denominator all terms involving Z_L.

3.6.5 • Possibility of Severe Resonance

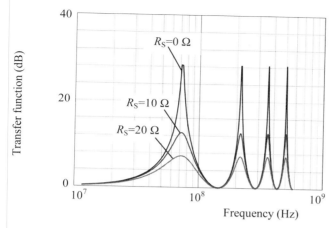

Figure 3.22—An LC line driven with a low-impedance source exhibits resonance at odd multiples of $1/(4v_0)$ Hz.

$$G_{\text{OPEN}} = \frac{1}{\left(\dfrac{H^{-1}+H}{2}\right)+\left(\dfrac{H^{-1}-H}{2}\right)\dfrac{Z_S}{Z_C}} \qquad [3.96]$$

In the case of a very low source impedance ($Z_S \approx 0$) the main difficulty with this equation is the cancellation that occurs within the term $H^{-1}+H$. Figure 3.23 shows how this cancellation can occur. The figure plots the trajectories of H and H^{-1} on the complex plane as a function of frequency. For a line with a lot of phase delay but not much attenuation, the trajectory of H remains at almost unit magnitude as it spirals clockwise around the origin. The trajectory of H^{-1} describes a similar path, spiraling counterclockwise just outside the unit circle. As H approaches $-j$, and H^{-1} at the same frequency approaches $+j$, the average value of the two functions passes close to zero. In the absence of any significant source impedance Z_S, when the first term in the denominator of G_{OPEN} passes close to zero, the function G_{OPEN} spikes upward, forming a sharp resonant peak.

In the RC transmission mode such a resonance cannot happen. In the RC mode the real and imaginary parts of the propagation coefficient are always equal so that a phase of $-\pi/2$ must always be accompanied by an attenuation of $e^{-\pi/2} = 0.208$. The attenuation of H (and amplification of H^{-1}) prevents the average of H and H^{-1} from passing close to the origin.

In the LC mode, it is the contribution of Z_S that pushes the denominator at $f = 1/(4v_0)$ up and away from the origin, damping the resonance.

If a current-source driver (high source impedance) is used, similar resonance effects occur but at even multiples of $f = 1/(4v_0)$.

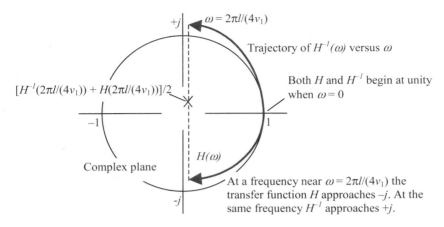

Figure 3.23—As the trajectories of H^{-1} and H approach $+j$ and $-j$, the average value $[H^{-1} + H]/2$ passes close to the origin.

3.6.5.1 Alternate Interpretation of Equation [3.17]

Equation [3.17] may be algebraically manipulated into the following form:

$$G = \frac{1}{2}\frac{(1-\Gamma_1)(1+\Gamma_2)}{1-H^2\Gamma_1\Gamma_2} \qquad [3.97]$$

where G is the overall system gain from the open-circuited output of the driver v_1 to the load v_3,

$\Gamma_1 = \dfrac{Z_S - Z_C}{Z_S + Z_C}$ is the reflection coefficient at the source end of the transmission structure,

$\Gamma_2 = \dfrac{Z_L - Z_C}{Z_L + Z_C}$ is the reflection coefficient at the load end of the transmission structure, and

H is the one-way propagation function of the raw, unloaded transmission structure [3.14].

This formulation makes clear the relation between the one-way propagation function H, the reflection coefficients Γ_1 and Γ_2, and the height of the resonant peaks in the overall system gain G. Assume for the moment that the reflection coefficients are large quantities (near unity) that do not vary strongly with frequency. Under that assumption the overall gain is determined mostly by action of the denominator $(1-H^2\Gamma_1\Gamma_2)$. If H can acquire a large phase shift without much attenuation, then the possibility exists that it will (at some frequency) pass very close to zero, causing a large spike in the overall system gain G.

Appropriate remedies for the resonance include

1. Design a load impedance Z_L such that Γ_2 is at all frequencies limited in magnitude, or
2. Design a source impedance Z_S such that Γ_1 is at all frequencies limited in magnitude, or
3. Do both (1) and (2).

These remedies correspond to the end, source, and both-ends termination strategies, respectively, outlined in Section 3.6.6.

3.6.5.2 Practical Effect of Resonance

In all cases please remember that any LC resonance affects only signals whose bandwidth extends into the resonant region. As long as the knee frequency [3.19] of your digital logic falls well below the first resonance, the LC-resonance effects won't significantly distort your signal. This condition is effectively the same as condition [3.27] for lumped-element operation.

POINTS TO REMEMBER

> In the LC region a signal can accumulate a substantial amount of phase delay without suffering much attenuation. This property indicates that a transmission line can act as an extremely high-Q resonant circuit.
> LC-mode resonance affects only signals whose bandwidth extends into the resonant region.

3.6.6 Terminating an LC Transmission Line

As explained in the previous section, the overall circuit gain G of a long transmission-line configuration can differ substantially from H, where H represents the one-way attenuation of signals moving from end to end across the transmission line. Even in cases where H is relatively well-behaved, the overall circuit performance can sometimes suffer from terrible resonance.

There are three classic ways to stabilize the propagation function of an LC transmission line, eliminating the undesirable resonance. Each case uses a resistive termination to deliver a circuit gain proportional to the one-way transmission line propagation function $H(\omega)$ (see [3.14]). In the case of a relatively short LC transmission line (such as normally used on a printed-circuit card for high-speed digital signals) the propagation function H is practically flat with linear phase, indicating the line itself acts as nothing more than a time-delay element with a minor amount of attenuation.

Presuming that the attenuation (and possibly distortion) of $H(\omega)$ falls within acceptable bounds, a properly terminated LC-mode transmission line may be operated satisfactorily at any length.

The three classic termination styles appear in Figure 3.24. The overall circuit gain G for each is derived from [3.17], repeated here for convenience.

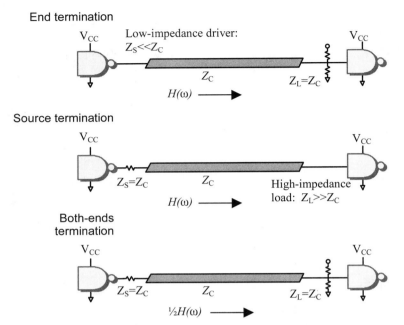

Figure 3.24—The end, source, and both-ends termination schemes each deliver a system response nearly proportional to $H(\omega)$.

$$G = \frac{1}{\left(\dfrac{H^{-1}+H}{2}\right)\left(1+\dfrac{Z_S}{Z_L}\right)+\left(\dfrac{H^{-1}-H}{2}\right)\left(\dfrac{Z_S}{Z_C}+\dfrac{Z_C}{Z_L}\right)} \qquad [3.98]$$

3.6.6.1 End Termination

Assuming a load impedance[24] Z_L reasonably close to Z_C, you may substitute 1 for the expression Z_C/Z_L in [3.17]. Further assuming a source impedance Z_S much less than Z_C, you may ignore the terms Z_S/Z_L and Z_S/Z_C in [3.17]. With these simplifications made, the expression for the overall circuit gain of an end-terminated transmission line reduces to

$$G \approx \frac{1}{\left(\dfrac{H^{-1}+H}{2}\right)(1+0)+\left(\dfrac{H^{-1}-H}{2}\right)(0+1)} = H \qquad [3.99]$$

---

[24] The load is assumed connected to ground. Connection of the load impedance to any other voltages requires a superposition of the solution given here with a second static solution related to the applied termination voltage.

3.6.6.2 Source Termination

Assuming a source impedance Z_S reasonably close to Z_C, you may substitute 1 for the expression Z_S/Z_C in [3.17]. Further assuming a load impedance Z_L much larger than Z_C, you may ignore the terms Z_S/Z_L and Z_C/Z_L in [3.17]. With these simplifications made, the expression for the overall circuit gain of a source-terminated transmission line reduces to

$$G \approx \frac{1}{\left(\dfrac{H^{-1}+H}{2}\right)\left(1+\dfrac{1}{\infty}\right)+\left(\dfrac{H^{-1}-H}{2}\right)\left(1+\dfrac{1}{\infty}\right)} = H \qquad [3.100]$$

3.6.6.3 Both-Ends Termination

Assuming source and load impedances to be the same, you may substitute 1 for the expression Z_S/Z_L in [3.17]. Further assuming both impedances to be reasonably close to Z_C, you may substitute 1 for both expressions Z_S/Z_L and Z_C/Z_L in [3.17]. With these simplifications made, the expression for the overall circuit gain of a source-terminated transmission line reduces to

$$G \approx \frac{1}{\left(\dfrac{H^{-1}+H}{2}\right)(1+1)+\left(\dfrac{H^{-1}-H}{2}\right)(1+1)} = \frac{H}{2} \qquad [3.101]$$

3.6.6.4 Subtle Differences Between Termination Styles

The three termination styles react differently to the presence of line resistance. Figure 3.25 shows those differences. All cases depict a hypothetical pcb trace having a fixed DC resistance of 6.4 Ω/m but no skin effect or dielectric loss. The trace is a 50-ohm stripline constructed of 1/2-oz copper (17.4 micron thickness) on FR-4 with a width of 150 microns (6 mils, $v_0 = 1.446 \cdot 10^8$ m/s, $L_0 = 346$ nH/m, $C_0 = 138$ pF/m). The trace length $l = 0.5$ meters. The response of the both-ends terminated circuit has been artificially multiplied by two, raising its output to a level comparable with the other two plots for display purposes.

In all cases the initial step amplitude is the same, corresponding to the circuit gain at high frequencies given by [3.93]. At DC, however, the responses differ.

In the source-terminated case there is no load at the far end of the line to draw any DC current, so the final value of current within the transmission line is zero, and the steady-state voltage drop from end to end across the transmission line must also be zero. The final value of the circuit output at the far end of the line therefore begins at a point defined by [3.93] and then drifts up to the full open-circuit voltage of the source (in this case, 1.0 volts).

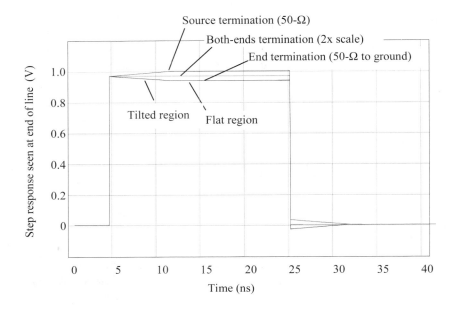

Figure 3.25—Different termination styles react differently to a fixed line resistance.

In the end-terminated case the load at DC draws a substantial current. The attenuation at DC depends on the relation between the load impedance and the total series impedance of the line:

$$\lim_{\omega \to 0} G_{END} = \frac{Z_L}{Z_L + l \cdot R} \qquad [3.102]$$

where G_{END} is the overall circuit gain of an end-terminated transmission line with one-way propagation function H,

Z_L is the load impedance applied to the end of the line,

the source impedance Z_S is assumed much smaller than the characteristic impedance of the line Z_C,

the line length is l (m), and

R is the transmission line series resistance per meter (Ω/m).

The value [3.102] is always *less* than the circuit gain predicted by [3.93]. The circuit output at the far end of the line therefore begins at a point defined by [3.93] and then drifts *down* to the value indicated by [3.102].

In Figure 3.25 the total resistance of the trace ($l = 0.5$ meters times 6.4 Ω/m) is 3.2 Ω, which amounts to 6.4% of the trace impedance. The steady-state amplitude of the end-terminated waveform is therefore reduced by a factor of 6.4%.

The both-ends terminated case is least sensitive to internal resistance. To first order, internal series resistance induces no tilt to the circuit output. The initial step amplitude and the final value of step amplitude of a both-ends terminated circuit are the same.[25]

3.6.6.5 Application of Termination Equations to Other Regions

The LC, skin-effect, and dielectric-loss-limited regions all share the same asymptotic high-frequency value of characteristic impedance Z_0. The same termination approaches therefore work for all three regions, delivering in each case an overall circuit gain G equal to the one-way propagation function H applicable within each region.

Because the propagation function H is flat with frequency within the LC region, a well-terminated line operated within this region delivers a minor amount of attenuation but no distortion in the shape of the signal (i.e., no dispersion of the rising edge).

In the skin-effect and dielectric-loss-limited regions the propagation function H is not flat. It acts in these regions as a low-pass filter, attenuating and dispersing the edges of all signals passing through the transmission line.

POINTS TO REMEMBER

> ➤ The source, end, and both-ends terminations can all be used to eliminate LC-mode resonance.
>
> ➤ The end-termination is least sensitive to the series resistance of the transmission line.

3.6.7 Tradeoffs Between Distance And Speed (LC Region)

Given a fixed receiver architecture, and presuming that you remain within the LC region, the maximum speed of operation is not directly limited by transmission-line length. The tradeoff between length and speed disappears within the LC region because the attenuation (in dB) does not vary with frequency. A 10% increase in transmission-line length therefore increases the attenuation (in dB) by 10%. Such an increase in the attenuation of an LC-mode transmission line cannot be offset by a reduction in system operating speed.

POINT TO REMEMBER

> ➤ The speed of operation achievable within the LC region is not directly limited by transmission-line length.

---

[25] In a both-ends termination structure the difference between the upwards tilt of the TDR response at the head of the line and the progressive attenuation of an initial step edge as it propagates down the same line is just enough to ensure that by the time your initial step edge reaches the end, a satisfactory steady-state condition has been reached with the far end voltage a little less than 50% [3.106] and the near-end voltage a little greater than 50%.

3.6.8 Mixed-Mode Operation (LC and RC Regions)

Figure 3.26 illustrates the transmission line modes spanned by two systems, each represented by rays. System (A) operates at a length below the critical length l_{RC} (see [3.54], under Section 3.5.1, "Boundary of RC Region"). At that length its frequency range of operation (0 to 20 MHz) spans only the LC and lumped-element regions. This system should be terminated with an impedance equal to the high-frequency characteristic impedance asymptote, Z_0. Any such terminator will likely continue to do an adequate job as the frequency of operation descends down into the lumped-element region. This is the normal arrangement for pcb traces—before you get down to a frequency low enough to observe RC effects, the whole system acts as a big lumped-element circuit anyway and (provided you don't have reactive loads) within the lumped-element region any old termination scheme will work.

System (B) is quite different. It operates at a length greater than l_{RC}. Its frequency range of operation is the same, but because of its greater length, its operational modes span the LC, RC, and lumped-element regions. The frequency response of this system will be a strong function of frequency (in the RC zone), so some form of equalization may be required. The ideal termination impedance for this system may be a frequency-varying network.

POINT TO REMEMBER

> Transmission lies operated at a length greater than l_{RC} may display characteristics of both LC and RC behavior.

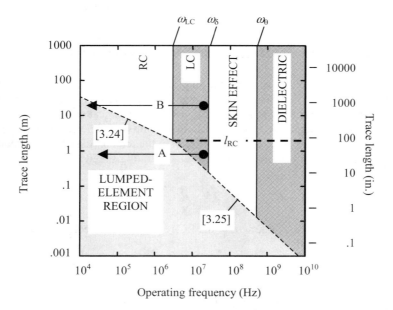

Figure 3.26—The range of performance for some applications may span multiple regions.

3.7 SKIN-EFFECT REGION

Beyond the LC transition the internal inductance of the conductors (a mere fraction of the total inductance) becomes significant compared to the DC resistance. This development forces a redistribution of current within the bodies of the conductors. The redistribution of current heralds the arrival of the skin-effect region (Section 3.2).

3.7.1 Boundary of Skin-Effect Region

As illustrated in Figure 3.2, the skin-effect region extends to all frequencies above that point ω_δ where the real part of the skin-effect resistance R_{AC} [2.44] equals the DC resistance R_{DC} [2.37].

$$\omega_\delta = \frac{2}{\mu\sigma}\left(\frac{k_a p}{k_p a}\right)^2 \qquad [3.103]$$

where ω_δ is the onset of the skin-effect region (rad/s),

k_p is the proximity factor (see Section 2.10.1, "Proximity Factor"),

k_a is the DC resistance correction factor (see Section 2.4, "DC Resistance"),

μ is the magnetic permeability of the conductors (H/m),

σ is the conductivity of the conductors (S/m),

p is the perimeter of the conductors (m), and

a is the cross-sectional area of the signal conductor (m^2).

NOTE: For differential configurations the parameters p and a are interpreted as properties of only one of the signal conductors. The constants k_a and k_p both assume values near two, accounting for the doubling of both DC and AC resistance due to the differential configuration. The result is that the frequency of skin-effect onset for a differential configuration occurs very nearly at the same frequency as the skin-effect onset for a single-ended configuration using one of the same physical signal conductors in juxtaposition with a solid reference plane located halfway between the two differential conductors.

If you have already computed parameters R_{DC} and R_0 (see Table 3.1), formula [3.103] reduces to

$$\omega_\delta = \omega_0 \left(\frac{R_{DC}}{R_0}\right)^2 \qquad [3.104]$$

where R_{DC} is the DC resistance of the conductors, Ω/m [2.37],

ω_0 is the frequency at which the AC line parameters are specified, rad/s, and

R_0 is the real part of AC resistance at frequency ω_0, Ω/m [2.43].

At all frequencies above ω_δ (including the dielectric and waveguide regions) the skin effect plays a significant role and must be taken into account in system modeling.

The distinguishing feature of the skin-effect mode is that within this region the characteristic impedance remains fairly flat, while the line attenuation in dB varies in proportion to the square root of frequency.

POINTS TO REMEMBER

> - The skin effect region starts when the internal inductance of the signal conductor becomes significant compared to its DC resistance.
> - Within the skin effect region the characteristic impedance remains fairly flat, while the line attenuation in dB varies in proportion to the square root of frequency.

3.7.2 Characteristic Impedance (Skin-Effect Region)

The full expression for characteristic impedance (neglecting the conductance G) includes contributions from both the external inductance L_0 and resistance R of the line.

$$Z_C = \sqrt{\frac{j\omega L_0 + R(\omega)}{j\omega C}} \qquad [3.105]$$

In the limit as you proceed to frequencies far above ω_δ (which by definition exceeds ω_{LC}), the contribution of the term R becomes negligible, leading to this approximation for characteristic impedance in the heart of the skin-effect territory.

(skin-effect assumption) $\quad Z_0 \triangleq \text{Lim}_{\omega \to \infty} \sqrt{\dfrac{j\omega L_0 + R(\omega)}{j\omega C}} = \sqrt{\dfrac{L_0}{C}} \qquad [3.106]$

Even though in the skin-effect region $R(\omega)$ grows in proportion to the square root of frequency, the term $j\omega L_0$ grows faster, so that once past the crossover ω_{LC}, the contribution of $R(\omega)$ to [3.106] is of strictly diminishing importance.

The only difference between [3.106] and the previous equation [3.85] is that here I explicitly provide for the possibility that the resistance $R(\omega)$ may change with frequency, and the inductance term is clearly marked L_0 to represent only the *external inductance* of the transmission configuration (see Section 2.7, "Skin-Effect Inductance" and Section 2.8, "Modeling Internal Impedance"). External inductance is the value of series inductance, in Henries per meter, computed by a 2-D field solver under the assumption that current rides on the surface of each conductor without penetrating its bodies.

3.7.2 • Characteristic Impedance

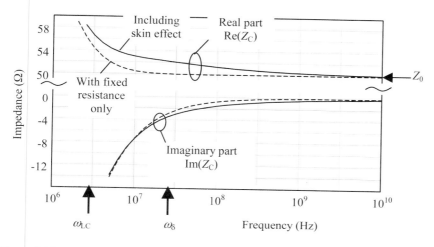

Figure 3.27—The characteristic impedance of a transmission line with skin effect (solid lines) does not approach Z_0 quite as rapidly as a line without (dashed lines).

The lumped-element, RC, and LC operating analyses presented previously assume that both L and R are constants, so it is therefore not generally necessary in those presentations to distinguish between the internal and external inductance, only lumping together all the inductance you have of either type under the umbrella term "L." Here the term $L_0 \triangleq Z_0/v_0$ from Equation [3.5] is used to represent the external inductance L_e. The *internal inductance* of the signal conductors is represented by variations in the imaginary part of $R(\omega)$.

As your operating frequency crosses upward from the LC to the skin-effect region, the variations in $R(\omega)$ change (slightly) the precise shapes of the asymptotically convergent curves in Figure 3.17, but not the asymptotic points of convergence (Figure 3.27). As long as your circuit does not depend on the precise shapes of the curves, the changes have little practical significance.

The plots starting with Figure 3.27 and going through Figure 3.30 all depict a hypothetical pcb trace having a fixed DC resistance of 6.4 Ω/m plus the skin effect, but no dielectric loss. The trace is a 50-ohm stripline constructed of 1/2-oz copper (17.4 micron thickness) on FR-4 with a width of 150 microns (6 mils, $v_0 = 1.446 \cdot 10^8$ m/s, $L_0 = 346$ nH/m, $C_0 = 138$ pF/m). The trace is 0.5 meters long.

The general relationship between the characteristic impedance and input impedance is given by [3.33] and discussed in Section 3.5.2 "Input Impedance (RC Region)" and in Section 3.2.1 "A Transmission Line Is Always a Transmission Line."

POINTS TO REMEMBER

> At frequencies far above ω_δ the characteristic impedance of a skin-effect limited transmission line asymptotically approaches $Z_0 \triangleq \sqrt{L/C}$.

➢ The asymptotic convergence is not quite as fast as for an LC transmission structure with fixed (non-frequency-varying) resistance.

3.7.3 Influence of Skin-Effect on TDR Measurement

As the skin effect increases the series resistance of a transmission line, it must also increase the degree of slope in the first TDR plateau as described under Section 3.6.3, "Influence of Series Resistance on TDR Measurements." Figure 3.28 depicts two TDR plots, the first having only a DC resistance of 6.4 Ω/m and the second having the same DC resistance but in addition showing the skin effect.

For a transmission line gripped by the skin effect, the sloping effect in the first TDR plateau begins with a slope steeper than would be predicted by the DC resistance alone. The steeper slope corresponds to the large AC (skin-effect) resistance of the line at frequencies associated with the rising edge. As time progresses the slope abates to a more moderate value corresponding to the DC resistance of the line at frequencies less than ω_δ.

Within the skin-effect region the slope over any particular scale of time t is roughly indicated by [3.89], substituting for R_{DC} the skin-effect resistance R_{AC} [2.44] at a frequency corresponding roughly to $0.35/t$ Hz [3.20].

For any open-circuited transmission line,[26] the second plateau of the TDR response shows a signal which has traversed the line twice, going to the far end and then reflecting

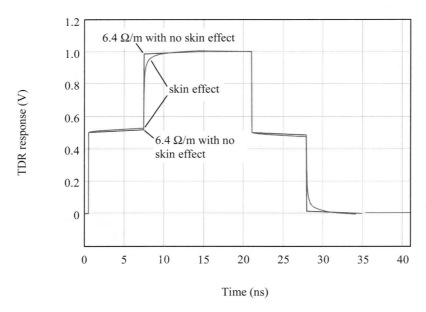

Figure 3.28—The AC resistance associated with the skin effect increases the slope of the initial TDR plateau.

---

[26] The usual condition of measurement for checking the impedance of a pcb test coupon.

back. This reflected signal displays the marks of its round-trip venture in the form of a noticeable dispersion of both rising and falling edges.

POINT TO REMEMBER

> ➤ The skin effect induces an upward tilt in the TDR measurement with a steep initial slope, gradually tapering to a more gentle rise.

3.7.4 Propagation Coefficient (Skin-Effect Region)

Figure 3.29 shows the propagation coefficient for the same hypothetical pcb trace as Figure 3.21, but this time in addition showing the skin effect (but not dielectric loss). This figure plots the real and imaginary parts of the propagation coefficient versus frequency. The plot is drawn with log-log axes to highlight the polynomial relationships between portions of the curves.

In the RC region below ω_{LC} both the real part of the propagation coefficient (log of attenuation) and the imaginary part (phase in radians) rise together in proportion to the square root of frequency. Above ω_{LC} the imaginary part (phase) grows linearly with increasing frequency, while the real part (attenuation) inflects to the right at a lesser slope. Because the skin-effect onset is so close to the LC-mode onset, the real part doesn't have sufficient room to flatten out in a true constant-loss profile. In the skin-effect region above ω_δ the real part once again turns upward, indicating an attenuation (in dB) proportional to the square root of frequency. In this example the LC region is quite narrow, a typical result for pcb traces.

In the skin-effect region, as in the LC region, the decoupling of phase and attenuation makes it possible to construct a line with an enormous phase delay and yet very low attenuation. Terminations are used within the skin-effect region to abate resonance on long transmission lines in precisely the same manner as used in LC region.

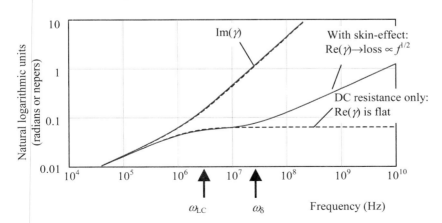

Figure 3.29—Above ω_δ the signal loss (real part of the propagation coefficient) for a skin-effect-limited transmission line grows in proportion to the square root of frequency.

You can predict the phase and amplitude response of a skin-effected transmission line beginning with the definition of the propagation function [3.13], assuming operation at a frequency well in excess of ω_δ so that $R_{AC} \gg R_{DC}$.

$$\gamma(\omega) = \sqrt{(j\omega L_0 + R_{AC})(j\omega C)} \qquad [3.107]$$

where L_0 represents the external inductance and where information about the internal inductance is coded into the imaginary part of R_{AC}.

Factor out the common $j\omega L_0$ and $j\omega C$ terms.

$$\gamma(\omega) = \sqrt{j\omega L_0 \cdot j\omega C} \sqrt{\left(1 + \frac{R_{AC}}{j\omega L}\right)} \qquad [3.108]$$

Assuming you are dealing at a frequency well in excess of ω_{LC} so that $|j\omega L_0| \gg |R_{AC}|$, you may linearize the second square-root:

$$\gamma(\omega) \approx \sqrt{j\omega L_0 \cdot j\omega C} \left(1 + \frac{1}{2}\frac{R_{AC}}{j\omega L_0}\right) \qquad [3.109]$$

and distribute the first term into the second:

$$\gamma(\omega) \approx \sqrt{j\omega L_0 \cdot j\omega C} + \sqrt{j\omega L_0 \cdot j\omega C}\,\frac{1}{2}\frac{R_{AC}}{j\omega L_0} \qquad [3.110]$$

Combine all the L_0 and C elements in the right-hand term, and make the substitution $\sqrt{L_0/C} \to Z_0$:

$$\gamma(\omega) \approx \sqrt{j\omega L_0 \cdot j\omega C} + \frac{1}{2}\frac{R_{AC}}{Z_0} \qquad [3.111]$$

On the right, substitute your assumption [2.50] about R_{AC}, and on the left factor $j\omega$ out of the square root:

$$\gamma(\omega) \approx j\omega\sqrt{L_0 C} + \frac{(1+j)}{2}\frac{R_0}{Z_0}\sqrt{\frac{\omega}{\omega_0}} \qquad [3.112]$$

3.7.4 • Propagation Coefficient

What remains is a linear-phase term $j\omega\sqrt{L_0C}$, representing the bulk transport delay of the transmission-line medium, and a low-pass filter term whose attenuation in dB grows in proportion to the square root of frequency.[27]

The linear-phase term implies that the one-way propagation function H of a skin-effect-limited transmission line, for frequencies above ω_δ, acts as a large time-delay element. The bulk propagation delay in units of seconds per meter is given by

$$t_p \triangleq 1/v_0 = \sqrt{L_0 C} \quad \text{s/m} \qquad [3.113]$$

Equation [3.113] may also be stated in terms of the effective relative permittivity ϵ_{re} associated with the transmission line configuration $t_p = \sqrt{\epsilon_{re}}/c$ s/m, assuming the speed of light $c = 2.998 \cdot 10^8$ m/s and further assuming the relative magnetic permeability μ_r equals unity (see [2.28]).

The overall bulk transport delay for a transmission line varies in proportion to its length. Doubling the length doubles the bulk transport delay.

The transfer loss in nepers per meter is given by the real part of the propagation function. This value is called the *skin-effect loss coefficient*:

$$\alpha_r \triangleq \mathrm{Re}[\gamma(\omega)] = \frac{1}{2}\frac{R_0}{Z_0}\sqrt{\frac{\omega}{\omega_0}} \quad \text{neper/m} \qquad [3.114]$$

$$\alpha_r \triangleq \mathrm{Re}[\gamma(\omega)] = 4.34\frac{R_0}{Z_0}\sqrt{\frac{\omega}{\omega_0}} \quad \text{dB/m} \qquad [3.115]$$

The low-pass filtering action of the skin effect implies that the step response will be dispersed, slurring crisp rising edges into edges of limited rise and fall time, according to the properties of the filter.

The magnitude of H is determined from the real part of the propagation function (see [2.16]):

$$|H(\omega,l)| = e^{-l\cdot\mathrm{Re}(\gamma(\omega))} = e^{-l\cdot\frac{1}{2}\frac{R_0}{Z_0}\sqrt{\frac{\omega}{\omega_0}}} \qquad [3.116]$$

where $|H(\omega,l)|$ is the magnitude of the low-pass filter function implied by [3.112],

l is the length of the transmission line in meters,

ω_0 is the frequency at which AC line parameters are specified, rad/s,

R_0 is the AC resistance of the line at frequency ω_0, and

Z_0 is the characteristic impedance of the line at frequency ω_0.

---

[27] Remember that the real part of the propagation coefficient shows the attenuation in nepers. *Increasing* attenuation at high frequencies indicates a low-pass filtering effect, *decreasing* the propagation function H (see [3.14]).

The signal loss represented by H, measured in nepers (or decibels), varies in proportion to the length of the transmission line and in proportion to the square root of frequency. Double the distance yields twice the loss in neper or decibel units (one neper equals 8.6858896 dB). Doubling the frequency multiplies the loss (in neper or dB units) by the square root of two.

Binary signaling tolerates a tilt of no more than about 3 dB (and certainly never more than 6 dB) in the channel attenuation over the band occupied by the coded data. Any tilt greater than that amount must be flattened out with an equalizer (see Section 3.14, "Linear Equalization: Long Backplane Trace Example").

Be careful with your units conversions when calculating attenuation, as the data rate f for most systems is expressed in Hertz, while the frequency argument for the function $H(\omega)$ is specified in radians per second. You must multiply f by 2π to get ω.

The properties in Table 3.6 hold only for transmission lines operated at frequencies well above the skin-effect onset ω_δ but below the onset of the dielectric loss region. Subsequent sections discuss the implications of dielectric loss, which further increases line attenuation.

Table 3.6—Summary of Skin-Effected-Limited Transmission Line Properties at Frequencies Well Above ω_δ

Property	Formula	Ref. equation
Asymptotic value of characteristic impedance	$Z_0 = \sqrt{\dfrac{L_0}{C}}$ Ω	[3.106]
Bulk transport delay per meter[1]	$t_p \triangleq 1/v_0 = \sqrt{L_0 C} = \sqrt{\epsilon_{re}}/c$ s/m	[3.113]
Attenuation in nepers per meter (1 neper = 8.6858896 decibels)	$\alpha_r \triangleq \dfrac{1}{2}\dfrac{R_0}{Z_0}\sqrt{\dfrac{\omega}{\omega_0}}$ neper/m	[3.114]
Attenuation in decibels per meter	$\alpha_r \triangleq 4.34 \dfrac{R_0}{Z_0}\sqrt{\dfrac{\omega}{\omega_0}}$ dB/m	[3.115]
Transfer gain at length l	$\|H(\omega,l)\| = e^{-l \cdot \frac{1}{2}\frac{R_0}{Z_0}\sqrt{\frac{\omega}{\omega_0}}}$	[3.116]
Inductance per meter	$L_0 = Z_0/v_0$ H/m	[3.5]
Capacitance per meter	$C = \dfrac{1}{Z_0 v_0}$ F/m	[3.6]

NOTE (1) —TIA/EIA 568-B specifications suggest following more detailed approximation for phase delay on long cables operated within the skin-effect limited region: $-\dfrac{\beta}{\omega} \approx \sqrt{L_0 C} + \dfrac{1}{2}\dfrac{R_0}{Z_0}\sqrt{\dfrac{1}{\omega_0 \omega}}$. This expression may be derived from [3.116]. A related expression exists for group delay within the skin-effect limited region: $-\dfrac{d\beta}{d\omega} \approx \sqrt{L_0 C} + \dfrac{1}{4}\dfrac{R_0}{Z_0}\sqrt{\dfrac{1}{\omega_0 \omega}}$.

POINTS TO REMEMBER

> The attenuation (in dB) within the skin-effect region grows in proportion to the square root of frequency.
> Doubling the length of a skin-effect-limited transmission line doubles the attenuation.

3.7.5 Possibility of Severe Resonance within Skin-Effect Region

Transmission lines operated in the skin-effect region fall prey to the same resonance difficulties that afflict structures operated in the LC region. The termination means used to conquer resonance are the same as those described in Section 3.6.6, "Terminating an LC Transmission Line."

One difference, however, between the severity of resonance associated with a pure LC structure and the severity of the resonance associated with a skin-effect-limited structure has to do with the additional attenuation present due to the skin effect. In the equation for overall system gain G, the height of the resonant peaks is determined by the depth of the cancellation in the denominator. If the magnitude of the one-way propagation function H evaluated at the frequency of resonance indicates a substantial amount of attenuation, then the magnitude of the term $H^2\Gamma_1\Gamma_2$ can never approach unity, and the resonance will fail to materialize.

In the frequency domain the skin effect causes a lessening of the heights of the resonant peaks in Figure 3.22, with the heights (in dB) of higher resonant modes falling off inversely with the square root of frequency.

In practice it is impossible to incorporate into a transmission structure sufficient internal loss to satisfactorily limit the resonance (i.e., reflections) while also providing a reasonably clean, full-sized signal at the far end of the structure. The skin-effect losses do, however, have the beneficial effect of reducing (somewhat) the requirements for efficacy of the intentional termination placed in the circuit.

Example of Limitation on Size of Resonance

Suppose on an open-circuited line of length 0.1 meter (3.9 in.) the first resonance $f_1 = 1/(4v_0)$ occurs at 754 MHz. Further assume that at that frequency the magnitude of $H(\omega)$ equals 1/2 (i.e., –6 dB.). The magnitude of $H^2\Gamma_1\Gamma_2$ therefore cannot exceed 1/4, and the worst-case resonance therefore will not exceed $1/(1-1/4) = 4/3$, or about 33% overshoot.

I should caution you against the wisdom of designing any system whose proper operation depends upon some "minimum attenuation" within the interconnections. It is quite possible that while your system as it is implemented today may enjoy the benefits of a minimum amount of attenuation within every interconnection, this property may not be true in the future, particularly if the technology used to construct the interconnections is upgraded. One excellent class of examples of systems that have been designed assuming some minimum attenuation, and then rendered *less* robust by the use with a lower-attenuation class of cabling, would be those LAN systems designed for use with category-III unshielded twisted-pair cabling. This cabling infrastructure was upgraded in many companies beginning in

1995 to a newer, lower-attenuation form of supposedly backwards-compatible cabling called category-5 unshielded twisted-pair cabling.

Another good example would be transceivers designed for FR-4 pcb substrates that may someday be used in systems with low-dielectric loss substrates like Getek.

As a final example, the original T1 connections deployed by the telephone company depended upon a minimum attenuation property. In the original version of that system each line was padded by hand, using special circuitry to guarantee a consistent line loss (and frequency response) independent of line length on every wire pair. The receivers therefore could incorporate a large fixed gain without concern for overloading. That sort of hand tweaking and testing on every individual wire pair is no longer economically feasible.

3.7.5.1 Subtle Differences Between Termination Styles

Figure 3.25 is repeated here showing the inclusion of skin-effect losses (Figure 3.30). All three traces clearly show dispersion of the rising and falling edges caused by the low-pass filtering properties of the skin-effect resistance.

3.7.5.2 Application of Termination Equations to Other Regions

The LC, skin-effect, and dielectric-loss-limited regions all share the same asymptotic high-frequency value of characteristic impedance Z_0. The same termination approaches therefore work for all three regions, delivering in each case an overall circuit gain G nearly equal to the one-way propagation function H applicable within each region.

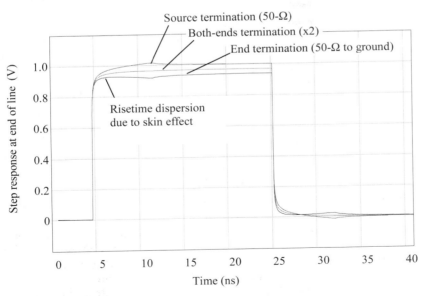

Figure 3.30—Different termination styles react differently to skin-effect resistance.

POINT TO REMEMBER

> Transmission lines in the skin-effect region fall prey to the same resonance difficulties that afflict the LC region and respond to the same means of termination.

3.7.6 Step Response (Skin-Effect Region)

This section presents the step response of a skin-effect-limited transmission line. The applicability of these results hinges on two crucial assumptions:

1. The risetime of the step input to the system must be substantially faster than the skin-effect risetime (meaning that the shape of the output waveform is determined primarily by the system itself and not strongly influenced by the precise risetime of the input step), and
2. The impedance of the line's series inductance greatly exceeds its series resistance (so that you have a true low-loss transmission media and not an RC dispersion line).

The solution requires knowledge of these parameters:

> Z_0, the nominal characteristic impedance of the transmission line at high frequencies (excluding the effect of the internal inductance of the wires), in ohms,

> ω_0, some particular frequency at which the skin-effect resistance is specified (rad/s),

> R_0, the series skin-effect resistance (Ω/m) at frequency ω_0, and

> l, the length of the line (m).

Beginning with [3.112], substitute $1/v_0$ for the radical $\sqrt{L_0 C}$ and make the substitution $(1+j) \to \sqrt{2j}$.

$$\gamma(\omega) \approx \frac{j\omega}{v_0} + \frac{1}{2}\frac{R_0}{Z_0}\sqrt{\frac{2j\omega}{\omega_0}} \qquad [3.117]$$

Multiply the result by the (negative of the) transmission line length l and exponentiate (as in [3.14]) to find H.

$$H(\omega) \approx e^{-l \cdot \frac{j\omega}{v_0}} e^{-l \cdot \frac{1}{2}\frac{R_0}{Z_0}\sqrt{\frac{2j\omega}{\omega_0}}} \qquad [3.118]$$

The first exponential term in [3.118] represents a linear-phase bulk transport delay. This delay, for the purposes of investigating attenuation and dispersion of the propagated signals, may be ignored. The second exponential term encodes the low-pass filter function associated with the skin effect. Call this filter $H'(\omega)$. Normalize $H'(\omega)$ by substituting into it the arbitrary constant $\tau = \dfrac{2}{\omega_0} \dfrac{l^2 R_0^2}{4 Z_0^2}$.

$$H'(\omega) \approx e^{-\sqrt{j\omega\tau}} \qquad [3.119]$$

To compute the *step response* of the skin effect numerically, multiply [3.119] by $1/j\omega$ and then apply an inverse Fourier transform. Multiplying by $1/j\omega$ converts [3.119] from an impulse-response to a step-response. Evaluating the inverse Fourier transform of the result produces a time-domain waveform that represents the step response of the skin effect.

The following well-known Fourier transform pair accomplishes the calculation of the skin-effect step response.[28,29]

$$\mathcal{F}^{-1}\left(\dfrac{1}{j\omega} e^{-\sqrt{j\omega\tau}}\right) = u(t)\,\mathrm{erfc}\left(\dfrac{1}{2}\sqrt{\dfrac{\tau}{t}}\right) \qquad [3.120]$$

where $u(t)$ is the unit step function, equal to one for all $t \geq 0$ and zero otherwise,
erfc( ) is the complementary error function, defined in Appendix E.
the symbol $\mathcal{F}^{-1}$ represents an inverse Fourier transformation, and
τ may take on any positive value.

The normalized skin-effect time-and-frequency-response transform pair illustrated in Figure 3.31 conforms exactly to equation [3.120] with parameter $\tau = 1$. The top two figures show the magnitude of the propagation function (in dB) of a long, perfectly terminated skin-effect-limited channel. The bottom two figures show the step response of the channel, one with a logarithmic time scale, the other with a linear timescale.

Assuming you have calculated the value τ used in [3.119], you may read from these charts directly the frequency response (magnitude) and step response (assuming a normalized 1-volt step input with zero risetime) for any skin-effect-limited system.

Table 3.7 presents selected values from the charts in Figure 3.31. Two rules of scaling apply to the quantities in this table:

---

[28] The author has been informed that a trivial variant of this transform pair is found in the 10th edition of the *CRC Handbook of Chemistry and Physics,* in the table of transforms, item number 84, although not in more modern editions of this book. The transform pair is also found in the appendix of R.V. Churchill's text *Operational Mathematics,* 3rd edition, also denoted 84. A more modern reference is found in the *Table of Integrals, Series and Products,* 5th ed., 1994 (translated from Russian) by Gradshteyn and Ryzhik, although in their listing they call out the use of the function erf( ) instead of erfc( ), which appears to be a mistake.

[29] The identification of the step response associated with the skin effect was first made by R. L. Wigington and R. S. Nahman, "Transient analysis of coaxial cables considering skin effect," Proc. IRE, vol 45, pp. 166–174; Feb. 1957. Their work was conducted in the context of the development of nuclear instrumentation for test-blast monitoring. The specific problem at hand had to do with whether signals from the blast sensors could travel down a coaxial cable to the test shack faster than the radiant wave of heat from the blast vaporized the cable.

3.7.6 • Step Response

1. The relative ratios among the time-domain measurements listed apply equally to any skin-effect-limited channel.
2. The product of any time-domain measure and any frequency-domain measure (for example, $T_{10\text{-}90\%} \cdot F_{3dB}$) remains constant for any skin-effect-limited channel.

Notice that the skin effect acts over a very wide scale of times. At the beginning of the step response, the slope is very sharp. Towards the end, the step response has a long, slow-moving tail. Any equalizing circuit built to undo the skin effect must be capable of producing subtle variations in its propagation function over a correspondingly wide range of frequencies.

The long settling time of the skin-effect phenomenon has serious implications for testing. Any measurement setup used to research skin-effect behavior must provide enough time for the skin-effect response to fully settle to 99% or more of its final value. Experiments involving truncated step-response waveforms will be missing significant portions of the overall step response waveform and will not reproduce the same risetime ratios as predicted in Table 3.7.

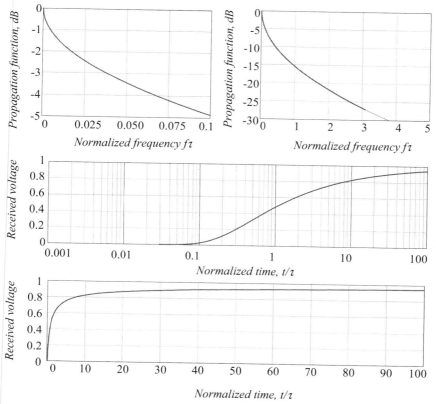

Figure 3.31—The step response associated with the skin effect has a quick rise and a long, sloping tail.

Table 3.7—Selected Values of Skin-Effect Step Response

Percentile point	Time to achieve (s)
T_{10}	$0.1848\,\tau$
T_{20}	$0.3044\,\tau$
$T_{47.9500}$	$1.0000\,\tau$
T_{50}	$1.0990\,\tau$
T_{80}	$7.7900\,\tau$
T_{90}	$31.6643\,\tau$
T_{95}	$127.159\,\tau$
T_{97}	$353.52\,\tau$
T_{98}	$795.63\,\tau$
T_{99}	$3183.03\,\tau$
Attenuation point	**Frequency at which point is reached (Hz)**
$F_{-3\text{dB}}$	$0.0379/\tau$
$F_{-6\text{dB}}$	$0.1518/\tau$

Similarly, bit-error-rate measurements taken with a pseudorandom pattern insufficiently long to exercise the full range of skin-effect behavior will return overly optimistic results.

Example of Insufficiently Long Pseudorandom BER Test

The skin-effect band for Belden type 8237 coaxial cable extends from 24 KHz to 5.9 GHz. Suppose you are working with a section of this cable whose length creates a skin-effect risetime of $T_{50} = 1$ ns. How long does the step response take to reach 99% of full value? In Table 3.7 the ratio T_{99}/T_{50} is $3183/1.099$; therefore, the time to reach 99% is $(3183/1.099)\cdot(1\text{ ns})$, which equals 2896 ns.

If the system bit time is 100 ns (10 Mb/s), then it takes (in theory) 29 bits for the skin-effect step response to reach 99% of its full value. A good pseudorandom BER test should therefore include a pattern with at least 29 bits in a row held low, followed by a transition, and at least 29 bits in a row held high, followed by a transition. This would seem to indicate a need for at least a 29-bit pseudorandom pattern generator.

A 7-bit (128-length) pattern generator holds at most 7 bits in a row high or low, at which point the skin-effect response progresses to only 97% or 98% of its final value, leaving untested the final 2% to 3% of intersymbol interference (ISI).

The value τ associated with [3.119] and [3.116] varies with the square of distance; therefore, the risetime of a skin-effect-limited channel scales with the square of its length. To see the relation directly, substitute the definition of τ into the expression in Table 3.7 for T_{50}.

$$T_{50} = 1.099\left[\frac{2}{\omega_0}\frac{l^2 R_0^2}{4Z_0^2}\right] \quad [3.121]$$

where ω_0 is the particular frequency at which the skin-effect resistance is specified (rad/s),

l is the length of the line (m),

R_0 is the series skin-effect resistance (Ω/m) at frequency ω_0, and

Z_0 is the nominal characteristic impedance of the transmission line at high frequencies (excluding the effect of the internal inductance of the wires), (Ω).

The square of length is a very fast-changing function. If you cut in half the length of a skin-effect-limited transmission channel, you decrease its risetime fourfold. Cutting the length to 1/10 shortens the risetime by a factor of 100.

The skin-effect risetime also scales with the square of R_0, which is inversely proportional to the perimeter of the signal conductor. Therefore, a conductor twice the diameter (or width) has 1/2 the AC resistance and 1/4 the skin-effect risetime. This scaling principle explains the tremendous bandwidth advantages of large coaxial cables. It also makes clear that the difference in skin-effect risetime between a 75-micron (3-mil) trace and a 150-micron (6-mil) trace is a factor of four.

Beware that the skin-effect and dielectric-loss effects mix together over a *very* broad range of frequencies. It's rare in pcb problems that you see skin-effect losses without also having to take into account dielectric dispersion. The best general rule for combining dispersion due to different sources is the sum-of-squares rule, which works as follows.

$$t_{10-90\%}^{\text{skin effect}+\text{dielectric loss}} \approx \sqrt{\left(t_{10-90\%}^{\text{skin effect}}\right)^2 + \left(t_{10-90\%}^{\text{dielectric loss}}\right)^2} \qquad [3.122]$$

If you seek a more exact answer, take the skin-effect and dielectric-loss impulse responses outlined in this and the next section and convolve them together. Or better yet, use the complete frequency-domain model from the previous chapter, sample it on a dense grid of frequencies, and inverse-FFT to get the complete time-domain step response.

POINT TO REMEMBER

> The step response associated with the skin effect has a quick rise and a long, sloping tail.

3.7.7 Tradeoffs Between Distance and Speed (Skin-Effect Region)

Given a fixed receiver architecture, and assuming the BER of the receiver is dominated by errors due to intersymbol interference, the speed of operation within the skin-effect region scales inversely with the square of transmission-line length. This tradeoff between length and speed arises because the distributed skin effect produces attenuation (in dB) proportional to the square root of frequency. For example, a 10% increase in length increases the attenuation (in dB) by 10%. To regain a normal signal amplitude at the end of a transmission line limited by skin-effect losses, the 1% increase in attenuation may be offset by a 20% reduction in the system operating speed.

POINTS TO REMEMBER

- The risetime of a skin-effect-limited channel scales with the square of its length.
- The speed of operation within the skin-effect region scales inversely with the square of transmission-line length.
- A conductor twice the diameter (or width) has 1/2 the AC resistance and thus 1/4 the skin-effect risetime.
- It's rare in pcb problems that you see skin-effect losses without also having to take into account dielectric dispersion.

3.8 DIELECTRIC LOSS REGION

Dielectric losses are present at all frequencies, growing progressively more severe at higher frequencies. These losses become noticeable only when they rise to a level comparable with the resistive losses, a point after which the line is said to operate in the dielectric-loss-limited region (Section 3.2).

3.8.1 Boundary of Dielectric-Loss-Limited Region

For typical transmission media used in macroscopic digital applications (i.e., not on-chip), the frequency ω_δ falls low enough that the losses due to dielectric absorption *at that frequency* are swamped by skin-effect losses. As the frequency is increased, however, the skin-effect loss grows only in proportion to the square root of frequency, while the dielectric loss grows at a faster rate in direct proportion to frequency. Above some frequency ω_θ the dielectric loss equals, and then exceeds, the skin-effect loss.

The derivation of ω_θ begins with simple approximations for the skin effect α_r and dielectric loss α_d, in units of nepers per meter.

$$\alpha_r \triangleq \frac{1}{2}\frac{R_0}{Z_0}\sqrt{\frac{\omega}{\omega_0}} \quad \text{neper/m} \qquad \text{see equation [3.114]}$$

$$\alpha_d \triangleq \frac{1}{2}\frac{\theta_0 \omega}{v_0}\left(\frac{\omega}{\omega_0}\right)^{-\theta_0/\pi} \quad \text{neper/m} \qquad \text{see equation [3.131]}$$

Equating the preceding two expressions while neglecting the slowly varying term $(\omega/\omega_0)^{-\theta_0/\pi}$ generates the following formula for the crossover frequency ω_θ.

$$\omega_\theta = \frac{1}{\omega_0}\left[\frac{v_0}{Z_0}\frac{R_0}{\theta_0}\right]^2 \qquad [3.123]$$

3.8.1 • Boundary of Dielectric-Loss-Limited Region

where ω_b is the frequency beyond which dielectric losses exceed skin effect losses, rad/s,

ω_0 is any arbitrary frequency at which the AC resistance may be specified, rad/s,

v_0 is the velocity of propagation at ω_0, m/s,

Z_0 is the characteristic impedance of the transmission line at ω_0, Ω,

R_0 is the series AC resistance at ω_0, Ω/m, and

$\tan \theta_0$ is the loss tangent of the dielectric material at frequency ω_0.

NOTE: For differential configurations, define R_0 to represent the sum of the AC resistances of both outbound and returning conductors, and Z_0 as the impedance of the differential transmission line thus formed.

In coaxial and twisted-pair transmission lines constructed from modern, low-loss dielectric materials the dielectric absorption at frequencies below 1 GHz accounts for only a small portion of the overall loss, but at some point in the range 1 to 10 GHz it can become quite significant. Radiation losses from such conducting structures, while very important to the radiated emissions problem, are generally so tiny that for signal integrity purposes they may also be ignored.

In the pcb realm dielectric losses can emerge as a significant problem at frequencies below 1 GHz due to the use of poor dielectric materials (like FR-4).

Beware the width of the mixing zone between the skin-effect and dielectric-effect regions. It is extremely broad, because the relative difference between skin effect and dielectric losses (in dB/m) changes only as fast as the square root of frequency, and also because the effects add directly "in phase." At a frequency 10 times lower than ω_b, skin-effect losses exceed dielectric losses by only a factor of $\sqrt{10} = 3.16$. The dielectric loss at this frequency contributes fully 24% of the overall attenuation. At a frequency 10 times higher than ω_b, the situation reverses. The dielectric loss at this frequency contributes 76% of the overall attenuation. Assuming the transmission performance is dominated by only resistive (i.e., skin-effect) and dielectric losses, Table 3.8 lists the percentage of total signal loss contributed by each factor.

The distinguishing feature of the dielectric-loss region is that within this region the characteristic impedance remains fairly flat, while the line attenuation in dB varies in direct proportion to frequency.

POINT TO REMEMBER

➢ Skin-effect loss grows only in proportion to the square root of frequency, while the dielectric loss grows in direct proportion to frequency. Above some frequency ω_b the dielectric loss equals, and then exceeds, the skin-effect loss.

3.8.2 Characteristic Impedance (Dielectric-Loss-Limited Region)

The full expression for characteristic impedance (neglecting the conductance G) includes contributions from the external inductance L_0, resistance R, and capacitance of a transmission line.

$$Z_C = \sqrt{\frac{j\omega L_0 + R(\omega)}{j\omega C(\omega)}} \qquad [3.124]$$

In the limit as you proceed to frequencies far above ω_{LC}, the contribution of the term R becomes negligible, leading to this approximation for characteristic impedance at any frequency above the skin-effect onset.

$$\text{(skin-effect assumption)} \quad Z_C \triangleq \lim_{\omega \to \infty} \sqrt{\frac{j\omega L_0 + R(\omega)}{j\omega C(\omega)}} = \sqrt{\frac{L_0}{C(\omega)}} \qquad [3.125]$$

The only difference between this and the previous equation [3.85] is that here I explicitly provide for the possibility that the resistance $R(\omega)$ and capacitance $C(\omega)$ may each change with frequency, and the inductance term is clearly marked L_0 to represent only the *external inductance* of the transmission configuration (see Section 2.7, "Skin-Effect Inductance" and Section 2.8, "Modeling Internal Impedance"). External inductance is the value of series inductance, in Henries per meter, computed by a two-dimensional field solver under the assumption that current rides on the surface of each conductor without penetrating its bodies.

Table 3.8—Relative Skin and Dielectric Losses

Relative operating frequency: ω/ω_θ (rad/s)	Percentage of loss contributed by skin effect	Percentage of loss contributed by dielectric effect	Loss slope[1]
.0001	99	1	.505
.001	97	3	.515
.01	91	9	.545
.1	76	24	.62
1	50	50	.75
10	24	76	.88
100	9	91	.955
1000	3	97	.985
10000	1	99	.995

NOTE (1)—The loss slope is defined as the slope of a curve showing the log of dB attenuation along the vertical axis and the log of frequency along the horizontal, as in Figure 3.1. A loss slope of α indicates the transmission loss in dB is growing proportional to f^α at that frequency (see Appendix B).

3.8.2 • Characteristic Impedance

The lumped-element, RC, and LC operating analyses presented previously assume that both L and R are constants, so it is therefore not generally necessary in those presentations to distinguish between the internal and external inductance, only lumping together all the inductance you have of either type under the umbrella term "L." Here the term $L_0 \triangleq Z_0/v_0$ from Equation [3.5] is used to represent the external inductance L_e. The *internal inductance* of the signal conductors is represented by variations in the imaginary part of $R(\omega)$.

The capacitance model [3.7] expresses the dielectric properties of the insulating medium surrounding your conductors as a single complex-valued function $C(\omega)$. The real part of $C(\omega)$, when multiplied by the operating frequency $j\omega$, represents displacement current flowing within the insulator. The total displacement current (in unit of A/m) is proportional to $j\omega \operatorname{Re}(C(\omega))V$, where V is the voltage impressed upon the transmission line at any one point. The displacement current flows in quadrature with the applied electric field. The displacement current effect acts like what you normally think of as capacitance.

The imaginary part of $C(\omega)$, when multiplied by the operating frequency $j\omega$, generates a conduction current that flows in phase with the applied voltage. It is a peculiar coincidence that the magnitude of conduction current in the sorts of insulators used for high-speed digital applications happens to vary in almost direct proportion to frequency—that's what makes it reasonable to approximate the conduction current using a model like $j\omega(j\cdot\operatorname{Im}(C(\omega)))V$. Those persons accustomed to seeing a conductance parameter G in the transmission equations will recognize that the term $j\omega(j\cdot\operatorname{Im}(C(\omega)))$ plays the role of G in [3.124] and [3.125]. Because the imaginary part of $C(\omega)$ from [3.7] is always negative, the term $j\omega(j\cdot\operatorname{Im}(C(\omega)))$, representing the conductance in S/m, is positive.

As your operating frequency crosses upward from the skin-effect region to the dielectric region, the variations in $C(\omega)$ change (slightly) the precise shapes of the asymptotically convergent curves in Figure 3.17, but not the asymptotic points of convergence (Figure 3.27). As long as your circuit does not depend on the precise shapes of the curves, the changes are of little practical significance.

If the precise shape of the impedance curve is of importance to your design (for example, in the construction of ultra-accurate terminations it would be crucial), then you should know that the skin-effect losses and dielectric losses have opposite effects on the shape of the characteristic impedance curve.

The plots starting with Figure 3.32 and going through Figure 3.35 all depict a hypothetical pcb trace having a fixed DC resistance of 6.4 Ω/m plus the skin effect, and with a dielectric loss tangent of 0.025. The trace is a 50-ohm stripline constructed of 1/2-oz copper (17.4 micron thickness) on FR-4 with a width of 150 microns (6 mils, $v_0 = 1.446 \cdot 10^8$ m/s, $L_0 = 346$ nH/m, $C_0 = 138$ pF/m). The trace is 0.5 meters long.

Figure 3.32 illustrates the relative influence of skin effect and dielectric losses on the characteristic impedance of a lossy line. The chart depicts the characteristic impedance of a trace with only skin-effect and DC resistive losses (assuming a perfect dielectric), a trace with only dielectric losses (assuming zero resistance), and a combination of both.

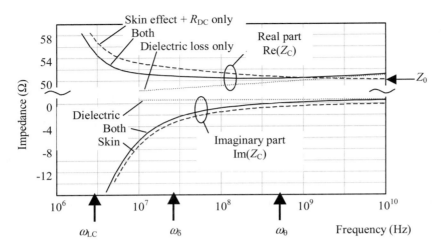

Figure 3.32—In the vicinity of the skin-effect onset, skin-effect and dielectric losses affect the characteristic impedance in opposite directions.

The topmost three curves in each case show the real part of impedance, and the bottom three curves, the imaginary part. Compared to a base case that includes no losses of any type, skin-effect losses *increase* the real part of the impedance curve in the vicinity of the skin-effect onset at ω_δ, while the dielectric losses *decrease* the impedance in the same area. The two effects almost cancel each other in the band near ω_δ, causing the characteristic impedance in this region to approach the asymptotic value marked Z_0 more rapidly and more completely than when either effect is present alone. The cancellation is nothing more than a grand coincidence. Do not depend on the cancellation effect to stabilize the impedance of a practical design, as the values of dielectric loss in most materials vary substantially with temperature and water content.

At frequencies above the onset of the dielectric-loss-limited mode ω_θ, dielectric losses ultimately force the characteristic impedance back up above Z_0. This happens because of the slow deterioration in line capacitance with increasing frequency caused by the dielectric loss.

The general relationship between the characteristic impedance and input impedance is given by [3.33] and discussed in Section 3.5.2 "Input Impedance (RC Region)" and in Section 3.2.1 "A Transmission Line Is Always a Transmission Line."

POINTS TO REMEMBER

> In the vicinity of the skin-effect onset ω_δ the skin effect *increases* characteristic impedance while dielectric loss *decreases* it.

> At frequencies above the onset of the dielectric-loss-limited mode ω_θ, dielectric losses ultimately force the characteristic impedance back up above Z_0.

3.8.3 Influence of Dielectric Loss on TDR Measurement

Dielectric losses progressively diminish the available line capacitance as you move to higher and higher frequencies, thereby causing an upward tilt to a plot of characteristic impedance versus frequency. In the time domain, the tilt suggests that the effective impedance measured over very short scales of time should exceed that measured at larger scales of time. Figure 3.33 illustrates precisely that effect.

The detailed blow-up of the first plateau in the TDR response for a lossy transmission line shows three waveforms: the response of a trace with only skin-effect and DC resistance losses (assuming a perfect dielectric), a trace with only dielectric losses (assuming zero resistance), and a combination of both.

The dielectric losses produce a negative slope in the first plateau. The resistive losses create a positive slope. Working together, the two effects almost cancel, in this particular example, creating a slope less steep than when either effect is present alone—the same peculiar coincidence mentioned in the previous section. It cannot be depended upon to stabilize the impedance of a practical design, as the values of dielectric loss in most materials vary substantially with temperature and water content.

Dielectric loss distorts the slope in the first plateau, obliterating your ability to accurately infer the resistance of the line from a single TDR plot.

For any open-circuited transmission line,[30] the second plateau of the TDR response shows a signal which has traversed the line twice, going to the far end and then reflecting

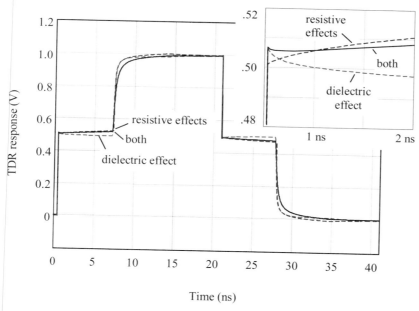

Figure 3.33—The dielectric effect starts with a higher impedance and then trends lower, while the resistive effects do the opposite.

---

[30] The usual condition of measurement for checking the impedance of a pcb test coupon.

back. This reflected signal displays the marks of its round-trip venture in the form of a noticeable dispersion of both rising and falling edges.

POINT TO REMEMBER

> Dielectric losses cause an upward tilt to a plot of characteristic impedance versus frequency. Resistive losses create a neagtive slope. Working together, the two effects can sometimes almost cancel, creating a TDR slope less steep than when either effect is present alone.

3.8.4 Propagation Coefficient (Dielectric-Loss-Limited Region)

Figure 3.34 shows the propagation coefficient for the same hypothetical pcb trace as Figure 3.32, but this time in addition showing dielectric and skin effect loss. This figure plots the real and imaginary parts of the propagation coefficient versus frequency. The plot is drawn with log-log axes to highlight the polynomial relationships between portions of the curves.

In the RC region below ω_{LC} both the real part of the propagation coefficient (log of attenuation) and the imaginary part (phase in radians) rise together in proportion to the square root of frequency. Above ω_{LC} the imaginary part (phase) grows linearly with increasing frequency, while the real part (attenuation) inflects to the right at a lesser slope. The dotted line extending horizontally to the right illustrates the effect on the real part of the propagation coefficient of DC resistance alone. The DC resistance produces a loss coefficient that is constant with frequency. Combining the DC resistance with skin-effect losses produces the dashed line marked "With skin effect."

The chart splits out the real and imaginary parts of the dielectric effect separately, showing them as dashed lines with a slope of +1. These curves represent the phase and magnitude coefficients that would accrue to a hypothetical transmission line with dielectric losses only, but no resistance. Comparing the dashed line for pure dielectric loss (marked "Dielectric loss") with the dashed line representing pure resistive losses (marked "With skin

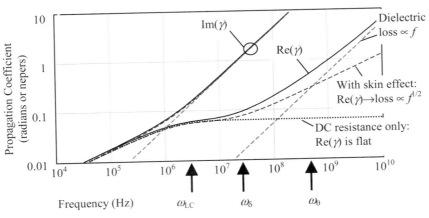

Figure 3.34—Below ω_θ the resistive losses dominate. Above ω_θ the dielectric losses matter most.

3.8.4 • Propagation Coefficient

effect"), you can see that below the frequency marked ω_θ the losses are dominated by resistive effects, and above ω_θ by dielectric effects.

In the skin-effect region the overall line attenuation (solid line marked "Re(γ)") grows proportional to the square root of frequency. In the absence of dielectric losses the attenuation would continue with the same slope forever. Once dielectric losses enter the picture, however, the slope becomes steeper. Above the point ω_θ where the dielectric losses take effect the attenuation grows in direct proportion to frequency.

In this example all the regions are crammed fairly close together with the result that the attenuation curve does not show explicit regions with perfect slopes. Instead, it describes a smooth arc with graduated changes in slope between regions.

In the dielectric region the decoupling of phase and attenuation is less complete than in the LC and skin-effect regions. Above ω_θ the ratio of phase (radians) to attenuation (nepers) never exceeds $2/\theta$, limiting the maximum Q of any resonator you might try to build. Because microwave designers like to make high-Q circuits, this limitation renders high-loss materials, like FR-4, practically useless for their applications. Digital designers, on the other hand, who strive to build low-Q, nonresonant circuits, discover that FR-4 is a very useful material even up to frequencies as high as 10-GHz, provided that their traces are kept sufficiently short that the line attenuation does not become a problem.

Terminations are used within the dielectric region to abate resonance on long transmission lines in precisely the same manner as used in LC regions.

You can predict the phase and amplitude response of a dielectric-loss-limited transmission line beginning with the definition of the propagation function [3.13]. To make this calculation, you must assume operation at a frequency so far in excess of ω_θ that the dielectric losses dominate the performance of the circuit so that the resistive (skin-effect and DC) losses may be safely ignored. In practice such a situation is created when a large, fat conductor is combined with a high-loss dielectric material. For example, a microstrip trace 5000 microns wide (200 mils), as might be used in a microwave application to mitigate skin-effect loss, if implemented on an FR-4 substrate, would enter the dielectric region at a frequency just below 10 MHz. The propagation coefficient above 1 GHz would be 91% dominated by the dielectric effect.

Rewriting [3.13] to ignore the resistive effects, and substituting [3.7][31] for the capacitive term,

$$\gamma \approx \sqrt{(j\omega L_0)\left(j\omega C_0 (j\omega/\omega_0)^{-\frac{2\theta_0}{\pi}}\right)} \qquad [3.126]$$

Group the ordinary inductive and capacitive terms, substitute $\sqrt{L_0 C_0} \rightarrow 1/v_0$, and remember to divide the exponent $(2\theta_0/\pi)$ by two as it is pulled out from under the radical.

$$\gamma \approx \frac{j\omega}{v_0}\left(\frac{j\omega}{\omega_0}\right)^{-\theta_0/\pi} \qquad [3.127]$$

---

[31] This equation makes the assumption that the loss tangent is constant with frequency.

Next isolate the exponential j term.

$$\gamma(\omega) = \frac{j\omega}{v_0}\left(\frac{\omega}{\omega_0}\right)^{-\theta_0/\pi}(j)^{-\theta_0/\pi} \qquad [3.128]$$

Assuming θ_0 to be a small positive constant, you may represent the term $(j)^{-\theta_0/\pi}$ by a linear expression. For θ_0 less than 0.1 the error in this approximation is less than one part in 800.

$$\gamma(\omega) \approx \frac{j\omega}{v_0}\left(\frac{\omega}{\omega_0}\right)^{-\theta_0/\pi}\left(1 - j\frac{\theta_0}{2}\right) \qquad [3.129]$$

The imaginary portion of [3.129] is a linear-phase term very nearly equal to $j\omega/v_0$. This term represents the bulk transport delay of the transmission-line medium. The real portion of [3.129] is the same but a factor of $\theta_0/2$ smaller. It represents a low-pass filter whose attenuation in dB grows in direct proportion to frequency.[32]

The linear-phase term implies that the one-way propagation function H of a dielectric-loss-limited transmission line, for frequencies above ω_0, acts primarily as a large time-delay element. The bulk propagation delay in units of seconds per meter for frequencies near ω_0 is given by

$$t_p \triangleq 1/v_0 = \sqrt{L_0 C_0} \qquad [3.130]$$

Equation [3.130] may also be stated in terms of the effective relative permittivity ϵ_{re} associated with the transmission-line configuration $t_p = \sqrt{\epsilon_{re}}/c$ s/m, assuming the speed of light $c = 2.998 \cdot 10^8$ m/s and further assuming the relative magnetic permeability μ_r equals one (see [2.28]).

The bulk transport delay varies in proportion to the length of the transmission line and slightly with frequency (see Section 3.8.6, "Step Response (Dielectric-Loss-Limited Region)"). Doubling the length doubles the bulk transport delay.

The transfer loss in nepers per meter is given by the real part of the propagation function. This value is called the *dielectric loss coefficient*:

$$\alpha_d \triangleq \mathrm{Re}[\gamma(\omega)] = \frac{1}{2}\frac{\theta_0 \omega}{v_0}\left(\frac{\omega}{\omega_0}\right)^{-\theta_0/\pi} \quad \text{neper/m} \qquad [3.131]$$

---

[32] Remember that the real part of the propagation coefficient shows the attenuation in nepers. *Increasing* attenuation at high frequencies indicates a *decreasing*, or low-pass filtering, propagation function H (see [3.14]).

3.8.4 • Propagation Coefficient

If you are operating in a fairly narrow band near ω_0, you may omit from your calculation the term $(\omega/\omega_0)^{-\theta_0/\pi}$ without much loss of accuracy. This is the standard practice for RF designers, who then convert the attenuation in nepers to dB, leading to this expression for the attenuation per meter:

$$\alpha_d = -20 \log |H(\omega)| = 4.34 \frac{\theta_0 \omega}{v_0} \quad \text{dB/m} \qquad [3.132]$$

The low-pass filtering action of the dielectric-loss effect implies that the step response will be dispersed, slurring crisp rising edges into edges of limited rise and fall time, according to the properties of the filter.

The magnitude of H is determined from the real part of the propagation function (see [2.16]):

$$|H(\omega,l)| = e^{-l \cdot \text{Re}(\gamma(\omega))} = e^{-l \cdot \frac{1}{2} \frac{\theta_0 \omega}{v_0} \left(\frac{\omega}{\omega_0}\right)^{-\theta_0/\pi}} \qquad [3.133]$$

where $|H(\omega,l)|$ is the magnitude of the low-pass filter function implied by [3.129],

l is the length of the transmission line in meters,

ω_0 is the frequency at which AC line parameters are specified, rad/s,

v_0 is the velocity of propagation at frequency ω_0, m/s, and

$\tan \theta_0$ is loss tangent of the dielectric material at frequency ω_0.

The signal loss represented by H, measured in nepers (or decibels), varies in proportional to the length of the transmission line and in proportion to the square root of frequency. Double the distance yields twice the loss in neper or decibel units (one neper equals 8.6858896 dB). Doubling the frequency multiplies the loss (in neper or dB units) by two.

Binary signaling tolerates a tilt of no more than about 3 dB (and certainly never more than 6 dB) in the channel attenuation over the band occupied by the coded data. Any tilt greater than that amount must be flattened out with an equalizer (see Section 3.14, "Linear Equalization: Long Backplane Trace Example"). Be careful with your units conversions when calculating attenuation, as the data rate f for most systems is expressed in Hertz, while the frequency argument for the function $H(\omega)$ is specified in radians per second. You must multiply f by 2π to get ω.

The properties in Table 3.9 hold only for dielectric-limited transmission lines operated at frequencies well above ω_θ and for which the dielectric loss tangent remains constant with frequency.

POINTS TO REMEMBER

> ➤ The attenuation (in dB) within the dielectric-loss-limited region grows in direct proportion to frequency.
> ➤ Doubling the length of a dielectric-loss-limited transmission line doubles the attenuation.

Table 3.9—Summary of Dielectric-Limited Transmission Line Properties at Frequencies Well Above ω_0

Property	Formula	Ref. equation
Asymptotic value of characteristic impedance	$Z_C = \sqrt{\dfrac{L_0}{C(\omega)}}\ \Omega$	[3.125]
Bulk transport delay per meter (for frequencies near ω_0)	$t_p \triangleq 1/v_0 = \sqrt{L_0 C_0} = \sqrt{\epsilon_{re}}/c$ s/m	[3.130]
Attenuation in nepers per meter. NOTE: For use only over a narrow range of frequencies, you may omit the term $(\omega/\omega_0)^{-\theta_0/\pi}$.	$\alpha_d \triangleq \dfrac{1}{2}\dfrac{\theta_0 \omega}{v_0}\left(\dfrac{\omega}{\omega_0}\right)^{-\theta_0/\pi}$ neper/m	[3.131]
Attenuation in decibels per meter. (1 neper = 8.6858896 decibels) NOTE: For use only over a narrow range of frequencies, you may omit the term $(\omega/\omega_0)^{-\theta_0/\pi}$.	$\alpha_d \triangleq 4.34\dfrac{\theta_0 \omega}{v_0}\left(\dfrac{\omega}{\omega_0}\right)^{-\theta_0/\pi}$ dB/m	[3.132]
Transfer gain at length l	$\|H(\omega,l)\| = e^{-l\cdot\frac{1}{2}\frac{\theta_0 \omega}{v_0}\left(\frac{\omega}{\omega_0}\right)^{-\theta_0/\pi}}$	[3.133]
Inductance per meter	$L = Z_0/v_0$ H/m	[3.5]
Capacitance per meter (varies slightly with frequency)	$C(\omega) = \dfrac{1}{Z_0 v_0}\left(\dfrac{\omega}{\omega_0}\right)^{-2\theta_0/\pi}$ F/m	[3.7]

3.8.5 Possibility of Severe Resonance within Dielectric-Loss Limited Region

Transmission lines operated in the dielectric region fall prey to the same resonance difficulties that afflict structures operated in the LC-region. The termination means used to conquer resonance are the same as those described in Section 3.6.6, "Terminating an LC Transmission Line."

Dielectric losses have the beneficial effect of reducing (somewhat) the requirements for efficacy of intentional terminations placed in a circuit, in the same way as skin-effect

3.8.5 • Possibility of Severe Resonance

losses (see Section 3.7.5, "Possibility of severe resonance within dielectric-loss limited region"), with the exception that in the dielectric region the heights of the resonant peaks of the transfer gain (measured in dB) fall off inversely with frequency.

3.8.5.1 Subtle Differences Between Termination Styles

Figure 3.25 is repeated here as Figure 3.35, this time showing the inclusion of skin-effect losses. All three traces clearly show dispersion of the rising and falling edges caused by the low-pass filtering properties of the skin-effect resistance. In comparison to the results in Figure 3.30, the source- and end-terminated circuits shown here benefit from a partial cancellation of the corrective terms applied to the characteristic impedance function by the resistive and dielectric loss effects respectively. The both-ends terminated circuit provides the best performance, independent of any requirement for balance between the resistive and dielectric loss mechanisms.

3.8.5.2 Application of Termination Equations to Other Regions

Although the LC, skin-effect, and dielectric-loss-limited regions all share the same asymptotic high-frequency parameter for characteristic impedance Z_0, the dielectric effect induces a gradual rise in the actual characteristic impedance Z_C in proportion to the log of freqeuncy. This gradual rise suggests that the best value of termination for a dielectric-loss-limited structure is probably $\sqrt{L/C(\omega)}$, where $C(\omega)$ is calculated near your anticipated maximum frequency of operation.

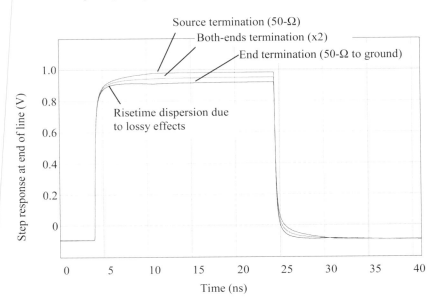

Figure 3.35—The source- and end-terminated circuits shown here benefit from a partial cancellation of the corrective terms applied to the characteristic impedance function by the resistive and dielectric loss effects respectively.

The same general termination approaches therefore work in all three regions, delivering in each case an overall circuit gain G nearly equal to the one-way propagation function H applicable within each region.

POINTS TO REMEMBER

> - The dielectric effect induces a gradual rise in the characteristic impedance Z_C in proportion to the log of freqeuncy.
> - Transmission lines in the dielectric-loss-limited region fall prey to the same resonance difficulties that afflict the LC region and respond to the same means of termination.

3.8.6 Step Response (Dielectric-Loss-Limited Region)

This section presents the step response of a dielectric-effect-limited transmission line. The applicability of these results hinges on two crucial assumptions:

1. The risetime of the step input to the system must be substantially faster than the dielectric-effect risetime, meaning that the shape of the output waveform is determined primarily by the system itself and not strongly influenced by the precise risetime of the input step, and
2. The dielectric losses greatly exceed the resistive losses so that you have a true dielectric-limited transmission media and not a resistively limited media.

The solution requires knowledge of these parameters:

> - $\tan \theta_0$, the loss tangent of the dielectric material at frequency ω_0,
> - v_0, the velocity of propagation at frequency ω_0, m/s,
> - l, the length of the line (m), and
> - ω_0, some particular frequency at which the high-frequency line parameters are specified (rad/s).

The derivation of step response begins with [3.127], into which you may substitute the following infinite series.[33]

$$a^x = 1 + x\ln(a) + \frac{(x\ln(a))^2}{2!} + \frac{(x\ln(a))^3}{3!} + \cdots \qquad [3.134]$$

Making the identifications $a = j\omega/\omega_0$ and $x = -\theta_0/\pi$ produces this expression for the propagation coefficient [3.127]:

---

[33] Burnigton, *Handbook of Mathematical Table and Formulas*, Handbook Publishers, Inc., 3rd ed., 1948, p 44.

3.8.6 • Step Response

$$\gamma = \frac{j\omega}{v_0}\left[1 - \frac{\theta_0}{\pi}\ln(j\omega/\omega_0) + \frac{\left(\frac{\theta_0}{\pi}\ln(j\omega/\omega_0)\right)^2}{2!} - \frac{\left(\frac{\theta_0}{\pi}\ln(j\omega/\omega_0)\right)^3}{3!} + \cdots\right] \quad [3.135]$$

In cases where the constant (θ_0/π) is much smaller than unity and ω lies within a reasonable factor of ω_0 you may safely ignore all but the first two terms of the substitution.

$$\gamma(\omega) \cong \frac{j\omega}{v_0}\left[1 - \frac{\theta_0}{\pi}\ln(j\omega/\omega_0)\right] \quad [3.136]$$

The next step both adds and subtracts a peculiar logarithmic term inside the square brackets: $(\theta_0/\pi)\ln(\omega_0\tau)$. The purpose of this operation is to mold the equation into a separable form so that the fixed bulk delay may be isolated from a canonical low-pass filter function. The constant τ is defined

$$\tau \triangleq l\theta_0/v_0\pi \quad [3.137]$$

The new version of [3.136] with the peculiar terms added and subtracted is

$$\gamma(\omega) \cong \frac{j\omega}{v_0}\left[1 + \frac{\theta_0}{\pi}\ln(\omega_0\tau) - \frac{\theta_0}{\pi}\ln(\omega_0\tau) - \frac{\theta_0}{\pi}\ln\left(\frac{j\omega}{\omega_0}\right)\right] \quad [3.138]$$

Combine the arguments of the right-most two logarithmic terms, and then distribute the term $j\omega/v_0$ into the result.

$$\gamma(\omega) \cong \frac{j\omega}{v_0}\left[1 + \frac{\theta_0}{\pi}\ln(\omega_0\tau)\right] - \frac{j\omega}{v_0}\frac{\theta_0}{\pi}\ln(j\omega\tau) \quad [3.139]$$

Multiply [3.139] by the (negative of the) transmission line length l and exponentiate (as in [3.14]) to find H.

$$H(\omega,l) \cong e^{-j\omega\frac{l}{v_0}\left[1 + \frac{\theta_0}{\pi}\ln(\omega_0\tau)\right]} e^{j\omega\tau\ln(j\omega\tau)} \quad [3.140]$$

The first exponential term in [3.140] represents a linear-phase bulk transport delay. The amount of delay $(l/v_0)[1+(\theta_0/\pi)\ln(\omega_0\tau)]$ depends not only on the nominal line delay l/v_0 but also on the dielectric loss and other variables. This delay, for the purposes of investigating attenuation and dispersion of the propagated signals, may be ignored. The second component $e^{j\omega\tau\ln(j\omega\tau)}$ represents a frequency-varying low-pass filter that accounts for the peculiar shape of the dielectric loss step response.

To compute the *step response* of the dielectric effect, multiply [3.140] by $1/j\omega$ and then apply an inverse Fourier transform. Multiplying by $1/j\omega$ converts [3.140] from an impulse-response to a step-response. Evaluating the inverse Fourier transform of the result produces a time-domain waveform representing the step response $g(t)$ associated with the dielectric effect.

$$g(t) \triangleq \mathcal{F}^{-1}\left(\frac{1}{j\omega} e^{-j\omega\frac{l}{v_1}\left[1+\frac{\theta_0}{\pi}\ln(\omega_1\tau)\right]} e^{j\omega\tau\ln(j\omega\tau)}\right) \qquad [3.141]$$

There is no recognized closed-form expression for this inverse Fourier transformation. It is best evaluated by sampling [3.141] on a dense grid of frequencies and using an inverse FFT to determine the step response.

Figure 3.36 compares the magnitude of the propagation function (in dB) of a long, perfectly terminated skin-effect-limited channel with that of a perfectly dielectric loss-limited channel. The –3 dB points on the two curves are very close. The bottom part of the figure compares the step responses of the two channels. The skin-effect step response displays a sharper initial rise due to the larger magnitude of its high-frequency content, but a longer, more slowly-evolving tail due to the unusual behavior of its frequency response near DC.

The normalized dielectric-loss step response $g_1(t)$ in Figure 3.36 is derived from

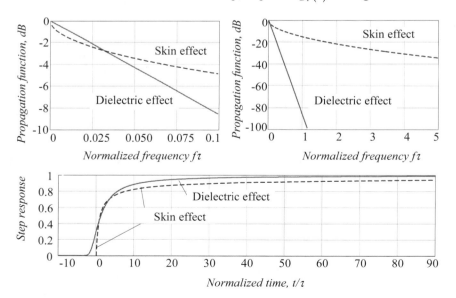

Figure 3.36—As a function of time, the dielectric step response begins more slowly than the skin-effect response, but finishes sooner.

3.8.6 • Step Response

[3.141] by ignoring the fixed bulk delay and then setting $\tau = 1$.

$$g_1(t) \triangleq \mathcal{F}^{-1}\left(\frac{1}{j\omega}e^{j\omega \ln(j\omega)}\right) \qquad [3.142]$$

From the normalized step response $g_1(t)$ you may determine the complete step response $g(t)$ by first scaling the time axis by a factor of τ and then delaying by amount $(l/v_0)[1+(\theta_0/\pi)\ln(\omega_0\tau)]$.

$$g(t) = g_1\left(\frac{t - \frac{l}{v_0}\left[1+\frac{\theta_0}{\pi}\ln(\omega_0\tau)\right]}{\tau}\right) \qquad [3.143]$$

Assuming you know the value τ defined in [3.137], you may read directly from Figure 3.36 the frequency response (magnitude) and step response (assuming a normalized 1-volt step input with zero risetime) for any dielectric-effect limited system.

As impressive as this chart may appear, please do not make the mistake of assuming it accurately predicts real-world dielectric effects. The skin effect has a precise physical model and behaves in a very predictable way, but the dielectric approximations discussed here depend for their accuracy on the assumption of *constant dielectric loss* across a very wide band. Although this may be the form of the specification for many materials, if you look at the actual dielectric loss curves, you will find the loss is *not* constant, but undulates slowly between inflection points, coming near the specification at perhaps more than one location but (hopefully) never crossing it. The charts in Figure 3.36 represent the behavior of a *worst-case* system with constant dielectric loss set at the maximum value θ_0 across all frequencies of interest.

In Figure 3.36 the dielectric step response begins before time zero. This happens because at very high frequencies above ω_0, as the dielectric of the insulating material deteriorates [3.7], the propagation velocity of the transmission line slightly exceeds v_0. Some portions of your signal therefore arrive just before you might have predicted based upon consideration of v_0 alone. This behavior does not constitute a violation of causality. All that has happened is that the bulk delay subtracted from the normalized step-response curve presented in Figure 3.36 is (apparently) a little too generous. Correcting Figure 3.36 by the amount of delay stipulated in [3.140] places the step-response waveform in the correct position.

Beware that the skin effect and dielectric loss effects mix over a *very* broad range of frequencies. It's rare in pcb problems that you see dielectric losses without also having to take into account skin-effect dispersion. The best general rule for combining dispersion due to different sources is the sum-of-squares rule, which works as follows.

$$t_{10-90\%}^{\text{skin effect}+\text{dielectric loss}} \approx \sqrt{\left(t_{10-90\%}^{\text{skin effect}}\right)^2 + \left(t_{10-90\%}^{\text{dielectric loss}}\right)^2} \qquad [3.144]$$

If you seek a more exact answer, use the frequency-domain model defined in Section 3.1, "Signal Propagation Model," sample it on a dense grid of frequencies, and inverse-FFT the result to get a full time-domain step response (see also Chapter 4).

POINT TO REMEMBER

> Given two systems with the same −3dB loss at frequency f_1, one system having only dielectric losses and the other having only skin-effect losses, the dielectric step response begins more slowly than the skin-effect response, but finishes sooner.

3.8.7 Tradeoffs Between Distance and Speed (Dielectric-Loss Region)

Given a fixed receiver architecture, and assuming the BER of the receiver is dominated by errors due to intersymbol interference, the speed of operation within the dielectric-absorption zone scales inversely with the transmission-line length. This tradeoff between length and speed arises because of the unit slope of the dielectric-absorption attenuation curve. For example, a 10% increase in transmission-line length increases the attenuation (in dB) by 10%. To regain a normal signal amplitude at the end of a transmission line limited by dielectric-absorption, the 10% increase in attenuation must be offset by a 10% reduction in system operating speed.

POINTS TO REMEMBER

> The risetime of a dielectric-loss-limited channel scales directly with its length.
> The speed of operation within the dielectric-loss region scales inversely with transmission-line length.
> A dielectric medium with twice the loss tangent incurs twice the loss (in dB) and induces a settling time twice as long.
> It's rare in pcb problems that you see skin-effect losses without also having to take into account dielectric dispersion.

3.9 WAVEGUIDE DISPERSION REGION

At frequencies so high that the wavelength of the signals conveyed shrinks to a size comparable with the cross-sectional dimensions of a transmission line, strange non-TEM modes of propagation appear. These modes do not by themselves portend a loss of signal power, but they can create objectionable phase distortion (i.e., dispersion of the rising and falling edges) that limits the maximum speed of operation (Section 3.2).

3.9.1 Boundary of Waveguide-Dispersion Region

If you attempt to operate a transmission line at such a high frequency that the wavelengths of the signals conveyed approach the dimensions of your conductors, strange modes of propagation begin to appear. These modes have to do with the possibility of signal power bouncing back and forth between two interfaces within the transmission structure. These bouncing modes are called non-TEM modes (see Section 5.1.5, "Non-TEM Modes").

In a coaxial cable the *critical dimension* of interest is the diameter of the shield. In a stripline configuration it's the spacing between the planes. In a microstrip it's the thickness of the dielectric.

At frequencies high enough that the signal wavelength becomes comparable with the critical dimension, a full-wave analysis of the situation predicts received waveforms that have what looks like severe overshoot and ringing, even if the line is perfectly terminated.

The frequency at which fully developed non-TEM modes may exist within a transmission structure is

$$\text{Stripline:} \quad \omega_c = \frac{\pi c}{b\sqrt{\epsilon_r}}$$

$$\text{Microstrip:} \quad \omega_c = k\frac{\pi c}{h\sqrt{\epsilon_r}} \qquad [3.145]$$

$$\text{Coaxial:} \quad \omega_c = 0.586\frac{\pi c}{d_2\sqrt{\epsilon_r}}$$

where ω_c appears in units of rad/s,

b is the interplane spacing of a stripline, m,

h is the dielectric thickness of a microstrip,

k is a constant in the range of 1/10 to 1/6,

d_2 is the inner diameter of a coaxial shield, m,

c is the speed of light, $2.998 \cdot 10^8$ m/s, and

ϵ_r is the relative dielectric constant of the insulating material, as measured in the vicinity of frequency ω_c.

Microstrips suffer more than other configurations from non-TEM modes because the bouncing modal power does not get a clean bounce off the dielectric-to-air interface. The properties of this interface introduce a significant phase shift into the modal equations with the result that non-TEM distortion appears in a noticeable way for microstrips at frequencies much lower than for other configurations. This peculiar form of non-TEM behavior is called *microstrip dispersion*.

For ordinary digital signaling on FR-4 printed circuit boards at 10 Gbps you may use microstrip trace heights up to 20 mils without encountering significant microstrip dispersion. At lower frequencies you can use correspondingly bigger traces. Above 10 Gbps, you must use correspondingly smaller ones.

POINTS TO REMEMBER

> ➤ If the wavelengths of the signals conveyed approach the dimensions of your conductors, strange modes of propagation begin to appear.
> ➤ For ordinary digital signaling on FR-4 printed circuit boards at 10 Gbps you may use microstrip trace heights up to 20 mils without encountering significant microstrip dispersion.

3.10 SUMMARY OF BREAKPOINTS BETWEEN REGIONS

Example of 100-ohm Differential Stripline

Length, $l = 0.6$ m (23.6 in.) (backplane application)
Conductor parameters: $w = 152$ μm (6 mil), $t = 17.4$ μm (1/2-oz. Cu), perimeter $p = 2(w + t) = 339$ μm (13.35 mil)
Conductivity of signal conductor: $\sigma = 5.98 \cdot 10^7$ S/m
The critical dimension for non-TEM considerations is the separation between the planes $b = 508$ μm (20 mil)
Specification frequency for AC parameters: $\omega_0 = 2\pi \cdot 10^9$
Characteristic impedance at ω_0: $Z_0 = 100$ ohms
Effective dielectric constant: $\epsilon_R = 4.3$
Effective loss tangent for FR-4 dielectric: $\tan \theta_0 = 0.025$
Proximity factor (see chapter on *printed circuit traces*): $k_p = 3.2$

Computed Values

Propagation velocity above RC region: $v_0 = c/\sqrt{\epsilon_R} = 1.4457 \cdot 10^8$ m/s ($t_p = 175.7$ ps/in.)
Differential inductance per meter: $L = Z_0/v_0 = 691$ nH/m (17.6 nH/in.)
Differential capacitance per meter: $C = 1/(Z_0 v_0) = 69.1$ pF/m (1.76 pF/in.)
DC resistance: $R_{DC} = 2/(\sigma w t) = 12.64$ Ω/m (0.321 Ω/in.)
AC resistance: $R_0 \triangleq \dfrac{k_p}{p}\sqrt{\dfrac{\omega_0 \mu}{2\sigma}} = 76.74$ Ω/m

Lumped-Element Region

First check the critical distance test specified in [3.30] and [3.31]:

$$critical\ length = (.25/R_{DC})\sqrt{L/C} = 1.97 \text{ m } (77.6 \text{ in.}) \qquad [3.146]$$

3.10 • Summary of Breakpoints

The trace length of $l = 0.6$ m falls short of the critical length. The exit from the lumped-element region will therefore fall along the curve prescribed by [3.27], proceeding directly into the LC region. There will be no observable RC region in this configuration.

Transitions into all the major regions are presented in Table 3.10 in units of Hz, which requires an adjustment by a factor of $(1/2\pi)$ from the formulas presented previously in this chapter.

Table 3.10—Onset of Various Transmission Regions (Pcb Trace Example)

Onset of Region	Formula for lower band edge	Value	Source
LC	$\omega_{LE} = (\Delta/l)(LC)^{-1/2}$	9.58 MHz	[3.27]
Skin-effect	$\omega_\delta = \omega_0 (R_{DC}/R_0)^2$	27.1 MHz	[3.104]
Dielectric	$\omega_\theta = (1/\omega_0)\left[(v_0 R_0)/(Z_0 \tan\theta_0)\right]^2$	498 MHz	[3.123]
Waveguide (for stripline)	$\omega_c = (\pi v_0 / b)$	142 GHz	[3.145]

Operation of this differential stripline trace at frequencies less than 9.58 MHz may be successfully modeled using a one-stage pi network. Above 9.58 MHz a distributed LC model applies until the skin effect takes hold at 27.1 MHz. Above the skin-effect transition the loss grows (theoretically) in proportion to the square root of frequency until you reach the onset of the dielectric effect at 498 MHz. Because the spacing between the skin-effect onset and dielectric-loss onset is not very great in this example (a factor of only 18.4), you may expect that even at the lower band edge of the skin-effect region the dielectric effect will still be exerting noticeable influence. Figure 3.1 illustrates this stripline example as the topmost waveform in the figure. From the skin-effect onset at 27 MHz upwards, there is no clearly defined region with a loss slope of precisely 1/2. Instead, the skin and dielectric effects mush together over a broad band, gradually increasing the loss slope from 1/2 to 1 over the range from 27 to 1000 MHz.

Provided that the trace is shortened to reduce the overall trace loss to a manageable size, this stripline trace geometry may be successfully operated at frequencies up to nearly 142 GHz without fear of exciting any unusual non-TEM modes of propagation.

Example showing Belden type 8237 (RG-8) Coaxial Cable

Length, $l = 2000$ m (6561 ft)
Center cond. 7x#21AWG stranded Cu
DC resistance (including shield) $R_{DC} = 0.0103$ Ω/m (3.14 mΩ/ft)
Critical dimension for non-TEM considerations = Shield diameter $d_2 = 7.239$ mm (0.285 in.)
Conductivity, $\sigma = 5.8 \cdot 10^7$ S/m
Specification frequency for AC parameters: $\omega_0 = 2\pi \cdot 10^7$
Characteristic impedance at ω_0: $Z_0 = 52$ ohms
Effective dielectric constant: $\epsilon_R = 2.29$
Effective loss tangent for solid polyethylene dielectric: $\tan\theta_0 = 0.00052$
Proximity factor for 7-way stranded center conductor: $k_p = 1.07$
Surface roughness factor for braided copper shield: $k_r = 1.8$ (applies to frequencies above 10 MHz).

Computed Values

Effective diameter of center conductor: $d_1 = d_2 \exp\left(-\left(Z_0\sqrt{\varepsilon_R}\right)/60\right) = 1.947$ mm (0.077 in.)

$$R_{AC,CENTER}(\omega_0) \triangleq \frac{k_p}{\pi d_1}\sqrt{\frac{\omega_0 \mu}{2\sigma}} = 0.1443 \text{ }\Omega/\text{m}$$

$$R_{AC,SHIELD}(\omega_0) \triangleq \frac{k_r}{\pi d_2}\sqrt{\frac{\omega_0 \mu}{2\sigma}} = 0.0653 \text{ }\Omega/\text{m} \quad [3.147]$$

$$R_0 = 0.1443 + 0.0653 = 0.2096 \text{ }\Omega/\text{m} \text{ }(0.0638 \text{ }\Omega/\text{ft})$$

Propagation velocity above RC region: $v_0 = c/\sqrt{\varepsilon_R} = 1.979 \cdot 10^8$ m/s (t_p=1.54 ns/ft)

Inductance per meter: $L = Z_0/v_0 = 253$ nH/m (77.1 nH/ft)

Capacitance per meter: $C = 1/(Z_0 v_0) = 101$ pF/m (30.8 pF/ft)

Lumped-Element Region

Check the critical distance test specified in [3.30] and [3.31]:

$$critical\ length = (.25/R_{DC})\sqrt{L/C} = 1262 \text{ m (4140 ft)} \quad [3.148]$$

The cable length of l = 2000m well exceeds the critical length. The exit from the lumped-element region will therefore fall along the curve prescribed by [3.26], and the cable will at higher frequencies then proceed to the RC region. Had the length been less than 1262 meters, the cable would have exited the lumped-element region along the curve prescribed by [3.27], transitioning directly into some higher region and bypassing the RC region altogether.

Transitions into all the other major regions are presented in Table 3.11 in units of Hz, which requires an adjustment by a factor of $(1/2\pi)$ from the formulas presented previously in this chapter.

Table 3.11—Onset of Various Transmission Regions (Coaxial Cable Example)

Onset of region	Formula for lower band edge	Value	Source
RC	$\omega_{LE} = (\Delta/l)^2 (R_{DC}C)^{-1}$	2.48 KHz	[3.26]
LC	$\omega_{LC} = (R_{DC}/L)$	6.24 KHz	[3.53]
Skin-effect	$\omega_\delta = \omega_0 (R_{DC}/R_0)^2$	24.1 KHz	[3.104]
Dielectric	$\omega_\theta = (1/\omega_0)\left[(v_0 R_0)/(Z_0 \tan\theta_0)\right]^2$	5.9 GHz	[3.123]
Waveguide (for coax)	$\omega_c = 0.586(\pi v_0/d_2)$	8 GHz	[3.145]

Operation of this cable at frequencies less than 2.48 KHz may be successfully modeled using a one-stage π network. Above 2.48 KHz a distributed RC model applies until you reach the onset of the LC region at 6.24 KHz. From that point until the skin effect takes hold at 24.1

3.11 • Equivalence Principle

KHz, the cable response remains fairly flat. Above that the skin effect loss grows in proportion to the square root of frequency until you reach the onset of the dielectric effect at 5.9 GHz. Remember that the transition into the dielectric-loss-limited region is very broad, extending a factor of 100 either side of ω_0. Well above 5.9 GHz, the loss slope gradually increases until the loss in dB is growing in direct proportion to frequency.

The cable may be successfully operated at frequencies up to nearly 8 GHz without fear of exciting any unusual non-TEM modes of propagation.

3.11 EQUIVALENCE PRINCIPLE FOR TRANSMISSION MEDIA

Figure 3.37 shows the transfer function and step response of six selected Belden coaxial cable types, at a fixed length of 25 meters. Eye diagrams for all six cable types appear in Figure 3.38. Of the six types, the 8237-type cable delivers the best transfer function (least loss) and the best step response (fastest risetime).

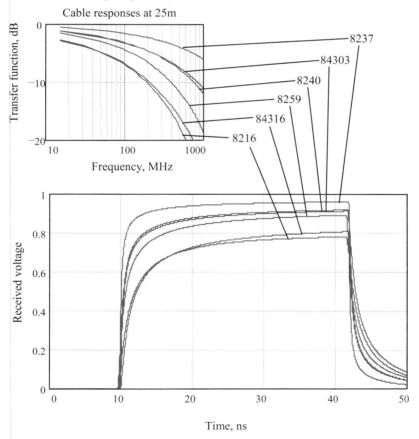

Figure 3.37—Frequency-response and step-response graphs for selected Belden coaxial cable types.

Suppose you are planning to use these cables in a communication system operating at a distance of 25 m and a speed of 500 Mbaud. You can determine the likelihood of success by a simple examination of the transfer function. The key item to observe is the gross attenuation in dB at 1/2 the data rate (250 MHz).

The 8237-type cable at a length of 25 m clearly attenuates less than 3 dB at 250 MHz, meaning that the opening at the center of the eye when transmitting a continuous 101010... pattern will likely be larger than 70%. Keep in mind that this is not an exact calculation because your data pattern consists of square waves, not sine waves, but for planning purposes it is quite useful. Figure 3.38 shows the calculated eye patterns, taking into account the limited risetime of the driver and the complete frequency response of the cable.

The next two cable types (84303 and 8240) have more attenuation at 250 MHz and so deliver more eye-pattern distortion. The next cable type (8259) is marginal, and the last two cable types (84316 and 8216) exhibit far too much attenuation at 250 MHz. The eye diagrams for these last two cable types are practically closed. Apparently, the performance of these coaxial cable types varies tremendously as a function of the construction.

Even though the performance varies widely, the cables still bear remarkable similarity to one another. You can see the similarity best in Figure 3.1, which shows the same six coaxial-cable curves plotted on a logarithmic vertical axis (log-of-dB). Over the frequency range from 10 to 1000 MHZ the slopes of all the cable responses are the same, because all six cables happen to operate in the same mode (the skin-effect limited mode) over this range.

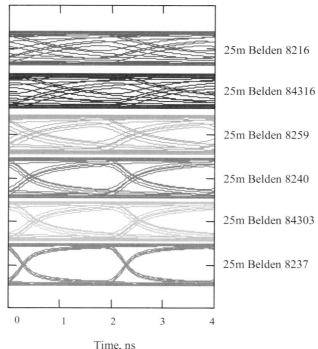

Figure 3.38—The eye quality for six selected Belden coaxial cable types at a data rate of 500 Mb/s and a fixed cable length of 25 m varies dramatically.

3.11 • Equivalence Principle

Over the frequency range 10 to 1000 MHz the six curves are just shifted copies of each other. That is, if you shift one curve vertically, it will overlay almost perfectly on top of any of the other curves. Of course there are limits to the application of this rule. For example, over the range between 0.1 and 10 MHz the various cables transition into the skin-effect region at different points. Over the more limited range from 10 MHz to 1000 MHz, however, all six curves have a similar slope.

You can use this similarity to derive an equivalence relation between all six cables. This equivalence depends on the fact that the cable attenuation in dB scales linearly with distance. That is, if you take one particular cable and scale its length by a factor k, then the attenuation of that cable, as measured in dB, also scales by the same factor k. Since the vertical axis in Figure 3.1 is a logarithmic axis, scaling the attenuation in dB by k will raise (or lower) the attenuation curve by a fixed amount equal to $\log(k)$. Scaling the length does not change the slope in Figure 3.1; it just raises or lowers the whole curve. Over the range of frequencies dominated by the skin effect, you can use this length-scaling principle to cause any cable to mimic the performance of any other cable.

For example, suppose you adjust the length of each cable from Figure 3.37 to guarantee attenuation in each case at 250 MHz of, say, 3.5 dB. This adjustment slides the transfer curves in Figure 3.1 vertically until they all line up, passing through the same point at 250 MHz. Since they have the same slope over the range 10 to 1000 MHz, they will then each produce precisely the same transfer function in every way, with the exception of the bulk transport delay, which will be less on the shorter cables (see Figure 3.39).

The selection of 3.5 dB as the point of similarity is arbitrary. The same theory works at any point. I chose 3.5 dB at 250 MHz because that is an aggressive specification for non-return-to-zero (NRZ) random binary signaling at 500 Mbps.

A *DC-balanced* data code having a strict guarantee of equal numbers of ones and zeros over some finite interval can get by with slightly more attenuation. In that case one typically strives to achieve no more than a 3.5 dB attenuation *difference* over the range between the lowest to the highest possible data frequencies.

Figure 3.40 shows the resulting eye-patterns with each cable adjusted to a length that provides 3.5 dB of attenuation at 250 MHz. As you can see, the performance graphs for all cables at the new adjusted lengths are the same. For example, 45 meters of type 8237 cable gives the same performance as 23 meters of type 8240 cable over the frequency range from 10 to 1000 MHz. In a sense, all skin-effect-limited channels do the same thing. Almost any copper cable, if its length is adjusted properly, will generate the same eye pattern.

Also note that the adjusted lengths fall almost in direct proportion to the diameter of the center conductor (Figure 3.41). For its size, the type-8259 cable delivers the worst performance. This cable has a tinned center conductor that artificially raises the AC skin-effect resistance of the cable, thus lowering its bandwidth.

POINT TO REMEMBER

> ➢ Over the range of frequencies dominated by the skin effect, you can scale the length of one coaxial cable type to cause it to mimic the performance of any other type.

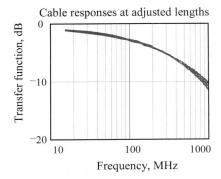
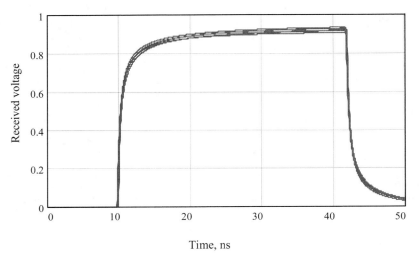

Figure 3.39—Frequency-response and step-response graphs for specific lengths of selected Belden coaxial cable types, the length in each case chosen to create precisely 3.5 dB attenuation at 250 MHz.

3.12 SCALING COPPER TRANSMISSION MEDIA

Lengthening a transmission line increases its delay while reducing its bandwidth. These two properties are inextricably interrelated. Longer almost always means *slower*.

Let me pause here to explain that I am writing to you about the optimal use of transmission lines—specifically, the use of properly terminated transmission lines with drivers that put out clean, fast, full-sized signals and receivers that do not excessively load the far-end terminus of the line. In that case the behavior of the transmission line is governed by its propagation coefficient [3.13]. This coefficient prescribes the attenuation at each frequency in units of dB per meter (or nepers per meter). The *overall attenuation* of such a physical link scales linearly with distance.

3.12 • Scaling Copper Transmission Media

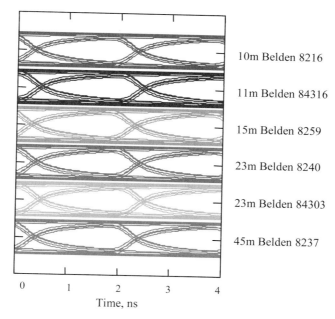

Figure 3.40—Eye diagrams for adjusted lengths of six selected Belden coaxial cable types at a data rate of 500 Mb/s.

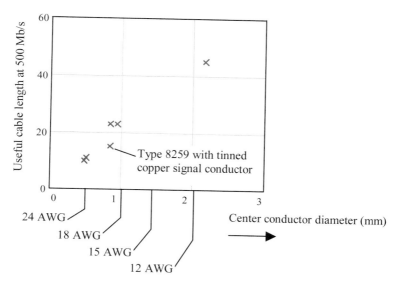

Figure 3.41—The 8259 cable has a tinned-copper center conductor, which raises the skin-effect resistance, limiting the distance at which it may be used.

For example, suppose you have one transmission line that has an attenuation of 3 dB at 100 MHz. A second line of twice the length (but otherwise the same) would have precisely 6 dB of attenuation at 100 MHz. Doubling the length doubles the attenuation.

Any exaggeration of the attenuation (by lengthening or any other method) must necessarily lower the 3-dB point. This happens because the attenuation function for real-world transmission lines is always monotonic and decreasing. *Increasing the length always reduces the bandwidth.*

For robust performance on unequalized receivers with binary data, one normally requires that the 3-dB bandwidth of the transmission media exceed 70% of the bit rate; in this case, increasing the length reduces not only the bandwidth but also the maximum rate of digital transmission that can occur over that transmission system.

If you know the slope of the attenuation function, you can predict the precise relation between the maximum communication rate and distance. Here nature helps us in a profound way: Almost all practical transmission lines have about the same shape to their attenuation function. Most transmission lines used for high-speed digital work display a smooth, rounded knee in their attenuation function, with a profile something like this:

$$a(\omega) \propto \omega^\eta l \qquad [3.149]$$

where a is the attenuation in decibels,

the constant of proportionality depends on the materials and geometry of the cabling,

ω is the frequency of operation, rad/s,

η is a slowly varying constant between 1/2 and 1, and

l is the length of the transmission line, m.

A value of $\eta = 1/2$ is typical for transmission lines that are limited primarily by the skin effect. Good examples of skin-effect-limited media would include any of the transmission lines in Figure 3.1 taken over the frequency range from 10 to 1000 MHz.

In the skin-effect-limited range the $\omega^{1/2}$ dependence creates an interesting property of scaling: Doubling the length but cutting the frequency by 1/4 produces precisely the same attenuation. In the time domain,

for skin-effect-limited media, doubling the length while slowing down to 1/4 the bit rate produces the same eye pattern.

Horrible, isn't it! The penalty for doubling the line length is a reduction in bandwidth by a factor of four.

Turn that around the other way and you see the other side: Cutting the length in half speeds up the system by a factor of four. This is what made 10BASE-T Ethernet so popular. At one time, prevailing wisdom suggested that telephone-style unshielded twisted-pair cabling had an inherent bandwidth of only 3KHz to 4 KHz. This reasoning was based on an assumption of length, namely, that every system had to be able to operate at distances sufficient to reach the nearest telephone central office, which could be as much as 5000 meters distant. Once people in the LAN business recognized that interoffice LAN communications needed only to go 100 meters, the bandwidth assumption could be boosted by a factor of $(5000/100)^2$, resulting in easily sufficient bandwidth to operate at 10 Mbps.

For many high-speed systems the skin effect is the most significant bandwidth-limiting factor. At extremes of frequency, however, the skin effect is superceded by dielectric loss. In the dielectric-loss-limited region the constant η asymptotically approaches a value of 1. In the dielectric-limited region the relation between speed and distance becomes merely inverse, not inverse-squared.

> For dielectric-effect limited media, doubling the length while slowing down to 1/2 the bit rate produces the same eye pattern.

Fortunately for high-speed digital designers, the bandwidth of a typical pcb trace is pretty incredible. Figure 3.42 illustrates the performance of a 152-μm (6-mil) stripline trace implemented on Getek. The trace is 0.3-m long. The 3-dB attenuation point for this trace occurs at 5 GHz.

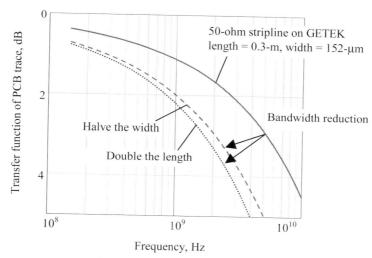

Figure 3.42—Either doubling the line length or halving the line width cuts bandwidth by a big factor.

If this sort of performance is not enough for your application, let me explain how to get even higher bandwidth. These ideas build on a simple approximation for transmission-line attenuation:

$$a(\omega) = 4.34 \cdot l \left[\frac{R_0}{Z_0} \sqrt{\frac{\omega}{\omega_0}} + \frac{\omega \theta_0}{v_0} \left(\frac{\omega}{\omega_0} \right)^{-\theta/\pi} \right] \text{ dB/m} \qquad [3.150]$$

where a is the attenuation in decibels,

ω is the frequency of operation, rad/s,

l is the length of the transmission line in meters,

ω_0 is the frequency at which AC line parameters are specified, rad/s,

R_0 is the AC resistance of the line at frequency ω_0,

Z_0 is the characteristic impedance of the line at frequency ω_0,

v_0 is the velocity of propagation at frequency ω_0, m/s, and

$\tan \theta_0$ is loss tangent of the dielectric material at frequency ω_0.

This expression contains many terms; there should therefore exist many ways to reduce the attenuation, thus increasing the 3-dB bandwidth.

1. **Use more copper.** If you want higher bandwidth, try using wider traces (for cables, use bigger signal conductors). At the same time you make the trace wider, also raise it up further away from the nearest solid plane. This change results in a new geometry with the same impedance as the original, but less resistance R_0. The resistance R_0 varies inversely with trace width.

    NOTE: Above 1 GHz the dielectric losses become rapidly more significant than skin-effect losses. Monkeying around with skin-effect loss in a system that is dominated by dielectric problems makes progressively less and less sense as you go to frequencies far above ω_θ.

2. **Don't go as far.** In the skin-effect zone the bandwidth varies inversely with the square length; in the dielectric zone it's an inverse relationship. Either way, longer traces have less bandwidth. If you must go a long way, consider using repeaters.

3. **Use a higher-impedance trace.** Moving your signal conductor farther from its nearest return path (without increasing its trace width, or diameter) increases the Z_0 while leaving R_0 mostly unchanged. This adjustment lowers the ratio R_0 / Z_0, lowering the skin-effect attenuation, thus raising the bandwidth. Unfortunately, this method has the side effect of rendering the trace more susceptible (percentage-wise) to lumped capacitive loads.

4. **Do something so the attenuation doesn't matter.** Fixed equalization can extend the operating distance by at least 50%. This approach is used in the popular 10BASE-T Ethernet standard at 10 Mb/s over category-3 unshielded twisted-pair wiring (see Section 8.2, "UTP Signal Propagation Example: 10BASE-T"). Fixed equalization may be incorporated into the driver, the receiver, or a combination of both. Adaptive equalization is a more powerful technique, although more difficult to design. It can in some cases more than double the operating distance. An adaptive equalization approach is used in many 100BASE-X Ethernet chips at 100 Mb/s over category-5 unshielded twisted-pair wiring. In systems with very low levels of background noise, and where the complexity of the receiver is not an objection, multilevel coding can provide even greater benefits. The 1000 Mb/s Ethernet standard for category-5 unshielded twisted-pair wiring (1000BASE-T) combines adaptive digital equalization, multilevel coding, and adaptive digital cancellation of near-end crosstalk to obtain a signaling rate of 250 Mb/s per pair on 100-m lengths of category-5 unshielded twisted-pair cabling.

5. **Use a better dielectric material.** The lower the loss tangent of the material, the less dielectric loss your signals will endure. Less dielectric loss translates to higher bandwidth. A lower dielectric constant, even with the same loss tangent, also helps because that increases the propagation velocity v_0, lowering dielectric losses.

As technology advances, more options become available. You can look at recent LAN standards to get a glimpse of what may someday become commonplace in ordinary digital logic families. For example, fixed equalization (10BaseT), adaptive equalization (100BaseTX), and multilevel coding with digital adaptive filtering and near-end crosstalk cancellation (1000BaseT) are fast becoming mass-market realities.

Many designs have not yet reached the point at which trace bandwidth becomes a serious limitation, but just you wait. When typical trace widths go down to 0.002 in. and typical clocks reach 1 GHz, you'll be there.

POINT TO REMEMBER

> Five ways to improve the performance of a copper transmission channel: use more copper, don't go as far, use a higher characteristic impedance, add equalization, or use a better dielectric material.

3.13 SCALING MULTIMODE FIBER-OPTIC CABLES

Just for fun, let's compare the theory of scaling for fiber with the theory of scaling for copper conductors.

In the fiber-optic case there are two predominate bandwidth-limiting effects: modal dispersion and chromatic dispersion. Both bandwidth-limiting effects vary inversely with distance. If you go twice as far,[34] you get half the bandwidth.

Fibers are also afflicted with a fixed transmission loss. The transmission loss in dBmW varies directly with length. The further you go, the less signal comes out the far end of the cable.

In a practical optical transmission system, as the cable is stretched further and further, one of two things eventually goes wrong. In some systems the bandwidth becomes a limiting factor, in which case the received signal has plenty of power, but the bits are slurred into each other. In other cases the power is a limiting factor, meaning that at great distances the signal looks okay, but simply becomes too small to reliably detect. In either case a 10% reduction in length results in a 10% improvement in signal quality.

For a skin-effect-limited copper medium, a 10% reduction in length generally results in a 20% improvement in signal quality, because copper bandwidth in this zone varies with the length squared. Copper systems are generally more sensitive to length than are fiber systems.

---

[34] Theoretically, once you exceed the mode-coherence length for a fiber-optic cable, the bandwidth descends only with the half-power of length, but since no cable manufacturers specify the mode-coherence length, this fact is not useful to designers of fiber-optic links.

Fiber cabling exhibits enormous variations in bandwidth and loss. For example, a typical length of cable with a bandwidth-distance specification of 500 MHz-km may have an actual bandwidth two or four times higher than the specification. The same holds for attenuation. The experience of technicians in the field is that fiber systems often go much further than advertised.

Not so with copper. The performance of a metallic interconnection is heavily affected by its physical construction, which is comparatively well controlled in the manufacturing process. Metallic transmission systems have a relatively hard, fixed upper limit on distance that should never be exceeded.

POINT TO REMEMBER

> ➢ The performance of a metallic interconnection is heavily affected by its physical construction, which is comparatively well controlled in the manufacturing process. Metallic transmission systems have a relatively hard, fixed upper limit on distance that should never be exceeded.

3.14 LINEAR EQUALIZATION: LONG BACKPLANE TRACE EXAMPLE

Figure 3.43 depicts the propagation function for four 50-Ω striplines with different lengths. The longest trace (1.5 m) corresponds to the maximal configuration you might encounter in a large backplane if you include long daughter card traces at each end.

The propagation functions in this collection exhibit low-pass filter traits that are characteristic of a combination of skin and dielectric loss effects: The longer the trace, the lower the 3dB attenuation frequency.

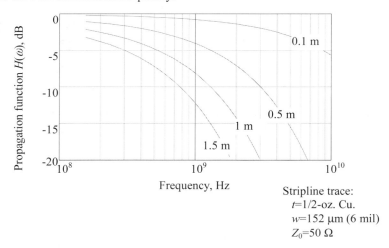

Figure 3.43—The propagation function for a long pcb trace varies strongly with trace length.

3.14 • Long Backplane Trace Example

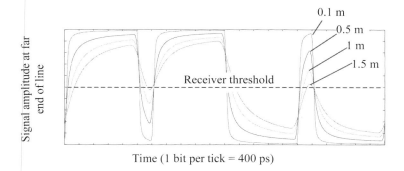

Figure 3.44—The received signal amplitude varies as a function of line length.

Figure 3.44 illustrates the signal as received at the far end of each trace. The signal pattern used to generate this figure is 1111101111100000100000, transmitted at a bit rate of 2.5 Gb/s (400-ps bit intervals). The assumed transmit rise/fall time is 20% of a bit interval. This particular pattern (a maximal run length of ones interrupted by a single zero, and vice versa) is a good worst-case eye-opening test for many systems whose transfer function declines monotonically like the curves shown in Figure 3.43.

The shortest trace (0.1 m) conveys the signal with almost perfect fidelity. As the trace is lengthened, the quick transition (101) is attenuated by successively greater amounts until, at 1.5 meters, this transition fails to adequately cross the receiver threshold. The attenuation is caused by dispersion of the rising and falling edges of the signal—in a quick 101 pattern the signal does not have adequate time to completely cross the threshold before it must reverse course.

Note that a sine wave purely composed of the alternating 101010.... pattern would of course cross the threshold admirably in both directions. A long run of ones, however, followed by one quick pop in the downward direction (or the inverse of this pattern), presents the greatest difficulty.

Figure 3.45 details the exact amplitudes of the received signal at several points. The peak-to-peak amplitude of a large (slow-speed, or long run-length) transition is marked as a_1. The peak-to-peak amplitude of a quick 010 transition is marked as a_2. The magnitude by which the peak of the 010 sequence manages to top the receiver threshold is given as

$$a_3 = -(a_1/2) + a_2 \qquad [3.151]$$

Equation [3.151] is derived by noting that the signal amplitude beginning at $-(a_1/2)$ is increased by amount a_2 at its peak. The ratio in dB by which the quick transition a_3 fails to attain the same peak value as the slow transition $+(a1/2)$ is called the dispersion penalty p_{d1}:

$$p_{d1} = 20\log\left(\frac{a_3}{+(a_1/2)}\right) \qquad [3.152]$$

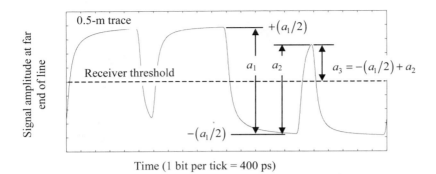

Figure 3.45—The dispersion penalty is a function of the ratio of the largest to the smallest transition amplitudes (a_1/a_2).

The dispersion penalty may be expressed in terms of the ratio r of the amplitudes of the low- and high-frequency transitions:

$$p_{d1} = 20\log\left(\frac{-(a_1/2) + a_2}{+(a_1/2)}\right) = 20\log(2r - 1) \quad [3.153]$$

where a_1 equals the amplitude of the largest (slow-speed, or long run-length) transition,

a_2 equals the amplitude of the smallest transition (usually a quick transition like 010),

r is the ratio a_2/a_1, and

p_{d1} is the noise-tolerance penalty suffered at the receiver in exchange for having to deal with signal dispersion above and beyond low-frequency attenuation.

Figure 3.46 charts the dispersion penalty p_{d1} as a function of r, with both quantities in units of dB. Due to the form of the dispersion-penalty equation, a received-amplitude ratio of 3dB causes a lot more than 3dB worth of degradation in the received amplitude. A received-amplitude ratio of 6 dB causes total receiver failure.

The dispersion penalty p_{d1} as defined here represents the worst-case received amplitude degradation above and beyond consideration of the attenuation of the lowest frequency (slow-speed, or long run-length) patterns present in the data waveform.

For noncoded random data, amplitude a_1 is taken to be the peak-to-peak received amplitude at DC. For coded data waveforms having zero meaningful DC content (like Manchester coding, or 8B10B coding), amplitude a_1 is taken to be the peak-to-peak received amplitude of the lowest possible frequency long-run-length sequence produced by the code.

The dispersion penalty could alternately be defined as the ratio in dB by which the quick transition amplitude a_3 fails to attain the peak amplitude of a repeating fast transition sequence 101010.... The amplitude of this repeating fast-transition sequence is $+(a_2/2)$. In this case the dispersion penalty calculation is modified:

3.14 • Long Backplane Trace Example

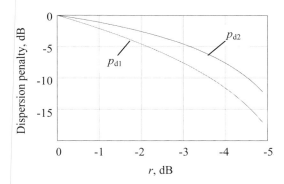

Figure 3.46—The dispersion penalty is a function of the ratio r of the amplitudes of the highest and lowest-frequency signals on the line.

$$p_{d2} = 20\log\left(\frac{-(a_1/2)+a_2}{+(a_2/2)}\right) = 20\log\left(\frac{2r-1}{r}\right) \quad [3.154]$$

This dispersion penalty p_{d2} represents the worst-case received amplitude degradation above and beyond consideration of the attenuation of the highest frequency (quickest, or shortest run-length) patterns present in the data waveform.

Either penalty calculation can be used, as long as you remember the underlying basis of the definition. The difference between p_{d1} and p_{d2} equals precisely $20\log(r)$.

An improvement in the dispersion penalty can be made in binary systems by a simple fixed-linear equalizer. Any monotonic, slowly changing equalization function that improves the ratio r, either by lowering the amplitude of the low-frequency transitions a_1 or boosting the amplitude of the high-frequency transitions a_2, will improve performance. The simple equalizer depicted in Figure 3.47 adjusts the transmitted signal magnitude during each bit interval to one of two values: a regular magnitude when repeating a bit, but double-magnitude when changing states. The combination of two magnitudes and two polarities results in a total of four possible transmit amplitudes that must be generated at the transmitter. By this scheme the received amplitude at the quick-transition locations is improved. This simple transmit-based equalization scheme can reliably boost the operating distance of a digital channel by 50% compared to the maximum reliable operating distance of an unequalized system. More elaborate equalization schemes can, if properly implemented, deliver even more impressive benefits.

The ratio r may be *estimated*, but not determined precisely, from a plot of the transfer function of the digital channel. For example, in Figure 3.43 the signal attenuation for a 0.5-m trace at a frequency of 1.25 GHz (half the data rate) amounts to roughly 5 dB, whereas the attenuation at 1/10th that rate (corresponding to a repeating pattern of 10 zeroes and 10 ones) amounts to roughly 1 dB, a difference of 4 dB. This difference is indicative, but not exactly equal to, the ratio a_2/a_1 in Figure 3.45. The reason such calculations are not exact is that the transfer function depicts the response to sine wave excitation at each particular frequency, whereas the desired variables a_1 and a_2 result from square wave excitation. Still, the transfer-function response ratio is often quite useful. My rule of thumb is that a system

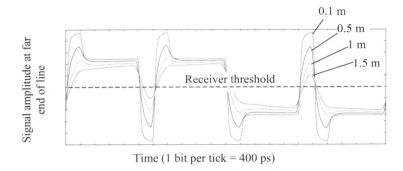

Figure 3.47—A modest amount of fixed equalization improves the received eye pattern in a skin-effect limited or dielectric-effect limited binary signaling channel.

with less than 3 dB of difference in attenuation between the lowest relevant frequency and half the data rate will work reasonable well unequalized. Systems with a difference of more than 6 dB are marginal, and would benefit from fixed linear equalization. Systems with a difference of more than 12 dB generally require adaptive equalization, especially if you expect them to operate reliably at all channel lengths from zero to the maximum.

3.15 ADAPTIVE EQUALIZATION: ACCELERANT NETWORKS TRANSCEIVER

Figure 3.48 illustrates the data waveforms produces by a sophisticated and highly adaptable equalization scheme first introduced for 5 Gb/s point-to-point pcb applications by Accelerant Networks. This system uses PAM-4 data coding, meaning that at each transition the transmitter sends one of four discrete levels. In that way it communicates two bits of information to the receiver on each transition.

Because the amount of information conveyed per transition is twice that in ordinary binary signaling, PAM-4 coding reduces the number of transitions required by a factor of approximately two, thus halving the bandwidth of the data stream.[35] The reduction in bandwidth allows a system using the Accelerant transceiver to accommodate stubs, connector artifacts, layout imperfections, and other packaging flaws physically twice as large as would be permitted in a binary system operating at the same overall bit rate.

In addition to the use of PAM-4, the Accelerant transceiver incorporates an adaptive transmit-based equalizer. The equalizer is automatically tuned based on measurements taken at each receiver, which are communicated back to that receiver's local transmitter using a hidden control channel superimposed on the reverse data channel. This scheme assumes the user has implemented a symmetrical duplex link with one differential pair running in each direction.

---

[35] I say approximately a factor of two, and not exactly a factor of two, because a small amount of coding overhead is taken from the data stream to provide clocking transitions, DC balance, and control functions.

3.15 • Accelerant Networks Transceiver

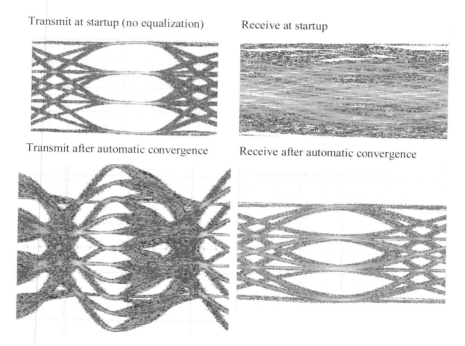

Figure 3.48—At startup (without equalization) the received data eye is closed. After automatic convergence the transmitter pre-distorts (equalizes) the transmitted PAM-4 signal so that the received signal is easily recovered. *Figure courtesy of Accelerant Networks, manufacturer of digital adaptive transceivers for communication at 5 Gb/s and beyond.*

During start-up, a reliable low-frequency code initiates the equalization sequence. After start-up, a continuous flow of control information passes between the transceivers, keeping both ends of the link properly adapted to changes in the transmission properties of the channel. This allows the system to respond to changes in temperature (see Section 5.1.3.5, "Variations in Dielectric Properties with Temperature").

All data on the link is scrambled at the transmitter, and unscrambled at the receiver, resulting in optimal radiation and crosstalk characteristics.

This code successfully operates on ordinary FR-4 differential pairs at distances up to 30 inches or more, traversing multiple connectors.

The top-left portion of Figure 3.48 shows a PAM-4 transmit signal with no pre-distortion (no equalization). On the top right the received signal after an aggregate 30 inches of FR-4 trace, composed of two paddle cards, two connectors, and a backplane, is barely perceptible as a mushy, closed eye.

The bottom-left portion of the figure shows the same PAM-4 transmit signal, but with pre-distortion (equalization) after the link has converged. On the bottom right the received signal after the same 30 inches of FR-4 trace is easily recovered. Adaptive equalization works wonders.

POINTS TO REMEMBER

➢ Intersymbol interference may be characterized by a dispersion penalty.
➢ The dispersion penalty can only be circumvented by equalization.

For further study see: www.sigcon.com

CHAPTER 4

FREQUENCY-DOMAIN MODELING

Some operations that are very complicated in the time domain (like linear filtering, also called *convolution*) become simple in the frequency domain. That's one good reason for learning about frequency-domain analysis.

The other reason to use frequency-domain simulation is because it gives you incredible control over the exact form of frequency-dependent losses, like the skin effect and dielectric-loss effect. Because the frequency-domain method may be easily programmed in any software spreadsheet application (like MatLab, Mathematica, or MathCad), you can control any aspect of the simulation, including searching for optimum and worst-case parameter values. For the analysis of a single, long, serial high-speed digital interconnection, especially if the link is terminated at both ends so it adheres to the properties of linearity and superposition, the frequency-domain method may be your best alternative.

4.1 GOING NONLINEAR

Article first published in *EDN Magazine*, May 16, 2002

SPICE is grand for solving circuits that have complicated topologies or include nonlinear elements. If, however, you are working with a simple, linear circuit, you might question whether SPICE is best.

For example, suppose you are designing a 2.5-Gbps serial-communications channel terminated at both ends. Your source-terminated, 100-Ω driver is a highly linear creature. So is your load. Simulations of this circuit therefore produce the same output whether you use a linear-analysis tool or SPICE.

The linear-analysis method is the same as the frequency-domain method (Figure 4.1). To perform linear analysis, you first transform your excitation waveform $x(t)$ into the frequency domain. You accomplish the transformation using a canned FFT algorithm. It produces a vector of frequency-domain values, X_k, sampled on a very dense grid of frequencies. Next, you compute the frequency response of the communication channel. If your system comprises a transmission line with frequency-varying values of $R, L, G,$ and C combined with some terminators, you can directly write down an expression for the frequency response.[36] Then, for every frequency on the dense grid, multiply X_k by the frequency response of the communication channel, H_k, and apply an inverse FFT to convert the resulting spectrum back into a time-domain waveform. The FFT method handles reflections at the source; reflections at the load; impedance mismatches within the communications channel; linear filters at the transmitter, receiver, or both; connector effects; and extended data patterns.

The FFT shines as an efficient computational tool for long transmission channels.

The FFT shines as an efficient computational tool for long transmission channels. Whenever the bulk transport delay of the channel exceeds the risetime of the driver, a time-step simulator must break the channel into many tiny sections, independently simulating each section. A simulation with N channel sections and M time steps requires $N \cdot M$ separate simulation activities. A frequency-based simulation, on the other hand, computes the performance of the communication channel as a whole, proceeding with only one computation (albeit a very complex one) for each frequency on the dense grid. When the number of simulated cable sections in the time-step method becomes large, the FFT-based approach computes much more rapidly.

The FFT method is far from perfect. It assumes the underlying system has a simple topology and is both linear and time-invariant. That means, for example, that you cannot analyze a diode termination (a nonlinear circuit) using the FFT method. Fortunately, most transceivers on long communication channels adhere to the linear-time-invariant assumption, so FFT-based simulation of serial-communications links usually delivers excellent results.

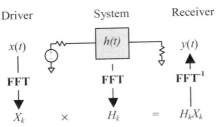

Figure 4.1—Canonical form of linear systems analysis.

---

[36] See Chapter 3, and Appendix C.

A final advantage of FFT-based simulation is really not an advantage of the method per se, but more an advantage of the mathematical spreadsheet tools used to do the computations. MatLab, Mathematica, and MathCad (my favorite) easily perform the required FFT operations and at the same time provide a highly flexible environment in which to optimize your design. All three tools provide automatic optimizing routines that can adjust the circuit parameters to improve just about any objective criteria. The criteria can include both time-domain and frequency-domain constraints, such as "adjust my transmit-equalizer component values to maximize the eye opening of the received data subject to minimizing the radiated emissions above 1.5 GHz." You can't come close to that kind of flexibility with SPICE.

Next time you face a challenging high-speed serial-communications problem, consider the advantages of frequency-based simulation before you go totally nonlinear.

POINTS TO REMEMBER

- Frequency-domain simulation gives you incredible control over the exact form of frequency-dependent losses, like the skin effect and dielectric-loss effect.
- Frequency-domain simulators may be easily programmed in any software spreadsheet application (like MatLab, Mathematica, or MathCad), giving you control over every aspect of the simulation, including searching for optimum and worst-case parameter values.
- Frequency-domain simulation applies only to linear systems.

4.2 APPROXIMATIONS TO THE FOURIER TRANSFORM

The frequency-domain model discussed in this chapter is based on the Fourier transform. The Fourier transform is one of many transformations that establish relations between the time domain and the frequency domain (Figure 4.2). With it, you can turn time-domain functions into frequency-domain functions, and vice versa. The most important aspect of the Fourier transform is that it changes time-domain convolution (difficult to compute) into frequency-domain multiplication (easy to compute).

Given a signal $x(t)$ and the impulse response of a filter $h(t)$, you can compute the output $y(t)$ by first Fourier-transforming both x and h to the frequency domain. Then, for all possible frequencies ω, form the product $X(\omega)H(\omega)$ and transform the resulting frequency-domain function $Y(\omega)$ back to the time domain using an inverse Fourier transform.

The Fourier transform is defined as

Fourier transform $$A(\omega) \triangleq \int_{-\infty}^{\infty} a(t) e^{-j\omega t} dt$$ [4.1]

Inverse Fourier transform $\qquad a(t) \triangleq \dfrac{1}{2\pi} \int_{-\infty}^{\infty} A(\omega) e^{j\omega t} dt \qquad$ [4.2]

Unfortunately, evaluation of the Fourier transformation requires the calculation of continuous-time integrals, which, unless a closed-form solution is available for your particular signal, is generally impossible to perform. Therefore, a shortcut has been developed for approximating Fourier transformations. The shortcut is called the DFT (discrete Fourier transform).

The DFT is a discrete-time approximation to the Fourier transform. It operates on a vector x_n with $n \in 0,1..(N-1)$, translating x_n into a discrete-frequency domain according to the following finite sum.

DFT $\qquad \mathbf{X}_k \triangleq \dfrac{1}{N} \sum_{n=0}^{N-1} \mathbf{x}_n e^{-j\frac{2\pi k}{N} n} \qquad$ [4.3]

Inverse DFT $\qquad \mathbf{x}_n \triangleq \sum_{k=0}^{N-1} \mathbf{X}_k e^{j\frac{2\pi k}{N} n} \qquad$ [4.4]

The DFT [4.3] is closely related to the Fourier transform [4.1] with the exception that both time and frequency have been rendered in a discrete form, changing the continuous integrations of [4.1] into discrete summations. With a suitable mapping between the continuous-time domain and the discrete-time domain, the DFT can successfully approximate the Fourier transform.

The fast-Fourier transform (FFT) is nothing more than a very clever implementation of equation [4.3]. It works only for particular values of N. The most common variety is the Cooley-Tukey FFT, which works only for N equal to a power of two. The FFT algorithm arranges the calculation so that the total effort required to accomplish the sum grows not in proportion to N^2 (as would a simple-minded matrix multiplication) but in proportion to

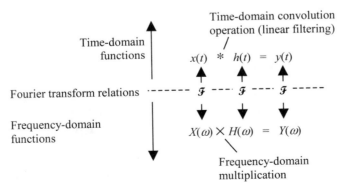

Figure 4.2—The Fourier transform $\mathcal{F}$ maps time-domain functions to frequency-domain functions, turning time-domain convolution into frequency-domain multiplication.

$N \log_2 N$. Its efficacy is so great that the DFT practically never appears in its direct form—the FFT universally has replaced it. Except for some very subtle differences involving coefficient quantization and sensitivity to rounding errors, the mathematical properties of the FFT and DFT are identical ([32], [33], [34] and [35]).

NOTE: There is some disagreement in the literature about whether the factor $1/N$ in equation [4.3] belongs in the forward DFT, in the inverse DFT, or perhaps split equally as $1/\sqrt{N}$ in both transformations. If you want the forward and inverse transformations to match so that for any time-domain vector **x**, you get $DFT^{-1}\left[DFT(\mathbf{x})\right] = \mathbf{x}$, then a factor of $1/N$ has to go *somewhere,* but it doesn't really matter where. Here I have shown the factor of $1/N$ in the forward transform as implemented in the MathCad function FFT( ). If your tool defines the FFT otherwise, then small adjustments will have to be made in the normalizing constants used with your FFT algorithm (see Section 4.5, "Normalizing the Output of an FFT Routine").

POINTS TO REMEMBER

➢ The DFT is a discrete-time approximation for the Fourier transform.

➢ The popular Cooley-Tukey FFT algorithm is a clever, highly efficient implementation of the DFT that works only for N equal to a power of two.

4.3 DISCRETE TIME MAPPING

The FFT requires that you identify the mapping between the continuous-time world and the discrete-time world. This is done with two parameters, ΔT and N.

The parameter ΔT defines the time interval between successive samples of the waveform to be transformed. The samples are evenly spaced. The spacing ΔT must be small enough to fairly represent the complete signal waveform without loss of significant information. This implies that on a scale of time commensurate with one sample, the signal waveform had better look smooth, not changing very much between samples.

For example, if you are working with a waveform having a 10% to 90% percent rise and fall time of t_{10-90}, then ΔT should be set anywhere from 4 to 100 times smaller than t_{10-90}. Gaussian-shaped rising and falling edges require only a 4x oversampling. Parabolic edges (as created by some IBIS simulators) require about 10x. Linear-ramp edges require much higher sampling rates. Plain square edges, no matter how fast you sample them, create special sampling problems that require fantastic sampling speeds.

The requirement for a small ΔT is roughly equivalent to a requirement that the *sampling rate* $1/\Delta T$ exceed twice the highest frequency contained in the sampled waveform. Precisely defining the highest frequency can be somewhat problematic. The general rule is that if the spectral content of your signal is down to –60 dB below the meat of the signal spectrum at frequency f_1, and if the spectrum above f_1 plummets at a rate of at least –40 dB from that point onward, then a sampling rate of $2f_1$ should provide a discrete-time conversion accuracy on the order of one part in one thousand.

One simple test of sampling efficacy is to halve the sampling interval and look for any observable change in the computed spectrum. A noticeable change indicates the original sampling rate was inadequate. Factors that dictate very high oversampling rates include outrageous requirements for time-domain accuracy, investigation of high-frequency EMC phenomena occurring at large multiples of the signal bandwidth, and inadequate low-pass filtering in the system you are trying to model. If you see results that change markedly with a small shift in the sampling phase, that's another sign of sample-rate inadequacy.

Another test of sampling sufficiency is made by observing the magnitude of your FFT output. If the magnitude as a function of frequency is not quickly diminishing toward zero at 1/2 the sampling rate, then your output will most certainly suffer noticeable amounts of aliasing (see Section 4.7, "Effect of Inadequate Sampling Rate").

When working with simulated signals that have linear ramps or square edges, I always include in the simulation a model of a scope probe with limited bandwidth. The scope probe model limits the spectral content of the observed signals (just as in real life), so you can safely use a finite sampling rate to process the simulated waveforms.

The parameter N defines the length of the time-domain sample vector. In conjunction with ΔT it defines the total window of time ($N \cdot \Delta T$) available within which the FFT must do its work. Always provide an N large enough to allow your simulated system to come to a steady-state condition before the end of the FFT time window. This condition comes about because FFT time, unlike time in the real world, wraps back around onto itself in a loop. Any residual tail in your waveform that "falls off the end of time" doesn't really fall off, but wraps back onto the beginning of time, superimposing onto and distorting your view of the initial circuit conditions.

In rare cases the wrapping helps. For example, when analyzing a clock waveform, if you establish the parameters such that $N \cdot \Delta T$ equals a precise multiple of the clock period, then the tail of the last clock cell overlaps with the beginning of the first period in a manner reminiscent of how the circuit would actually work in continuous time. Most often, though, the wrapping is a bothersome artifact of the FFT approximation and must be dealt with by providing sufficient additional time at the end of the FFT time window to allow your system to fully stabilize to within a negligible tolerance. When working with channels subject to the skin-effect the full stabilization time can be quite long.

When you plot discrete-time waveforms, it is handy to precompute a horizontal axis vector $\mathbf{t}$. The horizontal time axis $\mathbf{t}_n$ associated with each point $\mathbf{x}_n$ in the sampled time-domain vector depends on the index n and the step interval ΔT.

$$\mathbf{t}_n = n \cdot \Delta T \qquad [4.5]$$

POINTS TO REMEMBER

> ➤ The FFT requires two parameters: a sample interval ΔT and a sample vector length N.
> ➤ The spacing ΔT must be small enough to fairly represent the complete signal waveform without loss of significant information.

> Always provide an *N* large enough to allow your simulated system to come to a steady-state condition before the end of the FFT time window.

4.4 OTHER LIMITATIONS OF THE FFT

The FFT requires that your signal waveform have the same value at start and finish. This requirement is a consequence of FFT time wrapping, which makes the endpoint and beginning point actually *adjacent* in FFT time.

Another limitation of the FFT has to do with the maximum allowable rate of change for signals represented in discrete time. The Nyquist sampling theorem says that no band-limited signal can transition perfectly from one value to another in one time step without causing ripples in the adjacent samples. For example, if you try to work with a signal that incorporates a step change in signal amplitude at the end of the time window, it will cause ripples in both directions, distorting both the signal at the end of the time window and the samples near the beginning. To solve this problem, make sure the signal amplitude has *by design* the same amplitude (usually zero) at start and finish.

The normal way of harmonizing the beginning and ending of your excitation signal is to drive the system with a pulse, which first steps up, holds longs enough for you to see the resulting waveform, and then steps back down. After a suitable waiting interval, the system will have stabilized and the FFT time window may come to a close with zero excitation. The total FFT time window must therefore exceed the pulse duration plus one system stabilization time. I usually make the pulse duration half the width of the FFT time window and ensure that the total FFT time window exceeds twice the system stabilization time.

POINT TO REMEMBER

> The FFT requires that your signal waveform have the same value at start and finish.

4.5 NORMALIZING THE OUTPUT OF AN FFT ROUTINE

The DFT expression [4.3] doesn't have a clue about the original sampling rate used to form the discrete-time vector; therefore, it can't really compute the correct value of the Fourier transform without auxiliary scale factors. If you want your Fourier transformations to produce results with the correct amplitude, you must apply the scale factors every time you use either the forward or inverse transformation.

The mathematical spreadsheet package I use, called MathCad, provides a forward-transform FFT and its inverse IFFT defined as in [4.3] and [4.4]. When using these transformations in an environment with a sample interval of ΔT and a discrete-vector length of N, the appropriate scale factors work like this:

$$MyDFT(\mathbf{x}) \triangleq (N\Delta T) FFT(\mathbf{x}) \qquad [4.6]$$

$$MyInverseDFT(\mathbf{X}) \triangleq \frac{1}{(N\Delta T)} IFFT(\mathbf{X}) \qquad [4.7]$$

Other tools may define their FFT routines differently, requiring different scaling constants (see Section 4.10, "Checking the Output of Your FFT Routine").

MathCad routines FFT( ) and its inverse IFFT( ) are specialized for working with real-valued data sequences (as opposed to sequences with complex values). A real-valued data sequence $\mathbf{x}_n$ has the property that the FFT of that sequence, $\mathbf{X}_k$, is complex-conjugate-symmetric around the origin, which means in practical terms that if you know the output point $\mathbf{X}_i$, then you can trivially compute the output point $\mathbf{X}_{N-i}$. The MathCad functions FFT( ) and IFFT( ) therefore don't bother to generate, store, or use values of $\mathbf{X}_k$ for k greater than $N/2$. Since the top half of each frequency-domain vector associated with FFT( ) and IFFT( ) is unnecessary, the length of $\mathbf{X}$ is truncated to $(N/2) + 1$. The length of the time-domain vector $\mathbf{x}$ remains N.

4.5.1 Deriving the DFT Normalization Factors

When used as an approximation to the Fourier transform, the DFT requires a scaling factor. This scaling factor derives from [4.1], substituting for the integral an approximate summation carried out using the available samples taken at points in time $\mathbf{t}_n = n \cdot \Delta T$.

$$A(\omega) \approx \sum_{n=0}^{N-1} a(n \cdot \Delta T) e^{-j\omega n \cdot \Delta T} \Delta T \qquad [4.8]$$

Presuming that you wish only to compute values of the output on a dense grid of frequencies $\omega_k = 2\pi k/(N \cdot \Delta T)$, where $k \in 0,1..(N-1)$,

$$A\left(\frac{2\pi k}{N \cdot \Delta T}\right) \approx \sum_{n=0}^{N-1} a(n \cdot \Delta T) e^{-j\frac{2\pi k}{N \cdot \Delta T} n \cdot \Delta T} \Delta T \qquad [4.9]$$

The sampled values of $a(t)$ and $A(\omega)$ may be represented in vector notation:

$$\mathbf{x}_n \triangleq a(n\Delta T) \quad \text{for } n \in \{0,1..N-1\} \qquad [4.10]$$

$$\mathbf{X}_k \triangleq A\left(\frac{2\pi k}{N\Delta T}\right) \quad \text{for } k \in \{0,1..N-1\} \qquad [4.11]$$

which simplifies the appearance of [4.9]:

$$A\left(\frac{2\pi k}{N \cdot \Delta T}\right) \approx \mathbf{X}_k = \Delta T \sum_{n=0}^{N-1} \mathbf{x}_n e^{-j\frac{2\pi kn}{N}} = (N\Delta T) DFT[\mathbf{x}] \qquad [4.12]$$

Going in the other direction, suppose you have an expression for the Fourier transform $A(\omega)$ of a waveform $a(t)$. To approximate $a(t)$ using the inverse DFT, you first sample the frequency-domain function $A(\omega)$ according to [4.11], apply the inverse DFT to **X**, and then multiply by the scaling factor $1/(N \cdot \Delta T)$.

$$a(n\Delta T) \approx \mathbf{x}_n = \frac{1}{N\Delta T} DFT^{-1}[\mathbf{X}_k] \qquad [4.13]$$

The scaling factors apply to a DFT of any length, although when using the Cooley-Tukey FFT algorithm the length N will always be a power of two.

POINT TO REMEMBER

> Most FFT routines require external scale factors that depend on the sample interval ΔT and sample vector length N.

4.6 USEFUL FOURIER TRANSFORM-PAIRS

Use the standard functions in Table 4.1 to form test signals, data patterns, and pulses, or to feather the edges of fast-changing signals.

The first three entries reveal the DFT sequences necessary to create time-domain pulses and clock waveforms. If the pulse width in each case is set to an integral number of time samples, the time-domain transform will display no aliasing. That's nice for creating good-looking charts and graphs, but doesn't really hide the aliasing problem. If you shift the waveform by a nonintegral number of sample points (using the shift operator), the aliasing, also called Gibbs phenomenon, comes back. The only true solution to aliasing is to apply somewhere within your simulation a heavy-duty smoothing filter and sample on a dense grid of points with a spacing at least 4x finer than the risetime of the smoothing filter. If nothing else, you should be simulating the risetime of your oscilloscope probe or the bandwidth of your receiver using a Gaussian filter.

The next three entries depict various edge-smoothing filters. The linear filter leaves a square-edged signal with an unrealistically large amount of high-frequency content. The high-frequency content generates aliasing that must be fixed by either (1) further filtering the signal to limit its bandwidth, or (2) using an outrageously high sampling rate.

For general work I use the Gaussian filter the most. Provided the Gaussian risetime is set to a minimum of four sample points, it practically eliminates aliasing.

The quadratic filter lies between the extremes of the linear and Gaussian filters. If used as the only band-limiting filter in a simulation, it requires a sampling rate greater than the Gaussian filter but not as great as the linear-ramp filter. Quadratic responses are popular with some IBIS model proponents because the time-domain result is parabolic and therefore easy to compute. This filter synthesizes the same exact parabolic response.

If you must simulate ramp waveforms in your circuit, then I suggest you use the linear-ramp filter with the necessary risetime followed by a Gaussian filter with a much smaller (10x smaller) risetime, and then sample the composite signal on a dense grid 4x finer than the Gaussian filter risetime.

Table 4.1—Useful Fourier Transform Pairs

Fourier Transform	DFT
Pulse of width b	
$PulB(\omega) = \dfrac{1-e^{-j\omega b}}{1-e^{-j\omega \Delta T}} \Delta T$	$\textbf{PulB}_k = \begin{cases} b & \text{if } k = 0 \\ \dfrac{1-e^{-j\frac{2\pi k}{N}\frac{b}{\Delta T}}}{1-e^{-j\frac{2\pi k}{N}}} \Delta T & \text{otherwise} \end{cases}$
Pulse of width $(N/2)\Delta T$	
$PulN(\omega) = \dfrac{1-e^{-j\omega(N/2)\Delta T}}{1-e^{-j\omega \Delta T}} \Delta T$	$\textbf{PulN}_k = \begin{cases} (N/2)\Delta T & \text{if } k = 0 \\ \dfrac{1-e^{-j\pi k}}{1-e^{-j\frac{2\pi k}{N}}} \Delta T & \text{otherwise} \end{cases}$
Clock waveform with M complete cycles	
$ClkM(\omega) =$ $\dfrac{1}{1-e^{-j\omega \Delta T}} \dfrac{1-e^{-j\omega N \Delta T}}{1+e^{-j\omega \frac{N}{2M}\Delta T}} \Delta T$	$\textbf{ClkN}_k = \begin{cases} (N/2)\Delta T & \text{if } k = 0 \\ \dfrac{2M}{1-e^{-j\frac{2\pi k}{N}}} \Delta T & \text{if } \mathrm{mod}(k,2M)=M \\ 0 & \text{otherwise} \end{cases}$
Delay operator (delays by amount τ)	
$Dly(\omega) = e^{-j\omega \tau}$	$\textbf{Dly}_k = e^{-j\frac{2\pi k}{N\Delta T}\tau}$
Linear-ramp LPF with 10-90% rise-fall time r (set $q = 1.25 \cdot r$)	
$Lin(\omega) = \dfrac{\sin(\omega q/2)}{(\omega q/2)}$	$\textbf{Linr}_k = \begin{cases} 1 & \text{if } k = 0 \\ \dfrac{\sin(\pi k q/N\Delta T)}{(\pi k q/N\Delta T)} & \text{otherwise} \end{cases}$
Quadratic LPF with 10% to 90% rise/fall time r (set $q = 0.9045084972 \cdot r$)	
$Quad(\omega) = \left[\dfrac{\sin(\omega q/2)}{(\omega q/2)}\right]^2$	$\textbf{Quad}_k = \begin{cases} 1 & \text{if } k = 0 \\ \left[\dfrac{\sin(\pi k q/N\Delta T)}{(\pi k q/N\Delta T)}\right]^2 & \text{otherwise} \end{cases}$
Gaussian LPF with 10% to 90% rise/fall time r (set $q = .275 \cdot r$)	
$Gaus(\omega) = e^{-\omega^2 q^2}$	$\textbf{Gaus}_k = \begin{cases} 0 & \text{if } \left(\dfrac{2\pi k}{N\Delta T}\right) q > 7 \\ e^{-\left(\frac{2\pi k}{N\Delta T}\right)^2 q^2} & \text{otherwise} \end{cases}$

POINT TO REMEMBER

> ➢ Table 4.1 shows how to form FFT frequency vectors for test signals, data patterns, pulses, and feathered edges.

4.7 EFFECT OF INADEQUATE SAMPLING RATE

Aliasing is the word for the class of problems that result from an inadequate sampling rate. In old Western movies you will sometimes see an extreme form of aliasing cause wagon wheels to turn backwards. This happens because the movie camera takes discrete samples of the wheels at an inadequate rate. If the wheels turn almost one full spoke between each sample, it creates the illusion that the wheels are precessing slowly backwards in successive frames.

In simulations of high-speed digital systems an inadequate sampling rate causes a waveform to "wiggle around" as a function of precisely where it is sampled. For example, suppose you are sampling a continuous-time function with a step at time 1 and a small glitch near time 2.50.

$$x(t) = \begin{cases} 0 & \text{for } t < 1 \\ 1 & \text{for } t \geq 1 \\ \text{except } 0.5 & \text{for } 2.49 < t < 2.51 \end{cases} \quad [4.14]$$

Let the sampling interval ΔT be 1 second. As the samples t_n progress through the continuous-time waveform, you get different results depending on where the samples start. For example, samples starting at $t_0 = 0$ look like this:

$$t_n = \{0 \quad 1 \quad 2 \quad 3...\}$$
$$x_n = \{0 \quad 1 \quad 1 \quad 1...\} \quad [4.15]$$

A train of samples taken starting near $t_0 = 0.5$, however, would produce this totally different result:

$$t_n = \{0.5 \quad 1.5 \quad 2.5 \quad 3.5...\}$$
$$x_n = \{0 \quad 1 \quad 0.5 \quad 1...\} \quad [4.16]$$

Which is correct? The answer, of course, is neither. The waveform defined in [4.14] contains transitions far too quick to be sampled at such a pedestrian rate as 1 sample per second.

Figure 4.3 illustrates the kind of aliasing you are likely to encounter with the frequency-domain simulation if your sampling rate is too slow. This figure shows two rising edges. One is a linear ramp, the other a Gaussian rising waveform. The rise time in both

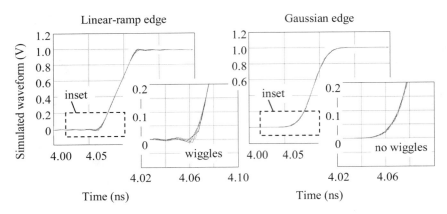

Figure 4.3—The Gaussian rising edge displays less wiggling (Gibbs phenomena) than the linear-ramp edge.

cases is set to 50 ps. The sample interval is 12.5 ps, providing four samples during each rising edge.

Four frequency-domain simulations are shown. In each simulation the time-domain waveform is sampled and converted to the frequency-domain. The frequency-domain vector is then delayed (using the frequency-domain delay operator) by 1/4, 1/2, and 3/4 of a sample time. The inverse FFT is then computed for each case, normalized, and the time-domain displays are then successively offset back in time to overlay the original signal. If the sampling rate were infinite, the time-domain shifting would precisely cancel the frequency-domain delay operator, and the waveforms would overlay perfectly. As you can see, they do not.

The linear-ramp edge shows a classic pattern of frequency-domain truncation *(Gibbs phenomena)*. The limited sampling rate imposes a maximum value on ω_k, a value which is apparently too small to capture all of the significant high-frequency information present within the signal. The missing high-frequency content induces extraneous wiggles in the signal before and after the rising edge indicating a sensitivity to precisely how the time-domain samples align with the waveform. If you've ever used a scope with an inadequately fast sampling rate, you have probably seen a similar effect—the sampled waveform appears to "jiggle" on the screen even when you know it contains no jitter.

The figure also shows a Gaussian-rising edge simulated using the same oversampling rate and the same delay procedures. The Gaussian rising edge clearly displays less Gibbs phenomena than the linear-ramp edge. With Gaussian edges an oversampling of 4x the risetime generates a sufficiently dense grid to produce a good representation of the underlying continuous-time signal regardless of the sampling phase alignment.

An FFT theorist would say that the frequency content of the Gaussian-filtered edge is sufficiently suppressed at the Nyquist rate ($0.5/\Delta T$ Hz) to avoid aliasing.

POINT TO REMEMBER

➢ In simulations of high-speed digital systems an inadequate sampling rate causes a waveform to "wiggle around" as a function of precisely where it is sampled.

4.8 IMPLEMENTATION OF FREQUENCY-DOMAIN SIMULATION

The example code in Table 4.2 uses MathCad syntax, although it could easily be rewritten into any mathematical spreadsheet notation. The MathCad symbol := means the variable on the left is assigned the value of the expression on the right.

The example in Table 4.2 simulates a pulse of length $N/2$, with Gaussian rising and falling edges having 10% to 90% rise/fall time equal to 40 ps ($4\Delta T$), and a delay equal to 4096 ps (($1/10$)$N\Delta T$). These factors were implemented using definitions **PulN**, **Gaus**, and **Dly** from Table 4.1. Such a specification might well represent the differential output of a driver with 1-V amplitude and 40-ps rise/fall time (see Figure 4.4).

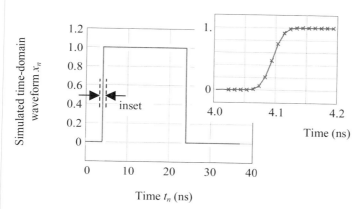

Figure 4.4—The time-domain signal x_n shows a pulse of length $(1/2)N\Delta T$, offset by a delay of 4096 ps = $(1/10)N\Delta T$. The inset reveals a Gaussian rising edge with a 10% to 90% risetime of 40 ps (four samples).

Assuming the differential driver is connected in a system configuration as shown in Appendix C, "Two-Port Analysis," the system gain G may be computed, sampled on the dense grid of frequencies ω_k to produce a frequency-domain vector $\mathbf{G}_k$, and then multiplied point-by-point times the vector $\mathbf{X}_k$. The frequency-domain result, once inverse-transformed to the time domain, represents the response of system $\mathbf{G}$ to the stimulus $\mathbf{X}$. The time-domain vector $\mathbf{y}$ will show the effects of all resistive losses, dielectric losses, bulk transport delay, and reflections within the transmission environment defined by $\mathbf{G}$.

$$\mathbf{Y}_k := \mathbf{X}_k \mathbf{G}_k \qquad [4.17]$$

$$\mathbf{y} := \frac{1}{N\Delta T} IFFT(\mathbf{Y}) \qquad [4.18]$$

Table 4.2—Example Code Showing FFT Simulation

Item	Expression	Units
Sampling resolution	$\Delta T := 10^{-11}$	sec
Length of sample vector	$N := 4096$	a power of two
Index to time points	$n := 0, 1 .. (N-1)$	integer
Horizontal axis for time-domain plots	$t_n := n\Delta T$	sec
Index to frequency points	$k := 0, 1 .. (N/2)$	integer
Horizontal axis for frequency-domain plots	$f_k := k/(N\Delta T)$	Hertz
Frequencies used to sample Fourier transform functions	$\omega_k := 2\pi f_k, \; k \in 0,1..(N/2)$	rad/sec
Pulse of width $(N/2)\Delta T$	$\mathbf{PulN}_k = \begin{cases} (N/2)\Delta T & \text{if } k = 0 \\ \dfrac{1 - e^{-j\pi k}}{1 - e^{-j\frac{2\pi k}{N}}} \Delta T & \text{otherwise} \end{cases}$	vector
Delay operator (delays by amount τ)	$\tau = (1/10)\,N\Delta T$ $\mathbf{Dly}_k = e^{-j\frac{2\pi k}{N\Delta T}\tau}$	vector
Gaussian LPF with 10% to 90% rise/fall time equal to $4\Delta T$	$q = 0.275 \cdot (4\Delta T)$ $\mathbf{Gaus}_k = \begin{cases} 0 & \text{if } \left(\dfrac{2\pi k}{N\Delta T}\right) q > 7 \\ e^{-\left(\frac{2\pi k}{N\Delta T}\right)^2 q^2} & \text{otherwise} \end{cases}$	vector
Example definition of signal in the frequency domain	$\mathbf{X}_k := \mathbf{PulN}_k \cdot \mathbf{Gaus}_k \cdot \mathbf{Dly}_k$	vector
Inverse transformation of frequency-domain vector $\mathbf{X}$ to produce time-domain vector $\mathbf{x}$ (see Figure 4.4)	$\mathbf{x} := \dfrac{1}{N\Delta T} IFFT(\mathbf{X})$	vector

4.9 EMBELLISHMENTS

4.9.1 What If a Large Bulk-Transport Delay Causes the Waveform to Slide Off the End of the Time-Domain Window?

The waveform doesn't really fall off the end; it wraps around to the beginning (remember, FFT time is circular). The circular-shifting effect is particularly troublesome on lines operated above the LC-mode boundary, where it is possible to generate enormous bulk delays (many hundreds of bits) without much signal attenuation. In those cases it is possible for the signal to have wrapped around your FFT time window many times without your having noticed. The position of the observed waveform then becomes extremely sensitive to the exact bulk delay (very confusing).

To fix this problem, remove the bulk delay by applying (multiplying point-by-point) a comparable but opposite delay. The inverse-delay function is computed using the delay operator **Dly** with parameter τ set to a negative value.

4.9.2 How Do I Transform an Arbitrary Data Sequence?

If the sampling rate is an exact multiple of the data bit time t_b, then you have an easy job ahead of you.

$$\frac{1}{\Delta T} = \frac{K}{t_b} \qquad [4.19]$$

where t_b is the data bit (or baud) time.

In this case the constant K is called the *oversampling ratio*.

First read the data pattern into a vector $\mathbf{x}_u$, where $u \in 0,1..(length(\mathbf{x})-1)$.

Now define a new vector $\mathbf{x}'$, which will hold a modified version of the input data sequence. The input sequence is expanded into the modified sequence by the method of repeating each value K times:

$$\mathbf{x}'_{u \cdot K + v} \triangleq \mathbf{x}_u$$
$$u \in 0,1..(length(\mathbf{x})-1) \qquad [4.20]$$
$$v \in 0,1..(K-1)$$

Use the normalized FFT to turn sequence $\mathbf{x}'$ into a frequency-domain entity $\mathbf{X}$ and then apply edge filtering $\mathbf{Gaus}_k$ and amplitude scaling ΔV in preparation for further processing.

$$\mathbf{X} = (N\Delta T) FFT(\mathbf{x}')$$
$$\mathbf{Y}_k = \mathbf{X}_k \mathbf{Gaus}_k \Delta V \qquad [4.21]$$

If the oversampling rate is not an exact multiple of the data rate, then you may construct the frequency-domain vector directly as a summation of variously delayed pulses. Take care that the full length of the data sequence fits within the FFT time window $N\Delta T$.

$$\mathbf{X}_k = \Delta V\, \mathbf{Gaus}_k\, \mathbf{PulB}_k \sum_{u=0}^{length(\mathbf{x})-1} \mathbf{x}_u e^{-j\frac{2\pi k}{N\Delta T} t_b u} \quad [4.22]$$

where parameter b for the pulse length **PulB** is set equal to the bit (or baud) interval t_b. In this example the data train has been given Gaussian rising and falling edges and an amplitude ΔV.

The data sequence $\mathbf{x}_u$ may be either binary (1–0) *or multivalued*.

4.9.3 How Do I Shift the Time-Domain Waveforms?

Occasionally you will want to directly shift a time-domain waveform in order to line it up with some other signal for comparison. The most general-purpose method for accomplishing a shift is to use modulo arithmetic on the index variable of the time-domain vector $\mathbf{x}$.

$$\mathbf{y}_n = \mathbf{x}_{\mathrm{mod}(n-m+N,N)} \quad [4.23]$$

The new vector $\mathbf{y}$ is a shifted copy of the old vector $\mathbf{x}$, delayed by m samples. The mod function avoids references to negative index values. It wraps the negative time points back around to index values near N, which is exactly how FFT time works. Values in FFT time that occur just *before* sample N will, if delayed by a few samples, wrap around to the front of the vector and show up just *after* time 0.

The time-domain circular shift is very handy for producing eye patterns, which are nothing more than multiple copies of $\mathbf{x}$, delayed by various integral numbers of bit times and overlaid on top of each other.

4.9.4 What If I Want to Model a More Complicated System?

Two-port analysis handles some pretty complicated situations. See the comments at the end of Appendix C.

4.9.5 What About Differential Modeling?

You can model your differential-mode signal as a single-ended signal with characteristic impedance Z_{DIFF}. A load of Z ohms connected across your differential line is modeled as a load of Z ohms to ground in the single-ended model. This approach models the differential signal, but not the common-mode signal, in your transmission environment.

Alternately, you could choose to model the odd-mode signal of your differential pair. The odd-mode signal equals the signal on one side of a purely differential transmission line, as opposed to the model mentioned previously, which models the *difference* between the two signals of a differential pair. The voltages in the odd-mode model are generally half those in the differential-mode model. An odd-mode model appears as a transmission line with an impedance equal to $\tfrac{1}{2}Z_{DIFF} = Z_{ODD}$. A load of Z ohms connected across your differential line is modeled as a load of $Z/2$ ohms to ground in the odd-mode model. This approach models the odd-mode signal, but not the even-mode signal, in your transmission environment.

The two-port analysis presented in Appendix C is not capable of modeling differential-to-common mode conversion problems or coupled transmission lines, but the frequency-domain approach can be made to work in that environment by using four-port matrices (two voltages and two currents on either side of each four-port matrix). A four-port transmission matrix, having four inputs and four outputs, is described by 16 internal cross-terms.

POINT TO REMEMBER

➤ Frequency-domain simulation handles some pretty complicated situations.

4.10 CHECKING THE OUTPUT OF YOUR FFT ROUTINE

Before using an unfamiliar FFT routine, you should always check the transform of a simple impulse at time 0 and also the transform of the same impulse delayed by one sample. Table 4.3 shows the results you should obtain. These checks will establish that you have used the correct scaling factor. It is common to find FFT routines that scale the forward transformation by $1/N$, $1/\sqrt{N}$, or 1, a simple problem to correct.

The checks also establish the direction of frequency rotation in the complex plane. Some FFT routines pervert the kernel of the transformation by using a plus j instead of a minus j in the exponent. This change inverts the frequency axis in the forward transform and reverses the time axis in the inverse direction, as compared to the transformation described in [4.3].

If your transform uses a plus-j kernel, you can fix it by forming the complex conjugate of the frequency domain result after running the perverted transform. The conjugation gives you the answer you should have gotten with a proper transform. In the reverse direction you must remember to conjugate each correct frequency-domain vector before submitting it to the perverted inverse transform for calculation of time-domain results. Alternately, you could use the perverted reverse transform for the forward transform, and vice versa, with attention paid to the scaling factor of N in both directions.

Table 4.3—FFT Test Cases

Object	Time-domain	Frequency-domain
impulse	$\mathbf{x}_n = \begin{cases} 1/\Delta T & \text{if } n=0 \\ 0 & \text{otherwise} \end{cases}$	$\mathbf{X}_k = 1$
shifted impulse	$\mathbf{x}_n = \begin{cases} 1/\Delta T & \text{if } n=1 \\ 0 & \text{otherwise} \end{cases}$	$\mathbf{X}_k = e^{-j\frac{2\pi k}{N}}$

POINT TO REMEMBER

➢ Before using an unfamiliar FFT routine, check the transform of a simple impulse at time 0 and also the transform of the same impulse delayed by one sample.

For further study see: www.sigcon.com

CHAPTER 5

PCB (PRINTED-CIRCUIT BOARD) TRACES

Pcb traces are used for extremely high-frequency connections within a single pcb and on backplanes connecting one or more pcbs. Figure 5.1 illustrates the microstrip, embedded microstrip, centered stripline, offset stripline, and coplanar waveguide structures as commonly employed in digital products.

The microstrip trace has a solid reference plane on only one side. The stripline lies between two solid reference planes. A coplanar waveguide has grounded metal on either side of the trace, with or without a solid reference layer underneath.

Analysis of pcb performance generally relies on three assumptions about the pcb-trace geometry.

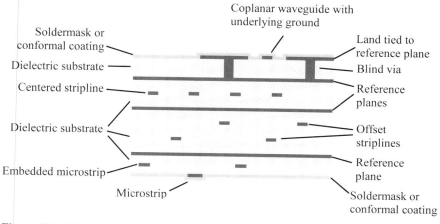

Figure 5.1—Digital products employ a rich variety of pcb transmission-line structures.

1. There exist well-defined uniform paths for the flow of both signal current and returning signal current.
2. The conductors are long compared to the spacing between the signal and return paths.
3. The conductors are shorter than the critical RC length l_{RC} (see [3.56] in Section 3.5.1, "Boundary of RC Region").

Conditions 1 and 2 make possible the use of the telegrapher's equations. In high-speed digital designs these conditions generally imply the existence of a solid conducting reference plane underlying the signal conductors, although it is also possible to satisfy the conditions using a differential pair routed without a reference plane.

In the absence of a well-defined return-current path—for example, in a two-layer pcb that may incorporate a grid of power and ground tracks instead of a solid reference plane— one is relegated to analyzing each circuit as a large lumped-element structure (or a series of lumped-element structures) with randomized values of inductance and capacitance. In this situation the performance of individual traces at high speeds will be highly variable. Terminations may improve performance, but will not be a reliable means of controlling ringing, as the trace performance will change every time the trace is moved in successive layout passes.

Solid reference planes, if adjusted to the correct distance from each trace layer, endow all traces with a uniform impedance regardless of their routing, making possible the use of terminations as a reliable strategy for controlling ringing. Solid planes also have the side benefit of greatly reducing crosstalk between signals.

Condition 3 ensures that the total DC resistance of the trace remains a small fraction of the characteristic impedance. A trace under this condition never exhibits any noticeable RC behavior. It transitions directly from the lumped-element region (which applies at low frequencies) to the LC region (at higher frequencies) without passing through the RC region. Such traces have a fairly constant value of characteristic impedance at all frequencies above the lumped-element region, a property which simplifies the analysis. Figure 3.2 illustrates the critical RC length for a typical 50-Ω pcb trace of 150-μm (5.9-mil) width at about 2 meters (78 in.).[37] Analysis of longer traces must take into account the varying nature of characteristic impedance and attenuation in the RC region.

On-chip interconnections do not meet condition 3. They suffer from very high resistance in the metallization layer, due to the use of extremely thin metallic plating, and even higher resistance in the poly layers. As a result, most on-chip interconnections do in fact operate in the RC region.

When analyzing pcb traces shorter than 25 cm (10 in.) operating at edge rates no faster than 500 ps (1 GHz bandwidth), you can generally ignore trace losses. They just aren't significant enough to worry about. On longer traces or at higher speeds, the skin and dielectric losses can become quite significant.

---

[37] The critical RC length is that length at which the outer boundary of the lumped-element region intersects the left edge of the LC region.

Point to Remember

> Analysis of pcb performance generally assumes:
>
> Well-defined uniform paths for both signal current and returning signal current.
>
> Conductors long compared to the spacing between the signal and return paths.
>
> Conductors shorter than the critical RC length l_{RC}.

5.1 PCB Signal Propagation

5.1.1 Characteristic Impedance and Delay

At this point in time, all major pcb CAD packages incorporate 2-D field solvers. These tools are extremely effective at calculating transmission-line impedance and delay, so much so that a listing of the old transmission-line impedance approximations would be of little use to the average reader. The values computed for impedance and delay correspond to the variables Z_0 and v_0 in the signal propagation model of Chapter 3.

If you don't already have a 2-D field solver, get one. If you absolutely cannot afford an automated tool, at least get a copy of Wadell [41]. His book lists all the closed-form approximation formulas for impedance and propagation delay of transmission-line configurations that were known as of the time of publication, including microstrips, striplines, some coplanar structures, and various differential arrangements. Failing that, this author modestly recommends the formula sets included in the appendices of [51].

In a practical sense, what you need to know about pcb-trace impedance is that it varies approximately with the square root of trace height, inversely with the square root of width, and inversely with the square root of the effective dielectric constant. These variations are not exact, but indicate generally the sensitivity of trace impedance to the parameters of width and height. The trace thickness is a secondary parameter.

For striplines, the effective dielectric constant (and thus the delay) is entirely determined by the dielectric constant of the surrounding substrate material. For microstrips, the effective dielectric is somewhat less (see 5.1.3.1, "Estimating the Effective Dielectric Constant for a Microstrip"). Examples are given in Table 5.1.

Because parameters for skin-effect resistance and dielectric loss are less broadly disseminated, I shall endeavor to provide some guidance on those matters.

Point to Remember

> If you don't already have a 2-D field solver, get one.

5.1.2 Resistive Effects

5.1.2.1 DC Resistance of Pcb Trace

The formula for the nominal DC resistance of a rectangular pcb trace is given in [2.41] (Section 2.4, "DC Resistance").

POINT TO REMEMBER

> ➤ A 1/2-oz copper pcb trace with 100-μm (3.9 mil) width has a DC resistance of 9.6 Ω/m. The DC resistance scales inversely with the width and inversely with the copper plating weight.

5.1.2.2 AC Resistance of Pcb Trace

The proximity effect for pcb traces takes hold at rather low frequencies on the order of a few megahertz. Below that frequency the magnetic forces due to changing currents in the traces are too small to influence the patterns of current flow. Low-frequency current in a pcb trace therefore follows the *path of least resistance*, filling the cross-sectional area of the trace, flowing uniformly throughout the signal conductor. That same current, as it returns to its source through the power-and-ground planes, tends to spread out in a wide, flat sheet, occupying as much of the surface area of the planes as possible on its way back to the source. That is the least resistive path through the planes.

Above ω_{LC} (a few Megahertz for most pcb traces) the magnetic forces surrounding a trace become quite significant, so much so that they change the patterns of current flow. The current is forced into a new distribution dependent not on the trace resistance, but on the trace inductance. This new distribution follows the *path of least inductance,* which is by definition that distribution of current that minimizes the total stored magnetic field energy surrounding the trace. Since the path of least inductance differs by definition from the path of least resistance, you would expect effective trace resistance to be increased when current is forced to flow in this high-frequency mode.

The magnetic fields constrain the high-frequency current in two ways. First, the current is confined to a shallow band of depth δ around the perimeter of a conductor, increasing the apparent resistance of the trace. This increase in apparent resistance is called the skin effect.

Next, the magnetic fields distribute the current around the perimeter of the conductor in a non-uniform manner (the proximity effect). The relative increase in the apparent resistance of the conductor due to this non-uniformity, above and beyond what you would expect from the action of the skin effect alone, is represented by the proximity factor k_p.

The proximity effect draws signal current towards the side of a microstrip facing the reference plane, or that side of a stripline that faces the nearest reference plane. It simultaneously concentrates the returning signal current on the reference plane, forming it into a narrow band that flows on the plane, staying mostly underneath the signal trace.

The increase in resistance of a typical high-speed digital signal conductor due to the proximity effect (above and beyond simple consideration of the skin depth and trace

5.1.2 • Resistive Effects

circumference assuming a uniform current distribution) typically ranges from 25% to 50%. This fraction remains fixed at all frequencies high enough for the effect to have taken hold (typically somewhere in the vicinity of ω_{LC}).

Another similar-sized increase in resistive dissipation occurs due to the nonuniform distribution of current on the reference plane. At DC, you can assume that current spreads widely in all directions, so the DC reference-plane resistance should be (nearly) zero. At high frequencies, however, the returning signal current in the planes flows most heavily just underneath your trace, with wide tails of residual current falling off (at least) quadratically as you move away from the trace on either side. The current in this reference-plane distribution, because it remains well concentrated in a narrow band underneath the signal conductor, dissipates a noticeable amount of power, typically on the order of 25% to 50% of the signal-trace dissipation. This fraction remains fixed at all frequencies high enough for the effect to have taken hold.

The proximity factor takes into account the additional resistance due to redistribution of current on both the signal conductor and the reference planes.

Table 5.1 lists values of the AC resistance R_0 and proximity factor k_p for typical trace geometries used in high-speed digital designs. The value of R_0 already incorporates k_p; I'm just showing the k_p in case you want to use it to work out other similar configurations later and so you can see how gradually it changes with trace geometry.

In Table 5.1, parameter w is the trace width, h the height of the bottom surface of the trace above the lower reference plane, and b (striplines only) the separation between the planes (Figure 5.2). The microstrips listed in Table 5.1 assume 1-oz copper (including plating) with a conformal coating (soldermask) consisting of a 12.7-μm (0.5-mil) layer having a dielectric constant of 3.3. The stripline examples assume 1/2-oz bare copper. The resistance data was developed using a method-of-moments magnetic field simulator with 120 segments equally spaced around each pcb trace, with the current linearly interpolated across each segment, and three points at each corner rounded to fit a circular arc. The author estimates the accuracy of the data generated by this simulator at approximately +/− 2%.

Table 5.1 assumes a dielectric constant of 4.3. If you are working with a different dielectric constant, you should know that what matters most for determining the proximity factor is the ratio w/h. The ratio t/h, assuming the trace is thin compared to its width, is of secondary importance. Traces with similar ratios of w/h inherit similar values of k_p regardless of the dielectric constant.

Table 5.1 does not take into account either surface roughness or the inevitable random variations in trace width that occur in real boards.

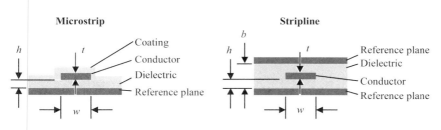

Figure 5.2—Printed-circuit trace configurations are defined by parameters w, t, h, and b.

Table 5.1—AC Resistance and Skin Effect Loss at 1 GHz for Selected Single-Ended Microstrips and Striplines

h mil	w mil	b mil	R_{AC} Ω/in.	R_{AC} Ω/m	k_p	α_r dB/in.	α_r dB/m	Z_0 Ω	ϵ_{re}
	50-ohm microstrips, 1-oz copper with soldermask								
10	17.9	n/a	0.420	16.5	1.953	.036	1.44	50	3.24
9	16.0	n/a	0.460	18.1	1.928	.040	1.57	50	3.24
8	14.0	n/a	0.513	20.2	1.903	.045	1.75	50	3.24
7	12.1	n/a	0.578	22.7	1.878	.050	1.98	50	3.24
6	10.2	n/a	0.663	26.1	1.851	.058	2.27	50	3.25
5	8.3	n/a	0.780	30.7	1.820	.068	2.67	50	3.25
	60-ohm microstrips, 1-oz copper with soldermask								
10	12.5	n/a	0.531	20.9	1.776	.038	1.51	60	3.16
9	11.1	n/a	0.583	23.0	1.754	.042	1.66	60	3.16
8	9.7	n/a	0.648	25.5	1.731	.047	1.85	60	3.16
7	8.3	n/a	0.730	28.8	1.704	.053	2.08	60	3.16
6	6.9	n/a	0.839	33.0	1.675	.061	2.39	60	3.16
5	5.6	n/a	0.981	38.6	1.650	.071	2.80	60	3.16
	70-ohm microstrips, 1-oz copper with soldermask								
10	8.8	n/a	0.665	26.2	1.631	.041	1.62	70	3.10
9	7.8	n/a	0.728	28.7	1.609	.045	1.78	70	3.10
8	6.7	n/a	0.813	32.0	1.584	.050	1.99	70	3.10
7	5.8	n/a	0.905	35.6	1.566	.056	2.21	70	3.10
6	4.8	n/a	1.036	40.8	1.542	.064	2.53	70	3.10
5	3.8	n/a	1.215	47.8	1.516	.075	2.97	70	3.10
	50-ohm striplines, 1/2-oz bare copper								
5	3.5	10	1.498	59.0	1.513	.013	5.12	50	4.30
5	5.1	15	1.144	45.0	1.597	.099	3.91	50	4.30
7	5.8	15	1.014	39.9	1.586	.088	3.47	50	4.30
5	5.7	20	1.074	42.3	1.655	.093	3.67	50	4.30
7	7.2	20	0.865	34.0	1.646	.075	2.96	50	4.30
10	8.0	20	0.782	30.8	1.637	.068	2.67	50	4.30
5	6.2	30	1.033	40.7	1.714	.090	3.53	50	4.30
7	8.5	30	0.782	30.8	1.732	.068	2.67	50	4.30
10	11.1	30	0.609	24.0	1.729	.053	2.08	50	4.30
15	12.7	30	0.533	21.0	1.722	.046	1.82	50	4.30

NOTES—AC parameters R_{AC}, α_r, Z_0, ϵ_{re}, and FR-4 dielectric $\epsilon_r = 4.30$ are all specified at 1 GHz. The microstrip examples assume copper traces of 1-oz thickness (including plating) with $\sigma = 5.98 \cdot 10^7$ S/m plus a conformal coating (soldermask) consisting of a 12.7 μm (0.5-mil) layer having a dielectric constant of 3.3. The stripline examples assume copper traces of 1/2-oz thickness with no plating.

5.1.2 • Resistive Effects

The other columns listed in Table 5.1 are

- α_r—resistive trace loss (dB/m 1 GHz)
- Z_0—real part of characteristic impedance (Ω at 1 GHz)
- ϵ_{re}—dielectric constant (real part of effective relative electric permittivity) at 1 GHz

The propagation velocity v_0 (m/s) and delay t_p (s/m) are found from ϵ_{re}, where $1/t_p \triangleq v_0 \triangleq c/\sqrt{\epsilon_{re}}$ and $c = 2.998 \cdot 10^8$ m/s. The DC resistance constant k_a is unity for all the single-ended pcb trace examples in Table 5.1.

Modern 2-D electromagnetic field solvers automatically calculate the proximity effect when reporting skin-effect resistance.

POINTS TO REMEMBER

- Low-frequency current in a pcb trace therefore follows the *path of least resistance*, filling the cross-sectional area of the trace.
- The skin effect confines high-frequency current to a shallow band of depth δ around the perimeter of a conductor.
- The proximity effect draws signal current towards the side of a microstrip facing the reference plane, or that side of a stripline that faces the nearest reference plane.
- The increase in resistance of a typical high-speed digital signal conductor due to the proximity effect (above and beyond simple consideration of the skin depth and trace circumference assuming a uniform current distribution) typically ranges from 25% to 50%.
- Another similar-sized increase in resistive dissipation occurs due to the nonuniform distribution of current on the reference plane.
- Traces with similar ratios of w/h inherit similar values of k_p regardless of the dielectric constant.

5.1.2.3 Calculation of Perimeter of Pcb Trace

When using [2.43], [2.44], or [2.63] to estimate the total series resistance R_{AC} of the conductor, taking into account both skin effect and proximity effect, the perimeter of the signal conductor is defined

$$p = 2(w+t) \quad [5.1]$$

where p is the perimeter of the conductor (m), and

w and t are the width and thickness of the signal conductor respectively (m).

5.1.2.4 Very Low Impedance Pcb Trace

In the limit as the trace height h approaches zero (and the impedance descends towards zero), the trace concentrates all its current on the reference-plane facing side of the trace, with none on the reverse side. This doubling of the current density on one side doubles the apparent resistance. In addition, the trace suffers from an identical distribution of current on the reference plane, doubling the result again, for a total of $k_p \approx 4.0$.

5.1.2.5 Calculation of Skin-Effect Loss Coefficient for Pcb Trace

The skin-effect loss in units of nepers per meter is given by [3.114]:

With R_0 a fixed parameter:
$$\alpha_r \triangleq \frac{1}{2}\frac{R_0}{Z_0}\sqrt{\frac{\omega}{\omega_0}} \quad \text{neper/m} \qquad [5.2]$$

where R_0 is the skin-effect resistance (Ω/m) at the particular frequency ω_0.

NOTE: This formula applies only at frequencies above the onset of the skin effect.

If the line resistance $R(\omega)$ is specified as a function of frequency, the resistive loss coefficient in units of nepers per meter equals,

with R as a function of ω,
$$\alpha_r \triangleq \frac{1}{2}\frac{R(\omega)}{Z_0} \quad \text{neper/m} \qquad [5.3]$$

For microstrips, Pucel et al. [45] (and [52], which is reproduced in [41]) provide empirical closed-form approximations for the calculation of α_r. A spot-check comparison of values from Pucel's closed-form expressions with the values listed for resistive loss in Table 5.1 reveals a corroboration error of about 2%.

5.1.2.6 Popsicle-Stick Analysis

Article first published in *EDN Magazine*, March 7, 2002

You can simulate the magnetic field surrounding a pc-board stripline using a rubber sheet and a Popsicle stick. First, stretch the rubber sheet over a rectangular frame. Then square off the end of the Popsicle stick and push it into the rubber (Figure 5.3). Your simulation shows the magnetic-field potential surrounding the pc-board trace.

    A close look at the precise forces bearing on the end of the stick teaches you a lot about stripline behavior. First, note that the force required to support the rubber sheet is not uniform at all points at the end of the stick. Because the stick is flat across its end face and because the rubber sheet is assumably lightweight, the forces acting on the stick are practically zero everywhere except around the edges of its end face. As a result, you don't really need any meat in the center of the stick; a hollow stick would do just as well.

5.1.2 • Resistive Effects

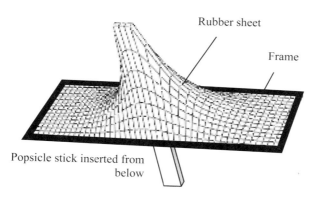

Figure 5.3—A rubber sheet simulates the magnetic field structure surrounding a stripline.

Next, check the distribution of force around the edges of the end face. It's not uniform either. With a round, hollow stick, you might expect a uniform distribution of force where the edges of the stick meet the rubber, but when a rectangular stick presses against the rubber, the corners bear a disproportionate share of the weight. You know this if you have ever erected a big tarp for a picnic shelter. Once you put the corner posts in place and stretch the tarp so it is taut, your work is nearly complete. The corners do most of the work of holding up the tarp, and all you must do to complete the job is provide minor supports every so often to keep the sides and middle from sagging (or leave them to sag). In Figure 5.3, the wire frame lines near the tip of the protrusion show the same effect—the sheet is stretched hardest at the corners.

Now to the point of this article, which is the correspondence between the rubber-sheet world and the electrical world. The flat end of the stick represents the cross section of a pc trace, the rectangular frame represents reference planes above and below the trace, and the protrusion of the rubber sheet mimics the magnetic potential. The slope of the rubber sheet at any point indicates the magnetic-field intensity. Contour rings drawn on the rubber sheet at constant heights above the frame show familiar "magnetic lines of force" typically used to depict magnetic fields (Figure 5.4).

> **The greatest concentrations of current prevail on the inside-facing surfaces of opposing conductors.**

A direct correspondence exists between the forces acting on the end of the Popsicle stick, which induce curvature in the rubber sheet, and the density of electrical currents flowing in a pc-trace of a similar cross-section profile. These currents likewise induce curvature in the magnetic potential. Where the rubber sheet is flat (no curvature), there is no force, and in the electrical world, there is no current. Where the rubber sheet is most tightly curved (at the corners), the force acting on the stick and also the density of current are greatest.

From this simple analogy you may conclude that high-frequency current flows only around the periphery of a trace, not in the middle. Such is the case at frequencies that are sufficiently high that the skin depth shrinks to much less than

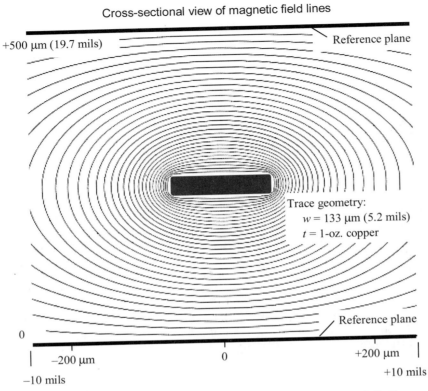

Figure 5.4—The magnetic field is most intense near the ends, and especially the corners, of a stripline trace.

the conductor thickness, a condition required for the analogy to hold. Some other conclusions that flow easily from Popsicle-stick analysis include the following:

1. The density of current near the corners of a trace profile, such as the forces acting near the corners of a Popsicle stick, exceeds the density of current elsewhere. Because traces tend to be thin and wide, a concentration at the corners has the same general effect as a concentration on either side of the trace. This effect is called *edge-current concentration,* the mathematics of which are predicted by rubber-sheet analysis.

2. If you make a differential pair by placing a signal conductor and a return conductor in close proximity, with the signal stick pressing up on the rubber sheet while the return conductor presses down (Figure 5.5), the greatest concentrations of current will prevail on the inside-facing surfaces of the two conductors where the magnetic potential traverses from peak to valley in the least distance (Figure 5.6). The concentration of current at the inside-facing edges is called the *proximity effect.*

5.1.2 • Resistive Effects

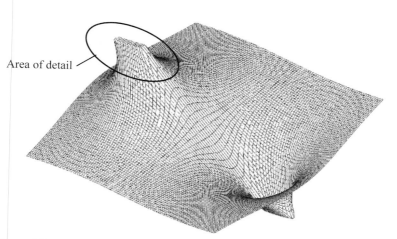

Figure 5.5—The magnetic potential surrounding a differential stripline is elevated at one conductor and depressed at the other.

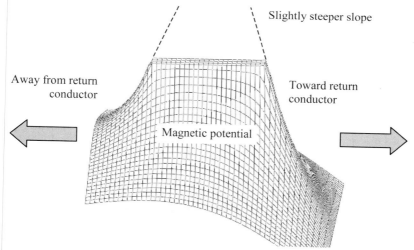

Figure 5.6 (Detail from Fig. 5.5)—A steeper slope prevails on the side facing the other conductor.

3. If a stripline trace is offset towards one reference plane, the slope of the rubber sheet on that side (and thus the force) must necessarily increase, indicating a larger concentration of current on the side of the trace facing the reference conductor than on the opposing side. This effect increases the apparent resistance of a signal trace above the value predicted, assuming a uniform distribution of current around the periphery of the conductor.

I know people who have spent years studying the mathematics behind these three conclusions but still don't understand *why* they happen. You've just grasped the whole subject in a few minutes, and you got to enjoy a Popsicle!

POINT TO REMEMBER

➤ You can simulate the magnetic field surrounding a pc-board stripline using a rubber sheet and a Popsicle stick.

5.1.2.7 Nickel-Plated Traces

High-Speed Digital Design Online Newsletter, Vol. 5, Issue 6
Paul Greene writes

I am an engineer with Rexcan Circuits, a pcb manufacturer in Ontario. We are evaluating various processes for the use of immersion gold. We are depositing approximately .00012 in. nickel followed by .000005 in. gold over the copper pads. I am investigating the effect of two alternate procedures:

1. Coating all the traces/pads with nickel and then applying soldermask, versus
2. Applying soldermask over the traces first and then coating only the pads with nickel and gold.

We have been advised that due to the changes to the skin effect caused by the Ni/Au on the traces, we could be building in a problem for high-frequency RF designs. Can you find time to comment? I have run some characteristic impedance tests at around 65 Ohms using a TDR tester and not seen much difference. Is nickel a real problem?

I hope you don't mind my contacting you, but your Web page has lots of good stuff in this area!

Reply

Thanks for your interest in *High-Speed Digital Design*. I've received a number of letters since writing "Steel-Plated Power Planes," pointing out that nickel is magnetic too, and a well-established chemistry exists for plating it onto copper. That's an interesting point. Of course the magnetic permeability of nickel is nowhere near that of steel, so you won't get nearly as dramatic an effect, but it might be worth investigating.

Regarding your specific question, I've often wondered about the same issue. I presume you are familiar with the concept of the skin effect and how current at high frequencies flows only on the outer surface (skin) of your conductor, not in the middle. Because of the high magnetic permeability of nickel, the skin-effect resistance on the nickel-plated side of your conductor will be considerably higher than that on the bare-copper side (core side).

You might be tempted to think that this will be okay because even if one side of the trace is messed up because of the nickel plating, you've still got a good copper surface on the bottom of the trace. The bottom-side surface acts in parallel with the nickel-plated side, so even if the nickel-plated resistance were infinite, the overall resistance of the configuration you might think would be no more than twice as bad as an all-copper trace. Unfortunately, that is a *very* poor analogy. At high frequencies the current distributes itself around the periphery of your trace in a

5.1.2 • Resistive Effects

pattern that minimizes the total inductance of the trace configuration *without regard to the skin resistance of the trace*. That is, if you change the skin-effect resistance on one side (by plating), then in the limit at very high frequencies you hardly change the distribution of current around the periphery of trace at all.

That action is contrary to the way current behaves at DC. At DC, current distributes itself to minimize the total dissipated power. For example, imagine two resistors R_t and R_b in parallel, both with resistance 2 Ω. The overall (parallel) resistance is 1 Ω. If you double the value of R_t (changing its value to 4 Ω), the DC resistance of the parallel combination becomes $(4\Omega \cdot 2\Omega)/(4\Omega + 2\Omega) = (4/3)\Omega$. Driven with a constant DC current, you get less current in R_t, but more in R_b, in a combination that minimizes the total dissipation. No matter how high you make the value of R_t, the resistance of the parallel combination never gets bigger than 2 Ω.

At high frequencies, the same effect does not prevail. At high frequencies, the current distributes itself in a way that minimizes the overall inductance (this minimizes the energy stored in the magnetic field surrounding the circuit). In a pcb trace, this means the ratio of currents on the top and bottom of the trace are fixed by the inductance and will not respond to modest changes in the surface resistivity of the two surfaces.

Going back to the example of resistances R_t and R_b, suppose R_t represents the resistance of the top surface of your microstrip trace and R_b the bottom surface, both with resistance equal to 2 Ω. In response to a DC current of 1A, each resistor initially dissipates 0.5 W, for a total dissipation of 1 W. If you double the resistance of the top surface without changing the distribution of current, the power dissipated in R_t doubles while the power in R_b remains at 0.5 W. The total dissipation jumps to 1.5 W. You could look at this and say that the effective resistance of the combination went up by 50% (to 1.5 times the original value).

If you multiply the resistivity of the top surface in the present example by 10, the effective resistance goes up to $(0.5) \cdot 10 + (0.5) = 5.5$. Continued increases in surface resistivity of the top side make almost unlimited increases in the overall effective resistance.

Let's calculate how bad the effect can get for a nickel-plated microstrip.

- The resistivity of nickel exceeds that of copper by a factor of $k = 4.5$.
- The relative magnetic permeability of nickel at 1 GHz is in the range of 5 to 20 (take $u_r = 10$ as a nominal value).
- The increase in surface resistance of nickel at 1 GHz (above and beyond that of copper at 1 GHz) equals the square root of $k \cdot u_r$, which works out to about 6.7.
- The current density on the top side of a 50-ohm FR-4 pure-copper microstrip contributes about 1/3 the total dissipation. If you increase that 1/3 of the dissipation by a factor of 6.7, then I would expect an overall increase in resistive trace loss by a factor of $(1/3) \cdot 6.7 + (2/3) = 2.9$, roughly tripling the resistive trace loss.
- The effective length at which you may use a high-speed pcb trace varies almost inversely with resistive trace loss.

At frequencies on the order of 1 GHz, the nickel-plating cuts in third the effective useful length of your traces.

I checked the skin depth of nickel at 1 GHz (see equation [2.42]) and found it's about 1.4 μm (0.055 mil), much thinner than your nickel plating. If you could make the nickel plating as thin as the gold, it won't raise the trace resistance, but I bet it doesn't work as a barrier layer if it's that thin.

In a TDR waveform, any series resistance present in a conductor causes an upward tilt to observed waveform. You can think of this as the trace showing slightly less impedance at first (high frequencies), then gradually transitioning to a larger value of impedance as time goes by (lower frequencies). The amount of tilt is related to the amount of series resistance. I predict your nickel-plated traces will show a greater upward tilt than your bare-copper traces. That is one way you can determine the extent of the effect (ultimately, this test measures the magnetic permeability, and thus the purity, of your nickel).

If you look carefully at the TDR step edge that returns from the far end of a long (perhaps 10-in.) line, you should see a noticeable degradation in the risetime. This degradation will be worse on the nickel-plated traces than on the bare copper traces.

The effect of nickel plating is real and commonly understood in the microwave community.

POINT TO REMEMBER

> At frequencies on the order of 1 GHz, nickel-plating the top surface of a microstrip cuts in third the effective useful length of the trace.

5.1.3 Dielectric Effects

Microwave designers worry a lot about dielectric loss. The dielectric-loss problem is particularly acute when constructing high-Q circuits intended to ring without signal amplitude loss for many cycles. As a result, microwave designers often choose ceramic substrates, like alumina, that have excellent dielectric loss properties in the gigahertz regime. Digital designers typically avoid high-Q circuit topologies and so are not nearly as sensitive to dielectric loss.

For FR-4 digital circuit board applications at operating frequencies below 1 GHz (corresponding to the highest frequency of interest in the rising and falling edge spectra of a system with rise and fall times of 500 ps), at distances up to 10 inches, you may safely ignore dielectric losses. At longer distances or at higher speeds, dielectric losses can become quite significant.

POINTS TO REMEMBER

> For FR-4 digital circuit board applications with risetimes of 500 ps or slower, at distances up to 10 inches, you may ignore dielectric losses.

> At longer distances or at higher speeds, dielectric losses can become quite significant.

5.1.3.1 Estimating the Effective Dielectric Constant for a Microstrip

A stripline trace contains all (or most of) its electric fields within the dielectric cavity formed between two solid conducting reference planes. The dielectric properties of stripline traces therefore depend only on the dielectric constant and loss tangent of the pcb material.

Microstrips, on the other hand, contain only part of their electric field within the dielectric layer. The remainder spews up into the air. The microstrip configuration therefore constitutes a type of *mixture dielectric* with properties intermediate between the properties of the dielectric substrate and air. The effective permittivity for a microstrip depends on the geometry of the trace and the proportion of total field energy stored in each of the two regions.

Many empirical closed-form approximations exist for the calculation of effective permittivity. The book by Gupta [39] provides numerous references for such approximations. The particular form of Gupta's equations reported here comes from a paper by I. J. Bahl and R. Garg [50].

$$F \triangleq \begin{cases} (1+12(h/w))^{-1/2} + 0.04(1-(w/h))^2 & (w/h \leq 1) \\ (1+12(h/w))^{-1/2} & (w/h \geq 1) \end{cases} \quad [5.4]$$

$$C \triangleq \frac{\epsilon_r - 1}{4.6} \frac{t/h}{\sqrt{w/h}} \quad [5.5]$$

$$\epsilon_{re} \approx \frac{\epsilon_r + 1}{2} + \frac{\epsilon_r - 1}{2} F - C \quad [5.6]$$

where w, h, and t are the width, height, and thickness respectively of a pcb microstrip (m),

ϵ_r represents the real part of the relative electric permittivity (dielectric constant) of the substrate material,

ϵ_{re} represents the real part of the *effective* relative electric permittivity (also called effective dielectric constant) of the complete transmission configuration,

C and F are empirically derived constants, and

no limits to accuracy or applicability were supplied with this equation.

A spot-check comparison of values from Bahl and Garg's closed-form expression with the values for effective dielectric constant listed in Table 5.1 reveals a corroboration error of about 3%.

For striplines, $\epsilon_{re} = \epsilon_r$.

POINT TO REMEMBER

> A microstrip has dielectric properties intermediate between the properties of the dielectric substrate and air.

5.1.3.2 Propagation Velocity

From the effective dielectric constant ϵ_{re}, you may determine the propagation velocity.

$$v_0 = \frac{c}{\sqrt{\epsilon_{re}}} \quad [5.7]$$

where c is the velocity of light in vacuum, $2.998 \cdot 10^8$ m/s,

ϵ_{re} is the *effective* relative electric permittivity (also called effective dielectric constant) of the complete transmission configuration, and

v_0 is the velocity of propagation (m/s).

Examples showing the variation in ϵ_{re} (and thus trace delay) for various microstrip trace geometries appear in Sections 5.1.3.6, "Passivation and Soldermask," 12.4, "Differences Between Stripline and Microstrip Delay," and 6.10.1, "Differential (Microstrip) Trace Impedance." The effective relative permittivity for a stripline always equals the relative permittivity of the substrate material.

5.1.3.3 Calculating the Effective Loss Tangent for a Microstrip

For striplines, the loss tangent of the transmission configuration equals the loss tangent of the substrate material.

For microstrips, you may determine the effective loss tangent of the transmission configuration given three pieces of information:

$\tan\theta$ The loss tangent of the substrate material,

ϵ_r The real part of the relative electric permittivity (also called dielectric constant) of the substrate material, and

ϵ_{re} The real part of the effective relative electric permittivity (also called effective dielectric constant) of the complete transmission configuration.

Assume you know $\tan\theta$ and ϵ_r, but not ϵ_{re}. Here are several ways to glean ϵ_{re}.

> From knowledge of the effective propagation velocity, presumably given by a 2-D electromagnetic field solver, you may invert equation [5.7],
> From the direct output of a 2-D electromagnetic field solver,
> By interpolation from the values of ϵ_{re} listed in Table 5.1, or
> By the closed-form approximation [5.6].

5.1.3 • Dielectric Effects

Once all three parameters are known, the effective loss tangent of the microstrip configuration is determined (see [2.87]):

$$\tan \theta_e = \frac{\epsilon_r}{\epsilon_{re}} \frac{\epsilon_{re} - 1}{\epsilon_r - 1} \tan \theta \qquad [5.8]$$

where ϵ_r is the real part of the relative electric permittivity (also called dielectric constant) of the substrate material,

ϵ_{re} is the real part of the *effective* relative electric permittivity (also called effective dielectric constant) of the transmission configuration,

$\tan \theta$ is the loss tangent of the substrate material, and

$\tan \theta_e$ is the *effective* loss tangent of the transmission configuration.

Microstrips do have a small advantage over striplines in regards to the loss tangent, which, due to the partial propagation in air, is somewhat less than what would be experienced by a stripline. For example, using values from Table 5.1, $\tan \theta_e$ for a 50-ohm microstrip implemented in FR-4 ($k = 4.3$) is approximately 90% of the native $\tan \theta$ of the underlying substrate.

5.1.3.4 Dielectric Properties of Laminate Materials (Core and Prepreg)

Core and prepreg laminate materials are now available in a staggering array of types and variations. A *core* laminate is a piece of insulating material with copper sheet bonded to both sides, ready for etching. A *prepreg* laminate is a sheet of insulating material, often in a partly cured state, that is inserted between etched cores to build up a finished multilayer pcb. The prepreg sheet, once heated and pressed in the stacking operation, cures into a material very similar to the insulation used in the cores.

Table 5.2 lists generic properties of selected materials used in pcb construction. Please remember that these are gross ballpark numbers for materials in pure form. All manufacturers add fillers to their resin mixture to improve properties such as toughness, foil adhesion, pot life, and so forth. The filler materials modify the dielectric and mechanical properties. The purpose of the table is to indicate in a general way the magnitude of the improvement to be had from switching between various materials.

In most high-speed digital products, the core or prepreg laminate comprises a fabric of fine threads embedded in a solidified resin. The threads may be made of glass, quartz, Kevlar (a trademark of DuPont), or some other durable material. Names associated with those materials are e-glass, s-glass, quartz, and aramid (Kevlar) fibers. The fabric imparts a great deal of mechanical stability to the finished core but creates a fundamentally inhomogeneous material with an uneven surface.

Table 5.2—Properties of Selected Materials in Pure Form

Material	ϵ_r 1 MHz	tan θ 1 MHz	Specific gravity g/cm³	Source
Natural materials				
Hard vacuum	1.0000	—	0	
Air	1.0005	—	0.0012	at STP, *CRC Handbook*
Water	80		0.998	at 20 °C, *CRC Handbook*
Pcb resins				
Epoxy (as used in FR-4)	3.8	0.049	1.54	ϵ_r is extrapolated from Harper [36], p. 8.46; tan θ and specific gravity inferred from Park/Nelco specification of N4000-2 laminate
Cyanate ester	3.	0.023	1.29	ϵ_r is extrapolated from Harper [36], p. 8.45; tan θ and specific gravity inferred from Park/Nelco specification of N8000 laminate at 1 GHz; values as low as 0.005 have been reported for tanθ by IBM [37]
Polyimide	3.3	0.029	1.33	ϵ_r comes from *www.zeusinc.com*, a manufacturer of polyimide resin; tan θ inferred from Park/Nelco specification of N7000 laminate at 1 GHz; specific gravity from *www.maropolymer.com*
PTFE	2.1	< .0002	2.14 to 2.20	*www.mdmetric.com*, a manufacturer of PTFE resin

The resin fills the spaces between fibers, forming a sturdy matrix with a smooth surface for supporting traces, vias, and other pcb structures. Popular resin systems include various epoxies, cyanate ester, polyimides, and polytetrafluoroethylene (PTFE, also known by the trade name Teflon).

Fiber materials are selected on the basis of strength, coefficient of thermal expansion, dielectric constant and loss, hardness (which affects the lifetime of your drill bit), and cost.

Resin systems are selected on the basis of strength, glass transition temperature T_g, dielectric constant and loss, and cost. T_g is the temperature near which a resin begins to soften and become gooey, and its mechanical properties undergo substantial change. For example, above T_g most pcb resins undergo substantial z-axis expansion. A human wouldn't particularly notice the expansion, but since it exceeds the thermal expansion of copper, it can easily rip apart pcb vias. The pcb resin materials in Table 5.2 are listed in order of increasing T_g, ranging from ordinary FR-4 epoxies in the range of 125 °C to 150 °C to BT/epoxy (not listed) at 185 °C, cyanate ester at 245 °C, and polyimide at 285 °C [44]. The PTFE material doesn't have a glass transition temperature, it just melts, but only at very high temperatures (so high that it makes a great cooking surface).

There are other categories of laminates, including ceramic structures used in multichip modules and some space applications. Ceramics are mechanically stable and can have an exceptionally low dielectric-loss tangent combined with a high dielectric constant. These

5.1.3 • Dielectric Effects

Table 5.2 (continued)—Properties of Selected Materials in Pure Form

Material	ϵ_r 1 MHz	tan θ 1 MHz	Specific gravity g/cm³	Source
Pcb fibers				
e-glass	5.8	.0011	2.54	Harper [36], p. 8.26
s-glass	4.52	.0026	2.49	Harper [36], p. 8.26
Quartz	3.5	.0002	2.20	Harper [36], p. 8.26
Aramid (Kevlar)	4.0	.001	1.40	Harper [36], p. 8.26
Other substrates				
Alumina 99.5%	10.1	0.0001 to 0.0002		T. C. Edwards [38]
96%	9.6	0.0006		
85%	8.5	0.0015		
Si (high resistivity)	11.9	0.001 to 0.01		T. C. Edwards [38]
GaAs	12.85	0.0006		T. C. Edwards [38]
Single-crystal sapphire	9.4, 11.6[1]	0.00004 to 0.00007		T. C. Edwards [38]
Fused quartz	3.8	0.0001		T. C. Edwards [38]

NOTE (1)—Sapphire is an anisotropic medium and so has different dielectric constants in different directions.

properties make ceramics ideal for flip-chip mounting of microwave circuits. Microwave engineers like the high dielectric constant of alumina ($\epsilon_r \approx 10$) because it shrinks the size of the printed structures they use to produce $\frac{1}{4}\lambda$ resonators and filters. The biggest disadvantage of ceramics is their incredibly high cost.

Inhomogeneities in a fabric-resin laminate ultimately limit the size of the thinnest dielectric that can be produced from that combination of materials. Obviously, the laminate can be no thinner than one fiber diameter. In practical implementations the laminate can be no thinner than a few fiber diameters to account for crossings of the fibers within the weave and to prevent fibers sticking up out of the surface. Laminates as small as 75 microns (3 mils) seem the practical limit today, although considerable research is ongoing to reduce this limit [36].

The fiber size also limits the minimum practical trace dimensions useful in high-speed applications. For example, a trace laid directly on top of one individual fiber, as compared to a trace laid on top of the resin-filled space between two fibers, would experience a different value of effective electric permittivity and thus inherit different values of impedance and delay. Visible evidence of this effect is observed in the present technology when using 3-mil cores with 3-mil trace widths. There is some hope that future all-polymer materials might alleviate the need for fibers and permit the construction of ever-smaller circuitry.

The resin systems in Table 5.2 have dielectric constants generally smaller than the listed pcb fibers. Raising the proportion of resin in a mixture therefore generally *decreases* the dielectric constant of the resulting laminate. Exceptions to this rule include mixtures of epoxy or polyimide resin with quartz fibers, for which the effective dielectric constant would hardly change with the proportions of the mixture.

The specifications for resin/glass mixtures usually list the proportion of resin by *weight*, whereas the formulas for computing the effective permittivity of the resulting mixture rely on the proportion of resin by *volume*. To convert from one system of units to the other, first assume you are given two proportions by weight, w_a and w_b, with the proportions adding to unity. Further assume the densities of the two material are known (d_a and d_b, in g/cm³). The volume of material a equals w_a/d_a, the volume of material b is w_b/d_b, and the volumetric proportion v_a of material a, out of the total volume of material present, equals

$$v_a = \frac{\frac{w_a}{d_a}}{\frac{w_a}{d_a} + \frac{w_b}{d_b}} \quad [5.9]$$

Table 5.3 lists the dielectric properties of a few representative practical laminates. These materials may be used to construct either core or prepreg layers. The resin constitutes 50% by weight of the resin/glass mixture in all but the PTFE laminate examples. Other percentage mixtures are possible. For example, the construction of very thin cores (perhaps 2 to 3 mils) sometimes calls for a uniaxial glass fabric with a higher than normal resin content. Such a layer would most likely display a smaller than normal dielectric permittivity. At some frequencies (notably 1 GHz) the material specifications call out a dielectric constant but no corresponding value of loss tangent.

Table 5.3 indicates for some materials a variation in ϵ_r versus frequency. In an ideal constant-loss-tangent material the slope of ϵ_r relates precisely to the loss tangent (tan θ) according to [2.86] and [2.88]. Such a material displays a smooth and continuous reduction

Table 5.3—Dielectric Properties of Practical Laminates

Trade name	1 MHz		1 GHz		10 GHz	
	ϵ_r	tan θ	ϵ_r	tan θ	ϵ_r	tan θ
Park/Nelco N4000-2 epoxy-resin/e-glass	4.4	0.027	4.1			
Park/Nelco N7000 polyimide-resin/e-glass			3.9		3.8	0.016
Park/Nelco N8000 cyanate ester/e-glass			3.7		3.5	0.011
Neltec NH9318 PTFE					3.18	.0024
Rogers RT/duroid 5580 PTFE					2.2	.0009

5.1.3 • Dielectric Effects

in ϵ_r with increasing frequency. For every decade of increase in frequency, the reduction equals $10^{-\frac{2}{\pi}\theta}$.

This discussion might lead you to believe that Park/Nelco 4000-2 should display a per-decade reduction in ϵ_r equal to a factor of $10^{-\frac{2}{\pi}0.027} = 0.96$. Over the three-decade range from 1 MHz to 1 GHz, the reduction would amount to a factor of $1000^{-\frac{2}{\pi}0.027} = 0.89$, reducing the dielectric constant from 4.4 at 1 MHz to $4.4 \cdot 0.89 = 3.91$ at 1 GHz.

Note, however, that the specification calls out a dielectric constant of 4.1 at 1 GHz, a value indicating less degradation in the dielectric constant than that suggested by the dielectric-loss tangent specification. The discrepancy is resolved by observing that of course the dielectric-loss tangent is not a constant value across all frequencies, but varies. The number 0.027 is a *worst-case* specification for the dielectric-loss tangent, not an average value. Apparently, the loss tangent over the broad range from 1 MHz to 1 GHz averages something more like 0.016 (even though it may somewhere attain the peak value of 0.027).

Since you never know how close to the limit the loss tangent may lie, or at what frequencies it may approach its peak value, the most straightforward option in simulation is to use the worst-case value for loss tangent at all frequencies.

POINTS TO REMEMBER

> - Core and prepreg laminate materials are now available in a staggering array of types and variations.
> - The core or prepreg laminate comprises a fabric of fine threads embedded in a solidified resin.
> - Inhomogeneities in a fabric-resin laminate ultimately limit the size of the thinnest dielectric that can be produced from that combination of materials.

5.1.3.5 Variations in Dielectric Properties with Temperature

Email correspondence received March 5, 2002

We recently made some measurements of the "S" parameters of an FR-4 backplane with VHDM connectors and paddle cards. We repeated the measurement at several temperatures. The magnitude of the variations with temperature surprised us (see Figure 5.7). What causes these kinds of changes?
—Jeff Sonntag

Reply

Here are descriptions of three effects I know about.

Changes in ϵ_r. The dielectric constant of FR-4 changes, if I remember correctly, by as much as +/– 10% (20% total variation) over an extended –20 °C to

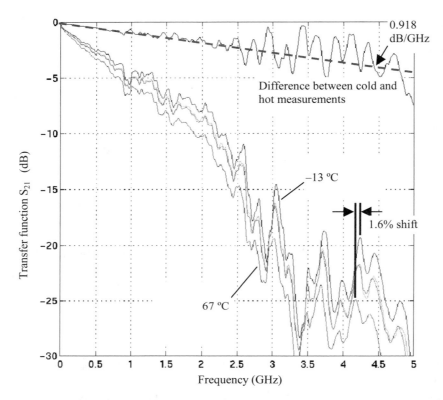

Figure 5.7—The difference in trace loss from –13 °C to 67 °C is a straight-line function of frequency for this backplane. *(Figure courtesy of Accelerant Networks, a manufacturer of digital adaptive transceivers for communications at 5 Gb/s and beyond.)*

80 °C temperature range. You can corroborate this using a simple capacitance meter and a bare (double-sided copper) board in your temperature chamber.

An increase in capacitance, if the loss tangent stayed fixed, would decrease the line impedance and increase the delay. Since the dielectric loss equations have to do with the ratio of line delay to risetime, more delay would by itself give you more dielectric loss even if the loss tangent stayed the same, with the loss increasing in proportion to the square root of the change in ϵ_r.

The skin-effect loss has to do with the ratio of series resistance to line impedance, so it too will change with temperature, getting worse as the line impedance drops in proportion to the square root of the change in ϵ_r.

From your plots, it looks like all the resonant effects shift slightly *lower* at high temperatures, corroborating the concept of a bigger ϵ_r and larger delay when running hot. The shift in your case looks only like about 1.6%, indicating a total change in ϵ_r of 3%, which is less than I would have expected, meaning that perhaps the resonances I am seeing are an artifact of some part of your test setup that doesn't change much with temperature. Anyway, a bare-board capacitance

5.1.3 • Dielectric Effects

measurement will clear up precisely what portion of your result is due directly to changes in ϵ_r.

Changes in the resistivity of copper. The thermal coefficient of resistance in copper is 0.0039, meaning that if the resistance of a copper wire at room temperature is R, then the resistance at a temperature 1 C higher would be R(1 + .0039). Over the range 0 °C to 70 °C, the resistance of copper wires should vary 28%. Of course, this variation changes only the skin-effect portion of your loss, which is not the lion's share of the problem above 1 GHz.

Changes in the loss tangent with frequency. There are few references to the magnitude of this effect. Synoptics presented data that looks very much like yours for PVC-insulated category-3 cables to the Ethernet 10BASE-10 committee. Its data prompted recommendations that non-plenum-rated category-3 cables never be used in attics because when they get too hot the dielectric loss skyrockets and the system would no longer meet its loss budget. I've not seen equivalent curves for FR-4.

You can see in your curves that the difference between the low- and high-temperature measurements (in units of dB) is a straight-line function of frequency, roughly .918 dB/GHz. Such straight-line behavior is the hallmark of an increased dielectric loss.

Conclusions: I would say your biggest problem is the change in loss tangent. A higher-T_g material should show a markedly smaller change. Higher-T_g materials tend to be more stable in all their mechanical and electrical properties.

I'm glad you designed continuous adaptation into the equalizer for your backplane transceiver, as it is apparently much needed.

POINT TO REMEMBER

> ➢ The dielectric loss of a backplane may change substantially with temperature.

5.1.3.6 Passivation and Soldermask

Copper corrodes quickly when exposed to oxygen. Copper traces on outer layers may be protected from corrosion by passivation or by coating them with an inert material.

Passivation refers to any treatment of the metal surface that renders it less reactive. For example, plating traces with tin or nickel produces a surface less susceptible to oxidation. Also, there are some organic catalysts that have been proposed for passivation (see box "How Passivation Works").

Any form of plating increases the thickness and width of a trace, lowering somewhat its characteristic impedance. Pcb fabrication shops normally compensate for the decrease in impedance by reducing the width of the etched trace prior to plating. As a result, for the same impedance, a composite plated structure will never achieve as low a value of DC resistance as an unplated, bare-copper trace.

The AC resistance of the trace may be adversely affected as well, particularly if the plating material used is a magnetic nickel alloy. All materials with high magnetic permeability suffer from a shallow skin depth and correspondingly large sheet resistivity at

> **How Passivation Works**
>
> At a microscopic level, passivation actually works not by retarding oxidation but by promoting it. Materials like copper or steel oxidize at a slow pace, allowing the oxygen to permeate deep into the metal, corroding and ultimately destroying the entire structure. A material like zinc oxidizes much more rapidly. Zinc quickly forms a microscopically thin, but highly impermeable, oxide layer that prevents further penetration of oxygen. Zinc is the basis of the *galvanizing* process used to protect steel from oxidation.
>
> Some organic molecules can act as catalysts to promote the quick buildup of an impermeable oxidation layer on copper. These catalysts may be applied to an entire board surface after etching. They don't require plating and don't produce dangerous metal residues.

high frequencies. The increased sheet resistivity exaggerates resistive trace losses. The amount of increase is not predicted by commonly available 2-D electromagnetic field solvers. Those tools assume you have enough sense to stay away from nickel plating on long, high-speed traces.

Inert protective coatings serve a variety of purposes [44].

Protect the copper traces from oxidation. Every exposed piece of metal that can't be protectively coated must be passivated. Examples of structures requiring passivation include solder pads and edge-connector fingers.

Protect the substrate from humidity. Most substrate materials readily absorb water vapor from the air. In epoxy-glass materials the water vapor can enter microscopic interstitial voids between the glass fibers and the epoxy matrix, traveling great distances into the interior of the substrate. The water reduces the dielectric strength of the substrate and greatly increases its dielectric loss.

Increase insulation resistance between traces. The inevitable accumulation of dirt and water degrades the insulation resistance between exposed metallic contacts on the surface of a board. In high-voltage circuits it can cause arcing (sparks). An insulating coating covering the traces prevents conduction from one trace, through the dirt, to the next trace.

Decrease spurious shorts between traces. During the reflow soldering process, any water trapped within the solder paste explodes into steam, spewing tiny solder balls all over your board. If the traces are covered, the solder balls are less likely to fall where they short anything out. Wave-solder processes similarly benefit from protective coatings, which prevent solder being dragged out of the pot onto the wrong places, causing shorts.

Control movement of solder during soldering. During reflow soldering, a via placed too close to a solder pad tempts the solder paste to slide over to the via and disappear down the hole. A thin line of solder mask drawn between the solder pad and the via acts as a solder dam, preventing the solder from moving to the via.

5.1.3 • Dielectric Effects

If the inert coating is applied prior to soldering, it is usually called a *soldermask*. Coatings applied after assembly are called *conformal coatings*. Inert coatings may be applied in either sheet (dry) form or as a liquid.

Any coating increases the total mass of dielectric material near the trace, decreasing somewhat its characteristic impedance. Pcb fabrication shops normally compensate for the decrease in impedance by reducing the width of the etched trace prior to plating, with a corresponding adverse impact on the DC and AC resistance.

All coatings that are screened, curtain-coated, or dipped will flow in a viscous manner after application. Even if initially applied in a uniform thickness, the material will sag off the tops of traces, leaving a thinner coating on the top of each trace and a pool of material bunched along the side. The exact geometry of the coating, and therefore its effect on trace impedance, is impossible to precisely calculate.

Table 5.4 illustrates the effect of either having or not having a soldermask overlay applied to a typical pcb trace. The coating is assumed 0.5-mils thick with a dielectric constant of 3.3. The trace widths in the table have been adjusted to compensate for the presence (or absence) of the coating. In each case the coating slows the signal propagation by a small amount and increases the resistive trace loss (because of the trace width adjustment).

Comparing the traces at $h = 10$ mils to the ones at $h = 5$, the larger traces seem less affected by the presence of the coating. This example illustrates a general principle: The larger the trace, the easier it is to maintain precise control over impedance.

Soldermask properties vary considerably. Many have not been investigated for high-

Table 5.4—Effect of Soldermask Coating on Single-Ended Microstrip Traces

Trace type	h	Coating	w	R_{AC} Ω/m	α_r dB/m	ϵ_{re}
50-ohm microstrip	10	yes	17.9	16.54	1.44	3.24
	10	no	18.2	16.35	1.42	3.17
	5	yes	8.3	30.71	2.67	3.25
	5	no	8.8	29.66	2.58	3.08
60-ohm microstrip	10	yes	12.5	20.91	1.51	3.16
	10	no	12.9	20.49	1.48	3.07
	5	yes	5.6	38.63	2.80	3.16
	5	no	6.1	36.80	2.66	2.96
70-ohm microstrip	10	yes	8.8	26.17	1.62	3.11
	10	no	9.3	25.26	1.57	2.96
	5	yes	3.8	47.83	2.97	3.10
	5	no	4.3	44.75	2.98	2.86

NOTE (1)—AC parameters R_{AC}, α_r, Z_0, ϵ_{re}, and FR-4 dielectric $\epsilon_r = 4.30$ are all specified at 1 GHz.

NOTE (2)—These microstrip examples assume copper traces of 1-oz thickness (including plating) with $\sigma = 5.98 \cdot 10^7$ S/m. When the conformal coating (soldermask) is present, it consists of a 12.7-μm (0.5-mil) layer having $\epsilon_r = 3.3$ at 1 GHz.

NOTE (3)—The propagation velocity v_0 (m/s) and delay t_p (s/m) are found from ϵ_{re}, where $1/t_p \triangleq v_0 \triangleq c/\sqrt{\epsilon_{re}}$ and $c = 2.998 \cdot 10^8$ m/s.

frequency use. A suggested range of dielectric constants from 3.3 to 4.2 appears in the reference text by Brian Wadell [41].

POINT TO REMEMBER

> ➢ Copper traces on outer layers may be protected from corrosion by passivation or by coating them with an inert material.

5.1.3.7 Dielectric Properties of Soldermask Materials

There is a dizzying array of soldermask materials available, including some new organic polymers that quickly catalyze a very thin, but highly impermeable, oxide layer. Table 5.5 illustrates some of the better-known choices. Good sources for further information about soldermask coatings are [36], [41] page 450, [46], and [47].

5.1.3.8 Calculation of Dielectric Loss Coefficient for Pcb Trace

The dielectric loss in units of nepers per meter is given by [3.131]:

$$\alpha_d \triangleq \frac{1}{2} \frac{\theta_0 \omega}{v_0} \left(\frac{\omega}{\omega_0} \right)^{-\theta_0/\pi} \text{ neper/m} \qquad [5.10]$$

At frequencies near ω_0 the term $(\omega/\omega_0)^{-\theta_0/\pi}$ is very nearly unity and may be ignored.

Table 5.5—Dielectric Properties of Representative Soldermask Layers

Trade name	Finished thickness (mil)	ϵ_r (1 MHz)	tan θ (1 MHz)
Dry-film photo-imageable			
Dynamask KM (epoxy)	3	4.23	0.031
Vacrel 8000	3 or 4	3.6	0.033
Vacrel 8100	3 or 4	3.8	0.042
Liquid photo-imageable			
Lea Ronal OP SR 550 (epoxy/acrylate)	0.15–0.50	3.5	.042
WR Grace AM-300 (epoxy/acrylate)	0.10–0.55	3.5	0.011
M and T Photomet 1001 (acrylate)	2–6	4.3	0.007
Ciba-Geig Probimer 52 (epoxy)	0.05–0.17	3.7	0.002
Dynachem EPIC SP-100 (epoxy)	0.05–0.17	4.07	0.035
NOTE—Data adapted from Brzozowski [36], Tables 8.27 and 8.28.			

5.1.4 Mixtures of Skin Effect and Dielectric Loss

Long, high-speed pcb traces operate in a zone influenced by both skin effect and dielectric losses. Both mechanisms attenuate the high-frequency portion of your signals, but in slightly different ways. Figure 5.8 illustrates 36 mixtures of varying amounts of skin and dielectric effects, showing the step response in each case.

The top row of the chart, working from left to right, shows step response waveforms corresponding to successively increasing amounts of dielectric loss. The dielectric loss in each case is rated in nepers of attenuation at frequency ω_0 (one neper equals 8.6858896 dB).

The left side of the chart, working from top to bottom, shows step-response waveforms corresponding to successively increasing amounts of skin-effect loss. The middle of the chart displays various combinations of both types of loss.

The horizontal axis for each waveform is calibrated in units of $2\pi/\omega_0$ seconds per division. For example, if the specification frequency for skin effect and dielectric-loss calculations is set to 1 GHz ($\omega_0 = 2\pi \cdot 10^9$ rad/s), then the horizontal axis reads out at 1 ns/div. If your loss calculations fall outside the range of values listed in the chart, try recalculating your loss coefficients assuming a different value of ω_0. By scaling ω_0, you can always ensure that your total loss numbers land somewhere within the values listed on the chart.

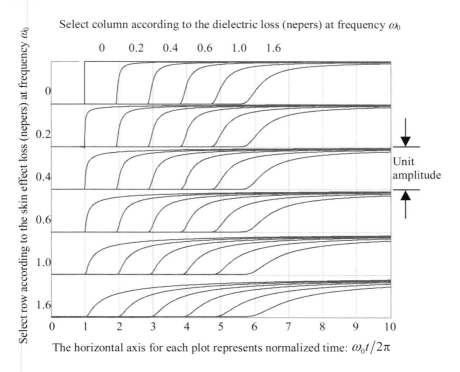

Figure 5.8—Given the same degree of attenuation at frequency ω_0, the skin effect produces a step response with a sharper initial rise and longer tail than does the dielectric effect.

The vertical axis for each waveform is one unit (full-scale response) per division. Waveforms in successive columns are each offset horizontally by one unit of time purely for convenience in reading the display. These waveforms were calculated using the technique of frequency-domain modeling.

Both the skin effect and the dielectric-loss effect degrade digital signals in the same fundamental way—by smearing out the rising and falling edges—but there are some differences. For the same degree of loss at frequency ω_0, the skin effect produces a steeper initial rise, but a longer, more lingering tail, than does the dielectric effect.

In the frequency domain, the differences in the slope of the initial rise imply that dielectric effect induces a steeper, more severe roll-off at high frequencies than does the skin effect, consistent with the general principle that dielectric attenuation (in nepers or dB) varies directly with frequency, while skin-effect attenuation varies only in proportion to the square root of frequency.

POINT TO REMEMBER

> ➢ The skin-effect step produces a sharper initial rise, but a longer, more lingering tail, than does the dielectric effect.

5.1.5   Non-TEM Modes

The way to reduce skin-effect loss is to use big, fat traces held at a great distance above the reference planes. The width of the trace provides a greater surface area for the flow of current, lowering the AC resistance. The great height is required to provide a reasonable impedance given the enormous trace width. The big-fat-trace approach commonly appears in microwave designs in combination with a very low-loss dielectric substrate. This is the standard approach in the microwave industry for controlling trace loss; however, for digital applications *there is a catch*.

First, you should know that adjacent-trace crosstalk scales *quadratically* with trace height. That implies that a very generous spacing must be applied to large parallel traces.

Second, you should know that very large traces, while they do not lose any signal power, sometimes fail to deliver all the power in a step edge coincident in time. For example, a trace may support multiple modes of propagation, called non-TEM modes. Each non-TEM mode has its own propagation velocity, so that even if the input power is coupled coincidently into the various modes at the near end of the line, it will not emerge in phase at the far end. The resulting *all-pass filter* response disperses the rising edge and can induce severe overshoot and ringback.

The next article discusses the general nature of the non-TEM mode problem and suggests an upper limit to the trace height for multi-gigabit digital designs.

5.1.5.1   Strange Microstrip Modes

Article first published in *EDN Magazine*, April 26, 2001

By now, many of you are probably used to looking at the electromagnetic-field-pattern pictures that most signal-integrity simulators produce. These simulators

5.1.5 • Non-TEM Modes

show the pattern of electric and magnetic fields surrounding a microstrip trace (Figure 5.9). From these field patterns, a simulator can determine the quasistatic values of capacitance and inductance per unit length and, from those values, the characteristic impedance and line delay.

Wait a minute. The quasi-what?

The quasistatic values of capacitance and inductance. *Quasistatic* are the values of capacitance and inductance that you get at low frequencies, near DC.

Do quasistatic values differ from real-world values?

At low frequencies, where the signal wavelength looms far larger than the trace height, the quasistatic values are the correct values.

What about higher frequencies? Is there a place where quasistatic calculations don't work?

Yes there is, and our industry is rushing madly toward it. For example, a 10-Gbps serial data stream with a rise and fall time of 35 psec has a maximum bandwidth (−6 dB) of approximately 15 GHz. The wavelength in FR-4 corresponding to 15 GHz is approximately 0.37 in. If you place this signal on a tiny trace, 10 mils wide and 5 mils above the nearest reference plane, the ratio of signal wavelength (0.37) to trace height (0.005) is a comfortable value of 74 to 1. No problem. If, on the other hand, in an attempt to mitigate skin-effect loss, you place the same signal on a huge, fat trace, 120 mils wide and 60 mils high, the ratio of signal wavelength (0.37) to trace height (0.060) is only 6 to 1. Such a small ratio causes big trouble.

What goes wrong?

Whenever the wavelengths of the signals conveyed approach the dimensions

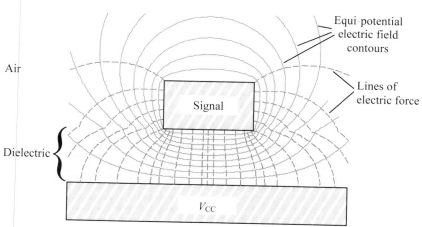

Figure 5.9—Contours of constant voltage (solid lines) encircle the signal conductor; electric lines of force (dotted lines) connect the signal conductor to the reference plane.

of your traces, strange modes of propagation begin to appear. Reference [39] provides a simple proof of the inevitability of these modes. You can't escape them. The idea that electromagnetic fields propagate straight down a microstrip structure in a simple and uniform manner is incorrect. To use a ray-tracing analogy, if the trace floats sufficiently high above the reference plane, a portion of the high-frequency signal power actually bounces up and down between the top surface of the dielectric layer (where the trace lies) and the underlying reference plane.

So it bounces. Doesn't all the signal power end up at the far end of the line anyway?

Yes, it does, but with different timing from what you may have anticipated. Part of the signal propagates in the normal TEM (transverse electric-and-magnetic) mode straight down the trace, arriving all together at one time. Another part bounces up and down along the way, spending part of its time interacting with the air above the trace and arriving at a different time. This difference in delay for various parts of the signal is called *microstrip dispersion*.

> **For normal digital signaling on FR-4 pc boards at 10 Gbps, you may use any trace height up to 20 mils without encountering significant microstrip dispersion.**

If different parts of a signal arrive at different times, doesn't the received waveform look distorted?

Yes. That's the whole problem with the microstrip configuration. At frequencies high enough that the signal wavelength becomes comparable with the trace height, the microstrip introduces unavoidable distortion. Full-wave analysis of the microstrip predicts received waveforms that have what looks like severe overshoot and ringing, even if the line is perfectly terminated.

Is this effect similar to multimode dispersion in fiber optics?

Yes. As an optical signal bounces around within a multimode fiber, various modes of propagation spend different proportions of the time in the core of the fiber versus in the cladding. Because the core and cladding have different dielectric constants, the different modes arrive at slightly different times. Manufacturers of low-dispersion multimode fibers gracefully taper the dielectric constant of the glass near the core-cladding boundary to mitigate this effect.

Can I ever use microstrips again?

Of course. For normal digital signaling on FR-4 pc boards at 10 Gbps, you may use any trace height up to 20 mils without encountering significant microstrip dispersion. At lower frequencies, you can use correspondingly bigger traces. At frequencies higher than 10 Gbps, you must use correspondingly smaller ones.

Is there an optimum trace height?

5.1.5 • Non-TEM Modes

Probably, but how to calculate it remains unclear. Skin-effect resistance varies inversely with the trace width. A fat trace, high above the planes, has little skin-effect dispersion, encouraging designers of high-speed systems to use huge, fat traces. Microstrip dispersion, on the other hand, compounds quadratically with trace height, discouraging the use of big, fat traces. There's a happy medium somewhere (I think) around a trace height of 20 mils for minimum-dispersion 10-Gbps signaling on copper microstrips (Figure 5.10).

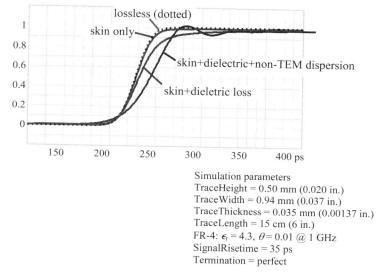

Simulation parameters
TraceHeight = 0.50 mm (0.020 in.)
TraceWidth = 0.94 mm (0.037 in.)
TraceThickness = 0.035 mm (0.00137 in.)
TraceLength = 15 cm (6 in.)
FR-4: ϵ_r = 4.3, θ = 0.01 @ 1 GHz
SignalRisetime = 35 ps
Termination = perfect

Figure 5.10—At a trace height of 0.50 mm (0.020 in.), the microstrip dispersion effect induces a small degree of overshoot with 5% ringback (simulation). The severity of microstrip dispersion varies in proportion to the square of trace height.

Microwave designers cringe at the thought of limiting trace height to only 20 mils, because they like gigantic, wide traces on enormous substrates 100 mils or more thick. The purpose of such large traces is to control skin-effect loss. Digital designers, on the other hand, regard 20 mils as an absurdly great height, so the microstrip dispersion limit does not impose on them much of a penalty.

Do the same things happen on a stripline trace?

Yes, but they don't emerge until you get to such high frequencies that we digital folks rarely notice them. The non-TEM modes in a stripline (either TE or TM modes) bounce up and down within the stripline cavity from reference plane to reference plane. Reference [40] explains that the lowest order non-TEM mode for a stripline occurs at a cutoff frequency of

$$f_c = \frac{c}{\sqrt{\epsilon_r}} \frac{1}{2b} \quad [5.11]$$

where f_c is the cutoff frequency above which the first non-TEM mode appears (Hz),

c is the speed of light, $2.998 \cdot 10^8$ m/s,

ϵ_r is the relative electric permittivity of the substrate, and

b is the interplane spacing.

Example: A 0.5-mm (0.020 in.) cavity with $\epsilon_r = 4.3$ exhibits $f_c = 141$ GHz.

Below the cutoff frequency f_c, non-TEM modes in a stripline cannot propagate, so they aren't a problem today, and it's going to be a while before we have to worry about 141 GHz.

The problem with microstrips is that there isn't a clean cutoff frequency for non-TEM modes. On a microstrip, when you approach within even 1/10 of the cutoff frequency f_c (as computed with b set equal to the trace height), the effective permittivity of the structure begins to change, creating microstrip dispersion.

Will my simulator show me non-TEM effects?

Probably not, unless you are using specialized tools crafted for the microwave industry. These tools usually work in the frequency domain and don't produce time-domain waveforms. You will need a 3-D electromagnetic field simulator with time-domain output to observe microstrip dispersion.

How do I fix microstrip dispersion?

First, don't let it sneak up on you. If you are planning a 10-Gbps system, get (or borrow) a full-wave 3-D electromagnetic-field simulator. Don't use megasized traces, and watch out for the extra resistive loss due to current concentration. You might try a lower-dielectric-constant substrate; it exhibits less of the effect. Alternatively, try a stripline. As a last resort, you can glue an extra piece of dielectric on top of the offending microstrip. The dielectric cap will somewhat reduce, but not completely cure, the microstrip dispersion.

POINT TO REMEMBER

➤ For normal digital signaling on FR-4 pc boards at 10 Gbps, you may use any trace height up to 0.5 mm (0.020 in.) without encountering significant microstrip dispersion.

5.1.5.2 Simulation of Non-TEM Behavior

References [38], [39], and [40] provide formulas for estimating the effect of non-TEM dispersion on microstrip impedance and propagation velocity. These formulas detail the changes in the real part of dielectric permittivity (the dielectric constant) but not the loss tangent. The simplest of their approximations is an analytic combination of the relative permittivity of the substrate and the effective relative permittivity (at low frequencies) of the microstrip configuration:

5.1.5 • Non-TEM Modes

$$\epsilon_{re}(f) = \epsilon_r - \frac{\epsilon_r - \epsilon_{re}(0)}{1 + G \cdot (f/f_p)^2} \quad [5.12]$$

where ϵ_r is the complex relative permittivity of the substrate,

$\epsilon_{re}(0)$ is the complex effective relative permittivity (at low frequencies) of the microstrip configuration,

$\epsilon_{re}(f)$ is the complex effective relative permittivity (at all frequencies) of the microstrip configuration,

f is the frequency of operation (Hz),

f_p is a constant approximately equal to $Z_0/(2\mu_0 h)$,

μ_0 is the magnetic permeability of free space ($4\pi \cdot 10^{-7}$ H/m),

h is the height of the trace above the reference plane (m),

Z_0 is the trace impedance in the skin-effect region, Ω, and

G is a constant approximately equal to either ϵ_{re}/ϵ_r according to [40] or $0.6 + 0.009 \cdot Z_0$, "...dependant mostly on Z_0 but also to a lesser extent upon h" according to [38]. G is reasonably near 1.0 for 50-ohm microstrips as used in digital circuits.

At a frequency less the 1/10 of f_p one would expect the effects of microstrip dispersion to be negligible. In the region below $f_p/10$ the effective permittivity is constant, but as you approach f_p it begins to rise, the change quickening quadratically with frequency, heading ultimately at very high frequencies toward an asymptotic value equal to the permittivity of the substrate.

Provided that the functions ϵ_r and $\epsilon_{re}(0)$ are both causal, the simple analytic combination of functions [5.12], if evaluated in the complex domain, produces a causal result with the correct overall loss tangent. Unfortunately, the more accurate approximations reported in [39] and [38] for the dielectric constant are not so simple, and it isn't obvious how to extract from them a good value for the loss tangent. To use the more accurate expressions, I suggest you follow a procedure very similar to that described in Section 2.12.6, "Finding $|\epsilon_r|$ to Match a Measured Loss Tangent," with the exception that at step [2.94], you must fill vector H with the known values of the real part of the log-magnitude of ϵ_r, the point of the procedure then being to synthesize a matching function for the phase (i.e., loss tangent).

$$H_k = \ln\left(\left|\epsilon_r(f_k)\right|\right) \quad [5.13]$$

The simple approximation [5.12] produces the result shown in Figure 5.10 for the step response of a very wide (and very high) pcb trace running 15 cm (6 in.) on a dielectric with loss tangent 0.01.[38] The non-TEM mode dispersion induces a ringing-like pattern in the step response. For operation at 10 Gb/s, this author therefore recommends not exceeding a trace height of 0.5 mm (20 mils).

The severity of non-TEM effects scales with the square of trace height. For example, a 15-cm (6 in.) trace implemented at the outlandish height of 1.5 mm (0.060 in.) would create severely objectionable ringing roughly nine times that shown in Figure 5.10. Improving the terminations would not affect the oscillatory behavior, because it is induced by the peculiar phase response of non-TEM propagation.

5.2 LIMITS TO ATTAINABLE DISTANCE

In the pcb environment two factors limit the distance at which you can send reliable information: sensitivity and dispersion.

As you stretch the channel length to extreme distances, sensitivity-limited systems (also called loss-limited systems) fail due to insufficient signal amplitude at the receiver. This definition infers that at the limit of operating distance, the received signal *would* be reliably detectable if amplified. Such systems display an eye pattern with a good eye opening and a reasonable amount of jitter at the receiver, but too small a signal level to properly activate the receiver circuit.

Common causes of poor sensitivity include

1. Poor control over the receiver switching thresholds. This problem is shared by many single-ended logic families, which therefore cannot tolerate any signal loss much larger than about one dB.
2. Thermal or shot noise within the receiver. This difficulty is often encountered in the design of high-speed fiber-optic receivers, but not in pcb transceivers. The voltage levels used in copper-based pcb transceivers usually reside several orders of magnitude above the thermal noise floor.
3. Crosstalk from alien sources.

As you stretch the channel length to extreme distances, dispersion-limited systems do not fail due to insufficient sensitivity. They fail due to signal distortion in the form of severe deterministic jitter, also called intersymbol interference (ISI). This definition implies that at the limit of operating distance, the received data eye pattern is not open, or not open sufficiently far to permit a reasonable window of operation for the sampling circuits.

Amplifying the signal does not change the performance characteristics of a dispersion-limited system, because it is not the *size* of the received signal that matters, but the *shape*. The only fix is to adopt some form of equalization or echo cancellation. Common causes of signal distortion include

1. Bandwidth limitations within the transmitter circuit or transmitter package.

---

[38] In order to mix together dielectric and non-TEM effects, this author first computes ϵ_r according to [2.88], and then $\epsilon_{re}(0)$ according to the approximation given in [5.6]. Both ϵ_r and $\epsilon_{re}(0)$ will then vary as a mild functions of frequency. The complex values of ϵ_r and $\epsilon_{re}(0)$ are then used in [5.12]. This procedure assigns a reasonable (and causal) loss tangent to the final result of [5.12].

2. High-frequency losses within the communications channel.
3. Echoes and other undesirable features in the step response of the communications channel.
4. Bandwidth limitations within the receiver circuit or receiver package.

A simple chart appropriate for estimating signal loss as a function of frequency, trace geometry, and dielectric material appears in Figure 5.11. The chart applies to 50-ohm single-ended striplines and 100-ohm loosely coupled differential striplines in materials with a dielectric constant of approximately 4 at 1 GHz.

The vertical axis in the chart shows the signal lost in units of percent-signal-lost per inch. For example, the skin-effect loss for a 6-mil trace operating at a frequency of 1 GHz equals 1% of the signal amplitude for every inch traveled. The amplitude a remaining at the end of 18 inches would then be

$$a = 0.99^{18} = 0.83 \qquad [5.14]$$

For small amounts of attenuation, you may approximate the exponential operator in [5.14] by simply *adding* the percentage signal lost in each inch. In the above example the approximate method estimates a signal loss of 1% per inch times 18 inches, for a total estimated loss of 18%, leaving a remaining signal amplitude of 0.82, fairly close to the value computed in [5.14]. For any percentage loss less than 25%, the correspondence between the exact method [5.14] and the simpler method of addition is better than 1 part in 20 (5% error).

To convert units of percentage-signal-lost per inch to units of dB per meter, multiply times 3.44.

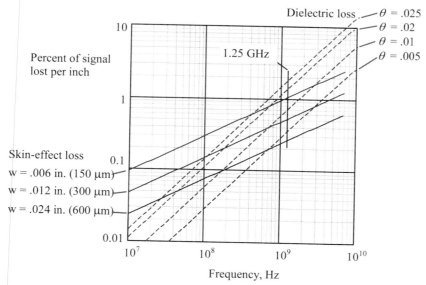

Figure 5.11—This chart estimates skin effect and dielectric losses for 50-ohm single-ended striplines and 100-ohm loosely coupled differential striplines using the approximations developed in Chapter 3.

Skin-effect loss curves are included for trace widths of 6, 12, and 24 mils. The trace thickness is a second-order parameter not considered in the calculation (all traces are assumed to be 1/2-oz copper striplines). The trace loss scales approximately with the inverse of trace width. Microstrips perform similarly. The same values of skin effect and dielectric loss apply equally well for 50-ohm single-ended striplines and 100-ohm loosely coupled differential striplines.

Dielectric-loss curves are included for dielectric loss tangents of 0.005, 0.01, 0.02, and 0.025. Other values of dielectric loss may be linearly interpolated on the chart from the figures shown. The skin effect and dielectric-loss effects must be summed to determine the total signal loss.

Example showing 2.5-Gb/s Serial Link

The operational characteristics of a 2.5 Gb/s link are best evaluated at a frequency of 1.25 GHz, corresponding to the maximum alternation rate for binary data operating at that speed. The loss values at this particular frequency will be highly indicative of the eye-opening loss in an actual system.

At 1.25 GHz, a 6-mil stripline trace operating on FR-4 accumulates a signal loss of roughly 1% per inch due to skin-effect loss. At an assumed dielectric loss tangent of 0.025. it accumulates another loss factor of 2% per inch due to dielectric loss. Operating at an overall distance of 8 inches, the total loss amounts to roughly 24%, a workable value provided the receiver thresholds are well centered. A modest degree of pre-emphasis would likely extend the distance at which such a link could operate.

Ordinary single-ended CMOS or bipolar TTL transceivers require that the received signal exceed V_{IH} (or fall below V_{IL}) at the instant the signal is sampled. In many logic families the difference between the voltage V_{OH} that all drivers are guaranteed to produce and the voltage V_{IH} that all receivers are guaranteed to accept provides for a loss budget of only about 10% of the peak-to-peak signal swing. When using such a simple, unadorned transceiver, the total percentage of the signal lost, as read from Figure 5.11, must not exceed 10%. This limitation imposes a severe distance limitation on the operation of multi-gigabit data links.

The means at your disposal to improve upon any system with a sensitivity limitation include

1. Select a transceiver with a larger spread between V_{OH} and V_{IH} (and also between V_{OL} and V_{IL}). Any improvement thus obtained in the noise margin translates directly into an increased budget for signal loss.

2. Select a transceiver with greater sensitivity. This strategy is the same as (1), just couched in analog terminology rather than in the terminology of ordinary digital logic.

3. If the system is limited by crosstalk or other interference, then reduce the interference. Determining the influence of crosstalk on sensitivity is a matter of selectively disabling adjacent channels while measuring the sensitivity of the receiver.

4. If the system is limited by self-generated noise within the receiver, then further improvements depend upon the use of a less-noisy receiver

architecture, or a reduction in temperature (which tends to reduce internal thermal noise), or an increase in the transmitted signal amplitude.

Systems limited by dispersion may sometimes be improved by a change in data coding. For example, run-length limited data coding schemes that enforce DC balance produce a data spectrum with very little low-frequency content. The reception of such a coding scheme therefore depends mostly upon the spectral properties of the communications channel between some low-frequency limit (the low-frequency cutoff) and the Nyquist frequency (1/2 the data rate). To the extent that this range is smaller than the range required by ordinary binary coding (NRZ), whose spectrum extends from DC to the Nyquist frequency, the expected maximum variation in attenuation from the low-frequency limit to the Nyquist frequency is improved. Since it is the *variation* in attenuation across the data frequency band that determines the maximum and minimum eye height at the receiver, excising the low-frequency part of the spectrum reduces this maximum variation, resulting in a cleaner eye with less distortion. Systems limited primarily by the skin effect respond well to a change in coding because this change essentially truncates the long, lingering tail in the skin-effect step response.

In cases where the signal amplitude lies far above the thermal noise floor, but is severely distorted by AC-coupling within the communications channel, a DC-restoration circuit can completely restore the appearance of the eye. This method constitutes a type of nonlinear equalization. One type of DC-restoration circuit is discussed in the following article "SONET Data Coding."

Means of linear equalization are discussed in Section 3.14, "Linear Equalization: Long Backplane Trace Example," and in Section 8.2, "UTP Transmission Example: 10BASE-T."

POINTS TO REMEMBER

> ➤ As you stretch the channel length to extreme distances, sensitivity-limited systems fail due to insufficient signal amplitude at the receiver.
>
> ➤ Dispersion-limited systems fail due to signal distortion, also called intersymbol interference (ISI).
>
> ➤ Amplifying the received signal does not change the performance of a dispersion-limited system. Equalization is what helps.
>
> ➤ Systems limited by dispersion may sometimes be improved by a change in data coding.

5.2.1   SONET Data Coding

High-Speed Digital Design Online Newsletter, Vol. 5, Issue 5
Bhavesh Patel writes

I read your article on fiber-optic encoding where you explain about DC balance in different encoding schemes, and in the latter half you explain SONET data which is scrambled and how many long runs of zeros and ones it can have.

I use SONET scrambled data on my backplane. Statistically, the maximum number of 0s or 1s should be limited to 72. In my SONET framer, however, the J0/Z0 byte is not scrambled. Unless the user intentionally stuffs the J0/Z0 byte with transitions, the run length can be even higher. Even if I treat the J0/Z0 byte properly, I think a run length of 72 used with AC coupling will be a problem.

Why? Because the DC level could drift depending on the run length of zeros and ones, and this could cause eye closing at the receiver and hence increase BER. Is there a solution to this if a user has to use AC coupling? Also, does the capacitor need to be of microwave quality, which has resonance above the f_{knee} frequency?

Reply

Thanks for your interest in *High-Speed Digital Design*. Regarding your inquiry, the DC balance of SONET can be terrible. If I understand your description of the coding correctly, you can get a run of 72 ones, followed by a few transitions, followed by 72 *more* ones, and so on. There's no good, cheap way to AC-couple such a system. Any linear circuit you try is subject to the statistical possibility that the average DC level might approach 1 (or 0) and stay there long enough to defeat your circuit. You are therefore relegated to using some form of nonlinear DC-level restoration.

Figure 5.12 shows one way to build a nonlinear DC restorer. This circuit assumes the signal has been AC-coupled at some point prior to the terminating resistor R_T, stripping the signal of all low-frequency information. The purpose of the circuit is to restore the lost low-frequency components of the signal. This circuit requires that you first terminate your signal at R_T and buffer it with a low-impedance linear driver.

The circuit accomplishes two goals, behaving as both a *signal filter* and a *feedback filter* at the same time. The signal $x(t)$ from the low-impedance buffer feeds through capacitor C_1 into the high-impedance input of the data slicer. From the perspective of this signal, the data slicer input appears loaded by resistance R_1, which leads to the low-impedance output of the sampling register. The signal pathway for $x(t)$ therefore comprises a high-pass filter with transfer function

$$H_{SIGNAL} = sR_1C_1/(1+sR_1C_1).$$

The signal from the sampling-register output feeds through resistor R_1 into the high-impedance input of the data slicer. From the perspective of this signal, it sees the data slicer input loaded by capacitance C_1. This pathway therefore comprises a low-pass filter with transfer function

$$H_{FEEDBACK} = 1/(1+sR_1C_1).$$

The signal at the input to the slicer is the sum of the two filtered signals:

$$z(t) = H_{SIGNAL}[x(t)] + H_{FEEDBACK}[y(t)]$$

5.2.1 • SONET Data Coding

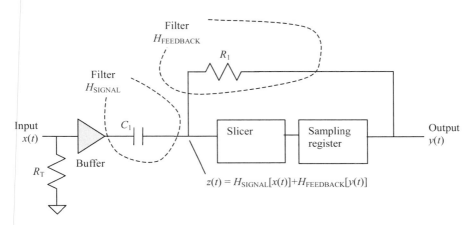

Figure 5.12—Low-frequency information lost passing through C_1 is restored by the action of R_1.

The clever part of this architecture is that both filters have *precisely* the same time constant, so the sum of the filter transfer functions H_{SIGNAL} and $H_{FEEDBACK}$ is *exactly* unity. That's important because, if you suppose for a moment the circuit does function such that the sampling register output $y(t)$ equals the input $x(t)$, then the slicer input $z(t)$ must also precisely equal the input $x(t)$, as shown:

Assume $y(t) = x(t)$.

$$z(t) = H_{SIGNAL}[x(t)] + H_{FEEDBACK}[x(t)]$$

Linearly combine the filters.

$$z(t) = (H_{SIGNAL} + H_{FEEDBACK})[x(t)]$$

Recognize that the sum of these filters is a unity-gain operation.

$$z(t) = x(t)$$

Provided that $y(t) = x(t)$, whatever low-frequency information H_{SIGNAL} removes from $x(t)$ is unerringly restored by the operation of $H_{FEEDBACK}$ on $y(t)$.

Even if the output $y(t)$ is delayed by a couple of bits, as it might be in a practical slicer circuit, the DC-restoration circuit still works as long as the amount of delay is small compared to the time constant of the low-pass filter $H_{FEEDBACK}$.

This circuit makes use of two pieces of information to restore your AC-coupled signal:

1. The high-frequency content of $x(t)$
2. The fact that $x(t)$ is binary-valued

This circuit does *not* use the low-frequency content of $x(t)$. That information is implied by the aforementioned two conditions. Therefore, if the low-frequency content of $x(t)$ is missing, because you AC-coupled $x(t)$ at some prior point in the system, the DC-restoration circuit doesn't care—it still works.

As long as the time constant R_1C_1 is *shorter* than the time constant of any AC-coupling network that precedes the buffer, the circuit will properly restore your signal at the input to the slicer.

This clever circuit does not require high tolerances on the AC-coupling capacitors that precede R_T, nor on C_1 or R_1. It is a very old circuit concept. It is used in (among many other modern systems) some Fast Ethernet 100 Mb/s transceivers.

I like this circuit better than systems that differentiate the incoming signal and then fire "set" and "reset" operations based upon the appearance of either positive or negative pulses in the differentiated signal. A differentiation circuit is extremely susceptible to high-frequency noise and requires multiple slicer levels. The approach in Figure 5.12 adds only a clean, low-frequency signal to your received data and therefore has almost no impact on noise performance.

POINT TO REMEMBER

➢ A non-linear DC restoration system can un-do the effects of AC coupling.

5.3 PCB NOISE AND INTERFERENCE

There are two major sources of noise and interference in a high-speed pcb communications channel: reflections and crosstalk.

Radio-frequency interference rarely afflicts digital logic signals in a direct way if the system is reasonably well shielded, the traces are kept close to a solid reference plane, and the digital logic levels are reasonably large (1 volt or more).

5.3.1 Pcb: Reflections

When you blast a high-speed step edge into a reasonably well-terminated pcb trace, most of your signal power reaches the far end at the same time, duplicating the size and shape of the original transmitted waveform. The received signal may be distorted or reduced in amplitude, but in a good system most of the signal power at least arrives reasonably coincident in time.

Any portions of the original transmitted signal that bounce off imperfections with the transmission structure, arriving later at the far-end load, are called *far-end reflections*. If the reflections arrive at the far end during the current baud period, they are considered a form of settling problem. Echoes that arrive during the reception of some later baud constitute a source of random noise to the receiver at the far end of the cable. In either case, the peak amplitude of the reflections must be limited to a value compatible with the noise margin budget of your logic family. The following sections detail a number of transmission-line imperfections that can generate substantial reflections.

Pcb traces terminated at *both ends* enjoy a great advantage in immunity to reflections as compared to their singly terminated cousins (either the end-terminated or source-

5.3.1 • Pcb: Reflections

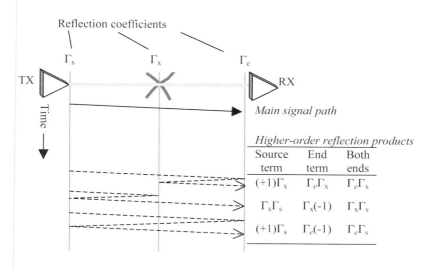

Figure 5.13—Second-order reflection products for a serial link are greatly attenuated by the use of both-ends termination.

terminated variety). Figure 5.13 illustrates this point with a simplified time-space reflection diagram estimating the magnitude of all second-order reflections reaching the far end of a generic serial link. This figure assumes that an imperfection in the transmission structure at point x creates a small reflection of magnitude Γ_x. In a source-terminated structure the magnitude of the reflection coefficient Γ_e is near unity, so the end-products $\Gamma_e\Gamma_x$ and $\Gamma_e\Gamma_s$ will be roughly the same size as Γ_x and Γ_s respectively. Similarly, the magnitude of the reflection coefficient Γ_s in the end-terminated structure is near unity (the driver has to have a low impedance or else the transmitted waveform wouldn't be full sized), so in that case the products $\Gamma_x\Gamma_s$ and $\Gamma_e\Gamma_s$ will be roughly the same size as Γ_x and Γ_e respectively. The both-ends-terminated structure provides a good attenuation of the reflected signal at both ends of the transmission line, substantially attenuating *all* second-order reflection products. To first order, therefore, the reflection at Γ_x has little effect on signal quality in a both-ends-terminated structure.

The relative immunity to reflections enjoyed by both-ends-terminated structures extends to reflections generated due to poor terminations, vias, connectors, or any other type of impedance mismatch.

5.3.1.1 Both Ends Termination

Article first published in *EDN Magazine*, January 18, 2001

Terminations exist to control ringing. Ringing (sometimes called overshoot or resonance) is the tendency for signals within a distributed transmission environment to slosh back and forth, bouncing from end to end and creating oscillatory ripples in the received digital data.

The best ways to control ringing on very long transmission lines are source termination, end termination, and both-ends termination. The both-ends termination is supremely tolerant of imperfections within the transmission system and within the terminators themselves.

Figure 5.14 depicts a time-space analysis of the both-ends termination. The graph depicts the evolution of one step edge from the time the driver injects it into the transmission line until it dissipates, bouncing back and forth.

The horizontal axis represents various physical positions along the transmission line from the source position (at the far left) to the load (at the right). The vertical dimension represents the flow of time, beginning at time t_0, when the driver first impresses onto the line a rising step edge of amplitude A.

As the step edge interacts with various obstacles along the way, each encounter spawns a new reflected signal. The time-space diagram tracks the magnitude of all the reflection products. Each arrow is labeled according to the attenuation factors (reflection coefficients) it encounters. The four reflection coefficients Γ_1 to Γ_4 are schematically defined at the top of the figure. Assume for this simple example that all four coefficients Γ_1 to Γ_4 are small, meaning that the line is well-terminated at both ends ($|\Gamma_1| \ll 1$ and $|\Gamma_4| \ll 1$) and that the obstacle in the middle, whatever it is, generates only mild reflections ($|\Gamma_2| \ll 1$ and $|\Gamma_3| \ll 1$).

In general, the amplitude of any step passing through obstacle n is multiplied by a factor $(1 + \Gamma_n)$. For simplicity, the figure leaves out these $(1 + \Gamma_n)$ terms under the assumption that, in this discussion, Γ_n is always small, so $(1 + \Gamma_n)$ must be reasonably close to 1.

The first thing you should notice about the diagram is that all the first-order products (solid arrows), having bounced one time, are heading from right to left.

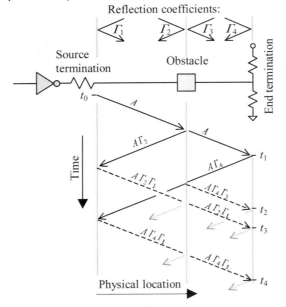

Figure 5.14—Reflection coefficients govern how each step edge bounces back and forth throughout a transmission line.

5.3.1 • Pcb: Reflections

None of these products reach the endpoint. Only second-order products (dashed arrows) and higher order even-numbered products can reach the endpoint. Because each reflection attenuates the signal, the higher order products are very small. In Figure 5.14 the higher-order products appear with successively faint lines, denoting that they are too small to worry about.

The next thing to notice is that each of the second-order products has been attenuated by two small coefficients. For example, both Γ_4 and Γ_3 attenuate the product arriving at time t_2. Both Γ_2 and Γ_1 attenuate the product arriving at time t_3. In both cases, the surviving signal has been double-attenuated. That's the beauty of a double-end-terminated net. All second-order reflection products are attenuated twice. It hardly matters what kind of obstacle lies in the middle; the terminators always get a chance to damp out the second-order-reflected products.

Contrast that behavior with what would happen on a plain end-terminated line. In that case, the magnitude of the coefficient Γ_1 would equal almost unity. (A powerful, low-impedance driver creates a reflection coefficient at the source of approximately −1.) The second-order term at time t_3 would then loom much larger.

> **The both-ends termination attenuates all second-order reflection products.**

Similarly, on a plain source-terminated line, the reflection coefficient Γ_4 would be practically +1, enlarging the second-order term at time t_2.

The both-ends termination attenuates all second-order reflection products, improving signal quality over any single termination. Mathematically, reducing the magnitude of both Γ_4 and Γ_1 renders your design impervious to variations in Γ_2 and Γ_3.

Of course, the big disadvantage of the both-ends termination is the half-amplitude received signal. The driver (whose source impedance matches the characteristic impedance of the transmission line) produces only a half-sized step. This half-sized step remains half-sized at the end-terminated endpoint. It takes an especially sensitive receiver to work with a both-ends-terminated transmission line.

The both-ends termination is an excellent choice for very high-speed serial links in which you anticipate encountering connectors, vias, or other impedance discontinuities in the middle of the line and for which you can afford a super-sensitive receiver.

POINT TO REMEMBER

> Pcb traces terminated at *both ends* enjoy a great advantage in immunity to reflections as compared to their singly terminated cousins.

5.3.1.2 Pcb: Lumped-Element Reflections

A shunt capacitance connected in the middle of an otherwise long, uniform transmission structure will distort any signal that passes by. In response to an incoming step edge the distortion takes two specific forms—a backwards-propagating reflection and an impairment of the forward-propagating signal.

The reflected signal is a brief pulse with a polarity opposite the polarity of the incoming step. The polarity is opposite because a capacitor, when interacting with a fast-moving step edge, draws at first a large surge of current. The surge of current required to charge the capacitor has the same effect as the temporary connection of a low-impedance shunt across the transmission line; namely, it creates an inverted reflection. In the steady-state condition after the traveling pulse has passed, the reflection effect ceases because capacitors have no effect on an unchanging signal. For small reflections, the duration of the reflected pulse approximately equals the rise or fall time t_r of the incoming step edge.

The peak amplitude a of the reflected pulse is given approximately by

$$a \approx \Delta V \frac{(1/2) Z_C C_L}{t_r}, \text{ or}$$

$$a \approx \Delta V \frac{\tau}{t_r}$$

[5.15]

where a is the peak magnitude of the reflected signal, volts,

ΔV is the amplitude of the incoming step, volts,

τ is a time constant computed from the lumped-element capacitance C_L and the transmission-line impedance Z_C, having the value $\tau = Z_C C_L / 2$, and

t_r is the 10% to 90% risetime of the incoming step, in seconds.

Equation [5.15] derives from the Fourier transform expression for the reflection coefficient $\Gamma(\omega) = -j\omega\tau/(1 + j\omega\tau)$ that occurs when a signal traveling in a line of impedance Z_C encouters a load formed by the lumped-element capacitor C_L in parallel with another section of transmission line having impedance Z_C. Equation [5.15] assumes that the time constant τ is small compared to t_r, so that the term $j\omega\tau$ in the denominator of the reflection expression may be ignored at all frequencies within the bandwidth of the incoming waveform, leaving the denominator equal to approximately unity. It further assumes that the maximum value of the derivative of the incoming step approximately equals $\Delta V/t_r$.

Figure 5.15 compares approximation [5.15] to the signal magnitude computed from time-domain simulations. This figure assumes the delay of the transmission line segments on either side of the lumped-element load each exceed $t_r/2$. Under that condition, the fully developed reflection amplitudes listed in Figure 5.15 apply. The same fully developed reflection amplitude applies if either side of the lumped-element load is well-terminated (e.g., if the lumped load is located coincident with an end termination).

For small reflections less than 25%, the simple approximation predicts the reflected signal amplitude remarkably well. Any time the simple approximation indicates a reflection larger than 25%, you should conclude that your system probably won't work. At that point it hardly matters the extent of the accuracy of the simple approximation.

5.3.1 • Pcb: Reflections

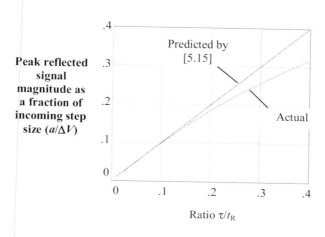

Figure 5.15—For reflections less than 25%, the simple approximation [5.12] works remarkably well.

If the length of the transmission segment leading up to the load falls short of $t_r/2$, then the source begins to interact with the lumped-element load directly. This interaction happens before the conclusion of the initial rising or falling edge from the source, with the result that the transmission impairments are either exaggerated or ameliorated, depending upon whether the impedance of the source is greater or less, respectively, than the line impedance.

The magnitude of the reflected pulse subtracts from the magnitude of the forward-propagating signal. The forward-propagating signal therefore does not immediately rise to a full amplitude; instead, it displays a somewhat degraded risetime. The rise or fall time t_{FWD} of the forward-propagating signal is estimated as

$$t_{FWD} \approx \sqrt{t_r^2 + (2.2 \cdot \tau)^2} \qquad [5.16]$$

where τ is a time constant computed from the lumped-element capacitance C and the transmission-line impedance Z_C, having a value $\tau = Z_C C_L / 2$, and

t_r is the 10% to 90% risetime of the incoming step, in seconds.

Provided that the peak magnitude of the reflected signal is no larger than 25% of the incoming step height, the correspondence between [5.16] and actual results computed for a Gaussian input step is better than 1 part in 40.

The group delay of the low-pass filter created by the lumped-element load represents the delay of the signal at approximately the mid-level and is the value that must be added to your propagation-delay calculations. The group delay created by a shunt capacitance in the middle of an otherwise long, uniform transmission structure equals τ.

The above approximations may be used to estimate the magnitude of the disturbance caused by any lumped capacitive load, including the parasitic capacitances associated with gate inputs, connectors, through-hole mounting vias, and surface-mount pads.

The same general discussion applies to any series inductance L_S with the provision that the time constant τ equals $L_S/(2Z_C)$ and the polarity of the reflected signal is positive.

POINTS TO REMEMBER

> ➢ A small lumped-element capacitance shunting a transmission line creates a backwards-propagating reflection.
> ➢ A small lumped-element inductance in series with a transmission line does the same, but with the opposite polarity.

5.3.1.3 Potholes

Article first published in *EDN Magazine*, November 11, 1999

Driving home from the Spokane, Washington, airport one clear night, a steaming cup of coffee cradled in my hand, I took a shortcut across the Colville Indian Reservation. Almost immediately—bam!—my truck hit a giant pothole. Hot java flew in every direction. I stopped the truck to see what I'd hit.

The pothole was about a foot across. It was filled with water, so it was difficult to see. It looked like it would be a hazard to other motorists, so I scrounged around for a big rock and dropped it into the hole.

The rock wasn't a perfect fit. It bulged in the center, but it seemed to be the right overall size for the hole. I backed up and tried driving over the hole again (no coffee this time). It was much better. Satisfied with my good deed, I continued the drive homeward.

> **Adjustments to transmission-line width can partially compensate for one isolated capacitive load.**

This incident reminded me of a similar treatment used in transmission-line design. You can improve a big imperfection in a transmission line (such as a capacitive load) by adding a compensating imperfection to the line. One imperfection partially cancels the other. Going back to the driving analogy, as long as the residual imperfection is smaller than your wheels, you won't feel it.

Figure 5.16 illustrates the scenario. Adjustments to the transmission-line width on either side of the load partially compensate for the capacitive load. The load adds extra capacitance to the line, but the extra-skinny trace takes away a compensating amount of capacitance (and adds some inductance). The negative reflection from the capacitive lump is counteracted by a positive reflection from the skinny trace segment.

The skinny-line adjustment in the figure can substantially reduce the reflected wave height of any incoming edge whose rise or fall time is slower than the effective delay of the adjusted segment.

Given a fixed value of $k = Z_1/Z_0$ corresponding to the skinniest transmission line (highest Z_1) you can reliably produce, select the length of the adjusted segment (x)

5.3.1.3 • Potholes

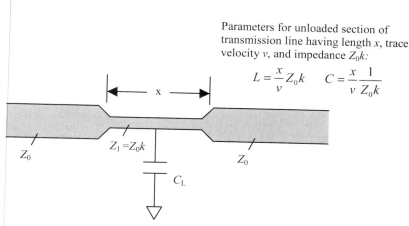

Figure 5.16—A short section of a skinny pc-board trace partially compensates for a lumped capacitive load.

so that the ratio of overall inductance and capacitance in the adjusted segment, including the effect of the lumped load, produces an effective impedance of Z_0. You mathematically represent this impedance condition as

$$Z_0 = \sqrt{\frac{(x/v)(Z_0 k)}{\frac{(x/v)}{Z_0 k} + C_L}} \quad [5.17]$$

where Z_0 is the impedance of the surrounding transmission medium, Ω,
v is the unloaded trace velocity of the adjusted segment, m/s,
$k = Z_1/Z_0$ defines the unloaded impedance of the adjusted segment,
$(x/v)(Z_0 k)$ is the total inductance of the adjusted segment, H,
$(x/v)/(Z_0 k)$ is the unloaded capacitance of the adjusted segment, F, and
C_L represents the lumped capacitive load, F.

Solving for the adjusted trace length x tells you how long (in meters) to make the skinny segment for a best compensating match. You may notice in this next formula that the skinnier you make the adjusted trace (the higher the Z_1 and thus the greater the k), the shorter you can make the adjusted segment:

$$x = Z_0 C_L v \left(\frac{k}{k^2 - 1} \right) \quad [5.18]$$

The skinny-trace compensation technique works only when the rise or fall time of the incoming edge is significantly slower (three to six times slower) than the *effective delay* of the adjusted structure. The effective delay t_{LOADED} of the adjusted structure (including load) is

$$t_{\text{LOADED}} = Z_0 C_{\text{L}} \left(\frac{k^2}{k^2 - 1} \right) \qquad [5.19]$$

Equation [5.19] assumes you have implemented the trace length prescribed by [5.18]. Equation [5.19] tells you whether the pothole-filling technique will be effective. Namely, when the time-constant $Z_0 C_{\text{L}}$ is much less than the signal rise time, it's easy to find a reasonable value of k for which t_{LOADED} remains acceptably small.

Example of pothole calculation

Let $Z_0 = 50\,\Omega$,

the pothole $C_{\text{L}} = 3$ pF,

the pcb effective dielectric constant $\epsilon_r = 4.3$, so $v = 1.446 \cdot 10^8$ m/s, and

suppose the skinniest line you can make yields $k = 2$.

$$x = \left((50\,\Omega)(3\,\text{pF})(1.446 \cdot 10^8 \text{ m/s}) \right) \left(\frac{2}{2^2 - 1} \right) = 1.446\,\text{cm}$$

$$t_{\text{LOADED}} = (50\,\Omega)(3\,\text{pF}) \left(\frac{2^2}{2^2 - 1} \right) = 200\,\text{ps}$$

The resulting structure remains practically invisible to any signal with a rise or fall time slower than 600 psec.

On the other hand, if $Z_0 C_{\text{L}}$ is comparable with or larger than the signal rise or fall time, you won't be able to adequately compensate for such a large C_{L}. To fix *that* problem, you need a smaller C_{L}, a smaller Z_0, or a slower rise and fall time.

In the automotive world, a similar effect applies: Potholes bigger than your wheels are not easily filled with a single rock.

POINT TO REMEMBER

> ➢ Adjustments to transmission-line width can partially compensate for one small, isolated capacitive load.

5.3.1.4 Inductive Potholes

An inductive pothole (i.e., a connector or via with too much inductance) embedded in the middle of an otherwise long, uniform transmission structure creates a positive reflection. The compensation for this sort of imperfection follows from the same theory given in the article "Potholes," but with slightly different formulas.

Given a fixed value of $k < 1$ corresponding to the fattest transmission line you can reliably produce, select the length of the adjusted segment (x) so that the ratio of overall inductance and capacitance in the adjusted segment, including the effect of the lumped load, produces an effective impedance of Z_0. The impedance condition for a series-connected lumped inductor L_S is

$$Z_0 = \sqrt{\frac{(x/v)(Z_0 k) + L_S}{\frac{(x/v)}{Z_0 k}}} \qquad [5.20]$$

where Z_0 is the impedance of the surrounding transmission medium, Ω,
         v is the unloaded trace velocity of the adjusted segment, m/s,
         $k = Z_1/Z_0$ defines the unloaded impedance of the adjusted segment,
         $(x/v)(Z_0 k)$ is the unloaded inductance of the adjusted segment, H,
         $(x/v)/(Z_0 k)$ is the total capacitance of the adjusted segment, F, and
         L_S represents the series inductance, H.

Solving for the adjusted trace length x tells you how long (in meters) to make the fat segment for a best compensating match. You may notice in this next formula that the fatter you make the adjusted trace (the smaller the Z_1 and thus the smaller the k), the shorter you can make the adjusted segment:

$$x = (L_S/Z_0) v \left(\frac{k}{1-k^2} \right) \qquad [5.21]$$

The fat-trace compensation technique works only when the rise or fall time of the incoming edge is significantly slower (three to six times slower) than the *effective delay* of the adjusted segment. The effective delay t_{LOADED} of the adjusted structure (including load) is

$$t_{\text{LOADED}} = (L_S/Z_0) \left(\frac{1}{1-k^2} \right) \qquad [5.22]$$

POINT TO REMEMBER

> ➤ Adjustments to transmission-line width can partially compensate for one small, isolated series inductance.

5.3.1.5 Who's Afraid of the Big, Bad Bend?

Article first published in *EDN Magazine*, May 11, 2000

Right-angle bends in pc-board traces perform perfectly well in digital designs in speeds as fast as 2 Gbps.

In most digital designs, the right-angle bend is electrically smaller than a rising edge. For example, the delay through a right-angle bend in an 8-mil wide, 50-Ω microstrip trace in FR-4 is on the order of 1 psec. That's less than 1% of a 100-psec risetime. For any object of this tiny physical scale, a lumped-element model should suffice. Years ago, Terry Edwards reproduced in his book [38] (also see[39]) a good lumped-element model for a right-angle bend. His model indicates that a right-angle bend has two primary effects: a slight delay plus some excess lumped capacitance. You might imagine that as a signal traverses a right-angle corner, the trace appears to grow wider at the corner. This simple idea explains why you see an excess capacitance (lower impedance) near the corner.

For an 8-mil wide, 50-Ω microstrip transmission line in FR-4, the excess lumped capacitance works out to 0.012 pF. Assuming that you are using 100-psec rise and fall times, the size of the reflected signal that bounces off this capacitive discontinuity is 0.30% (that's 0.003) of the incoming step amplitude. I conclude from this analysis that the reflection from a single corner is too small to worry about. (The reflected signal size scales in proportion to the trace width and inversely with rise and fall times.)

Some people worry that conduction electrons are traveling so fast that they won't be able to make it around a square corner. Perhaps they might reflect back or fly off into space. Such arguments are ridiculous. Sure, individual electrons move at high speeds, but their aggregate drift velocity is less than 2.5 cm/s (1 in./s) as they bounce from atom to atom. Your average electron smacks into something and changes directions billions of times in a length of 10 mils. Electrons don't have any trouble banging around a corner.

Might the electric-field concentration at a sharp, pointy corner create a lot of radiation? Hogwash. As a trace rounds a corner, it stays a constant distance from the underlying reference plane the whole way. The electric field intensity from trace to plane doesn't radically vary at any point along this track except for a modest perturbation in the vicinity of the actual pointy tip of the corner. It's true that a microscopic electric-field probe directly adjacent to the corner would detect this field concentration. However, measurements taken from farther away aggregate radiation from the whole trace, not just the corner. The corner, because it is so small, cannot noticeably affect the far-field radiation.

Layout professionals often point out that modern layout systems already round off all the outside corners, assuming that this rounding eliminates the square-corner effect. It doesn't. Rounding the corners removes 21% of the copper in the corner. Edwards shows that you must remove 70% to 90% of the copper from a right-angle bend to neutralize (to first order) the excess capacitance. Rounding removes only a small fraction of the required amount of copper. Rounded-corner right-angle bends

5.3.1.3 • Potholes

work well in digital designs not because they are rounded, but because the corners are too tiny to cause significant problems in the first place.

Today, only microwave designers need to worry about right-angle bends. At microwave speeds, roughly 10 times the rate of most digital designs, parasitic capacitance presents 10 times more of a problem. Additionally, microwave designers often use big, fat, 100-mil traces to reduce skin-effect losses, so their corners appear electrically 10 times bigger. Lastly, they also tend to linearly cascade multiple stages. Cascading sums the imperfections in each stage, making microwave designs about 10 times more sensitive to tiny imperfections. Overall, contemporary microwave designs can be 1,000 times more sensitive to right-angle bends than are digital designs.

As digital designs push toward higher speeds, you may eventually reach a point where the right-angle bends begin to matter. For example, corners are just beginning to affect the design of 10-Gbps serial connections, and they also contribute perceptibly to skew in certain poorly routed differential pairs. If you accumulate oodles of corners, as in a serpentine delay structure, you may begin to see a little extra delay. Other than these extreme applications, right-angle bends remain electrically transparent.

> **Microwave designs are far more sensitive to right-angle bends than digital designs.**

Some manufacturing engineers complain about the use of right-angle bends when using wave-soldering equipment. They worry that wayward solder balls or solder flux will get trapped in the inside corners. With reflow soldering and good soldermasking, neither should be a problem. I have heard no other credible negative comments about the manufacturability of right-angle bends, but I am always happy to hear from others whose experience may differ.

POINT TO REMEMBER

➢ Right-angle bends in pc-board traces perform perfectly well in digital designs in speeds as fast as 2 Gbps.

5.3.1.6 Stubs and Vias

High-Speed Digital Design Online Newsletter, Vol. 2, Issue 25
Gary Griffin writes

I would like to know the effects of short pcb trace stubs and pcb vias, and how to eliminate or reduce them. The stubs could be short traces needed to hook a bus up to a socket, or something like that.

Vias are a necessary evil, but how can they be made more transparent to the high-speed signal? The speeds I am talking about range from 400MHz to 3GHz (digital and analog).

Reply

Thanks for your interest in *High-Speed Digital Design*.

Regarding socket stubs, there isn't much you can do except to not have them (i.e., don't use sockets). I know that's impractical in many cases.

About vias,

1. Blind or buried vias are smaller and have less effect than full-sized vias.
2. When your trace is adjacent to one power (or ground) plane, and then jumps through a signal via to run adjacent to a different power (or ground) plane, place a bypass capacitor in the vicinity to help returning signal currents follow along (any bypass connecting the reference planes helps returning signal currents jump from plane to plane), or
3. Better yet, route your trace so that it pops back and forth between two planes which carry the same DC potential. You can then use a plane-to-plane via near the signal via to help returning signal currents jump from plane to plane, or
4. Even better than that, route your trace so that it pops back and forth between the two sides of a single plane. In this case no special return current provision is necessary.
5. If your via can be modeled as a simple lumped-element capacitive discontinuity, consider addressing it with the technique discussed in Section 5.3.1.3, "Potholes". If your via looks like an inductive discontinuity, the same mathematical approach applies, only you will want to make Z_1 less than Z_0 (see 5.3.1.4, "Inductive Potholes").
6. The use of both-ends termination (series and source termination used together), often combined with differential signaling, will produce a circuit with a great deal of immunity to the reflections caused by vias. Simulate this one to see the difference. To make use of this idea, you will of course need a receiver with sufficient sensitivity to work with the half-sized signal that results from using both-ends termination.

Good luck!

POINT TO REMEMBER

➢ Blind or buried vias are smaller and have less effect than full-sized vias.

5.3.1.7 Parasitic Pads

Article first published in *EDN Magazine*, August 17, 2000

I'm designing a 2.5-Gbps, OC-48 transceiver card. Between two chips on my board I have several discrete capacitors and resistors to slow the edges and pad the signal. It seems that the very short 2.5-cm (1-in.) trace that I'm using is covered more with part pads than with 50-Ω trace. My software calculates the [trace]

5.3.1.3 • Potholes

impedance but does not consider these parts. What effect does the pad size have on my trace impedance, and can I neglect it?
—Code Cubitt

The component pads have a big impact on trace impedance, and at 2.5 Gbps, you will really notice it. The basic effect is that each pad contributes a little extra parasitic capacitance C to the trace. A single 1206 pad contributes about 0.72 pF, estimated at 31 pF/cm^2 for a 0.0127-cm thickness of FR-4, (200 pF/in.^{2} for a 0.005-in. thickness of FR-4), ignoring fringing fields at the edges.

If you space the pads equally along the line, and if the spacing is a small fraction of the signal rise and fall time, then the parasitic capacitance of the pads merely reduces the line impedance. A general rule for the line impedance Z_0 is

$$Z_0 = \sqrt{\frac{L}{C}} \quad [5.23]$$

where L is the total inductance of the line, H, and
C is the total capacitance, F.

Adding parasitic capacitance increases C. According to the formula, when you increase C, you decrease Z_0. By whatever ratio you increase the apparent capacitance of the trace, by the square root of that ratio you reduce the effective loaded impedance of the structure.

The exact amount of reduction in the apparent impedance of a loaded trace is proportional to the square root of the ratio of how much capacitance would have been distributed along a raw, unloaded transmission line of the same dimensions, divided by how much capacitance you end up with after you add the equally spaced loads.

> **Capacitive loading decreases the effective trace impedance.**

A good formula for the total capacitance C_{LINE}, distributed along an unloaded transmission line, is

$$C_{\text{LINE}} = \frac{T}{Z_0} \quad [5.24]$$

where T is the one-way delay of an equivalent unloaded trace, and
Z_0 is the characteristic impedance of an equivalent unloaded trace.

If the total load capacitance added to the trace is C_{LOAD}, you can write the impedance-reduction-ratio formula as

$$\frac{Z_{\text{LOADED}}}{Z_0} = \sqrt{\frac{C_{\text{LINE}}}{C_{\text{LINE}} + C_{\text{LOAD}}}} \quad [5.25]$$

If the loads are equally spaced, at least you still have a transmission line with a defined (albeit rather low) impedance. You might consider shrinking the line width to get the impedance back up to your target value of Z_0.

Example of heavily loaded transmission line

Assume a transmission line with nominally 1.2 pF/cm of shunt capacitance. Add to that a collection of loads at 0.72 pF each, with a spacing of one load per cm.

$$\frac{Z_{\text{LOADED}}}{Z_0} = \sqrt{\frac{1.2}{1.2 + 0.72}} = 0.79$$

The loaded impedance in this example will be 79% of the target design value.

If the loads are not equally spaced or if their spacing is too great, the signal bounces back and forth between the various capacitive discontinuities in a disagreeable manner. The worst reflection coefficient Γ that you can get from an isolated capacitive discontinuity in the middle of an otherwise perfect transmission line is

$$\Gamma = \frac{1}{2} \frac{Z_0 C}{t_r} \qquad [5.26]$$

where Γ is the reflection coefficient, interpreted as the ratio of the peak height of the reflected pulse to the height of the incoming step,

Z_0 is the transmission-line impedance,

C is the capacitance of the lumped load, pF,

t_r is the signal rise (or fall) time, ps, and

Closely spaced loads generate much smaller reflections.

If one section of your line has many loads, you should decrease the line width in that section to compensate but leave the line width at its normal size over the long unloaded sections.

Remember that loads sufficiently heavy to decrease the line impedance also increase the line delay. By whatever ratio the loads reduce the trace impedance, they increase the trace delay. The best way to reduce the effect of parasitic loading is to use smaller parts. The smaller the parts, the better (for example, 0603 is much better than 1206).

On a related subject, I've been told that the metallic film on most surface-mounted resistors is generally placed on only one face of the package. Such one-sided parts exhibit more parasitic capacitance to ground, and less inductance, when turned "face down" rather than "face up." If you want these parts to appear with a particular orientation on your reels, you have to ask for it.

Some engineers cut a little hole in the solid reference planes underneath the pads to reduce their parasitic capacitance. I am unaware of any inexpensive software or tools to help you determine the amount of cutting required, but I know

5.3.1.8 How Close Is Close Enough?

Article first published in *EDN Magazine*, April 9, 1998

Let's say you can't fit your series termination in the ideal location, next to the driver. There isn't room. You have to place it a little further away than you'd like. Will it still work?

A series termination resistor is supposed to absorb high-frequency energy, damping reflections on the net. To perform at its best, it must be directly connected to a very low impedance source, presumably your driver. Anything placed in series with the termination resistor changes its value, making it less effective. That includes the short pcb trace, or connection stub, that hooks the driver to the termination resistor. Applications that need very accurate termination (like clock lines) should take this effect into account. Fortunately, we can easily calculate the degradation due to a connection stub. As long as the stub delay is less than 1/3 of the signal risetime, the approximations given below will be accurate to within ± 25%.

The connection stub (Figure 5.17), because it connects at one end to a low-impedance driver, acts like a little inductor L_{STUB}. This stub inductance acts in series with the termination resistor R_1, adding to the impedance of the termination. If you add the impedance of the stub $j\omega L_{STUB}$ to the resistor value R_1, you get a reasonable model for the combined termination impedance. To this termination impedance you must also add the natural output resistance and inductance of the driver. I'll assume you have set up the value of R_1 so that it, plus the natural output resistance of the driver, together match the line impedance Z_0, so the overall model of the termination impedance looks like this:

> **Place a series terminator no more than a small fraction of one risetime away from the driver.**

$$Z_T = j\omega\left(L_{STUB} + L_S\right) + Z_0 \qquad [5.27]$$

where Z_T is the effective termination impedance in ohms,
Z_0 is the line impedance in ohms,
L_S represents the output inductance of the driver in H,
L_{STUB} represents the inductance of the stub in H, and
The inductance of the resistor package is assumed much less than either L_S or L_{STUB}.

Series termination with stub

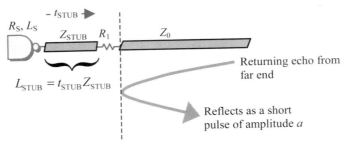

Figure 5.17—The returning echo from the far end encounters a composite load formed by the resistor R_1, the inductance L_{STUB} of the stub, plus the R_S and L_S of the driver.

The returning echo from the far end of the line, when it encounters termination network Z_T, reflects a short pulse. This amplitude of the reflected pulse (as observed at the open-circuited far endpoint of the line) will be approximately

$$a = \Delta V \left[\frac{1}{2} \frac{(L_{STUB} + L_S)/Z_0}{t_{10\text{-}90}} \right] \quad [5.28]$$

where a is the peak height of the reflected pulse,
ΔV is the open-circuit step size of the driver,
Z_0 is the line impedance in ohms,
L_{STUB} represents the stub in pH, equal to $(t_{STUB})(Z_{STUB})$,
t_{STUB} is the one-way propagation delay of the stub (ps),
Z_{STUB} is the characteristic impedance of the stub (Ω),
L_S is the series output inductance of the transmitter (pH), and
$t_{10\text{-}90}$ is the 10% to 90% risetime of the driver (ps).

The relative accuracy of this formula is about ± 25 percent.

Hint: If the stub inductance is significant compared to L_S, you've unnecessarily increased the size of the reflected pulse. The best results are obtained by making Z_{STUB} very *low* (fat trace), reducing its effective inductance.

5.3.1.3 • Potholes

That's the theory, except for one embellishment—the stub affects the risetime of the first incident waveform by a tiny amount. Keep the stub delay less than 1/3 of the risetime and you will hardly see this risetime degradation. (Thanks to Tom Giovannini and Joe Cahill for reminding me to mention this).

Example estimation of reflection from series terminator

Assume a BGA package with L_S = 6000 pH,

Trace impedances Z_0 and Z_{STUB} equal to 70 Ω,

An ideal series-termination resistor located 72 ps from the driver, and

A 3.3-v driver with a 1-ns risetime.

$$L_{STUB} = (72 \text{ ps})(70 \text{ Ω}) = 5040 \text{ pH}$$

$$L_{STUB} + L_S = (5040 \text{ pH} + 6000 \text{ pH}) = 11040 \text{ pH}$$

$$a = 3.3 \left[\frac{1}{2} \frac{(11040 \text{ pH})/(70 \text{ Ω})}{(1000 \text{ ps})} \right] = 260 \text{ mV}$$

The series termination in this case, even though it is adjusted for an ideal value of *resistance,* fails to completely damp the reflections because of the associated package and stub inductance. As a result, you may need to wait for the ringing to decay in this example before sampling the signal.

Pay close attention to the length of your connection stub. Stub delays less than 1/3 of the signal risetime create residual reflections that can be approximated by [5.28]. Stub delays in excess of 1/3 of the signal risetime can create a significant resonance that grows rapidly with increasing trace length. Don't stretch your luck. If you want 20dB or more of reflected-wave attenuation, use a stub delay of no more than 1/6 the risetime, a very good low-inductance package, and an accurate carbon-composition or low-inductance metal film resistor.

POINT TO REMEMBER

> ➢ Place a series terminator no more than a small fraction of one risetime away from the driver.

5.3.1.9 Placement of End Termination

High-Speed Digital Design Online Newsletter, Vol. 2, Issue 7
Bob Haller writes

I agree that series termination is a very effective way to eliminate SI problems on networks, and generating a simple expression to handle effectiveness based on placement is a great idea. Can you please address the stub length of parallel, terminated nets [for] both inline and downstream parallel termination?

When I am performing parallel termination, it is often difficult to sequence the termination in the proper order (driver, long line, load, short stub, and termination). I have found that if the stub length is kept very short, and edge rates are not excessively fast, terminating in the alternate sequence (driver, long line, termination, short stub, and receiver) can be as effective and save significant routing channels when surface-mounted components are utilized (especially high pin count BGAs).

Reply

Thanks for your interest in *High-Speed Digital Design*.

Great idea! When working with very fast edge rates, the sequencing of the end terminator and its associated load can make a measurable difference in signal quality. We can use "short stub" analysis to predict the effect. If you hook up a net in this sequence: driver, long line, terminator, short stub, and receiver, the additional short stub (which is *open*-circuited at both ends), will act as a small lumped-element capacitor (Figure 5.18).

This small capacitance of the stub, along with the parasitic input capacitance of the receiver pin, creates an imperfection in the termination network. When the first incident wave arrives from the driver, part of that wave, a small pulse, bounces off the imperfection and returns to the driver.

The small reflected pulse travels backwards along the line to the driver, where it bounces again (off the low impedance of the driver output) and returns, one round-trip time later, to the receiver. What we observe at the receiver is an initial rising edge, followed one round-trip time later by a secondary pulse. If the initial

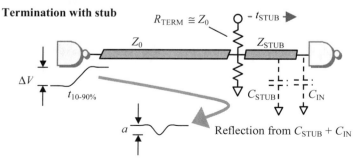

Figure 5.18—A dangling stub adds to the total load capacitance at the end of the trace, increasing the size of the reflected pulse.

5.3.1.3 • Potholes

reflected pulse is sufficiently small, all tertiary and subsequent events will be of negligible amplitude.

Assuming the delay of the short stub is less than 1/3 of a risetime, you can model the amplitude of the reflected pulse:

$$a \approx \Delta V \left[\frac{1}{2} \frac{Z_0 (C_{STUB} + C_{IN})}{t_{10\text{-}90}} \right] \quad [5.29]$$

where a is the peak height of the reflected pulse,

ΔV is the height of the incoming step,

Z_0 is the line impedance in ohms,

C_{STUB} is the stub impedance in pF, equal to $(t_{STUB})/(Z_{STUB})$,

t_{STUB} is the one-way propagation delay of the stub (ps),

Z_{STUB} is the characteristic impedance of the stub (Ω),

C_{IN} is the parasitic input capacitance of the receiver (pF), and

$t_{10\text{-}90}$ is the 10% to 90% risetime of the driver (ps).

The relative accuracy of this formula is about +/− 25 percent.

Hint: If the stub capacitance is significant compared to C_{IN}, you've unnecessarily increased the size of the reflected pulse. The best results are obtained by making Z_{STUB} very *high* (skinny trace), reducing its effective capacitance.

Fly-by termination

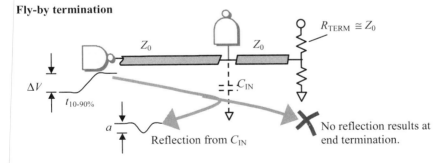

Figure 5.19—A fly-by termination suffers the capacitance of the receiver, but no additional reflection from the termination.

A stub whose length equals 10% of the signal rise (or fall) time, and whose impedance equals the impedance of the main signal trace, contributes a reflection 5% the size of the incoming step. A stub of 1/3 the length of the signal rise (or fall) time reflects on the order of 20% of the incoming step, destroying the effectiveness of the termination.

A signal routed first to the receiver, then on to a parallel termination at the bitter end of the line, suffers no additional load capacitance other than the receiver itself (Figure 5.19).

POINT TO REMEMBER

> Place an end terminator no more than a small fraction of one risetime from the end of the line.

5.3.1.10 Making an Accurate Series Termination

High-Speed Digital Design Online Newsletter, Vol. 4, Issue 14
Bill Daskalakis writes

I am aware that the driver output impedance or source impedance of a device may vary depending on whether it is in a high or low state. I have found that the source impedance may be as high as 200 ohms for the logic-1 state and as low as 20 ohms for a logic-0 state. How are you supposed to calculate an appropriate series termination when you have such a large variance in source impedance?

Note: I assume you are supposed to calculate the series resistance from the equation $R_T = Z_0 - R_S$,

where R_T = value of external series termination resistor,

Z_0 = impedance of the transmission line, and

R_S = natural output resistance of the driver.

Reply

Thanks for your interest in *High-Speed Digital Design*. Not only is there a wide variation in impedance from the *high* state to the *low* state, but there is an even wider variation from chip to chip, and between manufacturers of the same chip, and over the allowed operating temperature range, and over the allowed power-supply voltage range.

The on-state output impedance of a partially turned-on FET is very difficult to control. It depends quadratically on the exact value of the gate-switching threshold, which varies wildly depending on everything else. Of course, you get huge variations in the output impedance.

If you had access to +/− 20V supply rails that you could use to overdrive the FET gates in your I/O circuit (as is commonly done in switching-power-supply circuits), each FET would then turn on completely, producing an output resistance dependent on nothing but the bulk resistivity of your silicon and the size of the FET. As it is, most digital designs underdrive the gates, barely turning on the transistors, leaving the circuit quite sensitive to changes in its environment.

You also are fighting the tendency of most chip designers to make the pull-up side of the totem-pole output circuit fundamentally weaker than the pull-down side. I'm not a chip design expert, but I believe this has something to do with the superior carrier mobility available within the N-channel FET on the bottom of the totem pole as compared to the P-channel FET on the top. A larger topside FET could ameliorate the problem, but only at the expense of significant additional output capacitance (which becomes a problem in the tri-state condition).

As you have noticed, it is impossible given the specifications you have quoted to construct a series-terminated transmission line with sufficiently good termination to ensure first-incident-wave switching with a full-amplitude output signal.

If you can afford to wait a few round-trip times, however, your gate performs admirably. Assuming you use transmission lines with a 65-ohm impedance, the gate output impedance will be mismatched by a ratio of no worse than 3.25:1 in either direction (either 200/65 or 65/20 equals about 3.2), producing a reflection coefficient no greater than 53%. After five round trips, the residual reflection will die down to less than 5%, at which point you can safely clock the line. That's the way you are supposed to use this gate. If you can't afford to wait, you need a more accurate series termination.

What you must do to construct an accurate series-terminated configuration is use a driver with a much smaller output impedance. For example, consider the case of a driver whose output impedance varies from 1 to 10 ohms. Even though 10:1 is a huge variation in percentage terms, it is a small variation in absolute terms compared to 50 ohms. I may therefore place 45 ohms in series with this driver to produce an output structure whose impedance varies from 46 to 55 ohms, a pretty darn good match to a 50-ohm transmission line.

Alternately, you could use a current-source output circuit having an output impedance much *greater* than 50 ohms and then place an accurate resistor in parallel with the output to control the source impedance.

Either way, you end up using a good resistor to provide your well-controlled output impedance.

In the bipolar world you have other options available. For example, an emitter-follower output circuit biased with a small but constant output current exhibits a fairly well-controlled output impedance. ECL drivers (if properly biased) make use of this property to synthesize an output impedance very close to 10 ohms.

POINTS TO REMEMBER

> ➤ A low-impedance driver combined with a tight-tolerance resistor in series makes an accurate series termination.
>
> ➤ A high-impedance current-source driver combined with a tight-tolerance resistor in shunt across the driver also makes an accurate series termination.

5.3.1.11 Matching Pads

Article first published in *EDN Magazine*, December 21, 2000

Suppose you are connecting a 75-Ω cable to a piece of 50-Ω test equipment, or perhaps you are hooking up a pc-board trace to an unusual cable. If the transmission lines on either side of the connecting junction are long (compared with the signal rise or fall time) and if the shift in impedance is significant, reflections from the junction may degrade your signal. To fix the degradation problem, you can add circuitry at the junction.

The objectives for junction-matching circuitry vary according to your needs. Sometimes you want to cleanly pass signals in just one direction, the other direction, or both. Whatever the direction of signal flow, you want the signals to traverse the junction with minimal distortion, attenuation, and reflections. You can configure the circuit in Figure 5.20, called a resistive matching pad, to accomplish all of these objectives. The same basic circuit works for either single-ended or differential configurations.

In AC-coupled applications, in which no meaningful DC content exists, you can use a transformer to modify the circuit impedance. Examples of wideband applications with no meaningful DC content include audio, video, and some data signals specially coded to enforce an equal number of ones and zeros (such as Manchester data coding or 8B10B coding). The transformer is a good component to use for impedance translation, because by winding different numbers of turns on the primary and secondary of the core, you can amplify (or attenuate) the voltage at the expense of an opposite change in current. Unfortunately, transformers don't work at DC.

In narrowband applications, such as carrier-based AM or FM radio, you can sometimes use resonant-circuit tricks to accomplish impedance transformation. A classic example is the resonant pi filter. It can accomplish voltage amplification (or attenuation) over a narrow band of frequencies but not over a wide band.

> **The only passive circuits that guarantee good impedance translation for wideband signals are resistive pads.**

Random digital data, whose spectrum spreads across a vast range from DC to daylight, renders useless all standard narrowband and AC-coupling tricks. The only passive circuits that guarantee good impedance translation for wideband signals are resistive pads.

You can configure the matching pad shown in Figure 5.20 for left-to-right transmission, right-to-left transmission, or both. Table 5.6 presents the required component values and lists the performance for each application.

Each row of Table 5.6 shows values for R_1 and R_2. The signal gain G (never bigger than unity) and reflection coefficient Γ (bounded by ±1) are then given for signals traveling in either direction. Without loss of generality, the table assumes that $Z_1 < Z_2$. (If your circuit is the other way around, then look at Figure 5.20 in a mirror.)

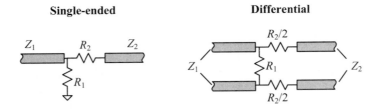

Figure 5.20—A resistive pad matches two circuits with characteristic impedances Z_1 and Z_2.

5.3.1.3 · Potholes

Table 5.6—Matching Pad Performance

Pad type	Component values		Performance in L→R direction		Performance in R→L direction	
	R_1	R_2	G	Γ	G	Γ
None	∞	0	$\dfrac{2Z_2}{Z_2+Z_1}$	$\dfrac{Z_2-Z_1}{Z_2+Z_1}$	$\dfrac{2Z_1}{Z_2+Z_1}$	$-\dfrac{Z_2-Z_1}{Z_2+Z_1}$
Optimal L→R	$\dfrac{Z_1 Z_2}{Z_2-Z_1}$	0	1.000	0	$\dfrac{Z_1}{Z_2}$	$-\dfrac{Z_2-Z_1}{Z_2}$
Optimal R→L	∞	Z_2-Z_1	1.000	$\dfrac{Z_2-Z_1}{Z_2}$	$\dfrac{Z_1}{Z_2}$	0
Both ways	$\dfrac{Z_1}{k}$	$Z_2 k$	$\dfrac{1}{1+k}$	0	$1-k$	0
NOTE— $k \triangleq \sqrt{1-Z_1/Z_2}$, Z_1 must be less than Z_2						

When you configure the circuit for optimal operation in one direction, the reflection coefficient in the opposite direction is not very good. In fact, it's worse than with a raw, unmatched junction. Sometimes such lopsided performance is acceptable. For example, with a good source-terminated network (source on the left), when the driver emits a fresh edge, you don't really care what bounces off the junction. The first bounce merely returns to the source termination and dies. You do, however, care about the signals that reach the end and then bounce off the massive open-circuited endpoint. These signals, on their return trip, take a second pass across the junction, and the reflection coefficient from right to left mostly determines the performance of the system in this circumstance. For a source-terminated application with $Z_1 < Z_2$, choose the optimal right-to-left pad. For an end-terminated driver on the same line, the optimal left-to-right pad works best. The both-ends termination (using both source and end-termination) is the least sensitive of all configurations to reflections at the junction. With both ends terminated, your circuit may not need a matching network at all.

To match two differential circuits with differential impedances Z_1 and Z_2 respectively, place R_1 directly between the two conductors of impedance Z_1. Then split R_2 into two resistors, each of value $R_2/2$, and put one in series with each conductor of impedance Z_2.

Postlog: Several readers pointed out that any signal having no DC component can be converted from one impedance domain to another using a transformer. Signals having no DC component include a 50 percent duty-cycle clock, Manchester-coded data, and data subject to other specialized codes like 8b10b that have the property of generating equal numbers of 1s and 0s. Even signals that have a DC component may be passed through a transformer provided the receiver is equipped with DC level restoration circuitry (see Section 5.2.1, "SONET Data Coding").

POINT TO REMEMBER

> ➢ Impedance translation over any band that includes DC is accomplished using a resistive pad.

5.3.2 Pcb Crosstalk

5.3.2.1 Purpose of Solid Plane Layers

Crosstalk in pcb applications is dramatically reduced by the presence of a solid reference plane. That's the primary reason for using solid planes in high-speed digital products. The reference planes can carry any DC voltage, including ground, V_{CC}, or anything else. Because crosstalk is an AC effect, the DC voltage on the reference plane makes no difference. What does matter, however, is that the plane nearest the signal trace must be continuous everywhere underneath the trace. Where a trace jumps through a signal via from layer to layer, changing reference planes, the planes must be interconnected with a suitably low impedance near the signal via.

The planes act to promote an eddy current that flows in opposition to, and directly underneath, the current flowing on every individual signal trace. The eddy current creates its own magnetic field that opposes the magnetic field from the signal current. The electromagnetic field from the signal current, combined with the equal-but-opposite field from the eddy current, creates an overall field pattern that falls off quite rapidly with increasing distance. The overall result is that crosstalk between two parallel traces varies strongly with trace separation and with the trace height above the plane. Either an increase in trace separation or a decrease in trace height will markedly reduce crosstalk.

Slots or cuts in the reference plane made perpendicular to a signal trace interrupt the formation of eddy currents, usually resulting in a marked increase in crosstalk, especially if the slot or cut passes directly underneath the trace.

POINT TO REMEMBER

> ➢ Solid reference planes exist to control crosstalk.

5.3.2.2 Variations with Trace Geometry

In a microstrip configuration with an underlying solid reference plane the crosstalk varies approximately quadratically with both trace separation and trace height. This means that a 10% increase in separation (or a 10% decrease in trace height) will decrease crosstalk by roughly 20%. A doubling of distance (or halving of height) decreases crosstalk by a factor of about four.

In stripline configurations the variation can be even stronger. Separation and height are your two greatest weapons when it comes to fighting crosstalk.

For example, suppose you have two microstrips, each 100-µm (4 mils) above a solid reference plane. Separating the two microstrips by 10 mm (400 mils) of white space (a separation-to-height ratio of 100:1) limits the crosstalk to something on the order of 1 part in

10,000 (−80 dB). This is more crosstalk immunity than needed for almost any ordinary digital logic purpose. Digital traces of similar amplitudes, even at extremely high speeds, do not need to be isolated from each other by cuts or moats in the reference plane. They just need to be routed with an appropriate spacing.

A two-dimensional field solver does a terrific job of estimating crosstalk between traces routed over a common solid reference plane. Using a field solver is the best way to estimate crosstalk for general digital purposes. Unfortunately, the assumptions behind a two-dimensional solver break down at any holes, slots, or gaps in the continuity of the planes, especially if those holes cross the path of either the victim or aggressor trace. Any time you see a trace crossing hole or gap in the plane you can expect a tremendous upsurge in crosstalk and radiation.

POINTS TO REMEMBER

> Crosstalk varies strongly with trace separation and with the trace height above the reference planes.

> A field solver is the best way to estimate crosstalk for general digital purposes, provided that no holes, slots, or gaps in the planes cross the path of either the victim or aggressor trace.

5.3.2.3 Directionality

Crosstalk in pcb traces is highly directional. A fast-edged signal propagating in a microstrip trace produces less crosstalk in the forward direction than in the reverse (see following sections about NEXT and FEXT). When measured as shown in Figure 5.21, the forward and reverse crosstalk waveforms differ noticeably.

Whether initially headed in the forward or reverse direction, crosstalk reflects and

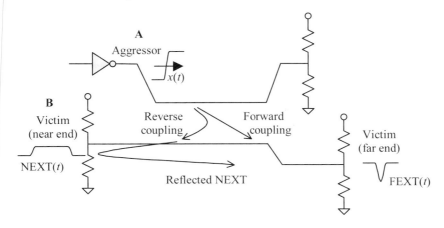

Figure 5.21—Crosstalk is directional—the forward and reverse crosstalk waveforms differ (voltage vs. time shown at each end).

bounces off any imperfections in the transmission structure, often ending up at both ends of the line. It is therefore difficult (and meaningless) in some configurations to bother distinguishing between the two forms of crosstalk. What matters in most cases is that you can exert powerful control over the coupled crosstalk amplitude by changing either the trace separation or height.

Due to the differences in NEXT and FEXT, and the way crosstalk bounces off the source and load, the timing and polarity of coupled crosstalk depends on whether the lines run in the same direction, or opposite directions.

POINTS TO REMEMBER

> ➤ Crosstalk is highly directional.
> ➤ Whether initially headed forward or backward, crosstalk reflects and bounces off any imperfections in the transmission structure, often ending up at both ends of the line.

5.3.2.4 NEXT: Near-End or Reverse Crosstalk

A fast-changing signal propagating along a transmission line couples some portion of its power into the adjacent traces, inducing signals that travel in both directions along the victim traces. The portion of the coupled power that flows backwards along the victims in a direction opposite to the forward progress of the aggressive signal is called *reverse crosstalk*. Reverse crosstalk is also called near-end crosstalk, or NEXT.

A pair of parallel traces having zero length exhibit no crosstalk of any kind. As the length of the parallel overlap is gradually lengthened, however, the amplitude of reverse crosstalk at first grows in proportion to length. The growth continues until the trace delay associated with the parallel overlap *exceeds* half the signal risetime t_r. Beyond this point the reverse-crosstalk amplitude saturates at a maximum step height. Lengthening the line beyond $t_r/2$ lengthens the *duration* of the reverse-crosstalk pulse, but not its amplitude. In response to a positive step input at the aggressor, the near-end crosstalk waveform appears as a long, low rectangle with a flat top. The initial rise time and final fall time of the near-end crosstalk waveform are the same as the rise time of the source. The duration of the near-end crosstalk waveform equals twice the delay of the parallel overlap between aggressor and victim plus one source risetime.

The maximum peak height of the reverse-crosstalk waveform at saturation, expressed as a fraction of the amplitude of the aggressive step, is called the *reverse-crosstalk coupling coefficient,* or *NEXT coefficient*.

Once you reach a length sufficient to saturate the reverse crosstalk, further extension of the length of overlap does not affect the amplitude of the reverse crosstalk. Management of crosstalk by controlling the length of parallelism does not work in systems where the line delay exceeds $t_r/2$.

In typical pcb applications involving sub-nanosecond signals, and especially on serial links, where the line delay greatly exceeds the signal risetime, the reverse crosstalk almost always achieves its maximum, saturated value.

The value of the NEXT coefficient is not determinable by analytic formula (although it has been approximated in closed-form expressions applicable for limited conditions). It is best found by 2-D field simulation. Examples showing single-ended to differential NEXT for a stripline configuration appear in Section 6.11.6, "Reducing Local Crosstalk."

POINTS TO REMEMBER

> - For parallel traces shorter than half the signal risetime, near-end crosstalk varies in proportion to the length of parallelism.
> - For parallel traces longer than half the signal risetime, near-end crosstalk saturates at a maximum level. The ratio of crosstalk to aggressive step-size at saturation is the NEXT coefficient.
> - Saturated NEXT looks like a long, low rectangle with a flat top and a duration equal to twice the trace delay plus one source risetime.

5.3.2.5 FEXT: Far-End or Forward Crosstalk

The portion of the coupled power that flows forward along a victim trace in a direction coincident with the forward progress of the aggressive signal is called *forward crosstalk*. Forward crosstalk is also called far-end crosstalk, or FEXT.

Forward crosstalk grows with trace length, saturating only when the amplitude of the coupled signal grows to a magnitude comparable with the aggressor. The shape of the coupled pulse is proportional to the derivative of the aggressor signal. In response to a positive step input on the aggressor, FEXT looks like a short pulse with duration equal to the signal risetime t_r.

The FEXT coupling coefficient may be expressed as a dimensionless fraction, where it is expected that you multiply that fraction times the line delay (accounting for length) and then divide by the risetime t_r (accounting for differentiation). Alternately, given a fixed risetime t_r (or for sine wave analysis, a fixed frequency f), you may choose to work with FEXT in units like percent crosstalk per meter.

In a microstrip trace having a delay equal to half the signal risetime ($t_r/2$) the peak height of the FEXT waveform is substantially less than the NEXT waveform.

The value of the FEXT coefficient is not determinable by analytic formula. It is best found by 2-D field simulation.

In a microstrip configuration the mutual capacitive coupling between adjacent traces is generally weaker than the mutual inductive coupling, driving the FEXT coefficient negative. Negative FEXT means that in response to a positive rising edge the FEXT appears as a negative pulse (and in response to a negative falling edge the FEXT appears as a positive pulse).

In a stripline trace (or any parallel configuration embedded in a homogeneous dielectric medium) the fine balance between inductive and capacitive coupling produces almost no observable forward crosstalk. In a stripline trace the coupled crosstalk initially flows almost entirely in the reverse direction, although it may quickly bounce off imperfections in the transmission structure and end up generating interference at the far end of the line anyway.

POINTS TO REMEMBER

> Far-end crosstalk varies in proportion to the trace length.

> FEXT looks like a short pulse with a duration equal to the source risetime.

5.3.2.6 Special Considerations

You can arbitrarily reduce FEXT or NEXT by separating the aggressor and victim traces, or by reducing their height above the planes (and making a proportional reduction in trace width to keep the impedance constant). Unfortunately, boards often lack sufficient space to achieve the desired reduction by spacing alone. In that case you'll want to know how to cheat.

Suppose serial-link connections **A** and **B** are oriented in the same direction, both driven by low-impedance sources (Figure 5.21). A fast-changing signal on trace **A** induces NEXT on trace **B**. The NEXT coupled into **B** travels backwards along trace **B** towards its driver (represented by resistors on the left). When the NEXT encounters the driver at the left end of **B** it will reflect, eventually ending up at the far end of trace **B**. The reflected signal is called *reflected NEXT*. The size of the reflected NEXT is determined by the product of the original NEXT amplitude times the reflection at driver **B**. At the right end of **A** you see both the FEXT and the reflected NEXT, superimposed.

A series termination at driver **B** would establish a reflection coefficient of zero at driver **B**, eliminating the reflected NEXT just described. Although this sounds helpful, the series termination would not by itself eliminate the complete effect of NEXT. To see why you must follow the progress of the aggressive signal on **A** all the way to its conclusion.

The main body of the aggressive signal on **A** is driven towards the right. Depending on the termination scheme employed by net **A**, this signal may bounce off the far endpoint of **A**, returning to the left. Along the way back towards the source the leftward-moving reflected signal creates both forward and reverse crosstalk on **B**. The forward crosstalk travels along the same direction as the leftwards-moving signal, while the reverse crosstalk travels in the opposite direction—that is, back to the right! This is the second form of reflected NEXT that can cause a disturbance at the far end of the victim net.

To prevent the first form of reflected NEXT you need a series termination at source **B**. To prevent the second form of reflected NEXT you need an end termination on net **A**. The end termination prevents the generation of the leftward-moving reflection, cutting off the possibility of further crosstalk. The combination of source-termination on the victim net with end-termination on the aggressor produces a pair of traces that, to first order, are immune to NEXT.

Unfortunately, the reciprocal combination (source-terminated aggressor and end-termination victim) does not work to reduce NEXT.

In the context of a number of parallel lines the only feasible way to eliminate all NEXT combinations is to equip every line with both near-end and far-end terminations, a combination burdened with the disadvantage of halving the received signal amplitude.

The FEXT situation is different. To gain immunity to FEXT, you must use stripline traces. The FEXT coefficient for a pair of raw, unloaded striplines is always zero (see Section 5.3.2.5, "FEXT").

From the above argument you may conclude that of all common pcb structures the both-ends-terminated stripline is the least susceptible to crosstalk.

In a practical circuit you will never *completely* eliminate crosstalk, because even the tiniest imperfections in either aggressor or victim create reflected signals that eventually flow towards endpoint **B**. For example, a small reflection on the aggressor trace produces a reverse-flowing waveform. The NEXT from this reverse-flowing signal then flows on the victim in a direction headed, once again, towards the far end.

Also, the cancellation of FEXT depends on a delicate balance between inductive and capacitive crosstalk within the transmission structure. This fine balance applies to any transmission structure constructed in a homogeneous dielectric environment and having a uniform cross section. If the transmission parameters are artificially modified due to surface roughness or other imperfections, the ideal balance between voltage and current is disturbed, creating nonzero FEXT. Vias, connectors, and other loads that disturb the trace impedance all create FEXT.

Crosstalk in a bus can never be eliminated, but it can be substantially abated through the use of both-ends terminated striplines.

Now suppose links **A** and **B** proceed in *opposite* directions. In this case the NEXT from **A** appears directly at the input to **B**, an effect that cannot be eliminated by fancy footwork with terminations. Accordingly, for best performance (i.e., maximum packing density on a backplane), one might select both-ends-terminated striplines, grouping all tracks with the same orientation together and providing extra spacing between traces with opposite orientations.

I should point out that in a synchronous bus FEXT has almost no impact, as it dissipates quickly after each rising edge, so that in practice a both-ends-terminated microstrip bus performs equally as well as a both-ends-terminated stripline bus.

In most cases it is the crosstalk measured at the far end of a trace that matters. If you are producing a full-duplex link (utilizing a hybrid circuit), however, the near-end reflections and NEXT issues discussed in Section 8.3, "UTP Noise and Interference," will also apply to your situation.

POINT TO REMEMBER

> ➤ The both-ends terminated stripline architecture greatly reduces, but does not completely eliminate, both FEXT and NEXT.

5.3.2.7 Directionality of Crosstalk

Article first published in *Electronic Design Magazine*, August, 1997

Crosstalk is a fact of life in modern digital systems. We can't eliminate it, but it's our job to figure out how to control it, manage it, and just plain live with it.

Consider the circuit in Figure 5.22. In the terminology of crosstalk, the gate at position **A** is the aggressor, and the gates at positions **D** and **F** are the victims. Gates **C** and **E** remain stuck at zero for the duration of this discussion.

Whenever aggressor **A** changes state, we observe a characteristic crosstalk waveform at both victims. Those of you doing dense, high-speed designs probably recognize this all-too-familiar scenario.

One of the fascinating things about crosstalk is its directionality. Crosstalk waveforms are a function of the orientation of the driver and receiver. For example, in Figure 5.22 the two victim circuits have opposite orientations. In response to a rising edge on the aggressor, the waveforms at **D** and **F** display opposite polarities.

> **Crosstalk is directional.**

The differing polarities suggest that we are not dealing with capacitive crosstalk. Many digital engineers assume that crosstalk is primarily a capacitive effect. It isn't. Mutual capacitance acting alone would cause the *same* polarity of crosstalk at both endpoints.

The differing polarities indicate that the interference is due (at least in part) to mutual inductive coupling. That's the same kind of coupling you get in a transformer. Everyone knows that reversing the leads on the primary winding of a transformer will reverse the polarity of the voltage on the secondary. Coupled pcb traces act in much the same way. If you think of each pcb trace as a little loop of current, you can see how the "crosstalk" transformer works.

First, imagine current from the gate at position **A** flowing out through the aggressor trace to the load at **B**. From there the current returns, along the power and ground system, back to the source at **A**. The aggressive current thereby makes a loop. Think of this loop as the primary winding of a transformer.

Driver:
  $R_S = 28\ \Omega$
  rise time=500 ps
Receiver:
  $C_{IN} = 8$ pF each
PCB Microstrip:
  $h = 125\ \mu m$
  $w = 125\ \mu m$
  1/2-oz Cu.
  $Z_0 = 65\ \Omega$
Crosstalk overlap region:
  length=7.5 cm
  spacing=125 μm

Simulation by
  Hyperlynx Linesim

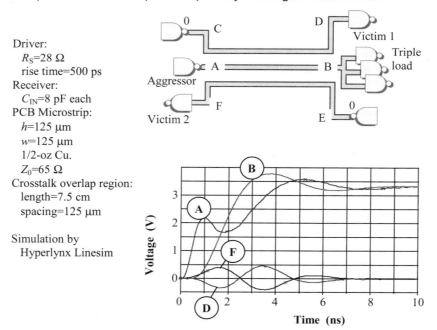

Figure 5.22—Crosstalk for short, heavily-loaded lines is highly directional.

One secondary winding of that same transformer lies nearby. It is the loop formed starting with the gate at position **C**, moving out along the victim trace to the load, and back along the power and ground system, returning to the gate at **C**.

The primary and secondary loops acting together behave almost exactly like a weakly coupled, single-turn transformer.

Because the orientation of the bottom circuit **E–F** opposes the top circuit **C–D**, the mutual inductive crosstalk captured at endpoint F is exactly opposite that captured at **D**. This behavior corresponds to the action of a transformer having two secondaries wound in opposite directions. One receives a positive signal, the other, negative.

The existence of transformer-type mutual inductive coupling between traces has profound implications for digital designs. For one thing, it implies that crosstalk varies depending on the applied load.

For example, Figure 5.22 assumes a *short pcb trace* such that the driver and load behave as if directly coupled. In this case the aggressor current varies strongly as a function of the applied load. The heavier the load, the more aggressor current the circuit draws and the more crosstalk it generates. The triple-loaded network in the figure therefore generates more crosstalk than would a similar net, with a similar topology, having only one load.

In contrast, a simple mutual-capacitance coupling model incorrectly predicts a *decrease* in crosstalk as loads are added to the circuit due to the smaller dv/dt present in the aggressive circuit when heavily loaded.

The loading effect is particularly acute when driving banks of SIMM memory modules. Such traces tend to be very short, but heavily loaded. When you insert the second SIMM module, the drive current rises markedly, creating noticeably more crosstalk.

If you are trying to debug a crosstalk problem on a dense multilayer board, knowledge of how trace loading affects crosstalk can help you understand, and fix, crosstalk problems.

If you are trying to manage crosstalk from first principles so it comes out right on the first spin, look into the new crosstalk prediction tools that feature IBIS I/O modeling. Many of these new tools are capable of calculating crosstalk, including the loading effects, in an automated, highly efficient manner.

POINT TO REMEMBER

> ➢ Both voltage and current affect crosstalk.

5.4 PCB CONNECTORS

5.4.1 Mutual Understanding

Article first published in *EDN Magazine*, January 1, 1998

"Whenever we execute this piece of code, the processor goes crazy." Sound familiar? This problem may involve crosstalk in your connectors. In large systems, especially those comprising multiple circuit cards, wide, fast bus structures must traverse connectors at many points. As bus signals pass through the connectors, the driven signals, or *aggressors,* couple some of their energy onto the other signals, or *victims.*

In a good connector the crosstalk is small even for adjacent pins, and it decreases rapidly as the victim pin is moved further from the aggressor. You can directly observe this effect with the following simple test setup. The test setup stimulates the connector with an aggressive signal that precisely mimics the in-circuit conditions of risetime, voltage, and current:

1. Turn off the system or hold it in reset.
2. Substitute a pulse generator for one of the bus drivers.
3. Adjust the rise time and voltage swing to simulate your real driver.
4. On the far side of the connector, apply a load calibrated to draw the same amount of current as your real in-circuit load.

Your test setup must receive the crosstalk voltage in a manner similar to the real system:

1. Cut each victim trace on both sides of the connector.
2. On one side of the connector, short all the victim pins to ground. This simulates the action of having a low-impedance bus driver holding each of the victim signals in a low state.
3. Measure the crosstalk received on the other side of the connector.

This approach pinpoints crosstalk generated inside the connector, eliminating trace-to-trace crosstalk on the pcb.

To understand what causes crosstalk in this configuration, remember that the aggressor current always flows in a loop. It goes to the other board and it also comes back. It flows to the other board through the signal pins on the connector, and it returns to its source back through the nearest power/ground pins. The current for every line on the bus flows in this kind of loop.

Now, here's the important part: When several bus lines are forced to share power/ground pins, their current loops overlap. These overlapping current loops form a single-turn, loosely coupled transformer with multiple inputs and outputs. Any signal on one loop couples, as in a transformer, to all the others.

A perfect example of this type of coupling happens on an open pin-field connector (i.e., a connector that uses ordinary signal pins for power and ground).

5.4.1 • Mutual Understanding

On a connector of this type, when you interchange the driver and its load in the test setup, thereby reversing the direction of signal flow on the aggressor signal pin, the crosstalk measured on the original victim circuit changes polarity. This polarity change proves that the crosstalk in that type of connector results primarily from mutual inductance (the transformer effect) rather than parasitic capacitance. This result may run counter to your intuition about crosstalk, but the evidence is irrefutable. Crosstalk in most connectors results primarily from mutual inductance rather than parasitic capacitance. Because the coupling is transformer-like, reversing the direction of current flow on the primary circuit inverts the voltage on the secondary.

When dealing with a connector configuration whose coupling is dominated by parasitic capacitances, it doesn't matter from which side you inject the aggressor signal. The received polarity stays the same. All

> **Crosstalk in connectors often results from mutual inductance rather than parasitic capacitance.**

that matters for capacitive crosstalk is the voltage you impose on the aggressor pin, not the current flowing through it. That's the effect you observe in high-impedance systems, such as low-level audio circuits. It makes sense that in this case the capacitance would matter most, because a high-impedance circuit deals with large voltages and small currents. The voltage-mode coupling (mutual capacitance) exerts a large influence on the circuit, but the current-mode coupling (mutual inductance) doesn't. Low-impedance digital circuits are the other way around—they have low voltage but high current—and so are more heaviliy influenced by mutual inductance.

Connectors designed for high-speed digital operation often have a solid ground shield adjacent to each signal pin. These connectors generate a mix of inductive and capacitive crosstalk that looks reminiscent of the NEXT and FEXT waveforms generated by adjacent parallel traces on a pcb.

Since connector crosstalk in open-pin-field connectors acts mostly through a transformer-like principle, anything you do to separate the current loops, such as providing private power or ground pins for each signal, will reduce the coupling between signals. Anything you do to reduce the magnitude of current in the aggressive circuit, such as using fewer loads on the destination side of the connector, also helps.

A connector configured with too few power and ground pins, or with too many heavy loads, generates a lot of crosstalk, easily enough to disturb edge-sensitive signals on the bus. The resulting flaky effects, like phantom interrupts, unexpected resets, and double clocking, are guaranteed to drive a processor crazy.

POINTS TO REMEMBER

> ➤ Connector crosstalk in open-pin-field connectors acts through a transformer-like principle.

> Separating the loops of signal current within a connector by providing private power or ground pins for each signal reduces crosstalk.
> Reducing the current in the aggressive circuit reduces crosstalk.

5.4.2 Through-Hole Clearances

Article first published in *EDN Magazine*, July 8, 1999

November 22, 5:45 a.m.
Ernie awoke with a start. It was still dark outside. His muscles tensed. Then, slowly, as his mind returned to consciousness, he began to relax. He always felt this way the morning after an all-night session in the lab—especially a session in which he'd discovered something important. He always feared that he would forget everything before he had a chance to write it all down and that his insight, his brilliant flash of inspiration, would melt away with the sun's first rays. It didn't. The ugly truth was still with him. Nothing was left to do now but craft his final message. Ernie stumbled into the kitchen to put on some coffee. Grumbling, he logged onto the main email server and began....

November 22, 6:22 a.m.
SENDER: Ernie
TO: Messrs Ulrich (VP Eng), Dagbottom (VP Mktg), Blumpf (Pres)
RE: Daily status report—Day 39
I have isolated our product failures to the layout of the daughterboard connector. Whenever the main data bus switches from mostly zeros to mostly ones, crosstalk within the connector activates the "write" strobe on the main system EEPROM circuits. This failure is repeatable. It explains the slow degradation of our system performance, especially the failures to reboot (due to corruption of the EEPROM configuration data).

Last night, I finally pinpointed the source of the problem. I used a pulse generator to transmit some test signals through the connector, simulating the same voltage and current conditions that would exist during an actual bus transition. Assuming that my measurements are correct, in actual use, the connector will induce an aggregate crosstalk of more than 2V on the EEPROM write line.

> **Never *assume* your board is built according to the layout specification.**

This amount of crosstalk was not supposed to happen. It is the result of a monumental grounding error. As you know, the daughterboard connector includes many ground pins. During the layout phase, I inspected the layout artwork and film to ensure that these pins were properly connected to the ground plane beneath this connector (Figure 5.23, part A).

Unfortunately—for reasons I do not totally comprehend—the fabrication shop did not build our boards according to the layout specification. I can demonstrate the problem by cutting the board along the dotted lines cut 1 and cut 2 in Figure 5.23, part B. If our boards were any good, the ground would remain electrically connected

5.4.3 • Measuring Connectors

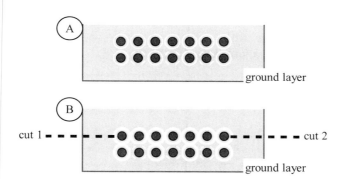

Figure 5.23—Ernie designed the ground plane with appropriate clearances and plenty of ground pathways between connector pins (A). The fabrication shop enlarged the ground plane's clearances, cutting off some of the ground pathways and creating massive amounts of crosstalk (B).

through the multiple ground pathways between pins. It doesn't. The clearance holes are too big. We'll have to order new boards.

My recommendation is that we immediately order a rush quantity of new boards and race to salvage at least a portion of our Christmas-deadline shipments. I regret that it has taken so long to resolve this problem. Because I couldn't directly see the inner layers, I assumed that the fabrication shop had built the board the way we asked.

November 22, 7:21 p.m.
SENDER: Ulrich
TO: Ernie

Our board vendor says that his company has always adjusted the trace widths, pad sizes, and clearances to meet our trace-impedance and finished-yield targets. He suggests that in the future you ask to check the finished film that they actually *use* after panelizing, instead of the film from our layout department. Alternatively, you could ask to see some preliminary unlaminated panels, which show the finished etching on the inside. X-ray services are also available to help you see the inner layers.

I hope this information will be useful to you in your next job. Because we will now surely miss the Christmas deadline, our board of directors has voted to shut down your project. I hope you don't take this action personally. Our pcb vendor says that mistakes like this one happen all the time.

POINT TO REMEMBER

> Never assume your fabrication shop will build the board the way you ask. Always check.

5.4.3 Measuring Connectors

Article first published in *EDN Magazine*, May 10, 2001

I would like to replace one connector type with a different, less expensive model. How do I prove the two connectors have the same electrical characteristics? Also, how will the power and ground-pin assignments within the connector affect its performance?
—John Lin

Three basic measurements will do the job. All three measurements use a mated pair of connectors hand-soldered to a solid ground plane on each side (Figure 5.24).

On either side of the connector, ground all the pins that you will use for power *or* ground connections. Leave the other pins unconnected but accessible to your test equipment.

> **Ground-shift voltages generated by connectors drive many common EMI failure mechanisms.**

First, test signal fidelity using a standalone signal generator to transmit a digital signal through the connector. Use the voltage and risetime that will be present in the finished system. Load, or terminate, the signal on the far side of the connector as you would normally so that you get realistic currents through the connector as well as realistic voltages. See if the signal looks okay after it passes through. This test is the easiest for a connector to pass.

If your signal generator has a 50-Ω output, and the coaxial cable is also 50 Ω, the connector under test will react as if a 50-Ω source is driving it. If you want to simulate a source impedance other than 50-Ω, use an impedance-matching pad. The most common difficulty associated with this test is a failure to appreciate the importance of keeping the hand-soldered connections extremely short. A hand-

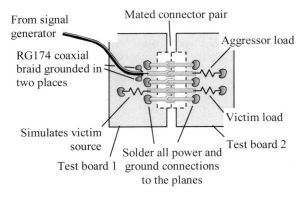

Figure 5.24—A simple test setup can characterize signal transmission, crosstalk, and EMI.

5.4.3 • Measuring Connectors

soldered test jig works fine up to about 100 Mb/s. For connectors operating at 500 Mb/s and above, the connector should be mounted with realistic vias and clearances using a layer stack designed for your application.

The signal-fidelity test tells you whether the connector impedance matches your transmission-line impedance. If it's a good match, the signal will shoot through unscathed. If it's a poor match, the initial signal edge may come through degraded, and you may also see subsequent residual reflections, depending on how you terminate the line.

Next prepare some victim pins for a crosstalk measurement. Terminate both ends of each victim signal pin with impedances that approximate the actual impedances that your system uses. For example, if low impedance sources drive your signals, then ground the source side of each victim signal pin as if a low impedance source is holding it in a zero state.

Make several measurements, and plot crosstalk versus the distance between aggressor and victim. From this plot you can estimate the worst-case aggregate crosstalk that might affect any particular victim.

The third test measures one form of EMI. Using the crosstalk measurement setup, tie a 6-ft wire onto the solid ground plane on one side of the connector. Stretch the wire horizontally across your (preferably wooden!) lab bench. Next, tie another 6-ft wire onto the solid ground plane on the *other* side of the connector. Stretch this wire horizontally in the other direction. You've just made a dipole transmitter.

Using a calibrated antenna and a sensitive spectrum analyzer, have an EMI engineer plot the received signal power as a function of frequency while you blast simulated data through one signal pin. If you can't detect the spectrum of the emissions from your connector against the fabric of your local background noise, increase the size of your transmitted signal, and then de-rate the measured results accordingly. Alternatively, do the measurement in an anechoic chamber. This measurement evaluates the *ground-transfer impedance* of the connector. When you pump high-frequency signal currents through a connector, the currents return to their sources through the ground (or power) pins of the connector. The returning signal currents passing through the ground-transfer impedance of the connector create tiny voltage shifts between the ground on one side of the connector and the ground on the other, driving the dipole antenna. These same tiny ground shifts also drive many common EMI failure mechanisms, which is why this test is a good way to measure EMI-shielding effectiveness.

Changing the number of power and ground pins in your layout will affect all three measurements. For open pin-field connectors, EMI changes inversely in proportion to the number of power and ground pins. Aggregate crosstalk changes inversely with the *square* of the number of power and ground pins. Signal fidelity improves when the configuration of power and ground pins immediately surrounding each signal pathway matches the correct trace impedance.

POINT TO REMEMBER

> ➤ Three primary measures of connector performance are signal fidelity, crosstalk, and EMI.

5.4.4 Tapered Transitions

Article first published in *EDN Magazine*, October 11, 2001

Consider the problem of adapting a straddle-mount SMA connector for a 10-Gbps digital application (Figure 5.25). The microwave guru who designed this particular SMA connector configured it to optimally launch into a gigantic microstrip measuring 1.52 mm (0.060 in.) wide suspended 0.81 mm (0.032 in.) above the nearest ground plane. Microwave circuits often incorporate such giant microstrips to curb skin-effect losses.

Your digital system probably uses much smaller microstrips with a much tighter trace-to-plane spacing. If you solder the straddle-mount SMA directly onto a small microstrip, it won't work properly. A small microstrip requires a ground plane much closer to the signal trace than does a large microstrip. With a tight signal-to-ground spacing, the parasitic capacitance between the SMA signal pin and your closely spaced ground plane will be too great, producing significant reflections. Assuming h = 150 μm (6 mils) for your digital microstrip, the excess parasitic capacitance of an SMA signal pad measuring 1.52 mm (0.060 in.) on a side is 0.69 pF, creating reflection coefficients at 1 and 5 GHz of 0.096 and 0.43 respectively. These reflection coefficients are unacceptably large. To make the circuit work at high speeds, you must reduce the parasitic capacitance of the SMA signal pin.

To reduce the parasitic capacitance, cut back the ground plane in the vicinity of the SMA signal pin. Figure 5.25 cuts back the ground plane in a beautifully tapered pattern reminiscent of a coplanar waveguide (CPW). It's not quite coplanar

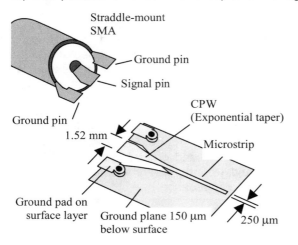

Figure 5.25—An SMA connector couples to a digital microstrip with an exponential CPW taper.

5.4.4 • Tapered Transitions

because the ground tracks on either side of the signal are offset down by one layer, but the effect is similar. For the purposes of this article, I'll call it a nearly-planar waveguide (NPW). The trace on layer 1 mimics the exponential taper of the cut in the ground plane. This tapered NPW converts the large geometry of the SMA footprint to the small geometry of a 250-μm (0.010-in.) microstrip trace while maintaining a constant 50-Ω impedance along its length. A tapered NPW defies analysis with 2-D quasistatic tools because its cross section changes along the length of the structure.

If your 2-D field solver computes ordinary CPW configurations, you can use it to design some appropriate trace and cut widths for various cross sections along the exponential taper, but don't expect those answers to be completely accurate. Your 2-D solver will improperly model the influence of the exponential taper on the impedance of the structure, and it may not understand that the signal and ground conductors are on different layers. Stretching the taper into a long, slowly evolving shape reduces the rate of change at any point, thus improving the accuracy of the 2-D solver. However, a long, gentle taper defeats your purpose—you need a short taper. To obtain a short taper, you have to build a few topologies or simulate them with a 3-D solver and then adjust the design to correct its impedance.

> **To reduce parasitic capacitance, cut back the ground plane in the vicinity of the SMA signal pin.**

A great example of a constant-impedance taper is the Eisenhart SMA connector (Figure 5.26 and [39]). This connector uses a long, tapered metal cone inside the body of the connector to form a constant-impedance transition from the 7-mm connector body diameter to a microstrip measuring 1.2 mm (0.047 in.) wide. The Eisenhart connector uses a linear taper approximately 25 mm (1 in.) long. At frequencies as great as 18 GHz, the connector provides a reflection coefficient no worse than 7%.

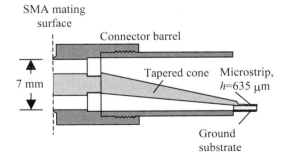

(Figure adapted from K. C. Gupta et. Al., *Microstrip Lines and Slotlines*, 2nd ed., Artech House, 1996 ISBN 0-89006-766-X)

Figure 5.26—An Eisenhart coaxial connector incorporates a conical taper.

The relative dimensions in the digital straddle-mount SMA layout are similar to those in the Eisenhart connector, so you should be able to use a similar rate of taper. Even better, the exponential transition, which etching technology enables, should provide even better performance than Eisenhart's linear taper. These factors indicate that a 1-in. exponential transition from the 1.52 mm SMA signal pad to a 250-µm trace should provide startlingly good performance from DC to 10 GHz.[39]

The values in Table 5.7 were generated using the coupled-line feature of Hyperlynx. They should provide a good starting place for your design of an NPW taper. Remember that these values won't be perfect, depending on the rate of taper adapted in your circuit. You'll have to build the structure (or simulate it with a 3-D field solver) and then tweak the geometry to get the TDR response just right.

Table 5.7—Fifty-Ohm NPW Configurations

Trace width w (mils)	Cut width s (mils)	Trace height h (mils)
10	0	6
14	4	6
15	8	6
17	12	6
24	24	6
32	36	6
39	48	6
48	64	6
58	96	6
60	128	6

The NPW table assumes the structure is made from a center trace of width w on layer 1, overlaid on top of a pair of big, fat ground traces on layer 2 (Figure 5.27). Each ground trace is 5 mm (0.200 in.) wide. The exact ground width on layer 2 doesn't much matter as long as it is at least 5 mm. The spacing between the ground traces on layer 2 (the ground-plane cut width) is s. The primary ground plane on layer 2 lies 150 µm (0.006 in.) beneath the surface.

There is also a second ground plane in the circuit. The second plane is solid with no cut, lying 0.81 mm (0.032 in.) beneath the surface. The second ground helps reduce EMI from the taper structure and is required for my 2-D quasistatic field simulator to do its job.

All the ground features should be tied to each other and to the SMA ground lugs with copious quantities of vias.

The assumed dielectric constant of the board is 4.3. The traces are coated with a 13-µm (0.5-mil) conformal coating having a dielectric constant of 3.3.

---

[39] The NPW technique usefully compensates for a signal-mounting pad that is too large given the signal-to-ground spacing of your board. The opposite problem (signal pad too small) is solved by adding capacitance to the signal pin (i.e., enlarging it).

5.4.5 Straddle-Mount Connectors

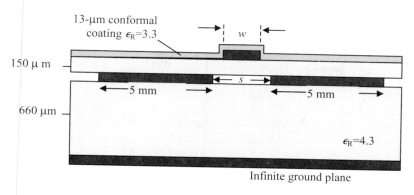

Figure 5.27—Table 5.7 lists the impedance of an NPW configuration involving two ground traces (5 mm wide each, grounded at both ends) and one signal trace.

POINT TO REMEMBER

> A tapered transition maintains constant impedance when interconnecting transmission lines having widely different physical scales.

5.4.5 Straddle-Mount Connectors

High-Speed Digital Design Online Newsletter, Vol. 4, Issue 18

In response to my previous *EDN* column, (Section 5.4.4) "Tapered Transitions," I received numerous messages asking, What is a *straddle-mount connector?*

A straddle-mount connector is one designed to hang over the edge of a pcb, as opposed to a connector that sits up on one side of the board only.

Connectors that sit up on one side of the board are convenient for mechanical reasons, as they absorb almost no mechanical headroom below the plane of the board (beyond that required to clear the back sides of the through-hole pins, if they have them). The disadvantage of a one-sided connector is that the connector pins are necessarily displaced up above the plane of the board (and the plane of the circuits), meaning that your signals must travel *up* from the board, then across through the connector, and then back *down* on the other side. In doing so the signals will encounter some inevitable parasitic inductance and capacitance.

A straddle-mount connector locates all its pins close to the plane of the pcb. It is used only at the edge of a board, as some part of the mating assembly of the connector is usually designed to hang off the edge of the board, protruding down below the plane of the board. This style of connector requires mechanical headroom below the plane of the board.

It is called a straddle-mount connector because, when viewed from the side, the connector assembly appears to straddle the edge of the board, as opposed to "riding on top" of the board.

The main advantage of a straddle-mounted connector is the reduction in the distance your signal must flow when traversing the connector. Because no long pins are required to bring your signals up out of the plane of the board, the straddle-mount connector may achieve a much more intimate connection between the grounds of the two mated circuits than is otherwise possible using a connector that sits up on one side of the board only.

An example of a straddle-mounted (also called edge-mounted) SMA connector—and I don't mean to imply it's the best one; it's just an example[40]—is the Tyco Electronics P/N: 449692-1.

Another question I got was, What's an Eisenhart connector? The Eisenhart connector was a very old N-type connector documented in Gupta and Garg's book [39]. I mentioned it as an example of where someone successfully used the tapering principle. I didn't intend for anybody to rush out and buy one...and I can't find a supplier that has any. If you locate one that's still available, let me know. It may now go by a different name.

POINT TO REMEMBER

➢ A straddle-mount connector locates all its pins close to the plane of the pcb.

5.4.6 Cable Shield Grounding

High-Speed Digital Design Online Newsletter, Vol. 2, Issue 2

I received numerous messages about cable grounding during my tenure as chief technical editor for the IEEE 802.3 Gigabit Ethernet standard. One part of that standard, the 1000BASE-CX link, runs at a data speed of 1.25 Gb/s ($1.25 \cdot 10^9$ bits per second) over two-pair, 150-ohm, balanced cabling. The link uses one pair for the transmit direction and another pair for the receive direction. The 150-ohm balanced cabling has an overall shield.

Some of the correspondents questioned the need for grounding the cable shield to the equipment chassis at both ends, suggesting that it should be grounded at one end only. Others suggested capacitively coupling the cable shield to the chassis, or coupling through parallel R-C networks. These suggestions were made to allay concerns about large AC power currents that might flow through the shield, should the shield become connected to differing 60-Hz potentials at each end. Here's how I replied.

Reply

In high-speed digital applications, a low impedance connection between the shield and the equipment chassis *at both ends* is required in order for the shield to do its job. The shield connection impedance must be low in the frequency range over which you propose for the shield to operate. The measure of shield connection

---

[40] At the time of publication, the Web link for this part was
http://catalog.tycoelectronics.com/TE/docs/pdf/1/22/208221.pdf.

5.4.6 • Cable Shield Grounding

efficacy for a high-speed connector is called the ground-transfer impedance, or shield-transfer impedance, of the connector, and it is a crucial parameter.

In low-speed applications involving high-impedance circuitry, where most of the near-field energy surrounding the conductors is in the electric field mode (as opposed to the magnetic field mode), shields need only be grounded at one end. In this case the shield acts as a Faraday cage surrounding the conductors, preventing the egress (or ingress) of electric fields.

In high-speed applications involving low-impedance circuitry, most of the near-field energy surrounding the conductors is in the magnetic field mode, and for that problem, only a magnetic shield will work. That's what the double-grounded shield provides. Grounding both ends of the shield permits high-frequency currents to circulate through the shield, providing a magnetic shielding effect.

For a magnetic shield to operate properly, you must provide means for current to enter (or exit) at both ends of the cable. As a result, a low-impedance connection to the chassis, operative over the frequency range of our digital signals, is required at *both* ends of your shielded cable [49].

How low an impedance is necessary? In the 1000BASE-CX cable the signal wires couple to the shield through an impedance of 75 ohms. That's another way of saying that the common-mode impedance of the cable is roughly 75 ohms. The standard requires that the shield be tied to the local chassis ground through an impedance of 0.1 ohm or less.

In such a configuration if you drive the signal wires with a common-mode voltage, you would expect to measure on the shield a voltage equal to (0.1/75) = 0.0013 times the common-mode voltage driven on the signal wires. The shield in this case should deliver a 57 dB shielding effectiveness. These are the specifications that our IEEE 802.3z 1000BASE-CX copper cabling group feels are necessary to meet FCC/VDE radiated emission regulations. In summary, the impedance between the shield and the chassis at the frequency of operation (about 1 GHz) must be less than 0.1 ohm.

To achieve such performance with a capacitively coupled shield, the effective series inductance of the capacitor would have to be limited to less than about 16 pico-Henries. That small an inductance cannot be implemented in a leaded component—you would need a very low-inductance distributed capacitance, possibly implemented as a thin gasket distributed all the way around the connector shell, insulating the connector shell from the chassis. We have seen proposals for this type of connector, but have not seen one work in actual practice.

The BERG MetaGig shielded connector exceeds the requirement for a shield transfer impedance. It does so by providing a direct metallic connection between chassis and shield that goes all the way around the connector pins, completely enclosing the signal conductors. A direct metallic connection is the only way we have found to beat the radiated emissions limits.

Keep in mind that the short copper link we are discussing (P802.3z clause 39) is intended for use inside a wiring closet. It only goes 25 meters. It will be used between pieces of equipment intentionally tied to the same ground (we call out in the specification that this must be the case). Between such pieces of equipment there will be no large circulating ground currents. For longer connections, we provide other links types that do not require grounding at either end (multimode fiber, single mode fiber, and category-5 unshielded twisted pairs). Direct grounding of the shield at both ends is the correct choice for a short, high-speed connection.

POINT TO REMEMBER

➢ A high-frequency shield needs direct metallic contact with the product chassis, completely surrounding the signal conductors.

5.5 MODELING VIAS

I have observed a great variety of models proposed for the analysis of vias. Everything has been tried from a zero-order model (ignore it) to a complicated S-parameter characterization painstakingly crafted from three-dimensional field simulations. Applying a via model is only half the battle, deciding how elaborate a model to use is equally important.

For general-purpose digital work using binary signaling you will need only three echelons of modeling.

None	Ignores the via, assuming it has no significant effect.
First order	The via acts as a single lumped-element reactance, either a shunt capacitance or a series inductance.
Pi model	The via exhibits the characteristics of both capacitance and inductance, including the possibility of resonance.

Before discussing these three models in detail, we must define the incremental capacitance and inductance of a via.

5.5.1 Incremental Parameters of a Via

The low-frequency (static) mutual capacitance between a via and its surrounding reference planes may be easily measured or simulated with the via standing in isolation, having no traces attached. When making this measurement the reference planes should be tied to each other, but not to the via under test.

Although such a measurement is well defined, there arise several difficulties with the test setup, as illustrated in the following thought experiment.

Suppose you measure the capacitance of a certain length of pcb trace. Then separately you measure the capacitance of a via, in isolation from the trace. If you now move the trace (by re-laying out the board) to connect it to the via, the total capacitance of the combination is less than the sum of the individual capacitances of the trace and via. The capacitances add in a non-linear way. This illustrates a general rule of capacitive combinations.

The capacitance to ground of any compound structure is always less than the sum of the capacitances to ground of its parts.

5.5.1 • Incremental Parameters of a Via

This rule applies to any bits of metal near a common reference plane, or sandwiched between two reference planes. It applies whether the planes carry ground or power. In the context of our via discussion, the rule says it is not enough to separately measure the capacitance of a trace and a via and then add them together. The properties of a via are modified by the trace to which it is attached.

The second difficulty with defining the properties of a via in isolation from its trace is that such measurements cannot be used to infer inductance. The inductance of a via depends critically upon the shape of the via *and also the path of returning signal current*.

In the evaluation of via capacitance you would never think of calculating the capacitance of a via floating in space, totally isolated from the planes. Obviously, it is the interaction between the via and the planes that matters. To talk about the *partial capacitance* of a via doesn't make much sense.

For similar reasons, this author dislikes very much the concept of "partial inductance." That concept has misled generations of engineers into believing that a via (or a wire) has, by itself, a well-defined value of inductance. It doesn't. Inductance becomes measurable if and only if you specify how and where current returns from the far end of the via (or the wire) back to the source.

To be fair, the partial inductance method does stipulate that you must take into account the partial inductance of the via, the partial inductance of the return path, and the mutual inductance between the two paths. Properly computed, the partial inductance method will reveal the correct answer, but only when you follow the method to completion. Along the way many engineers drop out of the process, omitting the crucial mutual-inductance correction terms and leaving themselves convinced that a via (or wire) by itself has a well-defined inductance.

I would have you remember that *inductance is a property only of an entire current pathway* (a loop of current). Changing the shape of the return path affects inductance just as much as changing the signal path.

It makes no sense to define, or to attempt to measure, the inductance of a via without also specifying how the attached traces bring current through it, and how the planes carry the returning signal current.

The foregoing difficulties with the definition of an isolated via have convinced me that via measurements should always be undertaken with the traces connected, and with the planes performing their job of carrying the returning signal currents in a realistic way.

To properly measure the *incremental shunt capacitance* of a via, C_V, first measure the static capacitance to the reference planes of a configuration that includes an input trace of length x, the via, and an output trace of length y, where both x and y greatly exceed the clearance-hole diameter. The lengths x and y are measured to the center of the drilled via hole. Then separately measure the static capacitance of a similar trace of length $x + y$ (with no via and no clearance hole). The incremental capacitance contributed by the via is defined as the difference between your two measurements (see Figure 5.28).

The purpose of extending the traces well beyond the clearance hole is to separate the fields near the via, which you are trying to measure, from the fields near the ends of the traces, which cancel when you take the difference between the two measurements. The planes must be electrically connected during this measurement.

The *incremental series inductance* of a via, L_V, is defined similarly, but with each trace shorted to the reference plane at its far end. Arrange your equipment to detect the loop

1. Measure the capacitance of a configuration including the via.

2. Separately measure the capacitance of an ordinary trace having the same total length as in the first measurement, but no via.

3. The difference between these two measurements is the incremental capacitance contributed by the via.

Figure 5.28—The measurement process cancels the effect of fringing-field capacitances due to the cut ends of the trace at A and B with equal fringing-field capacitances at C and D.

inductance of the path entering the trace at **A** (or **C**), passing through the short-circuit at the far end at **B** (or **D**), and returning through the reference planes to the equipment at the negative terminal shown under **A** (or **C**).

All return paths associated with the via must be in place during this measurement. For example, if bypass capacitors or ground vias connect the planes, those capacitors or vias must be present during measurement. The incremental inductance contributed by the via is defined as the difference between your two measurements.

The inductance measurement should be taken at a frequency sufficiently high that the via falls under the influence of the skin effect. This ensures that current flows realistically on the surfaces of the conductors, and in the correct minimum-inductance distribution. For 1/2-oz copper, a suitable measurement frequency would be at least 10 MHz and preferably 100 MHz.

When operating above the onset of the skin effect, whether a via is filled or hollow makes no significant difference to either the capacitance or the inductance measurements.

POINTS TO REMEMBER

- The properties of a via are modified by the trace to which it is attached.
- Inductance is a property of an entire current pathway (a loop of current). Don't use partial inductance values by themselves.
- A via contributes incremental shunt capacitance and incremental series inductance to a trace.

5.5.2 Three Models for a Via

Section 5.3.1.2, "Pcb: Lumped-Element Reflections," calculates the extent of the reflection produced when a transmission line is impaired by either a shunt capacitance or a series inductance. If the full effect of either the incremental capacitance or incremental inductance of your via is not sufficient to cause an objectionable reflection, then no model is required for your via. Simply ignore it. Keep in mind that the reflection calculations in Section 5.3.1.2 depend crucially on the risetime of the incoming signal, with shorter risetimes always making larger reflections.

If either the capacitance or the inductance alone is sufficiently large to cause a noticeable reflection, then you should know that the combined reflection generated by both effects working together will be *less* than the sum of the amplitudes of the individual effects. This happens because the reflections generated by shunt capacitance and series inductance have opposite polarities, leading to partial cancellation of their reflections when they appear together.[41]

The simple first-order model shown here elegantly handles the partial cancellation of capacitive and inductive effects. This first-order model reduces the configuration to either a single value of excess shunt capacitance, or a single value of excess series inductance, according to which effect creates the greatest reflection. The mathematics behind the reduction involves the concept of via impedance.

The impedance of a via is determined by the balance between its shunt capacitance and series inductance.

$$Z_V = \sqrt{\frac{L_V}{C_V}} \qquad [5.30]$$

where Z_V is the impedance (Ω) of the via,
L_V is the incremental static inductance (H), and
C_V is the incremental static capacitance (F)

When encountered by a rising edge, any via having an impedance less than the characteristic impedance Z_C of your transmission line creates a negative reflection. Such a via responds very much like a shunt-connected lumped capacitance. Conversely, a via with an impedance higher than Z_C creates a positive reflection, reacting very much like a series-connected lumped inductance.[42] In either case the reflection lasts only as long as the rise (or fall) time of the signal. A via with an impedance very close to Z_C creates almost no reflection at all.

For a via with impedance less than Z_C, the *excess capacitance* is defined as that capacitance above and beyond the amount required to balance the inductance L_V of the via. The amount of capacitance that would naturally balance the via inductance is $C_N = L_V / Z_C^2$, leading to the following expression for excess capacitance.

---

[41] This statement assumes that the signal current traverses the via, as opposed to the case of a hanging via stub which carries no current.
[42] When encountered by a falling edge, shunt capacitors make positive reflections and series inductances make negative ones.

$$C_{\text{EXCESS}} \triangleq C_V - C_N = C_V - L_V/Z_C^2 \qquad [5.31]$$

The completed via model in this case consists of a short section of transmission line having impedance Z_C and delay L_V/Z_C, loaded in the middle with a shunt capacitance of size C_{EXCESS}.

For a via with impedance greater than Z_C, the *excess inductance* is defined as that inductance above and beyond the amount required to balance the capacitance C_V of the via. The amount of inductance required that would naturally balance the via capacitance is $L_N = C_V Z_C^2$, leading to the following expression for excess capacitance.

$$L_{\text{EXCESS}} \triangleq L_V - L_N = L_V - C_V Z_C^2 \qquad [5.32]$$

The completed via model in this case consists of a short section of transmission line having impedance Z_C and delay $C_V Z_C$, interrupted in the middle by a series inductance of size L_{EXCESS}.

The first-order model presented here applies to any via that is electrically short compared to the risetime of your signals. Under that assumption, such a via can be reasonably modeled as a single lumped circuit element.

The peak amplitude of the reflection, in comparison to the height of the incoming step edge, is given in Section 5.3.1.2, "Pcb: Lumped-Element Reflections," where the value of capacitance or inductance used in the computation depends on the *excess* capacitance or inductance represented by the via.

The simple first-order model works well as long as the risetime remains at least three times bigger than the total delay through the via. The total via delay may be estimated (to within an order of magnitude) as $t_v = \sqrt{L_V C_V}$, or determined from consideration of the via length and the dielectric constant of the surrounding material.

To handle slightly shorter risetimes, you could represent the via with a pi model, similar to that shown in Section 3.4.2, "Pi Model." Place half the capacitance C_V on each side of the pi circuit, use no resistance, and place the full inductance L_V in the middle branch. In applications where the signal current passes through the via, the first-order model tends to be overly pessimistic, while the pi-model can produce somewhat optimistic results. In "dangling via" applications the pi-model performs better than the first-order model (see Section 5.5.3, "Dangling Vias").

If your signal risetime shrinks to a value comparable with the via delay the signal behavior becomes extremely complex. At this point I can tell you three things. First, to precisely predict the behavior you will need highly detailed model. Second, whatever you do it won't work very well. Third, you should circumvent both problems by using smaller vias.

Narrowband applications sometimes employ frequencies so high, and vias so large, that elaborate multistage models of the via structure become necessary to predict the exact phase and amplitude response. Such applications make use of the via at frequencies well beyond the useful band for digital applications.

Narrowband microwave applications can use huge vias at ridiculously high frequencies because carrier-based applications need only achieve a good, flat frequency response in the vicinity of the carrier. If a via introduces some terrible frequency-response

5.5.3 • Dangling Vias

trait, that imperfection need be flattened only over a narrow band of frequencies centered around the carrier frequency.

To improve the frequency response in one narrow band a microwave engineer might introduce a small reactive component, such as a patch of copper or a short transmission-line stub, *anywhere in the circuit*. Such games can be used to create a flat spot in the frequency response at any arbitrary frequency.

Unfortunately, narrowband response-flattening tricks work only over a limited range of frequencies. They are not applicable to digital (wideband) signals, for which the frequency response must be flattened at all frequencies (not just near the carrier).

If your via is so large compared to the signal risetime that you require anything more than a simple pi-model for the via, then it probably isn't going to work very well for a digital application. The requirements for good digital operation are that the via performance must be uniform across the entire frequency band of interest, from DC up to the knee frequency of the digital signals.

POINTS TO REMEMBER

➢ If the incremental capacitance or inductance of your via is not sufficient to cause an objectionable reflection, then no model is required.

➢ A first-order model reduces the via to either a single value of excess shunt capacitance, or a single value of excess series inductance, according to which effect creates the greatest reflection.

➢ If your via is so large compared to the signal risetime that you require anything more than a simple pi-model for the via, then it probably isn't going to work very well for a digital application. Use a smaller via.

➢ Narrowband applications sometimes use large vias at frequencies well beyond the useful band for digital applications.

5.5.3 Dangling Vias

Through-hole vias that transition to and from the inner layers of a multi-layer board may leave vestigial sections hanging above or below the path of current (Figure 5.29).

If a via is short compared to the signal risetime, you can still use a first-order lumped-element model, most likely an excess shunt capacitance. The entire capacitance of the via becomes C_V, while the inductance only of the inner section, where signal current actually flows, makes up L_V. From these values you may calculate C_{EXCESS}.

If the via is too long, however, the dangling section can develop a resonance, exacerbating the effects of its capacitance. The pi-model is helpful in understanding and predicting this effect, although I will admit it is not clear precisely how to apportion the capacitance and inductance of the pi model among the three sections of the via (the central body section that carries signal current, an upper dangling section, and a lower dangling section).

The clearest example of via resonance involves a thick FR-4 backplane with press-fit connectors. Suppose the backplane is 6.35 mm thick (0.250 in.), and imagine a signal traversing the connector routes onto the backplane on the top layer (nearest the daughter

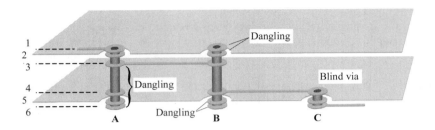

Figure 5.29—Through-hole vias that transition to and from the inner layers may leave vestigial sections dangling above or below the path of current.

card). The entire length of the press-fit via, with the connector pin embedded in it, forms a dangling stub connected to the signal line at the point where the signal enters the backplane (see **A** in Figure 5.29). The minimum transit time, t_v, through the stub structure approximately equals

$$t_v = (6.35\,\text{mm})\frac{\sqrt{\epsilon_r}}{c} = (6.35\,\text{mm})\frac{\sqrt{4.3}}{2.998 \cdot 10^8\,\text{m/s}} = 44\,\text{ps} \qquad [5.33]$$

The dangling stub acts as an unterminated transmission line, generating a quarter-wave resonance (a zero) in impedance at a frequency corresponding to four times t_v, or roughly 5.7 GHz. Any signal power near this resonant frequency will be seriously attenuated by the dangling via.

A pi-model of the same dangling stub would have shown a similar resonance at a frequency related to the combination of series inductance L_V and the capacitance $C_V/2$ at the far end of the structure, approximately ten percent lower (more conservative) than the quarter-wave stub theory would predict, with the exact value depending on how you calculate L_V. In any structure where the current returns from the far end of the stub through the action of distributed displacement current the determination of L_V is particularly difficult. Although neither model is perfect, they both draw your attention to the dangers of unterminated stubs.

If you find yourself worrying about precise models of stub resonance I suggest that you spend your time instead figuring out how to build smaller vias, or changing from binary signaling to a multi-level signal code that operates in a lower frequency band. Don't waste your life engineering the details of each and every via.

A blind via truncates the dangling stubs, producing a more compact structure with substantially less overall capacitance.

POINT TO REMEMBER

> A long, dangling via can develop a resonance, exacerbating the effects of its capacitance.

5.5.4 Capacitance Data

The following sections present three studies of via capacitance.

5.5.4.1 Three-Layer Via Capacitance

Brock J. LaMeres, in his BSEE thesis for Montana State University, simulated, using the Avant! Raphael 3D Field Simulator, a number of simple 3-layer via configurations (see Table 5.8, and the illustration in Figure 5.32). His simulations were then corroborated with physical measurements.

The 3-layer configuration represents the simplest form of via transition. There are only three layers, the top signal layer, the reference plane layer, and the bottom signal layer. At the position where the signal via penetrates the plane the returning signal current merely pops from one side of the solid reference plane to the other, passing through the clearance hole. In the three-layer configuration the series inductance of the via is negligible compared to its capacitance, suggesting the use of a simple lumped-element model of the excess-capacitance, a conclusion well supported by LaMeres' findings.

The capacitances were calculated in each case with the via standing in isolation, with no traces attached (see Table 5.8). The simulation software approximated each via pad, via body, and clearance hole with octagonal cylinders.

The dielectric spacing between layers 1 and 2, and also layers 2 and 3, was 18.9 mils. The thicknesses of the conductors on layers 1, 2, and 3 were all the same (2.2 mils), yielding an overall via length of 44.4 mils (from the top surface of layer 1 to the bottom surface of layer 3). The values of capacitance should be treated as approximate. The capacitance of a shorter via would be considerably less.

LaMeres also reports values of inductance for his vias, but these values are suspect, as they were developed using a model of one via standing in isolation and therefore do not take into account the presence of magnetic fringing fields from the associated traces. Inductance should be defined (and measured) using the technique of Section 5.5.1, "Incremental Parameters of a Via."

Table 5.8—Simulated Three-Layer Via Capacitance (LaMeres)

Drilled hole dia., mil	Clearance dia., Mil	Pad dia., mil	Via length, mil	Via capacitance C_V pF
18	46	20	44	0.23
18	46	30	44	0.31
18	46	40	44	0.46
12	36	20	44	0.21
12	36	30	44	0.31
12	36	40	44	0.46
8	28	20	44	0.19
8	28	30	44	0.30
8	28	40	44	0.46

NOTE—Data summarized from Brock J. LaMeres, *Characterization of a Printed Circuit Board Via*, B.S.E.E., Montana State University Technical Report EAS_ECE_2000_09, 1998 [48]

The data in Table 5.8 may be extrapolated to other via sizes (like buried vias). If all the specifications for a via (drilled hole, pad, clearance, length, and interplane spacing) are scaled by a factor k, the capacitance and inductance of the resulting configuration also scale by the same factor k. The capacitance also scales in proportion to the dielectric constant of the pcb material. The dielectric constant of the substrate assumed in Table 5.8 is 4.3.

5.5.4.2 Effect of Back-Drilling

In the event that your signal happens to traverse only layers 1 and 3, you might wonder if it is possible to cut off the dangling bottom section of the via, thus reducing its capacitance. This is possible in three ways:

> The *buried-via process* drills and plates various sublayers of a pcb before final board lamination, making possible the existence of miniature vias penetrating only partway through the board stack.

> The *micro-via process* laser-ablates through only a very thin dielectric separating the outermost two layers of a pcb. The finished structure is equivalent to a buried-via process with holes penetrating from the surface layer inward. Micro-vias may also be called *blind vias*.

> Finished plated-through-hole (PTH) vias may be *back-drilled* after plating. This secondary process uses a drill bit slightly larger than the original drill bit used to form the via hole. The back-drilling bit penetrates from the back side of the board partway through. This process cuts away the metallized surface of the via wall from the back side of the board, leaving a conductive structure that penetrates from the top surface only partway through the board.

Table 5.9 lists the capacitances of large through-hole vias that have been back-drilled. These vias are used with press-fit connectors in very thick, high-speed backplanes with multiple solid reference planes. The total backplane thickness in these examples was 250 mils. The via length was modified by progressively back-drilling the via to shorten the plated length of the hole. Your results vary according to the configuration of reference planes within the board.

The data in Table 5.9 may be extrapolated to other via sizes. If all the specifications for a via (drilled hole, pad, clearance, length, and interplane spacing) are scaled by a factor k, the capacitance of the resulting configuration also scales by the same factor k. The capacitance also scales in proportion to the dielectric constant of the pcb material. The

Table 5.9—Change in Via Capacitance Due to Back-Drilling

Drilled hole dia., mil	Plated hole dia., mil	Clearance dia., mil	Pad dia., mil	PTH length, mil	Via capacitance pF
26	22	52	38	250	2.4
26	22	52	38	200	2.0
26	22	52	38	225	1.8
26	22	52	38	150	1.5
26	22	52	38	125	1.3
26	22	52	38	100	1.0
NOTE—(The data in this table were adapted from Teradyne [43])					

5.5.4 • Capacitance Data

dielectric constant of the substrate assumed in Table 5.9 is approximately 4.

Example estimation of via capacitance

Assume a via with these dimensions (all values in mils):

9	finished hole size (irrelevant to electrical properties)
13	drilled hole size (this sets the diameter of the outer wall of the via)
19	pad diameter
26	clearance diameter
63	via length

The ratio of via length to pad diameter for this example via most closely matches row 5 of Table 5.9. It differs from the via in row 5 by a scale factor of 1/2.

The capacitance listed in row 5 is 1.3 pF. The estimated capacitance of the example via should therefore be reduced from row 5 by your scale factor of 1/2 to a value of 0.65 pF.

To more accurately estimate the capacitance of a via, you should use a 3-D field solver.

5.5.4.3 Effect of Multiple Planes

The total capacitance of a via depends on the geometry of the via, the surrounding reference planes, the trace width used to connect to the via, and the dielectric constant of the substrate material. The capacitance does *not* depend (much) on which particular layers are used to interconnect traces leading to and from the via.

For example, suppose in a 14-layer board a certain through-hole via accepts a signal coming in on layer 1 and leaving on layer 3. An identical adjacent via accepts a signal coming in on layer 1 and leaving on layer 14. Even though the signal current traverses different sections of the two vias, the voltages on the vias are the same in both cases, so the total incremental capacitance added to the traces is the same. Central to this analysis, of course, is the assumption that the full length of the via remains electrically short compared to the signal risetime and that the pads on all layers bear similar relationships to their respective reference planes.

Factors that do matter a great deal in the calculation of via capacitance include the number and positions of the reference planes.

The analysis of via capacitance presented in Tables 5.10A through 5.10D was by performed by Matt Hudale of Ansoft Corporation using the Ansoft Q3D Extractor Version 5. The via-capacitance data was extracted from field simulations using the technique described in Section 5.5.1.

The traces were 5-mils wide (approx. 70 ohms). Hudale measured the total capacitance of the combination of via and trace, and then subtracted out the capacitance of a trace of equivalent length (and with equivalent fringing fields at its start and finish points) to arrive at a final figure for the incremental capacitance added to the line by the presence of the via structure.

The simulated area of the planes was a block 0.216 in. on a side with the via located in the center. The traces were 0.108-in. long on the top layer, and 0.108-in. long on the bottom layer, extending in each case to the edge of the simulated planes. The via barrels, via pads, and clearance holes were simulated with octagonal cylinders as shown in Figure 5.30. The percentage error for convergence was set to 1% in all scenarios with the exception of the 2-

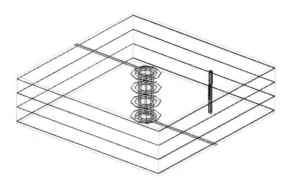

Figure 5.30—Matt approximated the parts of each via with octagonal cylinders.

plane geometry for which it was set to 0.5%. The dielectric constant was for all calculations was set to 4.4.

The first three tables show computed results for pad diameters of 24, 30, and 36 mils, and clearance diameters of 24, 30, and 36 mils. Examination of the data indicated that the computed data formed a fairly flat surface when plotted against the pad and clearance diameters, so I interpolated this data to finer gradations, showing interpolated results at steps of 2 mils. The interpolation doesn't add any new information; it just makes it easier to use the tables.

All the vias listed in Tables 5.10A through 5.10D penetrate completely through the board. In each case the board includes top and bottom layers for signals, plus from two to six additional solid reference layers. Signal pads are always present on the top and bottom

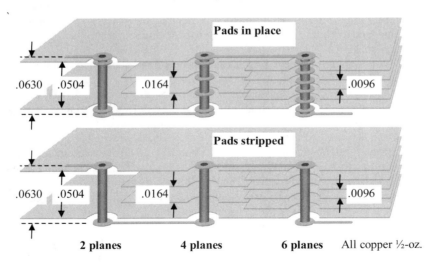

Figure 5.31—Stripping pads from the reference plane layers reduces the capacitance of a via, but increases the risk of a short between the via pad and the plane.

5.5.4 • Capacitance Data

Table 5.10A—Via capacitance data for 0.063-in. board with TWO reference planes

	Pad dia. (mil)	Clearance diameter (mil)						
		24	26	28	**30**	32	34	**36**
Pads stripped	**36**	**535**	510	485	**461**	438	416	**394**
	34	500	476	451	427	406	385	364
	32	466	442	417	393	373	353	333
	30	**432**	408	383	**358**	340	322	**303**
	28	398	375	352	328	312	295	279
	26	365	342	320	298	283	269	254
	24	**331**	310	289	**268**	255	242	**229**
Pads in place	30							429
	28						407	380
	26					385	358	331
	24				**362**	335	308	**281**

NOTE—HOLE DIA=12 mil, BOARD THICKNESS=63 mil
NOTE—Items in bold were simulated, others are interpolated from the simulated points.
NOTE—All capacitances in fF

Table 5.10B—Via capacitance data for 0.063-in. board with FOUR reference planes

	Pad dia. (mil)	Clearance diameter (mil)						
		24	26	28	**30**	32	34	**36**
Pads stripped	**36**	**639**	599	560	**521**	485	449	**413**
	34	603	565	527	489	454	418	383
	32	567	530	494	457	422	388	353
	30	**531**	496	460	**425**	391	357	**323**
	28	494	462	430	399	365	332	299
	26	456	428	401	373	340	307	275
	24	**419**	395	371	**347**	315	283	**251**
Pads in place	30							706
	28						664	613
	26					622	571	520
	24				**580**	529	478	**427**

NOTE— HOLE DIA=12 mil, BOARD THICKNESS=63 mil
NOTE—Items in bold were simulated, others are interpolated from the simulated points.
NOTE—All capacitances in fF

Table 5.10C—Via capacitance data for 0.063-in. board with SIX reference planes

Pad dia. (mil)		Clearance diameter (mil)						
		24	26	28	30	32	34	36
Pads stripped	36	**694**	649	604	**560**	521	482	**444**
	34	654	610	566	522	485	447	410
	32	614	571	528	484	448	412	376
	30	**575**	532	489	**447**	412	377	**342**
	28	544	503	462	422	389	355	322
	26	513	474	435	397	365	334	302
	24	**482**	445	409	**372**	342	312	**282**
Pads in place	30							**910**
	28						852	780
	26					794	722	650
	24				**737**	665	592	**520**

NOTE— HOLE DIA=12 mil, BOARD THICKNESS=63 mil
NOTE—Items in bold were simulated, others are interpolated from the simulated points.
NOTE—All capacitances in fF

Table 5.10D—Via capacitance data for 0.096-in. board

		Drilled hole dia., mil	Pad dia., mil	Clearance dia., mil	Ref. planes	Via cap. (fF)
Pads stripped		26	38	52	4	639
		26	38	52	6	684
		26	38	52	8	721
Pads in place		26	38	52	4	784
		26	38	52	6	934
		26	38	52	8	1081

layers. The pads may be either present or stripped on the plane layers, as indicated in the tables. Stripping pads from the reference plane layers (see Figure 5.31) reduces the capacitance of a via, but increases the risk of a mechanical short between the via pad and the plane due to capillary movement of the plating solution between the dielectric layers. In cases where the pads are present on plane layers the size of the clearance holes obviously imposes a restriction on the maximum pad diameter. If the pads on reference layers are stripped, as one might do when squeezing the clearance holes to maintain continuity of the reference planes in a very dense design, the pad diameter may exceed the clearance diameter.

5.5.5 • Inductance Data

When only two reference-plane layers are used, they are located 5 mils below the top surface and 5 mils above the bottom surface of the board. When more reference-plane layers are used, the first two remain in the positions just described, and the others are spaced equally throughout the remaining interior of the board.

POINTS TO REMEMBER

> ➢ The incremental capacitance of a via is affected by the geometry of the via, the surrounding reference planes, the trace width used to connect to the via, and the dielectric constant of the substrate material.
>
> ➢ Via capacitance varies in proportion to the overall size of the via.

5.5.5 Inductance Data

5.5.5.1 Through-Hole Via Inductance

The inductance of a signal via depends on the location of the return path associated with that signal via. A signal via that traverses only one plane keeps the returning signal current close at hand all along the signal pathway (Figure 5.32). Where the signal via dives through the plane on a signal via, the return current dives through the clearance hole around the via onto the back side of the reference plane (at high frequencies the reference plane thickness exceeds the skin depth). The returning signal current flows only on the top or bottom surface of the reference plane; it does not penetrate the plane. The clearance hole around the signal via provides a portal through which the returning signal current can pop between top and bottom surfaces. To the left of the via the return current flows on the top surface of the reference plane, to the right it flows on the bottom surface. Such a signal via encounters almost no parasitic inductance; the effect of the via is primarily capacitive [48]. See also typical values in Table 5.8.

The signal in Figure 5.33 traverses two reference planes. The returning signal current therefore must flow through the nearest available interplane connection. In Figure 5.33 the interplane connection is a via. The inductance of the overall configuration depends on the location of this interplane connection.

Keep in mind that at high frequencies the plane layers are many skin depths thick—so thick that the returning signal current cannot penetrate the plane. It requires a hole in the plane to pop from one side to the other. Because the wall of a via is similarly thick, returning current does not generally flow down the hole on the inner surface of a via. It chooses to flow instead on the outside surface. The precise pathway of the returning current, including the exact place where it switches from the one surface of each plane to the other, is indicated in Figure 5.33. In a more complex configuration the returning signal current would spread out, penetrating several nearby clearance holes and using multiple nearby vias to accomplish the jump between layers 2 and 3.

If the planes are interconnected with one or more vias, the inductance becomes a function of the positions of all the nearby vias (see examples in Figure 5.34). If the signal changes reference planes from a ground plane to a power plane (or vice versa), the interplane path will include at least one capacitor. The inductance of the bypass capacitor

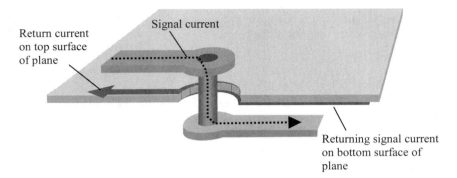

Figure 5.32—Returning signal current at a via pops between top and bottom surfaces of a solid reference plane by passing through the clearance hole.

above and beyond the topmost plane layer must be added to the interplane inductance computed per Figure 5.34. In a realistic situation with randomly placed interplane vias, the closest vias exert the lion's share of influence upon the overall inductance.

If you suspect that via inductance will play a significant role in your design, your choices are to (1) buy and learn to use a 3-D field solver, (2) prototype a stock set of geometries and catalog their performance for later use, or (3) build something (anything) and then adjust its geometry in successive passes until you get it right. The most common via adjustments are as follows:

1. For large vias, back-drill the holes. This removes part of the plated metal from the inside of the finished via, starting from the back side of the board and working back in. If your signal goes only partway through the board, this method can be effective. Back-drilling directly reduces C_V, raising Z_V.

2. Change the size of the drilled hole. This works if you are dealing with a signal via that doesn't have to accommodate a fixed size of connector pin. Shrinking the hole size while leaving the clearance and pad unchanged will reduce C_V and increase L_V.

3. Change the size of the clearance hole. This works only to the extent that you can maintain continuity of the planes in and among your field of vias. Increasing the clearance diameter decreases C_V and increases L_V.

4. Change the pad diameter. A smaller pad decreases C_V but doesn't have much effect on L_V.

5. Reducing the hole size, clearance size, and pad diameter all together in proportion doesn't much change anything. You need to change the *ratio* of the drilled hole to the clearance hole, or the pad to the clearance hole, to get much of an effect.

6. Enlarge the clearance hole in a noncircular manner. For example, elongating the clearance (anti-pad) on the side facing the incoming trace creates an egg-shaped or elliptical hole. The enlargement decreases C_V, thus raising Z_V. The advantage of using an elongated clearance is that it helps reduce the via

5.5.5 • Inductance Data

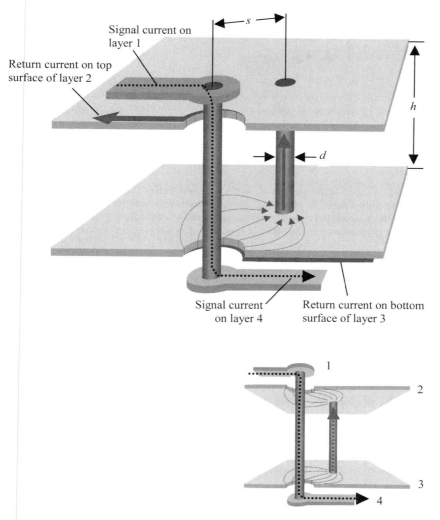

Figure 5.33—Returning signal current from the bottom surface of plane layer 3 pops through the lower clearance hole to the top surface of plane layer 3, then flows across the plane to the bottom of the interplane via, goes up that via to plane layer 2, then across the lower surface of plane layer 2 to the upper clearance hole, and finally pops through the upper clearance hole to the top surface of plane layer 2.

capacitance without affecting the spacing between adjacent vias on either side. In a tightly spaced row of vias this helps preserve the continuity of the reference planes between clearance holes [53].

In all cases check with your pcb fabricator to ensure that they comply with your precise adjustments of the hole, pad, and clearance sizes. Pcb vendors, often without telling you,

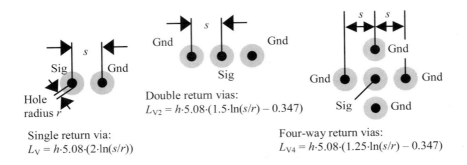

Single return via:
$L_V = h \cdot 5.08 \cdot (2 \cdot \ln(s/r))$

Double return vias:
$L_{V2} = h \cdot 5.08 \cdot (1.5 \cdot \ln(s/r) - 0.347)$

Four-way return vias:
$L_{V4} = h \cdot 5.08 \cdot (1.25 \cdot \ln(s/r) - 0.347)$

Figure 5.34—The inductance contributed by an interplane separation depends on the location of the nearest interplane connections. In all cases L_V is the via interplane inductance (nH), r is the hole radius (in.), h is the spacing between the planes (in.), and s is the separation between via centers (in.). If the via traverses multiple plane layers, then h equals the aggregate distance traversed.

make last-minute changes to these parameters in an attempt to improve their finished board yield.

POINTS TO REMEMBER

> The inductance of a signal via depends on the location of the return path associated with that signal via.

> A signal via that traverses only one plane keeps the returning signal current close at hand all along the signal pathway.

> A signal via that traverses two reference planes forces returning signal current through the nearest available interplane connection.

> If a signal changes reference planes from a ground plane to a power plane (or vice versa), the interplane return path must traverse bypass capacitors.

> Pcb vendors, often without telling you, make last-minute changes to hole, pad, and clearance sizes in an attempt to improve their finished board yield.

5.5.5.2 Via Crosstalk

This section concerns crosstalk generated by signals that traverse a stripline cavity (i.e., the space between any two solid reference planes).

Figure 5.35 illustrates a typical scenario for two signal vias (aggressor and victim) that traverse the same cavity. The height of the cavity (spacing between the planes) is marked h.

Signal current on the aggressor via creates patterns of magnetic fields within the cavity that look like concentric circles. The field intensity is constant in the z-axis direction (up and down in the figure), but drops off with $1/x$ at distances removed from the aggressive trace. For a signal current of 1 amp, the magnetic field intensity B at a point within the cavity removed by distance x away from the signal via is

5.5.5 • Inductance Data

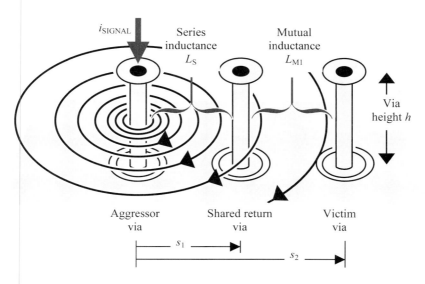

Figure 5.35—Magnetic fields from the aggressive signal cause crosstalk.

$$|B(x)| = \frac{\mu}{2\pi x} \quad [5.34]$$

where μ is the magnetic permeability of the dielectric substrate (H/m),
 x is the horizontal distance from the signal via, and
 for nonmagnetic substrate materials, $\mu = 4\pi \cdot 10^{-7}$ H/m.

The mutual inductance L_{M1} from the signal current to the victim via is determined by integrating [5.34] over a range of x from s_1 (the distance from the signal via to the shared-return via) to s_2 (the distance from the signal via to the victim via), and also over the range of z from 0 to h. Taking into account the polarity of the coupling,

$$L_{M1} = -\int_0^h \int_{s_1}^{s_2} \frac{\mu}{2\pi r} dr\, dz = -\frac{\mu h}{2\pi} \ln\left(\frac{s_2}{s_1}\right) \quad [5.35]$$

where μ is the magnetic permeability of the dielectric substrate (H/m),
 s_1 is the horizontal distance from the signal via to the shared-return via (m),
 s_2 is the horizontal distance from the signal via to the victim via (m),
 h is the separation between the reference planes (m), and

for nonmagnetic substrate materials, $\mu = 4\pi \cdot 10^{-7}$ H/m.

The mutual inductance L_{M2} from the aggressive return current to the victim via must also be taken into account (Figure 5.36). This inductance is determined by integrating [5.34] over a range of x from r (the radius of the shared-return via) to s_3 (the distance from the shared-return via to the victim via), and also over the range of z from 0 to h.

$$L_{M2} = \int_0^h \int_r^{s_3} \frac{\mu}{2\pi r} \, dr \, dz$$
$$= \frac{\mu h}{2\pi} \ln\left(\frac{s_3}{r}\right) \qquad [5.36]$$

where μ is the magnetic permeability of the dielectric substrate (H/m),
s_3 is the horizontal distance from the shared-return via to the victim via (m),
r is the radius of the shared-return via (m),
h is the separation between the reference planes (m), and
for nonmagnetic substrate materials, $\mu = 4\pi \cdot 10^{-7}$ H/m.

The overall mutual inductance between the aggressive path (signal and return) and the victim path is the sum of [5.35] and [5.36].

$$L_M = \frac{\mu h}{2\pi}\left[-\ln\left(\frac{s_2}{s_1}\right) + \ln\left(\frac{s_3}{r}\right)\right] \quad \text{H}$$
$$= \frac{\mu h}{2\pi} \ln\left(\frac{s_1 s_3}{s_2 r}\right) \quad \text{H} \qquad [5.37]$$
$$= h \cdot 200 \cdot \ln\left(\frac{s_1 s_3}{s_2 r}\right) \quad \text{nH}$$

where μ is the magnetic permeability of the dielectric substrate (H/m),
s_1 is the horizontal distance from the signal via to the shared-return via (m),
s_2 is the horizontal distance from the signal via to the victim via (m),
s_3 is the horizontal distance from the shared-return via to the victim via (m),
r is the radius of the shared-return via (m),
h is the separation between the reference planes (m), and
the last equation assumes for nonmagnetic substrate materials $\mu = 4\pi \cdot 10^{-7}$ H/m.

5.5.5 • Inductance Data

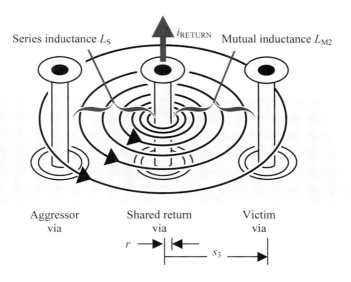

Figure 5.36—The return current, being closer to the victim than the signal path, causes even more crosstalk than does the signal current.

In the configuration of Figure 5.36 the return current via, being closer to the victim than the signal via, causes the majority of the crosstalk.

In English units, equation [5.37] is rewritten

$$L_M = h \cdot 5.08 \cdot \ln\left(\frac{s_1 s_3}{s_2 r}\right) \quad \text{nH} \qquad [5.38]$$

where s_1 is the horizontal distance from the signal via to the shared-return via (in.),

s_2 is the horizontal distance from the signal via to the victim via (in.),

s_3 is the horizontal distance from the shared-return via to the victim via (in.),

r is the radius of the shared-return via (in.),

h is the separation between the reference planes (in.), and

this equation assumes a nonmagnetic substrate.

Various simplifications of [5.38] are listed in Figure 5.37 and Figure 5.38. For example, the case L_{MC} in Figure 5.37 sets $s_1 = s_3 = s$, and $s_2 = 2s$.

Example calculation of mutual inductance of vias

Parameters:
20-mil pad
5-mil gap between pads
$h = 0.025$

$r = 0.003$
$s = 0.025$

Self-inductance of via with single return path:
$L_V = .539$ nH $= 3.38\ \Omega$ at 1 GHz

Mutual inductances from Figure 5.37 and Figure 5.38:
$L_{MA} = .357$ nH $= 2.24\ \Omega$ at 1 GHz
$L_{MB} = .269$ nH $= 1.69\ \Omega$ at 1 GHz
$L_{MC} = .181$ nH $= 1.14\ \Omega$ at 1 GHz

The crosstalk voltage induced in a victim circuit equals the rate of change of current in the aggressor times the mutual inductance, L_M, shared between the two circuits. With a step change of aggressive current having a maximum rate of change equal to $(\Delta V/Z_C)/t_r$, the peak voltage in the victim circuit equals

$$v_{\text{peak,step}} = \frac{\Delta V}{t_r} \frac{L_M}{Z_C} \quad [5.39]$$

For a sinusoidal excitation of amplitude a the maximum rate of change in the aggressor current equals $(a2\pi f)/Z_C$, for a peak victim voltage of

$$v_{\text{peak,sinusoidal}} = a\frac{2\pi f L_M}{Z_C} \quad [5.40]$$

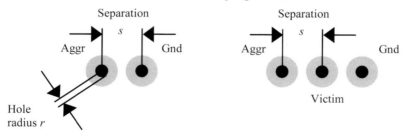

Figure 5.37—Self-inductances and mutual inductances are easily computed for simple via configurations.

5.6 • Future of On-Chip Interconnections

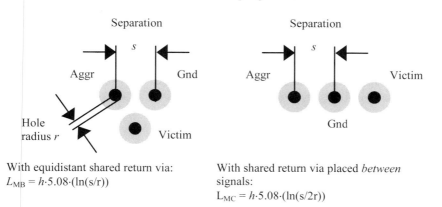

With equidistant shared return via:
$L_{MB} = h \cdot 5.08 \cdot (\ln(s/r))$

With shared return via placed *between* signals:
$L_{MC} = h \cdot 5.08 \cdot (\ln(s/2r))$

All dimensions in inches
All inductances in nH

Figure 5.38—Mutual inductances are easily computed for simple via configurations.

A mutual impedance of $2\pi f L_M = 1\ \Omega$ shared between two 50-Ω circuits therefore induces a crosstalk voltage in the victim of approximately 2%.

In a both-ends-terminated architecture, half of the inductively coupled crosstalk amplitude propagates towards either end of the victim circuit.

Within a group of signal vias that all share one common return path, the crosstalk from all the nearby aggressors aggregates.

POINTS TO REMEMBER

> Vias that traverse a common stripline cavity (i.e., the space between two reference planes) create crosstalk.
> The crosstalk voltage induced in a victim circuit equals the rate of change of current in the aggressor times the mutual inductance, L_M, shared between the two circuits.

5.6 THE FUTURE OF ON-CHIP INTERCONNECTIONS

Article first published in *EDN Magazine*, February 3, 2000

The specialized branch of mathematics known as topology studies obscure relationships between seemingly unrelated items. For example, a topologist might point out that a coffee cup and a doughnut are topologically equivalent, both being solid objects with a single hole.

Topological equivalence between two objects means that if the first object were made from an infinitely stretchy, rubber-like substance, you could stretch or mold the first object into the shape of the other object without tearing any new holes or closing off any existing holes.

You can extend the concept of equivalence relations to the field of digital communications. By changing the impedance, stretching the distance, or adjusting the bandwidth, you can often mold one problem into a form that looks a lot like another problem.

Consider the relationship between the telegraph and the telephone. Both use copper-based transmission media but at varying distances and speeds. In the early 1860s telegraph systems operated with simple circuitry at distances of hundreds of miles and an average bandwidth of only a few baud. Later developments boosted the speed considerably, using high-speed paper-tape recording and playback machines to multiplex messages from many human operators onto a single telegraph wire [54].

> If you want to know what will happen on-chip tomorrow, look at a pcb today.

The telephone uses the same type of wiring as a telegraph, but uses the wiring at a greatly reduced distance and correspondingly higher bandwidth.

Plain-old-telephone-service connections with simple circuitry operate at maximum distances of 1 mile and maximum bandwidths of 3 kHz. If you want your telephone line to operate at digital-subscriber-line speeds, the required circuitry gets more complex—just like in the telegraph problem. The type of wire, the distance, and the sophistication of your circuitry determine the achievable speed of applications such as the telegraph and the telephone. The shorter the wire, the greater its natural bandwidth. The more sophisticated the circuitry, the closer you can push toward full use of the ultimate bandwidth of your channel. The physical principles controlling the behavior of the wires are the same in both telephone and telegraph applications; they just operate at different points on the speed-distance curve. You can usually scale up the speed of any copper-based communications system by scaling down its length.

The speed-distance scaling principle applies equally well to wiring used in LAN applications. In the early 1980s several companies noticed that although telephone service was required to operate at distances as great as 1 mile, in-building data wiring needed to go only about 100 m. The available bandwidth of unshielded twisted-pair-type telephone wiring at that short distance is enormous, as demonstrated by the ever-popular 10-Mbps Ethernet 10BaseT, 100-Mbps Fast Ethernet 100BaseTX, and 1000-Mbps Gigabit Ethernet 1000BaseT. Shrinking the wire distances from a telephone-centric 1-mile requirement to a data-centric 100 m gets you easily from telephone speeds to 10 Mbps. To get to 100 Mbps, you use better cabling (category 5). To get to 1000 Mbps, you use four pairs of category-5 cabling and advanced adaptive-equalization techniques.

Going further, you can apply the speed-distance-scaling principle to pcb traces. They are transmission lines too, just as in the telegraph, telephone, and LAN applications. Being shorter, they of course work at higher speeds. For applications

5.6 • Future of On-Chip Interconnections

that reach about 1 GHz at 25-cm distances (10 in.), simple CMOS totem-pole switching circuitry works fine. Applications faster than 1 GHz or longer than 25 cm require advanced circuitry that takes into account trace impedance, terminations, crosstalk, high-frequency losses, multilevel signaling and adaptive equalization.

Transmission-line theory will next become important on-chip. Today's chip-layout software takes into account the RC propagation delays of major bus structures and clock lines. In tomorrow's designs, at even higher speeds, the full RLC nature of the on-chip transmission channels will emerge. Ringing, terminations, and adaptively equalized multilevel signaling will all eventually appear on-chip, just as they do on-board. It's inevitable. If you want to know what will happen on-chip tomorrow, look at a pcb today; the problems are rapidly becoming equivalent.

For further study see: www.sigcon.com

CHAPTER 6

DIFFERENTIAL SIGNALING

This chapter explores the essential elements of differential signaling. It illustrates the many benefits of differential operation and lays out the conditions under which those benefits may be obtained. It also gives some practical advice about how to best manage differential connections.

I'll begin with a detailed, start-from-the-basics look at *single-ended signaling* architectures, then extend the analysis to cover *differential signaling*. The point of this presentation is to break down differential signaling into its component parts to show how precisely the various benefits are obtained.

6.1 SINGLE-ENDED CIRCUITS

The scissors in Figure 6.1 have cut the circuit, interrupting communications. With the wire cut, regardless of the position of the switch, no current flows through the bulb. To light the bulb, this elementary circuit needs the top switch closed *and* the bottom wire connected.

It takes two wires to efficiently convey electrical power.

This easy example illustrates an important principle of lumped-element circuit operation—namely, that current flows only in a *complete* loop. Another way of expressing the same loop idea is to simply say that if current comes out of the battery, it must return to the battery. That is the way circuits operate in practice. Electrons can't just come out of the battery, pass through the bulb, and then pile up somewhere. The movement of electrons requires a complete loop path.

The current-loop principle may be stated in general mathematical terms:

> *The total sum of all current going in and out of the battery is zero.*

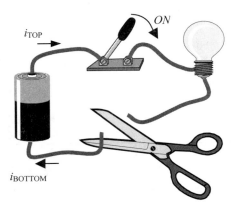

Figure 6.1—Whether cut by the switch or the scissors, this circuit is open and no current flows.

This principle is equivalent to Kirchoff's current law, named after its discoverer, Gustav Robert Kirchoff, 1824–1887, a German physicist. It applies to all lumped circuit elements (not just batteries), including nonlinear circuits, time-varying circuits, passive circuits, active circuits, and circuits with more than two wires. It remains true even in the presence of distributed parasitic capacitances provided one is careful to include the displacement currents.

Applying this principle to Figure 6.1, cutting the bottom wire prevents current from entering the bottom of the battery; therefore, current is also prevented from leaving at the top.

I hope this discussion makes it clear that you cannot propagate electrical current on just one wire. The propagation of electrical current requires a system of at least *two* conductors.

Now let's apply this same reasoning to high-speed digital systems. I'll restrict my attention to *electrical* logic families such as TTL, CMOS, and ECL that have electrical inputs (as opposed to optical or telepathic inputs). First please note that any electric input requires *current* to operate. Specification sheets may emphasize the performance in terms of voltage specifications, but current is still required. Even on CMOS parts, for which the input current is practically zero, it takes a fair amount of current on every rising edge just to charge the parasitic input capacitance.[43] Every electrical input requires current. As a result, the propagation of every (electrical) logic signal requires a system of at least two conductors. Even though the "second wire" does not appear on a schematic circuit diagram, its presence is required.

As you probably know, the two-conductor requirement is rather inconvenient. Most digital logic designers would rather use a single wire for each signal, or at least *make believe that they are doing so*. Toward that end, we designers have adopted a certain convention for the handling of the second wire necessary in all circuits. What we do is hook one side of all transmitters and all receivers to a common reference voltage.[44] In high-speed designs this voltage is usually distributed throughout the system as a ground plane or a pair of power and

---

[43] It takes 1 mA to charge one picofarad to 1 volt in 1 nanosecond.

[44] A global reference voltage may be distributed by means of a solid ground plane, a solid V_{CC} plane, or the Earth itself. Anything can work as long as it ties everything together through a reasonably low impedance.

6.1 • Single-Ended Circuits

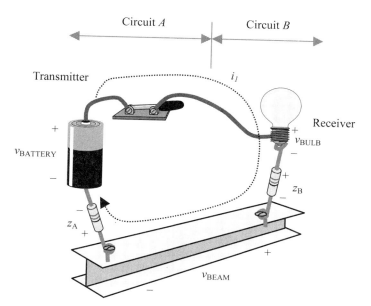

Figure 6.2—Digital systems use a common ground reference for all returning signal currents. Components z_A and z_B represent the ground connection impedances of a semiconductor package.

ground planes. All circuits share this same "second conductor" for the conveyance of returning signal currents. Figure 6.2 illustrates the arrangement for a single transmitter (key switch), receiver (light bulb), and reference system (I-beam). The use of a shared reference voltage for all circuits is called *single-ended signaling*. Single-ended systems require only one apparent wire for each signal. What one needs to keep in mind about this arrangement is that the second wire is still physically there and that it still physically carries the returning signal currents for each transmitter; it's just implemented as a big, common shared connection.

Here's how the circuit in Figure 6.2 works. The transmitter (circuit *A*) includes a battery and a switch. One side of the battery connects to a big, solid, shared-reference system (I-beam). This connection between the transmitter circuit and the reference system passes through impedance z_A, representing the finite impedance of the package pins or balls used to accomplish this connection.

The receiver (circuit *B*) is merely a light bulb connected to the beam through impedance z_B. The bulb brightness indicates the condition of the transmitter (switch *ON* or *OFF*). As with the transmitter, the bulb's connection to the common reference beams has a finite impedance.

Provided that the reference-connection impedances are sufficiently low and that there is no significant voltage drop across the steel I-beam, the circuit functions well. When the switch is depressed, the bulb lights.

Let's take a moment to investigate how the finite-impedance reference connections affect the voltage sensed by the bulb. This effect is controlled by the following physical principle:

The battery voltage equals the sum of the voltages across series-connected loads.

This principle is equivalent to Kirchoff's voltage law. It applies to all voltage sources (not just batteries), including nonlinear sources and time-varying sources.

Applied to the circuit in Figure 6.2, Kirchoff's voltage law predicts the following relation:

$$v_{\text{BATTERY}} = v_{\text{BULB}} + z_A i_1 + v_{\text{BEAM}} + z_B i_1 \quad [6.1]$$

Here I have substituted the expression $z_A i_1$ and $z_B i_1$ for the voltages across the two resistors, with the variable i representing the current flowing around the loop. Presuming that the battery voltage remains constant, if either of the voltages $z_A i_1$ or $z_B i_1$ go up, the voltage across the bulb must go down. Let me state that again in a slightly different way: Anything that affects the voltage drop across z_A, z_B, or the beam also affects the voltage received at the bulb. This statement uncovers a great weakness of single-ended signaling: The reference voltages at transmitter and receiver must match. Unfortunately, as anyone with an oscilloscope knows, noise exists between every two points on a ground plane (or power plane). If the local reference voltages at the receiver and at the transmitter differ by too great an amount, single-ended signaling can't work.

The difficulty here is that the light bulb has no way of knowing what is the *true* voltage coming out of the battery. It has no magic connection to the center of the earth with which to measure the *true* earth potential. It sees only what remains of the transmitted battery voltage after subtraction of the various voltage drops around the signaling loop. In this sense the receiver is a *differential receiver*. Is responds only to the *difference* between the voltage on its input terminal and the voltage on its reference terminal (the reference terminal is the ground pin in TTL logic or the most positive power supply pin in ECL logic). All digital receivers operate the same way. Adding 100 mV of noise to the reference pins of a single-ended IC accomplishes precisely the same thing as adding 100 mV of noise directly to every receiver input. This extra noise detracts directly from the available *noise margin*[45] in your logic family and must be incorporated into any voltage margin analysis for the system.

There is no way around this limitation. If you wish to use single-ended signaling, you must limit the voltage differences within the reference system to a small fraction of the signal amplitude. The impedance of the reference system itself must be low enough to absorb all returning signal currents from all sources without producing objectionable voltage drops at any point.

Figure 6.3 depicts another problem inherent to single-ended signaling. This figure incorporates a second transmitter. Assume the second transmitter is part of circuit *B*. It resides within the same physical package as the receiver; therefore, they share a common ground connection internal to the package. In the figure this shared connection appears as a solid wire connecting the bottom of the light bulb to the bottom of the second battery. This shared internal ground then connects through the impedance of the ground pins or balls on the package (represented by impedance z_B) to the circuit board ground (the I-beam). The

---

[45] The *noise margin* for a logic family is the difference between V_{OL} and V_{IL}, or V_{OH} and V_{IH}, whichever is less.

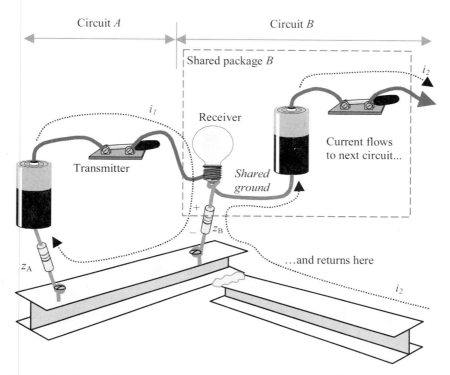

Figure 6.3—*Ground Bounce:* **In circuit B, a transmitter and receiver share the same ground connection z_B. As current i_2 passes through component z_B, it creates voltage $-i_2 z_B$, which interferes with reception.**

boundary of the package containing both the first receiver and the second transmitter is depicted in Figure 6.3 as a box drawn with dashed lines.

Consider what happens when current i_2 flows in the second transmitter circuit. First current i_2 exits the shared package (dotted region). After circulating through its load, this same current must therefore find its way back inside the shared package. There are only two paths available for current returning to the shared package—either through impedance z_B or through the wire leading to the input port of the receiver (light bulb). In a digital system the impedance z_B is much lower (hopefully) than the impedance of the path through the input port, so most of the current reenters the package through z_B. So far, so good. Next let's examine the implications of this returning current.

As returning current i_2 reenters the package, it induces voltage $-z_B i_2$ across the impedance z_B. Because impedance z_B is shared with the receiving circuit, any voltages appearing across it disturb the apparent voltage received by the light bulb (equation [6.1]). This form of interference goes by the name of *common impedance coupling* [60]. It happens any time there is any overlap between the paths of current flow for a transmitter and a receiver.

In a high-speed digital application the inductance of the ground connection z_B usually causes more difficulties that its resistance. This inductance, multiplied times the *di/dt* of the returning current, can generate voltages easily sufficient to disturb normal receiver operation. The problem of noise generated by returning signal current acting across the finite impedance of a common ground connection within an IC package is called *ground bounce* [61], or more generally *simultaneous switching noise* [59].

The same general problem also happens wherever the reference system is weakened or necked-down. Returning signal currents from many drivers, flowing through the finite inductance of the necked-down region, generate differences in the reference potential (with respect to true earth ground) at different points in the system. These voltages can disturb receiver operation. This problem is often called a *ground shift*, or *noisy ground*. Noticeable ground shifts often happen across places, like connectors and cables, where the integrity of the solid ground plane has been violated.

Power and ground distributions networks are both susceptible to returning signal currents in the same way. Currents flowing through a power distribution net can perturb the power-supply voltages at various points within the system in the same way that currents in the ground network perturb the ground voltages. Whether you care more about power-supply noise or ground noise depends on whether your single-ended circuits use the power or the ground rail as their internal reference for the discrimination of logic signals. TTL integrated circuits and most high-speed digital CMOS circuits use the ground terminal as the designated reference voltage. ECL circuits powered from ground and -5.2 volts use the ground terminal as the designated reference voltage. ECL circuits powered from a positive supply and ground (sometimes called PECL) use their positive supply terminal as the designated reference voltage.

POINTS TO REMEMBER

- The big advantage of single-ended signaling is that it requires only one wire per signal.
- Single-ended signaling falls prey to disturbances in the reference voltage.
- Single-ended signaling is susceptible to ground bounce.
- Single-ended signaling requires a low-impedance common reference connection.

6.2 TWO-WIRE CIRCUITS

Two-wire signaling cures many noise problems at the cost of a second signal trace. As shown in Figure 6.4, a two-wire transmitter sends current on two wires: a first wire, which carries the main signal, and a second wire, which is provided for the flow of returning signal current. As drawn, the *currents* on the two wires will be equal and opposite, but the voltages will not be. This architecture provides three important benefits.

First, it frees the receiver from requiring a global reference voltage.[46] In effect, the second wire serves as a reference for the first. The receiver need merely look at the difference between the two incoming wires. Two-wire signaling renders a system immune to

---

[46] The reference voltage for TTL, most high-speed CMOS, and ECL is ground; for PECL (positively-biased ECL), it is the power voltage.

6.2 • Two-Wire Circuits

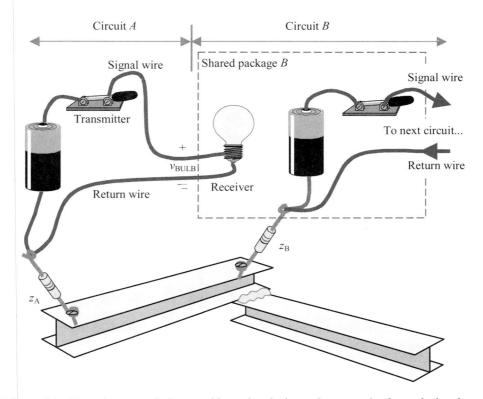

Figure 6.4—*Two-wire* transmission provides a signal wire and return wire for each signal.

disturbances in distribution of global reference voltages, provided the disturbances do not exceed the power-supply noise tolerance of the logic family or the common-mode input range of the receivers.

Second, the two-wire architecture eliminates shared-impedance coupling between a receiver and transmitter in the same package. In Figure 6.4, returning associated with transmitter B flows through the return wire back to the battery at B without traversing z_B, and therefore *without disturbing the receiver*. By eliminating the shared-impedance coupling between circuits A and B, two-wire signaling conquers ground bounce locally generated within the package.

Third, two-wire signaling counteracts any type of interfering noise that affects both wires equally. A good example would be the *ground shifts* encountered in a high-speed connector. When two systems are mated by a connector, the net flow of signal current between the systems returns to its source through the ground (or power) pins of the connector. As it does so, tiny voltages are induced across the inductance of the connector's ground (or power) pins. These tiny voltages appear as a difference between the ground (or power) voltage on one side of the connector and the ground (or power) voltage on the other side. This problem is called a *ground shift,* and it is yet another form of common impedance coupling. Two-wire signaling fixes this problem.

These three benefits do not depend on the use of any *changing* voltage on the second wire. As shown in Figure 6.4, the return wire merely carries the local reference voltage (ground, in this case) from the transmitter to the receiver, where it may be observed. This simple circuit renders the system immune to local disturbances in the power and ground voltages, ground bounce generated within a package, and ground bounce generated within a connector. That's pretty good.

The performance of a two-wire signaling circuit hinges on the assumption that no current flows through impedances z_A and z_B. Under this assumption the receiver at B can directly observe (on the return wire) the local reference voltage at transmitter A, and the next receiver C can observe the local reference voltage at transmitter B. Any currents flowing through z_A or z_B change the references voltages on the return wires, interfering with reception. The two-wire circuit must be arranged so that it limits the current through z_A and z_B to innocuous levels.

Unfortunately, in a high-speed system all wires couple to the surrounding chassis and other metallic objects, whether you want them to or not. In Figure 6.4 you can model this coupling as a collection of parasitic lumped-element connections connected from each wire to the reference beam. Current transmitted on the signal wire therefore has a choice of returning pathways. It can return to the source along the return wire (the intended path), or it can flow through the parasitic connection to the reference beam and from there return to the transmitter through impedance z_A. The current that flows through the parasitic pathway is called *stray returning signal current*. At high speeds the stray returning signal current is often significant enough to impair the effectiveness of a two-wire signaling system.

Does this impairment defeat the utility of two-wire signaling for high-speed circuits? Not necessarily, provided that you pick a particular, unique signal for the second wire. The second wire must carry a signal equal in amplitude to the first, but opposite in polarity (an *antipodal*, or *complementary*, signal). If you do that, everything still works.

POINTS TO REMEMBER

> Two-wire signaling renders a system immune to disturbances in distribution of global reference voltages.
> Two-wire signaling counteracts any type of interfering noise that affects both wires equally.
> Two-wire signaling counteracts ground bounce (also called simultaneous switching noise) within a receiver.
> Two-wire signaling counteracts ground shifts in connectors.
> Two-wire signaling works when there is no significant stray returning signal current.

6.3 DIFFERENTIAL SIGNALING

The transmission of two complementary signals over identical, matched traces is a special case of two-wire signaling. It is called *differential signaling* (Figure 6.5). A differential

6.3 • Differential Signaling

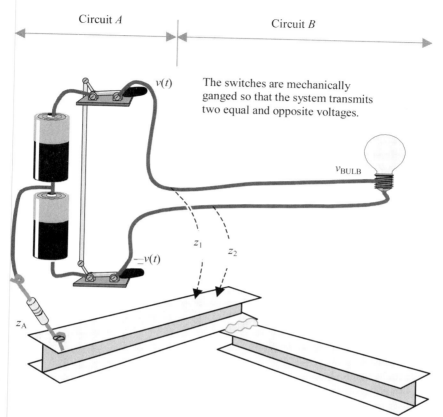

Figure 6.5—*Differential transmission:* As long as parasitic impedances z_1 and z_2 remain well balanced, no net current flows into the reference system.

signaling system delivers equal but opposite AC voltages *and currents* on the two wires. Under these conditions, and assuming the layout is symmetrical so that both wires have equal coupling to the reference system, any AC currents induced in the reference system by one wire are counteracted by equal and opposite signals induced by the complementary wire.

This effect is illustrated in Figure 6.5, where parasitic impedances z_1 and z_2 represent the impedance from one wire to the surrounding reference system (chassis or other bits of metal) and from the other wire to the reference system respectively. In a well-balanced differential system these two impedances are equal. As long as the AC voltages on the two wires are complementary, the stray currents through z_1 and z_2 will cancel, resulting in *no net flow of stray returning signal current in the reference system*.

The performance of such a system does not hinge on the particular value of impedance z_A (since no current flows through it). Conversely, the system enjoys a measure of immunity from other circuits that do induce currents in the reference structure, because the differential receiver does need the voltages on the reference system to be the same everywhere. A differential receiver requires only that the disparity between reference voltages at either end of a link not cause the received signal to exceed the common-mode operating range of the receiver.

If the impedances z_1 and z_2 are not well balanced, or if the transmitted voltages are not precisely complementary, some amount of current will flow in the reference system. This current is called *common-mode current*. Assuming that the transmitted voltages are equal, I'd like to point out two physical means for reducing the magnitude of common-mode current: the *weak-coupling* approach and the *precise-balance* approach.

High-performance twisted-pair data cabling (Figure 6.6) uses the *weak-coupling* approach, whereby the parasitic coupling through z_1 and z_2 to Earth is generally weakened. The weakening is accomplished by thickening the plastic jacket on the cable to keep other wires and objects outside the jacket relatively far away from the signal conductors. At the same time, the cable holds the wires of each pair closely together in a 100-ohm differential configuration. This geometry increases the magnitude of z_1 and z_2 relative to the impedance of the load. As a result, regardless of where the cable is laid, the fraction of current that can possibly be conducted through z_1 or z_2 cannot be very great. Further reductions in the common-mode current are obtained by tightly twisting the wires so that any nearby objects are approached an equal number of times by each wire. The tight twists produce a better balance between z_1 and z_2. The combination of a weakened coupling magnitude and good balance is used in category 5, 5e, 6, and 7 unshielded twisted-pair cabling to attain spectacular common-mode rejection.

Lest you doubt that impedances z_1 or z_2 could carry enough signal current to represent a significant radiation problem, let me suggest the following experiment. Beginning with a working LAN adapter, strip back a section of the outer jacket of the twisted-pair cable. Untwist the wires of one of the active pairs, and tape one wire of that pair against a solid metal part of the product chassis for a distance of about two inches. Run in this configuration, the product will not pass its FCC or EN mandated radiation tests. The amount of excess common-mode current flowing through the parasitic capacitance of the one wire taped against the chassis causes enough radiation to exceed FCC or EN mandated limits. Unbalanced stray current to ground makes a big difference in twisted-pair transmision systems and must be strictly limited.

Differential traces on a pcb must take a different approach to the control of common-

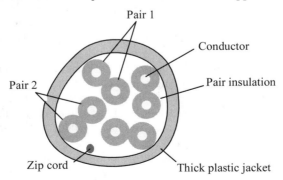

Figure 6.6—Construction of high-performance 4-pair unshielded twisted-pair data cable. The zip cord shown in this figure is a tough string that technicians pull to rip open long sections of the outer insulation.

6.4 • Differential and Common-Mode Voltages

$$b = c - d/2 \qquad [6.5]$$

In a good differential system one usually strives to limit the AC component of the common-mode signal. This is done because the common-mode portion of the transmitted signal does not enjoy any of the noise-canceling or radiation-preventing benefits of differential transmission. The common-mode and differential signals also propagate differently in most cabling systems, which can lead to peculiar skew or ringing problems if the common-mode component is an appreciable fraction of the overall signal amplitude, especially if those common-mode currents are accidentally converted into differential signals (see Section 6.8, "Differential to Common-Mode Conversion"). Intercabinet cabling, particularly, is extremely sensitive to the presence of high-frequency common-mode currents, which radiate quite efficiently from unshielded cabling.

Another decomposition of the two-wire transmission problem defines odd-mode and even-mode voltages and currents. These are similar to, but slightly different from, differential and common-mode voltages and currents.

An odd-mode signal is one that has amplitude $x(t)$ on one wire and the opposite signal $-x(t)$ on the other wire. A signal with an odd-mode amplitude of $x(t)$ has a differential amplitude of $2x(t)$. If the signal $x(t)$ takes on a peak-to-peak range of y, then the peak-to-peak odd-mode range is simply y, but the *peak-to-peak differential amplitude* is $2y$.

An even-mode signal is the same on both wires. An even-mode signal with a peak-to-peak range of y also has a peak-to-peak common-mode range of y. The even-mode amplitude and common-mode amplitude are one and the same thing.

Two-wire transmission systems sometimes send a signal voltage on one wire, but nothing on the other. In this case the differential-mode amplitude equals the signal amplitude on the first wire. The common-mode amplitude is half that value. In this case the odd-mode and even-mode amplitudes are the same and both equal to half the signal amplitude on the first wire.

Here are the translations between odd-mode and even-mode quantities. The same decomposition applies to currents.

Odd-mode signal:
$$o \triangleq \frac{a-b}{2} \qquad [6.6]$$

Even-mode signal:
$$e \triangleq \frac{a+b}{2} \qquad [6.7]$$

Signal on first wire:
$$a = e + o \qquad [6.8]$$

Signal on second wire:
$$b = e - o \qquad [6.9]$$

where a and b represent the voltages on the two wires with respect to a common reference.

The differential-and-common-mode decomposition and the even-and-odd mode decomposition share very similar definitions. The discrepancy between the two models has

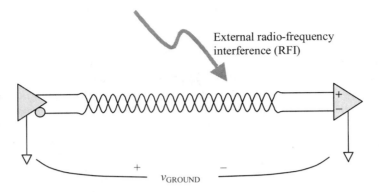

Figure 6.8—A good differential receiver cancels any noise that affects both wires equally, such as external RFI.

to do with the definition of the differential mode. The differential voltage is what you read with an electrical instrument when you put it across two wires. The odd-mode voltage is a mathematical construct that simplifies the bookkeeping in certain situations.

$$o = d/2 \qquad [6.10]$$

$$e = c \qquad [6.11]$$

Any noise like external RF interference that equally affects both wires of a differential pair will induce a common-mode (even-mode) signal, but not a differential-mode (odd-mode) signal (Figure 6.8). A good differential receiver senses only the differential signal and is therefore immune to this type of noise.

POINTS TO REMEMBER

- Differential and common-mode signals are used to describe the voltages and currents on a two-wire transmission system.
- Odd-mode and even-mode signals are yet another way to describe the voltages and currents on a two-wire transmission system.
- Differential receivers cancel common-mode noise.

6.5 DIFFERENTIAL AND COMMON-MODE VELOCITY

In a configuration with a homogeneous dielectric (like a stripline) the propagation velocities for the differential and common-mode (also read odd-mode and even-mode) signals are equal.

Configurations with inhomogeneous dielectrics (like microstrips), however, support slightly different propagation velocities for the two modes. The impact of this difference is not very great as long as there is not much coupling between the modes (see Section 6.8) and as long as the receiver remains sensitive only to the differential component of the signal. It is theoretically possible to observe deleterious effects from the difference in velocities under the following circumstances:

> A differential signal is created.
> Part of the signal is inadvertently converted to a common-mode signal.
> The differential-mode and common-mode signals propagate independently, and with slightly different velocities, to the far end of a long transmission line.
> The common-mode signal is inadvertently converted back to a differential-mode signal.

In this case the receiver will perceive a superposition of two incident waveforms with slightly different timing and amplitudes (the double-converted signal presumably being smaller).

POINT TO REMEMBER

> Microstrips support slightly different propagation velocities for the differential and common modes. The impact of this difference is not very great.

6.6 COMMON-MODE BALANCE

Common-mode balance is a term that applies to a differential transmission system. Common-mode balance is the ratio of common-mode to differential-mode signal amplitudes within the system. A perfectly differential system with no common-mode component is *perfectly balanced* (i.e., the signals on the two wires are exactly antipodal, or opposite).

When calculating the common-mode balance ratio, it is common to refer only to the AC component of the common-mode signal. For example, a perfectly complementary pair of TTL 3.3-V signals may share a common-mode DC offset of 1.65 volts, but the common-mode AC voltage (the changing part) might still be very small.

The common-mode balance ratio is often expressed in decibels. A common-mode balance of 1 part in 10,000 (–80dB) is exceptionally good. Digital logic parts sold as differential transmitters may exhibit common-mode balance of perhaps only –30 dB or even –20 dB.

Some authors express the common-mode balance in terms of the even-mode to odd-mode signal amplitude. The even-to-odd mode amplitude ratio is half the common-to-differential mode amplitude ratio, making it 6dB less impressive-sounding. An older term used to describe the common-mode to differential-mode amplitude ratio is *longitudinal balance*.

POINT TO REMEMBER

➢ Common-mode balance is the ratio of common-mode to differential-mode signal amplitudes.

6.7 COMMON-MODE RANGE

Every digital receiver comes with a specification for its common-mode input range. You are expected to ensure that input signals remain within this range at all times. As long as both inputs stay within the common-mode operating range, the component will meet or exceed its specification for the input switching threshold. Beyond that, manufacturers give few clues as to how the component will operate. It may operate normally. On the other hand, it may reverse its outputs, it may saturate and take a long time to recover, it may lock up into a brain-damaged state until power-cycled, or it may permanently fail. You never know which [55]. Don't violate this specification (even for a brief period).

The common-mode range specification is definitely useful, but I'd like to know more about a receiver. For example, let's say you are receiving a 500-mV differential clock input. Add to each signal a common-mode noise voltage of 1-V p-p. How much jitter will come out of the clock? You can't figure that out from the specifications.

Another source of clock jitter is noise from the power supply. Suppose there is a 100-mV AC ripple on V_{CC}. How much jitter will you get? You can't figure that out from the specifications, either.

Both examples deal with the issue of common-mode rejection, which is one measure of how much the input switching threshold changes in response to a defined noise input. For linear amplifiers, it's common to see a common-mode rejection ratio (CMRR) specification for common-mode noise at the input terminals and also for noise at the power terminals. For example, a CMRR of –50 dB for V_{CC} means that a 100-mV ripple on V_{CC} will have the same effect as an equivalent differential noise source of 0.3 mV (that's 100-mV less 50 dB). The CMRR translates each type of noise into an equivalent differential noise level at the input. You can then add up all the equivalent input noise figures to determine the overall signal-to-noise ratio, or jitter performance, of your system.

The common-mode range specification used in digital comparators and receivers doesn't break down the sources of noise, so there's no way to do jitter analysis.

POINT TO REMEMBER

➢ Don't violate the common-mode input range specification for a receiver (not even briefly).

6.8 DIFFERENTIAL TO COMMON-MODE CONVERSION

Any imbalanced circuit element within an otherwise well-balanced transmission channel creates a region of partial coupling between the differential and common modes of transmission at that point. The coupling can translate part of a perfectly good differential signal into a common-mode signal, or vice versa.

Such differential-to-common mode conversion problems frequently arise in the design of LAN adapters. For example, assume the output winding of the transformer in Figure 6.9 has equal capacitances C_1 connected from point a to ground and from b to ground. If the capacitances are exactly equal (and the cable and transformer perfectly symmetrical), the differential signal present on the cable forces equal but opposite currents through these two capacitances. In the product chassis, the two currents perfectly cancel. The perfect cancellation implies that no current circulates between the twisted-pair cable and the surrounding chassis. In actual practice, however, one capacitance is always a little larger than the other.

Let capacitor C_2 in Figure 6.9 represent the small amount of physical imbalance (2 pF) between the parasitic capacitances associated with circuit nodes a and b. Let's calculate the current flowing through this capacitor, see where it flows, and then decide if it causes any problems.

Using Ethernet 10BASE-T for this example, the drive amplitude is approximately 2V p-p on each wire, at a switching time of 25 ns. The current forced through capacitor C_2 is

$$i(t)_{\text{PEAK}} = C_2 \frac{dv}{dt} = (2\,\text{pF})\frac{2\text{V}}{25\,\text{ns}} = 160\,\mu\text{A} \qquad [6.12]$$

This current flows *through* capacitor C_2 to the product chassis. It couples from the product chassis to the Earth (either through the green-wire ground or through the capacitance between the product chassis and the Earth). From the Earth it couples capacitively to the cabling, along which it travels as a common-mode signal riding on the twisted-pair cable

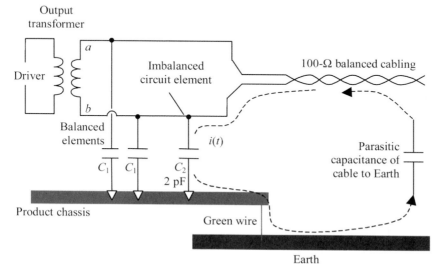

Figure 6.9—An imbalanced circuit element within the transmitter causes current to circulate through the external cabling and the product chassis.

back to the transformer, completing the loop.

A *balanced* load composed of equal-valued capacitors from *a* to ground and from *b* to ground would *not* generate any common-mode currents, because the currents through the two capacitors cancel, leaving nothing to exit the system in common-mode format. In this example the *imbalance* in capacitive loading generates the common-mode current.

A capacitive imbalance even as small as 2 pF causes a big problem in this example, because 160 µA of high-frequency common-mode current on an exposed cable easily violates U.S. and international emissions regulations.

POINT TO REMEMBER

> An imbalanced circuit can translate part of a perfectly good differential signal into a common-mode signal, or vice versa.

6.9 DIFFERENTIAL IMPEDANCE

What is differential impedance? Differential impedance is the ratio of voltage to current on a pair of transmission lines when driven in the differential mode (one signal positive and the other negative).

For example, the circuit in Figure 6.10 drives a signal $x(t)$ differentially into a pair of uncoupled transmission lines.[48] Because the lines are symmetrical, the voltage splits evenly and you see voltage $\frac{1}{2}x(t)$ on the top line and $-\frac{1}{2}x(t)$ on the bottom. The current through each load must therefore equal $\frac{1}{2}x(t)/Z_0$. This current flows through both lines, and through both loads, in the direction shown.

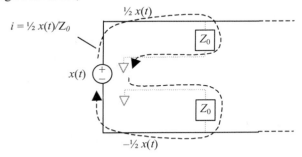

Figure 6.10—The differential input impedance of a pair of noninteracting transmission lines is $2Z_0$.

---

[48] Traces separated by more than four times the trace height *h* interact so slightly that for purposes of impedance analysis, one usually ignores the effect of one on the other. Such traces are said to be *uncoupled*. The exact degree

6.9 • Differential Impedance

The differential impedance (ratio of differential voltage to current) is

$$Z_{\text{DIFF}} = \frac{x(t)}{\left[\tfrac{1}{2} x(t)/Z_0\right]} = 2Z_0 \qquad [6.13]$$

where Z_{DIFF} is the differential impedance of a pair of uncoupled transmission lines (Ω),

Z_0 is the characteristic impedance of either line alone, also called the uncoupled impedance (Ω),

$x(t)$ is the differential voltage applied across both lines (V), and

$\tfrac{1}{2} x(t)/Z_0$ is the current driven into each line by the source (A).

The differential impedance of two matched, noninteracting transmission lines is double the impedance of either line alone.

If the lines are *coupled*, the situation changes. As an example of *coupling*, think of two parallel pcb traces. These traces will always exhibit some (perhaps very small) level of crosstalk. In other words, the voltages and currents on one line affect the voltages and currents on the other. In that way the two transmission lines are coupled together.

With a pair of coupled traces, the current on either trace depends in part on crosstalk from the other. For example, when both traces carry the same signal, the polarity of crosstalk is positive, which reduces the current required on either trace. When the traces carry complementary signals, the crosstalk is negative, which increases the current required on either trace. Figure 6.11 illustrates the situation. The characteristic impedance of each line to ground (including consideration of the adjacent trace) is represented as a single lumped-element component Z_1. The impedance coupling the two transmission lines is depicted as a single lumped-element device Z_2.

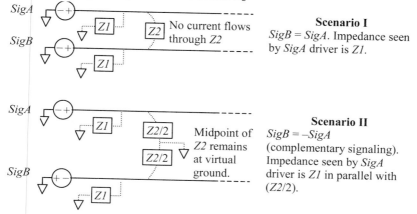

Figure 6.11—Driving point impedance of two different two-wire scenarios.

---

of separation required to produce an inconsequential interaction of course depends upon what you consider to be inconsequential.

In the first scenario *SigB* is driven identically with *SigA*. The voltages on the two wires are therefore the same at all points in time, so (conceptually) no current flows through the coupling impedance Z_2. The impedance (ratio of voltage to current) measured for each wire under this condition equals Z_1. This is called the *even-mode impedance* of the transmission structure.

In the second scenario *SigA* and *SigB* are driven with complementary signals. This is called differential signaling, or sometimes antipodal signaling. Under these conditions the midpoint of impedance Z_2 remains at a virtual ground. An AC analysis of this circuit reveals that the impedance measured for each wire under this condition equals Z_1 in parallel with half of Z_2. This value of impedance is called the odd-mode impedance. *The odd-mode impedance is always less than the even-mode impedance.*

The terms even-mode impedance and odd-mode impedance are closely related to the terms *common-mode impedance* and *differential-mode impedance*. Common-mode impedance is measured with the two wires driven in parallel from a common source. Common-mode impedance is by definition half the even-mode impedance.

Differential-mode impedance is twice the odd-mode impedance. It is measured using a well-balanced source, computing the ratio of the differential voltage (twice the odd-mode voltage) to the current on either line.

Combining your knowledge about common and differential mode impedances, you should be able to prove that the common-mode impedance always exceeds one-fourth of the differential-mode impedance.

Just as an example, suppose we have two 50-ohm, uncoupled transmission lines. As long as the lines are far enough apart to remain uncoupled, the even-mode and odd-mode impedances for these two lines will be the same, and equal to 50 ohms. Should you connect these two lines in parallel, the common-mode impedance would be 25 ohms. Should you connect these two lines to a differential source, the differential input impedance of the pair of lines would be 100 ohms.

POINTS TO REMEMBER

- Differential impedance is the impedance measured *between* two conductors when they are driven in the differential mode.
- Odd-mode impedance is the impedance measured *on either of* two conductors when they are driven with opposite signals in the differential mode.
- The value of differential-mode impedance is twice the value of odd-mode impedance.
- The differential impedance of two matched, uncoupled transmission lines is double the impedance of either line alone.
- The odd-mode impedance of two matched, uncoupled transmission lines equals the impedance of either line alone.
- Coupling between two parallel pcb traces decreases both differential and odd-mode impedances.

> Common-mode impedance is the impedance measured on two wires in parallel when they are driven together.
> Even-mode impedance is the impedance measured *on either of* two wires when they are driven with identical signals in the common mode.
> The value of common-mode impedance is half the value of even-mode impedance.

6.9.1 Relation Between Odd-Mode and Uncoupled Impedance

I should now like to discuss the relation between the odd-mode impedance and the *uncoupled impedance*. The uncoupled impedance Z_C is what you would measure if the same transmission lines were widely separated, so they couldn't interact. What you need to know is simple: The odd-mode impedance of a coupled transmission line is always *less* than the uncoupled impedance. The even-mode impedance is always *greater*. The closer you place the line, the more coupling you will induce, and the greater a discrepancy you will see between the odd-mode, uncoupled-mode, and even-mode impedances.

Let's codify this into a differential impedance principle:

> *Coupling between parallel pcb traces **decreases** their differential (or odd-mode) impedance.*

When implementing tightly coupled differential traces on a pcb, one normally reduces the width of the lines within the coupled region in order to compensate for the expected drop in differential impedance.

6.9.2 Why the Odd-Mode Impedance Is Always Less Than the Uncoupled Impedance

The proof relies on the construction of a thing called an equipotential plane midway between two differential traces. Due to symmetry, all the electric fields in the odd-mode situation will lie perpendicular to this plane. Therefore, the potential everywhere along the equipotential plane will be zero. If the potential everywhere along the plane is zero, I could replace the imaginary equipotential plane with a real, solid copper wall and it wouldn't make any difference. The odd-mode characteristic impedance is not affected by the wall. What's really neat about this construction is that *once the wall is in place,* the problem is partitioned into two noninteracting zones. This gives us a way to evaluate the odd-mode impedance using a single trace and a solid copper wall instead of two traces. If you think about the impact of the wall on the characteristic impedance of a single trace, it's pretty obvious that the wall will add capacitance and decrease the impedance. The net result of this argument is that the odd-mode impedance of a coupled structure is always less than the uncoupled impedance Z_0.

6.9.3 Differential Reflections

High-Speed Digital Design Online Newsletter, Vol. 2, Issue 21
John Lehew writes

In the *High-Speed Digital Design* book and in a few other places it states the fractional reflection Γ coefficient caused by a mismatch in impedance is

$$\Gamma = \frac{Z_2 - Z_0}{Z_2 + Z_0} \qquad [6.14]$$

where Z_0 is the characteristic impedance of a primary transmission line, and

Z_2 is the characteristic impedance of a mismatched section to which the primary line is coupled.

This formula is typically used to calculate reflections that happen in a single transmission line that is referenced to a ground plane. Does this formula also apply to differential or balanced lines?

Reply

Thanks for your interest in *High-Speed Digital Design*.

Aside from the complications introduced by unbalanced modes, differential transmission lines behave pretty much like single-ended ones. Equation [6.14] applies to both.

Assume I have a section of differential transmission line with a differential impedance of Z_1. Assume I couple that into a load with differential impedance Z_2 (it doesn't matter whether Z_2 is a lumped-element load or another section of differential transmission line with characteristic impedance Z_2). The size of the signal that bounces off the joint, in comparison to the size of the incoming signal, is given by your equation for the reflection coefficient Γ.

Let's do an example using unshielded twisted-pair cabling (UTP). Suppose I couple a section of category 5, 100-Ω (nominal) UTP cabling to another section of category 4 120-Ω (nominal) UTP cabling (such cable is available only in France). The reflection off the joint will be of (nominal) size:

$$\Gamma = \frac{120 - 100}{120 + 100} = 0.09 \qquad [6.15]$$

Now, what could go wrong with this simple example? If the cable is inherently *un*balanced (i.e., more capacitance from one side to ground than on the other), then you have a more complicated situation. In general, there are four modes of propagation involved in the problem, one differential mode and one common mode for each of the two cables. The complete problem is described by a 4x4 coupling matrix whose entries vary with frequency.

Imperfections in the balancing of the cable result in cross-coupling between the differential modes and the common modes at the joint, which is one of the things that creates EMI headaches.

POINT TO REMEMBER

> ➤ Aside from the complications introduced by unbalanced modes, differential transmission lines behave pretty much like single-ended ones.

6.10 PCB CONFIGURATIONS

The requirements for high-speed differential traces in a solid-plane pcb are these:

1. The two traces carry complementary voltages,
2. The two traces carry complementary currents—in conjunction with point 1 this implies that their characteristic impedances be the same,
3. The two traces have equal impedances to the surrounding reference system—ground planes, V_{CC} planes, or both, and
4. The two wires have equal propagation delay.

These requirements may be satisfied by many trace configurations. The most popular cases are the differential microstrip, the edge-coupled stripline, and the broadside-coupled stripline (Figure 6.12).

The author assumes you have access to a good 2-D field solver that can predict the differential impedance for various combinations of six variables: trace width, trace height, trace separation, trace thickness, configuration, and dielectric constant. If you don't have a 2-D field solver, then get one (it comes with any signal integrity package). Tell your

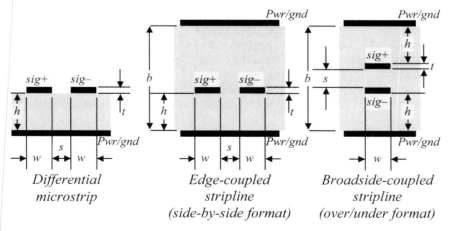

Figure 6.12—Differential pcb traces can be arranged in many different ways.

manager I said you have to have it. A 2-D field solver is absolutely the best way to compute the impedance of any pcb transmission structure.

The following sections present some limited data on differential trace impedance, but don't expect to find all possible combinations of the six primary variables. I will concentrate instead on helping you understand the meaning and purpose of the various adjustments you can make to trace geometry.

6.10.1 Differential (Microstrip) Trace Impedance

***High-Speed Digital Design Online Newsletter*, Vol. 5, Issue 2**
Mitch Morey of San Diego writes

I'm working on a board with 100-ohm differential signaling that I would like to design for microstrip routing. I've used the Polar Instrument calc, the ADS LineCalc software, and have got two additional stack-up constructions from our fabrication houses, and have talked to numerous people on this.

Here are the recommendations I have gathered so far. All configurations represent 100-Ω differential microstrips operating at 2.4Ghz speeds using a 5-mil FR-4 dielectric.

- .005" lines with .005" edge to edge (fab shop 1)
- .004" lines with .008" edge to edge (fab shop 2)
- .005" lines with .008" edge to edge (ADS LineCalc)
- .006" lines with .0065" edge to edge (ADS LineCalc #2)
- .016" lines with .016" edge to edge (engineer #1)

Why the discrepancies?

Reply

Thanks for your interest in *High-Speed Digital Design*.

What you need is a piece of software called a 2-D E&M field solver. This program calculates the magnetic and electric fields surrounding your traces, and from that data extracts the impedance and delay. This is the best way to do impedance calculations. The good field solvers allow you to specify the trace width, height, spacing, thickness, dielectric constant, *and* they allow you to overlay the trace with a soldermask layer.

I'm not sure what ADS LineCalc uses, but if it's not a 2-D field solver, you shouldn't trust its results. I have reason to distrust the accuracy of the examples you have provided.

First let me give you some general principles to help you understand what's happening, and then I'll rule out a couple of the solutions below.

The first thing you need to know is that the patterns of electric field lines in a dielectric medium follow the same shapes as patterns of current flow in salt water. This sounds pretty obtuse, but it's going to help you in a major way, because it will help you *see* what is happening when you change the trace geometry.

Follow me for a minute on this mental experiment. Imagine a microstrip of length *x*. I want you to mentally replace the dielectric medium surrounding this trace

6.10.1 • Differential (Microstrip) Trace Impedance

with a slightly conductive salt-water mixture. Now imagine that you connect an ordinary ohmmeter between the trace and the ground plane. The value of DC resistance you measure in this experiment will be exactly proportional to the *impedance* of the trace. I hope you can now imagine what would happen if you press the trace closer to the ground plane. Can you see that the impedance must go *down,* because there is now less water between the trace and the reference plane? If the trace is pressed down to the point where it nearly touches the reference plane, its resistance to ground (i.e., impedance) approaches zero.

What about doubling the width? This adjustment doubles the surface area of the trace, substantially lowering its resistance to the reference plane (i.e., impedance). I like this DC analogy because most engineers find it a lot easier to imagine simple patterns of DC current flow than they do high-frequency electromagnetic fields. The constant of proportionality isn't important—I just want you to see what's going to happen as you make various adjustments.

So far I've shown two things that decrease the impedance in microstrips:

1. Moving a trace closer to the reference plane decreases its impedance.
2. Fattening a trace (i.e., increasing its width) decreases its impedance.

The converse statements are also true:

1. Moving a microstrip further away from the reference plane increases its impedance.
2. Shaving down the trace width increases its impedance.

Stripline traces are a little more complicated in that you must account for the distance from your trace to both top and bottom reference planes. The general result for offset striplines is that whichever plane lies closest to the trace has the most influence on the impedance. Smack in the middle, the planes are both equally important.

Let's now imagine a differential configuration with two traces, still embedded within your slightly salty water. Connect the ohmmeter between the two traces (from one to the other). The resistance you read now will be proportional to the *differential impedance* of two-trace configuration. [Note: One-half the differential impedance is defined as the odd-mode impedance.]

If your two traces are set far apart, and they have the same dimensions as in the first experiment, your new differential measurement will be exactly *twice* as great as before. If you draw out the patterns of DC current flow, you can see why. For *widely* separated traces, the current flows mostly from one trace straight down to the nearest reference plane, then it shoots across the plane to a position underneath the second trace, and from there it leaks back up to the second trace. As this current flows, it encounters a resistance R when leaving the first trace, practically zero resistance flowing across the plane, and then another amount R as it flows back up to the second trace. The total resistance encountered is $2R$.

> *The differential impedance of two widely separated traces equals twice the impedance to ground of either trace alone.*

Now let's see what happens to the differential impedance as you slide the two traces towards each other. When they get close enough, significant amounts of current begin to flow directly between the traces. You still get the same old currents going to and from the reference plane, but in addition to those currents you have now developed a new pathway for current, directly from trace to trace. This additional current pathway acts like a new resistance in *parallel* with the original, widely spaced current pathways. The new parallel pathway lowers the differential impedance of the configuration. You may conclude that

> *The differential impedance of a tightly spaced pair is less than twice the impedance to ground of either trace alone.*

If the traces are moved so close that they nearly touch, the differential resistance (impedance) approaches zero. In general, the differential impedance is a monotonic function of the trace separation.

> *All other factors being equal, the tighter the intertrace spacing, the less the differential impedance.*

I view any decrease in impedance as an annoying side effect of close spacing. If I could redesign the universe, I'd try to make it not happen. Fortunately, you can counteract the annoying drop in impedance by shaving down the width of your traces. If you shave off just enough width, you can push the impedance back up to where it belongs. In this way, the trace separation and trace width become somewhat interchangeable.

> *To maintain constant impedance, a reduction in spacing must be accompanied by a reduction in trace width (or an increase in trace height).*

With these eight rules in mind, let's now look at the specific recommendations you have been given.

With your 5-mil dielectric, the individual impedance of a 16-mil trace on FR-4 already falls below 50 ohms, so the differential impedance will be less than 100 ohms regardless of what spacing you use. You can therefore rule out the 16-mil configuration. I suspect your engineer #1 may have been thinking about using a thicker dielectric than what you propose.

The two ADS LineCalc results conflict with each other. Starting from a pair 5-mil wide with an 8-mil space, *increasing* the trace width to 6 mils will lower the impedance, and *decreasing* the spacing to 6.5 mils will lower it even further. Therefore, one of these results must be wrong. They cannot both be 100-ohm solutions. Therefore, I suspect something is wrong with either your copy of ADS LineCalc or (dare I say it) your use of the tool.

Table 6.1 presents some data from another commercial 2-D field solver (HyperLynx). All these combinations should give you a 100-ohm differential microstrip impedance under the conditions listed in the table:

Each row lists the trace height h, the finished, plated trace width w, the finished, plated edge-to-edge separation s, the proximity factor k_p, the skin-effect resistance R_{AC}, the skin-effect loss coefficient α_r, and the effective permittivity ϵ_{re}. All

Table 6.1—AC Resistance and Skin-Effect Loss (at 1 GHz) for Selected 100-Ω Differential Edge-Coupled Microstrips

h mil	w mil	s mil	k_p	R_{AC} Ω/in.	R_{AC} Ω/m	α_r dB/in.	α_r dB/m	ϵ_{re}
5	8	30	3.48	1.54	60.7	0.067	2.63	3.17
5	7	11	3.17	1.57	61.7	0.068	2.68	2.97
5	6	7	3.01	1.69	66.5	0.073	2.89	2.85
5	5	5	2.91	1.89	74.4	0.082	3.23	2.78

NOTE (1)—AC parameters R_{AC}, k_p, α_r, ϵ_{re}, and FR-4 dielectric $\epsilon_r = 4.3$ are specified at 1 GHz.

NOTE (2)—These microstrip examples assume copper traces of 1-oz thickness (including plating) with $\sigma = 5.98 \cdot 10^7$ S/m, plus a conformal coating (soldermask) consisting of a 12.7-μm (0.5-mil) layer having a dielectric constant of 3.3.

NOTE (3)—The propagation velocity v_0 (m/s) and delay t_p (s/m) are found from ϵ_{re}, where $1/t_p \triangleq v_0 \triangleq c/\sqrt{\epsilon_{re}}$ and $c = 2.998 \cdot 10^8$ m/s.

AC parameters are specified at 1 GHz. The trace construction is assumed to be 1/2-oz etch plus 1/2-oz of plating, yielding a total 1-oz thickness, with a conformal coating (soldermask) consisting of a 12.7-μm (0.5-mil) layer having a dielectric constant of 3.3. If you select a different type of soldermask, your pcb vendor will adjust the trace width to compensate for the thickness of the dielectric material above the traces. A thicker soldermask will slightly reduce the finished propagation velocity.

The resistance data was developed using a method-of-moments magnetic field simulator with 120 segments equally spaced around each pcb trace, with the current linearly interpolated across each segment. The author estimates the accuracy of the data generated by this simulator at approximately ±2%.

Whatever you choose to do, insist that your board fabrication shop place differential impedance test coupons on your panels and test each one to verify that you are getting the correct impedance.

6.10.2 Edge-Coupled Stripline

In a pcb application, the differential impedance of closely spaced traces varies with their spacing. As the gap between traces narrows, the differential impedance goes down. In extreme cases, the widths of the traces may require adjustment in order to keep the differential line impedance within a specified target range. This pesky adjustment in width is the one key disadvantage of closely spaced differential lines.

Figure 6.13 plots the impedance of edge-coupled differential traces versus spacing and trace width. The data for this plot were generated using a method-of-moments magnetic field simulator with 120 segments equally spaced around each pcb trace, with the current linearly interpolated across each segment.

At small separations (less than 9 mils, as shown in Figure 6.13) the traces couple significantly, so the impedance varies with both trace width and spacing. When the intrapair

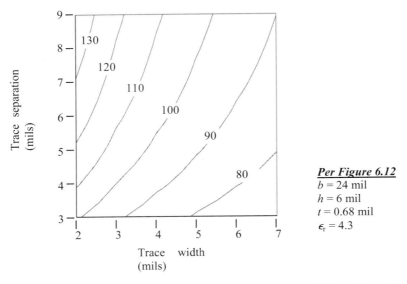

Figure 6.13—Differential impedance Z_{DIFF} (ohms) of edge-coupled differential striplines.

separation exceeds about four times the trace height h (a separation of 24 mils in this example), the coupling between the two traces of the pair usually becomes so weak that the traces hardly affect each other. The differential impedance then depends mostly on just the trace width. At great separations the contours become completely vertical.

Very widely separated traces comprise an *uncoupled* or *loosely coupled* differential pair. For an uncoupled differential pair, the even-mode and odd-mode impedances are the same, and the differential impedance is twice the impedance of either line alone.

Differential traces *can* be pushed really, really close together. Squishing them together saves board area. If you do so, you will need to compute a new trace width to compensate for the fact that the differential impedance goes down for closely spaced pairs. Widely spaced pairs are not subject to this picky, difficult-to-implement requirement.

Another disadvantage of closely spaced pairs has to do with trace routing. Once the signals are closely paired, they *cannot* be separated, or else you will mess up their impedance (unless you readjust their widths). This effect imposes a routing penalty on edge-coupled traces, because it is difficult to get closely spaced, edge-coupled pairs to go around obstacles without temporarily separating them (Figure 6.14).

Finally, the use of tight coupling decreases the trace width, exacerbating the skin-effect loss.

What do I do? Unless absolutely pressed for space, I normally set the trace separation at about four times the trace height h. This setting usually yields a less-than-6% reduction in impedance, a small enough value to simply ignore. All stripline traces, differential or not, then have the same width. I instruct my layout professional to keep differential pairs near each other, but allow them to separate from time to time as needed to go around obstacles. I also insist that the elements of each pair be equal in length, to within 1/20 of a risetime, which limits the common-mode signal contributed by trace skew to less than 2.5% of the single-ended signal amplitude.

6.10.2 • Edge-Coupled Stripline

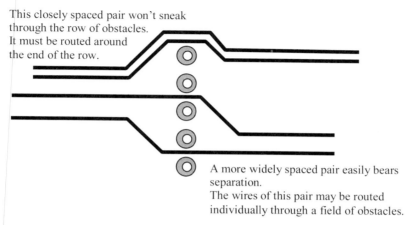

Figure 6.14—Closely spaced differential pairs can be difficult to route.

Figure 6.15 illustrates the magnetic lines of force surrounding an edge-coupled differential stripline pair. It shows intense concentrations of magnetic flux near the corners of the traces, indicating a substantial peaking of the current density at the corners. The proximity factor k_p for these traces takes into account the crowding of current at the corners of the traces, plus a slight current concentration on the inside-facing surfaces of the pair, plus an allotment for the current induced on the top and bottom reference planes of the stripline cavity. The value of the proximity factor k_p for the traces illustrated in Figure 6.15 is 3.08. Differential pcb traces with impedances between 100 to 150 ohms possess a proximity factor typically within the range of 2.5 to 3.5. Figure 6.16 illustrates the distribution of current around the periphery of the signal conductors, and also the lower reference plane. Current flows also on the upper reference plane, although that plane lies above the limits of the vertical scale in the drawing, so it is not shown.

Figures 6.15 and 6.16 were generated using a 2-D field solver. This same program also calculates skin-effect resistance and skin-effect loss. A listing of such calculations for selected edge-coupled stripline geometries appears in Table 6.2. These values provide a good starting point for planning a high-speed interconnection. The table shows that the most important determiner of skin-effect loss is simply the trace width. All other factors are

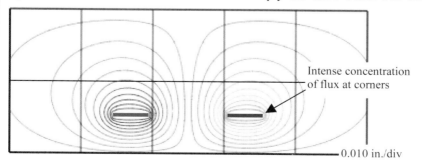

Figure 6.15—This cross-sectional view of the magnetic field in the vicinity of a differential edge-coupled stripline shows intense field concentrations near the corners of the traces.

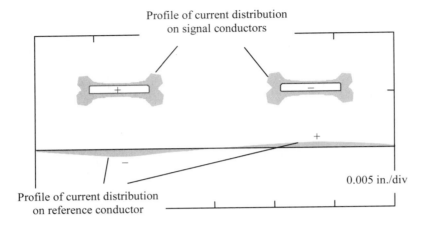

Figure 6.16—The distribution of current around the perimeter of a differential edge-coupled stripline takes on a dog-bone appearance, with the current peaking at the corners of the trace. The inside-facing corners of the pair have a slightly higher peak current than the outside-facing corners.

secondary.

If you don't find the configuration you need in the table, you may interpolate values of r_{SKIN} and α_R from the entries listed in the table. The interpolation process begins by first correcting the listed table values to account for your actual value of b. This is accomplished by scaling the trace geometry as follows:

- Select any row from the table with an interplane spacing b_2 that is reasonably close to your target b_1.
- Compute the scaling constant $k = b_1/b_2$.
- Multiply h, w, s and also the trace thickness t by the constant k, producing a new row for the table.
- The impedance of the new row will be identical to the original Z_{DIFF}.
- The skin-effect resistance (at 1 GHz) for the new row will equal the original r_{SKIN} divided by k.
- The skin-effect attenuation constant for the new row will equal the original α_r divided by k.

This scaling process is exact; however, it produces an unusual value of t (the thickness is usually standardized to either 1/2 or 1 oz of copper). Because t has only a second-order effect on characteristic impedance and resistance, you can usually ignore this change. Using the scaled values of b, h, w, and s while leaving t fixed will introduce some small inaccuracies into the scaling process, but still produce a new configuration with a characteristic impedance workably close to Z_{DIFF}.

Once you have corrected a few of the table entries to account for your actual value of b, you can then interpolate along the h and w dimensions to estimate the skin-effect resistance and skin-effect loss.

6.10.2 • Edge-Coupled Stripline

This principle of scaling and interpolation works with the results from any field solver.

Example showing scaling of edge-coupled differential stripline

Suppose you have implemented a differential layout with $b = 20$, $h = 7$, $w = 5$, and $s = 7$. The skin-effect loss (Table 6.2) is 0.0929 dB/in versus a hypothetical budget of 0.1 dB/in.

How much can you squeeze the trace width without exceeding the skin-effect loss budget?

First let's try changing w and s without changing b. The next smallest trace size listed in Table 6.2 shows $w = 4$, $s = 5.2$, yielding a resistive loss of 0.1089 dB/in. You can adjust w and s 58% of the way towards this next smallest size, where the fraction 58% is determined by the ratio of skin-effect loss numbers:

$$58\% = \frac{0.1 - 0.0929}{0.1089 - 0.0929} \qquad [6.16]$$

The interpolated values of w and s are

$$\begin{aligned} w &= 5 + 0.58(4 - 5) = 4.42 \\ s &= 7 + 0.58(5.2 - 7) = 5.96 \end{aligned} \qquad [6.17]$$

Next let's start the problem over, this time changing b. You have plenty of excess budget for skin effect loss in your initial configuration, so the new b can be smaller than the original, where the ratio of 93% is determined by the ratio of skin-effect losses:

$$93\% = \frac{0.0929}{0.1} \qquad [6.18]$$

Scaling all values by the factor 93% produces

$$\begin{aligned} b &= 0.93 \cdot 20 = 18.6 \\ h &= 0.93 \cdot 7 = 6.51 \\ w &= 0.93 \cdot 5 = 4.65 \\ s &= 0.93 \cdot 7 = 6.51 \end{aligned} \qquad [6.19]$$

This example has produced two new approximate configurations. The next step is to tweak the exact spacing s for each configuration to zero in on the exact impedance. This is done using your 2-D field solver. Then choose the exact interpair pitch to satisfy your crosstalk constraint. After that you will be in a position to determine which approach gives you the best layout density: just shrinking w and s while keeping b constant or shrinking b and everything else along with it.

POINTS TO REMEMBER

> Differential traces *can* be pushed really, really close together. If you do so, compute a new trace width to compensate for the fact that the differential impedance goes down for closely spaced pairs.

> Widely spaced (i.e., loosely-coupled) pairs are not subject to picky, difficult-to-implement spacing and width requirements.

> The most important determiner of skin-effect loss is the trace width.

> An interpair trace separation of four times h yields about a 6% effect on impedance, a small enough value in many cases to simply ignore.

> Matching the elements of each pair to within 1/20 of a risetime limits the common-mode signal contributed by trace skew to less than 2.5% of the single-ended signal amplitude.

Table 6.2—AC Resistance and Skin-Effect Loss (at 1 GHz) for Selected 100-Ω Differential Edge-Coupled Striplines

b	h	w	s	R_{AC} Ω/in.	R_{AC} Ω/m	α_r dB/in.	α_r dB/m	Z_{DIFF} (Ω)
10	3	3	40.0	3.50	137.8	0.152	5.99	99.0
10	4	3	7.0	3.24	127.6	0.139	5.47	100.4
10	5	3	7.0	3.22	126.8	0.137	5.40	101.2
10	5	4	40.0	2.76	108.7	0.126	4.95	94.6
14	4	3	5.5	3.19	125.6	0.136	5.36	101.0
14	4	4	12.0	2.74	107.9	0.118	4.63	100.3
14	5	3	4.5	3.14	123.6	0.135	5.31	100.1
14	5	4	7.5	2.60	102.4	0.112	4.39	100.5
14	5	5	40.0	2.33	91.7	0.101	3.98	99.5
14	7	3	4.5	3.11	122.4	0.132	5.19	101.8
14	7	4	6.5	2.56	100.8	0.110	4.31	100.6
14	7	5	13.0	2.24	88.2	0.096	3.77	100.9
14	7	6	40.0	2.01	79.1	0.091	3.59	95.0

NOTE—All values b, h, w, and s in mils.
NOTE—AC parameters R_{AC}, α_r, and Z_{DIFF} specified at 1 GHz with FR-4 dielectric $\epsilon_r = 4.3$.
NOTE—These stripline examples assume copper traces of 1/2-oz thickness with $\sigma = 5.98 \cdot 10^7$ S/m.

6.10.2 • Edge-Coupled Stripline

Table 6.2 (continued)—AC Resistance and Skin-Effect Loss (at 1 GHz) for Selected 100-Ω Differential Edge-Coupled Striplines

b	h	w	s	R_{AC} Ω/in.	R_{AC} Ω/m	α_r dB/in.	α_r dB/m	Z_{DIFF} (Ω)
20	5	3	4.4	3.14	123.6	0.134	5.28	101.0
20	5	4	6.5	2.59	102.0	0.111	4.37	100.7
20	5	5	11.0	2.27	89.4	0.097	3.84	100.6
20	5	6	40.0	2.09	82.3	0.092	3.61	98.4
20	7	3	3.9	3.13	123.2	0.134	5.26	101.1
20	7	4	5.2	2.55	100.4	0.109	4.29	100.9
20	7	5	7.0	2.17	85.4	0.093	3.66	100.7
20	7	6	10.0	1.91	75.2	0.082	3.24	100.3
20	7	7	19.0	1.75	68.9	0.075	2.96	100.7
20	7	8	40.0	1.61	63.4	0.072	2.85	96.3
20	10	3	3.7	3.14	123.6	0.135	5.30	100.6
20	10	4	5.0	2.54	100.0	0.108	4.25	101.6
20	10	5	6.5	2.15	84.6	0.092	3.60	101.4
20	10	6	8.5	1.88	74.0	0.081	3.17	100.7
20	10	7	12.0	1.68	66.1	0.072	2.85	100.5
20	10	8	25.0	1.56	61.4	0.067	2.65	100.3
30	5	3	4.3	3.15	124.0	0.135	5.30	100.7
30	5	4	6.3	2.60	102.4	0.111	4.38	100.7
30	5	5	10.0	2.27	89.4	0.097	3.83	100.7
30	5	6	22.0	2.09	82.3	0.090	3.54	100.2
30	6	3	4.0	3.14	123.6	0.134	5.27	101.0
30	6	4	5.4	2.56	100.8	0.110	4.33	100.7
30	6	5	7.5	2.20	86.6	0.094	3.71	100.7
30	6	6	11.2	1.96	77.2	0.084	3.31	100.6
30	6	7	20.0	1.81	71.3	0.078	3.07	100.4

NOTE—All values b, h, w, and s in mils.
NOTE—AC parameters R_{AC}, α_r, and Z_{DIFF} specified at 1 GHz with FR-4 dielectric $\epsilon_r = 4.3$.
NOTE—These stripline examples assume copper traces of 1/2-oz thickness with $\sigma = 5.98 \cdot 10^7$ S/m.

Table 6.2 (continued)—AC Resistance and Skin-Effect Loss (at 1 GHz) for Selected 100-Ω Differential Edge-Coupled Striplines

b	h	w	s	R_{AC} Ω/in.	R_{AC} Ω/m	α_r dB/in.	α_r dB/m	Z_{DIFF} (Ω)
30	7	3	3.8	3.14	123.6	0.134	5.29	100.9
30	7	4	5.0	2.56	100.8	0.109	4.31	100.9
30	7	4.5	5.7	2.35	92.5	0.101	3.96	100.7
30	7	5	6.5	2.17	85.4	0.093	3.67	100.6
30	7	6	8.8	1.91	75.2	0.082	3.22	100.7
30	7	7	12.5	1.73	68.1	0.074	2.92	100.8
30	7	8	21.0	1.60	63.0	0.069	2.72	100.5
30	8	3	3.7	3.15	124.0	0.134	5.29	101.0
30	8	4	4.7	2.56	100.8	0.110	4.33	100.6
30	8	5	6.0	2.17	85.4	0.093	3.66	100.7
30	8	6	7.7	1.89	74.4	0.081	3.20	100.6
30	8	7	10.2	1.69	66.5	0.073	2.85	100.9
30	8	8	14.0	1.55	61.0	0.066	2.62	100.6
30	8	9	23.0	1.45	57.1	0.062	2.45	100.5
30	8	10	40.0	1.36	53.5	0.060	2.37	97.8
30	10	3	3.6	3.16	124.4	0.135	5.30	101.2
30	10	4	4.5	2.57	101.2	0.110	4.32	101.0
30	10	5	5.5	2.17	85.4	0.093	3.67	100.7
30	10	6	6.8	1.89	74.4	0.081	3.19	100.7
30	10	7	8.4	1.67	65.7	0.072	2.82	100.9
30	10	8	10.5	1.51	59.4	0.065	2.55	100.8
30	10	9	13.5	1.38	54.3	0.059	2.34	100.6
30	10	10	19.0	1.29	50.8	0.055	2.18	100.6
30	10	11	34.0	1.22	48.0	0.053	2.08	100.2
30	10	12	40.0	1.15	45.3	0.052	2.03	96.4
30	15	3	3.5	3.17	124.8	0.135	5.33	101.0
30	15	4	4.3	2.58	101.6	0.111	4.36	100.7
30	15	5	5.3	2.18	85.8	0.093	3.65	101.2
30	15	6	6.3	1.89	74.4	0.081	3.19	100.9
30	15	7	7.5	1.67	65.7	0.072	2.82	101.0
30	15	8	9.0	1.50	59.1	0.064	2.53	100.9
30	15	9	11.0	1.36	53.5	0.058	2.30	101.0
30	15	10	13.0	1.25	49.2	0.054	2.14	100.0
30	15	11	17.0	1.17	46.1	0.050	1.98	100.3
30	15	12	25.0	1.10	43.3	0.047	1.87	100.5

NOTE (1)—All values b, h, w, and s in mils.
NOTE (2)—AC parameters R_{AC}, α_r, and Z_{DIFF} specified at 1 GHz with FR-4 dielectric $\epsilon_r = 4.3$.
NOTE (3)—These stripline examples assume copper traces of 1/2-oz thickness with $\sigma = 5.98 \cdot 10^7$ S/m.

6.10.3 Breaking Up a Pair

Article first published in *EDN Magazine*, November 9, 2000

The two traces comprising a differential pair, when routed close together, share a certain amount of cross-coupling. This coupling lowers the differential impedance between the traces. For example, when two traces with $Z_0 = 50\ \Omega$ are well separated (uncoupled), the differential impedance between them should be precisely $Z_{DIFF} = 2Z_0 = 100\ \Omega$. When you jam the same two traces close together (tightly coupled), as you might do to improve routing density, the differential impedance will be a lower value, perhaps something in the range of 70 to 90 Ω.

If the coupling effect lowers the differential impedance too much for your taste, you can fix it. Just reduce the trace widths, thus raising their impedance. The traces remain coupled, but you can (theoretically) always push the differential impedance back up to 100 Ω by making the traces skinnier.[49]

> **Model the mismatched region as a short transmission line of impedance Z_{DIFF} plus a lumped inductance L_{EXCESS}.**

What happens to a tightly coupled pair when it traverses an obstacle, such as a via (Figure 6.17)? If you have room to route both traces on the same side of the obstacle, maintaining their constant separation, no special problems arise. If, on the other hand, you separate the traces to pass by, then the differential impedance in the separated region reverts to its original, uncoupled value of $2Z_0$. If you have thinned the trace widths to produce exactly 100 Ω in the coupled state, then the reverted, uncoupled impedance with skinny traces will *exceed* 100 Ω.

To calculate (approximately) the effects of such a mismatch, let's assume you have a long, uniform differential transmission configuration with differential impedance Z_{DIFF}. Insert into the middle of this line a short section with differential impedance Z_2, having length (in time) t_d. Further assume that t_d is much less than

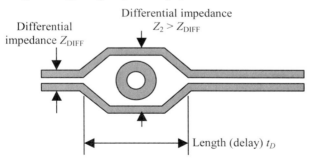

Figure 6.17—The impedance of a tightly-coupled differential pair changes when you separate the wires.

---

[49] You must avoid placing the traces so close together that the required trace width becomes unmanufacturable.

the signal risetime (or fall time) t_r, so the mismatched section acts as a simple lumped-element circuit.

Analysis begins by computing the values L_2 and C_2 corresponding to the mismatched section: $C_2 \triangleq t_d/Z_{\text{DIFF}}$ and $L_2 \triangleq t_d Z_{\text{DIFF}}$. Next, mentally break L_2 into two pieces, $L_2 \triangleq L_N + L_{\text{EXCESS}}$, with $L_N \triangleq Z_{\text{DIFF}}^2 C_2$.

The value L_N is the *natural inductance* you would expect to accompany capacitance C_2 in a differential transmission line with differential impedance Z_{DIFF}, as evidenced by the ratio $\sqrt{L_N/C_2} = Z_{\text{DIFF}}$.

Inductor L_{EXCESS} represents the *excess inductance* in the mismatch region above and beyond L_N. In other words, this procedure models the mismatch region as a short transmission line of impedance Z_{DIFF} (comprising L_N and C_2) plus a series inductance L_{EXCESS}. You will next model the reflection generated by component L_{EXCESS}.

When a fast step input hits any small inductive discontinuity, you get a reflected pulse. The pulse duration equals the rise (or fall) time of the incoming step. You can approximate the reflection coefficient Γ (ratio of reflected pulse height to the incoming step size) generated by L_{EXCESS} as follows:

$$\Gamma \approx \frac{1}{2t_r} \frac{L_{\text{EXCESS}}}{Z_{\text{DIFF}}} \quad [6.20]$$

Substituting the definition of L_{EXCESS} yields

$$\Gamma \approx \frac{1}{2t_r} \frac{L_2 - Z_{\text{DIFF}}^2 C_2}{Z_{\text{DIFF}}} \quad [6.21]$$

Further substituting your basic expressions for L_2 and C_2 gives

$$\Gamma \approx \frac{1}{2t_r} \frac{t_d Z_2 - Z_{\text{DIFF}}^2 (t_d/Z_2)}{Z_{\text{DIFF}}} \quad [6.22]$$

And consolidating the terms provides

$$\Gamma \approx \frac{t_d}{2t_r} \left(\frac{Z_2}{Z_{\text{DIFF}}} - \frac{Z_{\text{DIFF}}}{Z_2} \right) \quad [6.23]$$

That's about the best approximation you will find for the case of a short separation between the elements of a differential pair. If you want better accuracy, use a time-domain simulator.

In the example of Figure 6.17, supposing the ratio of Z_2/Z_{DIFF} to be $(122\,\Omega)/(100\,\Omega) = 1.22$, equation [6.23] reduces to

$$\Gamma \approx \frac{t_d}{2t_r}\left(1.22 - \frac{1}{1.22}\right) = 0.200\frac{t_d}{t_r} \qquad [6.24]$$

If this amount of signal degradation is troublesome, try thickening the traces in the separated region to match the impedance of the thinner, more highly coupled traces elsewhere.

If you've kept t_d/t_r less than 1/6, you can expect a proportional accuracy of a couple of percentage points from approximation [6.23]. If you try to stretch the approximation to a ratio of t_d/t_r as big as 1/3, expect the approximation to be good only to about 20%. Beyond that, at t_d/t_r = 1/2 it falls completely apart, delivering totally erroneous answers.

The same approximation works for BGA layouts, where signals escaping from the inner rows neck down to pass through the BGA ball field. The neck-down region raises the local trace impedance in a small region.

In the event your separated traces pass particularly close to the edges of a via pad, the parasitic capacitance between your trace and the via may add to the value of C_2 in a noticeable way, modifying the values of L_N, L_{EXCESS}, and the reflection coefficient Γ.

If the parasitic capacitance is large enough to produce a negative value of L_{EXCESS} (meaning that the effective impedance Z_2 within the mismatched region is less than Z_{DIFF}), then your analysis must begin with the known value of L_2 and then compute the *excess capacitance* above and beyond that necessary to balance out the inductance L_2. It so happens that the reflection coefficient for the case of $Z_2 < Z_{DIFF}$ is also given by equation [6.23]. The negative amplitude associated with the result in that case indicates that a positive-going step edge produces a negative reflected pulse.

POINT TO REMEMBER

➤ If you separate elements of a tightly-coupled pair the differential impedance reverts to twice the uncoupled value of Z_0.

6.10.4 Broadside-Coupled Stripline

Figure 6.18 plots the impedance of broadside-coupled differential traces versus spacing and trace width. The data for this plot were generated using a method-of-moments magnetic field simulator with 120 segments equally spaced around each pcb trace, with the current linearly interpolated across each segment.

As you can see in the figure, widening the traces always decreases the differential impedance. Making the traces skinnier always raises the impedance.

Changing the height invokes a more complicated behavior. The impedance is maximized by a trace height (to the centerline of the trace) of 25% of the interplane separation (6 mils, as drawn in the figure).

From the 25% maximal point, a reduction in *h* moves the traces nearer to the planes, increasing the trace capacitance to the planes. This maneuver reduces the characteristic impedance. The more tightly you press the traces against the planes, the less direct coupling exists between the traces themselves.

Going in the other direction from the 25% maximal point, an increase in *h* moves the traces nearer to each other (*s* gets smaller). This increases the direct capacitance between the two traces. This maneuver also reduces the characteristic impedance. The closer you bring the traces towards each other, the greater the coupling between traces.

At the 25% maximum-impedance-point the trace impedance is least sensitive to changes in width or height. Also, given a fixed interplane spacing, the 25%-point also maximizes the trace width, thus minimizing the skin-effect losses.

The only *disadvantage* to the 25%-point is that it maximizes crosstalk between adjacent broadside-coupled pairs; however, the advantages to be gained in trace pitch don't seem to this author to justify the penalties associated with using any trace height much different from 25%.

When converting from an edge-coupled pair on the surface layer of a pcb to a broadside-coupled pair on the inner layers, there is a subtle asymmetry built into the conversion. The asymmetry is illustrated in Figure 6.19.

In the figure, signal **A** starts on layer 1 and then proceeds through blind via **Y** to layer 4. The signal current has no difficulty changing layers at this position; it just passes through **Y**. The return current associated with signal **A** (dotted line) has a more difficult time managing the layer transition. At the left of the figure, the return current for **A** is shown flowing on the top surface of solid plane layer 2. In the center of the figure, most of the return current for this trace flows on solid plane layer 5 (returning signal current always

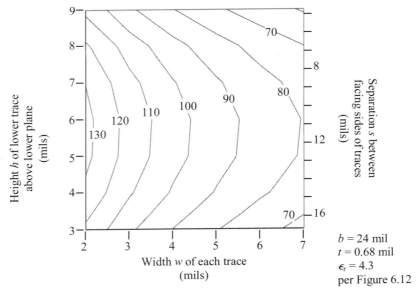

Figure 6.18—Differential impedance Z_{DIFF} (ohms) of broadside-coupled differential striplines.

6.10.4 • Broadside-Coupled Stripline

flows mostly on the nearest solid plane). In the vicinity of the blind via, the return current must find some path (other than the capacitance of the planes, which is not terribly significant) through which it can hop from layer 2 to layer 5. Wherever a signal changes reference planes, the return current must always find a way to follow along.

In the figure, because planes 2 and 5 happen to carry the same potential, they are connected with via **X**, which forms the path for returning signal current. If the planes carry different voltages, the return-current path must traverse a bypass capacitor. In either case the returning signal current temporarily diverts away from the signal current.

The return current associated with signal **B** displays no such difficulty. At the position of blind via **Z** the returning signal current (dotted line) merely needs to change from the top surface of solid plane layer 2 to the bottom surface of that same layer.[50] As you can see, there is a clearance hole located at blind via position **Z** on layer 2. The returning signal current easily pops through this hole from one side of the plane to the other. There is no significant diversion of signal and return current at this location.

The effect of the dissimilarity in return paths is that signal **A** experiences an extra delay as the return current finds its way from plane to plane. To minimize this additional delay, make sure you put a number of plane-to-plane connections near any points where the signals dive into an over/under configuration, where they change layers within the board, and again where they emerge.

Blind vias are used in Figure 6.19 for illustrative purposes; the same effect applies to through-hole vias.

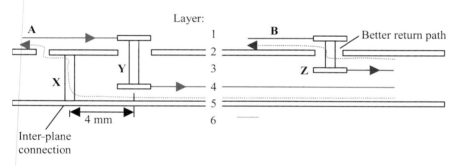

Figure 6.19—The diversion through **X** in the path of returning signal current for signal **A** creates more delay than for trace **B** (cutaway view).

Example showing asymmetry in broadside configuration

The broadside-coupled layout illustrated in Figure 6.19 includes two traces, **A** and **B**, that begin on the surface layer of the pcb and then pass through layer 2 into a broadside-coupled stripline configuration on layers 3 and 4. The plane-stitching via **X** is located 4 mm away from signal via **Y**. The return current for trace **A** therefore has to divert 4 mm out of its way to find the plane-stitching via, plus another 4 mm to get back, giving a total estimated additional delay of (4 mm)·2·(4 ps/mm) = 56 ps. If this same asymmetry exists at both ends of the broadside configuration, the total additional delay on trace **A** equals 112 ps.

---

[50] The reference planes in a pcb are many skin depths thick; high-speed currents do not penetrate the planes, but rather flow only on one surface or the other of the plane.

This crude estimate doesn't perfectly model what happens, as both the capacitance and the inductance of the via configuration are involved, but if an intrapair skew number anywhere near 100 ps matters to you, then either place the plane-stitching vias closer to the signal vias or don't use a broadside configuration.

In addition to possible asymmetry caused by imbalances in the return paths, the broadside configuration falls prey to any differences between the AC voltages on the two planes. Because the top trace is coupled more heavily to the top plane, and the bottom trace to the bottom plane, any differences in the voltages on these two planes induce a differential signal on the two traces. When using the broadside configuration, it pays to use the same power-supply voltage on both planes and nail them together with numerous vias on a tight grid. I like to use whatever plane voltage delivers the best common-mode noise rejection at the receiver (usually the ground plane).

If you use different power-supply voltages on the two planes, all the power-supply noise between them couples directly into the differential broadside-coupled configuration. In the side-by-side configuration, both traces naturally couple equally to the same nearby plane, so differential pickup of power-supply noise doesn't happen.

Broadside-coupled traces suffer from the dielectric layer-thickness tolerances on the layers separating the traces from their respective solid planes. For example, in a design with 5-mil separation from each trace to its respective plane, a layer-thickness tolerance of +/– 1 mil might result in one trace being 4 mils and the other 6 mils away from the reference planes. This arrangement impairs your ability to achieve good symmetry between the traces, which is after all the whole purpose of using a differential configuration. Edge-coupled traces, because they are etched at the same time under the same conditions on the same layer with the same layer thickness, are generally more symmetric.

The impedance of a broadside-coupled trace is affected by the mechanical registration tolerance of the two signal layers (3 and 4 in Figure 6.19).

The one possible area where broadside-coupled traces have an advantage over edge-coupled traces is routing density. For example, if you need to interleave a large bus through a succession of connector pin fields on a large backplane, the broadside configuration requires only single-track routing between pins, whereas an edge-coupled configuration might require double-track routing between pins to achieve the same layout density. The broadside configuration is also somewhat easier to lay out by hand, because both traces go everywhere together (except at the launch and recovery sites). I avoid broadside-coupled traces unless they are made necessary by routing considerations.

Figure 6.20 illustrates the magnetic lines of force surrounding a broadside-coupled differential pair. It shows intense concentrations of magnetic flux near the corners of the traces, indicating a substantial peaking of the current density at the corners. The proximity factor k_p for these traces takes into account the crowding of current at the corners of the traces, plus any current concentrations on flat surfaces of the pair, plus an allotment for the current induced on the top and bottom reference planes of the stripline cavity. The proximity factor for the traces illustrated in Figure 6.20 is 2.73.[51] Differential broadside-coupled pcb traces with impedances between 75 to 135 ohms possess a proximity factor typically within the range of 2.5 to 3.5.

---

[51] The proximity factor is the ratio of the actual AC resistance to the resistance one would compute assuming current distributed uniformly around the periphery of one signal conductor only, and taking into account the skin depth.

6.10.4 • Broadside-Coupled Stripline

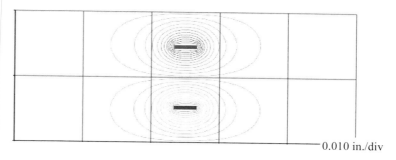

Figure 6.20—This cross-sectional view of the magnetic field in the vicinity of a broadside-coupled stripline shows intense field concentrations near the corners of the traces.

Table 6.3 presents the skin-effect resistance and skin-effect attenuation for selected broadside-coupled stripline configurations.

If you need results for some value b_1 that does not appear in the table, you can linearly interpolate the values in the table to accommodate an intermediate value of b, using the b-column as the x-axis and any other column as the y-axis in your interpolation.

POINTS TO REMEMBER

> Broadside differential trace impedance is maximized by a trace height equal to 25% of the interplane separation.
> The bottom trace of a broadside-coupled differential pair has some extra delay built in at the endpoints.
> Avoid broadside-coupled traces unless they are made necessary by routing considerations.

Table 6.3—AC Resistance and Skin-Effect Loss (at 1 GHz) for Selected 100-Ω Differential Broadside-Coupled Striplines

b mil	h mil	w mil	R_{AC} Ω/in. @1GHz	R_{AC} Ω/m @1GHz	α_r dB/in. @1GHz	α_r dB/m @1GHz
14	4	1.9	4.04	159.1	0.175	6.89
20	5	3.5	2.71	106.5	0.117	4.61
30	7	5.3	2.01	78.9	0.087	3.43
45	10	9.1	1.32	52.0	0.057	2.24

NOTE (1)—All values b, h, and w in mils.
NOTE (2)—AC parameters R_{AC}, and α_r specified at 1 GHz with FR-4 dielectric $\epsilon_r = 4.3$.
NOTE (3)—These stripline examples assume copper traces of 1/2-oz thickness with $\sigma = 5.98 \cdot 10^7$ S/m.

6.11 PCB APPLICATIONS

The following sections describe the main applications for differential signaling on pcbs.

6.11.1 Matching to an External, Balanced Differential Transmission Medium

Differential traces are often used to connect to balanced cabling. For this purpose, the tightness of coupling between the two traces making up the differential pair is irrelevant. What matters is that the differential characteristic impedance of trace configuration matches the differential characteristic impedance of the balanced cabling.

The most popular types of balanced cabling are 100-Ω twisted-pair cabling (ISO 11801 categories 3, 5, 5e, 6, and 7), and the old 150-Ω shielded twisted-pair cabling (IBM Type I). When connecting directly to these cabling types, one normally uses two 50-ohm traces (or two 75-ohm traces for 150-Ω cabling) to couple into the cable.

Figure 6.21 depicts a typical LAN coupling situation. The target cable is a 100-Ω unshielded twisted-pair cable. Your objective in this application is to generate a purely differential transmitted signal of standard size with as little high-frequency power and as low a common-mode content as practicable.

The low-pass filter formed by L1-C1 (and L2-C2) truncates any unnecessary power in the frequency range above the bandwidth of the digital signal. The natural balance of the transformer combined with the additional balancing properties of the common-mode choke together serve to limit the common-mode content of the transmitted waveform. The reason common-mode balance is so important is that common-mode radiation from an unshielded twisted-pair cable is many orders of magnitude more efficient than differential-mode radiation. Minimizing the common-mode current minimizes the cable emissions.

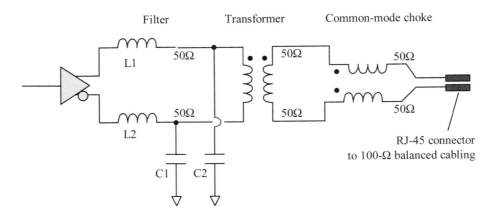

Figure 6.21—Ethernet 10/100BASE-T interface, showing use of 50-Ω transmission lines to match 100-Ω balanced load.

After passing through the common-mode choke you should make the two pcb traces as symmetrical as possible, with equal impedances to ground. The traces must be symmetrically positioned with respect to all nearby grounded objects, but they do not necessarily need to be tightly coupled to each other.

POINTS TO REMEMBER

> Match the differential characteristic impedance of two pcb traces to the differential characteristic impedance of a balanced cable.

> Make the two pcb traces as symmetrical as possible, with equal impedances to ground.

6.11.2 Defeating Ground Bounce

Differential signals arrive naturally at a receiver with a built-in reference voltage. The receiver of a differential signal need not rely on its own internal reference, which could be corrupted by ground bounce or other disturbances in the reference supply. Differential signaling defeats ground bounce.

For ground-bounce cancellation to work, the receiver must see two complementary signals with equal delays from the driver. Any ground shifts or disturbances along the way that affect both wires equally will be cancelled at the receiver.

Note that the two halves of a differential signal must arrive synchronously so that they will be equally influenced by noise, but they do not necessarily need to be tightly coupled to each other.

POINT TO REMEMBER

> Differential signaling defeats ground bounce.

6.11.3 Reducing EMI with Differential Signaling

Differential signals radiate less than single-ended signals. That's one of the benefits of differential logic. If the two complementary signals of a differential pair are perfectly balanced, the degree of field cancellation is determined entirely by the separation between traces.

If, however, the two complementary signals are not perfectly balanced, then the degree of attainable field cancellation will be limited to a minimum value determined not by the trace spacing, but by the common-mode balance of the differential pair. Because the common-mode balance of most digital drivers is not particularly good, it often happens that differential pairs radiate far more power in the common-mode than in the differential mode. In such a situation no radiation benefit remains to be gained from squeezing the differential traces more closely together.

Figure 6.22 plots the theoretical radiation gain attained by a differential microstrip pair as a function of trace separation. The figure assumes the measurement antenna is located in the plane of the board, removed in a broadside direction a distance $r = 10$ m away from the traces (worst case). The radiation from one trace of the differential pair is *supposed to be* cancelled by the equal and opposite radiation from the adjacent trace, resulting in a marked reduction in emissions. The depth of cancellation is related to the ratio $2\pi s/\lambda$, where λ is the free-space wavelength of the highest frequency of interest and s is the separation between traces. This ratio controls the relative phase relationship of the two near-complementary waves as they leave your board. The cancellation is also related to the ratio $r/(r + s)$, which speaks to the relative intensities of the two near-complementary waves as they reach the antenna. The formula in Figure 6.22 shows that differential cancellation improves as you reduce s.

The common-mode radiation from the two traces of a differential microstrip reinforces rather than cancels, so that common-mode radiation does not vary strongly with trace separation. You can adjust the differential-mode radiation by adjusting the trace spacing, but you can't do much about the common-mode radiation (except to install a driver with better common-mode balance).

Under FCC class B measurement conditions, the differential-mode radiation from a differential microstrip pair with 0.5-mm (0.020-in.) separation should theoretically yield a 40-dB radiation improvement at 1 GHz, compared to the radiation measured if the same signal were implemented as a single ended layout. Smaller separations should yield even more improvement. While that theory sounds appealing, in practice you will rarely if ever achieve as much as a 40-dB improvement in overall radiation because your gains will be limited by the degree of balance available on the two outputs of your differential transmitter. Unless the outputs are balanced to better than 1 part in 100, a *common-mode radiation component* of at least 1% of the differential amplitude will emanate from your differential pair anyway. Given a 1% common-mode imbalance, even a differential spacing of zero would not improve the total radiation by more than 40 dB.

Taking an example from the LVDS differential driver family, which prescribes a differential balance no better than 1 part in 16, even the most Herculean efforts at trace

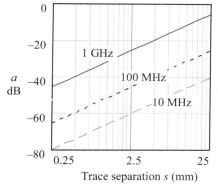

$$a = 20\log\left|1 - \frac{r}{r+s}e^{\frac{-j2\pi s}{\lambda}}\right|$$

where r is the distance to the receiver,
s is the trace separation, and
λ is the free-space wavelength of the highest frequency of operation.

Figure 6.22—Theoretical radiation improvement a for the differential portion of the far-field radiation from a microstrip, as a function of trace separation s.

balancing will never improve the overall radiation for that logic family by more than a factor of 16 (24 dB).

In plain terms, a differential trace spacing of 0.5 mm is close enough to deliver all the EMI benefit you are likely to ever achieve. Because radiation problems on digital pcbs are usually dominated by the common-mode radiation, you need not struggle to place ordinary differential digital traces any closer than 0.5 mm for any EMI purpose.

POINT TO REMEMBER

> ➢ You need not struggle to place ordinary differential digital traces any closer than 0.5 mm (0.020 in.) for any EMI purpose.

6.11.4 Punching Through a Noisy Connector

When two systems are mated by a connector, the net flow of signal current between the systems returns to its source through the ground (or power) pins of the connector. As it does so, tiny voltages are induced across the inductance of the connector's ground (or power) pins. These tiny voltages appear as a difference between the ground (or power) voltage on one side of the connector and the ground (or power) voltage on the other side. This problem is called a *ground shift,* and it is yet another form of common impedance coupling. In a single-ended communications system the ground shift voltages detract directly from the available noise margin for your logic family. In a differential signaling system, the ground-shift voltages don't matter, because they affect both wires of the differential pair equally. Subject to the limits of common-mode rejection, ground shifts generated within a connector are totally cancelled within the receiver.

Differential signaling usually reduces crosstalk generated by either mutual inductance or mutual capacitance within the connector itself. The exact gains available depend on the relative spacing of the differential signals as they pass through the connector and the distance to the nearest aggressive source. If the aggressive source is closer to one element of the pair, the crosstalk will not affect both pairs equally, and it will therefore not be cancelled in the receiver.

The cancellation of ground shifts generated by a connector and the cancellation of nearby aggressors within a connector both have a lot to do with the position of the signal pins within the connector, but very little to do with the pcb trace layout, or intertrace coupling.

POINT TO REMEMBER

> ➢ Subject to the limits of common-mode rejection, ground shifts generated within a connector are totally cancelled within a differential receiver.

6.11.4.1 Differential Signaling (Through Connectors)

High-Speed Digital Design Online Newsletter, **Vol. 3, Issue 12**
Sal Aguinaga writes

I have 16 differential pairs that go through a connector and terminate on a daughter card. What is the best signal-to-ground ratio and pattern I should consider?

In this case the connector is a high-density pin connector. If the differential impedance is 100 Ω, do I need a special ground pattern as the signals go through the connector to maintain the differential impedance close to 100 Ω?

Reply

Thanks for your interest in *High-Speed Digital Design*.

Regarding your correspondence, there is no general formula for the number of grounds required, as it depends on the spacing and sizes of the connector pins, and how they are bent.

Here are a few thoughts for you to consider.

- In a connector with an open field of pins, put the two elements of each differential pair on the same row of the connector. That will ensure that they get the same pin lengths and go through the same pattern of elbow bends.[52]
- On a synchronous bus, if you have enough time to wait for the crosstalk to settle, you may not need to isolate the differential pairs from each other.
- If isolation between pairs is required (for low crosstalk between nonsynchronized bus signals or for a clock or other asynchronous signal), place the pairs so that no signal wire from one pair falls adjacent to any wire from another pair. This implies that you will be using at least as many grounds as signal pins to separate the pairs, and probably more.

The differential impedance of most open-pin-field connectors is probably going to be a little higher than you want. You can measure this. You will need a pair of test boards on which you can mate the connector halves. The boards don't use any traces. They can be solid copper with holes drilled for the connector pins. Ground all the pins that will be grounded (or tied to a power plane) in your application. Use two RG-174 50-Ω coaxial cables to route a differential, 100-Ω signal into the designated signal pin pair. Let this signal go through the connector to the far side. On the far side of the connector, terminate the signal differentially with 100 Ω.

For any signal speed that will work with an open-pin-field connector, you will find that a 1/8-watt axial, 100-Ω resistor works fine as a terminator.

Blast in a differential signal from your 100-Ω source. Make a record of the resulting waveform as measured at the source. This should show the source waveform going out and a first reflection coming back (you've made a crude TDR

---

[52] Connectors with solid ground shields between columns of pins are usually designed to accommodate differential pairs collocated within each shielded cavity. This implies that the pin lengths may differ for the two elements of each pair. A perfect differential shielded connector would be designed to match the skew on each element of a differential pair as it traverses the connector, even if the elements were located on different rows. If your connector produces a known skew, it's up to you to cancel it somewhere else in the layout.

instrument). Use a step risetime commensurate with what you are going to be using in the real system.

Don't mess around with fancy 35-ps step edges on this type of connector. They will just show you a bunch of fine-structure detail that isn't going to matter in the real system.

Now disconnect the coaxial cables from the connector. Place the 100-Ω termination directly across the coaxial cable outputs, with the coaxial grounds tied together. Repeat the measurement. You should (ideally) see no reflection.

Looking at the difference between the first measurement and the second, if the reflected waveform bumps up in the positive direction (same polarity as the step input), the connector impedance is a little too high. If it bumps negative, the connector impedance is too low. If you don't see a bump then the impedance is just right.

Adding more ground pins around the signal pair lowers the impedance.

Spacing the signal pins further away from ground raises the impedance.

Bonus idea: Adding a little lumped-element capacitance from signal to ground on each side will lower the effective impedance. This may be implemented in the pcb layout by just using larger-than-normal via pads. Experimentation and remeasurement is required to get this idea to work. The "big-pad" concept works when the connector through-delay is less than 1/6 of the signal risetime and the connector is acting like a lumped-element inductor (too high an impedance).

6.11.5 Reducing Clock Skew

When a digital component receives a clock, the precise moment at which the clock is recognized depends upon the *switching threshold* for that component. For 5-V TTL logic, the switching threshold is defined to be somewhere between 0.8V (V_{IL}) and 2.4V (V_{IH}). The spread between V_{IL} and V_{IH} defines a window shown in Figure 6.23 within which the actual clock transition takes place:

$$t_{\text{UNCERTAINTY}} = \frac{V_{IH} - V_{IL}}{(dv/dt)} \quad [6.25]$$

where $t_{\text{UNCERTAINTY}}$ is the uncertainty (skew) in the clock switching moment,

V_{IH} and V_{IL} are the worst-case guaranteed high and low logic thresholds respectively, and

dv/dt is the rate of change of voltage on the clock input (roughly equal to the logic swing ΔV divided by the 10% to 90% risetime of the driver).[53]

---

[53] The risetime emanating from an unreasonably fast driver will be modified by the package parasitics of the receiver. The risetime of the signal measured at the input pads of the receiver die is therefore generally not quite as fast as the risetime measured external to the package. It is the risetime of the internal signal combined with the receiver switching thresholds that determines the actual amount of switching uncertainty. If the driver and receiver are implemented in similar technology, and the package is not deemed a significant impediment to reception, you may simply use the driver risetime for the calculation of dv/dt in equation [6.25].

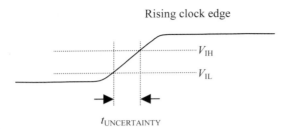

Figure 6.23—The input thresholds and the risetime of the input signal combine to create an uncertainty in the precise switching time for a clock.

The smaller you can make this uncertainty, the less clock skew you will have to recognize in your timing budget. In the differential world a similar effect takes place, but the specifications for V_{IH} and V_{OL} are replaced by a specification for the offset voltage of the receiver. The input offset voltage is the differential input voltage at which a particular receiver actually switches. An ideal receiver would switch precisely at zero (when the inputs are equal), regardless of common-mode voltage, temperature, power supply quality, age, and so on. Such a part would have an input offset voltage of zero. Practical receivers always switch at some finite, hopefully small, value. There is no way to predict the polarity of the offset.[54] The spec sheet for a differential receiver usually provides a worst-case upper bound for the magnitude of the offset. The effective spread between V_{IH} and V_{IL} for a differential receiver is twice the maximum offset magnitude.

A relevant figure of merit for comparing differential logic families in this regard is the ratio of the spread in differential input offset voltage to the peak-to-peak differential output voltage swing.[55] For single-ended logic the corresponding figure of merit is the spread between V_{IH} and V_{IL} divided by the peak-to-peak output voltage swing. Differential logic usually fares better on this measure of performance.

In a differential clock distribution system one usually attempts to match the delays of the two complementary signal traces as they traverse a pcb. The two traces need not follow the same path; they just need to have the same delay. If the delays of the two traces are unequal, if affects the switching time. For example, suppose the two complementary halves of a differential signal arrive at successive times t_1 and t_2. Let the separation between t_1 and t_2 be a small fraction (perhaps 1/10) of a rising edge. Under these conditions the receiver will switch very nearly at the average arrival time $(t_1 + t_2)/2$. In the worst case, if the separation is as great as a risetime, the receiver will switch no earlier than t_1 and no later than t_2. I normally match the delay of the two traces in a differential signal to within 1/20 of the signal risetime.

The clock skew contributed by a clock receiver is a function of the input risetime, the switching thresholds, and, for differential signaling, the degree of similarity in the times of arrival of the two complementary signals. Clock skew has little or nothing to do with trace spacing or geometry (other than delay).

---

[54] Unless the device is manufactured with a purposeful offset in one direction or another.
[55] In a differential signaling architecture, the peak-to-peak differential output voltage swing is twice as large as the peak-to-peak voltage swing on either of the two inputs.

POINTS TO REMEMBER

> Differential receivers often have more accurately specified switching thresholds than single-ended receivers.

> Uncoupled differential traces need not follow the same path; they just need to have the same delay.

6.11.6 Reducing Local Crosstalk

Differential traces on a pcb do a relatively poor job of reducing local crosstalk. As illustrated in Figure 6.24, when some local aggressive trace approaches a differential pair, the interference is not balanced. Interference couples much more strongly to the near side of a differential pair than to the far side.

In the figure, clock+ is twice as close to the aggressor as clock−, so you get a 4:1 difference in the crosstalk coupled into the two sides. Imbalanced crosstalk of this sort cannot be cancelled by a differential receiver. The receiver sees almost the full value of crosstalk from the aggressive trace to the clock+, with little or no cancellation from clock−.

The best way to prevent crosstalk onto a differential pair is to design a keep-out zone around the sensitive traces, forcing other traces to stay at a respectable distance. All modern layout systems support separation rules by net class that will allow you to keep big, dirty signal traces away from delicate, sensitive ones.

Cramming the traces of a differential pair closer together does yield marginal improvements in crosstalk reduction, but you don't get the *big* benefit you get from simply moving the whole pair further away from the problem. Figure 6.25 provides data to support this assertion. The data for this plot were generated using a sheet-conductance measurement method [56].

The figure plots crosstalk for three trace configurations. Each plot shows the measured near-end crosstalk coefficient in dB versus the separation x between traces.[56]

The first scenario shows how a single-ended aggressor affects a differential pair. The next two scenarios show how two differential pairs affect each other. The difference between scenarios II and III is that the intrapair separation changes from 8 to 4 mils. In all cases the separation between planes is 24 mil, and the height of all traces above the nearest plane is 6 mil. The widths of all traces have been adjusted to give a characteristic impedance

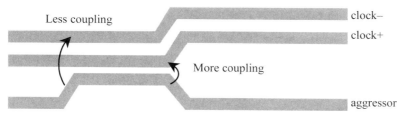

- The aggressor couples more strongly to clock+ than to clock−
- It is a good practice to enforce a bigger-than-normal spacing between clock signals and data signals

Figure 6.24—Differential signaling does *not* offer much help with nearby crosstalk.

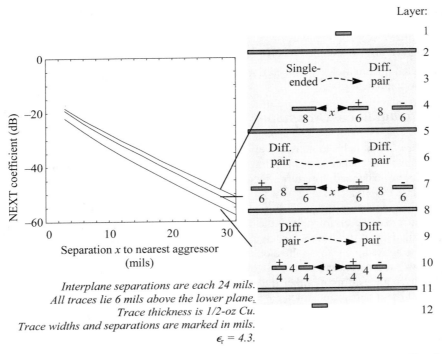

Figure 6.25—The crosstalk immunity of differential pairs is not much improved by tight coupling; mostly what matters is the interpair spacing.

of 100 ohms for all differential transmission-line pairs and 50 ohms for the single-ended aggressor.

In all cases the crosstalk falls off steeply as you increase the separation x between the aggressor and victim. Looking at the difference between scenarios I and II, the addition of a complementary signal to the aggressor (changing it into a differential pair) tends to cancel a little bit of the interference, but the complementary signal is too far away to have much of an effect. The net reduction in interference is less than 2 dB. When you go from scenario II to III, reducing the intrapair separation, the complementary signals is brought into play. This reduces crosstalk by an additional 4 dB, which is useful, but still nowhere near as significant as simply moving the aggressor further away.

Tightly coupling a differential pair delivers only a modest improvement in crosstalk, and therefore only a modest improvement in the achievable pair pitch (see also Section 6.10.2, "Edge-Coupled Stripline").

POINT TO REMEMBER

> ➢ Tightly coupling a differential pair delivers only a modest improvement in crosstalk.

6.11.7 A Good Reference about Transmission Lines

The *Transmission Line Design Handbook* by Brian C. Wadell [58] compiles handy approximations for transmission-line impedance, delay, skin effect loss, dielectric losses, and radiation losses. It is very comprehensive. The book addresses most of the popular transmission-line formats in use today, including microstrip, buried microstrip, offset stripline, and edge- or broadside-coupled differential striplines.

Wadell heavily references the original research articles and measurements. He doesn't pull any punches when the research is fuzzy or contradictory. He shows you just what is known and indicates what is not known. If you're looking for closed-form approximations, this is the best source.

P.S.: If you have an old copy of Wadell's text, check out his errata list on the Web.

6.11.8 Differential Clocks

High-Speed Digital Design Online Newsletter, Vol. 1, Issue 10
Fabrizio Zanella writes to the SI-List

I understand the benefits of using differential pairs for signals running at 100MHz and above. Can you speak about the impact of using differential clocks in a parallel bus? Do the differential clocks maintain the noise suppression characteristics when daisy-chained in a multidrop environment? Has anyone tried this and had positive experiences versus single-ended multidrop clocks?

Reply

Thanks for your interest in *High-Speed Digital Design*.

In general, I have found differential distribution to be a very effective means of combating ground bounce in the transmitting package, ground bounce in the receiving package, as well as the ground shifts that occur on either side of the connectors in high-speed systems. These benefits accrue in multidrop configurations as well as point-to-point configurations.

I have found differential distribution to be of little value in reducing the impact of crosstalk generated locally by other traces on the same pcb. This is because the crosstalk function from nearby traces falls off very steeply with distance. The impact of this is that differential pairs cannot be placed particularly close to any aggressive signal. For example, imagine a system that has one aggressive trace and a nearby victim trace that is receiving unacceptable amounts of crosstalk. Now I propose to protect the victim trace by splitting it into a differential pair and using a true differential receiver. Assuming that I don't want to affect the layout density, I plan to implement the centerline of the new pair right on top of the original victim trace. In other words, when we split the victim, one member of the pair will have to move closer to the offending source, while the other moves away. Unless the two traces of the pair are extremely close together (less than a third of the original separation between centerlines), the extra crosstalk we pick up from the nearby side of the differential pair overwhelms any "balancing" effect we might have hoped to gain from the far side of the differential system. To mitigate this effect, you have to separate the victim pair from the aggressive signal. In the final analysis, it's usually the extra separation that is providing most of the crosstalk benefit, not the fact of balanced signal distribution. When you want to battle crosstalk picked up on a pcb (over a solid ground plane), increasing the trace separation will probably result in a more dense design than using differential distribution.

For clocks, I see differential signaling used a lot, both in point-to-point distribution and in multidrop distribution. The multidrop aspect does not diminish the ground-bounce-canceling properties of differential signaling. There are only a few clock nets in a system (compared to the number of data nets) and it isn't that difficult to provide this extra measure of protection.

For parallel bus signals, I rarely see differential distribution used because it doubles the number of wires required. That will cause the phenomenon known as "routing headaches" among your pcb layout staff.

POINT TO REMEMBER

> The benefits of differential signaling apply to multidrop configurations.

6.11.9 Differential Termination

Article first published in *EDN Magazine*, June 8, 2000

I am designing a piece of equipment to interface to a digital tape recorder designed by another company. This recorder uses a differential ECL interface, and the user's manual recommends terminating the clock lines slightly differently from the data lines. Each clock line employs a split terminator (160-Ω to −5.2V and 100-Ω to ground), but the data signals simply use a single 120-Ω resistor between the wires of each pair. Because these two methods are Thevenin equivalents, why does the user's manual recommend different termination schemes? As far as I can tell, the transmitters and receivers for the clock and data lines are electrically equivalent, and all signals have 390-Ω pull-down resistors to −5.2V at the transmitter to properly bias their emitter-follower outputs. I have contacted the company that designed the circuit, but the original designers are unavailable. Does one termination scheme have any advantage over the other?
—Raymond Bullington

Engineers often have a difficult time figuring out why something was done. Sometimes there is no reason, sometimes there is a multitude of good reasons, and sometimes (as is common in the standards world) everyone wants it done the same way but all for conflicting and different reasons.

Anyway, differences do exist between the termination schemes you described. The single-resistor scheme (120 Ω across the two lines) terminates all differential-mode signals into 120 Ω but provides no termination for common-mode signals.

Your four-resistor scheme (independent terminations for each line) terminates all differential signals *and* all common-mode signals. The difference between these

6.11.9 • Differential Termination

two styles matters only if a common-mode signal is present. And where might a common-mode signal come from? It can come from any skew naturally present in the clock driver output plus any imbalances in the circuit that convert part of the output from the differential mode to the common mode.

Consider the single-resistor termination shown in Figure 6.26. Say that a positive-going edge $x(t)$ arrives first on trace A, and then, after a tiny skew interval Δt, the opposite signal $-x(t - \Delta t)$ arrives on trace B. During the tiny skew interval Δt, the single 120-Ω resistor R1 creates two tiny artifacts. First, the initial rising edge on line A shoots right through the resistor onto trace B, creating a little blob of crosstalk. The amplitude of the crosstalk compared to the amplitude of the incoming signal is 1/2. Second, coincident with the crosstalk, you get a small signal reflected back onto line A. The amplitude of the reflected signal compared to the amplitude of the incoming signal is also 1/2. Both artifacts have positive polarity, creating what amounts to a common-mode reflection.

> **Every long, differential link needs a differential termination for signal quality and *also* a common-mode termination to prevent common-mode resonance.**

After time Δt, the opposite signal $-x(t - \Delta t)$ arrives on line B. At this time you get a second set of crosstalk and reflection artifacts, but with negative polarities this time (because they originated on the negative half of the differential pair). The second set of artifacts partially cancels the first, with the degree of cancellation depending on the exact temporal alignment of the two signals. The two sets of artifacts perfectly cancel only when the signals on A and B arrive in perfect synchronism.

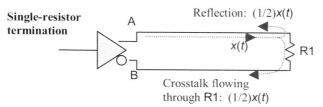

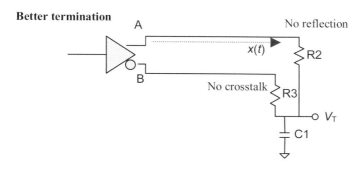

Figure 6.26—A network of two resistors terminates each half of the differential pair independently.

In this example, both the crosstalk and reflection amplitude coefficients equal 1/2. You may express the residual common-mode signal $y(t)$, induced on either trace by the single-resistor end-terminator, as $y(t) = (1/2)(x(t) - x(t - \Delta t))$. If the skew is less than the signal rise or fall time t_{10-90}, and you define the signal step height as ΔV, the peak amplitude of the reflected common-mode noise roughly equals $(1/2)\Delta V(\Delta t/t_{10-90})$.

Once created, the common-mode noise returns to the driver. In your case, the ECL driver presents a very low output impedance to the line, generating a big reflection. The reflected noise then proceeds to the receiver, where it once again encounters the single-resistor termination, but this time as a purely common-mode signal. Because a common-mode signal presents the same voltage on both traces, the single-resistor terminator draws zero current and acts as an open circuit. The open circuit generates another big reflection. After that point, the common-mode noise trapped on the line happily bounces back and forth between the driver (low impedance) and the receiver (open circuit) for a long time.

Terrible things happen to the common-mode noise if your trace delay equals one-quarter of the clock period. In that case, the little common-mode artifacts from each edge build and superimpose, cycle after cycle, magnifying the common-mode noise at the receiver and also magnifying the common-mode radiated emissions. This problem is called a *common-mode resonance*.

To avoid common-mode resonance, every long, differential link needs two terminations: first, a good differential termination at one end or the other to provide good differential signal quality, and second, a reasonable common-mode termination at one end or the other to prevent severe common-mode resonance. An ECL driver does not provide a good common-mode termination at the source; therefore, one is required at the load.

The four-resistor termination that was recommended to you for the end of the clock net independently terminates both lines, damping both differential and common-mode signals at that point.

An even better termination circuit appears in Figure 6.26. This circuit terminates both differential and common-mode signals but requires only two resistors (R2 and R3 are each set to half the differential-line impedance). The capacitor need be only large enough to hold its charge steady during the brief interval of skew Δt. In your case the ECL sources incorporate pull-down resistors, so you don't need to supply a special terminating voltage (V_T) to the capacitor.

Any driver that provides a reasonable common-mode termination at the source relieves you of the responsibility of providing one at the destination. For example, a source-terminated driver works fine with the single-resistor termination.

POINT TO REMEMBER

> ➢ Every long, differential link needs at least one good differential termination *and also* a reasonable common-mode termination to prevent severe common-mode resonance.

6.11.10 Differential U-Turn

Article first published in *EDN Magazine*, September 1, 2000

What is the effect of a split in a solid plane on the impedance of a coplanar differential pair? The differential pair passes over a solid plane (logic return) and then crosses a 50-mil void into an I/O area that has a solid plane of its own that is tied to the chassis.
—Boris Shusterman

Significant currents flow on the solid reference plane beneath your differential traces. Cutting the plane interrupts these currents.

Consider first a single-ended pcb trace. When a changing voltage propagates down a *single-ended* transmission structure, currents flow through the distributed capacitance of the trace to all nearby objects, especially the big, solid plane underneath the trace. This capacitive effect generates a returning (reverse) flow of current on the solid reference plane. The return current flows all the way back to the source along the reference plane, staying underneath the signal trace the whole way, making a complete circuit. (Current always makes a loop.)

Now, consider a differential-pcb-trace pair. Differential structures have capacitance from each trace individually to the reference plane and also *between* the traces. The between-trace, or mutual, capacitance of the differential pair induces returning current on the other

> **Counteract the U-turn by shrinking both the reference-plane gap and the spacing between traces.**

trace as well as on the solid reference plane. In a differential-pcb pair, most of the returning current from each trace still flows on the solid plane, not the other trace, because each differential trace couples much more strongly to the big, solid nearby plane than it does to its little, skinny differential buddy.

Try to visualize the propagation of a differential signal as a quad of four currents: two currents, i+ and i–, on the two signal traces, and the returning currents, r+ and r–, on the reference plane underneath the traces (Figure 6.27). In most cases the currents are almost as big as i+ and i–.

When a differential signal encounters the gap between reference planes, the two signal currents continue across the gap on the signal wires, but the gap blocks the two return currents r+ and r–. This situation forces the return currents to execute a U-turn maneuver, whereby each return current U-turns into the other position. On the new reference plane, a similar effect takes place with a second U-turn formation creating a pair of currents r′+ and r′– to complete the quad current formation.

In the space between the reference planes, current circulates clockwise on all sides of the U-turn zone. This current behaves like a small current-loop antenna, generating a substantial, fast-changing magnetic field within the U-turn zone.

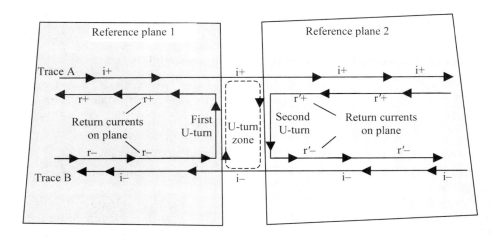

Figure 6.27—The U-turn zone at a reference-plane gap separates the return currents from the primary signal currents on traces A and B.

The magnetic field makes the circuit behave as if an inductor were in series with the signal path. The length and width of the U-turn zone determine the inductance. For example, a trace-to-trace separation of 0.100 in. and a plane-to-plane separation of 0.100 in. would generate an inductance on the order of 10 nH, which in a 100-Ω differential system would introduce a low-pass filter response with a time constant of 100 psec. If risetime exceeds 1 nsec, you probably won't even notice the effect. On the other hand, at very high speeds, a 100-psec risetime could mess up your signals.

The U-turn zone is more than merely a region of increased impedance due to the absence of the reference planes. It is an effect whose physical dimensions span both the gap between planes and the spacing between traces. You counteract the U-turn by shrinking both the reference-plane gap and also the spacing between traces. You can also practically eliminate it by providing continuous pathways adjacent to each signal trace for the conveyance of return currents from plane to plane, eliminating the need for a U-turn zone. If the planes carry different DC voltages, a bypass capacitor next to each trace isn't perfect, but it helps.

The magnetic fields within the U-turn zone induce crosstalk and EMI. The crosstalk couples to all the differential pairs that pass over the same gap. Both EMI and crosstalk vary in proportion to the size of the U-turn zone.

POINT TO REMEMBER

> Visualize the propagation of a differential signal as a quad of four currents.

6.11.11 Your Layout Is Skewed

Article first published in *EDN Magazine*, April 18, 2002

Passing through a sharp turn with an edge-coupled differential pair, the outside trace travels further than the inside trace. The difference in distance traveled contributes a small amount of skew to your differential signals. The skew acts as a mode converter, changing part of your differential signal power into common-mode power.

The pair-turning skew becomes noticeable only when the skew contributed by the turn rises to a level comparable with the natural skew already coming out of your driver. Therefore, before worrying about turns, first determine the skew of your driver. In many cases the driver skew is not specified, in which case you can assume the skew will be *at least* 10% of the signal risetime. Digital differential drivers just aren't balanced very well. Analog transceivers often are, which accounts for the importance of meticulous skew-matching in some analog applications.

For example, let's say an Ethernet 100BASE-TX LAN transceiver with a well-balanced output transformer and common-mode choke puts out differential signals balanced to 1 part in 1000—meaning that the common-mode output is 1000 times smaller than the differential signal. To avoid amplifying the common-mode signal on the wires (and thereby the radiation), the aggregate skew contributed by all components used with this transceiver must remain less than 1/1000 of 1 risetime. The risetime of a 100BASE-TX signal is approximately 8 nS, corresponding to roughly 94 inches of propagation in air, 1/1000 of which works out to 0.094 in., so the skew budget within cable connectors and board layout should be set somewhere around 0.1 inch. Worrying about little bitty skew effects much smaller than 0.1 inch doesn't buy you anything.

> **Chamfering or rounding of differential corners does not eliminate skew.**

To take a faster example, a 2.5 Gb/s serial link driver with a risetime of 200 ps has an output skew of probably no better than 20 ps (maybe a lot worse). In this case a skew budget of perhaps 20 ps seems reasonable.

Figure 6.28 illustrates the skew calculations for three alternative corner treatments, each with a trace pitch (centerline to centerline) of p. In each case, the two traces within the pair share the same number and type of sharp corners (diagonal-striped regions). The differences between the inside and outside traces are shaded dark with white lettering. The lengths added to the outside trace are $2p$, $1.65p$, and $1.57p$ respectively for the three corner styles. Apparently, chamfering or rounding of differential corners does not eliminate skew; it only makes at best a modest improvement.

If your trace separation p equals 20 mils, and the propagation delay of your media is 160 ps/in., the skew associated with a distance of p is (0.020 in. x 160 ps/in.) = 3.2 ps. According to Figure 6.28, the skew accumulated when rounding a corner of any of the three types shown ranges in that case from a

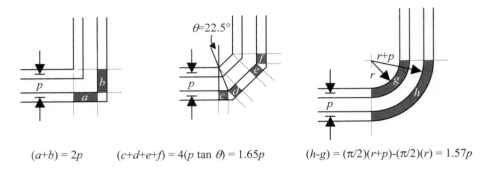

Figure 6.28—These three corner treatments all generate similar amounts of skew.

high of $2p \rightarrow 6.4$ ps to a low of $1.57p \rightarrow 5.0$ ps. If this amount of skew is superceded by the skew from your driver, then don't worry about the turns.

When skew becomes a problem, you can mitigate its impact in two ways. First, use a smaller spacing. The smaller you make *p*, the less skew you will get. This is one of the few benefits of tightly coupled pairs. Second, position your ICs so the traces leave the driver headed in the same direction that they enter the receiver. For example, a differential pair that starts out headed north and ends up headed north has by definition equal numbers of right- and left-hand turns no matter what happens in the middle (unless it makes a spiral), so the net skew accumulated is zero.

POINT TO REMEMBER

> Chamfering or rounding of differential corners does not eliminate skew.

6.11.12 Buying Time

Article first published in *EDN Magazine*, May 2, 2002

The previous article "Your Layout Is Skewed" concerned various styles of corners and bends normally used on differential edge-coupled pairs. It pointed out that all corners, whether chamfered or not, add extra length to the outside trace as it rounds the bend. The equivalent trace length added by 90 worth of bending ranges from one and one half to two times the intrapair trace pitch depending on how the corner is chamfered. The extra time added to the outside trace is a form of intrapair skew.

This article considers two strategies for minimizing the intrapair skew accumulated by a differential net. The six BGA chips within the dotted-line region in Figure 6.29 illustrate the first strategy.

Pair **A** exits the bottom chip heading north. It enters the receiver (top chip) also heading north. Along the way, this pair takes one right turn and one left turn. *The skew accumulated in the two successive turns cancels to zero.*

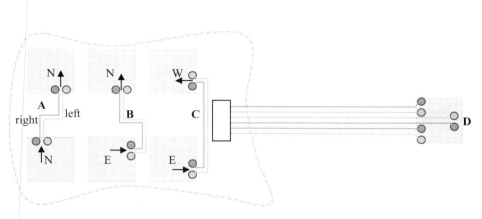

Figure 6.29—The positioning of entrances and exits affects the intra-pair skew.

In a general routing problem, the number and types of turns required depends on the relative orientations of the driver and receiver. Because pair **A** starts and ends going in the same direction, this pair will always make equal numbers of right-hand and left-hand turns no matter what happens in the middle (unless it makes a spiral). The net skew accumulated on any pair with a chip floor plan like **A** is zero.

Pair **B** doesn't fare as well. It exits the bottom chip headed east. It takes one left turn to get going north (the orientation of the receiver), after which the number of left and right turns remain balanced. The total skew accumulated by pair **B** equals the amount generated by one left-hand turn.

Pair **C** is the worst of all. It exits to the east and enters to the west. It therefore requires two extra left turns to achieve the correct orientation. If you are going fast enough so that every turn worth of skew matters, you should carefully plan your chip orientations so the accumulated skew naturally balances to zero.

The second strategy concerns the precise manner in which your pair enters or exits a ball grid array (or any field of connector pins, package pins, or vias). This strategy works best when the ball pitch exceeds the intrapair trace pitch. It works by offsetting the centerline of the pair as it enters (or exits) the BGA, as shown in the figure at **D**. By offsetting the top pair down half a position, extra time delay accrues to the topmost trace. Offsetting the bottom pair up half a position adds time to the bottommost trace. This strategy "buys some time," which you can use to pay for other floor planning inadequacies. It isn't perfect, but it delivers what you need—balanced skew.

> **A pair that starts and ends going north has by definition equal numbers of right-hand and left-hand turns.**

When you have to adjust the skew, I favor doing it near either the driver or the receiver, whichever has the poorest termination. That way the skew adjustment can't possibly affect the quality of the good termination at the other end of the line. If both ends have high-quality terminations, then you place the adjustment at either end. In an imperfect world, that's as well as you can do.

To those who yearn for a perfect layout with zero bends I say, with all due credit to the Rolling Stones, "You can't always get what you waaaannnnnnt... You can't always get what you waaaannnnnnnt... but if you buy some time, you just might find, you'll get what you need."

Oooh, yeah!

POINT TO REMEMBER

➤ A pair that starts and ends going north has by definition equal numbers of right-hand and left-hand turns.

6.12 INTERCABINET APPLICATIONS

The term used to describe the high-speed differential wiring often used between pieces of equipment is *balanced cabling*. This term has been adopted by the ISO building-wiring standards committee to describe any cable that provides one or more pairs of wire, each pair having a defined differential-mode impedance and each pair having a defined immunity to crosstalk from the other pairs within the same jacket.

There are two basic construction techniques used to produce balanced cabling: the *twisted pair* and the *quad configuration* (see Figure 6.30). Both arrangements hold the wires of each pair in a fixed arrangement with a uniform cross section. This stabilizes the differential impedance of the cable. Both guarantee low crosstalk among the pairs.

The twisted-pair cable guarantees low crosstalk by virtue of having a different rate of twist on all the pairs within the same jacket. The different rates of twisting are an essential part of the crosstalk cancellation process. This happens because of the way transmission-line coupling works between two adjacent differential pairs. The basic rule of thumb for pair-wise crosstalk is this: *When you flip one pair, the crosstalk reverses polarity.*

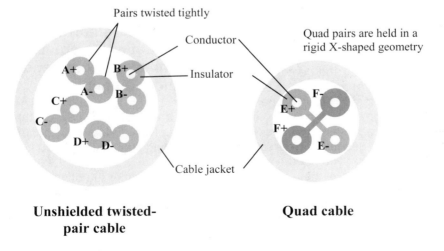

Figure 6.30—Construction of balanced cables.

A corollary to the flipping rule is this: *When you flip both pairs, the crosstalk retains the same polarity.* That might happen often if the twist rates were the same on two adjacent pairs. Every time both pairs flipped over, the crosstalk would remain the same. The crosstalk therefore might never cancel. To avoid this effect, the pairs in a multi-pair twisted cable are usually twisted at different rates. This randomizes the relation of one pair vis-à-vis its partner, nulling out the crosstalk.

On a well-constructed twisted-pair cable, one of the colored pairs will carry a noticeably tighter twist than the others. The crosstalk to and from this pair will be the best in the bunch, a nice property. Do not, however, be deluded into thinking that all cables will have the same hierarchy of twist performance. There are few, if any, standards concerning which of the pairs should carry the tightest twist. Manufacturers are free to change the twist pattern at will, including reassignment of the hierarchy of twist performance. The crosstalk standards for most cables specify only the worst-case crosstalk between any two pairs. They do not designate any particular pairs as having better performance than the others.

The quad cable guarantees low crosstalk by virtue of its unique geometrical alignment. Both capacitive and inductive coupling mechanisms between the pairs are cancelled by this construction technique. With regards to interpair crosstalk within the same jacket, quad cable, if carefully constructed, can exceed the performance of twisted-pair cabling. With regards to crosstalk with other objects and cables outside the jacket, quad cable performs less well than twisted-pair cable. Twisted-pair cable does a better job of canceling electromagnetic radiation from the cable and providing good common-mode rejection.

The following sections describe the main applications for balanced cabling between cabinets.

POINTS TO REMEMBER

> The twisted-pair cable guarantees low crosstalk by virtue of having a different rate of twist on all the pairs within the same jacket.

> Quad cable guarantees low crosstalk by virtue of its unique geometrical alignment.

6.12.1 Ribbon-Style Twisted-Pair Cables

Ribbon-style twisted-pair cables have the same twist pitch on all pairs and yet still deliver reasonable crosstalk performance. It seems counterintuitive that this would work, because when one pair twists (inverting its local field polarity) the adjacent pair twists as well (inverting its sensitivity). The crosstalk would seem to reinforce with the same polarity at every twist. How can it work?

This paradox is solved by looking closely at the exact variations in crosstalk as the wires turn about one another. Imagine an axis run horizontally through the centerline of both pairs. Now imagine you can continuously control the angle of rotation on each pair about their respective axes. Begin with a rotational phase of 0° (left view in Figure 6.31). The coupling for this configuration is dominated by the inside two wires, A– to B+, and so has a negative polarity.

Figure 6.31—In a twisted ribbon cable the crosstalk coupling polarity reverses every 90°.

At a rotational phase of 90° (right view in Figure 6.31), the coupling changes dramatically. In this case the A+ wire couples mostly to B+, and A– to B–, yielding a coupling amplitude almost exactly the same as in the previous case, but with opposite (positive) polarity.

The coupling reverts to the original (negative) polarity at 180° and inverts back to positive once again at 270°. As long as the rotational axes of each pair is held in a fixed position, the coupling averaged throughout the entire rotational cycle nulls to near zero. The wires of a twisted ribbon cable are varnished into place to ensure they maintain the correct geometry. Different arrangements of the starting phases and rotational directions are possible.

In a practical, multipair, jacketed cable it is not generally possible to hold all the wires in fixed positions. The wires in the 90° case are likely to slump towards each other, upsetting the cancellation. To circumvent this difficulty the manufacturers of multi-pair twisted cables resort to the ruse of varying the twist rate on each pair.

POINT TO REMEMBER

> ➤ Ribbon cables can use the same twist pitch on every pair because the wires are held in a rigid geometry.

6.12.2 *Immunity to Large Ground Shifts*

Differential signaling with unshielded twisted-pair cables does not require a direct ground connection between the two ends of the link. As long as the potential difference between the transmitter and receiver remains within the common-mode input range of the receiver, the system will function. In most cases, the existing green-wire ground connection implemented on most computer equipment keeps the product chassis at either ends of the link within an acceptable voltage range (see box). No additional grounding needs be added to the system.

High-frequency single-ended signaling, on the other hand, *does* require a direct ground connection between the two ends of the link. Because this ground connection carries high-frequency returning signal currents, it must follow closely along with the signal wires in a low-inductance, controlled-impedance structure. The green-wire ground is woefully inadequate for this purpose. If single-ended signaling is to be used between cabinets, additional grounding means (such as a coaxial cable shield) must be implemented.

6.12.2 • Immunity to Large Ground Shifts

These additional grounding means may violate one of the most sacred AC power safety principles:

Never introduce a metallic connection between any two frames powered by different AC power sources.

As explained in the box "Earth Potential," violation of this rule may draw significant currents through the green-wire connection. This is a problem because it upsets the sensitive green-wire current detectors built into the main electrical panel of most modern buildings. These detectors look for early warning signs of electrical malfunction. For example, a partial short between any *hot* wire and a product chassis will transmit green-wire currents back to the electrical panel where they may be detected. The circuit that detects these currents is called a ground-fault interrupter, or *GFI,* circuit breaker. When the detected current exceeds a critical threshold, power may be removed from that section of the building. Messing with the green wire is serious stuff. Don't do it.

If you must electrically connect the metallic frames of two systems, make sure that both systems are served by a green-wire ground connected to the same Earth potential. There are multiple ways to do this. For systems located within the same rack-mounted chassis, just screw all the frames to the same rack. For boxes located in the same room, but not in the same rack, provide a way to plug the AC power cord of one system into a convenience outlet on the other system. This arrangement daisy chains the green-wire connections, so you know they are all at the same potential. If daisy chaining is not possible, try to plug all the systems into the same outlet or power strip. For boxes located within different rooms, use differential signaling, fiber, or RF connections that don't require a metallic connection between frames.

> ### Earth Potential
>
> The potential across the surface of the Earth is not constant. Various mechanisms, including spurious power distribution currents, magnetic-field interactions, and lightening, induce large currents in the surface of the Earth. These currents, working across the surface resistance of the soil layers, produce noticeable potential differences. Between the ends of a typical building, one may observe several volts of potential difference.
>
> Large buildings are typically divided into several grounding domains. Each domain is powered by a local transformer, which typically sits near the center of the domain. At that location, the neutral wire of the transformer secondary, the green-wire ground, and a copper ground stake are all bonded together (see Figure 6.32). The green-wire grounds between domains do not touch. This arrangement limits the voltages between the machinery in your office (whose outer metallic skin is connected to the green-wire safety ground) and the actual local ground where you are standing to something just below the level of human perception (a few volts).
>
> If you connect together the metallic skin of a box in one domain with the skin of another box in another domain, several things will happen. First, you may see a noticeable spark. After that, several amps of current that used to be flowing in the Earth will now begin flowing through the connection. This current flows from the ground stake at position *A,* through a green wire to chassis *B,* through connection *C* to chassis *D,* and from there back to ground stake *E*.

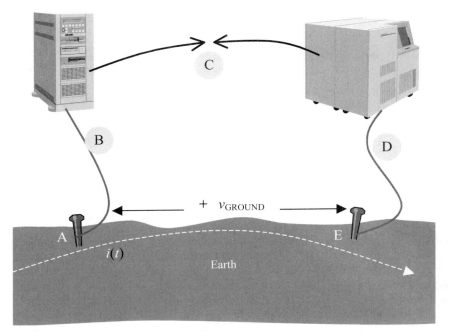

Figure 6.32—Huge currents *i(t)* circulating through the Earth's crust cause measurable differences in the electric potential at the Earth's surface.

POINTS TO REMEMBER

- Never introduce a metallic connection between any two frames powered by different AC power sources.
- If you must electrically connect two boxes, make sure that both boxes are served by green-wire grounds connected to the same Earth potential.
- Differential signaling with unshielded cables does not require a direct ground connection between the two ends of the link.

6.12.3 Rejection of External Radio-Frequency Interference (RFI)

External RF fields impinging on a twisted-pair cable tend to affect both wires equally. Any interference mostly appears as a common-mode signal on the cable, which is cancelled at the receiver. I say mostly because, as usual, a number of things can go wrong. As good as twisted-pair cables are for rejecting RF interference, here's what happens in the real world:

1. Part of the RF field energy is absorbed by the cable. This has to do with the efficiency of the complete cable structure as an electrical antenna, a subject outside the bounds of this book. For more information about antenna

6.12.4 • Superior Tolerance to Skin Effect

efficiency as it relates to RFI problems, see the excellent text by Clayton Paul [57].

2. The RF field energy absorbed is converted to a common-mode current. This happens according to the relation $i_{COMMON} = \sqrt{P/Z_{COMMON}}$, where P is the power received and Z_{COMMON} is the common-mode impedance of the cable, with respect to true Earth ground, including its common-mode terminations.

3. If the common mode of propagation is unterminated at both ends of the cable (as it would be using ordinary transformer-coupling at both ends), a significant resonance may occur that amplifies the received common-mode current.

4. Some fraction of the common-mode current flowing in the cable is converted to a differential-mode current. This conversion may take place due to a natural imbalance in the construction of the cable itself, an imbalance in the connectors, or an imbalance in the transmitter or receiver circuitry.

5. The receiver interprets the differential-mode current as a true differential signal.

POINTS TO REMEMBER

To get the best RF-rejection performance from your cabling,

> Use a tightly twisted, well-balanced cable. Twisted cables work better than quad cables in this respect.

> Don't scrimp on connectors. Buy and use connectors designed to go with the cable.

> Use well-balanced circuitry for both transmitter and receiver.

6.12.4 Differential Receivers Have Superior Tolerance to Skin Effect and Other High-Frequency Losses

Let's say you need to communicate one digital signal from box A to box B located 15.2 m (50 feet) away. You choose a single-ended 3.3-V 50-ohm line driver, running on RG-58 coax at 1000 Mbaud (one nanosecond per bit), with a rise/fall time of 250 ps.

The response of this system is shown in Figure 6.33. The figure shows the actual eye pattern, as predicted by simulation, using solid lines. The ideal transmitted waveform, assuming no skin-effect distortion or attenuation, is depicted with a dashed line. The transmitted data pattern is ...1111010001111....

The worst-case high and low receiver thresholds for single-ended 3.3-V JEDEC LVTTL logic are drawn in place at 2.0 and 0.8 volts respectively. Notice that the low-side threshold fails to catch the first negative-going excursion—causing a bit error. Even when the LVTTL receiver does properly interpret the data, you can expect a fair amount of jitter in the received waveform.

Differential receivers are commonly specified with more accurate switching thresholds than ordinary single-ended logic. Had you selected a differential receiver and a

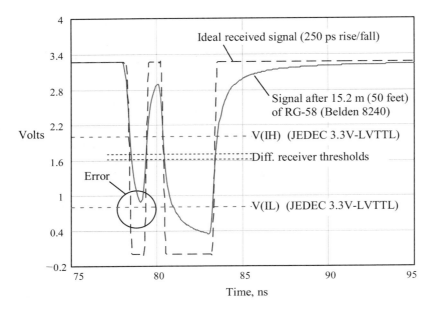

Figure 6.33—Fifty feet of Belden RG-58 distorts this 1-Gb/s signal.

differential cabling system, the effective receiver thresholds would have been more nearly centered in the middle of the data pattern.

For example, the chart shows the differential receiver thresholds for LVDS logic. These thresholds still properly discriminate the data even in the face of severe pulse distortion. In general, for the same amount of transmission-line distortion, a differential receiver generates less jitter than a single-ended receiver. This advantage follows from the generally better threshold tolerances available in differential receivers, not of the differential architecture itself.

In the example of Figure 6.33, you could improve the single-ended system performance by using a differential receiver with its negative input terminal tied to a stable and accurate source of 1.65 V. That simple change would create a single-ended receiver with much better control over the input threshold than indicated in the figure.

If the transmission cable is end-terminated, the end termination is best tied to some voltage halfway between V_{IH} and V_{IL} (or tied to a split terminator with a Thevenin equivalent voltage between V_{IH} and V_{IL}). That way the DC attenuation of the cabling symmetrically affects both high and low logic levels, keeping the received signal centered.

POINT TO REMEMBER

➢ Differential receivers have more accurate switching thresholds than ordinary single-ended logic.

6.13 LVDS Signaling

This section is not intended as a promotion of LVDS logic. It is a primer on how to interpret the specifications of any differential logic family.

Low-Voltage Differential Signaling (LVDS) is a good example of a high-speed differential logic family. The LVDS standard was generated in 1995 by the IEEE. [62]

The LVDS standard contemplates both *general-purpose* and *short-range* applications. Here I will confine my remarks to the *general-purpose* version of the standard. It defines a differential data path operating at data transfer rates in the range of 200 to 500 MHz with a source-synchronous clock and various bus widths up to 128 bits.

Table 6.5 lists some of the key performance specifications for LVDS general-purpose transceivers.

6.13.1 Output Levels

LVDS being a differential logic family, there are two (complementary) outputs per logic signal. The nominal steady-state operating conditions for these outputs are 1.0 and 1.4 volts,

Table 6.5—LVDS General-Purpose Link Specs (adapted from ANSI/IEEE P1596-3-1995)

Transmitter specifications					
Signal	Parameter	Conditions	Min	Max	units
V_{oh}	Output voltage high, either wire	$R_{load} = 100\Omega \pm 1\%$		1475	mV
V_{ol}	Output voltage low, either wire	$R_{load} = 100\Omega \pm 1\%$	925		mV
$|V_{od}|$	Output differential voltage	$R_{load} = 100\Omega \pm 1\%$	250	400	mV
R_0	Output impedance, single-ended	V_{cm}=1.0V and 1.4V	40	140	Ω
V_{os}	Output offset voltage		1125	1275	mV
ΔV_{os}	Change in VOS between 0 and 1 states (this specification defines the AC common-mode output voltage)	$R_{load} = 100\Omega \pm 1\%$		25	mV
t_{rise}, t_{fall}	V_{od} rise/fall time 20% to 80%	$R_{load} = 100\Omega \pm 1\%$	300	500	ps
Receiver specifications					
V_i	Input voltage range, either input	$|V_{gpd}| < 925$ mV	0	2400	mV
V_{idth}	Input differential threshold	$|V_{gpd}| < 925$ mV	−100	+100	mV
V_{hyst}	Input differential hysteresis	$V_{idthh}-V_{idthl}$	25		mV
R_{in}	Receiver differential input impedance	—	90	110	Ω
C_{in}	Not specified				
Implementation specifications					
	Pcb skew allocation	Worst case		50	ps

for the low and high states respectively. When one wire goes to 1.0 volts, the other goes to 1.4, and vice versa.

You may decompose this situation into a steady-state common-mode component of 1.2 volts, plus a changing differential voltage of ±0.4 volts.[57] The differential voltage varies from –0.4 V to +0.4 V, so we say that the differential peak-to-peak voltage is 800 mV. The peak-to-peak voltage on either wire alone would be 400 mV.

The standard requires that the steady-state differential voltage representing either a zero or one state be at least 250 mV, but no larger than 400 mV *when the outputs are loaded differentially by 100 ohms*. It also requires that no output ever fall below 925 mV or exceed 1475 mV under the same conditions.

Concerning output levels under other loading conditions, the standard provides little guidance. The only hints come in the form of stated constraints on output impedance. From the standard it is impossible to determine, for example, the output levels that would result from loading the outputs with a 75-ohm differential load. Nor is it possible to determine something else that the IBIS community has long sought in these sorts of documents: a precise specification of the shape of the rising and falling edge.

POINT TO REMEMBER

> Normal operating voltages for LVDS logic are 1.2 ± 0.2 V on each wire.

6.13.2 Common-Mode Output

The output offset voltage (DC bias) can change by 25 mV when switching from the one state to the zero state. This is a peak-to-peak change. The AC amplitude range of the common-mode voltage is therefore ±12.5 mV.

The AC amplitude of the differential output voltage lies somewhere in the range of ±250 to ±400 mV.

When considering certain radiation and crosstalk problems, it is handy to know the approximate ratio between common-mode and differential-mode emissions coming out of a driver. For LVDS, the ratio of common to differential amplitudes can be as poor as 12.5/250 = 5%.

POINT TO REMEMBER

> LVDS, like most digital transceivers, is not extraordinarily well balanced.

6.13.3 Common-Mode Noise Tolerance

Looking at the common-mode operating range of the receiver (called V_i in the specification), it is apparent that the receiver can tolerate inputs anywhere in the range of 0 to 2400 mV. From these numbers you can derive how much common-mode noise the

---

[57] Or an even-mode voltage of 1.2 volts plus an odd-mode voltage of ±0.2 volts.

system will tolerate. First assume a driver is at its lowest permissible level (925 mV). Now figure out how much common-mode noise you can add in the negative direction without the receiver input falling outside its guaranteed operating range. The answer is –925 mV. Next assume the driver is at its highest permissible level (1475 mV). Now figure out how much common-mode noise you can add in the positive direction without the input falling outside its guaranteed operating range. The answer is +925 mV. LVDS therefore tolerates a common-mode difference (V_{gpd}) between the ground potential at the driver and the ground potential at the receiver as great as ±925 mV.

If you exceed the common-mode operating range of the receiver, all bets are off. It could do anything (see Section 6.7, "Common-Mode Range").

POINT TO REMEMBER

> The common-mode noise tolerance for general-purpose LVDS logic is ±925 mV.

6.13.4 Differential-Mode Noise Tolerance

Next let's look at the differential noise margin. In the worst case the transmitter differential output may be as small as 250 mV. At the same time, the receiver threshold may be offset by as much as 100 mV. The difference between these two figures is the differential noise margin, which is 150 mV. As a percentage of the signal swing on either wire (400 mV p-p), the 150 mV figure represents 37%. A transmitter with a larger output swing would enjoy an even bigger percentage noise margin. These are excellent noise margins for digital logic. Most single-ended logic families have noise margin percentages on the order of only 10% to 15%.

POINT TO REMEMBER

> The high noise margin gives LVDS a built-in natural advantage in combating ringing, overshoot, and crosstalk from like devices.

6.13.5 Hysteresis

A receiver with hysteresis has two input switching thresholds, one used for positive-going signals and one used for negative-going signals. The positive-going threshold is always set a little higher than the negative-going threshold. Once the input crosses into positive territory, the receiver automatically switches to the negative-going threshold so that the signal must turn around and descend below the negative threshold before it can cause another switching event. Once the signal crosses below the negative threshold, the receiver flips back to using the positive threshold. A receiver equipped with hysteresis requires a certain amount of signal change before flipping to the other state.

The hysteresis feature is intended to avoid annoying self-oscillation that might happen with slowly changing, but very clean, inputs. The feature is usually implemented as a form of limited positive feedback from the receiver output back to own its input. With hysteresis, the inputs tend to switch quickly and firmly, once they reach an acceptable level, and then stay there. It's a good feature.

LVDS inputs have a guaranteed amount of hysteresis. You still, however, shouldn't supply any of these parts with a slowly moving input, because in that case as the input slowly sweeps through the transition region, any crosstalk that happens to exceed the hysteresis switching range will cause glitches (and therefore more noise) in the receiver.

POINT TO REMEMBER

> Always provide fast-edged inputs to LVDS logic.

6.13.6 Impedance Control

The LVDS specification goes to a lot of trouble to control ringing and reflections. This is one of the strongest provisions of the specification. Excellent control of ringing and reflections makes it possible to obtain first-incident-wave switching in most LVDS applications.

LVDS uses a both-ends termination strategy to control reflections. Each LVDS transmission line is terminated first at the source and again at the end of the line.[58]

The source impedance of the driver is constrained to the range of 40 to 140 Ω. That is a ratio of only 3.5:1 from highest value of allowed output impedance to the lowest. To a board-level analog designer, this sounds easy, because you can go out and buy very accurate lumped-element resistors. To a chip designer, it's a nightmare. You just can't control the absolute value of $R_{DS}(ON)$ or the absolute value of transconductance very accurately.

The achievement of a 3.5:1 output impedance specification represents a major accomplishment for the chip industry, one which I am sure will pay off in terms of higher volume, given the user-friendly advantages of both-ends termination (see box "Both-Ends Termination").

The worst-case reflection coefficient at the transmitter, assuming it is coupled to a perfect 100-Ω transmission line, will therefore be the worse of these two numbers:

$$\Gamma_{MAX\,RO} = \frac{140-100}{140+100} = +0.167 \quad\quad [6.26]$$

$$\Gamma_{MIN\,RO} = \frac{40-100}{40+100} = -0.428 \;\leftarrow\text{worst case} \quad\quad [6.27]$$

The input impedance of the receiver is constrained to the range 90 to 110 Ω. The LVDS specification recommends (but does not require) that this be implemented as a built-in terminator, placed inside the integrated chip package right at the die. From a signal integrity perspective, that would definitely be the best place to put it. Initial LVDS implementations,

---

[58] Variants of LVDS are available in which the drivers can enter a tri-state (high-impedance) mode, and the receivers do not incorporate terminations. These versions are suitable for building multidrop bus structures.

6.13.6 • Impedance Control

however, did not do this. Due in part to the difficulty of fabricating accurate on-chip resistances, early implementations of LVDS left the 100-ohm termination as an external component.

If you have to design with external terminations, use a 100-Ω ±10% external terminating resistor in a low-inductance package (0805 or smaller package) directly attached to the transmission line at the input terminals of the package, with very small pads (for low parasitic capacitance).[59]

The worst-case reflection at the terminator, assuming a line impedance of precisely 100 Ω, will be the worse of these two numbers:

$$\Gamma_{\text{MAX RIN}} = \frac{110-100}{110+100} = +0.047 \qquad [6.28]$$

$$\Gamma_{\text{MIN RIN}} = \frac{90-100}{90+100} = -0.053 \leftarrow \text{worst case} \qquad [6.29]$$

Multiplying together the worst-case transmitter and receiver reflection coefficients, $\Gamma_{\text{MIN RO}}$ [6.27] and $\Gamma_{\text{MIN RIN}}$ [6.29], shows that the amplitude of any residual reflections in the transmission structure (meaning anything that arrives after the initial step edge) can in no case exceed 2.25% of the initial signal amplitude. Therefore, you can expect solid first-incident wave switching performance from this system, assuming a perfect implementation with perfect 100-Ω differential transmission lines.

You may be wondering how far the line impedance may stray from the ideal value of 100-Ω while still guaranteeing first-incident-wave switching. Figure 6.34 reveals the answer. This figure shows the magnitude of the residual reflections remaining after the

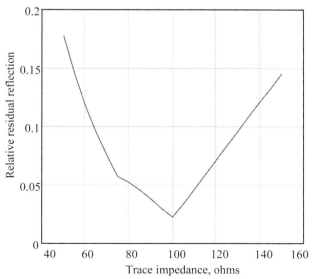

Figure 6.34—Residual reflection after arrival of initial step edge for terminated LVDS logic with worst-case transmitter and receiver impedances, as a function of trace impedance.

arrival of the first-incident waveform. The figure is a compilation of four different constraint lines corresponding to different combinations of worst-case high and low R_o interacting with worst-case high and low R_{in}. At various specific values of trace impedance, one constraint or another takes precedence, which accounts for the segmented appearance of the curve.

The chart indicates that a line impedance of 100Ω ±10 would produce an initial residual reflection no greater than 5% of the incoming step amplitude. A plus-or-minus 20-ohm tolerance would increase the initial residual to no greater than 7%. LVDS logic, because it uses a both-ends style termination, tolerates a fairly wide range of line impedances.

The existence of a significant residual reflection may not by itself endanger the performance of a particular link, depending on the polarity and timing of the arrival of the reflected signal power. Only time-domain simulation can tell.

This figure assumes the transmitter and receiver impedances are purely resistive. Any significant reactances at either the transmitter or receiver will further degrade the performance.

Both-Ends Termination

The both-ends-termination strategy uses a source termination at the transmitter and an end termination at the far end of the line. This is what I call the ax-murderer approach to reflection control. An ax murderer never takes just one whack at a problem. He keeps working and working on the problem until it is completely, finally solved.

That's how both-ends termination works. The main signal, once launched from the transmitter, proceeds toward the far end at full speed. Upon arrival, the end termination metes out one terrible, swift chop that reduces the incoming signal to little more than a small reflected bit of rubble. This surviving reflection retreats post-haste towards the safety of the transmitter, where the source terminator delivers another fatal blow. Only a very small remaining fraction of the original signal amplitude survives this double-chop to crawl back toward the far end a second time.

The second-incident wave amplitude is the product of the original transmitted signal size, the reflection coefficient at the far end, and the reflection coefficient at the near end. In a reasonably well-designed system this amplitude is small enough to simply ignore. That's the big advantage of both-ends-terminated systems, there are hardly any reflections to worry about.

Another advantage of both-ends termination is its high tolerance for obstacles in the middle of the transmission line (like vias). Single-end terminated transmission lines have a lower tolerance for obstacles. In a single-end terminated line, there is always some reflection pattern that can take one bounce off the obstacle, and one additional bounce off some unterminated end, and wind up at the receiver. In a both-ends-terminated design the reflection coefficient at each end of the line is small, so *all* reflection modes are damped, even ones that bounce off of obstacles in the middle of the line.

POINT TO REMEMBER

> ➢ LVDS works best with 100-Ω transmission lines.

6.13.7 Trace Radiation

In the LVDS specification the number ΔV_{OS} defines the degree of balance between the two complementary outputs. It calls for a peak-to-peak common-mode (or even-mode) content in the transmitted signal of no more than 25 mV. Compared to the peak-to-peak signal level on either of the two signal wires (400 mV), that's a relative common-mode content of 6.25%.

The common-mode content limits the degree of attainable radiated field cancellation to a value of –24 dB (=20log(0.0625)). You can easily achieve this amount of cancellation at all frequencies up to 1 GHz by placing the differential traces at any separation of 0.5 mm or less (see Section 6.11.3). Unless you need to save the circuit board space, it is not, in this author's opinion, worth the effort trying to cram LVDS traces closer together than 0.5 mm.

In individual circumstances with particularly well-balanced transmitters it is possible to get better cancellation, but you can't depend on always having parts that beat the specification.

POINT TO REMEMBER

> ➢ You need not struggle to place ordinary differential digital traces any closer than 0.5 mm (0.020 in.) for any EMI purpose.

6.13.8 Risetime

I'm glad to see a specification for the minimum risetime. That's a big help when dealing with all manner of high-speed phenomena, especially the calculation of crosstalk. I offer my sincere thanks to all the standards weanies who voted for this provision.

6.13.9 Input Capacitance

The last of the receiver specifications is the input capacitance. Sadly, this specification is lacking. The closest we get in the standard to addressing the input capacitance is a vague statement that the input capacitance "should not limit the high-frequency, 250-MHz operation of the receiver." That's nice, but it's not a specification. Standards like this leave open the possibility of receivers that meet the spec as written, but don't interoperate.

6.13.10 Skew

The LVDS committee did a lot of work on clock-to-data skew. They carved out an overall skew budget, defining a permissible amount of skew for each signal in an LVDS link. Their budget assumes a link architecture that includes two pcbs, each with a connector and each plugged into some sort of backplane media.

In this architecture the specification pcb designers need to worry about is the pcb skew number of 50 ps. If every data and clock signal in an LVDS link is matched to within this amount of delay, the timing for the link as a whole should work.

Keep in mind that skew accumulates as your signal progresses. If your signal must traverse more than two connectors, the skew budget for each is less than in a simpler system.

If you are using an FR-4 dielectric, the 50-ps delay number gives you an allowance for about 1/4 inch of line length imbalance between any two signals in an LVDS link. This figure is definitely achievable, but don't depend entirely on your autorouter—you need to take a close look at the final artwork to make sure you've stayed under the limit.

The LVDS specification does not make any specific reference to the degree of skew imbalance permitted between the two wires of an individual differential signal. My rule of thumb is that the skew imbalance in any differential pair should be kept to less than 1/10 of the risetime.

POINT TO REMEMBER

> Always double-check your final artwork to make sure you've met the specifications for skew.

6.13.11 Fail-Safe

LVDS components from National Semiconductor include a fail-safe circuit in the receivers. This feature shuts off the output in the event the input is disconnected (zero differential input). This feature is permitted by the standard, but not required, so check carefully if you will be mixing different vendors to make sure they all do it in a compatible fashion.

Figure 6.35 illustrates how fail-safe is implemented in the National LVDS logic family.

The figure depicts the mandated differential thresholds, V_{IH} and V_{IL}, for an LVDS receiver. The manufacturer of the receiver in Figure 6.35 has created a part with better control over the input threshold than the mandated minimum. The actual thresholds, $V_{TH}(+)$ and $V_{TH}(-)$, are specified at ±30 mV. The close tolerance of the actual thresholds is exploited to create the fail-safe feature.

The fail-safe feature is created by forward biasing the inputs. When the input is disconnected from any source (the transmitter is turned off or unplugged), biasing resistors R2 and R3 trickle enough current through the external end-termination resistor R1 to forward-bias the input by 50 mV. This level is above the actual component threshold, so the receiver output stays locked at 1.

When the input is connected to a source with a differential output impedance of 100 ohms, the current from resistors R2 and R3 forward-biases the input by only half as much, or only about 25 mV, shifting $V_{TH}(+)$ and $V_{TH}(-)$ to new worst-case values of +55 mV and −5 mV respectively. These values remain well within the mandated limits of ±100 mV.

Some applications require a greater margin of safety for the fail-safe feature. For example, let's say you are making a twisted-pair communication link. When the transmitter is powered off, you may expect more than 25 mV of differential noise. This can be

6.13.11 • Fail-Safe

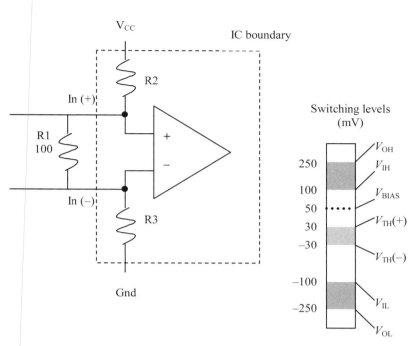

Figure 6.35—Switching levels for National LVDS logic family.

implemented by adding two new resistors, R4 and R5, in parallel with the existing bias resistors R2 and R3, but outboard of the IC package. The new resistors can be sized to enforce an arbitrary amount of offset in the case the transmitter is disconnected. One disadvantage of this technique, if taken to an extreme, is that the fail-safe bias current may be large enough to disturb normal operations.

The circuit in Figure 6.36 fixes this problem. In the event the transmitter is powered off or unplugged, the fail-safe resistors R4 and R5 provide a large amount of bias current. In the event the transmitter is connected and powered on, you can pick values for R6 and R7 in the transmitter that will source an equal and opposite amount of current, canceling the offset. Resistors R2 through R5 appear in parallel with the differential impedance of the termination network and must be taken into account when selecting values for R1 and trace impedance. The same applies for resistors R6 and R7 at the source.

POINT TO REMEMBER

> Fail-safe features are permitted by the LVDS standard, *but not required*.

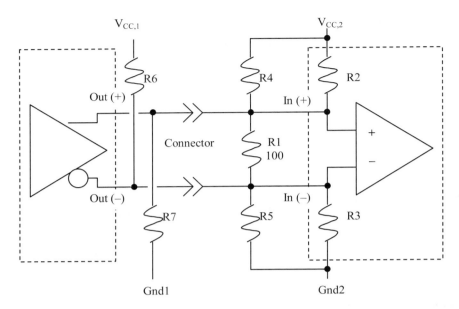

Figure 6.36—Combining external bias resistors at receiver and transmitter creates zero bias during operational mode.

For further study see: www.sigcon.com

CHAPTER 7

GENERIC BUILDING-CABLING STANDARDS

If you are planning any data communications link for use between pieces of equipment inside a building, or between buildings on a single campus, you need to know about generic building-cabling standards [64], [65]. In your application, these standards will

1. Save you money,
2. Shorten your development cycle, and
3. Make it possible for customers to install your product.

Compliance with generic building-cabling standards saves you money, because generic cabling components are *cheap*. These components are cheap because they are standardized and produced in high volumes. For example, there are today more than 100 million Ethernet nodes installed and operating on generic building cabling.[60]

Compliance shortens your development time because it limits your choices. For example, there are 74 different types of U.S. MIL-SPEC 50-ohm coaxial cables. Picking the right one is a nightmare. There are only a handful of popular varieties of unshielded twisted-pair generic building cabling.

Compliance makes it possible for customers to install your product. One hundred percent of MIS managers in Fortune 500 companies equip their buildings with standard, generic building cabling. Other types of cable are not generally available in these buildings and in some cases are *not permitted to enter the building*.

The international specifications for voice and data cabling are a boon not only to designers of LAN and telephone equipment, but to anyone contemplating the use of cables for any digital application, *as long as you stick with the standard parts*. This chapter introduces you to the building-cabling standards and shows you where to go for more information.

---

[60] This chapter is adapted from *Fast Ethernet: Dawn of a New Network* [63].

The world of building cabling has not always been clearly defined. Prior to 1985, each data communication standard mandated its own, unique type of cabling. When you bought new equipment, you put in new cabling. As a result, conduits overflowed with different kinds of coax, twin-ax, tri-ax, multiwire, quad, twisted-pair, and fiber-optic cables. Users accepted this sorry state of affairs by rationalinzing that the cost of new cables was only a small part of the cost of the new computers and communciation services they were buying.

Today, LAN interface cards sell for a fraction of the cost of cabling a new office. Installation of new cabling is no longer acceptable. Customers demand systems that use the existing cables already in the walls. The key buzzword here is *generic cabling*. Any system that connects from room to room, or from building to building, should use generic cabling.

Generic cabling is defined by two standards:

1. In North America: TIA/EIA 568-B.1-2001
2. In other countries: ISO/IEC 11801:2002

These standards specify the preferred cables, preferred methods of installation, and preferred topologies for three popular types of building cabling: 100-Ω balanced cabling, multimode fiber, and single-mode fiber. The scope of both standards includes data and telephone cabling, but not wiring for power, HVAC, or building control systems. Building architects around the world are currently designing new structures according to these standards, which provide much needed wiring-closet space within 100 m of every desktop location.

The best place to purchase cabling standards is from an independent technical-standards warehouse. Both these services take credit cards and ship overnight:

- Global Engineering, *http://global.ihs.com,* (800) 854-7179 (U.S. toll free).
- Techstreet, *http://www.techstreet.com*, (800) 699-9277 (U.S. toll free).

Each standard references a slew of other documents, all of which have been modified many times by various addenda and changes. I can't possibly list all the pieces here because building-cabling standards change so often. When ordering standards, always ask for the latest complete package, including all the detailed cabling specifications for both copper and fiber. Then ask if there are any outstanding drafts or revisions being prepared. The standards organizations that develop these documents are sometimes reluctant to inform you about anything other than complete, finalized standards, but the independent technical-standards warehouses are usually happy to ship copies of the latest drafts-in-progress. In some cases you may have to ask the standards organization how to contact or join the relevant technical working committee in order to obtain copies of drafts-in-progress. I can't overemphasize the importance of developing your own direct link into the standards process. That's how you predict the future.

The TIA/EIA and ISO/IEC standards are similar, but not identical. For example, definitions of critical terms in the two documents are inexplicably different. For the most part, they agree on these major points:

- A star-wiring topology is good,
- The maximum distance from each desktop to the telecommunications room should be no more than 100 m, and
- 100-Ω balanced cabling or fiber are the preferred means of connection.

7.1 • Generic Cabling Architecture

telecommunications room. It may include a short section of cabling with an intermediate transition point designed to facilitate layout of open-air offices using under-carpet cabling or cabling built into cubicle walls. People who install building cabling are expected to properly install and test all permanent links.

Channel—An entire connection, starting at the end of the telecommunications equipment cable and completing at the bitter end of the work area cable. People who use building cabling or who manufacture equipment that interfaces to building cabling care about the performance of the whole channel.

TIA/EIA and ISO/IEC cabling standards differ, but less and less as time passes. The most important point is that the horizontal channels for both standards include support for the most popular cable type—category 5e 100-Ω balanced cabling. Some of the remaining differences are called out in Table 7.1.

Table 7.1—Popular Cables Recognized for Use as Horizontal Cabling

TIA/EIA 568-B.1-2001		ISO/IEC 11801:2002	
Cable Type[1], [2]	Pairs or strands[3]	Cable Type	Pairs or strands
100-Ω category 5, 5e, or 6 balanced cabling	4	100-Ω category 5, 5e, 6, or 7 balanced cabling	2 or 4
62.5/125-μm multimode fiber	2	62.5/125-μm multimode fiber	2
50/125-μm multimode fiber	2	50/125-μm multimode fiber	2
NOTE (1)—Category 3 cables are still recognized by both standards but no longer widely available in the market. Category 5 and 5e cables perform better and cost no more. NOTE (2)—150-Ω STP-A is still recognized by both standards but no longer widely available in the market. It may be dropped from future versions of the generic wiring standards. NOTE (3)—Pairs of 100-Ω balanced cable or strands of fiber.			

POINTS TO REMEMBER

➢ Horizontal cabling is the most widely deployed, highest-volume element of the building-cabling architecture.

➢ New buildings in North America provide two outlets in every work area, with four-pair, 100-Ω UTP, category 5 or better cabling to both outlets.

➢ Backbone cables are mostly a mix of category 5 cables, multimode fiber (62.5-μm or 50-μm), and some single-mode fiber.

> A weird backbone cabling requirement is a sales obstacle to be overcome. A weird horizontal cabling requirement is a wooden stake in the heart of your project.

7.2 SNR Budgeting

After you've gained a little knowledge about cables, you will be sorely tempted to scratch out a rough signal attenuation budget on a napkin, add up the worst-case numbers, and declare that a certain system will or will not function. Be very careful with these calculations. Most communications product designers discover additional deleterious factors late in their design that weren't accounted for in the original SNR budget.

Always calculate the full amount of attenuation and crosstalk you expect from connectors, jumper cables, work area cables, equipment cables, chip packaging, board layout, receiver bandwidth, transmitter risetime, jitter, and anything else you can imagine. Then compute the reflections expected from all the connectors and cable junctures, including an allottment for structural return noise. After you've done all that, add another 2-dB margin for copper-based systems and a 3-dB power margin for fiber. The extra margin will cover you later when you discover other deleterious factors you forgot in the first-pass budget.

Point to Remember

> Don't underestimate the complexity of proper SNR budgeting.

7.3 Glossary of Cabling Terms

Cable	A continuous piece of balanced transmission media.
Cabling	A complete system of data delivery, including cables, jumpers, cords, and connecting hardware.
Category rating system	A rating system for the data-handling capacity of cables, connectors, work area cords, and jumpers. Category 1 is the poorest and 7 is the best.
Category 1, 2, and DIW	These cables are not sanctioned by international building-cabling standards. DIW is the classic 24-AWG twisted-pair phone wire traditionally installed in North America prior to the 1988 specification of category 3 cable. It was the original basis for the category 3 specifications. It should, in theory, meet category 3 specifications, but there are *no guarantees*. If you

7.3 • Cabling Terms

	are not sure, have each link professionally tested for compliance with category 3.
Category 3	Category 3 cabling, while no longer widely available in the market, remains in many buildings. Category 5e performs better and costs no more. TIA/EIA 568-B.2-2001 category 3 cable has a plastic (usually PVC) insulation surrounding four distinct pairs of wires. The wires in each pair twist gently around each other, reducing crosstalk between pairs.[62] The performance of this cable is specified to 16 MHz. Also listed in IEC 61156-2 (2001-09).
Category 4	This cable is no longer sanctioned by international building-cabling standards. At one time the French cable industry manufactured UTP cabling with a nominal characteristic impedance of 120 Ω. This cable interested only a relatively small community of users in France. Technically, it had somewhat lower attenuation than 100-Ω cable. Politically, it was hoped that the standardization of this cable as ISO/IEC 11801 category 4 would open new markets for the French cable industry. As it turns out, the U.S. standard TIA/EIA-568 never adopted category 4. It was retracted in ISO/IEC 11801-2002 in favor of category 5e and higher categories, all operating at 100 Ω.
Category 5	This cable is no longer sanctioned by international building-cabling standards. EIA/TIA 568-A-1995 category 5 specifies a four-pair balanced cable with performance to 100 MHz. Compared to category 3, the attenuation is much improved by using better dielectric materials, and the crosstalk and noise immunity is much improved by specifying tighter twists. Category 5 cabling has been superceded by Category 5e.
Category 5e	TIA/EIA 568-B.2-2001 category 5e supercedes the older category 5. It has the same general signal attenuation performance as category 5, but includes additional specifications for far-end crosstalk (ELFEXT) and a restating and improvement of the specification for characteristic impedance in the form of return loss. Also listed in IEC 61156-5 (2002-03).
Category 6	TIA/EIA 568-B.2-1-2002 category 6 improves marginally the attenuation requirements of category 5, improves substantially

---

[62] With every half twist, the polarity of crosstalk coupling reverses. After many twists, the alternating positive and negative crosstalk voltages tend to cancel.

	the noise characteristics, and raises the limit of specified performance to 250 MHz. Also listed in IEC 61156-5 (2002-03).
Category 7	IEC 61156-5 (2002-03) category 7 improves marginally the attenuation requirements of category 6, improves substantially the noise characteristics, and raises the limit of specified performance yet again to 600 MHz.
Class C, D, E, and F	ISO/IEC 11801-2002 specifies six classes of balanced cabling performance. Cabling classes C, D, and E are based on cable standards corresponding to TIA/EIA 568-B categories 3, 5e, and 6 respectively. Cabling class F is based on category 7 cables specified in IEC 61156-5.
Number of pairs	TIA/EIA standards stipulate four pairs at each work area outlet. ISO/IEC standards let users squeak by with just two. Four is better, because it leaves the user more options. In addition, the use of all four pairs avoids the problem of having to select which two pairs to equip at each four-pair wall outlet (Ethernet, FDDI, and Token Ring all use different combinations of two pairs). Lastly, new services like Gigabit Ethernet require all four pairs.
UTP	TIA/EIA name for 100-Ω balanced cables of categories 3, 5e, or 6 with no overall screen. This is the normal form of cabling used in North America.
ScTP	TIA/EIA name for 100-Ω balanced cables of categories 3, 5e, or 6 with an overall screen.
S/UTP, F/UTP, SF/UTP, S/FTP, F/FTP, SF/FTP	ISO/IEC nomenclature for various forms of shielded twisted-pair cables. The first part of the name (S, F, or SF) refers to the form of overall screen covering the entire cable. S = braid screen, F = foil screen, or SF = both. No marking indicates no overall screen. After the slash, the second part of the name (UTP or FTP) refers to the form of element screen used for each balanced element (a pair or a quad). UTP = unscreened, or FTP = foil screened.
150-Ω STP-A	IBM designed this cable in the early 1980s. Physically, the cable is massive, unwieldy, expensive, and difficult to terminate. Electrically, it is a very fine transmission medium. Some LAN architectures were first made available on 150-Ω STP-A because it's easy to get a transceiver up and

running on this incredible cable. Mainstream LAN technology, however, uses 100-Ω balanced cabling.

Although 150-Ω STP-A is still recognized it is no longer widely available in the market, and may be dropped from future versions of the generic wiring standards.

7.4 PREFERRED CABLE COMBINATIONS

TIA/EIA 568-B.1 and **ISO/IEC 11801** do more than simply define acceptable types of horizontal cabling. They define in what combinations such cables should be used. According to the standards, *two horizontal links serve each work area*.

The work area in Figure 7.3 shows a pair of links in typical usage. One serves the phone and the other serves the data device.

Table 7.2 shows cable combinations permitted by both standards. Here I have listed only the arrangements preferred by TIA/EIA 568-B.1-2001, not all the compliant possibilities.

ISO/IEC specifications are somewhat more permissive than TIA/EIA. Permissive standards proliferate unnecessary cable choices. For example, ISO/IEC permits substitution of category 3, 5e, 6, or 7 cables and substitution of practically any type of optical fiber for the items listed in Table 7.2.

Based on my belief that fiber-to-the-desk won't happen any time soon, I tell users to install category 5e four-pair 100-Ω balanced cabling to both outlets.

Table 7.2—Preferred Horizontal Cable Combinations

TIA/EIA 568-B.1-2001
First outlet: • Four-pair, 100-Ω UTP, category 5e or better
Second outlet, any one of   • Four-pair, 100-Ω UTP, category 5e or better   • Two-strand, 62.5/125 μm fiber   • Two-strand, 50/125 μm fiber

7.5 FAQ: BUILDING-CABLING PRACTICES

Question: How are most new office buildings cabled?

Answer: In North America each work area has at least two outlets. All outlets are cabled with category 5e four-pair 100-Ω UTP. Most users install a mix of backbone cables, including category 5e cables for voice and local data services plus multimode and single mode fiber.

Question: Do customers ever mix different categories of cabling in the same building?

Answer: Confucian philosophers say, "Everything that can happen has happened." That doesn't make it a good idea. When different categories are mixed in the same link, the guaranteed performance supposedly reverts to the lowest category. For example, if you install category 3 connectors (guaranteed to 16 MHz) with category 5e cabling (guaranteed to 100 MHz), the performance of the whole mess is only guaranteed to 16 MHz. If that bandwidth is adequate, it doesn't hurt anything. There are certain combinations, however, that don't mix. Category 6 and 7 connectors contain certain parasitic-coupling compensation features that may not interoperate properly with their lower-category counterparts. Plugging a category 6 jack into a category 5e receptacle, for instance, may degrade performance below the category 5e level.

When buildings are purposely wired with different categories of cabling to different offices, your installation people probably won't keep good enough records to remember which is which. It's best to keep the whole building at the same level to the greatest extent practicable.

Question: The customer doesn't have cable records. How can I find out what cable is installed?

Answer: Ask an installer to test the cable to see if it complies with any recognized category of cabling. A good installer should have equipment to verify compliance. If he or she does not have the equipment, find another installer.

Question: Why bother supporting older categories of cabling?

Answer: In certain ranges of speed (20 to 100 Mb/s, for example), the design of a category 3 transceiver can be challenging. Restricting your transceiver to operation only on cabling with category 5 or better performance would definitely simplify such a design. Regrettably, requiring category 5 or better cabling may also lock you out of that precentage of customer sites that are unwilling to upgrade their existing cabling plant from older category 3 cabling to the new standards. Your choice is therefore one of design difficulty versus the size of your total available market.

At present, customers installing new cabling choose category 5e (or higher). The proportion of significant users that retain category 3 cabling is dwindling; however, the practical life expectancy of building data cabling is 5 to 20 years, so it could take quite a while to change over the entire installed base.

Conventional wisdom among system developers holds that many, but not all, users who need bandwidth of 100 Mb/s or greater understand the need for category 5e cabling and will pay to have it installed. If your application demands the extra bandwidth, go ahead and specify good cabling. On the other hand, if your application needs to cover 99% of the installed base, and high bandwidth is not your first consideration, you are better off sticking with the older category 3.

The same considerations will apply at each successive transition as you work your way up from DIW through categories 3, 5, 5e, 6, and the new IEC category 7.

7.6 CROSSOVER WIRING

Many UTP star-wired LAN standards, like 10BASE-T and 100BASE-T, designate distinct pairs of wires for unidirectional transmission and reception. At the hub and at the client, these pairs occupy specific pin positions on the data connectors. In the preferred arrangement, the hub and client have complementary pin assignments so that the wiring may be accomplished straight through, from end to end, connecting pin 1 on the client to pin 1 on the hub, pin 2 to pin 2, and so forth. This is what you should do for copper. For fiber, you label the TX and RX connectors on your equipment and expect the user to cross them over for you.

The preferred wiring arrangement is illustrated in Figure 7.4. In this figure, the client has a normal pin assignment, and the hub a complementary one. The wires run straight through. Inside the hub, adjacent to the imaginary transmitter and receiver, is where you are supposed to implement the crossover. A wiring crossover is an essential function in every link. It must reside either in the client, the hub, or the wiring.[63] The preferred location for a crossover is in the hub.

The use of straight-through wiring simplifies installation and maintenance considerably. It eliminates any concern about whether a link might have a crossover at one end, the other, or both.

Certain exception conditions exist, like the connection between two hub ports. In this

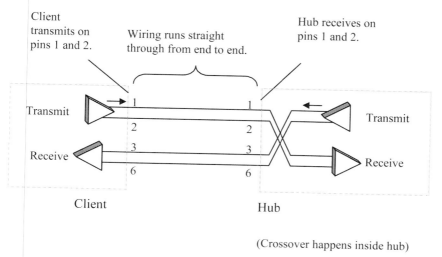

Figure 7.4—Example of straight-through wiring with internal crossover in hub.

---

[63] Technically, the requirement is for an odd number of crossovers.

case both pieces of equipment may already include an internal crossover. The installer is expected to provide an external crossover (that is, a crossover explicitly implemented in the wiring).

> *When necessary, an external crossover should be implemented in a short, clearly visible section of cabling and boldly labeled.*

Examples of crossover connections appropriate for Fast Ethernet (100BASE-TX) appear in Table 7.3 and Table 7.4.

Table 7.3—Wiring Crossover for 100-Ω Balanced Cables

	Pair 3→2		Pair 2→3		Pair 1→4 [2]		Pair 4→1 [2]	
From pin: [1]	1	2	3	6	4	5	7	8
To pin: [1]	3	6	1	2	7	8	4	5
NOTE (1)—Not used by 100BASE-TX.								
NOTE (2)—RJ-45 pin numbers.								

Table 7.4—Wiring Crossover for 150-Ω STP-A; References Are to DB-9 Pin Numbers

	Pair 1→2		Pair 2→1		Other Pins Not Used
From pin: [1]	5	9	1	6	2,3,4,7,8
To pin: [1]	1	6	5	9	
NOTE (1)—DB-9 pin numbers.					

POINTS TO REMEMBER

➢ Multi-pair building cables should be installed straight-through with no crossing of the pairs.

➢ When necessary, an external crossover should be implemented in a short, clearly visible section of cabling and boldly labeled.

7.7 PLENUM-RATED CABLES

This issue has to do with the type of air-conditioning system in your building. All modern central heating and air-conditioning systems pump conditioned air through enclosed ducts. The conditioned air flows through the building, where users receive the benefits of its coolness (or warmth), *and then it returns to the main blower*. At the main blower, most of the returning air is recirculated back through the fans. Only a little bit of fresh outside air is mixed into the return flow. This saves a lot of energy compared to continually cooling 100% fresh outside air. The plenum issue has to do with *how* the used air returns to the fans. Building architects have the option of letting air return to the main blower through the natural open space above the false ceiling in a multistory office building (or the building attic in a one-story structure). If the false ceiling (or attic) space is used for returning air, you have a *plenum-return system*. The false ceiling (or attic) space is called a *plenum*. The alternative is to return the air through separate, dedicated return ducts. If you have separate return ducts, the plenum is nothing but a big dead airspace.

When a fire happens in the plenum of a plenum-return building, smoke from flammable materials in the plenum are sucked directly into the main blower, which distributes the deadly smoke instantly throughout the building. As a safety measure, plenum-return systems are therefore required to use nonflammable materials in the plenum. The insulation used in most old category 3 cables is polyvinyl-chloride (PVC), which emits dangerous gases when burned. Such cables are not permitted in plenum-return air systems. Plenum-rated cables must be made of something other than PVC. Unfortunately, the materials used to make plenum-rated cables are heavy, stiff, and somewhat more expensive than PVC.

POINT TO REMEMBER

➢ The materials used to make fire-resistant plenum-rated cables are heavy, stiff, and somewhat more expensive than PVC.

7.8 LAYING CABLES IN AN UNCOOLED ATTIC SPACE

Even if the attic space is not used for air return, you may still need plenum-rated cables. When it gets hot in the summer, all the cables laid in the attic have to work at elevated temperatures. Unfortunately, the old category 3 PVC-insulated cables suffer from excessive signal attenuation at elevated temperatures. Do not use category 3 PVC cable at temperatures greater than 40 °C or 104 °F, a temperature easily attained in an enclosed attic. Instead, specify a less temperature-dependent cable, such as a FEP, PTFE, or PFA plenum-rated cable, or any category 5e or better cable.

With category 5e or better cables, the performance doesn't degrade as severely as with category 3 PVC, but the cable attenuation must still be de-rated to account for the increased resistance of the copper conductors at elevated temperature. See further information in Section 8.6, "Category-3 UTP at Elevated Temperature."

POINT TO REMEMBER

➢ Cable performance must be de-rated to account for operation at the elevated temperatures commonly found in building attics.

7.9 FAQ: OLDER CABLE TYPES

Question: What are TIA/EIA T568B connectors?

Answer: This question refers to two authorized ways of connecting twisted pairs to an RJ-45 style jack or plug. These methods are designated T568A (standard) and T568B (optional). These designations are not to be confused with the generic cabling standards TIA/EIA 568-A (older) and TIA/EIA 568-B (new and improved). The two connection methods swap the positions of the orange and green wires (pairs 2 and 3) on the connector.

According to the standard, pin assignments should be made according to the T568A style, or optionally the T568B style "if necessary to accommodate certain 8-pin cabling systems" (and if you want to permanently confuse everyone who works on your wiring). The U.S. Federal Government publication "NCS, TRP 109-1977" recognizes the only style T568A.

The technical issue here is that the worst crosstalk on a category 3 style T568A uncompensated connector occurs between pairs 1 and 2. Style T568B moves the worst-case crosstalk to pairs 1 and 3. Depending on what services you plan for which pair, one or the other style could be best.

Upgrading to cabling of category 5e or better renders the issue moot.

Question: My European customers have unusual old cables installed in their buildings. How do I support them?

Answer: If you can't work on their cable, then do them a big favor by insisting that they install ISO/IEC 11801-compliant cabling of at least class D (or whatever you need) to every outlet. The new cables will likely support all their older applications.

Question: Didn't a lot of buildings used to have 25-pair cables near the work area?

Answer: Yes. Such cables are relics left over from the era of mechanical key-system telephones. **TIA/EIA 568-B.1**, Annex C, states regarding the use of 25-pair horizontal cabling, "Although such an arrangement may provide installation efficiencies, it should not be used for the general case."

The sanctioned mode of installation runs individual 4-pair cables from each work area faceplate all the way to a switch room located within roughly 100 m. The use of individual cables from each faceplate provides terrific crosstalk isolation between devices connected to different work areas.

Question: My system runs on only two twisted pairs. Can I let my customers operate two systems through the same four-pair cable?

Answer: You will need to carefully check your crosstalk budget before supporting this feature. Even if you can support two systems in a single 4-pair cable, it likely won't increase your market. That's because modern buildings are cabled with a separate cable to each data faceplate. At the RJ-45 data faceplate it is not always easy to split out two of the pairs to run to a separate jack. Also, users don't expect to be able to put two systems on one cable, because the feature is not guaranteed by most LAN specifications.

Question: What is the difference between two-pair UTP and quad cable?

7.9 • Older Cable Types

Answer: The difference is in the twist. Two-pair UTP incorporates two pairs of wire, each of which twists round and round its mate, but not around the other pair. The low-crosstalk properties of this cable derive from the way alternating positive and negative crosstalk effects cancel as the wires twist.

High-quality quad cable, or star quad, is found mostly in Europe. It holds all four wires in a square configuration and then twists the whole bunch. The wires of each pair stay on opposite corners of the square at all times (Figure 7.5). The low-crosstalk properties of this cable derive from its exact geometrical symmetry. Quad cables are generally available only in two-pair configurations.

Persons familiar with telephone cable might remember the old telephone-grade quad cable, which also had four wires but no controlled symmetry. In the old telephone-grade quad cable, the wires of each pair were not twisted, and could flop into any position. With no controlled symmetry, such cabling has terrible crosstalk. Do not use it for data.

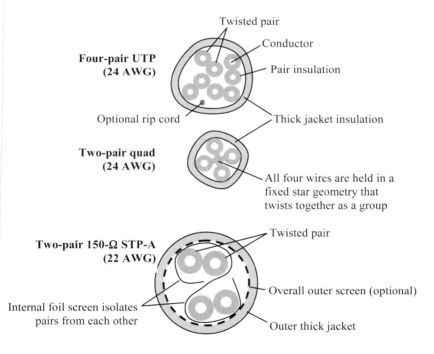

Figure 7.5—Types of twisted-pair cable (EIA definitions).

For further study see: www.sigcon.com

CHAPTER 8

100-OHM BALANCED TWISTED-PAIR CABLING

This Chapter uses the TIA/EIA nomenclature for twisted-pair cabling. TIA/EIA-568-B [70], [71], [72] calls out both unshielded and screened versions of 100-Ω balanced twisted-pair cabling.

Unshielded 100-Ω twisted-pair cabling (UTP) is composed of a number of balanced, twisted pairs. Each pair of conductors is wound tightly together. As defined for horizontal building wiring (see Chapter 7), UTP contains four pairs, for a total count of four or eight conductors respectively. Typical construction for four-pair UTP appears in Figure 8.1.

UTP is cheap, ubiquitous, and easy to handle. A 1,000-foot roll of four-pair UTP can be carried in one hand up a ladder by a lone technician. Judiciously hurling the reel through the space above the false-ceiling tiles in a modern high-rise, one technician can quickly

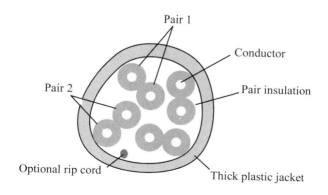

Figure 8.1—Construction of typical four-pair unshielded twisted-pair (UTP) LAN cable.

distribute UTP cabling among large number of offices.[64] The conductor is color-coded according to a well-recognized standard system. It may include a tough rip cord inside every jacket. Installation people yank the rip cord to quickly strip back the outer cable jacket prior to flaying out and terminating the individual conductors. An experienced technician can terminate four pairs of UTP in only a few seconds. These physical features contribute to the enormous popularity of UTP.

Screened 100-Ω twisted-pair cabling (ScTP) incorporates an electrically continuous shield just under the thick plastic jacket, enveloping the core of the cable. According to the standard, the shield consists of "plastic and metal laminated tape with one or more longitudinal, helical, or braided noninsulated solid tin-coated copper conductors(s) [drain wire(s)] of 26 AWG equivalent or larger that are in contact with the metal side of the tape." ScTP cables have the same interior construction as UTP and meet the same electrical specifications for categories 3, 5e, and 6. Since the signal propagation issues are the same in both cables, I shall refer subsequently to 100-Ω balanced twisted-pair cabling as UTP.

Operating in a parallel but slightly different universe, ISO/IEC standards for generic building wiring [73], [74] also specify category 3 horizontal wiring [75], [76], [77] and categories 5e and 6 horizontal wiring [78], [79] with performance essentially the same as that called out in TIA/EIA-568-B.2. In addition, [78] and [79] define another version of horizontal wiring called category 7 with performance slightly superior to category 6 and parameters guaranteed to 600 MHz. Although the North American TIA/EIA committees and the international ISO/IEC committees have worked hard to harmonize their standards, many differences still apply.

Beware that cable definitions are in constant flux. Cable manufacturers have determined it is to their advantage to promulgate new standards as often as possible, thus driving up customer demand for building rewiring. This works counter to the interests of communications systems vendors, who seek a consistent and reliable market for their products. The natural tension between these two groups produced this amusing statement in the forward to IEC 61156-5 [78], released in March 2002: "The committee has decided that the contents of this publication will remain unchanged until 2004." When first penned, that promise was good for only 21 months. It hints that some members of the committee would like the standard to change even sooner. If you expect to keep pace with the latest developments in the cabling industry, you must read all the latest standards, read all the amendments, and then check with the relevant committees to obtain copies of upcoming drafts.

POINT TO REMEMBER

➢ Cabling standards proliferate faster than bunnies.

---

[64] The wire unrolls as it flies thorough the false-ceiling space. I've done this, but if you break any ceiling tiles, please don't blame me.

8.1 UTP Signal Propagation

Electrical specifications for category 3, 5e, and 6 UTP, plus 150-Ω STP-A appear in TIA/EIA-568-B and are summarized in Table 8.1. Please note that these are cable specifications, not overall channel specifications. Channel specifications (which include connectors, jumper cables, work area cables, equipment cables, and other factors) are always worse (see note in Section 7.2, "SNR Budgeting").

Compared to category 3 cabling, categories 5e and 6 higher have progressively tighter twists and better plastic insulation with less dielectric loss at high frequencies. The resulting cables pick up less noise and have a superior frequency response (Table 8.1). Category 3 cabling is specified only up to 16 MHz, and therefore useful only up to about 25 Mbaud. Category 5e, 6, and 150-Ω STP-A carry the attenuation specifications up to higher frequencies.

Category 3 cabling, while no longer widely available in the market, is included because many buildings still have this wire installed. Category 5e is a better choice for new buildings because it performs better and costs no more.

Compared to any of the UTP categories, 150-Ω STP-A specifies larger conductors (22

Table 8.1—TIA/EIA-568-B UTP and 150-Ω STP-A Electrical Specifications

Item	Cat-3	Cat-5e	Cat-6	150-Ω STP-A	Notes
$|H(\omega)|$ (dB)					Maximum allowable attenuation in dB per 100 m (328 ft) at 20 °C
0.064 MHz	0.9	—	—	—	
0.256 MHz	1.3	—	—	—	
0.512 MHz	1.8	—	—	—	
0.772 MHz	2.2	1.8	1.8	—	
1.0 MHz	2.6	2.0	2.0	—	
4.0 MHz	5.6	4.1	3.8	2.2	
8.0 MHz	8.5	5.8	5.3	3.1	
10.0 MHz	9.7	6.5	6.0	3.6	
16.0 MHz	13.1	8.2	7.6	4.4	
20.0 MHz	—	9.3	8.5	4.9	
25.0 MHz	—	10.4	9.5	6.2	
31.25 MHz	—	11.7	10.7	6.9	
62.5 MHz	—	17.0	15.4	9.8	
100.0 MHz	—	22.0	19.8	12.3	
200 MHz	—	—	29.0	—	
250 MHz	—	—	32.8	—	
300.0 MHz	—	—	—	21.4	
Z_C Min	85	90	90	135	Characteristic impedance (ohms) as measured on a 100 m length
Max	115	110[1]	110[1]	165	
$t_p \triangleq 1/v_0$ Max, at 10 MHz	5.45	5.45	5.45	5.45	Permitted group delay (ns/m) varies slightly as a function of frequency.
NOTE (1)—As implied by return loss specification.					

AWG versus 24 AWG) and higher characteristic impedance (150-Ω versus 100-Ω). As a result, the 150-Ω STP-A bandwidth is much higher than either UTP category, and its performance is specified up to a higher frequency. Balancing these advantages are some severe practical disadvantages, discussed in Chapter 9. Although at the time of writing 150-Ω STP-A is recognized by TIA/EIA 568-B.2, it is not recommended for new cabling installations and may be dropped from future versions of that standard.

POINT TO REMEMBER

> ➢ Compared to category 3 cabling, categories 5e and 6 higher have progressively tighter twists and better plastic insulation with less dielectric loss at high frequencies. The resulting cables pick up less noise and have a superior frequency response.

8.1.1 UTP Modeling

What follows is a discussion of how to compute the model parameters for UTP cables (see Section 3.1, "Signal Propagation Model"). Since 150-Ω STP-A is so similar to UTP, this section also develops the 150-Ω STP-A parameters at the same time.

For 24-gauge or similar cabling, parameter ω_0 should be set to 10 MHz ($\omega_0 = 2\pi \cdot 10^7$ rad/s). Parameters Z_0 and v_0 appear directly in the specifications for UTP and 150-Ω STP-A cables.[65]

R_{DC} requires that you know the conductivity of annealed copper at 20 °C. The standards specify worst-case cable behavior at 20 °C. If your application operates at a significantly different temperature, increase the resistance by 0.39% per degree Centigrade. The formula for the DC resistance, per meter, of a UTP cable counts the resistance of *both* conductors (outbound and return). Hence, the factor of 2 appears on the left-hand side of this expression.

DC resistance of twisted pair $\qquad R_{DC} = 2\left(\dfrac{1}{\sigma \pi (d/2)^2}\right)$ Ω/m [8.1]

where σ is the conductivity of the conductors, S/m, and
d is the conductor diameter, m.

Sometimes the total DC resistance of a cable (outbound plus return path included) is specified directly on the datasheet, in which case you may use that value instead of calculating it from the conductor diameter. Datasheet values must always be used for cables with complex conductor construction, such as copper-coated steel conductors, stranded conductors, or tin-plated conductors.

---

[65] TIA/EIA 568-B presents a formula for the worst-case cable delay at 10 MHz from which you may calculate v_0 = 0.6116c, which is rounded down to 0.6c for all categories in Table 8.3.

8.1.1 • UTP Modeling

Example: Calculation of DC Resistance for Category 3 UTP

For annealed copper as prepared for ordinary cables at 20 °C, $\sigma = 5.80 \cdot 10^7$ S/m.
For AWG 24 copper conductors, the diameter $d = 0.508$ mm.

$$R_{DC} = 2 \left(\frac{1}{(5.80 \cdot 10^7 \text{ S/m}) \pi \left(\frac{0.000508 \text{ m}}{2} \right)^2} \right) = 0.1701 \ \Omega/\text{m}$$

The worst-case DC resistance for this grade of cabling is specified as 0.1876 Ω/m, suggesting that conductors with a diameter 5% smaller than nominal size AWG 24 are permitted under the standard.

To compute the skin-effect resistance R_0, you first need to compute the skin depth at frequency ω_0. The following equation shows the calculation of skin depth as it ordinarily appears [69].

$$\delta = \sqrt{\frac{2}{\omega \mu \sigma}} \qquad [8.2]$$

where δ is the skin depth, in meters,

ω is the frequency at which the skin depth is specified, rad/sec,

μ is the magnetic permeability of the conductor in H/m, and

σ is the conductivity of the conductor, S/m.

Next use the skin depth to calculate the AC resistance of one of the signal conductors at frequency ω_0. If the current distribution were uniform around the periphery of each conductor, the resistance would simply be $1/(\pi d \delta(\omega_0) \sigma)$. Unfortunately, life is not so simple.

The proximity effect distorts the pattern of current flow on the surface of the conductors, increasing the effective AC resistance by a fixed constant k_p dependent on the conductor geometry (see Section 2.10.1, "Proximity Factor"). The proximity factors listed in Table 8.2 take into account the existence of two conductors (thus the values are each bigger than 2).

$$R_0 = k_p / (\pi d \delta(\omega_0) \sigma) \qquad [8.3]$$

Expand $\delta(\omega_0)$ according to [8.2].

$$R_0 = \frac{k_p}{(\pi d \sigma)} \sqrt{\frac{\omega_0 \mu \sigma}{2}} = \frac{k_p}{\pi d} \sqrt{\frac{\omega_0 \mu}{2 \sigma}} \quad \Omega/\text{m} \qquad [8.4]$$

Table 8.2—Proximity Factors for Twisted-Pair Cabling

Cable type	Proximity factor k_p
cat-3	2.3
cat-5e	2.3
cat-6	2.3
150-Ω STP-A	2.06

Example: Calculation of Nominal Skin-Effect Resistance for Category-3 UTP

Assume $\omega_0 = 2\pi \cdot 10^7$ rad/sec (10^7 Hz).

Copper being a nonmagnetic conductor, the magnetic permeability $\mu = 4\pi \cdot 10^{-7}$ H/m.
For annealed copper as prepared for ordinary cables at 20°C, $\sigma = 5.80 \cdot 10^7$ S/m.
The proximity factor $k_p = 2.3$ (Table 8.2).
For AWG 24 copper conductors the nominal diameter $d = 0.508$ mm.

$$R_0 = \frac{2.3}{\pi(0.000508 \text{ m})} \sqrt{\frac{(2\pi \cdot 10^7 \text{ rad/s})(4\pi \cdot 10^{-7} \text{ H/m})}{2(5.80 \cdot 10^7 \text{ S/m})}} = 1.189 \ \Omega/\text{m}$$

The best-fit value in Table 8.3 for the worst-case specifications indicates a skin-effect resistance of 1.452 Ω/m, suggesting that the worst-case specifications permit a combination of a diameter smaller than nominal size AWG 24, plating or other impurities in the surface layer of the copper, surface roughness effects, and measurement error combining to increase the nominal skin-effect resistance by 22%.

Note about dielectric loss: For PVC used below 40 °C, assume a nominal effective dielectric loss of $\theta_0 = 0.02$ (Table 8.3 shows a best-fit, worst-case value of 0.01578 for category 3 UTP). Above 40 °C the PVC dielectric material typically used in category 3 cables exhibits a strong temperature dependence. Do not use PVC cable at temperatures greater than 40 °C, or 104 °F, a temperature easily attained in an enclosed attic. If you need to work at elevated temperatures, specify a less temperature-dependent cable, such as an FEP, PTFE, or PFA plenum-rated cable (see Section 8.6 "Category-3 UTP at Elevated Temperature").

8.1.2 Adapting the Metallic-Transmission Model

As you might expect, the many possible combinations of surface plating, types of shielding, and dielectric make it difficult to accurately predict the performance of all twisted-pair cables from the basic information provided on a datasheet. Fortunately, the copper-based cable propagation model is highly adaptable. By only adjusting two parameters, you can easily produce a model that mimics the performance of just about any twisted-pair cable.

To adapt a model, start with [8.1] and [8.4] for DC resistance R_{DC} and AC resistance R_0. That usually gets you pretty close to the worst-case frequency-domain specifications in the vicinity of ω_0. Then tweak R_0, which scales the attenuation at all frequencies above the onset of the skin effect, and θ_0, which increases the curvature of the attenuation graph at the very high end, until you get the best match to the worst-case specifications. That's how all the models in Table 8.3 were built. The "best match" criteria was a least-squares fit of the

8.1.2 • Metallic-Transmission Model

attenuation magnitude in dB. The models optimized for performance all the way down to DC match the worst-case specification to within 0.16 dB per 100 m at all specified frequencies.

The models optimized only over the frequency range above 1 MHz match the worst-case specification a little better, to within 0.03 dB at all specified frequencies. Modeling always works that way—the more limited the domain of a model, the better it can match any arbitrary specification. The main difference between the copper-propagation model and the TIA/EIA model (other than that the copper-propagation model includes the correct phase information and is therefore useful for time-domain simulation) has to do with the treatment of the transition from the skin-effect mode to the LC (constant-loss) mode. TIA/EIA models accomplish this transition over the limited range of frequencies near 1 MHz by incorporating a frequency-response term proportional to $1/\sqrt{f}$. The TIA/EIA approach, while enjoying some theoretical basis as an approximation in the range near 1 MHz, is unrealizable in a physical cable and poorly mimics cable behavior below 1 MHz.

To force the metallic propagation model to conform to the unnatural behavior of the TIA/EIA model near 1 MHz, parameter R_{DC} has been increased and R_0 lowered. This adjustment better matches the worst-case TIA/EIA specifications at frequencies above 1 MHz by sacrificing the correct value of DC resistance.

Figure 8.2 compares the signal-propagation model of Chapter 3 to the TIA/EIA-568-B specifications.

A perfect match is impossible to achieve. Standards often incorporate mysterious wobbles and bumps in their specifications, bumps that can't be matched with any rational modeling technique. No doubt these bumps are vestigial results of long-forgotten standards battles over compatibility with products that existed at the time. Also, remember that specifications represent a worst-case conglomeration of many effects. Each individual effect may cause the performance of a cable in the field to touch the specification limit at one particular frequency, but it may be impossible to construct a single cable that hugs the limit

Table 8.3—Worst-Case Transmission Line Parameters for TIA/EIA-568-B Cables

Cable type	Z_0 Ω	v_0/c	R_{DC} Ω/m	R_0 Ω/m	θ_0 rad	ω_0 MHz	Useful range MHz	Max. error dB/100 m
Models optimized to include DC performance									
cat-3	85	0.6	.1876	1.452	.01578	10	0–16	.012
cat-5e	85	0.6	.1876	1.253	.001153	10	0–100	.032
cat-6	90	0.6	.1876	1.257	.000447	10	0–250	.161
150-Ω STP-A	135	0.6	.1142	1.134	.000658	10	0–300	.164
Models optimized only from 1 MHz up									
cat-6	90	0.6	.2902	1.208	.000965	10	1–250	.028
150-Ω STP-A	135	0.6	.2855	1.061	.001136	10	1–300	.013

precisely at all frequencies. Lastly, keep in mind that cable installers rarely if ever calibrate their test equipment, and some of the tests aren't that accurate anyway. It is quite possible that you will see cables in the field certified for operation that don't quite meet the specifications.

Such is life. To respond to these problems, you should construct a rigorous cable

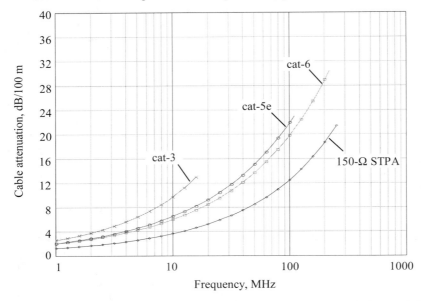

Figure 8.2—Comparison of attenuation computed by metallic transmission model (solid lines) to EIA-568B specifications (points).

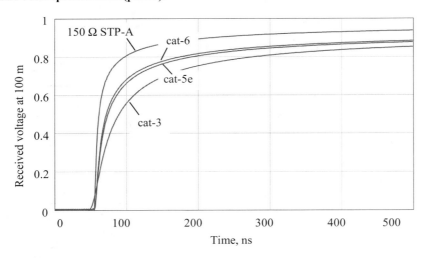

Figure 8.3—Comparison of step-response waveforms for 100 m of cat-3, cat-5e, cat-6, and 150-Ω STP-A cabling (computed).

model for system testing that either adds another 2 dB of fixed, flat loss to the standard or extends the simulated maximum cable length by another 10% to 20%. Transceiver designs that pass such testing will likely work well in the field.

Calculated worst-case (slowest) system step-response waveforms for 100 m of category 3, category 5e, category 6, and 150-Ω STP-A cabling appear in Figure 8.3. Longer or shorter cables scale the step response generally in proportion to the square of cable length. The step-response waveforms have been shifted horizontally to fit on the display.

POINTS TO REMEMBER

> ➤ The many possible combinations of surface plating, types of shielding, and dielectric make it difficult to accurately predict the performance of all twisted-pair cables from the basic information provided on a datasheet.

> ➤ The cable model you use for system simulation should either add another 2 dB of fixed, flat loss to the datasheet attenuation or extend the simulated maximum cable length by another 10% to 20%.

8.2 UTP TRANSMISSION EXAMPLE: 10BASE-T

The 10BASE-T system operates at 10 Mbaud on category 3 cable (Figure 8.4). It uses Manchester coding, which generates either one or two signal transitions per data bit. Let's do a set of system simulations of 10BASE-T signaling using cable lengths from 10 to 100 meters. These simulations highlight the effects of the cable transfer function. They use the worst-case category 3 cable model (at 20 °C), but no other system imperfections. The

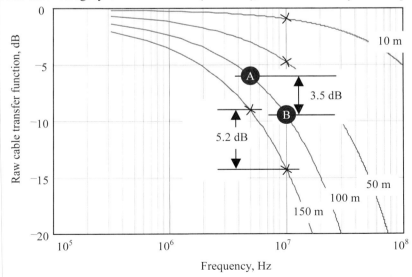

Figure 8.4—Raw transfer function of worst-case category 3 UTP at various lengths.

transmitted rise/fall time is 30 ns.

Figure 8.4 shows the raw cable frequency response for a variety of lengths. On each plot, I've marked the attenuation for each length at a frequency of 10 MHz, which corresponds to the maximum alternation rate for the 10 Mb/s Manchester-coded signal.

The 10-m cable gives an attenuation at 10 MHz of 1 dB. This produces in Figure 8.5 a near-perfect eye pattern. Note that I've set the transmitted signal rise/fall time to 30 ns, a value commonly used in the industry. This setting reduces the spectral content above 30 MHz, which helps to meet FCC and EN emissions regulations.

The 50-m cable gives an attenuation at 10 MHz of roughly 5 dB. The eye still looks okay. This is a good general lesson in cable equalization: 5 dB of attenuation at the maximum alternation rate doesn't distort a *Manchester-coded* signal very much.

The 100-m cable gives an attenuation at 10 MHz of almost 10 dB. The eye here shows some asymmetry, but still looks easily recoverable. Looking back at the attenuation curve, let's compare the attenuation at 10 MHz (the maximum alternation rate) and 5 MHz (the minimum alternation rate for Manchester coding). The difference is about 3.5 dB. This difference corresponds *roughly* to the difference in heights between the minimum-alternation-rate (maximum run length) eye height (A) and the maximum-alternation-rate amplitude (B) in Figure 8.5. This is about as far as you want to take a nonequalized

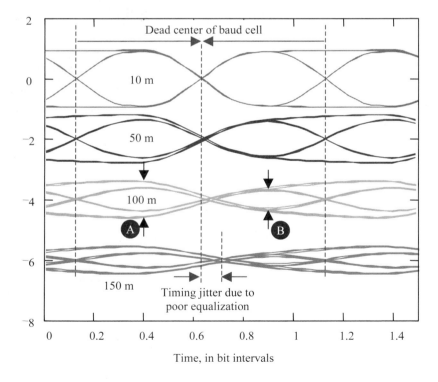

Figure 8.5—Eye patterns for unequalized Manchester coding. Data rate = 10 Mb/s, transmit rise/fall = 30 ns, cable = worst-case category 3 UTP at 20 °C.

8.2 • Example: 10BASE-T

Manchester system. For any coding system, when the attenuation difference between the maximum alternation frequency and the minimum alternation frequency[66] exceeds 3.5 dB, the eye pattern suffers.

The 150-m cable goes too far. At this distance the 10-MHz attenuation is roughly 15 dB, and the difference between the 5 and 10 MHz attenuation figures has grown to 5.2 dB. The zero crossings of the received data show noticeable amounts of timing jitter, and the noise margin is severely compromised.

From Figure 8.5 you can see that the received data amplitude at 150 meters is a function of past history. That is, the second 1 of a 1,1 pattern comes in larger than a 1 that follows a 0. This larger received amplitude then takes longer to come down on the next stroke, causing timing jitter. If the received amplitude could be made independent of data pattern history, the timing jitter would be improved. This is the principle of equalization:

> *The timing jitter is improved when all received amplitudes are independent of past history.*

To equalize the received amplitudes, we could ask the transmitter to please transmit a *smaller-than-normal* pulse when following a baud with the same polarity. One of the many possible circuits used to accomplish this trick appears in Figure 8.6. Any circuit that modulates the transmit amplitude in order to better equalize the signal at the end of a long cable is called a pre-emphasis circuit.

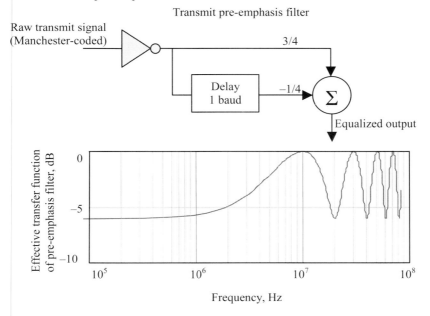

Figure 8.6—Effective transfer function of 10BASE-T transmit pre-emphasis filter.

---

[66] The minimum alternation frequency for uncoded random binary data is zero. The minimum alternation frequency for a code designed to equalize the DC balance (numbers of 1s and 0s) over each code groups of n baud intervals and having a baud period of b equals approximately $1/(2nb)$.

The circuit in Figure 8.6 assumes the Manchester-coded data takes on values of 1 or −1 in each successive baud.[67] The circuit takes 3/4 of the normal transmit signal and subtracts from that value 1/4 of the signal from one baud earlier. Whenever the raw transmit signal is the same in two successive 1/2-baud intervals, the net result is a signal of 1/2 size. Whenever the raw transmit signal executes a transition, successive 1/2-baud intervals will have opposite polarities, and the net result will be a signal having full value. The equalized output appears in Figure 8.7.

This circuit operates like a filter, with a frequency-domain transfer function equal to $\tfrac{3}{4} - \tfrac{1}{4}e^{-j2\pi f b}$, where b is the Manchester baud period. The transfer function magnitude is plotted in Figure 8.6. The transfer function attenuates all signals below 1 MHz by about 6 dB. At 5 MHz the attenuation is −2.5 dB, and at 10 MHz, the attenuation is zero. That looks just about like what we need. The difference between the 10-MHz attenuation and the 5-MHz attenuation is +2.5 dB, which should help correct the negative 3.5-dB slope of the 100 m category 3 UTP cable response. The peculiar behavior of the circuit at frequencies above 20 MHz is of little consequence, as the transmit edge-shaping heavily attenuates all signal power in that band.

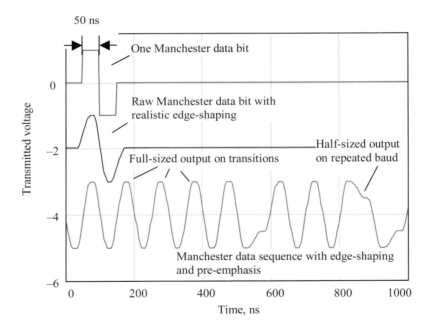

Figure 8.7—10BASE-T transmitted waveforms with edge-shaping and pre-emphasis.

---

[67] Manchester coding generates two bauds for each transmitted bit. The transmitted baud rate therefore equals twice the bit rate. To send a logical 1, the Manchester coder generates the pattern 1, −1. To send a logical 0, the Manchester coder generates a pattern −1, 1. This is one possible variation of Manchester coding that guarantees a transition in the middle of each bit cell. An alternate definition might guarantee a transition at the beginning of each cell, with a transition either present or missing in the center of the cell to indicate data values of 1 or 0.

8.2 • Example: 10BASE-T

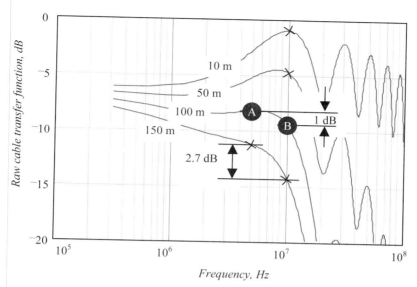

Figure 8.8—Equalized transfer functions for worst-case category 3 UTP at various lengths.

A family of composite (equalizer plus cable) frequency response curves is presented in Figure 8.8. The transmitted waveform contains very little power at frequencies above 20 MHz, so the lumps in the equalized transfer function at 30 MHz, 50 MHz, and beyond will not affect the received signal quality.

The equalized system eye patterns are shown in Figure 8.9. As you can see, the lower two eye patterns (100 m and 150 m) appear much better than in the unequalized system. In the lower two waveforms the mid-baud zero crossing occurs much closer to the dead center of the baud cell. This improvement in timing jitter makes life a lot easier on the clock recovery circuits. Also, the noise margin has improved. All around, it's a better system.

I should point out that all the eye pattern waveforms in this section show the signal at the end of a single segment of worst-case cable. This signal may not be the same as the signal present at the discriminator circuit in a practical receiver, due to the use of low-pass filtering in the receiver front end used to control noise. The receiver's low-pass filter in most cases introduces further pattern-selective degradation of the received amplitude and additional jitter. Other system imperfections, such as work-area cables, jumpers, reflections, near-end crosstalk (NEXT), and alien crosstalk, are similarly not incorporated into the figure.

The pre-emphasis circuit helps at long cable lengths but makes things worse at short cable lengths. For example, at 10 m the pre-emphasis causes noticeable overshoot of the transitions in Figure 8.9. In a two-level system this overshoot is of little consequence, because only the polarity of the received signal matters, not its amplitude.

At short lengths, the pre-emphasis also induces unwanted jitter on the transitions. Fortunately, any clock timing jitter contributed by the overshoot is more than compensated in the receiver noise budget by the presence of the absolutely enormous incoming signal level. At 100 m, where you need good jitter performance, the timing looks superb.

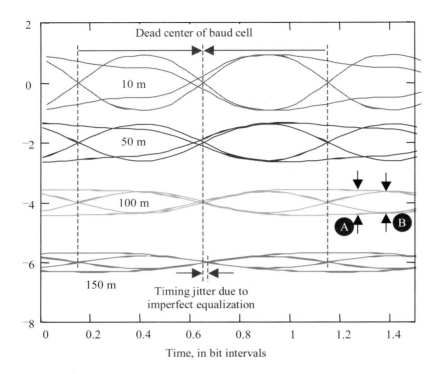

Figure 8.9—Equalized eye patterns for Manchester coding. Data rate = 10 Mb/s, transmit rise/fall = 30 ns, cable = worst-case category 3 UTP at 20 °C.

In a multilevel transmission system, overshoot can become a very serious issue. It can significantly confuse the receiver. The more levels that exist in a transmission scheme, the more accurately one must equalize. To achieve the requisite accuracy, equalizer circuits for multilevel coding usually incorporate some form of automated adaptation.

For a two-level system, the simple fixed pre-emphasis circuit presented in this section boosts the maximum operational cable length by at least 50%. I should point out that the Manchester equalizer is a particularly simple case, because with Manchester coding the majority of the signal energy in the data signal is contained within the band between 5 MHz and 10 MHz. The equalizer need only flatten the received signal spectrum over this relatively narrow band. Uncoded random binary data (NRZ coding), which produces a wider ratio between the high and low band edges of its data spectrum, demands equalization over a wider range of frequencies.

A more sophisticated adaptive equalizer, to the extent that it can maintain a constant gain over the range of frequencies contained in the data signal, will permit operation at even greater distances. Adaptively equalized systems commonly boost the maximum operational cable length by 100% or more.

Points to Remember

> - Timing jitter is improved when all received amplitudes are independent of past history.
> - Simple fixed pre-emphasis boosts the maximum operational cable length of a Manchester-coded link by at least 50%.
> - A more sophisticated adaptive equalizer can extend operation to even greater distances.

8.3 UTP Noise and Interference

There are six major sources of noise and interference in a high-speed UTP system:

1. Far-end reflections
2. Near-end reflections
3. Near end crosstalk (NEXT)
4. Alien crosstalk
5. Far-end crosstalk (FEXT)
6. Radio-frequency interference (RFI)

8.3.1 UTP: Far-End Reflections

When you launch power into a transmission system, three things can happen to it:

1. Part of the power is absorbed by the transmission medium as the signal propagates.
2. After a one-way propagation delay, part of the power is delivered to the load.
3. The remaining power reflects from the load or from intermediate imperfections within the transmission line, and then bounces around within the cable until it is dissipated by mechanism (1) or (2).

In any reasonable signal transmission channel, only a tiny fraction of the signal power leaks out of the cable system in the form of radiation. As far as the integrity of the received signal is concerned, this amount of lost power is insignificant. It may cause some significant EMI headaches, but it's not large enough to worry about when computing the amplitude or shape of the received signal.

Portions of the original transmitted power that bounce around within the cable, arriving later at the far-end load are called *far-end reflections* (Figure 8.10). In a modern UTP data transmission system, because the one-way cable delay greatly exceeds the data baud period, reflections from one baud typically arrive at the receiver during the reception of some later baud. The far-end reflections are therefore not correlated with the data in the current received baud period and constitute a source of random noise to the receiver at the far end of the cable. This section examines the magnitude of far-end reflected noise.

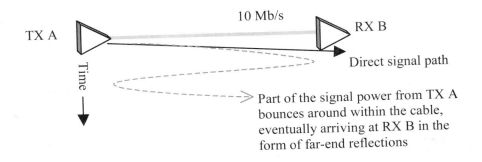

Figure 8.10—One-half of 10BASE-T link showing unidirectional data transmission on one pair of UTP.

The magnitude of the far-end reflected signal shown in Figure 8.10 depends in part on the characteristic impedance of the cable. The characteristic impedance of data-grade UTP is controlled to within 15% for category 3 and 10% for higher categories (Table 8.1).

At the bitter ends of the cabling where the wiring meets your transceiver, a reflection always occurs due to the inevitable mismatch between the cable impedance and the input (or output) impedance of your hardware. With some hard work, you can control the input (or output) impedance of your circuitry to within about ±5%. The worst-case reflections occur when your transceiver impedance is high and the cable is low (or vice versa). Assuming a +5% transceiver impedance and a −15% category 3 UTP cable impedance, the total impedance mismatch is on the order of 20%, which generates about a 10% reflection.[68] If you have terminated the cable at *both ends,* the worst-case round-trip reflection would in that case be only 10% of 10%, amounting to only 1% of the transmitted signal amplitude not counting the additional round-trip of cable attenuation suffered by the reflected signal. Such a scenario illustrates the big benefit of terminating a cable at both ends—each round-trip reflection is attenuated *twice*.

A complete reflections budget needs to take into account not only the reflections at the ends of the cable, but also those that occur in the middle. Wherever a transition occurs from one cable to another, there exists the chance that one cable will have a high impedance and the other a low impedance. When working with category 3 or 5 UTP, reflections from major cable transition points have a maximum amplitude of 15%,[69] provided that the lengths of cable on either side of the transition are sufficiently long to build up a full-scale reflection amplitude. The relation between the length of cable required to build up a full-sized reflection and the signal risetime is given approximately by

$$l_{full} = t_r v / 2 \qquad [8.5]$$

---

[68] A transceiver impedance of 105 ohms coupled to a cable impedance of 85 ohms generates a reflection of size (85 − 105)/(85 + 105) = −0.105 times the height of the incoming signal.

[69] A cable impedance of 85 ohms coupled to a cable impedance of 115 ohms generates a reflection of size (115 − 85)/(115 + 85) = 0.15 times the height of the incoming signal.

8.3.1 Far-End Reflections

where   cables of length l_{full} (m) generate a full-sized reflection,

t_r is the signal 10% to 90% risetime (or fall time) in s, and

v is the cable propagation velocity, m/s.

Cables shorter than the critical length l_{full} generate smaller reflections, in proportion to their length. For example, suppose you were working with a Nyquist band-limited system operating at 25 Mbaud on a cable with velocity $v = 0.6c$. The risetime for such a system approximately equals the bit period of 40 ns. The critical length l_{full} for this link is therefore $l_{full} = (40 \cdot 10^{-9} \text{ s})(0.6 \cdot 2.998 \cdot 10^8 \text{ m/s})/2 = 3.6 \text{ m}$. In this type of link, cables shorter than one meter won't generate much of a reflection even if their impedance mismatch is significant. Cables longer than 3.6 m, on the other hand, can easily produce noticeable reflections.

In a practical long-distance link, the worst-case magnitude of the far-end reflections must be limited by design. This is accomplished by establishing limits on the nominal cable impedance, the maximum variation of impedance within each cable, and the number of major transition points within each individual link. The number of transition points permitted in horizontal generic building wiring is four.

To compute the worst-case far-end reflection, you must sum all the second-order reflection modes. Figure 8.11 illustrates all the second-order reflection modes in a UTP connection with two transition points.[70] In the far-end received noise budget, all the worst-case second-order modes must be summed at the receiver.

When estimating the worst-case reflection modes by hand, just sum the absolute

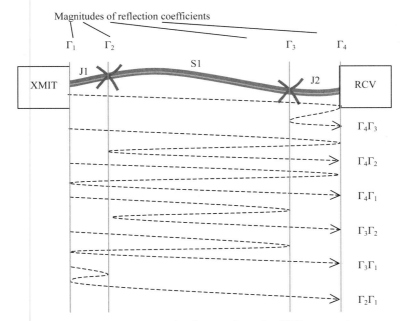

Figure 8.11—Second-order reflection products for UTP connection.

values of the worst-case reflection coefficients at each transition point.[71] In addition, assume zero cable attenuation between transition points. For a more accurate answer, try using a mathematical calculation tool like MathCad, Mathematica, or MatLab to grind through all possible combinations of worst-case high and low impedances in each section of cable. Program the tool to check the total reflection noise and received signal intensity for each combination.

The worst reflections usually result from transition points spaced apart by one-half data baud. Move the transition points closer, and the reflections tend to blur together, failing to attain peak worst-case values. Move the transition points further apart, and the reflected signals encouter additional round-trip cable attenuation. A spacing of about one-half data baud is the worst.

Connectors located at the cable transition points also generate their own reflections, which superimpose on the reflections due to changes in cable impedance. The magnitude of these reflections must be included in a good noise budget (see Section 8.4, "UTP Connectors").

Example Calculation of Far-End Reflected Noise Magnitude

Assume a system with cable impedances specified to within 100 Ω ± 10% and using both source and end terminations accurate to within 5%. Let there be one cable transition point located near each end of the cable (as in Figure 8.11).

The total cable attenuation between points is as follows:

J1 (short jumper)	−1 dB
S1 (long segment)	−10 dB
J2 (short jumper)	−1 dB

The assumed impedances are as follows:

Source	105 Ω
J1	85 Ω
S1	115 Ω
J2	85 Ω
End	105 Ω

The worst-case reflection coefficients are as follows:

$\Gamma 1$	−19.5 dB
$\Gamma 2$	−16.4 dB
$\Gamma 3$	−16.4 dB
$\Gamma 4$	−19.5 dB

---

[70] Transitions are usually accompanied by connectors.
[71] In a slow data transmission system one might assume that two successive reflections, one positive and one negative, might average together during the period of one data symbol, rendering a worst-case effect that is less than the sum of absolute values. In a fast system where the reflections are each separated by more than one data symbol, it is possible to construct a data sequence that actually produces the worst-case sum-of-absolute-values effect.

The worst second-order modes are as follows:

J1 S1 J2 Γ4 J2 Γ3 J2	−(1 + 10 + 1 + 19.5 + 1 + 16.4 + 1)	−49.9 dB
J1 S1 J2 Γ4 J2 S1 Γ2 S1 J2	−(1 + 10 + 1 + 19.5 +1 + 10 + 16.4 + 10 + 1)	−69.9 dB
J1 S1 J2 Γ4 J2 S1 J1 Γ1... J1 S1 J2	−(1 + 10 + 1 + 19.5 + 1 + 10+ 1 + 19.5 ... + 1 + 10 + 1)	−75.0 dB
J1 S1 Γ3 S1 Γ2 S1 J2	−(1+10+16.4+10+16.4+10+1)	−64.8 dB
J1 S1 Γ3 S1 J1 Γ1 J1 S1 J2	−(1 + 10 + 16.4 + 10 + 1 + 19.5 + 1 + 10 + 1)	−69.9 dB
J1 Γ2 J1 Γ1 J1 S1 J2	−(1 + 16.4 + 1 + 19.5 + 1 + 10 + 1)	−49.9 dB
Worst-case total (sum of magnitudes)		−42.2 dB

The signal-to-noise ratio (SNR) at the far-end receiver is as follows:

Signal: J1 S1 J2	−(1 + 10 + 1)	−12 dB
Noise		−42.2 dB
SNR		30.2 dB

POINTS TO REMEMBER

> A complete noise budget takes into account all reflections within a cabling system.
> Connectors generate reflections that superimpose on the reflections generated by changes in cable impedance.

8.3.2 UTP: Near-End Reflections

When you shout into a tunnel, it echoes back. The same thing happens when you blast a high-speed digital signal into a long UTP cable: You get an echo. The term *near-end reflection* encompasses two different types of echo effects. The first echo effect is due to reflections that occur at cable transition points, termination points, and connectors. The other echo effect is specified by cable *return loss* (RL). It has to do with reflections generated due to imperfections within the cable itself.

The return loss specifications are important in systems that attempt to transmit in both directions (full-duplex) at the same time on the same pair of conductors.

In a unidirectional link the near-end reflections are of no consequence, because when you are transmitting on one conductor pair, you are not also listening to that same pair. Bidirectional, full-duplex systems are different. In a bidirectional, full-duplex link you *are* listening on the same pair with which you are transmitting. In that case you hear your own echo.

Figure 8.12 shows an example of a bidirectional, full-duplex data link. This link was designed by ROLM in 1983 as the basis of its digital telephones. The ROLM link sends two-level PAM data at 256 Kbaud in each direction. Data are sent continually at all times, in

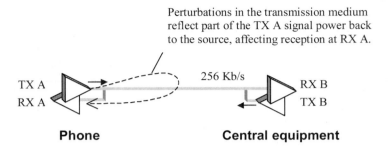

Figure 8.12—A ROLM digital telephone transceiver transmits bidirectionally on a single pair of category 3 cable (circa 1982).

both directions, on a single pair of a category 3 cable. Partly because the data rate was so slow, echo problems on this link were easily managed. The ROLM system used a hybrid circuit (see Section 8.3.3 "UTP: Hybrid Circuits") to enable reception of the far-end signal even while the local transmitter was broadcasting.

The noise reflected from cable transition points and connectors is qualitatively the same as *far-end* reflected noise, but quantitatively much more significant. The reason that *near-end* reflections matter so much in bidirectional systems is because the reflections in Figure 8.13 that arrive at RX A have bounced from only one reflecting body, as opposed to the reflections in Figure 8.11, which have each been attenuated by at least two bounces. The difference between the two situations is striking. The signal-to-noise ratio in the near-end case is much worse.

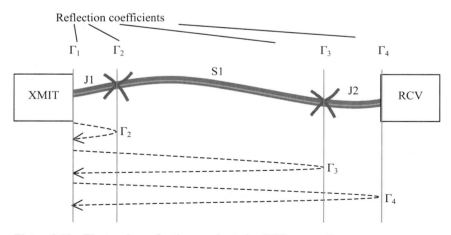

Figure 8.13—First-order reflection products for UTP connection.

8.3.2 • Near-End Reflections

Example: Near-End Reflected Noise Magnitude

Assume a system with cable impedances specified to within 100 Ω ± 10% and using both source and end terminations accurate to within 5%. Let there be one cable transition point located near each end of the cable (Figure 8.13).

The total cable attenuation between points is as follows:

J1 (short jumper)	−1 dB
S1 (long segment)	−10 dB
J2 (short jumper)	−1 dB

The assumed impedances are as follows:

Source	105 Ω
J1	85 Ω
S1	115 Ω
J2	85 Ω
End	105 Ω

The worst-case reflection coefficients are as follows:

$\Gamma 1$	−19.5 dB
$\Gamma 2$	−16.4 dB
$\Gamma 3$	−16.4 dB
$\Gamma 4$	−19.5 dB

The worst near-end reflection modes are as follows:

J1 Γ2 J1	−(1 + 16.4 + 1)	−18.4 dB
J1 S1 Γ3 S1 J1	−(1 + 10 + 16.4 + 10 + 1)	−38.4 dB
J1 S1 J2 Γ4 J2 S1 J1	−(1 + 10 + 1 + 19.5 + 1 + 10 + 1)	−43.5 dB
Worst-case total (sum of magnitudes)		−17.1 dB

The signal-to-noise ratio (SNR) at a near-end receiver is as follows:

Signal: J1 S1 J2	−(1 + 10 + 1)	−12 dB
Noise		−17.1 dB
SNR	(Signal − Noise)	5.1 dB

POINT TO REMEMBER

➢ Bidirectional links must tolerate near-end reflections.

8.3.2.1 UTP: (Structural) Return Loss

Compared to discrete reflections, the structural return effect is more complicated. *Structural return noise* results from minor variations in the characteristic impedance of a continuous cable. As a signal propagates down the cable, these minor variations create small reflections that propagate back towards the source. The composite of all the backwards-propagating signals is called structural return noise. This noise appears at the source as an almost Gaussian noise waveform. For sinusoidal inputs, the *expected magnitude* of the structural return noise for long cables varies in proportion to the 3/4-power of ω. The faster you go, the larger the structural return noise becomes.

At higher frequencies the increase in structural return noise complicates the design of hybrid circuits. For example, some of the very early Ethernet UTP transceivers made by 3Com used a bidirectional architecture very similar to the ROLM system depicted in Figure 8.12. These early Ethernet transceivers operated bidirectionally at 10 Mb/s over category 3 UTP. The data rate per pair was higher than in the ROLM system by a factor of (10 MHz/256 KHz) = 39, making structural return loss problems significantly more worrisome ($39^{3/4}$ = 15.6 times more worrisome).

Structural return noise shares many similarities with the noise generated by reflections from cable transitions. Both are self-induced effects, meaning that the magnitude of either is proportional to the transmitted signal. If the transmitter ever shuts off, both effects diminish to near zero within one cable round-trip time.

Structural return noise is contributed from many, many small sources, all of which blur together into a continuously evolving waveform. The effect looks like white noise (except that the step response of the noise is precisely the same every time you measure it). Cable transitions, on the other hand, generate much larger, distinct, individual reflected signals punctuated by long gaps (Figure 8.14). Both noise sources contribute to the overall near-end noise budget at the receiver.

Structural return noise is controlled for all generic building wiring types in one of two ways:

> A specification of *structural return loss* combined with a specification on the mean value of characteristic impedance (used for old category 3 cables), or

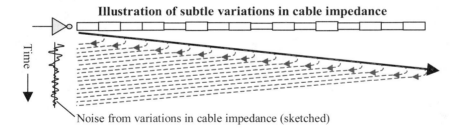

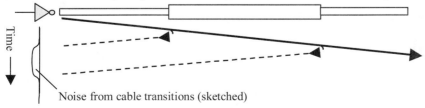

Figure 8.14—Time-space diagram showing qualitative difference between noise due to impedance variations and noise due to cable transition points.

8.3.2 • Near-End Reflections

> ➤ A single specification of cable *return loss* (used for newer category 5e and 6 cables).

To understand the difference between these two specifications, imagine that characteristic impedance is a function that fluctuates as you progress along the length of a cable. The old structural return loss specification decomposes the characteristic impedance function into a mean value plus fluctuations. The term *structural return loss* refers only to the noise (in dB) reflected due to the fluctuations in impedance. Reflections due to the mean value of impedance are counted separately. When this form of specification is used, limits on the mean value of impedance are explicitly called out.

The newer term *return loss* assumes the cable is terminated into a perfect 100-Ω differential load. The noise (in dB) reflected under this 100-Ω load condition includes reflections from the fluctuations in impedance (structural return loss) plus other reflections due to errors in the mean value of impedance. The return loss therefore includes both the structural return noise and the mean value of characteristic impedance simultaneously instead of with distinct specifications. When a return loss specification is indicated, the nominal value of characteristic impedance is implied by the value of load resistance used in the test, but no explicit limits on the mean value of characteristic impedance are called out.

The TIA/EIA 568-B standard for categories 5e and 6 specifies return loss, as opposed to structural return loss, in order to circumvent problems associated with accurately defining and measuring the mean value of characteristic impedance (see Table 8.4). An overall noise budget calculation taking into account all forms of near-end reflections including cable return noise should be undertaken for any bidirectional full-duplex system.

Table 8.4—(Structural) Return Loss for UTP and 150-Ω STP-A Cabling (Worst Pair)

Cable type	Freq. Range MHz	Return loss dB
cat-3 (structural return loss)	$1 \leq f \leq 10$	12
	$10 \leq f \leq 16$	$12 - 10\log(f/10)$
cat-5e (return loss)	$1 \leq f \leq 10$	$20 + 5\log(f)$
	$10 \leq f \leq 20$	25
	$20 \leq f \leq 100$	$25 - 7\log(f/20)$
cat-6 (return loss)	$1 \leq f \leq 10$	$20 + 5\log(f)$
	$10 \leq f \leq 20$	25
	$20 \leq f \leq 250$	$25 - 7\log(f/20)$
150-Ω STP-A	$0 \leq f \leq 20$	24
	$20 \leq f$	$24 - 10\log(f/20)$
NOTES—Values taken from TIA/EIA-568-B.2-2001 and TIA/EIA-568-B.2-1-2001. In all formulas, the frequency variable *f* is in MHz.		

POINTS TO REMEMBER

➤ A specification of *structural return loss* combined with a specification of the mean value of characteristic impedance is used for old category 3 cables.

➤ A single specification of cable *return loss* (as measured with the cable terminated in a 100-Ω load) simultaneously limits both the mean value and local perturbations in cable impedance.

8.3.2.2 Modeling Structural Return Loss

A transfer function representing the structural return effect may be modeled as a large number of separate little reflections, each with a frequency response proportional to ω. The general form of the model looks like this:

$$SRL(\omega) = \sum_{n=0}^{N} \alpha_n j\omega H(\omega, 2n \cdot \Delta x) \quad [8.6]$$

where $SRL(\omega)$ is a transfer function representing the structural return effect,

ω is the frequency of interest, rad/s,

α_n is a collection of independent, uncorrelated random variables

$H(\omega,l)$ is a transfer function representing a segment of cable l meters in length, as measured at frequency ω, and

Δx is the length of each segment in the model, m.

The model breaks the cable into N segments, each representing a small section of length Δx. Each element in the sum represents the reflections due to the structural return effect from portion n. The term α_n indicates the strength of the effect from segment n, which depends randomly on local impedance perturbations. When synthesizing a random SRL transfer function, a reasonable distribution to use for each α_n is a Gaussian distribution with zero mean and standard deviation σ.[72] In that case the standard deviation σ should be set to produce the correct nominal amount of overall gain A_0 at some frequency ω_0, according to the following formula:

$$\sigma = \frac{A_0}{\omega_0 \sqrt{\sum_{n=0}^{N} |H(\omega_0, 2n \cdot \Delta x)|^2}} \quad [8.7]$$

The term $j\omega$ in equation [8.6] indicates that the basic effect within each segment is a differentiating effect. This comes about due to the geometry of each segment. Imagine a single segment in isolation, coupled to an otherwise ideal, infinite cable. Let the segment in question have an impedance uniform over its length, but different from the characteristic

---

[72] The Greek letter sigma is used here to refer to the standard deviation of a random process, not conductivity.

8.3.3 • Hybrid Circuits

impedance of the remainder of the cable. The imperfect segment will induce two impedance discontinuities, one at either of its ends, with opposite reflection coefficients (that is, one reflection will be negative and the other positive). The essence of the reflection effect is this: As the main signal steps into the perturbed area, a reflection will be generated, and as the main signal passes back out, a second reflection *roughly equal in magnitude but opposite in sign* to the first will be generated. The result is a pair of small, closely spaced impulses with opposite polarities. If the delay represented by each segment in the above model represents much less than the rise (or fall) time of the signals involved, the structural return operation at each segment looks very much like a derivative.

The term $H(\omega, 2n \cdot \Delta x)$ indicates that the total path traversed by the reflection from segment n includes the distance $n \cdot \Delta x$ from the driver to segment n followed by an identical distance $n \cdot \Delta x$ on the way back.

The attenuation predicted by this model at any particular frequency ω is a random variable whose magnitude obeys the Rayleigh distribution. When the nominal gain is set to value A_0 (this sets the average reflected power to a value of A_0^2 times the incoming signal power), the actual gain may vary considerably from trial to trial above and below this value. If it is your intention to generate a model whose worst-case gain just touches some specified limit for SRL, you should de-rate A_0 by anywhere from 6 to 12 dB before using equation [8.7] to calculate an appropriate standard deviation. A de-rating factor of 8 dB was used to generate the synthetic NEXT curves in Figure 8.21.

Alternately, if you are casting random SRL functions for use in Monte-Carlo testing, you may individually adjust each function to guarantee that it just grazes the specified limit in at least one location.

If the function $H(\omega, l)$ represents a skin-effect-limited channel, then the SRL noise modeled by these equations displays a 3/4-power law dependence on frequency (i.e., the noise grows at a rate of 15 dB per decade).

POINTS TO REMEMBER

> ➤ Structural return noise is modeled as a summation of many noise sources with random amplitudes.
> ➤ Structural return noise grows at a rate of 15 dB per decade.

8.3.3 UTP: Hybrid Circuits

The circuit that makes possible bidirectional full-duplex transmission is called a *hybrid circuit*. Figure 8.15 illustrates a simple hybrid circuit appropriate for use on a single-ended transmission medium.

The waveform at $v(t)$ is a superposition of effects from both sides. Mathematically, it is a linear combination of the signal $x(t)$ and the signal $y(t)$. Setting $y(t)$ momentarily to zero, you may evaluate the gain from $x(t)$ to $v(t)$ using the voltage-divider equation.

In general these equations are pretty hairy, but they can be simplified in a major way by carefully designing impedance Z_S to match Z_C. Under that assumption the gain from $x(t)$ to $v(t)$ would be precisely 1/2. Similarly, the gain from $y(t)$ to $v(t)$ is also 1/2, but with an extra term included to account for the propagation loss in the transmission media.

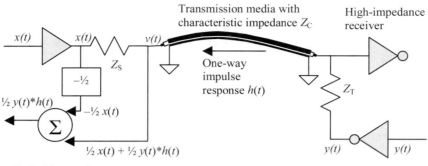

Hybrid receiver

Figure 8.15—A hybrid circuit subtracts half the transmitted signal from the received signal to form a remainder equal to the desired signal from the far end.

Mathematically, the propagation effect is the equivalent of convolving the signal $y(t)$ with the impulse response $h(t)$ of the transmission line, leading to a fairly simple composite form for the waveform at $v(t)$:

$$v(t) = \frac{1}{2}x(t) + \frac{1}{2}y(t)*h(t) \qquad [8.8]$$

where the symbol * represents the convolution (linear filtering) operator.

This last expression indicates that the signal at v is composed of two parts: one half the signal $y(t)$ you wish to receive, modified by its passage through the intervening cable segment, plus a pesky interfering signal $(1/2) \cdot x(t)$. The beauty of the hybrid concept is that you can predict *and therefore eliminate* the pesky interfering signal by simply subtracting from the received waveform $v(t)$ the known interfering quantity $(1/2) \cdot x(t)$. The summation block in Figure 8.15 subtracts $(1/2) \cdot x(t)$ from $v(t)$, producing a final waveform equal to $(1/2) \cdot y(t)*h(t)$. All hybrid circuits work on the same general principle—they subtract from the received signal the known (or measured) interference from the transmitter.

Notice that the circuit gain from $y(t)$ to $v(t)$ is in the best of cases no greater than 1/2. This loss of 6 dB is a consequence of using what amounts to a double-end termination scheme.

The hybrid doesn't rid the circuit of the effects of $h(t)$, but at least it eliminates interference from local transmissions. This technique is widely used to construct bidirectional full-duplex links.

The effects of $h(t)$, if objectionable, must be reversed by an equalizing circuit. The equalizing circuit may either predistort the waveform $y(t)$ to account for the assumed degree of distortion present on the cable or post-filter the resulting $v(t)$ to restore the original waveform, or a little of both.

Let's review the conditions necessary for the proper operation of the hybrid:

1. Foreknowledge of the transmission line impedance.
2. End-termination with impedance $Z_T = Z_C$ at all frequencies.

8.3.3 • Hybrid Circuits

3. Source-termination with impedance $Z_S = Z_C$ at all frequencies.

If the termination impedances do not match Z_C, then the gain from $x(t)$ to $v(t)$ will differ from 1/2 and the cancellation will not function properly. The same hybrid circuit is typically used at both ends of the line.

Hybrid circuits are rated in terms of the amount of the locally transmitted signal that leaks through the cancellation circuit to affect the hybrid output. This leakage is called the hybrid return loss. The hybrid return loss is reported in units of dB relative to the nominal gain of the received signal, which is usually 1/2. The hybrid loss is a function of the transmission-line impedance and the impedance of the terminations at both source and far ends of the transmission line.

It is instructive to examine the sensitivity of the hybrid circuit to various parameters.

A hybrid circuit's sensitivity to perturbations in the source impedance is approximately 1/2, meaning that a 10% change in the source impedance results in a 5% relative error at the output of the hybrid (26 dB return loss). The sensitivity to perturbations in the far-end termination impedance is similar.

The hybrid circuit in Figure 8.15 is designed to operate in a low-loss mode (LC mode or above), for which the characteristic impedance of the transmission line maintains a fairly constant real value Z_C over the frequencies of interest. What if the line were operated in the dispersive RC mode? In that mode the characteristic impedance changes substantially with frequency. Matching Z_S and Z_T to $Z_C(\omega)$ for a dispersive line can become a serious challenge.

Figures 8.16, 8.17, and 8.18 explore that challenge. Each figure plots the input

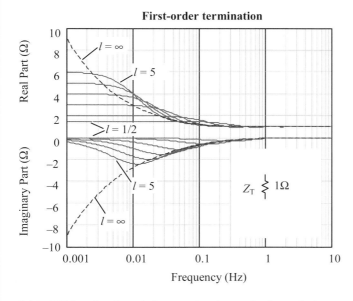

Figure 8.16—With a simple resistive termination at the far end, the precise input impedance of the terminated transmission line varies as a function of the transmission-line length. It is therefore impossible to build a perfect (nonadaptive) hybrid circuit that works at all lengths and all frequencies using a simple resistive termination.

impedance $Z_{IN}(\omega)$ of a loaded transmission line versus frequency for several situations. The dotted lines reveal the characteristic impedance $Z_C(\omega)$ of the transmission line. This is the input impedance you would measure if the line were of infinite extent (or perfectly terminated). The real part of Z_C appears in the top half of each figure, while the imaginary part (always negative) appears in the bottom half. The dotted-line plot for each figure is the same, representing the characteristic impedance of a normalized RC transmission line with per-unit-length parameters $R = 1$, $L = 1$, and $C = 1$.

The solid lines in each figure depict the actual input impedance of normalized transmission lines of various lengths with particular end terminations. The lengths selected are 1/2, 1, 2, 3, 4, and 5 units.

Figure 8.16 illustrates the consequences of terminating an RC line with a simple resistance, in this case $Z_T = 1\Omega$. This figure reveals an input impedance in the area above 1 Hz holding flat at 1Ω. That's because the line is operating at those frequencies mostly as a low-loss, or LC, transmission line, for which the characteristic impedance equals $\sqrt{L/C}$, a value that does not much change with frequency. Since the low-loss line is properly terminated in its characteristic impedance of $1\ \Omega$, the reflection coefficient at the far end is zero and the input impedance Z_{IN} does not vary with line length. It stays at $1\ \Omega$.

At low frequencies the input impedance of finite-length, 1-Ω terminated RC transmission line varies dramatically with both frequency and lne length. For example, at DC the input impedance equals the far-end termination resistance ($1\ \Omega$) *plus* the DC resistance of the line itself. In Figure 8.16 the series resistance of the line is $1\ \Omega$ per unit length, so for lengths of 1/2, 1, 2, 3, 4, and 5 units, the total DC resistance (line plus load) shown at the left side of the figure reads 1.5, 2, 3, 4, 5, and 6 ohms.

For this normalized transmission line, a hybrid constructed with source and load

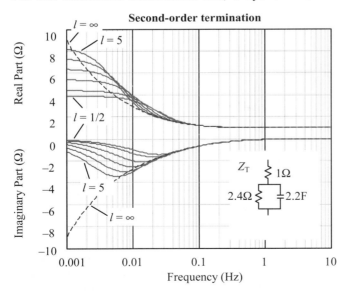

Figure 8.17—A second-order termination circuit improves the stability of the input impedance function at frequencies down to 0.1 Hz.

8.3.3 • Hybrid Circuits

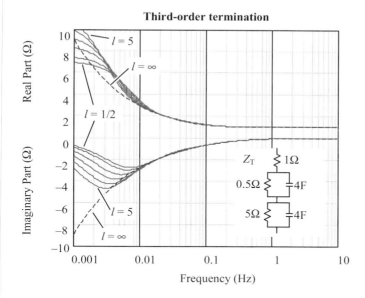

Figure 8.18—A third-order termination circuit stabilizes the input-impedance function at frequencies down to 0.02 Hz.

impedances of 1 Ω would function down to a frequency of approximately 1 Hz, below which the hybrid return loss would become unacceptably large.

You may generalize this rule for a resistively terminated hybrid, saying that the circuit only works above a cutoff frequency $f = R/L$ (Hz), where R and L are the per-unit-inch values of resistance and inductance in the transmission line.

One exception to this generalized rule happens if the line is short enough compared to the frequencies of operation to be considered a simple lumped-element circuit. In that case the hybrid circuit at the transmitter "sees through" directly to the receiver load so the circuit still functions regardless of Z_C.

Normally, though, if you want your hybrid to function below the RC-mode cutoff frequency, you have two choices:

1. Build a better terminating network, or
2. Use digital adaptive hybrid cancellation.

The principle behind choice 1 is that the more closely your terminations match the line impedance, the better your hybrid circuit will function. Figure 8.17 and Figure 8.18 illustrate the advantages of second-order and third-order termination networks. In each case the termination impedance Z_T matches the true characteristic impedance of the line Z_C to progressively lower frequencies. The second-order circuit works down to about 0.1 Hz, and the third-order circuit works down to 0.02 Hz, irrespective of line length.

The component values shown in the various termination networks may be scaled for use on other transmission lines by multiplying all the resistance times the factor $\sqrt{L/C}$ and multiplying all the capacitances by the factor $\sqrt{LC}/R$.

Note that the circuit of Figure 8.15 directly DC-couples the two endpoints. The DC operating point of the hybrid output therefore depends on, among other factors, the DC-average potential of signals $x(t)$ and $y(t)$. If the duty cycles of the two transmitters vary from 50% or the power-supply voltages available at the two ends of the link wander, DC voltages may be induced at the hybrid output. To eliminate these considerations, a hybrid output is generally AC-coupled, and then followed by DC level restoration should that be necessary.

Figure 8.19 illustrates a simple hybrid block diagram useful for balanced transmission in the low-loss mode. Assuming the differential signal coming out of the driver is $x(t)$, the differential signal developed across the primary (or secondary) of the transformer must be $\frac{1}{2}x(t) + \frac{1}{2}y(t)*h(t)$. The intermediate nodes of the bridge form weighted averages according to the ratio of resistances in the bridge legs. Assuming $R \gg Z_0$, this bridge forms a differential output voltage of $\frac{1}{3}y(t)*h(t)$.

The main factors that limit the performance of this circuit are parasitic capacitances within the transformer, imbalance in the resistor ratios, and the unavoidable uncertainty in the impedance of the cabling. A better hybrid circuit is formed by replacing the fixed fractional subtraction of $\frac{1}{2}x(t)$ with a variable gain or a digital adaptive filter. A sufficiently complex adaptive digital filter can compensate for cable roughness and also cable-transition reflections simultaneously. Such a filter is called an adaptive echo cancellation circuit.

Digital adaptive hybrid cancellation (also called echo cancellation) is beyond the scope of this book; however, a general overview of the echo cancellation problem appears in [67]. An excellent introduction to the theory behind digital adaptive echo cancellation

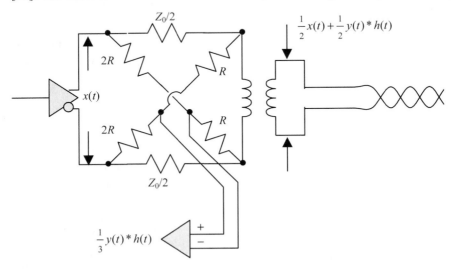

Figure 8.19—A simple balanced hybrid circuit subtracts a fixed fraction of the transmitted signal from the received signal.

appears in [68]. A detailed study of the mathematics behind adaptive filters in general and adaptive echo cancellation in particular is given in [66].

POINTS TO REMEMBER

> ➤ A hybrid circuit makes possible bidirectional full-duplex transmission through a single channel.
>
> ➤ A sufficiently complex adaptive digital filter can compensate for cable roughness and also cable-transition reflections simultaneously. Such a filter is called an adaptive echo cancellation circuit.

8.3.4 UTP: Near-End Crosstalk

Any data link that transmits on one pair while listening on a different pair is susceptible to *near-end crosstalk* (NEXT). Figure 8.20 illustrates the 10BASE-T Ethernet system, showing the general behavior of NEXT. The distinction between NEXT and alien crosstalk (see next section) has to do with the source of the interference. NEXT, since it comes from your own transmitter, is somewhat under your control. Alien crosstalk, which may come from another application operating on different pairs within the same cable jacket, is not.

The NEXT effect for UTP differs markedly from the simple NEXT effect observed on two parallel printed circuit board tracks X and Y. In the pcb case a fast step edge on trace X induces (ideally) a simple boxcar-shaped pulse of NEXT voltage on trace Y.

The UTP crosstalk waveform is complicated by the twisting pattern of the conductors. Every time one pair does a half twist, the adjacent NEXT reverses polarity. Successive twists tend, to first order, to cancel each other out. The crosstalk that is left over represents the degree to which the twisting was imperfect. The twisting imperfections comprise a classic second-order effect: a random generator of NEXT.

NEXT may be modeled as a large number of separate little crosstalk effects, each with a frequency response proportional to ω. The mathematical model used to generate synthetic NEXT transfer functions is identical with the one used to produce SRL transfer functions, so

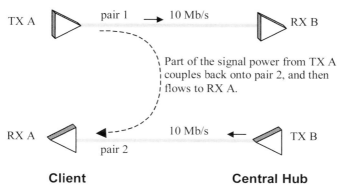

Figure 8.20—10BASE-T installation showing behavior of NEXT.

the NEXT effect has an overall frequency response proportional to the 3/4 power of frequency (Figure 8.21). NEXT must be included in the SNR budget for any full-duplex system.

If you are using 100-Ω UTP cable, TIA/EIA 568-B cable specifications limit the NEXT to the maximum values presented in Table 8.5. The table lists NEXT specifications

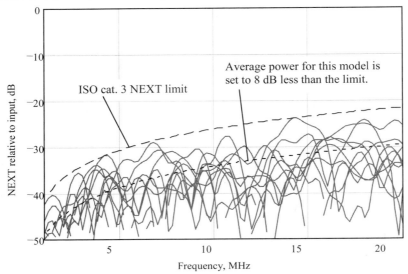

Figure 8.21—Synthetic NEXT curves for category 3 UTP generated by a random SRL/NEXT model.

Table 8.5—NEXT for UTP and 150-Ω STP-A Cabling (Worst Pair)

Cable	Freq. Range (MHz)	Min. allowed NEXT loss [1],[2] (dB)	Min. allowed NEXT loss at 10 MHz (dB)
cat-3	0.772–16	$11.26 - 15\log_{10}(f/100)$ [3]	26.3
cat-5e	0.772–100	$35.3 - 15\log_{10}(f/100)$ [4]	50.3
cat-6	0.772–250	$44.3 - 15\log_{10}(f/100)$	59.3
150-Ω STP-A	1–300	$38.5 - 15\log_{10}(f/100)$	53.5

NOTE (1)—Values from TIA/EIA-568-B.2-2001, as measured for cables greater than 100m in length.
NOTE (2)—In all formulas, the frequency variable f is in MHz.
NOTE (3)—The form of this equation has been modified to conform with the other NEXT equations in this table.
NOTE (4)—This value is 3 dB more stringent than the older EIA/TIA 568-A category 5.

8.3.4 • Near-End Crosstalk

using formulas exactly as shown in the relevant standards. For purposes of comparison, the table also shows the NEXT performance at the specific frequency of 5 MHz. Since all the curves vary with $15\log(f)$, you can extrapolate the entire high-frequency performance curve of each cable from that single value.

As shown in the table, 150-Ω STP-A has the best NEXT performance. That's part of what makes it easy to get a high-speed data link to work on 150-Ω STP-A. In comparison to category 5 UTP, the good NEXT performance of 150-Ω STP-A is obtained, in large measure, by increased physical separation and shielding of the conductors, both of which tend to increase the bulk and weight of the cable.

Whatever cable you use, an overall NEXT calculation, including NEXT from all adjacent pairs, must be included in the SNR budget for any bidirectional full-duplex system.

It is traditional when making multi-disturber NEXT calculations to assume that all pairs do not simultaneously suffer from worst-case NEXT. This argument implies that the worst-case sum of NEXT noise from N adjacent disturbers should be less than N times the disturbance from the worst pair. Such procedures have been in use (at least) since the inception of the 10BASE-T standard (see annex A of the IEEE 802.3 standard "Example Crosstalk Computation for Multiple Disturbers"). Take caution when using this procedure: Its proper use depends heavily on a precise knowledge of the statistics of NEXT in multiple-pair cables, something which is extremely difficult to pin down with any accuracy and which is not guaranteed by any known standard, specification, or measurement procedure.

One of the worst cases for NEXT happens when the far end of a short cable is left unplugged. With the far-end termination missing, the amplitude of the transmitted signal in most double-end-terminated architectures will double. For similar reasons, the receiver becomes twice as susceptible to noise. The overall effect is a quadrupling of crosstalk into the receiver. You may be thinking it doesn't matter how much crosstalk you have when the system is dysfunctional, but there is a subtle catch: If the crosstalk is sufficient to trip the carrier-detect threshold on the receiver, it will think it is listening to a system at the far end, when all it is doing is hearing its own crosstalk. This can cause some very strange system behavior. It is always a good idea to implement a hardware signal-detect function that disqualifies any strange-looking incoming signals. The signal-detect function should always reliably detect disconnected or powered-off conditions at the other end of the cable.

A circuit that fixes the NEXT problem is called an adaptive near-end crosstalk canceller. It takes as inputs the collection of all transmitted signals and produces one output for each received signal. The outputs attempt to predict the precise interference that will occur on each received signal line, given the past history of the transmitters. The circuit then subtracts the predicted interference from each incoming line. What remains is the desired signal from the other end. The design of good NEXT cancellation algorithms is a challenging digital signal processing problem (see [66], [67]).

POINTS TO REMEMBER

➢ NEXT is modeled as a summation of many noise sources with random amplitudes.
➢ NEXT grows at a rate of 15 dB per decade.

8.3.5 UTP: Alien Crosstalk

Alien crosstalk is noise that comes from other devices that use conductors within the same cable jacket as your system. Within the context of a 10BASE-T installation, an alien device could be a second 10BASE-T device or perhaps a telephone.

The 10BASE-T standards committee went through an exhaustive process of evaluating alien crosstalk for category 3 UTP. The committee determined that one primary source of alien noise would be telephone devices operating on an adjacent pair. The worst-case noise generated by this device happens when the telephone handset is raised while the phone is ringing, interrupting the ringing current. The differential noise spike coupled onto an adjacent pair by a ringing-interruption event can be as large as 264 mV, as measured using a 100-ohm receiver having a bandwidth B of approximately 15 MHz. The noise measured varies in proportion to the receiver bandwidth B.

While there was (and still is) some controversy about the rigor of the experimental procedure used to determine the 264-mV noise number, there is no question that Ethernet 10BASE-T systems that are designed to withstand 264 mV of differential noise prove extremely reliable in the field.[73] Other standards (notably 100BASE-TX) have addressed this issue by stipulating that any other unused pairs within the same jacket remain unused.

The importance of the distinction between alien crosstalk and NEXT is that NEXT can be cancelled by a suitably sophisticated circuit similar to a digital adaptive hybrid. Alien crosstalk, because your circuit does not have access to the driving function $x(t)$ within the alien device, cannot be cancelled.

The Gigabit Ethernet standard 1000BASE-T incorporates an elaborate form of multiway crosstalk cancellation. This is a four-channel full-duplex system operating at 125 Mbaud per pair. It uses a PAM-5 coding to achieve a transmission rate of 250 Mb/s on each pair. Each receiver is programmed to cancel the echo from its own transmitter in addition to NEXT from the other three active transmitters. Because this system uses all four pairs, installers are not tempted to install alien devices within the same jacket.

POINT TO REMEMBER

> Alien crosstalk comes from devices occupying unused pairs within your cable jacket.

8.3.6 UTP: Far-End Crosstalk

Figure 8.22 illustrates the far-end crosstalk (FEXT) scenario present in 100BASE-T2. The 100BASE-T2 data link uses two pairs, each transmitting 50 Mb/s in a bidirectional, full-duplex mode over 100 meters of category 3 UTP. FEXT couples part of the signal from transmitter A over to receiver D (and B to C, etc.). FEXT affects any link that utilizes multiple pairs *transmitting in the same direction* within the same cable jacket.

The FEXT parameters quoted in Table 8.6 are defined as *equal-level far-end crosstalk* (ELFEXT) measurements. ELFEXT is defined as the difference in dB between the signal

---

[73] By 1999 there were over 100 million Ethernet 10BASE-T nodes in operation worldwide.

8.3.6 • Far-End Crosstalk

Far-end crosstalk represents the extent to which signals traveling on one pair induce crosstalk *in the same direction* on an adjacent pair.

Figure 8.22—100BASE-T2 transmitter A induces FEXT measurable at receiver D.

levels measured at the inputs to transceivers C and D in Figure 8.22 when transmitting only from location A. The table includes information for categories 5e and 6 cabling.

ELFEXT for category 3 cables remains unspecified. This lack of specification hampered the development of standards for 100BASE-T2 and 100BASE-T4 transceivers, two variants of Ethernet built to operate at 100 Mb/s over category 3 cable. Both of these variants depended on the use of wire pairs transmitting in the same direction within the same jacket.

In the absence of concrete specifications for ELFEXT, the 100BASE-T2 and 100BASE-T4 LAN standards *assumed,* based on laboratory measurements, that ELFEXT on category 3 cables would never exceed $5.0 - 20\log(f/100)$ dB over the range of 2 MHz to 16 MHz. That assumption was used in both cases as part of the signal-to-noise budget. Unfortunately, the cable industry never adopted the assumption as a formal specification, nor were the sales of either transceiver sufficient to prove the assumption correct before

Table 8.6—ELFEXT Limits

Standard	Cable	Freq. Range (MHz)	Min. allowed ELFEXT loss (dB)
EIA/TIA 568-B	cat-5e	1–100	$23.8 - 20\log_{10}(f/100)$
EIA/TIA 568-B	cat-6	1–250	$27.8 - 20\log_{10}(f/100)$

NOTE (1)—Values listed apply to cable only. A complete link budget will differ (see EIA-568-B).
NOTE (2)—150-Ω STP-A incorporates only two pairs, one used in either direction, rendering the ELFEXT inapplicable.
NOTE (3)—Values of ELFEXT for category 3 cabling have never been standardized.

both variants were overtaken in the marketplace by the category-5 version, 100BASE-TX.

FEXT may be modeled as a large number of separate crosstalk effects in a manner similar to equation [8.6], but with major simplifications. Wherever on the cable the crosstalk occurs, the signal eventually must traverse one complete cable length.[74] Therefore, the term $H(\omega, 2n \cdot \Delta x)$ in [8.6] is replaced by $H(\omega, L)$, where L is the total length of the cable. Factoring out the product $j\omega H(\omega, L)$ from under the sum in [8.6] leaves you with nothing but a pure differentiation, multiplied by the unidirectional cable response, further multiplied by a single consolidated random variable (the summation of α's), which you may adjust to produce the appropriate worst-case behavior. ELFEXT noise grows in direct proportion to frequency and generally with the square root of the cable length.

FEXT must be included in the SNR budget for systems that transmits on multiple pairs in the same direction.

Whenever you encounter a situation where FEXT is important, you may also wish to consider the effect of *timing skew* between the pairs. Timing skew is the worst-case (maximum) difference between the propagation delay of any of the pairs in a multipair data link. Timing skew affects any data link that starts with a single transmitted data stream, parcels it out among more than one parallel path, and then reassembles the fragments at the far end. The fragments arriving at the receiver must be de-skewed before they are reassembled into a complete data stream. The de-skew logic realigns the arriving fragments by building out the delay of the quickest pairs to match the delay of the slowest pairs.

In UTP data links, the timing skew slowly varies over the period of a day with the changes in temperature of the cable. It is not sufficient to measure the skew once when the link is first powered on and then forever compensate with a fixed de-skew circuit. Data arriving at the receiver must be continuously de-skewed before assembling it into a complete data stream.

Your choice of fragment size has a dramatic impact on the complexity of the de-skew circuitry. If you select a fragment size whose duration is smaller than the timing skew (for example, a single bit), you will have to contend with the possibility that fragments from one data pair may be delayed by more than one full fragment, as compared to fragments from another pair. Therefore, the de-skew circuitry must incorporate means to detect and track which fragment follows which.

If, on the other hand, you select a fragment size whose duration is larger than the timing skew (for example, a 32-bit word), the order of reassembly for fragments is never in question. The de-skew circuitry merely realigns the fragments to begin on exact timing boundaries, using minor timing adjustments smaller than the fragment size. Larger fragments have the advantage of simplifying the de-skew circuitry, but the disadvantage of increasing overall latency.

POINTS TO REMEMBER

> FEXT is modeled as a single noise source with a random amplitude.
> FEXT grows at a rate of 20 dB per decade.

---

[74] The crosstalk signal generated by segment n travels along the aggressor pair from position 0 to position $n\lambda$, where the crosstalk happens, and thence along the victim pair from that point to the end of the cable.

8.3.7 Power Sum NEXT and ELFEXT

TIA/EIA 568-B specifies for four-pair horizontal wiring, categories 5e and 6, the worst-case aggregate power coupled into any victim pair as being no more than 3 dB worse than the noise power coupled from the most aggressive of the other three pairs within the same jacket.[75]

This number compares with the $10\log(3) = 4.77$ dB increase in total coupled power you would expect from three uncorrelated sources of equal strength. In effect, this specification says that while there may be one combination of pairs that press up against the limit for pair-to-pair NEXT or ELFEXT, not all combinations of pairs may do so.

POINT TO REMEMBER

➤ Within a single jacket there may be one combination of pairs that press up against the limit for pair-to-pair NEXT or ELFEXT, but not all combinations of pairs may do so.

8.3.8 UTP: Radio-Frequency Interference

Radio waves impinging on a twisted-pair cable from faraway sources influence both conductors equally, generating common-mode currents within the cabling. In some cases the common-mode currents can be quite large. Any imbalance in the connectors, transceivers, or cables will convert part of the common-mode induced power into a differential mode, which is picked up by differential receivers.

In a UTP system, the best antidote for RFI is good signal balance. To the extent that you can make the system perfectly symmetrical, the conversion from common to differential mode won't happen. Good transceivers that deliver excellent common-mode balance usually couple to the line through a transformer or common-mode choke (or both). Such transceivers are largely immune to external RFI as well as 60-Hz power shifts. Keep in mind that any system with a transformer will require the use of a DC-balanced code[76] or a DC-restoration circuit in the receiver.

Several authors have attempted to quantify the expected levels of RF noise on UTP cabling. The best presentation I've seen on this subject was delivered by W. Michael King of Costa Mesa, California, in a speech to the IEEE 802.3 Ethernet committee in September 1993. In that presentation, Mr. King emphasized the point that RF pickup is a four-stage process. To predict actual levels of interference in an indoor LAN-type application, you need to know

1. The intensity of RF fields in the outside environment,
2. The degree of RF shielding provided by a building,
3. The degree to which RF fields couple differentially into category 3 UTP, and

---

[75] In the case of category 6 NEXT the power-sum limit exceeds the pair-to-pair limit by only 2 dB.
[76] A DC-balanced data code is designed to equalize the numbers of ones and zeroes in the data stream, making it possible to transmit the code successfully through transformers and other AC-coupled circuits. An example of a DC-balanced data code is the well-known Manchester code.

4. The effect of input filtering present in the LAN receiver.

The intensity of RF fields in the outdoor environment (U.S. only) may be estimated based on the relevant power, frequency, and antenna height restrictions imposed on the AM, FM, and TV broadcast industry by the Federal Communications Commission. This data has been compiled by, among others, Don White Consultants and is summarized in Figure 8.23. Also shown in that figure are actual measurements of RF field intensity in the AM broadcast band taken by AT&T in the city of Chicago (deemed representative of a major U.S. metropolitan area). Potential emissions from mobile radios, which in contrast to major broadcast services can be moved quite close to the affected wiring, are depicted as well. The limits drawn for mobile radio emissions are those imposed by EN55024 specifications. The general conclusion from Figure 8.23 is that there are two major bands of interference, the AM band from .540 to 1.620 MHz and the Mobile/TV/FM from 27 to 200 MHZ and beyond. In each band the worst-case outdoor field strength expected at an ordinary commercial building site is 3 V/m. There are no licensed high-powered radio sources in the United States occupying the midrange from 1.620 to 27.0 MHz.

Bob Smith[77] (of Synoptics, later known as Bay Networks) in 1992 and 1993 measured FM and AM radio station pickup on category 3 cables. His measurements are reproduced in Figure 8.24. These measurements were performed using a single segment of well-terminated category 3 cable, 100-m in length, held in the horizontal polarization, and measuring the differential pickup. Mr. Smith's measurements were made using actual AM and FM broadcast signals located near San Jose, California. For each station, he carefully measured the local field strength using a calibrated loop antenna and then normalized his measured

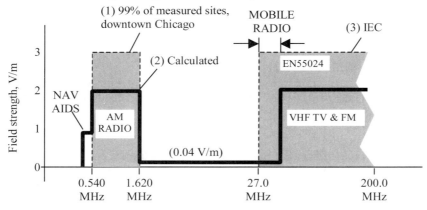

(1) D. N. Heirman, Bell Labs, National Telecomm. Conference, Nov. 29, 1976, Dallas, TX
(2) Don White Consultants, "Projected Electric Field Intensities From Licensed Services"
(3) IEC 801-3 (EN55024) specifications

Figure 8.23—Interoffice products must tolerate 3 V/m RF fields in the AM and Mobile/TV/FM bands.

---

[77] The same Bob Smith who patented the common-mode termination used on some FDDI-Copper and 100BASE-TX transceivers.

8.3.8 • Radio-Frequency Interference

differential pickup to predict what *would have resulted* from the presence of a 3 V/m local electric field. The measurements were made outdoors and so do not benefit from the building-attenuation effect anticipated in real-life LAN applications. In the absence of building attenuation you may conclude from this figure that RF pickup for horizontally polarized cables is less than 40 mV in the AM band and less than 200 mV in the Mobile/FM/TV band.

Mr. Smith's measurements represented cable pickup in a long, horizontal cable run. What about vertical cable runs, as often happen in tall buildings?

In the AM band most buildings efficiently attenuate the RF field intensity. Try it. With a portable AM radio, find out how many stations you can pick up outside your building, and then move the receiver indoors. The indoor reception is usually much worse. In a tall steel building the attenuation of AM radio stations ranges anywhere from 30 to 50 dB. This attenuation is so impressive that AM radios almost never work in high-rise buildings. RF pickup of AM radio stations on category 3 cable in high-rise buildings is therefore not much of a problem. Wooden or brick buildings don't provide as much attenuation as a steel high-rise, but neither are they as tall, so the horizontally-measured results in Figure 8.24 should apply.

In the FM radio band building attenuation is almost nonexistent, especially near the edges of the building. FM radio stations come through practically full-strength. Even worse, some mobile services may be literally located *inside* the building, where any putative building attenuation couldn't possibly help. You can't count on building attenuation to help you in the Mobile/FM/TV band. For category 3 cable systems, you must either design your system to work with 200 mV of RF noise or provide a low-pass filter in the receiver to filter out potential Mobile/FM/TV interference.

In a typical installation, only one RF source will produce the lion's share of the interference, so you don't have to worry about summing noise across multiple interference sources. The worst-case noise limits should be set at 40 mV in the AM band and 200 mV in the Mobile/FM/TV band.

If you limit the spectral content of your signals to the band below 27 MHz and provide

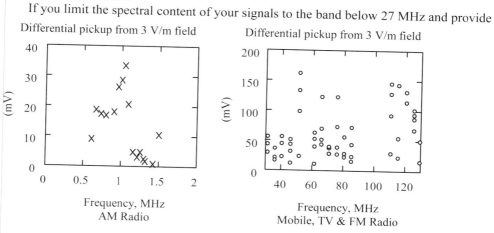

Figure 8.24—The susceptibility of typical category 3 UTP to an external 3 V/m RF field has been measured in the AM radio band (Bob Smith, Sept. 1993) and also in the Mobile, TV & FM band (Bob Smith, Oct. 1992).

good low-pass filtering in the receiver to cut off all interference above 27 MHz, you can eliminate most of the Mobile/FM/TV interference. That should result in a category 3 data link that picks up less than 40 mV of noise in most commercial situations.

Exceptions to this rule could include

> Installations immediately adjacent to high-powered commercial broadcast antennas,
> Installations immediately adjacent to high-powered military broadcast services, and
> Installations immediately adjacent to low-powered local transmitters if they are located within a few feet of the cabling.

Category 5e cabling, which sports a tighter twist and better overall balance of each pair, can be expected to pick up approximately 12 dB less RF interference than category 3 cabling at all frequencies. For category 6 cabling, the expected difference is approximately 17 dB.

POINTS TO REMEMBER

> The best antidote for RFI is good signal balance.
> A 27-MHz low-pass filter applied to category-3 horizontal cabling should cut RFI to less than 40 mV in most commercial situations.
> Categories 5e and 6 cabling pick up less RFI.

8.3.9 UTP: Radiation

UTP common-mode signals radiate much more effectively than differential mode signals. The key to obtaining good radiated performance is good common-mode balance. This problem is equivalent to the problem of hardening your system against RFI, and the same solutions apply.

Examples of systems that can be made to pass FCC Class-A radiated emissions, given a sufficiently well-balanced coupling network, appear in Table 8.7. For each link type, the table lists the number of simultaneously active pairs, the overall data transfer rate, the baud rate on each pair, and the maximum differential dV/dt on each pair. If you are planning a system with either a faster or larger signal swings than these known-good designs, now would be a good time to review your product plans.

The 100BASE-TX and 1000BASE-T systems are both *scrambled*, meaning that the outgoing data at the transmitter has been multiplied (X-OR'd) with a known, repetitive pseudorandom sequence. The inverse operation is then applied in the receiver to recover the original data. Scrambling accomplishes several purposes:

1. Scrambling spreads the spectral power density of the transmitted signal over a wide range of frequencies. This reduces the peak radiation in any one band.
2. Scrambling reduces unintentional correlation between successive bits in the outgoing data stream. This helps the adaptive equalizer circuits to converge quickly and efficiently.

3. Scrambling reduces unintentional correlation between the outgoing and received data streams. This makes the near-end noise cancellation systems work better.

Scrambling in the 100BASE-TX is applied after 4B5B data coding, with the unfortunate result that several properties of the 4B5B data code, like DC-balance and limited run-length, are destroyed. The 1000BASE-T standard takes the more reasonable approach of scrambling the data *prior* to coding.

Table 8.7—Systems That Pass FCC Class-A Radiated Emissions Limits

System	Cable	Active pairs within cable	Data rate Mb/s	Baud rate per pair	Max. diff. dV/dt
IEEE 802.3 10BASE-T	Cat. 3	2	10	20 Mbaud (10 MHz Manchester)	0.16 V/ns
IEEE 802.3 100BASE-TX	Cat. 5	2	100	125 Mbaud (3-level MLT-3) *scrambled*	0.33 V/ns
IEEE 802.3 1000BASE-T	Cat. 5	4	1000	125 Mbaud (5-level PAM with pre-distortion) *scrambled*	0.33 V/ns

POINTS TO REMEMBER

➢ The key to obtaining good radiated performance is good common-mode balance.

➢ Scrambling spreads the spectral power density of the transmitted signal, reducing the peak radiation.

8.4 UTP CONNECTORS

One outstanding feature of UTP cabling is the connector system. The UTP connector system is based on the pinching action of a single stamped-metal fitting that strips the wire, holds it in place, and establishes a gas-tight electrical connection all at the same time. The rather high insertion force required to make this connection is supplied by a handheld, mechanically activated *punch-down tool* (shown in Figure 8.25) or by the crushing action of the two halves of an RJ-45 connector shell as that connector is crimped onto the end of a UTP cable (Figure 8.26). A good technician with a punch-down tool can cut and terminate 24 pairs of UTP in about the time it takes to open the bag and sort out the parts associated with one coaxial connector.

These connectors are cheap, and the performance is outstanding. Category 5 UTP connectors based on punch-down technology are suitable for data transmission rates up to

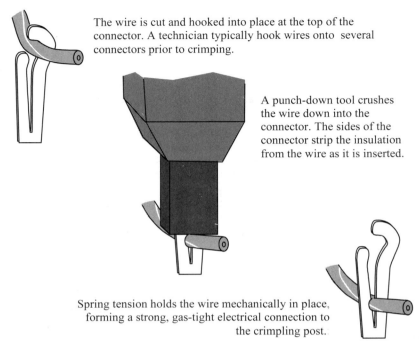

The wire is cut and hooked into place at the top of the connector. A technician typically hook wires onto several connectors prior to crimping.

A punch-down tool crushes the wire down into the connector. The sides of the connector strip the insulation from the wire as it is inserted.

Spring tension holds the wire mechanically in place, forming a strong, gas-tight electrical connection to the crimpling post.

Figure 8.25—One common variation of punch-down connection technology.

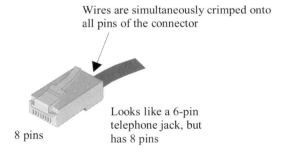

Wires are simultaneously crimped onto all pins of the connector

8 pins

Looks like a 6-pin telephone jack, but has 8 pins

Figure 8.26—RJ-45 UTP data connector.

125 Mbaud. A punch-down connector is also called an *insulation displacement connector* (IDC).

The RJ-45 connector goes by many names (see Table 8.8). In standard practice, work area cables use the same RJ-45 connector at the equipment and also at the wall outlet.

Table 8.9 lists the pin assignments used by Ethernet for two-pair UTP connections. This table provides two different pin assignment listings for devices with and without an internal crossover. As explained in Section 7.6, "Crossover Wiring," the customary LAN wiring arrangement presupposes a hub port equipped with internal crossover and a client

8.4 • UTP Connectors

Table 8.8—Different Names for the RJ-45 Connector

Name	Description
RJ-45	Common name for the traditional 8-pin data plug or jack. This is the name most used by data installers.
ISO 8877	An ISO connector specification to which RJ-45 conforms.
IEC 603-7 Detail Specification for Connectors, 8-Way	The latest IEC specification of the RJ-45 connector. It adds new mechanical characteristics to **ISO 8877**.

port without. In this case the wire is expected to connect the pins straight through, from pin 1 to pin 1, 2 to 2, and so on.

LAN ports equipped with an internal crossover should always be marked with an "X" symbol.

If the connectors at both ends of a link are marked with an "X", or neither is marked, then an external crossover may be required.

The customary correspondence between connector pins and the wiring color code, for North America, also appears in Table 8.9. On an RJ-45 connector, the physical wiring pairs are assigned to pins [4,5] and [3,6], and [1,2] and [7,8], in that order. Single-pair telephone systems most often use pair 1.

One of the most common installation errors involves accidentally swapping the two wires of a pair, for example, BLU/WHT and WHT/BLU. This mistake is called a *polarity reversal*. Some systems are designed to tolerate polarity reversal. Systems that tolerate polarity reversal greatly simplify installation.

As pointed out in TIA/EIA 568-B, a good cable, coupled with good connectors, is not enough to ensure good overall link performance. Other factors, such as the quality of the jumper cables, the total number of connections in a link, and the care with which the connecting components were installed, may affect link performance. Nevertheless, so you'll have some understanding of the magnitude of connector effects, Table 8.10 and Table 8.11 list the attenuation and noise performance of TIA/EIA 568-B connecting hardware.

Table 8.9—10BASE-T North American Contact Assignments for RJ-45

RJ-45 contact	10BASE-T No internal crossover	10BASE-T With internal crossover	UTP pair names	UTP conductor colors
1	—		Pair 3 +	GRN/WHT
2	—		Pair 3 –	WHT/GRN
3	RX +	TX +	Pair 2 +	ORG/WHT
4	TX +	RX +	Pair 1 +	BLU/WHT
5	TX –	RX –	Pair 1 –	WHT/BLU
6	RX –	TX –	Pair 2 –	WHT/ORG
7	—		Pair 4 +	BRN/WHT
8	—		Pair 4 –	WHT/BRN

> **A Warning About RJ-45 Wiring**
>
> You may violate the RJ-45 color code by consistently substituting one colored pair for another. Such a substitution has no impact on the electrical performance of the wiring. If your system supports polarity reversal, you may even swap the individual wires in any colored pair.
>
> But one thing you must never do is split the pairs apart. For example, a crossing of BLU/WHT and ORG/WHT at the connector will induce massive crosstalk between what is supposed to be pair 1 and pair 2. When the pairs get crossed, results are unpredictable. Service providers will charge a lot to debug a pair-crossing error.

Specialized connectors are available for use with 150-Ω STP-A cabling whose performance extends well beyond the basic frequency ranges indicated in the TIA/EIA 568-B standard.

In an ordinary, uncompensated connector the parasitic effects, which often involve a preponderance of either inductive or capacitive coupling, generate NEXT and FEXT noise with roughly the same magnitude. The asymmetry evident in the specifications below gives you a hint that the connectors are highly compensated for the purpose of reducing NEXT. This may be done by controlling the precise shapes of the metal electrodes and the proximity with which they approach each other within the shell of the connector.

Unfortunately, compensating elements that reduce crosstalk within one specified frequency band sometimes dramatically exaggerate it at other frequencies outside the band. Never extrapolate the crosstalk performance of a connector beyond its specified limits.

POINTS TO REMEMBER

> ➢ UTP connectors are cheap, and the performance is outstanding.
> ➢ Systems that tolerate polarity reversal greatly simplify installation.

Table 8.10—Insertion Loss of Connecting Hardware

Connector type	Freq. Range MHz	Max. allowed Insertion loss dB
cat-3	$1 \leq f \leq 16$	$0.1\sqrt{f}$
cat-5e	$1 \leq f \leq 100$	$0.04\sqrt{f}$
cat-6	$1 \leq f \leq 250$	$0.02\sqrt{f}$
150-Ω STP-A	$1 \leq f \leq 300$	$0.025\sqrt{f}$
NOTE (1)—All values taken from TIA/EIA-568-B.2-2001 and TIA/EIA-568-B.2-1-2001. NOTE (2)—In all formulas, the frequency variable f is in MHz.		

Table 8.11—Noise Budget Items for Connecting Hardware

Connector type	Freq. range MHz	Min. allowed loss dB
cat-3	$1 \leq f \leq 16$	NEXT $2.1 - 20\log_{10}(f/100)$
cat-5e	$1 \leq f \leq 100$	NEXT $43.0 - 20\log_{10}(f/100)$
	$1 \leq f \leq 100$	FEXT $35.1 - 20\log_{10}(f/100)$
	$31.5 \leq f \leq 100$	RL $\quad 20.0 - 20\log_{10}(f/100)$
cat-6	$1 \leq f \leq 250$	NEXT $54.0 - 20\log_{10}(f/100)$
	$1 \leq f \leq 250$	FEXT $43.1 - 20\log_{10}(f/100)$
	$50 \leq f \leq 250$	RL $\quad 24.0 - 20\log_{10}(f/100)$
150-Ω STP-A	$1 \leq f \leq 300$	NEXT $46.5 - 20\log_{10}(f/100)$
	$16 \leq f \leq 300$	RL $\quad 20.1 - 20\log_{10}(f/100)$

NOTE (1)—All values taken from TIA/EIA-568-B.2-2001 and TIA/EIA-568-B.2-1-2001.
NOTE (2)—In all formulas, the frequency variable f is in MHz.
NOTE (3)—RL stands for return loss. In each case, from 1 MHz to the specified lower limit, the RL specifications are flat.
NOTE (4)—FEXT and RL are not specified for category 3 connectors.

8.5 ISSUES WITH SCREENING

TIA/EIA 568-B.1 clause 4.6 requires that the drain wire of all screened cables (ScTP) be terminated at both ends. In the equipment room the drain wire is terminated to the telecommunications grounding bus bar. In the work area the drain wire is grounded to the chassis of the local equipment, which connects to the local power grounding system.

Unfortunately, as explained in Section 6.12.2, "Immunity to Large Ground Shifts," it is never a good idea to make direct ground connections between systems with separate AC power inputs.

To the credit of TIA/EIA 568-B.1, the standard does stipulate that prior to connection, the AC voltage between the two ends of each cable shall not exceed 1 V rms. That's a nice-sounding number, but in the field you can't expect technicians to check this specification. Even if your installation people do check, they aren't likely to observe the worst case because currents circulating in the ground fluctuate wildly during the day corresponding to the amount of electrical power consumed by major metropolitan areas.[78]

---

[78] For example, a surge of ground noise occurs during the Super Bowl at halftime when viewers all over America activate their blenders and refrigerators.

Point to Remember

> Even though screened cables are heavily favored in Europe, this author does not recommend their use.

8.6 CATEGORY-3 UTP AT ELEVATED TEMPERATURE

Both TIA/EIA and ISO/IEC specifications for horizontal cabling apply at a nominal inside-building operating temperature of 20 °C. When operating at elevated temperatures, the increased resistance of copper exaggerates the skin-effect losses. For category 5e and 6 cables, whose losses are dominated by the skin effect, the attenuation in dB scales directly in proportion to the increase in resistance of the base metal (becoming larger at elevated temperatures). TIA/EIA 568-B specifies a de-rating factor for categories 5e and 6 of 0.4% insertion loss (in dB) for each degree above 20 °C.

The performance of category 3 cable suffers more noticeably at high temperatures, because its de-rating factor depends on both skin effect and dielectric properties. Because the dielectric properties are not well specified, your performance will vary. TIA/EIA 568-B recommends a de-rating factor of 1.5% for each degree above 20 °C for category 3 cabling and also recommends that it not be used in hot environments.

Plenum-rated category 3 cables are made from plastic materials with a higher glass-transition temperature than ordinary PVC-insulated category 3 cables, rendering their electrical properties more stable at elevated temperatures. According to ISO 8802.3 clause 14 (Ethernet 10BASE-T), "The loss of PVC-insulated cable exhibits significant temperature dependence. At temperatures greater than 40° C (104° F) it may be necessary to use a less temperature-dependent cable, such as most plenum-rated cables."

What this means to you is simple: *Never use PVC-insulated category 3 cables in an uncooled attic space.*

Typical (not worst-case) attenuation measurements taken by Synoptics (later Bay Networks, later part of Nortel) were presented to the token-ring committee IEEE 802.5 in March 1993. These measurements are reproduced in Figure 8.27.

Point to Remember

> Never use PVC-insulated category 3 cables in an uncooled attic space.

8.6 • Elevated Temperature

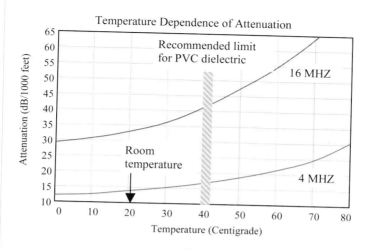

Synoptics, March 1993 presentation to IEEE
802.5 token-ring standards committee

Figure 8.27—The performance of PVC-insulated category 3 cable (non-plenum rated) varies dramatically with temperature.

For further study see: www.sigcon.com

CHAPTER 9

150-Ohm STP-A Cabling

In 1985 IBM rocked the world of integrated voice and data cabling with the introduction of the IBM wiring plan. So bold, so daring, so robust, so overbuilt, so incredibly heavy, so expensive, and so difficult to install, it landed with a deep blue thud on the doorsteps of MIS managers around the world. Few picked it up.[79]

What remains of the system today is 150-Ω twinax, also called 150-Ω STP-A, or IBM Type-I cable (see Figure 9.1). IBM managed at one point to get this twinax cable written into the EIA and ISO building wiring standards, although it is rarely used today and is not expected to be carried forward into future versions of international building-wiring standards. Nevertheless, the performance of the cable is so extraordinary it merits discussion.

One hundred-fifty-Ω STP-A generally contains two independent twisted pairs.[80] The

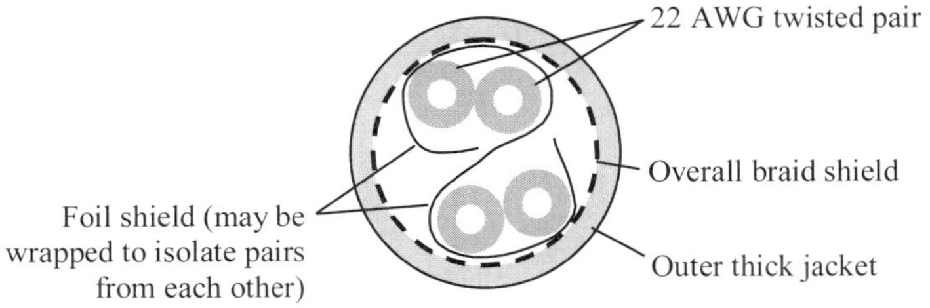

- 22 AWG twisted pair
- Overall braid shield
- Foil shield (may be wrapped to isolate pairs from each other)
- Outer thick jacket

Figure 9.1—Construction of two-pair 150-Ω STP-A.

---

[79] If you are one of the unfortunate customers who installed this cable, it's probably still in the walls. It's so heavy I don't think you can rip it out once it's installed. I've heard stories about old buildings in Manhattan that depend on IBM cables for structural support.

[80] Some of the initial IBM variants contained as many as two twisted data pairs, plus four additional twisted pairs for telephone usage.

cable incorporates an overall outer braid shield and also a plastic and metal-laminated foil shield. As shown in the figure, the metal-laminated foil shield may wrap individually around both pairs, isolating them from each other. The finished outside jacket diameter of this cable is 0.43 inch. Compared to a four-pair UTP diameter of 0.25 in., the cross-sectional area (and thus the cable reel volume and weight) for two-pair 150-Ω STP-A is three times larger, yet it contains only half as many pairs.

The time you want to think about specifying this cable is when you need a quick and dirty transceiver for a first product release (or beta-trial) and don't have time to design a complete, professional, UTP solution. The 150-Ω STP-A solution will be easier to design, because the cable bandwidth is so terrific, but it places a terrible installation burden on your customer.

POINT TO REMEMBER

➢ Think about 150-Ω STP-A when you need a quick and dirty transceiver for a first product release (or beta-trial).

9.1 150-Ω STP-A SIGNAL PROPAGATION

One hundred-fifty-ohm STP-A and UTP are geometrically and electrically very similar. You will find a discussion of 150-Ω STP-A signal propagation in Section 8.1, "UTP Signal Propagation."

9.2 150-Ω STP-A NOISE AND INTERFERENCE

There are three major sources of noise and interference in a high-speed 150-Ω STP-A system:

1. Far-end reflected noise
2. Near-end crosstalk (NEXT)
3. Skew

Before we look at the major noise and interference effects, I'd like to say a word about some effects that *don't* show up in this list: *near-end reflections, alien crosstalk,* and *far-end crosstalk*. These problems don't tend to arise in 150-Ω STP-A systems because of the way typical applications use 150-Ω STP-A.

All known 150-Ω STP-A LAN applications use the pairs unidirectionally: one pair for data traveling in one direction and the other pair for data traveling in the opposite direction. As a result, problems with near-end echoes don't arise, because when you transmit on one pair, you are not also listening on that same pair. Similarly, because both pairs are completely used, leaving no room for other alien applications within the same cable jacket, alien crosstalk doesn't matter. Lastly, far-end crosstalk doesn't arise because it requires two parallel transmission channels working in the same direction—something that doesn't happen with 150-Ω STP-A.

Why is 150-Ω STP-A always used in the unidirectional mode? The reason has to do with the typical development scenario for most LAN-type products. Early in their life, several high-speed LAN standards historically endorsed 150-Ω STP-A, because the bandwidth of the cable is so great that it is easy to make it work. The early adopters that purchase new LAN standards are often willing to undergo the pain of 150-Ω STP-A installation in order to gain the benefits of bleeding-edge LAN technology. Later, as the standard gains credibility and users want to deploy it in larger quantities, manufacturers invest in the development of UTP-style transceivers. The UTP transceivers are undoubtedly more complex and more costly, but the extra expense is more than compensated by the reduced installation cost associated with UTP. Examples of LAN standards that were originally available in 150-Ω STP-A format include 10 Mb/s Ethernet, Token Ring, FDDI, 100 Mb/s Ethernet, and 1000 Mb/s Ethernet. In all cases the UTP versions of these transceivers, once available, conquered the market.

The subjects of far-end reflections and near-end crosstalk are so similar to the UTP case that I have incorporated them into Section 8.3.2, "UTP: Near-End Reflections" and Section 8.3.4, "UTP: Near-End Crosstalk".

POINT TO REMEMBER

> When 150-Ω STP-A is used in a unidirectional mode it is not subject to near-end reflections, alien crosstalk, or far-end crosstalk.

9.3 150-Ω STP-A: SKEW

Inside a 150-Ω STP-A cable, the two wires of an individual pair couple more heavily to their shield than they do to each other. The net result is that the two wires act like a pair of weakly coupled transmission lines, with a near-end crosstalk coupling factor of only 3.7%.[81] If you have studied coupled transmission lines, you will know that such a system can support multiple modes of transmission and that the speed of propagation among the various modes need not be equal. In particular, the signal on one wire might, due to slight differences in the dielectric composition of the wires, arrive ahead of the signal on the other wire.[82] In extreme cases, the resultant skew could diminish the size of the eye opening of the far-end received signal. The skew problem in 150-Ω STP-A is particularly acute because of the short baud intervals involved in typical 150-Ω STP-A applications. For example, the Gigabit Ethernet 1000BASE-CX specification uses an 150-Ω STP-A-like cable with a baud

---

[81] Working with the two wires of a single pair, drive the (+) wire with respect to the shield and connect the (-) wire through 75 ohms to the shield. The near-end crosstalk induced on the (-) wire is 3.7%. The two wires are well balanced in that they bear equal impedances to the common shield, but they are hardly coupled to each other.

[82] Early pre-production versions of this cable used a colored dielectric insulation. The various colors of ink used to tint the wires had different dielectric constants that affected the speed of propagation on the two wires. Production cables available today generally use a thin painted coloration, partly to help control skew.

period of only 800 ps. Compared to such a tiny baud period, even a few tens of picoseconds of skew can have a measurable effect on the quality of the received signal.

One hundred-fifty-ohm cables with STP-A construction are available with a skew specification as tight as 150 ps in 25 m (this meets the requirements of Gigabit Ethernet 1000BASE-CX).

POINT TO REMEMBER

> ➢ Inside a 150-Ω STP-A cable, the signal on one wire of a pair might arrive ahead of the signal on the other wire.

9.4 150-Ω STP-A: RADIATION AND SAFETY

Radiation from high-speed 150-Ω STP-A data links is controlled by proper grounding of the shield at both ends of the link. The 150-Ω STP-A shield must be connected to ground at both ends to be effective. This connection must have a very low impedance at all frequencies within the data spectrum.

Unfortunately, this dual-grounding requirement runs directly counter to AC power-safety grounding considerations (see Section 6.12.2, "Immunity to Large Ground Shifts").

Any form of shielded twisted-pair cable should, in this author's opinion, be used only for short connections within a wiring closet or computer room between pieces of equipment intentionally tied to the same ground. Your specification should explicitly state that this must be the case. Between such pieces of equipment there will be no large circulating ground currents. For longer connections, you should provide other links types that do not require grounding at either end, such as multimode fiber, single-mode fiber, or UTP.

The susceptibility and radiation budget for Gigabit Ethernet requires a ground transfer impedance (that is the impedance between the chassis and the shield of the cable) of 0.1 Ω or less at 625 MHz. Connectors that meet this ground-transfer impedance specification provide a direct metallic connection between the chassis and the shield that goes all the way around the connector pins, completely enclosing the signal conductors. This specification is difficult (if not impossible) to achieve with any sort of pigtail connection of the shield to the chassis or with an AC-coupling capacitor in series with the shield.

A discrete capacitor can't be used to make an AC-coupled shield because the ground-transfer impedance requirement of 0.1-Ω at 625 MHz implies an effective series inductance of less than 25 pH. That small an inductance cannot be implemented in a leaded component. If you are dead-set on building an AC-coupled shield, try making a thin capacitive gasket of dielectric material distributed all the way around the connector shell, insulating the connector shell from the chassis, and then use plastic screws to hold the connector in place. I have seen proposals for this type of connector, but have not seen one working in actual practice.

CHAPTER **10**

COAXIAL CABLING

Coaxial cabling is the oldest and most venerable of transmission mediums. It has been in use since at least 1898, when Hertz made his original investigations of wave propagation on transmission lines. Despite its age, the electrical performance of coaxial cable is as good as anything else.

Physically, coax is a mess (see Chapter 7, "Generic Building-Cabling Standards"). The problem with coax is topological (Figure 10.1). To access the inner conductor, you must first peel back the outer jacket, outer shield, and inner dielectric. This error-prone procedure takes a long time in the field and often results in an unreliable connection. Pre-fabricated

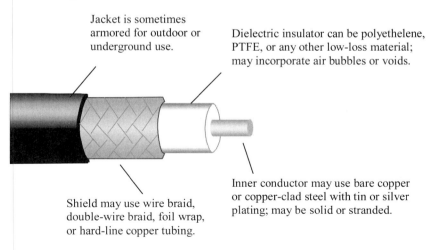

Jacket is sometimes armored for outdoor or underground use.

Dielectric insulator can be polyethelene, PTFE, or any other low-loss material; may incorporate air bubbles or voids.

Shield may use wire braid, double-wire braid, foil wrap, or hard-line copper tubing.

Inner conductor may use bare copper or copper-clad steel with tin or silver plating; may be solid or stranded.

Figure 10.1—A coaxial cable has four parts.

jumper cables made with good tools, under ideal manufacturing conditions, work just fine. Field-applied connectors don't. Coax connectors (compared to UTP) are extremely difficult to apply.

Coaxial cable suffers from an overabundance of standards. To begin with, the mechanical properties of coaxial cable are controlled by a series of RG specifications. These specifications were written in the early sixties by the American military. The electrical properties in the RG specifications are very loosely specified. These specifications leave open many questions having to do with conductor construction, stranding, surface roughness, surface plating, use of steel cores, and dielectric material selection, all of which affect the electrical properties. Most modern cables rated for a particular RG class of operation will far outstrip the capabilities of that class.[84] If you want to know how a particular cable will perform, you have to examine the vendor's specifications for that type of cable.

From among all the thousands of variants available, I've selected a particular set of cables manufactured by Belden *(www.belden.com)* for all the examples in this chapter. The cables are listed in Table 10.1.

Table 10.1—Selected Belden Coaxial Cable Types

Belden cable type	Class	Jacket O.D. (in.)	Conductor stranding and gauge (AWG)	Conductor composition	Dielectric
8216	RG-174/U	0.110	7 x 34	Bare copper-plated steel	Polyethylene
84316	RG-316/U	0.098	7 x 33½	Copper-plated steel with silver coating	TFE Teflon
8259	RG-58A/U	0.193	19 x 32	Tinned copper	Polyethylene
8240	RG-58/U	0.193	20 Solid	Bare copper	Polyethylene
84303	RG-303/U	0.170	18 Solid	Copper-plated steel with silver coating	TFE Teflon
8237	RG-8/U	0.405	7 x 21	Bare copper	Polyethylene

POINTS TO REMEMBER

➢ The electrical performance of coaxial cable is as good as anything else, but physically, coax is difficult to handle.

➢ Coaxial cable suffers from an overabundance of standards.

---

[84] By as much as a factor of two.

10.1 COAXIAL SIGNAL PROPAGATION

Coaxial cable is a single-ended transmission medium, meaning that the signal current flows through the main signal path (the inner conductor) and then returns to its source through a low-impedance, hopefully zero-voltage-drop return path (the outer shield). The dielectric exists to hold the center conductor in the middle of the structure, keeping it from flopping over to one side and shorting against the shield.

A good coaxial cable presents a nearly uniform impedance at all frequencies above the onset of the skin effect. The following equation shows the impedance obtained in the skin-effect region, with signal current flowing only on the facing surfaces of the inner and outer conductors.

$$Z_0 = \frac{60}{\sqrt{\epsilon_r}} \ln\left(\frac{d_2}{d_1}\right) \qquad [10.1]$$

where ϵ_r is the dielectric constant (real part of relative electric permittivity) of the insulating material,

d_2 is the inside diameter of the shield, m, and

d_1 is the effective diameter of the signal conductor, m.

The onset of the skin effect happens at about 100 KHz in RG-58 and related coaxial cables.

Mechanical perturbations in the cross-sectional geometry of the cable can affect its characteristic impedance. If, however, the perturbations are small and closely spaced compared to the wavelength of the signals used, and if the imperfections are uniformly distributed, the impedance remains unchanged. For example, the surface roughness on the inside face of a braided shield will increase the skin effect loss of the cable, but not the impedance.

Coaxial cables are readily available with characteristic impedances near 50, 75, or 93 ohms. Fifty ohms is most popular for test equipment applications; 75 Ω is most popular for audio-visual applications.

Electrical specifications for selected Belden coaxial cable types appear in Table 10.2. Coaxial cables are not subject to rigid standardization, as are UTP and some varieties of fiber cables, so you will find many, many types and varieties of coaxial cabling. I've selected only a small number of representative cables for study. Please note that these are cable specifications, not overall channel specifications. Channel specifications (which include connectors and patch cords) are always worse (see Section 7.2, "SNR Budgeting").

What follows is a discussion of how to compute the metallic-transmission model parameters for coaxial cables (See Section 3.1, "Signal Propagation Model"). For most coaxial cables a reasonable value for ω_0 is 10 MHz. Parameters Z_0, v_0, and R_{DC} generally appear directly in a coaxial cable datasheet. Parameters R_0 and θ_0 must be estimated.

R_{DC} is calculated as the sum of the DC resistance per meter of the center conductor and the shield. These parameters are generally listed on the cable datasheet. If there are two shields (triaxial cable), use the DC resistance of only the inner shield.

$$R_{DC} = DCR_{CENTER} + DCR_{SHIELD} \qquad [10.2]$$

where DCR_{CENTER} is the resistance of the center conductor, Ω/m,

DCR_{SHIELD} is the resistance of the center conductor, Ω/m, and

R_{DC} is the overall DC resistance of the cable, Ω/m,

If the center conductor is composed of a copper-plated steel core (as is commonly used), you will need to know the plating thickness and the worst-case variation in plating thickness to work out the DC resistance. These parameters are often difficult to obtain from cable manufacturers. In that case you can always resort to your ohmmeter to measure some sample cables.

You should also know that a composite cable composed of a steel core with copper plating (and perhaps also with a third layer of silver plating) undergoes a complicated series of skin-effect transformations. Current is first expunged from the steel at surprisingly low frequencies. This happens because the large magnetic permeability of steel imbues the material with an astonishingly small skin depth. This effect causes a ramp in the AC resistance at low frequencies, followed by a relatively flat plateau where the conductor acts essentially as a hollow copper tube. The tube resistance is fairly flat until one reaches a skin depth comparable with the plating thickness, after which the resistance again resumes rising at a rate proportional to the square root of frequency. The overall effect of a composite structure is to broaden the transition region associated with the onset of the skin effect.

The equation for the AC skin-effect resistance is modified to separately account for the center conductor and shield resistances, where σ_1 and d_1 are the conductivity (S/m) and diameter (m) of the center conductor respectively and where σ_2 and d_2 are the conductivity and diameter of the shield. First compute the depth of penetration of current (skin depth) for

Table 10.2—Electrical Specifications for Selected Belden Coaxial Cables

Item			8216 (RG-174/U)	84316 (RG-316/U)	8259 (RG-58A/U)	8240 (RG-58/U)	84303 (RG-303/U)	8237 (RG-8/U)
Attenuation dB/100 m at 20 °C	MHz	1	6.23	3.93	1.38	.98	1.11	.52
		10	10.82	8.85	4.92	3.61	3.61	1.84
		50	19.02	18.36	12.13	8.20	8.85	4.26
		100	27.5	27.2	17.7	12.5	12.8	6.2
		200	41.0	39.3	26.6	18.4	18.4	9.2
		400	62.3	57.4	40.7	27.5	26.9	13.8
		700	88.5	77.7	58.0	38.4	36.1	19.3
		900	101.6	89.5	69.2	44.9	41.0	22.6
Z_0 (Ω)	Min		48	48	48	49.5	48	48
	Max		52	52	52	53.5	52	52
v_0/c	Nom.		.66	.695	.66	.66	.695	.66

10.1 • Signal Propagation

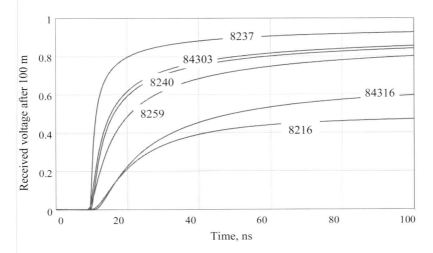

Figure 10.4—Comparison of worst-case step-response waveforms for six different Belden cable types. All cables 100 m length.

The raw size has more to do with high-frequency performance than any other single factor. The circumference of the conducting elements determines the skin-effect losses, which dominate the performance of the cable. Bigger cables almost always exhibit less loss.

The only limitation to the use of very large cables is the existence of non-TEM modes of propagation that may occur in circular waveguides. The lowest-order such mode is termed the cutoff frequency f_c for the waveguide.

$$\omega_c = .586 \frac{\pi}{d_2} \frac{c}{\sqrt{\epsilon_r}} \quad \text{rad/s}$$

$$f_C = \frac{.293}{d_2} \frac{c}{\sqrt{\epsilon_r}} \quad \text{Hz} \quad [10.7]$$

where d_2 is the inside diameter of the shield, m,

   ϵ_r is the relative dielectric constant of the dielectric material, and

   c is the velocity of light, $2.998 \cdot 10^8$ m/s.

In the case of RG-58 having a core diameter of $b = 2.95$ mm and $\epsilon_r = 2.3$, the lowest-order non-TEM mode appears at $f_c = 19.7$ GHz. Operation of this cable at or above 19.7 GHz begs for trouble. The bigger the cable, the lower the cutoff frequency.

Secondary differences between the cables involve the surface treatment on the conductors and the dielectric composition.

The main difference between the 8240 and 8259 cables is that the center conductor on the 8240 is bare solid copper, while the center conductor on the 8259 is both solder-tinned and stranded. Tinned conductors are easy to work with and don't corrode, but these features come at the cost of much increased cable attenuation. Stranding makes the center conductor more flexible and easy to work, but the surface roughness again increases the attenuation.

Plating the center conductor with silver works in the opposite direction. Silver plating helps *reduce* the attenuation. The resistivity of silver is about 8% less than copper. That's the approach taken in the 84303 cable. This cable also uses a TFE Teflon dielectric, which has a slightly lower dielectric constant compared to polyethylene (2.1 versus 2.6). For the same capacitance per foot, Belden can afford to fatten the center conductor, which again reduces the attenuation. This benefit must be traded off against the slight deterioration in dielectric loss of TFE Teflon versus polyethylene. The 84303 cable performs electrically a little better than the 8240. The same differences (silver-coated center conductor and TFE Teflon dielectric) distinguish the 84316 cable from the 8216.

If performance is super-critical in your application, ask for *sweep-tested* cable. That means the manufacturer actually tested the frequency response of the cable after it was made. This testing costs extra but roots out various manufacturing problems that can create subtle, annoying, repetitive defects in the cable structure. These repetitive defects can cause lumps in the frequency response. High-end video applications, for example, use sweep-tested cables.

POINTS TO REMEMBER

> - A good coaxial cable presents a nearly uniform impedance at all frequencies above the onset of the skin effect.
> - Coaxial cables formed from foamed, cellular, or helically-wrapped dielectrics exhibit a faster propagation velocity and less high-frequency loss than their solid-dielectric counterparts.
> - The step response duration for a coaxial cable scales roughly in proportion to the square of cable length.

10.1.1 Stranded Center-Conductors

To increase the mechanical flexibility of a cable, many manufacturers use a stranded center conductor. Patterns of either 7 or 19 wires are common (Figure 10.5). The effective diameter of a 7-way strand, for the purposes of computing the skin-effect resistance, is 2.63 times the diameter of each individual wire. The effective diameter of a 19-way strand, for the

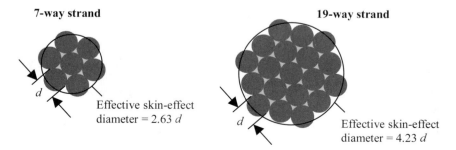

Figure 10.5—For the purpose of computing skin-effect losses, the effective diameter of a stranded configuration is less than the maximal outside diameter of the configuration.

purposes of computing the skin-effect resistance, is 4.23 times the diameter of each individual wire.

The effective diameter in each case was computed using a 2-dimensional method-of-moments field solver assuming a 50-ohm cable with a solid polyethelyne dielectric. A total of 120 line segments approximated the outer perimeter of the 7-strand or 19-strand configurations during this calculation. Other combinations of characteristic impedance and dielectric will return similar values of effective diameter.

The effective diameters for computing the DC resistances of these stranded-configuration wires are $\sqrt{7}$ and $\sqrt{19}$ times respectively, the diameters of the individual wires in each case.

10.1.2 Why 50 Ohms?

Article first published in *EDN Magazine*, September 14, 2000

Why do most engineers use 50-Ω pcb transmission lines (sometimes to the extent of being a default value for pcb layout)? Why not 60 or 70 Ω?
—Tim Canales

Given a fixed trace width, three factors heavily influence pcb-trace impedance decisions. First, the near-field EMI from a pcb trace is proportional to the height of the trace above the nearest reference plane; less height means less radiation. Second, crosstalk varies dramatically with trace height; cutting the height by half reduces crosstalk by a factor of almost four. Third, lower heights generate lower impedances, which are less susceptible to capacitive loading.

> **A characteristic impedance of 50 Ω minimizes the skin-effect losses in a solid-polyethylene coaxial cable.**

All three factors reward designers who place their traces as close as possible to the nearest reference plane. What stops you from pressing the trace height all the way down to zero is the fact that most chips cannot comfortably drive impedances less than about 50 Ω, so that's where one usually stops. (Exceptions to this rule include Rambus, which drives 27 Ω, and the old National BTL family, which drives 17 Ω).

Fifty ohms is not always best. For example, an old NMOS 8080 processor operating at 100 KHz doesn't have EMI, crosstalk, or capacitive-loading problems, and it can't drive 50 Ω anyway. For this processor, because very high-impedance lines minimize the operating power, you should use the thinnest, highest-impedance lines you can make.

Purely mechanical considerations also apply. For example, in dense, multilayer boards with highly compressed interlayer spaces, the tiny lithography that 70-Ω traces require becomes difficult to fabricate. In such cases, you might have to go with 50-Ω traces, which permit a wider trace width, to get a manufacturable board.

What about coaxial-cable impedances? In the RF world, the considerations are unlike the pcb problem, yet the RF industry has converged on a similar range of impedances for coaxial cables. According to IEC publication 78 (1967), the preferred values of coaxial cable impedance are 50, 75, and 100 Ω. The 75-Ω value is popular with radio engineers because you can easily match it to several common antenna configurations. The 50-Ω value is good for use with the most common dielectric (solid polyethylene) because it minimizes the skin-effect loss. The 100-Ω value was never widely adopted, although today 93-Ω cables are available for use in applications where you need a low value of capacitance per unit length.

You can prove the optimality of 50-Ω solid-polyethylene coaxial cable from basic physics. The skin-effect loss α_r (in decibels per unit length) is proportional to the total skin-effect resistance R_{AC} (per unit length) divided by the characteristic impedance Z_0 of the cable. The total skin-effect resistance R_{AC} is the sum of the shield resistance and center conductor resistance, both of which vary with the size of the conductors.

The skin-effect resistance of the coaxial shield varies inversely with its diameter d_2. The skin-effect resistance of the coaxial signal conductor varies inversely with its diameter d_1. The total series resistance R_{AC} therefore varies in proportion to $(1/d_2 + 1/d_1)$. Fixing the outer diameter of the shield d_2 and the relative electric permittivity of the dielectric insulation ϵ_r, you can minimize the skin-effect loss α_r as a function of d_1, starting with the following equation:

$$\alpha_r \propto \frac{\left(\dfrac{1}{d_2} + \dfrac{1}{d_1}\right)}{Z_0} \qquad [10.8]$$

In any elementary textbook on electromagnetic fields and waves, you can find the following formula for Z_0 as a function of d_2, d_1, and ϵ_r:

$$Z_0 = \frac{60}{\sqrt{\epsilon_r}} \ln\left(\frac{d_2}{d_1}\right) \qquad [10.9]$$

Substituting [10.9] into [10.8] and rearranging terms,

$$\alpha_r \propto \frac{\sqrt{\epsilon_r}}{60 \cdot d_2} \frac{\left(1 + \dfrac{d_2}{d_1}\right)}{\ln\left(\dfrac{d_2}{d_1}\right)} \qquad [10.10]$$

Equation [10.10] separates out the constant terms $\sqrt{\epsilon_r}/(60 \cdot d_2)$ from the operative terms $(1 + d_2/d_1)/\ln(d_2/d_1)$ that control the position of the minimum. Examination of the operative terms reveals that the position of the minima is a function only of the ratio d_2/d_1, and not of either ϵ_r or the absolute diameter d_2.

10.2.2 Coax: Radio Frequency Interference

Coaxial cables have fairly good natural immunity to external noise, due to the physical symmetry of the signal current conductor and the returning current conductor (the concentric shield). This symmetry cancels, to first order, all effects of external electromagnetic fields. Any residual susceptibility in a coaxial cable results from imperfections in its shield.

At frequencies up to a few megahertz coaxial susceptibility is proportional to the resistance of the cable shield. The end-to-end resistance of the shield, when excited by the large common-mode currents that can be induced by RFI, creates a small residual voltage from end to end across the cable shield. This residual voltage appears to the receiver as a source of noise. Susceptibility problems due to shield resistance happen most often in the below-30-MHz band. To conquer low-frequency susceptibility problems, use a thicker, lower-resistance outer braid or switch to a larger cable (which has a bigger, lower-resistance braid).

Higher-frequency electromagnetic fields can leak directly through the holes in the braid. To conquer high-frequency susceptibility problems, specify a cable with a heavy braid plus a solid foil shield. The solid foil shield is often wrapped around the dielectric, just underneath the heavy braid. The combination of foil shield and heavy, low-resistance braid works particularly well for combating external noise, although a thin aluminum foil will somewhat increase the skin-effect resistance of the shield, slightly worsening the high-frequency attenuation.

In all cases when working with fast digital systems, specify a good connector. Do not use a connector that has pigtails, pins, or little tabs that connect the coaxial shield to the chassis. Get a connector that makes 360-degree contact, all around the connector shell, with the chassis.

10.2.3 Coax: Radiation

The key to obtaining good radiated performance is to specify an adequate coaxial shield. This problem is equivalent to the problem of hardening your system against RFI, and the same solutions apply.

You will want a cable with a low value of *transfer impedance*. The transfer impedance for a coaxial cable is the ratio of the voltage generated longitudinally along the shield divided by the signal current flowing within the cable. This parameter is usually specified as a function of frequency. Quoting from ISO/IEEE 8802.3 (1996), "A [coaxial] cable's EMC performance is determined, to a large extent, by the transfer impedance value of the cable."

Bigger, heavier braids, or multiple braids, or a combination of foil wrap and braid, are approaches commonly used to reduce the transfer impedance. Above 100 MHz, data scrambling is often implemented to guarantee that ordinary cables will not radiate in excess of FCC or EN limits.[86]

---

[86] Unscrambled transmission systems radiate horribly because simple repetitive structures within the data stream, like the idle pattern, tend to concentrate all their radiated power at harmonics of the basic pattern repetition rate. These concentrated harmonics then leak from the coaxial cable, where they may be easily detected by FCC or EN test antennas. In contrast, scrambled transmission systems spread their radiated power across a wide frequency range, limiting the peak radiation in any one radio-frequency band.

POINT TO REMEMBER

➤ RF susceptibility and radiation in coaxial cables result from imperfections in the shield.

10.2.4 Coaxial Cable: Safety Issues

Wherever a coax link terminates on your equipment, you have two choices for the treatment of the coax ground: You may connect it to your equipment chassis, or not.

The treatment of the coax ground generally matches the treatment of the signal conductor. Figure 10.7 illustrates the direct-connection method.[87] If you direct-connect the signal, then you *must* also provide a direct, low-impedance path for the flow of returning signal current.

If you block the direct path of signal current with an isolating device, such as a transformer, optical isolator, or differential receiver, then you are free, as far as signal integrity is concerned, to disconnect the coax ground from your equipment ground (Figure 10.8), creating an *isolated cable*. In a unidirectional link, one traditionally directly-connects the transmitting end and isolates the receiving end. This is a good arrangement because, as explained in Chapter 6, Section 6.12.2, "Immunity to Large Ground Shifts," it is never a good idea to make direct ground connections between systems with separate AC power inputs.

A common-mode choke blocks the flow of intercabinet ground currents in another way. The common-mode choke is similar to a transformer, but connected differently (see Figure 10.9). The normal flow of signal current is in the forward direction through one winding and then in the reverse direction on the other. The magnetic fields from current that follows this path are exactly opposite and perfectly cancel. The choke therefore exerts *no net effect* on the normal flow of signal current.

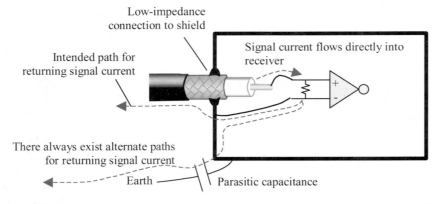

Figure 10.7—A directly-connected coaxial cable requires a low-impedance connection between the cable shield and the product chassis. If the impedance of the shield connection is too high, returning signal current will be encouraged to flow on alternate return paths. Alternate return-current paths often act as very efficient radiating antennas.

---

[87] For schematic clarity, the connector isn't shown, but I think you get the idea.

10.2.4 • Safety Issues

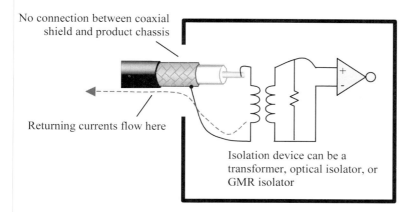

Figure 10.8—An isolated coaxial cable does not connect to the product chassis. It does not permit signal current to flow into the system. All signal currents are returned to the source on the incoming cable.

The choke *does* affect any current that enters the system through one winding and then attempts to leave on any path other than by the return winding. These currents are subject to and impeded by the full inductance of the choke. Given enough inductance in the choke (several Henries), you can attenuate the flow of intercabinet ground currents while still providing a good high-frequency path for digital signals. For this approach to work, the choke must possess a primary winding impedance of several thousand ohms at 60 Hz. It must also possess a leakage inductance small enough to pass your high-speed digital signals. Designing such a choke is a challenging project.

A DC-balanced signal (see box), leaves you more flexibility in your treatment of the cable shield. DC-balanced signals carry very little signal power at frequencies below some predefined cutoff frequency f_{DC}. Therefore, a DC-balanced coaxial transmission system does not require a ground at frequencies below f_{DC} because *there isn't any return current at those*

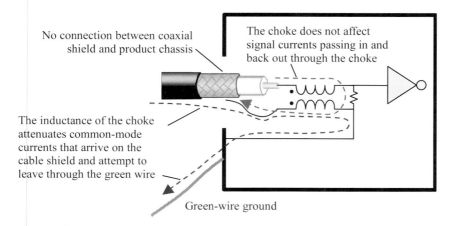

Figure 10.9—A common-mode choke attenuates intercabinet ground currents.

low frequencies. A 10-MHz Manchester-coded signal, for example, has a lower cutoff frequency on the order of 1 MHz. If the transmission system is interrupted below this frequency, it makes little difference to the received signal. With such a system, you might consider making a connection between the coaxial cable ground and the system chassis that is low impedance at high frequencies but high impedance at 60 Hz. This connection could conceivably be done with a capacitor, provided that the capacitor has a low enough series inductance for your application (see Section 9.4, "150-Ω STP-A: Radiation And Safety").

DC-balanced signals are perfectly suited for connection through transformers.

DC Balance

Any bit stream with equal numbers of ones and zeros has the property of *DC balance*. Examples of DC-balanced signals include a 50% duty-cycle clock, a Manchester-coded data signal, and an 8B/10B-coded data signal. The spectral power density of such signals is zero (or close to zero) at all frequencies below some critical cutoff frequency f_{DC}. The value of f_{DC} depends on the pattern of data and the length of the data-bit interval.

DC-balanced signals will pass relatively undistorted through any high-pass filter with a cutoff frequency less than f_{DC}.

POINTS TO REMEMBER

➤ If you block the direct path of signal current with an isolating device, such as a transformer, optical isolator, or differential receiver, then you are free, as far as signal integrity is concerned, to disconnect the coax ground from your equipment ground.

➤ A common-mode choke can also block the flow of intercabinet ground current.

➤ DC-balanced signals are perfectly suited for connection through transformers.

10.3 COAXIAL CABLE CONNECTORS

There are too many different families of coaxial connectors. This is the natural result of a broad, diverse industry operating for many years without coherent standardization. Table 10.5 summarizes a selected group of the more popular connectors used in digital applications.

Within each family, the basic choices you will need to make relate to the frequency response, the plating materials used, the style of attachment, the connection to the cable, and the quality of the springs.

Above 100 MHz, you should always match the characteristic impedance of the connector to the cable. A matched coaxial connector balances the parasitic series inductance L and the parasitic shunt capacitance C of the connector in such a way that the natural

10.3 • Connectors

Table 10.5—Selected Coaxial Connector Families

Connector Size	Nominal cable O.D.	Quick-disconnect (bayonet style)	Threaded style	Recommended operating frequency (max)
Standard	.060 to .425	C		4 GHz
Standard	.060 to .425		N	10 GHz
Miniature	.060 to .425	BNC		4 GHz
Miniature	.060 to .425		TNC	10 GHz
Subminiature	.060 to .141	SMB		4 GHz
Subminiature	.060 to .141		SMA, SMC	10 to 30 GHz

impedance of the connector $\sqrt{L/C}$ matches some particular value (perhaps 50 or 75 ohms). Even when this is done, the connector may still show imperfections at extremely high frequencies. Depending on the distribution of parasitics within the connector, there may be internal regions over which the matching is quite good and regions over which it is not so good. As the risetime of your signals shrinks to a value comparable with the propagation delay through the connector, the internal details begin to matter. In very high-speed applications, look for vendors that provide standing-wave ratio (SWR) data or reflection coefficient data to back up their claims of superior performance. Screw-on type connectors will always outperform quick-disconnect types because the threads accurately align the mated halves of the connector and provide a solid 360-degree ground connection concentric with the signal conductor. The electrical performance of the connector has more to do with the care taken during manufacturing to produce a uniform, properly dimensioned cavity than it does the physical dimension or other factors.

Here's a handy little table that relates the SWR, reflections, and return loss specifications for connectors (Table 10.6). *This table has been computed only for the case of sine-wave excitation.* In general, if you want your digital signal to pass through the connector 99% intact, select a connector with a return loss greater than 17 dB at all frequencies from DC up to the knee frequency of your logic (1/2 over the risetime). Also keep in mind that connector distortions aggregate across all the connectors in a particular data link.

The contact plating serves to stave off corrosion and eventual failure of the contacts. For applications that require multiple connector insertions, always look for connectors with gold or stainless steel mating surfaces.

The attachment style may be either quick-disconnect or threaded. This is a tradeoff of ease-of-use versus reliability. My rule of thumb here is simple: If it goes on a boat, a car, a plane, or anything that moves, it's got to be threaded. A quick-disconnect part will not survive the tough U.S. military-standard "500-hour salt-spray test" or the "2-minute Saturn-5 vibration, heat, and shock test," or for equipment mounted near a gasoline-tank, the even more excruciatingly difficult "Ford Pinto heat and flame trial."[88]

---

[88] OK, I made up the last one.

Regarding the choice of crimped versus soldered cable attachment, this choice is based on the facilities available at the point of assembly and the electrical performance of the connector. Crimping has two basic advantages: It doesn't require access to AC power (soldering does), and it's fast. On top of a telephone pole, down in a cable tunnel, or anywhere power may not be available, crimping is the way to go. In a factory environment, where speed matters, crimping wins again. Crimping is not appropriate in situations where your field technicians won't have access to the special crimping tools required to press the connectors onto the cable or access to any spare connectors. In those applications, like on a ship or a spacecraft, the solder-type connectors may be best. In all cases avoid the popular

Table 10.6—SWR and Return Loss Conversions for Connectors

SWR $\dfrac{1+\Gamma}{1-\Gamma}$	Return loss dB $20\log(\Gamma)$	Reflection coefficient Γ	Transmission coefficient $\sqrt{1-\Gamma^2}$
17.39	1	0.8913	0.4535
8.724	2	0.7943	0.6075
5.848	3	0.7079	0.7063
4.419	4	0.631	0.7758
3.57	5	0.5623	0.8269
3.01	6	0.5012	0.8653
2.615	7	0.4467	0.8947
2.323	8	0.3981	0.9173
2.1	9	0.3548	0.9349
1.925	10	0.3162	0.9487
1.785	11	0.2818	0.9595
1.671	12	0.2512	0.9679
1.577	13	0.2239	0.9746
1.499	14	0.1995	0.9799
1.433	15	0.1778	0.9841
1.377	16	0.1585	0.9874
1.329	17	0.1413	0.9900
1.288	18	0.1259	0.9920
1.253	19	0.1122	0.9937
1.222	20	0.1000	0.9950
1.196	21	0.0891	0.9960
1.173	22	0.0794	0.9968
1.152	23	0.0707	0.9975
1.135	24	0.0631	0.9980
1.119	25	0.0562	0.9984
1.106	26	0.0501	0.9987
1.094	27	0.0446	0.9990
1.083	28	0.0398	0.9992
1.074	29	0.0354	0.9994
1.065	30	0.0316	0.9995

10.3 • Connectors

"twist-on" style connectors. These seem to twist off just as easily as they twist on.

Crimp-type connectors tend to have the best dimensional control (because they don't have to accommodate blobs of solder) and so deliver the best control over impedance. This makes crimp-style connectors generally superior to the other types for high-frequency work, but always check the specifications for insertion loss and SWR at the frequency of operation.

Lastly, about the springs, always specify heat-treated beryllium-copper for critical contact springs. These maintain contact pressure for years, whereas ordinary copper or brass will soon deform and fail to connect.

Good practical information about coaxial connectors is found in [82] and [84]. General reference material concerning plating, crimping, and spring-loaded connector technology may be found in [83].

POINTS TO REMEMBER

- Above 100 MHz, you should always match the characteristic impedance of the connector to the cable.
- Contact plating serves to stave off corrosion and eventual failure of the contacts.
- If it goes on a boat, a car, a plane, or anything that moves, use threaded connectors.
- Crimp-style connectors generally superior to the other types for high-frequency work.
- Always specify heat-treated beryllium-copper for critical contact springs.

For further study see: www.sigcon.com

CHAPTER **11**

FIBER-OPTIC CABLING

Without question, the bandwidth-carrying capacity of modern fiber-optic cabling greatly exceeds that of any form of copper cabling. Unfortunately, the transceiver technology required to fully realize the performance advantages of fiber remains quite costly. Additionally, fiber suffers from some practical drawbacks that limit its utility. For example, fiber connectors are difficult to assemble and must be kept clean and free from scratches. Copper insulation-displacement connectors, on the other hand, are rugged and wipe clean upon insertion. Copper connections may be assembled anywhere, even in dirty environments.[89]

    I'll not dwell further on the cost or convenience problems you will encounter with fiber. That's something you'll have to evaluate in the context of your own application. Any discussion here would soon be moot anyway, as the cost and convenience of using fiber cabling are improving steadily with the passage of time. I look forward to the day when fiber-optic connections are as ubiquitous and easy to use as copper connections.

    What we discuss in this chapter are the physics of optical transmission. The following sections present the optical performance characteristics of both multimode and single-mode glass fiber data links. The cables used in the examples are selected from the TIA/EIA-568-B and ISO/IEC-11801 building wiring standards. Before we get to the detailed technical information, though, I should like to present a few pages of information about the general construction of fibers. If you're already familiar with fiber-optic cable construction, this might be a good time to step out for a cup of tea.

---

[89] This is not an issue to be taken lightly. Anyone who has spent time crawling around in attics and switch rooms knows that they are not clean, well-lighted places to work.

POINT TO REMEMBER

> The bandwidth-carrying capacity of modern fiber-optic cabling greatly exceeds that of any form of copper cabling, an advantage counterbalanced by the high costs and practical difficulties associated with fiber.

11.1 MAKING GLASS FIBER

Glass optical fiber is drawn as one continuous thread from a single cylinder of purified glass called a *preform*. The glass preform is manufactured by a process of *chemical vapor deposition* (CVD), whereby purified, gaseous glass vapor (silica) is accreted onto the surface of a uniform glass cylinder in thin layers under conditions of controlled heat and pressure. As the glass accumulates, impurities may be mixed into the glass vapor to modify the index of refraction of the finished product.

The two most common CVD processes for making a preform are *outside vapor deposition* (OVD) and *inside vapor deposition* [85], [90], [89]. The outside vapor deposition process deposits doped silica on the outside of a rotating glass mandrel, with growth taking place radially. The inside vapor deposition process works backwards, depositing doped silica onto the inner surface of a pure silica tube until it is almost filled. The resulting inside deposition preform has a hollow core, which is squeezed shut when the fiber is drawn.

Either process results in a large cylinder of pure glass with a carefully controlled radial variation in the index of refraction (Figure 11.1). A typical preform has a diameter in the range of 1 cm to 6 cm and a length of 1 m to 2 m [89]. The inside portion of the cylinder, where light will eventually flow, is called the core. The outer portion of the cylinder, which acts as a mirror to keep the light centered in the core, is called the cladding. Only a small percentage of the light power carried in a glass fiber travels in the cladding.

What happens next is, to me, the truly amazing part. The preform is heated in a drawing apparatus, and a finished fiber is pulled from the bottom (Figure 11.2). *As the preform sinks into the drawing funnel, the profile of the index of refraction is squeezed down*

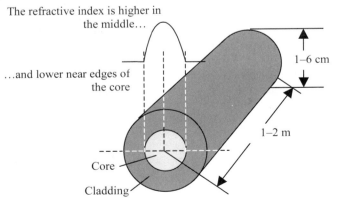

Figure 11.1—The ideal refractive-index profile for a graded-index multimode preform is almost perfectly parabolic.

11.2 • Finished Core Specifications

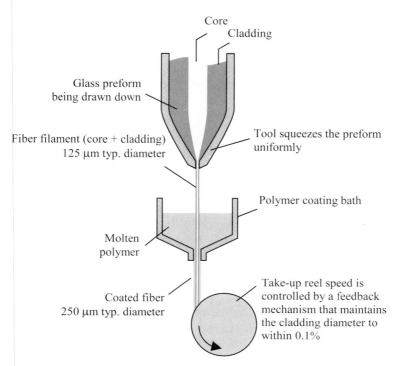

Figure 11.2—A finished fiber is pulled from the bottom of a heated preform.

to microscopic dimensions but retains its shape. A single fiber pulled from a preform may be as long as 100 km [89]. The finished fiber is given a protective polymer coating, reeled, and tested.

POINT TO REMEMBER

➤ Glass optical fiber is drawn as one continuous thread from a single cylinder of purified glass called a *preform*.

11.2 FINISHED CORE SPECIFICATIONS

A complete fiber-optic core is characterized by the diameter of its core, its cladding, its polymer coating, and the various mechanical and optical tolerances associated with those items (Figure 11.3). Of all the various parameters that define a fiber core, the key parameter that differentiates fiber in the marketplace is the core diameter. Fiber is segmented into different grades of electrical performance according to its core diameter.

As illustrated in Figure 11.4, large cores are classified as multimode fiber (MMF), while small cores as classified as single-mode fiber (SMF). See Section 11.5.1, "Multimode

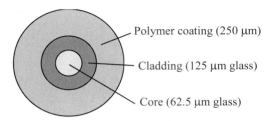

Figure 11.3—Construction of typical coated 62.5 μm multimode fiber core.

Signal Propagation," and Section 11.6.1, "Single-Mode Signal Propagation," for definitions of these terms. The distinction between MMF and SMF has to do with the relation between the size of the core and the wavelength of the light flowing through it.

The diameter of the core strongly affects both the cost and signal transmission bandwidth of a finished fiber. Enlarging the core renders a finished system less expensive but reduces the bandwidth. Larger cores reduce the overall system cost because they can accept light from less mechanically precise and less costly packages and connectors. Unfortunately, larger cores also reduce the bandwidth because they allow the light to bounce around more inside the fiber, dispersing the received optical energy over time. The smallest cores (about 10 μm diameter) deliver the greatest signal transmission bandwidth at the greatest finished system cost.

The largest cores are made from plastic, an inexpensive and easy-to-handle material. Plastic fibers have the same general optical characteristics as glass fibers; however, the optical attenuation of plastic is much higher, and the bandwidths are much lower. Plastic fibers are relegated mostly to relatively low-bandwidth, short-distance applications.

High-volume desktop LAN applications use multimode glass fiber. This fiber is produced in standard core diameters of 50, 62.5, 85, 100, and 140 μm. The most popular fibers for LAN applications are 50 and 62.5 μm.

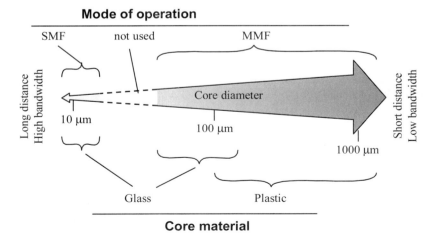

Figure 11.4—Segmentation of the fiber marketplace according to core diameter.

11.5 • Multimode Glass Fiber-Optic Cabling

amount.[93] Yet, their performance varies tremendously. The optical differences have to do with subtle variations in the index of refraction of the core and with the purity and clarity of the glass.

The two most popular standard core diameters for multimode glass fiber are 50 μm and 62.5 μm. Both cores operate as multimode cables for all transmission wavelengths within the wavelength range of 700 nm to 1600 nm. Larger core diameters are rarely used for high-speed applications because they admit too much modal dispersion (see Section 11.5.1). The core and cladding diameter specifications are usually given together, written as a pair like this: 62.5/125 μm.

Figure 11.8 illustrates the construction of a standard 62.5/125 μm graded-index multimode glass fiber core. The nominal index of refraction[94] of both core and cladding is approximately 1.5 with subtle variations in the central core region. The subtle variations in the index of refraction give fiber its unique properties.

The term *graded index* refers to the shape of the refractive index profile in the core area of the fiber.[95] Multimode fibers used in modern LAN applications have an inverted-parabolic refractive index profile (shown in Figure 11.8). This type of graded-index profile produces a very high-bandwidth fiber. Older multimode fibers have a stepped refractive index profile (Figure 11.9). The step-index profile is easier (and cheaper) to make, but does not produce as high a bandwidth, as explained in Section 11.5.2, "Why Is Graded-Index Fiber Better than Step-Index?".

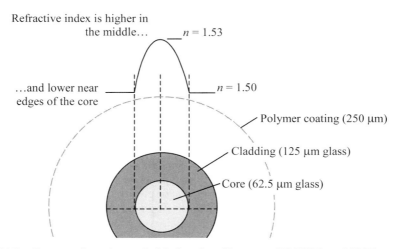

Figure 11.8—Construction of a graded-index glass fiber core (62.5/125 μm MMF).

---

[93] Yes, glass is flexible. When you push on a window, the glass flexes slightly. Thinner glass flexes more, in inverse proportion to the *square* of its thickness. Glass that is only 125 μm thick, like an optical fiber, becomes quite flexible.
[94] The velocity of propagation of light in a clear material is inversely proportional to the index of refraction. When light encounters a change in the index of refraction, the rays of light are bent or reflected.
[95] The index of refraction for a perfect graded-index core is circularly symmetric, being a function only of the radial distance from the center of the core. Looking at a cross-section of the core, if you draw a horizontal line across the center of the core and plot the refractive index as a function of position along this line, you will have measured the refractive index profile for the fiber.

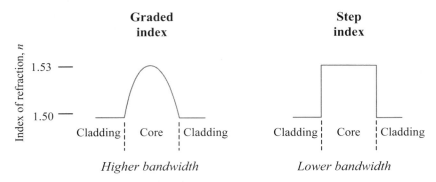

Figure 11.9—Refractive index profiles for two styles of multimode fiber.

POINTS TO REMEMBER

➤ The two most popular standard core diameters for multimode glass fiber are 50 μm and 62.5 μm.

➤ A graded-index multimode fiber higher bandwidth than a step-index multimode fiber of the same core diameter and quality.

11.5.1 Multimode Signal Propagation

The first violinist in a concert production emits sound waves in all directions. These different directions of sound propagation, or acoustic modes, scatter off the walls, ceiling, and floor, eventually reaching your ears. In a well-designed symphony hall, the various acoustic modes will all (hopefully) arrive closely grouped in time. That makes the music sound live, but not too reverberating. If the same concert were presented in a larger symphony hall, say, the New Orleans Superdome, the multiple disparate echoes would make it very difficult to properly interpret the music.[96]

A similar principle applies to fiber-optic signal propagation. Within a multimode fiber, there exist hundreds of different pathways, or modes of propagation. Some go straight down the center of the core, some bounce back and forth off the cladding walls, and others spiral around the center of the core as they make their way down the cable. What you need to understand about modal propagation in fibers is this: The modes don't interact very much as they travel. *Each mode proceeds separately at its own pace, with a unique value of attenuation and delay.*

What is the effect of this multimodal structure on a propagating signal? The primary effect is that a step transition in the transmitted power becomes gradually dispersed in time as its travels down the fiber.

---

[96] Unless you are listening to grunge music, which might be improved by the echo effect (anything would probably improve grunge music).

11.5.1 • Multimode Signal Propagation

Figure 11.10 illustrates the dispersion effect. This figure shows a broad-beam transmitting source on the left, like an LED. The source couples optical power into all the conducting modes of the fiber.[97] Such a launch condition is called an *overfilled launch*.

The independent propagating modes (only three are shown) appear temporally aligned near the source. As the modes progress, they each experience different values of attenuation and delay. Even though the modal delays in modern graded-index multimode fibers are matched to within better than 1 part in 1,000, the difference in arrival times at the end of a long fiber can become quite noticeable.

The receiver has no way to distinguish the arriving modes. It simply sums the power in all the arriving modal signals. On a rising edge, this summing action causes the receiver output to begin rising when the first mode arrives. The received waveform doesn't stop rising until after the arrival of the slowest, latest mode. If the spread in arrival times exceeds the natural risetime of the optical source, the incoming edge will be noticeably dispersed, or stretched out. The same effect happens on falling edges. The longer the fiber, the greater the modal separation, and the greater the rise/fall time dispersion.

Sufficiently large amounts of dispersion cause the tail of one edge to overlap with the beginning of the next. This type of eye pattern distortion degrades the receiver noise margin and must be included in the *optical power budget*. The power budget, and the way it accounts for dispersion, are the subjects of Section 11.5.5.2, "Multimode Attenuation Budget."

Specifications for dispersion in a multimode fiber are broken down into two categories, *modal dispersion* and *chromatic dispersion*. The modal and chromatic dispersion specifications separately account for dispersion due to the patterns of propagation of light within the fiber and the variations in refractive index with wavelength respectively.

Let's take a quantitative look at modal dispersion.[98] A multimode fiber illuminated

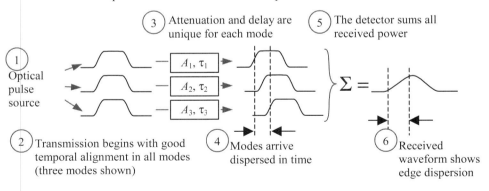

Figure 11.10—Each mode within a fiber propagates at a slightly different speed.

---

[97] An LED does not couple equal amounts of power into each mode; however, all modes are used.
[98] Here I use a ray-tracing analogy to describe the propagation of modes. This analogy, while not perfect, is a useful way to mentally visualize the modes. For a better analogy, you may recall from the study of ordinary differential equations Bessel's solution for the behavior of the surface of a round drum. The drum vibrates in a number of modes. Each mode is characterized by a number of concentric rings and a number of undulations as a function of angular position. The modes in a fiber-optic cable better resemble drumhead patterns than individual rays. For further study, see [90].

with only one wavelength of light (presumably this comes from a perfect LED having a zero spectral width) displays only its modal dispersion t_m. The various rays bang around inside the fiber, emerging with a total dispersion proportional to the length of the fiber.

The modal dispersion of a fiber may be estimated given its *modal bandwidth* and length:

$$t_m = (0.48/B_m) \cdot l \cdot 10^6 \qquad [11.1]$$

where t_m is the 10% to 90% modal dispersion risetime, in ps,

B_m is the 6-dB modal bandwidth of the fiber (FWHM bandwidth), in MHz-km, and

l is the length of the fiber, in km.

NOTE: The term B_m represents the 6-dB modal bandwidth of a 1-km fiber. In other words, it is the frequency in MHz at which the electrically detected signal amplitude from a sinusoidal optical source would be 6 dB below the detected signal amplitude at low frequencies. At this 6-dB *electrical loss* frequency, the received optical *peak-to-peak power* is down only by half (a 3-dB optical power penalty). The optical detector changes optical *power* to electrical *voltage,* changing a 3-dB optical power loss into a 6-dB electrical signal loss. The 6-dB loss point is the traditional specification method used for modal bandwidth.

Modal bandwidth is improved by manipulating the refractive index profile of the fiber as it is manufactured to create a structure that propagates all possible modes at more nearly the same speed. It is also improved by using a smaller core, which tends to restrict the number of possible modes.

Next let's examine chromatic dispersion. A fiber illuminated with a practical LED having a finite spectral width displays not only modal dispersion t_m, but also a degree of chromatic dispersion t_c, due to the fact that different wavelengths of light travel at different speeds in glass. As the spectral width of the source increases, so does the chromatic dispersion.

The chromatic dispersion of a fiber may be estimated given its *chromatic dispersion constant*, length, and the *spectral width* of the source:

$$t_c = D \cdot l \cdot \lambda_{RMS} \cdot 2.56 \qquad [11.2]$$

where t_c is the 10% to 90% chromatic dispersion risetime, in ps,

D is the chromatic dispersion constant, in ps/nm-km,

l is the length of the fiber, in km,

λ_{RMS} is the RMS spectral width of the source, in nm (the RMS spectral width is the standard deviation of the curve of power density versus wavelength), and

2.56 is a constant that converts standard-deviation-type pulse width measurements to 10% to 90% measurements.

11.5.1 • Multimode Signal Propagation

> *NOTE:* The term λ_{RMS} represents the RMS spectral width, in units of nanometers, of the source. It is defined as the standard deviation of a plot of source power versus wavelength. The RMS width is one traditional specification method used for spectral width (that is, you spread out the light using a diffraction grating or prism, then you put it into an electrical detector to convert power to voltage, then you plot the detector voltage versus wavelength, then you determine the standard deviation of that plot). Had the optical source been characterized by a FWHM spectral width (defined at the -3dB power, or -6dB electrical, points), the term λ_{RMS} would be replaced with λ_{FWHM}, and the conversion factor to 10% to 90% risetime would be changed to 1.09.

LED operation at 850 nm suffers markedly from chromatic dispersion. LED operation at 1300 nm suffers less, because 1300 nm lies near a natural null in the chromatic dispersion of glass. For LED-based systems, which have a source spectral width on the order of 150 nm, there is therefore a natural advantage to operation at 1300 nm.

A laser-based source produces a much narrower spectral width than an LED. Most laser sources are therefore less susceptible to chromatic dispersion at all wavelengths, although they may be affected by other peculiar factors (see Section 11.5.9, "Multimode Fiber with Laser Source").

Careful analysis reveals two primary factors that contribute to chromatic dispersion:

1. Variations in the refractive index of the glass as a function of wavelength (material dispersion), and
2. Variations in the penetration depth of each wavelength into the core/cladding interface (waveguide dispersion).

The first factor was discovered by Newton. Light travels at different speeds in glass, depending on its wavelength. Prisms used for spectrographic analysis work on this principle.

In an ideal fiber, the refractive index would not vary with wavelength, so that all wavelengths of light would travel at the same speed. In a practical fiber variations in the refractive index always occur. These variations arise due in part to the nature of amorphous silicon, and in part to the presence of trace impurities.

The second factor arises because different wavelengths of light tend to penetrate, or bleed over, to different depths within the core-cladding interface. Any change in the relative mix of signal power carried in the core versus cladding portions of the fiber affects the overall effective index of refraction experienced by waves at that particular wavelength and mode number; therefore, the speed of propagation varies with wavelength [90].

A natural null in chromatic dispersion occurs in glass near 1300 nm. Careful management of the index-of-refraction profile can modify the position and depth of this null. A fiber in which the dispersion curve has been intentionally changed is called a dispersion-shifted fiber, Dispersion-shifted fibers can be produced with a null anywhere in the range of 1300 to 1600 nm (Figure 11.11). Typical values for dispersion in the first window (800 nm) are on the order of 85 ps/nm-km. Shifting the dispersion null to the center of the third window improves operation in that window at the expense of worse performance in the second window.

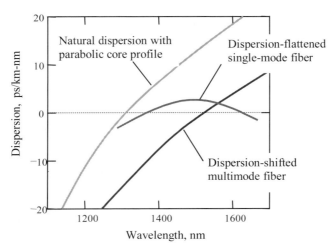

Figure 11.11—The dispersion curve may be shifted, or flattened, by manipulating the index-of-refraction profile within the core.

In the case of single-mode fiber intended for use *only* in the second or third window, sophisticated manipulation of the index-of-refraction profile within the core can produce a dispersion-flattened fiber, which has a low value of dispersion across a wide range of wavelengths within those windows.

Chromatic dispersion in typical installations is improved by the use of the 1300-nm operating wavelength (as opposed to 850 nm), by the use of narrow spectral-width sources like VCSEL laser diodes (as opposed to LEDs), and by limiting the link to short distances.

Modal and chromatic dispersion combine like this to produce the overall fiber dispersion t_f:

$$t_f = (t_m^2 + t_c^2)^{1/2} \qquad [11.3]$$

POINTS TO REMEMBER

- ➤ Within a multimode fiber, there exist hundreds of different pathways, or modes of propagation.
- ➤ The multiple modes cause a step input to gradually disperse in time as it travels down the fiber.
- ➤ Dispersion in a multimode fiber is divided into modal dispersion and chromatic dispersion.
- ➤ Modal bandwidth is a function of the refractive index profile of the fiber.
- ➤ Chromatic dispersion is a function of the material properties of the glass and also the refractive index profile of the fiber.

11.5.2 Why Is Graded-Index Fiber Better than Step-Index?

The core/cladding interface in a step-index fiber acts like an almost-perfect mirror. According to Snell's law [91], waves internal to a step-index fiber that graze the core/cladding interface at a sufficiently shallow angle will experience total internal reflection. No power is lost. Such waves may bounce back and forth between the cladding walls as many times as necessary as they travel down the fiber. This bouncing effect describes the essence of multimode fiber operation.

In the simple ray-tracing diagram (Figure 11.12), those modes that travel straight down the fiber experience the fewest reflections, have the straightest path, and therefore arrive with the least delay. Those modes that travel at the greatest angle with respect to the central core experience the greatest delay.[99]

The maximum dispersion that can occur in a step-index fiber has to do with the ratio of transit times between the slowest and fastest modes in the fiber. Here's where Snell's law comes in. It constrains the maximum angle at which light can propagate in the fiber. Using Snell's law you can predict that the maximum ratio of transit times, based on the ray-tracing analogy, equals n_2/n_1, where n_2 and n_1 represent the index of refraction in the cladding and core areas respectively.

For a typical step-index, the quantity n_2/n_1 equals approximately 1.01, or 1% greater than unity. In a 100-m cable with a natural delay of 400 ns, the step-index dispersion (worst-case difference in delays) would therefore be 1% of 400 ns, or 4 ns.

If you take a careful look at the ray-tracing picture, you may spot an easy way to improve this situation. Notice how the reflecting ray spends a lot of time near the edges of the fiber, but relatively little in the core. The central, straight ray is the other way around. It spends its time in the middle. If you artificially increase the index of refraction exactly in the center of the core, *the central ray slows down, but it doesn't much affect the reflecting ray*. This effect helps to balance the delay between modes. It also turns out that slightly reducing the index of refraction near the edges of the core speeds up the most broadly reflecting rays. A careful job of shaping the index profile can theoretically improve modal dispersion by orders of magnitude. The resulting fiber, which has gradual, controlled variations of the index of refraction within the core region, is called a graded-index fiber.

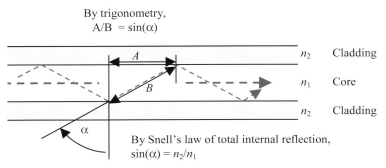

Figure 11.12—A reflecting ray takes a longer path than the straight ray.

---

[99] Ray-tracing is an imperfect description of how modes actually propagate in a circular fiber, but it's the best we can do without having to solve a bunch of Bessel's equations.

POINT TO REMEMBER

➢ Carefully grading the profile of the index of refraction greatly improves modal bandwidth.

11.5.3 Standards for Multimode Fiber

Internationally recognized specifications for 50 and 62.5 μm multimode optical fibers are provided by IEC 793-2 [86]. If you are planning to design systems that will be installed using existing in-building cabling, you should familiarize yourself with IEC 793-2. For each standard core diameter, IEC 793-2 identifies a large number of standard modal bandwidth and attenuation categories for operation at either 850 nm or 1300 nm. These are the two most popular wavelengths available for fiber-optic system design. Transceiver components for a third major wavelength range, at 1550 nm, are recently becoming widely available.

In the early days of fiber development, a fiber would be heavily optimized for use at 850 nm or for use at 1300 nm, but not both. Today the situation has changed. Today most fibers are specified for use at either wavelength. Fibers specified for use at either 850 or 1300 nm are called dual-window fibers.

System designers can call out from Table 11.1 minimum standard bandwidth categories for dual-window fibers that suit their application and can expect that users will be able to purchase and install these cables. Should the user change systems, possibly to a system that operates at a different wavelength, the dual-window property will protect the investment in the installed cabling.

Specifications for dual-window attenuation or bandwidth are usually written with a slash, like this: 62.5/125 μm 160/500 MHz-km. The first two figures specify a 62.5 μm core with a 125 μm cladding. The next two figures give the modal bandwidth, in MHz-km, at operating wavelengths of 850 nm and 1300 nm respectively.

If you are looking for "safe" values that will cover a larger percentage of the installed base, you can take a tip from the IEEE 802.3z Gigabit Ethernet committee. Based on customer surveys conducted in 1997, the committee picked minimum bandwidth figures of 160/500 MHz-km for 62.5/125 μm fiber and 400/400 MHz-km for 50/125 μm fiber (highlighted in bold below). If you design your system to work with these values, you will find a large installed base of cable already in the field that can support your application.

Specifications for fiber performance continue to evolve. For example, ISO 11801 standards stipulate a slightly higher bandwidth for 62.5/125 μm fiber of 200/500 MHz-km. In 1998, one representative from the predominant major manufacturer of 50/125 μm fiber cores indicated that they commonly ship a 500/500 MHz-km product. In 1999, presentations to the 802.3 committee regarding 10-Gigabit Ethernet suggested that 50/125 μm fiber with a bandwidth of 2500 MHz-km may be possible.

11.5.4 • Use of 50-micron Fiber?

Table 11.1—Standard Attenuation and Modal Bandwidth Specifications for Dual Wavelength Fiber, Adapted from IEC 793-3, with Additions

Fiber type	A1b (62.5/125 µm)		A1a (50/125 µm)		Units
Transmission wavelength	850	1300	850	1300	nm
Attenuation categories (max)	3.5	1.5	2.7	1.0	dB/km
	3.2	0.9	2.5	0.8	
	3.0	0.7	2.4	0.6	
Modal bandwidth categories	160	200	200	400	MHz-km
	160	500	200	600	
	200	200	400	400	
	200	400	400	600	
	200[1]	500[1]	400	800	
	200	600	400	1000	
	250	1000	400	1200	
	300	800	400	1500	
			500[2]	500[2]	
			600	1000	

NOTE (1)—Specified in the ISO/IEC 11801 generic cabling standard, but not part of IEC 793-3.

NOTE (2)—A common specification, according to the predominant manufacturer of 50/125 µm fiber, but not part of IEC 793-3.

Can you use these new cables in your design? That depends on your application. A sufficiently compelling application could cause a customer to rip out old wiring and install something new. More likely, though, you will need to work on the old, installed base of cabling.

You will occasionally hear the term *numerical aperture* used to describe fiber. This term refers to the light-gathering ability of the fiber. All other things being equal, a fiber with a greater numerical aperture will generally accept more light from an LED source. Mathematically, the value of numerical aperture is a function of the refractive index profile, roughly equal to $\sqrt{n_{CORE}^2 - n_{CLADDING}^2}$.

You will rarely have to specify the numerical aperture, because it is already specified in an international standard, IEC 793-2 (1992) [86]. Section A of IEC 793-2 calls out mechanical dimensions and numerical aperture values for several standard sizes of graded-index multimode fiber cores. Designers of high-speed systems need only ask for multimode optical fiber from IEC 793-2, category A1a (50 µm) or A1b (62.5 µm).

POINT TO REMEMBER

> Internationally recognized specifications for 50 and 62.5 µm multimode optical fibers are provided by IEC 793-2.

11.5.4 What Considerations Govern the Use of 50-micron Fiber?

Fifty-micron multimode fiber is generally available with a higher bandwidth and better transmission characteristics than 62.5-μm multimode fiber. This natural advantage is tempered by the fact that many data transmission products don't couple as well into 50-μm fiber as into 62.5-μm fiber. The coupling issue has to do with the construction of LED transmitting sources.

LED transmitters, which are normally used with multimode fiber (62.5 or 50 μm), produce divergent light beams. These transmitters may be classified into two broad categories: surface emitters and edge emitters.

A surface-emitting LED produces a beam spot generally large enough to overfill the front end of a 62.5-μm fiber (Figure 11.13). The beam power is calibrated to produce the correct transmitted light power when coupled to a 62.5-μm fiber. When the same transmitter is used with a 50-μm core, the smaller core intersects a smaller fraction of the transmitter's projected beam. It therefore picks up less power than the 62.5-μm core. A 50-μm fiber coupled to a surface-emitting LED suffers a power budget loss of approximately 2 dB to 5 dB, depending on the beam geometry.

An edge-emitting LED produces a beam spot that generally fits within either a 62.5-μm or 50-μm core. It couples approximately the same power into either fiber type. There is no 50-μm penalty associated with edge-emitting LED transmitters. The edge-emitting LED produces a superior beam, but, like many things in life, you gotta pay for it. Edge-emitting sources are generally more expensive than surface-emitting sources.

To better support the use of cheap 62.5-μm sources, many customers in North America install 62.5-μm fiber.[100]

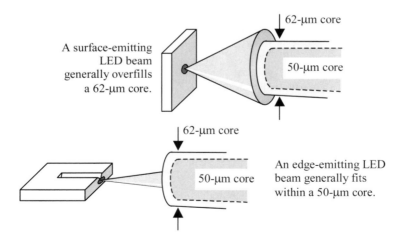

Figure 11.13—Surface-emitting and edge-emitting LED beam widths.

---

[100] The 50-mm versus 62.5-mm issue is part of a larger marketing war that rages constantly among cabling manufacturers.

POINT TO REMEMBER

> ➤ Fifty-micron multimode fiber has a generally higher bandwidth and less attenuation than 62.5-μm multimode fiber. These advantages are counterbalanced by the fact that some common LED sources can't couple efficiently into 50-μm core.

11.5.5 Multimode Optical Performance Budget

The *optical performance budget* for a fiber-optic link is a list of all known factors that affect link performance. It adds up the worst case for each effect and verifies that the system will still function under worst-case conditions.

The performance budget is one of the first places customers look when they want to lay blame for a system that doesn't work. If the customer's fiber meets the stated attenuation budget, but your system doesn't work, then in the mind of the customer any problems that arise will begin to seem like *your fault*. This issue cuts both ways. If you can prove that the customer's fiber doesn't meet its specification, you get to point the fickle finger of fate squarely at the cable installer. Make sure you have a performance budget and that it is accurate.[101]

A good optical performance budget should address three main areas: dispersion, attenuation, and jitter. Template budgets for all three areas appear in the following sections. These templates follow the same pattern used for the development of FDDI, Ethernet, and Fibre-Channel LAN standards.

These templates were developed for glass multimode fiber links operating at distances between 100 m and 1 km, and at speeds between 100 Mb/s and 1000 Mb/s. If your application is significantly different from that, you should spend some time carefully checking each of the assumptions used in these models to see which might not apply.

This brief chapter cannot possible hold enough information to satisfy the needs of an optical component designer. That is not my intent. Rather, I merely hope to define the most commonly used terms, show the relations between the various parameters, and prepare you as a digital designer to read an optical specification sheet, understand the terms used, and properly apply a fiber-optic transceiver in your next system design. Towards that end, let's look first at the causes of dispersion in fiber data links.

11.5.5.1 Multimode Dispersion Budget

An optically perfect, crisp rising-edge input to the front end of a long fiber emerges at the other end somewhat smaller and with a degraded risetime. The purpose of dispersion calculations is to determine the extent of risetime degradation and to estimate the impact that degradation will have on signal reception. I'm going to review the dispersion calculations very carefully, because they are riddled with assumptions that must be understood in order for the calculations to apply. You will need the following basic information (Table 11.2) to compute dispersion.

---

[101] I'm not kidding about this. Some customers will initiate legal action against your company for a failure to perform.

Table 11.2—Parameters Required for Dispersion Calculations

Name	Meaning	Units
t_s	Source rise/fall time	ps
t_b	Source baud interval	ps
t_w	Clock window at receiver	ps
B_m	Fiber modal bandwidth	MHz-km
l	Fiber length	km
λ_{RMS}	Source RMS spectral width	nm
D	Fiber chromatic dispersion constant	ps/nm-km
If you don't have access to D, you will need the next three parameters:		
λ_0	Zero-dispersion wavelength	nm
S_0	Dispersion slope	ps/nm^2-km
λ_c	Source center wavelength	nm

Figure 11.14 depicts four test points used to define dispersion. Point *TP1* represents the electrical interface just prior to the LED or laser driver. This interface is commonly a differential low-level signaling interface. The choice of signal levels is arbitrary.[102]

Point *TP2* represents the optical signal entering the fiber-optic cable plant. This is commonly defined at a point on the far side of the transmitter connector if the transmitter is directly connectorized or on the far side of the first connection placed at the end of the transmitter pigtail if the transmitter is a pigtail type. In some extraordinary circumstances—for example, when a special form of patch cord is required at the driver—point TP2 is defined at the end of a suitable patch cord.[103]

Point *TP3* represents the optical signal at the conclusion of the fiber-optic cable plant at the point where it enters the receiver. The receiver comprises an optically sensitive diode receiver (PIN diode or APD diode), a low-pass filter[104] (to cut down the level of white noise from the diode), and a limiting amplifier (a comparator).

Point *TP3b* represents a circuit internal to the receiver. This circuit is generally not available for direct testing, although its characteristic may be determined by indirect means. This node is after the low-pass filter but before the limiting amplifier.

Point *TP4* represents the electrical interface just after the limiting amplifier (comparator) in the receiver but before the sampling circuit (flip-flop). This interface is commonly a differential low-level signaling interface.

The specific problem addressed by the dispersion budget involves the degree of intersymbol interference present at point TP3b. The precise amount of intersymbol interference at point TP3b varies as a function of the optical risetime at TP2, the effects of

---

[102] Although ECL (or PECL) was for a long time the best choice, in 1999 it began being supplanted by LVDS and other low-voltage signaling schemes.
[103] That is the approach taken in Gigabit Ethernet.
[104] Most receivers comprise a chain of several limiting-amplifier stages, each with a finite bandwidth. The number of stages of amplification required to boost the signal to a fully saturated level is a function of the incoming signal amplitude. The low-pass filter in this diagram represents the composite bandwidth of those stages which, at the minimum signal amplitude, operate in the linear mode. The comparator then represents the operation of the remaining stages, which operate in a fully saturated mode. A well-designed receiver heavily restricts the bandwidth of the first stage, but leaves the others wide open, so that the effective receiver bandwidth of the resulting amplifier chain does not vary with signal amplitude.

11.5.5 • Multimode Optical Performance Budget

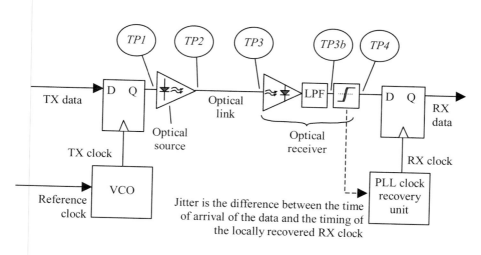

Figure 11.14—Block diagram showing critical points for jitter analysis.

the fiber as measured at TP3, and the effects of the low-pass filter inside the receiver. The intersymbol interference reduces the usable signal amplitude at point TP3b, rendering the system more susceptible to internal noise generated within the optical receiver. If the usable signal amplitude is sufficiently reduced, bit errors may result.

> *The purpose of the dispersion budget is to allocate some permissible degree of dispersion among the source, fiber, and receiver in a way that will provably limit the worst-case intersymbol interference at TP3b.*

Dispersion calculations are generally made using the following assumptions:

- ➤ The source risetime (or fall time, whichever is worse) at TP2 is known independent of other factors.
- ➤ The fiber performs a linear, time-invariant operation on the transmitted signal.
- ➤ The receiver detection diode is both linear and instantaneous in its operation, converting received optical power to an electrical voltage.
- ➤ The low-pass filter performs a linear, time-invariant operation on the transmitted signal.

Under these assumptions, we may conclude that the complete signal at point TP3b, and thus the intersymbol interference, may be completely predicted by convolving together

1. The source signal at TP2,
2. The impulse response of the fiber,
3. The gain of the diode, and
4. The impulse response of the low-pass filter.

Given a sufficiently powerful calculation tool, you could convolve together these waveforms and directly compute the intersymbol distortion. There is, however, a generally accepted shortcut that can save you a lot of time in the calculations. It is based on some additional assumptions, which I should be very careful to state clearly. This shortcut-calculation procedure does not work in all systems, only in those which conform to the following:

> The source signal rising edge is monotonic (so is the falling edge).
> The step response of the fiber is monotonic.
> The step response of the low-pass filter is monotonic.
> All rising and falling edge waveforms are Gaussian.

In fiber-optic systems, the aforementioned signal conditions are generally true, and under those conditions the following approximation describes the risetime at point TP3b.

$$(t_{TP3b})^2 = (t_{TP2})^2 + (t_{TP2 \to TP3})^2 + (t_{TP3 \to TP3b})^2 \qquad [11.4]$$

where t_{TP3b} is the 10% to 90% risetime at point TP3b,

t_{TP2} is the 10% to 90% risetime at point TP2,

$t_{TP2 \to TP3}$ is the 10% to 90% risetime of the fiber, and

$t_{TP3 \to TP3b}$ is the 10% to 90% risetime of the low-pass filter.

We'll use equation [11.4] to estimate the signal risetime, and from that the intersymbol interference, at TP3b.

I'd like to draw your attention for a moment to the use of the Gaussian assumption in equation [11.4]. The Gaussian assumption is also used to translate the signal risetime at TP3b into an equivalent amount of intersymbol interference. It is crucial that you understand that this assumption, while fairly reasonable for most fiber-optic systems, does not apply to copper-based transmission systems. In the copper-based world you are faced with a skin-effect impulse response which has long, significant, non-Gaussian tails. In a copper system, you really need to explicitly convolve together the whole system response to accurately predict what intersymbol interference will develop. In the fiber case, we can take the shortcut of making the Gaussian assumption.

Equation [11.4] is derived by first recognizing the relation between the risetime of a Gaussian step response and the area under the corresponding impulse response (see [88], and Figure 11.15).[105] The 10% to 90% points on the step response correspond in time to the 10% and 90% cumulative area points on the impulse response. That is, 80% of the area under the impulse response should lie between the 10% and 90% points on the step-response curve. If we knew the standard deviation σ of the impulse response waveform, and if the pulse were truly Gaussian, then the exact width which encompasses 80% of the area should equal $\pm 1.28\sigma$.

For every Gaussian waveform there exists this basic equivalence between risetime of the step response and the standard deviation of the corresponding impulse response: The 10% to 90% risetime equals 2.56 times the standard deviation.

---

[105] I should point out that the concept of RMS risetime makes sense only for signals that are monotonic, that is, lacking any form of ringing or other non-monotonic behavior.

11.5.5 • Multimode Optical Performance Budget

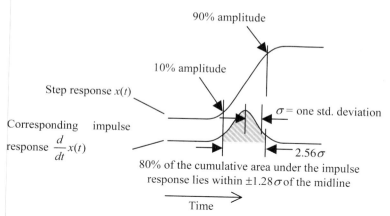

Figure 11.15—Relation of step response to impulse response (for a positive, integrable impulse response with finite variance).

If the risetime is proportional to the standard deviation, then the risetime squared should be proportional to the standard deviation squared (variance) of the impulse response. That's the form in which equation [11.4] may begin to make sense to you. It's a theorem about the variance of little impulses. It says, "When convolving impulse functions, their variances add."

The proof of this familiar fact appears in every elementary textbook on the subject of probability theory. You'll find that in probability theory this theorem applies to the determination of the variance of a sum of random variables, whereas here we look to determine the variance of a convolved chain of impulse responses, but it's the same theorem either way. Variances add.

To operate equation [11.4] you will need to know the risetimes of the component pieces. The risetime of the source t_{TP2} is easy: That's the 10% to 90% risetime called out on the datasheet for the optical driver.

The 10% to 90% risetime of the fiber is a little more involved, being a function of two types of fiber bandwidth: modal and chromatic. For a typical multimode fiber application the overall risetime is closely approximated by

$$\left(t_{TP2 \to TP3}\right)^2 = \left(\left(\frac{0.48}{B_m}\right) \cdot l \cdot 10^6\right)^2 + \left(D \cdot l \cdot 2.56 \lambda_{RMS}\right)^2 \quad [11.5]$$

where the assumed impulse of the fiber is approximately Gaussian,

$t_{TP2 \to TP3}$ is the 10% to 90% risetime contributed by the fiber, ps,

B_m is the 6-dB (electrical) modal bandwidth of the fiber (FWHM bandwidth), in MHz-km,

l is the length of the fiber, km,

D is the chromatic dispersion constant, ps/nm-km,

λ_{RMS} is the RMS spectral width of the source, nm (the RMS spectral width is the standard deviation of the curve of power density versus wavelength), and

2.56 is a constant that converts standard-deviation-type pulse width measurements to 10% to 90% measurements.

The value of the chromatic dispersion constant D varies with wavelength. For some fibers, you will find a chart that directly shows D as a function of wavelength. Other fibers are specified in terms of a zero-dispersion wavelength and dispersion slope from which you are expected to estimate D, like this:

$$D \approx \sqrt{\left((S_0/4)(\lambda_C - \lambda_0^4/\lambda_C^3)\right)^2 + (0.7 S_0 \lambda_{RMS})^2} \qquad [11.6]$$

where D is the chromatic dispersion constant, in ps/nm-km,

S_0 is the dispersion slope parameter at wavelength λ_0, in ps/(nm²-km),

λ_c is the source center wavelength, in nm,

λ_0 is the zero-dispersion wavelength, in nm, and

λ_{RMS} is the RMS spectral width of the source, in nm.

Equation [11.6] is the dispersion formula accepted for use by the IEEE 802.3z Gigabit Ethernet committee in 1998. At the time it was generally acknowledged by the working members of that committee to be the most accurate closed-form approximation available for the estimation of chromatic dispersion.

Equation [11.6] is appropriate for use with normal or dispersion-shifted fiber. It properly estimates the high-order dispersion effects that happen with wide-spectral-width sources near the dispersion null. It also works with narrow-spectral-width sources. It works in both 850 and 1300 nm windows. This formula is not appropriate for use with dispersion-flattened single-mode fiber.

The 10% to 90% risetime of the low-pass filter is a function of the bandwidth of low-pass filter and the various preamplifier circuits internal to the receiver. These bandwidths are typically specified in terms of a 3-dB roll-off frequency. For a typical multipole, critically damped roll-off, most receiver filters will have a risetime reasonably approximated by

$$t_{TP3 \rightarrow TP3b} = \frac{0.35}{B_{3dB}} \qquad [11.7]$$

where the equation assumes the impulse of the filter is not quite as good as a Gaussian response,[106]

$t_{TP3 \rightarrow TP3b}$ is the 10% to 90% risetime of the low-pass filter, s, and

---

[106] A perfect Gaussian filter would have a risetime equal to $0.338/BW_{3dB}$.

11.5.5 • Multimode Optical Performance Budget

B_{3dB} is the frequency (Hz) at which the filter gain falls 3 dB short of the gain at DC.

Now you have all the numbers necessary to apply equation [11.4].

Given t_{TP3b}, and assuming the overall response is Gaussian, you may use the following general expressions to represent the cumulative effect of source, fiber, and receiver filter, where the variable σ equals the standard deviation of the cumulative impulse response, equal to $t_{TP3b}/2.56$. The following expression takes into account the shape of the received waveform, but does not model the bulk transport delay of the fiber.

Impulse response
$$h(t,\sigma) = \frac{1}{\sigma\sqrt{2\pi}} \exp\left(-\frac{t^2}{2\sigma^2}\right) \quad [11.8]$$

Step response
$$g(t,\sigma) = \mathrm{erf}_2(t/\sigma) \quad [11.9]$$

Frequency response
$$H(\omega,\sigma) = \exp\left(-\frac{\omega^2\sigma^2}{2}\right) \quad [11.10]$$

where $h(t)$ is the system impulse response, as a function of time t,

$g(t)$ is the system step response, as a function of time t.

The function $\mathrm{erf}_2()$ is defined in Appendix E,

$H(\omega)$ is the transfer function from TP2→TP3b, as a function of frequency,

ω is the frequency of operation, rad/sec, and

σ is the RMS pulse width of the impulse response, equal in this case to $t_{TP3b}/2.56$, in units of seconds (NOTE: t_{TP3b} was previously computed in units of ps—here you must translate it to units of sec).

You may recognize the function $H(\omega)$ as a linear-phase transfer function, meaning that the step response will have equal length, symmetrical tails on the leading and trailing edges. This transfer function will not much affect the positions of the zero crossings of the received waveform, but it will create ISI whenever the tails from one transition have not been given sufficient time to dissipate before the center of the next bit.

Procedures for modeling the overall system response, given the function $H(\omega)$, appear in Chapter 4, "Frequency-Domain Modeling."

From the risetime T_{TP3b} you can predict the worst degradation in received amplitude likely to occur under any conditions due to *intersymbol interference* (ISI). This requires that you know the worst-case data pattern for intersymbol interference. That pattern, for any system with monotonic rising and falling step edges, is a single 1 preceded and followed by a long run of zeros (or the complement of this pattern).

Worst case ISI pattern: 000000010000000...   [11.11]

From knowledge of the system step response we may write an expression for the exact waveform $y_{BAD}(t)$ expected under the worst-case-ISI condition, namely, a single step going high at time t_0 followed by an opposite step at time $t_0 + t_B$, where the constant t_B represents the data link baud interval:

$$y_{BAD}(t) = g(t-t_0) - g(t-t_0-t_b) \qquad [11.12]$$

Figure 11.16 shows examples of various dispersion-induced ISI waveforms calculated in this manner. What counts is the degree to which $y_{BAD}(t)$ exceeds the switching threshold at the moment your sampling circuit makes its decision. In order to estimate the degrading effect of dispersion, separate from all other factors, one assumes the received waveform is of unit size, the sample is taken precisely halfway between the two step transitions at time $t_0 + t_b/2$, and the threshold is adjusted precisely midline (at 1/2). Under those conditions the sampled amplitude (above the threshold) would be

$$y_{SAMPLE} = y_{BAD}(t_0 + t_b/2) - 1/2 \qquad [11.13]$$

Plug in the definition of y_{BAD} [11.12].

$$y_{SAMPLE} = g(t_0 + t_b/2 - t_0) - g(t_0 + t_b/2 - t_0 - t_b) - 1/2 \qquad [11.14]$$

Simplify terms.

$$y_{SAMPLE} = g(t_b/2) - g(-t_b/2) - 1/2 \qquad [11.15]$$

Adapt the definition of $g(t)$ from [11.9].

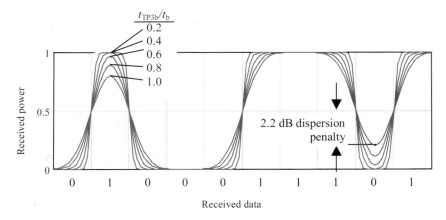

Figure 11.16—Received optical signal with varying amounts of Gaussian dispersion.

11.5.5 • Multimode Optical Performance Budget

$$y_{SAMPLE} = erf_2\left(\frac{t_b}{2\sigma}\right) - erf_2\left(-\frac{t_b}{2\sigma}\right) - 1/2 \qquad [11.16]$$

Apply the definition of σ.

$$y_{SAMPLE} = erf_2\left(1.28\frac{t_b}{t_{TP3b}}\right) - erf_2\left(-1.28\frac{t_b}{t_{TP3b}}\right) - 1/2 \qquad [11.17]$$

Equation [11.17] is usually converted to a dB penalty, expressing the ratio of two quantities: first, the sampled amplitude above the threshold in [11.17] assuming dispersion T_{TP3b}, and second, the nominal sampled amplitude that would have been received had there been zero dispersion (1/2 in this case).

The traditional approach for all penalty calculations is to express them as a received optical power penalty, using 10 times the logarithm of the received power ratio, recognizing that the received optical power is converted linearly by the optical receiver into a received voltage. An optical power penalty of 3 dB implies that the received voltage has been cut in half.

$$p_D = -10 \cdot \log\left(\frac{erf_2\left(1.28\frac{t_b}{t_{TP3b}}\right) - erf_2\left(-1.28\frac{t_b}{t_{TP3b}}\right) - 1/2}{1/2}\right) \qquad [11.18]$$

The above expression for the dispersion penalty p_D is plotted against the parameter t_{TP3b}/t_b in Figure 11.17, where t_{TP3b} is the total 10% to 90% dispersion in seconds and t_b is the baud interval in seconds (or both in ps). The dispersion penalty represents the loss of eye opening expected in the received data due to the interaction of one bit with the next. The ratio t_{TP3b}/t_b defines the dispersion in units of *unit intervals* where one unit interval is understood to be one baud interval.

In many systems a limit is placed on the worst-case magnitude of the dispersion penalty. Typical dispersion limits are set at 2 or perhaps 3 dB. With 3 dB of dispersion penalty, the eye at top dead center is half closed by intersymbol interference. Beyond 3 dB of dispersion, the system performance becomes extremely sensitive to dispersive effects, as the received eye pattern begins to completely close.

Any fiber-optic data link that conveys an acceptable amount of power to the receiver, but for which the degree of intersymbol interference (dispersion) prohibits reliable communication, is called a dispersion-limited link. Increasing the transmitter power in such a link will not improve performance. On the other hand, a link with an acceptable degree of intersymbol interference (dispersion), but which is otherwise limited by the total amount of received power, is called an attenuation-limited link.

After consideration of the dispersion penalty, any jitter or uncertainty in position of the clock further degrades the received amplitude at the precise moment of sampling. The additional degradation as a function of clock position could be computed by using an

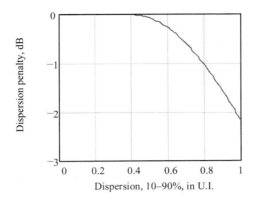

Figure 11.17—Dispersion penalty in dB versus the degree of dispersion, 10% to 90%, measured in units of bit intervals.

equation like [11.13] to compare the worst-case sampled power at one edge of the clock window (A_2 in Figure 11.18) to the sampled power at the center of the clock window (A_1). Alternately, you may assume that the received signal under worst-case conditions takes on the appearance of a half-sinusoid within each baud interval, leading to the following more commonly-used approximation for what is called the *clock window penalty*:

$$p_\mathrm{W} = -10 \cdot \log\left(\cos\left(\frac{\pi}{2}\frac{t_\mathrm{w}}{t_\mathrm{b}}\right)\right) \qquad [11.19]$$

where t_w is the full width of the clock uncertainty window (assumed to be symmetrically located about the baud center as in Figure 11.18),

t_b is the full width of each baud interval, and

p_W is the clock window penalty, in decibel units of received optical power, evaluated at the edges of the clock window.

The preceding assumption is a *pessimistic* assumption. It always *overestimates* the clock window penalty. The clock window penalty may be improved only by narrowing the window of clock uncertainty through the use of better clock-recovery circuitry.

To the extent that there is any *duty-cycle distortion* present in the transmitted signal, we must adjust the dispersion calculations. Duty-cycle distortion is caused by asymmetries in the turn-on and turn-off time of the optical source. It is a specification of the optical source. In cases where the duty-cycle distortion is significant it is customary to take the minimum ON or OFF duration permitted by the duty-cycle specification and use that value for t_b in the calculation of the dispersion and clock window penalties. For example, in FDDI the transmitter duty-cycle distortion is specified as no more than 1.00 ns peak-to-peak. That would mean that the worst-case baud duration must lie within the range of 8.00 ns +/− 0.50 ns, for a minimum of 7.50 ns. The value 7.50 ns should be used in all FDDI dispersion calculations.

11.5.5 • Multimode Optical Performance Budget

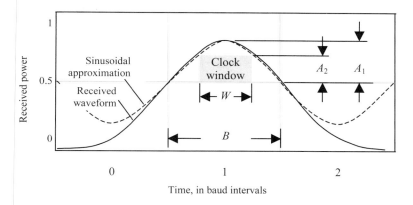

Figure 11.18—The received amplitude at the edges of the clock window is never as great as in the center of the eye.

The parameters p_D and p_W comprise the output of the dispersion-estimation process. These values feed into the signal attenuation budget discussed in the next section.

Example Calculation of FDDI Dispersion Penalties

NOTE: The following values represent typical datasheet parameters. These values do not reflect the worst-case parameters permitted by the FDDI standard.

Specifications		
Source rise/fall time	$t_s = 3500$	ps, 10–90%
Source baud interval (min)	$t_b = 7500$	ps
Clock window at receiver	$t_w = 2000$	ps
Fiber modal bandwidth	$B_m = 500$	MHz-km
Fiber length	$L = 2$	km
Source FWHM spectral width	$\lambda_{FWHM} = 140$	nm
Source RMS spectral width	$\lambda_{RMS} = \dfrac{\lambda_{FWHM}}{2.35}$	nm
Zero-dispersion wavelength	$\lambda_{0,min} = 1300$	nm
	$\lambda_{0,max} = 1350$	
Dispersion slope	$S_0 = 0.11$	ps/nm²-km
Source center wavelength	$\lambda_{c,min} = 1320$	nm
	$\lambda_{c,max} = 1360$	
LPF bandwidth	$B_{LPF} = 87.5$	MHz

Derived Quantities		
LPF risetime	$t_{LPF} = \dfrac{0.35}{B_{LPF}} \cdot 10^6$	ps 10–90%
Dispersion constant	$D(\lambda_0, \lambda_c) = \sqrt{\left(\dfrac{S_0}{4}\left(\lambda_c - \dfrac{\lambda_0^4}{\lambda_c^3}\right)\right)^2 + (0.7 \cdot S_0 \lambda_{RMS})^2}$	ps/nm-km
Fiber risetime	$t_{TP2 \rightarrow TP3} = \sqrt{\left(\dfrac{0.48}{B_m} \cdot L \cdot 10^6\right)^2 + (D(\lambda_0,\lambda_c) \cdot l \cdot \lambda_{RMS} \cdot 2.56)^2}$	ps 10–90%
Overall risetime	$t_{TP3b} = \sqrt{t_s^2 + t_{TP2 \rightarrow TP3}^2 + t_{LPF}^2} = 6119$	ps 10–90%
Dispersion penalty p_D	$-10 \cdot \log\left(\dfrac{\mathrm{erf}_2\left(1.28\dfrac{t_b}{t_{TP3b}}\right) - \mathrm{erf}_2\left(-1.28\dfrac{t_b}{t_{TP3b}}\right) - 1/2}{1/2}\right) = 1.154$	dB optical
Clock window penalty p_W	$-10 \cdot \log\left(\cos\left(\dfrac{\pi}{2}\dfrac{t_w}{t_b}\right)\right) = 0.393$	dB optical

POINT TO REMEMBER

➢ Dispersion calculations determine the extent of risetime degradation and estimate the impact that degradation will have on signal reception.

11.5.5.2 Multimode Attenuation Budget

The purpose of the attenuation budget (also called the power budget) is to allocate some permissible degree of attenuation among the long fiber cabling, the short fiber jumpers, and the connectors in a way that will provably limit the worst-case minimum received signal at TP3.

A proper attenuation budget takes the form shown in Table 11.3. The budget begins by listing the output power available from the transmitter and the input power required at the receiver. The difference between these two entries is the *optical power budget*, or *available power budget*, in the optical link. The other entries represent various attenuating factors present within the link. These entries are appropriate for evaluating a multimode glass-fiber link with an LED transmitter. (See also Section 11.5.9, "Multimode Fiber with Laser Source," and Section 11.6.1, "Single-Mode Signal Propagation.")

The sum of all attenuating factors should (hopefully) be less than the available power budget. The power margin represents the excess power budget remaining after subtraction of all known deleterious effects.

11.5.5 • Multimode Optical Performance Budget

The values in Table 11.3 represent typical datasheet parameters for FDDI components. These values do not necessarily reflect the worst-case parameters permitted by the FDDI standard.

Table 11.3—Multimode LED Optical Power Budget

Line	Item	Value	Total	Units
1	TX power (min)	–20		dBmW
2	Guaranteed RX sensitivity	–31		dBmW
3	Power budget *(line 1 less line 2)*		11	dB
4	Cable losses	4.0		dB
5	Connector and splice losses	2.0		dB
6	Dispersion penalty p_D	1.2		dB
7	Clock window penalty p_W	0.4		dB
8	Extinction ratio penalty	0.0		dB
9	Other penalties	n/a		dB
10	Total losses and penalties *(sum of lines 4–8)*		7.6	dB
11	Power margin *(line 3 less line 9)*		3.4	dB

NOTES

TX power. The guaranteed worst-case (minimum) output power specified under the worst combination of power supply voltage, temperature and aging conditions.

RX sensitivity. A level above which all receivers are guaranteed to operate with an acceptable BER, under worst-case conditions, including local self-generated crosstalk from any nearby transmitter circuits within the same package, VCC noise, etc.

Cable losses. Typically 2 to 10 dB per kilometer for glass fiber. This example assumes 2 km at 2 dB/km.

Connector and splice losses. Remember to provide for several connections so you can have patch panels and jumper cables. This example budget permits four connectors at 0.5 dB each.

Dispersion penalty. See "Example Calculation of FDDI Dispersion Penalties." In some systems the receiver input sensitivity is defined under a specific test condition which imposes a worst-case signal with artificially generated dispersion. In such a system, if the test conditions accurately reflect the real-life worst-case dispersion, you could assume that the dispersion penalty has already been accounted for within the receiver sensitivity specification.

Clock window penalty. See "Example Calculation of FDDI Dispersion Penalties."

Extinction ratio penalty. For systems that transmit equal numbers of ones and zeros (that's the normal case for a fiber-optic link), the peak-to-peak transmitted power is nominally twice the average transmitted power. The peak-to-peak power is what activates the receiver; the average power is what we read on a power meter during product testing or installation. If the transmitter does not go completely dark during the transmission of a logical zero, the peak-to-peak power will be somewhat smaller than twice the average power. The difference between the actual peak-to-peak power and the nominal value is expressed here as the *extinction ratio penalty*. The extinction ratio E expresses the ratio between the *zero*-state power and the *one*-state power, while the extinction ratio penalty in optical dB equals $10\log((1 - E)/(1 + E))$. The extinction ratio penalty for LED transmitters is often negligible.

Other penalties. Other penalties apply to laser-based links. See Section 11.5.9, "Multimode Fiber with Laser Source."

Power margin. The purpose of margin is to cover your posterior in case you forgot anything or were overly optimistic about any parameters. A 3-dB margin is highly desirable. Any finished system that incorporates a 3-dB optical margin will prove very robust in the field. Technicians will routinely be able to exceed its guaranteed link distances, eating into the margin without fear of catastrophic link failure. A big margin keeps your customers smiling.[107]

POINT TO REMEMBER

➤ An attenuation budget allocates attenuation among the long continuous runs of fiber cabling, the short fiber jumpers, and the connectors in a typical installation.

11.5.6 Jitter

Every practical communication system is affected by *jitter*. Simply put, jitter is the amount by which each individual rising or falling edge deviates from its ideal temporal position.

A large amount of jitter, something large enough to push individual data transitions into the clock window, will obviously cause bit errors. A smaller amount of jitter, while not enough to cause dramatic errors, can still induce system failure if it interferes with the operation of the PLL-based clock recovery subsystem.

In fiber-optic transmission systems it is common to divide the measured jitter into two components, the *deterministic jitter* and the *random jitter*. The deterministic jitter represents the misplacement of received edges due to any repeatable, data-dependent phenomenon. The most important property of deterministic jitter is that it is *bounded*. That is, there is a certain amount of it, and it never gets any worse.

One common measurement procedure for deterministic jitter uses a repeating data waveform that includes a variety of different run-lengths, both high and low. An ideal waveform would be something like this: [10101111000011110000 10101111000011110000...]. This waveform includes a rapid 1010... pattern, and also some slower 11110000... patterns. Such a repeating waveform must be observed, averaged over many repetitions to eliminate random effects, and analyzed to determine the precise position of each zero crossing. Such an analysis reveals the worst-case peak-to-peak deterministic jitter. Frequent causes of deterministic jitter include asymmetry in the rise/fall time of the

---

[107] In the initial stages of product planning, you should insist on an even greater margin (perhaps 6 dB). As your design progresses and you learn more about the behavior of optical components, there will be many heated arguments about the budget, and the margin will *always* diminish.

11.5.6 • Jitter

optical source, the non-ideal modal step response of the fiber, and phase nonlinearity in the receiver filter.

The random jitter is then calculated as the difference between the observed total jitter and the measured deterministic jitter. The reason we bother separating these two effects is that total jitter is usually measured at some nominal working point on the jitter histogram. A comfortable probability-of-occurrence level for jitter measurement in high-speed systems is about $p=10^{-6}$. In other words, the measurement gives us a jitter number and tells us that the jitter will exceed this number only one time out of 10^6. That's nice, but it isn't what we need to know. What we usually seek is a jitter number that will be exceeded only one time in, say, 10^{12}. To calculate that number, we can extrapolate the random jitter component, assuming it follows a Gaussian probability distribution from 10^{-6} to 10^{-12}, and then add back in the deterministic jitter to estimate the total jitter at a probability level of 10^{-12}. Extrapolation of the random jitter distribution is done according to the chart in Figure 11.19.

POINT TO REMEMBER

> Fiber-optic transmission systems commonly divide the jitter budget into *deterministic jitter* and *random jitter*.

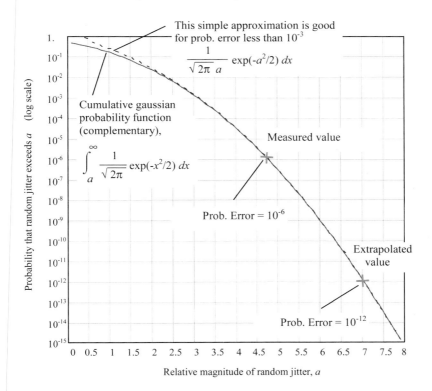

Figure 11.19—The extrapolated random jitter magnitude at a probability-of-error level of 10^{-12} exceeds the measured value at 10^{-06} by a factor of $7/4.8 = 1.46$.

11.5.7 Multimode Fiber-Optic Noise and Interference

Fiber-optic links use photons, not electrons. As a result, fiber-optic cables emit zero electromagnetic energy within the FCC or EN regulated bands. Fiber-optic links operate free from interference due to radio waves, power-line noise, lightning, and all other electromagnetic effects. Fiber links may run near power lines and other communication lines without fear of interference. They may be bundled tightly together without limit. Fiber is perfect for communication between buildings.

Any noise problems you encounter with fiber will come from the receiver circuitry itself, from local crosstalk within the receiver package or from the printed-circuit board or power system to which the receiver connects.

The self-noise generated within a high-speed fiber-optic receiver is true, wide-bandwidth white Gaussian noise (WGN). It's caused partly by thermal noise and partly by quantization noise (shot noise) within the detection diode and preamplifier circuits. In a good receiver this is the primary source of noise. Cooling down a receiver usually improves its sensitivity.

A low-pass filter in series with the detector output reduces the level of noise, but at the expense of additional signal dispersion. Most receivers attempt to set the low-pass filter cutoff somewhere around $0.7/t_b$, where t_b is the baud interval. Extremely high-speed receivers, because they cannot afford the dispersion that comes with low-pass filtering, have to operate in the presence of high levels of noise. Such receivers are generally less sensitive than lower-speed receivers.

Direct electrical crosstalk between the local optical transmitter and the optical receiver is another factor that can affect performance. This crosstalk is exaggerated when the local transmitter and receiver are located close together. The worst case for local crosstalk happens when the receive fiber in a two-fiber system is disconnected or when the far end is powered off. In either case the receiver sees no light. When that happens, the receiver's AGC circuit usually turns up the receiver gain as high as it will go, straining to hear any faint signals on the incoming fiber. If the gain goes up too high, the receiver picks up electrical crosstalk from the local transmitter. I've seen protocols that come alive and start talking to themselves when that happens, thinking there is a "live" device at the other end.

Appropriate solutions to the local crosstalk problems include

1. Limiting the maximum receiver gain,
2. Implementing a minimum-signal-level detection circuit to shut off the receiver in the event of a dark input, or
3. Reducing the crosstalk.

Regarding emissions, even though fiber itself doesn't radiate, beware of the fiber-optic transmitter. The transmitter is a fairly high-powered device mounted near a hole in your cabinet shielding. Potentially high levels of direct radiation can emanate from the optical transmitter.

11.5.8 Multimode Fiber Safety

POINT TO REMEMBER

> Fiber cabling may be immune to crosstalk and RFI, but your fiber-optic receiver is not.

11.5.8 Multimode Fiber Safety

One thing you should never do with a fiber is look into the end of it. Most fiber communication systems use infrared light. The human eye cannot see it, but it is very, very bright. The surface brightness over the tiny active cross section of the fiber can rival that of an oven heating element on full broil. While it is true that there are no internationally recognized safety guidelines for multimode LED emitters, it is still never a good idea to look directly into the end of a fiber.

This author will always remember the warning he found posted near the transmit port of one early fiber-optic transmitter: "Do not look into this orifice with your remaining good eye."

POINT TO REMEMBER

> Never look into the end of a fiber.

11.5.9 Multimode Fiber with Laser Source

LAN standards for ATM, Fibre Channel, and Ethernet all advocate the use of laser-diode transmitters on multimode fiber. These standards attain speeds of approximately 1 Gb/s, at distances ranging up to several hundred meters.

Laser-diodes enjoy many advantages over LED sources. Good laser-diodes have a faster risetime, a greater switching frequency, and a higher optical power. Laser-diodes also possess a narrower spectral width, which dramatically reduces chromatic dispersion. As far as traditional measures of optical performance are concerned, laser-diodes appear better in every respect.

The major areas where laser-diodes are *not* better than LED sources include cost and modal effects. The cost issue is, for the moment, pretty glaring. Some manufacturers hope that VCSEL technology (see Section 11.5.10, "VCSEL Diodes") will reduce the cost of their laser optics, but it will probably still never achieve the rock-bottom pricing of a simple surface-emitting LED.

The modal effects are very complex and deserve serious discussion. The problem here is that laser-diodes depend on some subtle, undocumented, and unspecified features of multimode fiber. That's not a good situation.

Multimode fibers are designed for illumination with a broad-beamed source like an LED. The LED induces an *overfilled launch condition*, meaning that all modes of propagation in the fiber are filled with light.[108] If the fiber manufacturer has done the job correctly, the propagation delays for most of the modes will be closely grouped. Notice that

---

[108] An overfilled launch condition does not carry equal amounts of power in each mode; however, all modes are used.

I did not say *all* the modes, but rather *most* of the modes. Since there are so many modes, and the power splits among them, no individual mode carries very much power. If there are a few **bad** modes with radically different delays, it really doesn't hurt anything. When confronted with all the modal power that does arrive together, the receiver will never notice a tiny amount of errant, **bad**-mode power that arrives late (or early).

This description may give you an idea about how to improve the bandwidth performance of multimode fiber. What if you could craft a source that broadcast power only into the really **good** modes, the ones with the best natural speed-matching? Such a miracle source might then enjoy the benefits of reduced modal dispersion. That would be wonderful, but no one knows how to do it reliably and cheaply with practical fibers.

What the industry does know how to do is how to craft a source that concentrates most of its power into a few very **bad** modes. This kind of source emphasizes the worst-than-worst case modal dispersion problems of those **bad** modes. A fiber illuminated with this kind of **bad**-mode source may exhibit modal dispersion in excess of the normal specified limits.

One way to create a really **bad**-mode source is to aim a very narrow-beam laser-diode straight down the center of a multimode fiber. This couples power into only a few modes of propagation, mostly centered in the middle of the fiber. It turns out that the manufacturing processes used to create multimode fiber often leave imperfections in the dead center in the fiber that are highlighted by this type of source.

Precisely this scenario happened to the Gigabit Ethernet committee in 1997. The committee at that time was trying to develop a multiuse 1300 nm source that would work on either single-mode fiber or multimode fiber. The SMF requirements mandated that the narrow-beam laser-diode source be very accurately centered. When this source was coupled to MMF, the pinpoint accuracy of the source constituted an almost perfect **bad**-mode source, with anomalous results. Eventually, the committee found ways around this difficulty [93].

Another problem with the use of laser-diodes on MMF relates to the modal structure of a laser source. Laser-diodes radiate fewer optical modes than do LED sources. For the purposes of this discussion you may imagine one mode as a beam of light shining in a particular direction with a particular wavelength.[109] Contrary to popular belief, laser-diodes do not produce a single beam of light at one wavelength. A laser-diode actually radiates in hundreds of modes at once, at a multitude of wavelengths, spewing out the front facet of the laser at a variety of angles. There are a lot of modes, but not an infinite number. The modal structure is discrete and measurable.

An LED, on the other hand, operates more like a light bulb. It emits light over a continuous range of wavelengths, with an even and continuous spatial distribution.

The modal structure of the laser-diode wouldn't really matter if all the modes were always present, all the time, with consistent amplitudes, but that isn't how they work. Some modes carry a lot of power, some very little. Also, sometimes the amplitude of one specific mode pops out of whack suddenly. This happens particularly in connectors if an air gap is present between the coupled fibers. It also happens in the fiber itself if the fiber is wiggled. When the amplitude of one mode changes, if you happen to have too many eggs in that one modal basket, the received signal glitches unpredictably. The term for this effect is *modal*

---

[109] Here I use a ray-tracing analogy for the modes. This analogy, while not perfect, is a useful way to mentally visualize the radiated modes.

noise. An LED is not subject to a modal noise penalty, because it uses a continuum of modes. With an LED source, the loss of any one particular mode is too insignificant to affect the remaining power in the optical signal.

As the laser-diode turns on, the distribution of modes changes. In the early stages of oscillation it may emit at one wavelength and later at a slightly different one. Also, the spatial distribution changes. If the fiber propagation delay for the early modes and the late modes are slightly different, you get another strange type of signal distortion. The term for this effect is *mode partition noise*. Gigabit Ethernet was among the first standards to incorporate mode partition noise to its power budget [93]. An LED source is not subject to a mode partition noise penalty.

The third weird effect associated with lasers is reflected input noise (RIN). A laser-diode depends on an intricate set of mirrored facets to do its work. When other mirrors are introduced into the system, it can mess up the lasing action. In a multimode system there may be connector facets downstream that reflect significant amounts of noise back to the source. With an LED source these reflections cause no difficulties. With a laser-diode source the reflections cause sporadic shifts in the laser's operation. These shifts and clicks have been characterized and a measurements procedure crafted to measure the susceptibility of a laser to reflections. The test procedure is specified in ANSI X3.230-1994 annex A, subclause A.5.[110]

My advice to you about the use of laser-diodes on MMF is simple: It can be done, but it can be dangerous. If you must go down this road, copy the specifications from an existing standard (preferably one that ships in high volumes).

POINT TO REMEMBER

> ➢ The use of laser-diodes on multimode fiber depends on subtle, undocumented, and unspecified features of the multimode fiber.

11.5.10 VCSEL Diodes

The term *VCSEL* stands for *vertical-cavity surface-emitting laser*. The key difference between this form of laser-diode and any other (like a typical CD-ROM laser) is the orientation of the emitted beam. In an edge-emitting laser the beam comes out the side of the die; in a VCSEL it comes out the top surface. The VCSEL beam orientation is the same as that used in an ordinary surface-emitting LED.

The edge-emitting configuration was the first approach used to make successful laser-diodes. With this approach, the diodes must be diced apart, polished, and tested to weed out the bad components (Figure 11.20). A certain amount of effort is wasted polishing and handling parts that turn out ultimately to be dead lasers.

---

[110] IEEE 802.3z: "For use with MMF applications the polarization rotator referenced in the ANSI X3.230-1994 should be omitted, and the single-mode fiber should be replaced with a multimode fiber."

The surface-emitting configuration eliminates some of the waste, because the diodes can be tested *en masse* as they sit on the wafer, prior to dicing (Figure 11.21). After testing, the diodes are dye marked to record the results of the test, and the parts may then be diced and packaged in the ordinary way. The difference in fabrication process is responsible for the claims of lower cost associated with VCSEL technology.

POINT TO REMEMBER

> ➤ A VCSEL shines perpendicular to its top surface, just like a surface-emitting LED.

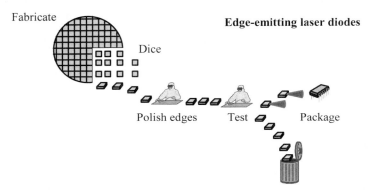

Figure 11.20—Edge-emitting diodes must be diced, polished, and individually tested to see if they are any good.

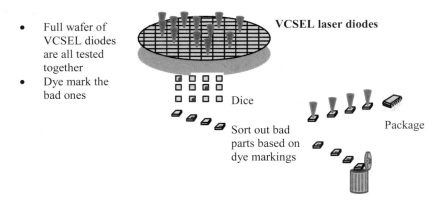

Figure 11.21—The VCSEL fabrication process eliminates the polish-and-test cycle.

11.5.11 Multimode Fiber-Optic Connectors

There continues a high degree of thrash in the fiber-optic connector arena, mostly because no one has yet designed a satisfactory, easy to install, inexpensive fiber-optic connector. Nor is anyone expected to do so in the near future.

Fortunately, if your customer's building site has the wrong connectors, you can always cut the fiber and put on different ones. Normal installation practices for fiber almost always leave a large loop of fiber at either end of a connection for just this purpose. That leaves you free to select just about any connector you want for your system. The three fiber-optic connectors most prevalent in building backbone wiring appear in Table 11.4, and Figure 11.22.

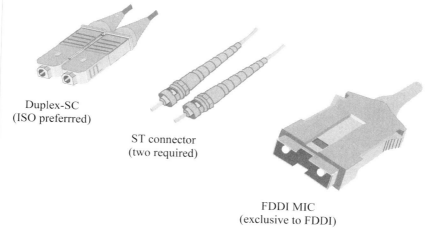

Duplex-SC
(ISO preferrred)

ST connector
(two required)

FDDI MIC
(exclusive to FDDI)

Figure 11.22—Widely-installed fiber-optic connectors.

Of the three connector styles in Table 11.4, the duplex SC has the smallest footprint.[111] It is a duplex connector, meaning that both transmit and receive fibers are terminated in a single keyed plastic connector assembly. At the time the Gigabit Ethernet LAN standard was finalized, ISO 11801 had designated the duplex SC as the connector of choice for future facilities designs. The duplex SC was selected as the only connector authorized for use with that standard. This is an excellent choice for all high-speed systems.

The ST connector remains popular in older Ethernet installation on 10 and 100 Mb/s equipment. Two ST connectors are required per port, one for the transmit circuitry and one

Table 11.4—Common Fiber-Optic Connectors

Duplex SC (also called duplex SCFOC/2.5)	EIA/TIA 604-3, and IEC 874-14	ISO-preferred connector for future facilities designs
ST (also called BFOC/2.5)	EIA/TIA 604-2	Separate connectors for TX and RX
FDDI fiber-MIC	ISO/IEC 9314-3	Unique to FDDI equipment

---

[111] Footprint is the amount of pcb space, or front panel space, consumed by the connector.

for the receive circuitry.

The FDDI MIC was developed exclusively for FDDI. The advent of 100 and 1000 Mb/s Ethernet backbone standards may obliterate the FDDI MIC.

Upcoming new fiber-optic connectors will focus on ease of assembly and reduced footprint. It's not clear which will turn out to be the better connector, but two connectors that show promise are the 3M Volition VF-45 (standardized by ANSI/TIA/EIA 604-7) and the Siemans MT-RJ series (see Figure 11.23).

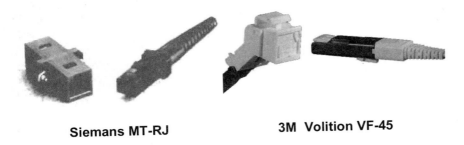

Siemans MT-RJ   **3M Volition VF-45**

Figure 11.23—New fiber-optic connectors are about the size of an RJ-45 data plug.

You have probably heard speculation about the use of optics for backplane connections. Someday that will become commonplace. Optical backplanes certainly have an EMC advantage, and it's true that the pricing is coming down. Unfortunately, the cost of a transceiver pair is still *stratospheric* compared to the cost of one pin on a Euro-connector. Within a system, optics just can't compete economically. Carefully check the pricing and availability for both *transceivers* and *optical connectors* before you proceed with any optical backplane architecture. Optics work well for intersystem connections, but I've not yet seen a cost-effective optical backplane.

POINTS TO REMEMBER

> No one has yet designed a satisfactory, easy to install, inexpensive fiber-optic connector.

> Optics work well for intersystem connections, but I've not yet seen a cost-effective optical backplane.

11.6 SINGLE-MODE FIBER-OPTIC CABLING

ISO-11801 recommends a standard core diameter for 1300 nm single-mode glass fiber of 9 μm to 10 μm. Figure 11.24 illustrates the construction of a standard 10/125 μm step-index single-mode glass fiber core.

At a diameter of 9 μm to 10 μm, any wavelength in the vicinity of 1300 nm or longer will propagate only as a single mode. The absence of additional propagating modes means

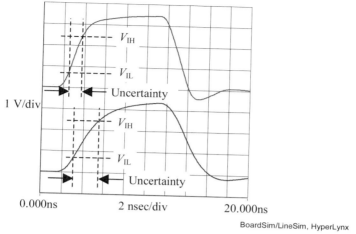

Figure 12.2—A faster clock transition (top waveform) generates less uncertainty in the receiver switching moment.

emissions are also improved by using a slower clock frequency, but that's not usually practical.

The *jitter* requirement becomes important when using your clock as a reference input for any kind of PLL circuit (Figure 12.3). To achieve low jitter, you must limit all forms of noise and crosstalk that impinge on the clock. Relevant sources of crosstalk may include other data traces on the same pcb, ground bounce within the clock driver package, and power supply noise that enters either the clock oscillator or any of its repeaters through their power terminals.

The required *fan-out* may be obtained with any of the topologies discussed later in this

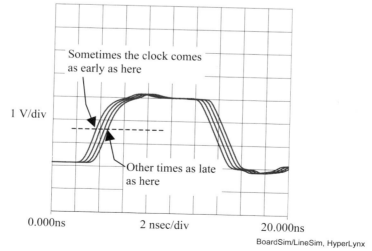

Figure 12.3—A jittery clock appears sometimes early and sometimes late.

chapter. In CMOS logic families fan-out is the most straightforward and easy to satisfy of the five requirements.

The last special issue for clocks is *clock skew*. For optimal system performance the clock transitions must arrive at all points precisely *on time*. The skew requirement constrains the tolerances on all terminations, line impedances, line delays, and load capacitances for every clock line. Let's look next at why *clock skew* is so important in high-speed architectures.

POINT TO REMEMBER

> ➢ Clock signals, because they are so fast, so heavily loaded, and so important for system timing, are subject to special requirements.

12.1 EXTRA FRIES, PLEASE

Article first published in *EDN Magazine*, January 7, 1999

Have you ever had one of those days when nothing seems to go right? My friend Bill Turner and I were debugging a high-speed, heavily pipelined, multiple-clock-phase hardware project one morning when we suddenly realized it would never work. The timing budget didn't add up. We were missing about 2 nsec of setup time at a key latch in the middle of the design.

> A specific, predictable clock skew, applied at a few crucial nodes, can optimize the overall timing budget.

At lunch, over burgers and fries, we put our heads together to find a solution. I marked out a crude timing chart on my place mat, with vertical lines representing the various clock phases. We then nibbled our french fries down to appropriate lengths and placed them horizontally on the chart to represent the setup-and-hold requirements around each clock edge. More french fries showed the minimum and maximum propagation delays out of each latch and through the surrounding logic. A blob of ketchup highlighted the main timing violation. When we finished, we had an accurate (and tasty) representation of the timing constraints for the whole system.

Next, we started sliding the french fries back and forth, adjusting the timing of each clock phase, seeking a combination of adjustments that would make the system function. We found that when phase 2 moved to the right about 1 in., it caused all of the other french fries to line up, and the system looked pretty solid. I made a drawing of this final configuration to take back to the office, and we ate the rest of the chart.

Timing adjustments like this one occur frequently in high-speed products. Designers often discover that a specific, predictable clock skew, applied at a few crucial nodes, can optimize the overall timing budget, increasing the maximum potential system-operating rate.

12.1 • Extra Fries, Please

Until recently, designers implemented intentional clock skew with either fixed-length physical transmission-line delay structures or lumped-element RC delay circuits. Pure semiconductor chain-of-gates delay approaches are unpopular for fixed delays because it's just impossible to produce MOSFET delays with sufficient accuracy. You can make a nice, electronically adjustable MOSFET delay pretty easily, but not an accurate fixed delay—at least, not if you are trying to produce a general-purpose delay line. For the intentional clock-skew application, however, the situation changes.

What's different about a clock signal is that you may use a phase detector to accurately measure the delay of one clock signal relative to another. For example, imagine a simple clock buffer with an electronically adjustable delay (see Figure 12.4). The clock buffer has IN, OUT, and control-adjustment (CNTL) terminals. Using a phase detector, measure the delay of OUT relative to IN. If the phase detector says the output is ahead of schedule, feed that signal back into the adjustment circuit to slow the output. When the output is a little behind, speed it up. Working in this manner, you can program the feedback circuit to obtain any arbitrary output phase to within the tolerance limits of the phase detector.

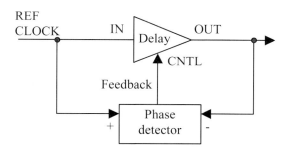

Figure 12.4—This clock buffer architecture uses a phase detector to precisely control delay

This architecture resembles a PLL. The basic feedback idea, combined with various internal ring-oscillator and divider-chain circuits, can produce a flexible array of precise output clock phases.

A number of vendors now offer clock repeater chips with just this sort of circuitry built in. The repeaters provide varying degrees of individual adjustability in the output clock phases. Just to mention a few, there are the Cypress RoboClock, the Quality Semiconductor TurboClock, and the AMCC S4402. You can use each to produce arbitrary, precise, intentional clock skew where and when you need it.

In the high-speed world, timing is everything, so I predict these types of components will be really hot. If you think these parts could help your design, place your order now and ask for a basket of fries to go with them.

POINT TO REMEMBER

➢ DLL or PLL technology can produce arbitrary, precise, intentional clock skew where and when you need it.

12.2 ARITHMETIC OF CLOCK SKEW

The circuit in Figure 12.5 is a two-bit *ring counter,* also called a *switch-tail counter.* When clocked at low speeds, the bit pattern at Q_1 repeats forever (...00110011...).

As you raise the clock frequency in Figure 12.5, the circuit continues to emit the same pattern until at some high frequency the circuit fails. The circuit fails because of a lack of setup time for flip-flop 2. At the failure frequency, the transitions at Q_1 emerge from gate G too late to meet the setup time requirement of D_2. Figure 12.6 diagrams this failure mode. When clocked at or beyond the failure frequency, the circuit no longer produces an 0011 output sequence. This type of failure I call a *setup-margin failure.*

In the parlance of some timing-verification tools, a setup margin failure may also be called a *setup-time violation,* a *critical-path failure,* or a *long-path failure.* Automated timing-verification tools uncover such failures by computing a max-delay analysis in which they assume a worst-case maximum delay through every data propagation pathway and check that the data arrives at each D input sufficiently in advance of the next clock.

Note that setup-time violations happen at (and above) some particular clock frequency. Slowing down the clock fixes setup problems. Speeding up the clock causes them. This behavior is in contrast to the other form of timing difficulty, the *hold-time violation.*

In popular parlance a hold-time violation may also be called a *short-path failure* or a *min-delay failure.* An automated min-delay analysis assumes a worst-case minimum delay through every data propagation pathway and then checks that the data remains valid at each D input sufficiently long after each clock edge to satisfy the receiving flip-flop's hold-time requirement.

In an ordinary synchronous state machine with only one clock phase a setup-time violation involves one clock that produces a data transition and a second clock that receives

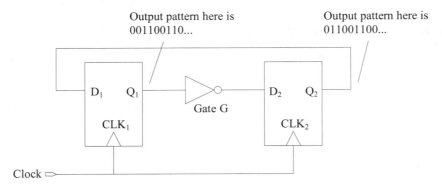

Figure 12.5—This two-bit ring counter fails at high frequencies due to a lack of setup time at D_2.

POINTS TO REMEMBER

➤ Timing margin measures the slack, or excess time, remaining in each clock cycle.

➤ Lowering the clock frequency fixes setup problems, but not hold problems.

➤ *Clock skew* affects operating speed as much as any other propagation delay.

12.3 CLOCK REPEATERS

A *clock repeater* can distribute clock signals on an individual pcb or between the boards of a more complex system. A wide variety of repeater architectures is available. As shown in Figure 12.8, the typical clock repeater chip has one input pin but multiple output pins. The outputs are all ganged together, replicating multiple copies of the input signal. Each output is usually endowed with a large, powerful driver. These high-powered outputs may be used to construct extremely well-terminated transmission lines, a key element in the battle for monotonic, square-edged, and perfectly damped clock signals.

The variety of clock repeater chips is practically unlimited. Repeater chips are available in almost every conceivable logic family, with a varying number of outputs and a broad selection of ancillary features. What these chips have in common are three things: multiple outputs, low-impedance drivers, and low skew (Figure 12.9).

The existence of multiple outputs facilitates high fan-out. The potential number of outputs is limited in theory only by the packaging and by marketing considerations. In practice, most clock repeater chips provide between 4 and 20 outputs.

The low-impedance driver outputs are *not* intended to help you hook up multiple loads to each output. Rather, they are provided so that you can attach an accurate external series-

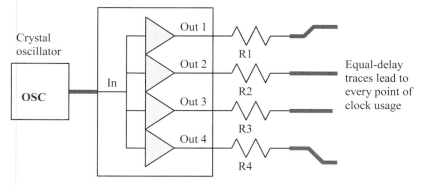

A clock repeater chip is particularly advantageous in a configurable system. The clock repeater isolates the various recipients of clock so they don't interact even as cards are plugged and unplugged from the system.

Figure 12.8—**A clock repeater with one input and multiple outputs.**

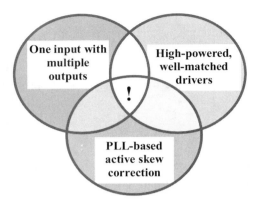

Figure 12.9—Clock driver chips can provide many combinations of features: multiple outputs, high-powered drivers, and active skew correction.

terminating resistor to precisely match your transmission line impedance. The output impedance of the driver, if it is sufficiently low, will then have little bearing on the performance of the termination.

Clocks should *always* be terminated regardless of line length. This helps insure that your clock signals enjoy a predictable relation between line length and signal delay.

Table 12.1 lists the timing performance of a selected sample of clock repeater chips. For each chip, the table lists the output-to-output skew and also the part-to-part timing uncertainty. The output-to-output skew is the magnitude of the worst-case skew from any output to any other output under all allowed operating conditions. The part-to-part uncertainty is the worst-case skew that would prevail in an application like Figure 12.10, which apparently requires more outputs than are available from a single clock repeater. The part-to-part uncertainty matters only in architectures that chain clock repeaters together into a multilayer clock distribution tree. In small applications served by a single clock repeater chip the part-to-part uncertainty specification is irrelevant.

Gallium-Arsenide and Bipolar chips show wonderfully skew; however, these parts also tend to be quite power-hungry. They may also provide in some cases non-TTL outputs which, if they must be translated to TTL levels before you can use the clock, will lose their wonderful low-skew properties as they pass through the translator.

Figure 12.10 illustrates one brute-force way to expand the number of low-skew outputs. If the oscillator output driver is sufficiently powerful, and if the clock repeater chips are located sufficiently close together, a single oscillator may directly drive multiple clock repeater inputs. If the oscillator cannot drive two loads, or if trace delay between the clock repeater inputs exceeds 1/6 of the clock risetime, the tree configuration shown in Figure 12.11 works better. In either case, any *predictable* delay through the clock repeaters is the same for both branches of the distribution and so has no impact on the delivered clock skew, but the input-to-output timing *uncertainty* affects the overall skew in a direct manner.

Beware that many clock repeaters, while they may have terrific output-to-output skew specifications, may not perform well in a chained application because of poor control over the input-to-output skew. Special considerations concerning chaining of clock repeaters

12.3 • Clock Repeaters

Table 12.1—Specifications for Selected Clock Repeaters

Part	Logic family	Intended use	Number of outputs	Output-to-output skew (±ps)	Part-to-part uncertainty (± ps)	Min. risetime (ps)	Max. operating frequency (MHz)
Vitesse VSC6110	GaAs	50-Ω diff.	4	50	—	—	1250
AZ10E111	5V-ECL	50-Ω diff.	9	75	150	250	500
Fairchild 100310	5V-ECL	50-Ω diff.	8	50	400	275	750
AMCC S3LV308	BiCMOS	65-75-Ω	20	350	500	1500	100
IDT 5T907	2.5V CMOS	±8 mA for short lines only	10	25	300	—	250
IDT CSPT857A	2.5V CMOS PLL	60-Ω diff.	10	75	100	1000	200
Cypress CY2300	CMOS zero-delay buffer	±8 mA for short lines only	4	200	400	—	66

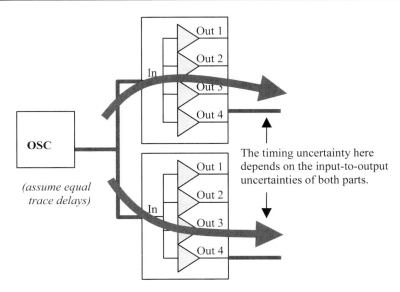

Figure 12.10—A sufficiently powerful oscillator can directly drive two or more proximate clock repeaters. The effectiveness of this approach is limited by the input-to-output timing uncertainty of the repeater chips.

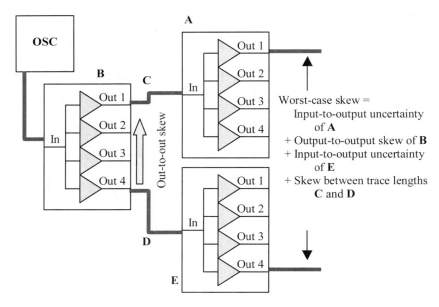

Figure 12.11—The performance of a tree structure depends heavily on the input-to-output uncertainty of the clock repeaters.

appear in Section 12.3.1, "Active Skew Correction," and Section 12.3.2, "Zero-Delay Clock Repeaters."

Keeping the clock repeater isolated in its own package is a good idea. When you use an independent clock repeater chip, there is no possibility that the clock outputs will become corrupted with ground bounce from other circuits within the same package. If you use instead a few extra outputs from, say, a 512-pin ASIC chip to source your clock signals, those outputs may become corrupted with substantial amounts of simultaneous switching noise (ground bounce) coming from other I/O cells switching within the same package. With the clock repeater safely isolated in its own, separate, isolated package, ground-bounce isn't a problem. A separate clock repeater package also makes it easy to maintain a healthy physical separation from other noisy signals on the board, keeping the clocks free from crosstalk.

POINTS TO REMEMBER

➢ The performance of a clock tree structure depends heavily on the input-to-output uncertainty of the clock repeaters.

➢ Keeping the clock repeater isolated in its own package is a good idea.

12.3.1 Active Skew Correction

No manufacturer can accurately predetermine the delay of a CMOS driver; however, anyone can construct a circuit that *measures differences in delay* with great precision. This idea suggests the use of phase-locked-loop (PLL) or delay-locked-loop (DLL) technology to automatically compensate for the natural propagation delays internal to a clock repeater.

Figure 12.12 illustrates one way that DLL technology might be used to null the output-to-output skew of a clock repeater. In the figure, one of the outputs (Y1) is designated as the reference output. Some fixed (but unpredictable) delay is built into the reference output circuit. The circuit compares the timing of the Y2 output with the timing of the reference (at pin REF), and then adjusts the internal delay on the Y2 driver until the Y2 output matches perfectly with the reference, to within the tolerance limits of the phase comparator. All the other channels include a similar auto-adjustment feedback loop. The automatic skew compensation architecture in Figure 12.12 is a form of *delay-locked loop*.

A skew-compensated structure delivers far better output-to-output skew than does a noncompensated structure, but does nothing to combat uncertainty in the overall input-to-output delay.

Beware that actively compensated repeaters, because they incorporate sensitive analog circuitry, are susceptible to power supply noise. Always follow the manufacturer's guidelines for power supply filtering and supply a clean, jitter-free reference clock.

POINTS TO REMEMBER

> - A skew-compensated clock repeater architecture does nothing to combat uncertainty in the overall input-to-output delay.
> - Actively compensated clock repeaters are highly susceptible to power supply noise.

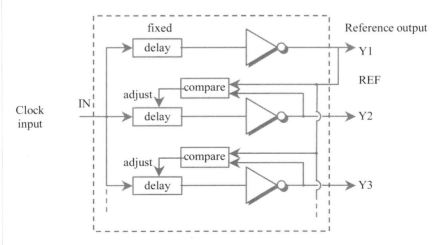

Figure 12.12—In a skew-compensated clock repeater, each output is automatically adjusted to match the reference output.

12.3.2 Zero-Delay Clock Repeaters

If it's the input-to-output uncertainty you wish to improve, you will need a part that regulates its outputs in comparison not to each other, but to the input. Such a part is called a zero-delay clock buffer. It's the ultimate part for big clock tree applications.

A PLL-type zero-delay repeater incorporates an ordinary clock repeater, a phase detector, and a voltage-controlled oscillator, or VCO (Figure 12.13). The DLL-style zero-delay repeater substitutes an adjustable delay chain for the VCO. In either case, the components are hooked up such that the phase detector can adjust the VCO or the delay chain to cause the output transitions to line up directly on top of the input transitions.

In a PLL-basead part, when first powered on, the phase comparator examines the frequency relationship between the input and output sections and works to slew the VCO into a rough frequency lock. From there, the phase comparator makes more delicate adjustments, eventually nudging the VCO into precise phase lock with the source. After a suitable warm-up period, the precise input-to-output delay should be very nearly zero. A DLL-based part undergoes a similar start-up sequence.

Beware that actively compensated repeaters, because they incorporate sensitive analog circuitry, are susceptible to power supply noise. Always follow the manufacturer's guidelines for power supply filtering and supply a clean, jitter-free reference clock.

Regarding the PLL versus DLL controversy, you will undoubtedly see many purported proofs of the inherent superiority of a DLL control loop as opposed to the PLL variety. It's commonly stated that a DLL, because it doesn't contain an oscillator, will produce less jitter and noise at its output. Such arguments are fallacious. While there may be certain architectural advantages to the DLL structure, the noise produced by all presently available integrated-chip PLL or DLL circuits is not dominated by subtle theoretical limitations of the system architecture. It is dominated by crosstalk from the power system, crosstalk from other circuits within the chip, and noise caused by switching between delay taps, operating charge pumps, and so forth within the circuit. The difference between parts is

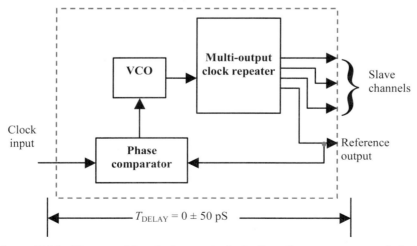

Figure 12.13—The zero-delay clock repeater locks the reference output precisely to the reference input.

not the architecture, but the talent and experience of the designer who built the part. Focus your decision-making process on externally observable specifications, like worst-case skew and RMS jitter, and forget about whether it's a PLL or a DLL inside.

POINT TO REMEMBER

> ➢ A zero-delay clock buffer directly controls the input-to-output uncertainty.

12.3.3 Compensating for Line Length

Low-skew and zero-delay repeaters have as their goal the creation of multiple clock waveforms with low skew. The skew is defined at the output pins of the repeater chip. What you really want, however, is something a little different. What you really want is low skew as defined *at the points of usage*. To obtain the latter from the former, you are forced to provide equal-delay traces to every load, with the same terminations and loading on every trace.

Wouldn't it be grand if the clock repeater could just automatically measure the trace delays to every point of usage? What if it could automatically compensate for the individual time-of-flight on every trace? Such a part would free you from having to worry about the clock trace lengths.

At one time, there was such a part (Microlinear ML6510—see Figure 12.14). This component made use of a special property of series-terminated lines. The figure illustrates a clock driver at point A, an external series-terminating resistor leading to point B, and a remote clock receiver at point C. *Assuming the line was perfectly terminated,* there would appear two transitions at point B for every clock edge. Each transition was half-sized, and the spacing between transitions was precisely one transmission-line round-trip delay.

If point B was made accessible to the clock repeater chip (Figure 12.15), the chip would have gained all the information necessary to determine when the clock edges were

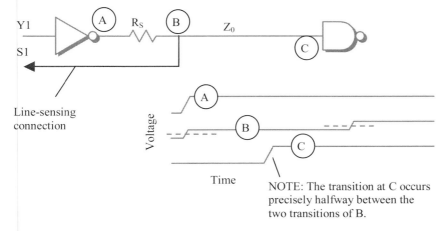

Figure 12.14—The reflections on a source-terminated circuit can be used to determine when each clock edge actually hits the far end.

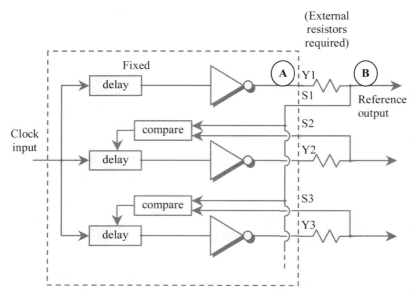

Figure 12.15—Each output requires two pins: a driver pin and a line-sensing pin.

actually reaching the end of each line. The clock repeater used two comparator circuits, one set to detect the first transition and the other set to detect the second transition, and then it averaged the two times to estimate the time-of-arrival for each clock edge.[115] The internal feedback loop then compensated the delay on each output such that the clocks were received at the correct times.

POINT TO REMEMBER

> ➤ What you really want is low skew as defined *at the points of usage*.

12.4 STRIPLINE VS. MICROSTRIP DELAY

A stripline is any trace separated from the air by a solid conductive reference plane on both sides. The electric fields from such a trace are totally contained between the two solid planes, so the speed of propagation for signals traveling on the trace is entirely determined by the dielectric constant of the pcb dielectric material. The speed of propagation for a stripline is independent of the trace geometry or impedance. A stripline need not be symmetrically located between the planes to qualify as a stripline. The term *offset stripline* is used to describe an asymmetrically positioned stripline (as shown in the pcb cross section in Figure 12.16).

---

[115] Circuits designed to perform this feat are very similar to circuits used in data-transmission systems to position the clock edges midway between successive data transitions.

12.6 • Effect of Clock Receiver Thresholds

The 25% threshold crossing reveals a much different picture. Near the 25% threshold the curves in Figure 12.19 bulge, displaying a greater-than-expected amount of variation with line length. Apparently, the absence of a series termination creates a volatile and nonlinear relationship between the line length, load capacitance, and delay.

If you are serious about controlling clock skew, you will carefully balance your clock distribution tree. Use the same clock drivers everywhere. Source-terminate every driver, and use the same length line with the same impedance and the same loading on each trace. Balance the loading on each line, even if you have to add dummy capacitors to one branch to balance out loads on the other branches. Pay close attention to the specifications for input-to-output delay on the drivers (not just the output-to-output skew), and check your results with a high-quality probe and high-bandwidth oscilloscope.

Tight control of clock skew can be accomplished only with a complete awareness of all the relevant circuit parameters. Merely balancing the trace lengths is not enough.

POINT TO REMEMBER

> ➤ For low skew, use the same clock drivers everywhere, source-terminate every driver, and use the same length line with the same impedance and the same loading on every trace.

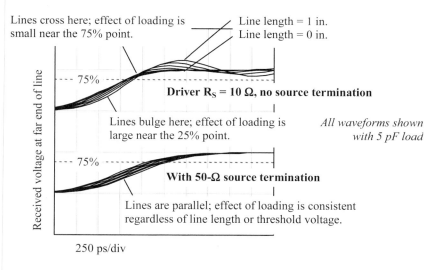

Figure 12.19—An unterminated transmission line exhibits a nonlinear relation between line length and delay.

12.6 EFFECT OF CLOCK RECEIVER THRESHOLDS

The spread between V_{IL} and V_{IH} creates an uncertainty in the exact moment at which a clock receiver will switch. This uncertainty is a normal part of any skew budget, although it may be accounted for in many different ways. For example, input setup times might be referenced to the instant the clock satisfies V_{IL}, while input hold requirements are referenced to the instant the clock satisfies V_{IH}.

Figure 12.20 illustrates the relationship between the signal risetime, signal amplitude, and the uncertainty (or skew) contributed by the effect of uncertain thresholds. Differential receiver families with tight control over V_{IL} and V_{IH} contribute very little skew due to threshold uncertainty.

While you cannot much improve the basic uncertainty of a logic family by using an exceptionally fast clock input (because the clock receiver won't respond any faster than normal anyway), you can disadvantage your system by using an overly slow clock waveform, exacerbating the effect of uncertain thresholds.

POINT TO REMEMBER

➢ The spread between V_{IL} and V_{IH} creates an uncertainty in the exact moment at which a clock receiver will switch.

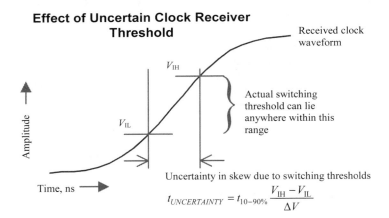

Figure 12.20—Always incorporate the clock receiver switching uncertainty in your skew budget.

12.7 EFFECT OF SPLIT TERMINATION

A resistive load attenuates the output of a digital driver, but does not change its rise (or fall) time. Figure 12.21 illustrates this point with a basic step-response test. A digital logic gate is represented in the figure by a step voltage source with an output impedance of 10 Ω. The response $y(t)$ is shown for four different values of R_L.

With R_L set to infinity (an open circuit) the output $y(t)$ duplicates $x(t)$, rising to a full open-circuit voltage of the driver. Both 10% and 90% crossings are marked on the chart, illustrating the 10% to 90% open-circuit risetime of the source, $t_{10-90\%}$.

12.7 • Effect of Split Termination

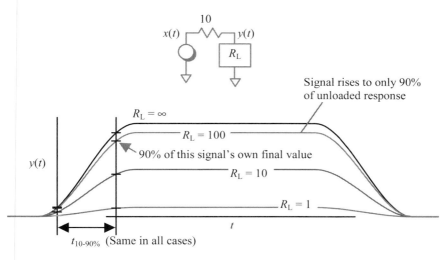

Figure 12.21—A resistive load leaves $t_{10-90\%}$ unchanged,

With R_L set to 100 ohms the signal does not rise quite so high. With a driver of 10 Ω and a load R_L of 100 Ω, you should expect to see only a 90% response.[116] The signal at its peak will therefore only just *approach* 90% of the open-circuit amplitude. If you define risetime by waiting to see when the signal *crosses* 90% of the open-circuit voltage, you could be stuck waiting for a very long time indeed. With an 89-Ω load, for example, you'd never make it—the risetime would be undefined.

To avoid such difficulties, the risetime of a digital signal is defined as the difference between the times at which the signal crosses 10% and 90% of *its own steady-state amplitude*. With this definition, the 10% and 90% crossings of the 100-Ω loaded waveform line up exactly with the 10% and 90% points on the open-circuit waveform.

Load the circuit with 10 Ω and it cuts the signal in half, yet the 10% and 90% crossings of the half-sized signal line up exactly with the other signals. A 1-Ω load produces a very small signal indeed, but with the same 10% to 90% risetime as in the other cases.

Resistive loads make the signal smaller but do not change how quickly it moves from one level to another.

What confuses many engineers about this subject is the way resistive loading changes circuit timing. This effect is illustrated in Figure 12.22. Here the signal $y(t)$ feeds a digital receiver, which has a fixed switching threshold. As the resistive load R_L is changed, the threshold remains fixed, thereby changing the point in time at which the loaded waveform crosses the receiver threshold.

In this example a load of 10 Ω to ground retards the switching time associated with the rising edge, but advances the switching time associated with the falling edge. A 10-Ω resistor to V_{CC} would have the opposite effect. Resistive loads have a definite effect on signal timing, but they do not change the 10% to 90% rise or fall time (or the signal bandwidth).

---

[116] The exact amplitude should be 100/(100 + 10) times the open-circuit amplitude.

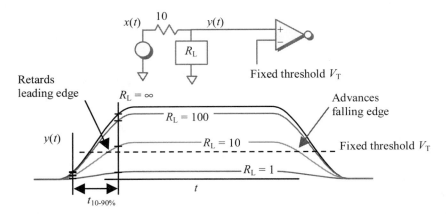

Figure 12.22—A resistive load pulling to ground retards the leading edge but *accelerates* the falling edge. Resistive loads pulling to V_{CC} have the opposite effect.

A split termination can provide an interesting timing adjustment. A split terminator acts in many ways just like a single resistor going to an adjustable battery. The setting of the battery is determined by the ratio R_1/R_2. This circuit (Figure 12.23) is called a Thevenin-equivalent circuit for the split termination.

The Thevenin-equivalent battery voltage may be adjusted up or down by changing the ratio of R_1/R_2. The effect of this adjustment is to bias the DC level of digital signal $y(t)$ up or down by a small amount. The degree of adjustment obtained depends on the ratio of the termination impedance (parallel combination of R_1 and R_2) as compared to the output resistance of the driver.

This DC-bias adjustment is normally made to ensure that the transmitting gate meets its V_{OH} and V_{OL} obligations; however, the same adjustment also is occasionally useful when making very small tweaks to system timing.

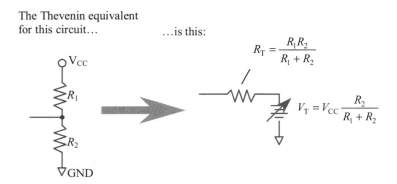

Figure 12.23—Changing the ratio R_1/R_2 biases a digital signal up or down by a small amount.

POINT TO REMEMBER

> ➤ Resistive loading attenuates the output of a digital driver, but does not change its rise (or fall) time.

12.8 INTENTIONAL DELAY ADJUSTMENTS

Sometimes a small positive (or negative) skew may be desirable. For example, a retarded (or advanced) clock usually improves timing margins for one part of the circuit but worsens them elsewhere. If the "elsewhere" part of the circuit already has a bounteous timing margin, intentionally skewing the clock to fix a local problem makes a lot of sense.

Once you realize that a purposeful, nonzero skew can be useful, you might change your objective for clock design. Instead of seeking to attain everywhere the lowest possible skew, you might view clock design as an exercise in reducing the *uncertainty* in the clock arrival time. Once the uncertainty is suitably reduced, you can always deal with the necessary fixed adjustments as a second step. Be careful, though. Use purposeful clock skew only when you have a good timing model for the whole circuit.

Adjustments to the clock timing are sometimes called adjustments to the *clock phase*. This term reminds us that the clock is a repetitive waveform, roughly sinusoidal.

12.8.1 Fixed Delay

The simplest form of clock adjustment is a fixed delay. A fixed delay provides a predetermined amount of clock delay that does not change after assembly. A fixed delay compensates for nominal delays elsewhere in the circuit. Because the delay is frozen at the time of design, a fixed delay cannot cancel variations in trace propagation delay or active component speed, neither of which is known until the circuit is built.

Fixed delays are built from three basic building blocks: transmission lines, chains of logic gates, and discrete circuits. Each has its advantages (Table 12.3). Transmission lines work well for short delays and can be very accurate. Gate delays use less board area than delay lines use but are considerably less accurate. A discrete-circuit delay element covers the widest range of possible delays. Its delay variation depends mainly on the quality of the analog components used in its construction.

Table 12.3—Common Fixed Delay Elements

	Practical amount of delay realizable (ps)	Approximate uncertainty in delay (percent)
Pcb trace (serpentine delay)	10–1000	±10%
Ordinary logic gate (each)	100–10,000	±50% or more
Discrete circuit	1000–1,000,000	±5% to ±20% depending on quality
NOTE—Gate delays on-chip are considerably less than these numbers.		

Delays implemented as transmission lines on pcb require space. For example, at a dielectric constant of 4.3, a 1-ns delay requires a trace length of 0.144 m (5.67 in.). Implemented with a 300-μm trace pitch (11.8 mil), each nanosecond of delay consumes approximately 0.43 square cm of board real estate (0.067 sq. in.).

When using a pc trace as a transmission line-delay, keep in mind the variation in relative permittivity of the trace with temperature. For FR-4 material, this variation results in a considerable change in propagation velocity over the temperature range of 0 to 70 °C. The variation is most easily measured for your material using a bare board with copper foil on both sides. Attach a handheld capacitance meter with long leads to the board to record the capacitance between the foil plates. If the board is .3 x .3m, with a core thickness of 1.5 mm (60 mils) and a dielectric constant of 4.5 at 1 MHz, the expected value of capacitance is

$$C = 8.854 \cdot 10^{-12} \frac{\epsilon_r wd}{h} = 2391 \, pF \qquad [12.5]$$

where   the dielectric constant $\epsilon_r = 4.5$ at 1 MHz,

the board width w is 0.3 m (12 in.),

the board depth d is 0.3 m (12 in.), and

the board thickness h is 1.5×10^{-3} m (60 mil).

Try the experiment with the leads alone and then with the leads touching the board so you can calibrate out the capacitance of the leads. Now place the board (but not the meter) in a temperature-controlled oven and measure the percentage change in capacitance over your range of temperatures. The percentage change in trace velocity, being related to the square root of the dielectric constant, will vary by half the percentage change in the dielectric constant.

Some commercial delay lines are built from a transmission line surrounded by a magnetically permeable material. The permeable material radically increases the delay per inch, shrinking the physical size of the delay line. These delay lines are available with or without buffering.

A spare gate makes an effective delay element. Such an approach is often used to guarantee compliance with a register hold time. The problem with using a gate for a delay element is that while all manufacturers specify maximum propagation delay, few talk about the minimum gate delay. The total variation in gate delay is so large that sometimes the use of a gate as a delay element hinders rather than helps clock skew. Unfortunately, inside a gate array or custom chip, there may be no choice but to use a gate for a delay element.

The discrete-circuit delay in Figure 12.24 produces clean, repeatable delays when used with CMOS gates. The slow risetime of the RC circuit retards the propagation of pulses from the first gate to the second gate by an amount approximately equal to the RC time constant.

If this circuit is built from bipolar gates (or any gates that require substantial amounts of DC input current), the signal at the input to the second stage may fail to rise or fall to a level sufficient to cause reliable switching if R is too large. This happens because of the voltage drop $I_{IH}R$ (or $-I_{IL}R$) across the resistor. An inductive bead or a wound inductor used as the series loss element has no loss at DC and so would pass the required input current without introducing a voltage drop. CMOS logic doesn't require any DC input current, and so just about any value of resistor *R* works (even 1 M Ω).

The accuracy of the delay circuit in Figure 12.24 depends mostly on the tolerance of components *R* and *C,* the parasitic input capacitance of the second gate, the switching threshold of the second stage, and the variability in delay of gates A and B. A differential receiver with a tightly controlled threshold (i.e., low offset voltage) would diminish the variability of this circuit. As you stretch the delay to amounts greatly in excess of the natural delay of gate B, the input at B becomes highly susceptible to noise due to the slow transition speed at that point. A Schmidt trigger at B will prevent oscillation near the threshold point, but will not improve the noise characteristics. A simple Schmidt-trigger feature ($C/10$) appears with dotted lines in Figure 12.24. If you need accurate, long-term delays, the circuit must be protected from crosstalk and provided with its own privately filtered power source.

Whatever form of fixed delay you choose, incorporate its delay uncertainty into your timing-margin calculations.

12.8.2 Adjustable Delays

An adjustable delay can compensate for actual as well as for nominal delays in a circuit. The adjustment, if properly set, reduces the uncertainty in clock skew caused by variations in board fabrication and active component delay. Unfortunately, the weakness in this approach is that technicians must make adjustments after assembly as part of the final test process.

Do not assume your manufacturing staff will understand the meaning of the adjustments provided. Write a test procedure for each adjustment, showing how to measure the clock delay at that point and indicating the limits of proper adjustment.

A delay line may be adjusted in quantized steps. The layout in Figure 12.25 illustrates a typical end-terminated adjustable delay. The transmission line has five adjustment taps. Provided that the five-way collection circuit on the right side of the diagram is short compared to the length of a rising (or falling) edge, the circuit works beautifully.

A more flexible arrangement appears in Figure 12.26, which produces 16 different delays with only 8 jumpers. The jumper sizes in Figure 12.26 are tuned to one, two, four, and eight times a basic delay T. The switches can select any combination of delay sections. Although the circuit in Figure 12.26 is technically more powerful, its complexity works against you, because complexity breeds mistakes.

A *shorting plug* makes a good adjustable tap at low frequencies. These tiny, removable plugs fit onto a pair of 0.025-in.-square posts separated by 0.100-in. (Figure 12.27). Some people call shorting plugs *software jumpers* because of their prevalence as

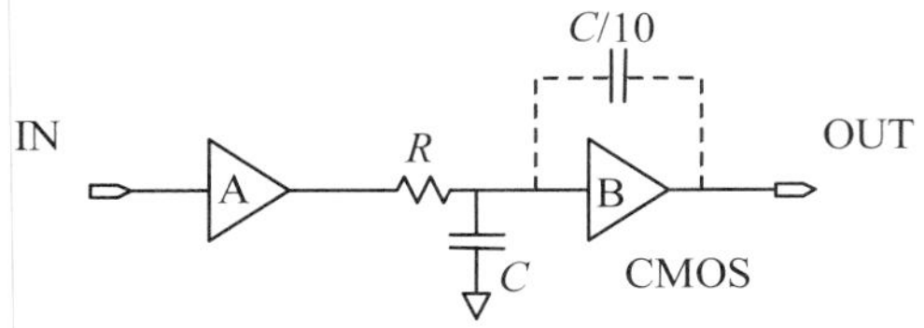

Figure 12.24—This circuit easily produces delays many times larger than an individual gate delay.

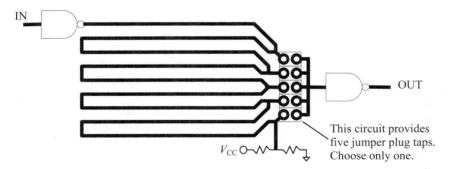

Figure 12.25—An adjustable delay line can be set to any of a number of fixed delay taps.

option jumpers on personal computer add-on cards. Above 100 MHz you may notice some side effects from the inductance of a shorting plug.[117] The inductance varies according to how far down the posts you have pressed the plug. For laboratory purposes, I've seen this effect used as a way to implement a simple adjustable delay (but it's not robust enough for use in field applications).

If the inductance of the jumper plugs is not acceptable, try *solder blob jumpers* (Figure 12.27). A solder blob jumper consists of two 0.50-in.-square pads separated by a 0.006-in. space. The 0.006-in. gap is wide enough to prevent solder bridging during assembly, yet narrow enough to be easily bridged by a technician or by a blob of solder paste deposited during manufacturing. A bridged solder blob clears quickly and cleanly with solder wick.

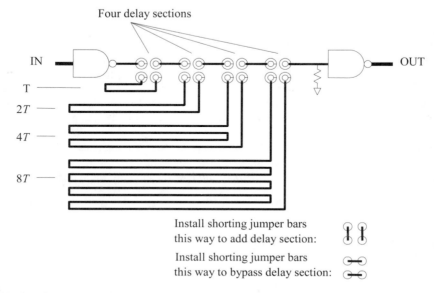

Figure 12.26—This adjustable delay provides 16 settings with only 8 jumpers.

---

[117] A plug with 0.1-in. pins on a 0.1-in. spacing produces an inductance somewhere in the range of 1 to 3 nH, depending on the shape of the conductors within the plug housing.

12.8.3 • Programmable Delays

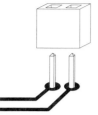
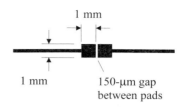

The shorting jumper plug seats onto two pins, thus shorting them together

The solder blob jumper may be bridged with solder and later cleared using solder wick

Figure 12.27—The solder blob jumper, being smaller, has better high-frequency properties than a shorting plug.

Compared to shorting jumper bars, solder blob switches take very little circuit board area. Another advantage of solder blob jumpers is that they don't fall off or move after assembly.

Gate delays may also be adjusted in quantized steps. A chain of gates tapped at discrete points makes a usable delay line. Delay circuits built from gates suffer from the basic inaccuracy of each gate delay. Otherwise, they behave much like a tapped transmission line.

A discrete-circuit delay may be adjusted by varying either R or C. Continuously variable resistors are cheaper and easier to get than variable capacitors. With either type, provide some mechanism for clamping or gluing the adjustment after setting it. Adjustable components are particularly susceptible to vibration.

Some step-variable passive components are available that incorporate several component values plus a tiny set of solder blob jumpers all on a 1206 surface-mount body. Such parts would allow quantized tuning of R-C delay circuits.

12.8.3 Automatically Programmable Delays

An ideal delay circuit would be continuously variable, would be stable over a wide temperature range, and would adjust itself in production. Sound impossible? Read on.

First let's see how to make a continuously programmable delay. Two approaches show promise in this arena. The oldest involves a varactor diode. The varactor diode is a diode whose parasitic capacitance varies as a function of applied reverse bias. Normally a hindrance to design, parasitic capacitance in the varactor is its primary selling point. The difficulty with using varactor diodes is that to obtain a wide range of capacitance, the device must be biased at a reverse-bias voltage significantly higher than the forward voltage drop across the diode. This typically requires at least a 12-volt (preferably 24-volt) power supply.

A second programmable delay approach uses a chain of gates. If all the gates are internal to one integrated circuit, the chain can be very long. A tapped version of the gate chain, with a giant multiplexer to select between taps, forms a useful digitally programmable delay. The design of the multiplexer must prevent glitches, which might occur when switching from one tap to the next.

Several manufacturers now produce clock repeater chips that can be programmed to intentionally skew their outputs. This trend will likely continue, with even more flexibility and granularity introduced in each successive generation of clock repeaters.[118]

---

[118] For example, see the Cypress CY7B991V "Roboclock" and the IDT IDT5T9950/A "Turboclock Jr."

Modulating the power-supply voltage to the entire chain is an interesting and effective way to change the total delay. This works because CMOS gates switch much more slowly when starved for power. The power-starvation effect, normally viewed as a troublesome source of undesirable variation, can be used to create continuously adjustable chains of delay.

With any adjustable delay circuit, you can store a table listing proper adjustment settings as a function of temperature. That improves the circuit's temperature stability. Better yet, set up a second dummy delay and tune its performance using a DLL until it matches an external standard (such as one clock period from a crystal oscillator). Then adjust your main delay using the same parameters. This technique gives you some hope of compensating for temperature, aging, and process variations.

If you want your chain of N delays to add up to precisely one clock period (making a multiphase clock generator), just run the output of the last delay along with the clock into a phase detector. Use the phase detector output to adjust the whole chain until you achieve a total delay of precisely one clock period. Presuming the delays of each stage are equal, the outputs so generated will precisely divide the clock into equal-sized intervals. For this purpose a chain of inverters works better than a chain of buffers. The inverters tend to better maintain a 50% duty cycle as you go down the chain.

Finally, consider how a clock-phase adjustment circuit used on a bus might automatically tune itself. As your clock skews out of adjustment in either direction, your system will likely show a marked increase in its error rate. You can detect that increase and then center the clock between the error-prone zones.

Alternately, you could directly sense the switching times of data signals on your bus. The receiving clock could then be automatically adjusted to match the transition times in the data waveform. This method is directly analogous to the clock recovery architectures used in serial data transmission.

POINTS TO REMEMBER

> Delay elements are built from three basic building blocks: transmission lines, logic gates, and passive lumped circuits.

> A fixed delay cannot cancel variations in board fabrication or active component delay.

> An adjustable delay compensates for actual delays, not just nominal delays, elsewhere in the circuit.

> Whatever form of delay you choose, incorporate its *uncertainty* in delay into your timing margin calculations.

12.8.4 Serpentine Delays

Article first published in *EDN Magazine*, February, 15, 2001

If you are using some form of delay line to match clock delays at all points of usage within a pcb, here's a short list of the items you need to match:

12.8.4 • Serpentine Delays

1. Trace length,
2. Trace configuration (microstrip or stripline, to match the delay per inch),
3. Trace width and impedance (to match high-frequency losses),
4. Dielectric constant (variations affect delay),
5. Trace loading (more capacitance slows down the rising edge),
6. Clock-receiver thresholds (higher thresholds switch later),
7. Terminations, and
8. Serpentine layout.

A tight design process calls out explicit tolerances on all of the above items. Simulations will usually show the slowest results with the longest trace on the slowest layer, with the narrowest line (most skin effect), the greatest dielectric constant, the greatest capacitive load, the highest receiver threshold (for a rising-edge clock), and the termination with the least overshoot. Conversely, the fastest results appear with the shortest trace on the fastest layer, with the widest line, the lowest dielectric constant, the least capacitive load, the lowest receiver threshold, and the most overshoot. The difference between the slowest and fastest results for your system is the *clock distribution skew*.

When selecting a serpentine layout for your system, you should avoid long, coupled switchbacks. The term *switchback* refers to the commonly used U-turn format, in which a trace goes out and then comes back parallel to the outbound path. If the outbound and returning traces pass too close to each other, crosstalk coupling between the two traces may distort the output.

For example, a 50-Ω microstrip layout with 8-mil traces and 5-mil spaces set 5 mils above a solid reference plane produces NEXT (near-end crosstalk) of approximately 10%. If the round-trip delay of each switchback is comparable with or greater than the signal risetime, each switchback translates the NEXT into a 10% distortion of the received signal. Any simulator capable of computing coupled transmission lines can show you this effect.

> **Short, coupled switchbacks produce smaller delays than the total trace length would indicate.**

If, on the other hand, your switchback delay is much less than the signal risetime, the NEXT distortion blends into the overall shape of the rising edge in a special way. The NEXT distortion for short switchbacks doesn't affect the *shape* of the rising edge, but it *advances the time of arrival*. That is, short, coupled switchbacks produce smaller effective delays than the total trace length would indicate. Long, coupled switchbacks distort signals in even more horrible ways.

The reduction in delay for a single, short, coupled switchback can be as much as twice the NEXT coefficient. When you place multiple switchbacks together in a serpentine configuration, the net reduction in delay can be as great as four times the NEXT coefficient.

The boundary between short and long coupled switchbacks is fuzzy. When the round-trip delay of a heavily coupled switchback far exceeds one-third of the rise time, you get seriously distorted signals; when it's much less than one-third, you get advanced timing. A 1-nsec risetime used on an FR-4 dielectric thus limits the

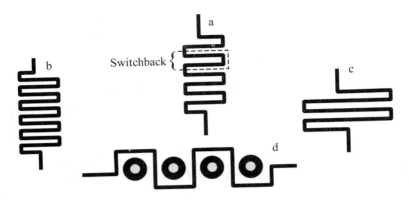

Figure 12.28—Serpentine layout affects signal quality and delay.

maximum useful size of one switchback section to about 1 in. (2 in. round trip). A 100-psec risetime limits the maximum coupled-switchback length to about 0.1 in.

Figure 12.28 illustrates some of the tradeoffs in serpentine design. Assume that Figure 12.28a produces a standard amount of delay. To save space, try squashing the traces closer together (Figure 12.28b).

If the reduction in delay due to NEXT coupling requires the use of more sections (as shown), the layout in Figure 12.28b may not actually save space at all. Rearranging the serpentine to make it shorter and fatter (Figure 12.28c) may distort the received signal if the delay of each section becomes too great (and if the structure is significantly coupled). The layout in Figure 12.28d stretches out the serpentine to eliminate the coupling issue. The stretched-out layout does not suffer from delay reduction or distortion, nor does it wipe out big blocks of space for vias on other layers.

12.8.5 Switchback Coupling

Figure 12.29 illustrates three microstrip delay-line layouts with identical total trace length, but substantially different performance.

The first layout is a single switchback, also known as a two-section serpentine. The two elements used in this design are each 150 mm long, for a total of 300 mm. The traces are laid out using a 200-130-200 µm pattern (8-5-8 mils), meaning that a plan view of the board would show one 200-µm trace, one 130-µm space, and another 200-µm trace. The trace pitch (distance between trace centerlines) is 330 µm. The trace height is 130 µm, yielding a trace impedance of approximately 50 Ω.

The second layout is a more convoluted serpentine built from 24 sections of 0.5 inches each, again using the 200-130-200 µm trace-width pattern.

The third layout is a straight 50-Ω trace.

Figures 12.30 and 12.31 illustrate the nature of serpentine coupling on all three layouts.[119] In the simulations all trace losses have been neutralized and perfect terminations

---

[119] Computed using HyperLynx LineSim v.5.01.

12.9 • Driving Multiple Loads with Source Termination

edge at the receiver won't be full sized. Note that in the series-terminated case the driver does not have to meet V_{OH} at the stipulated peak current, but you *must know* with some degree of precision what voltage your driver is guaranteed to produce when sourcing i_{PEAK}. Given the voltage produced at the required i_{PEAK}, the external series-terminating resistor is then sized to produce a voltage drop precisely equal to the difference between the driver output voltage at current i_{PEAK} and the initial required voltage on the line, which is half of V_{CC}. If the resistor is sized properly, the voltage-doubling effect at the unterminated far end of the line will ultimately bring the initial rising edge of the received signal up to exactly full value.

Some driver circuits can easily source enough current to drive two source-terminated lines. Is it possible to drive two or more source-terminated lines from such a driver? Yes, but only under the limited conditions diagramed in Figure 12.34.

The trick to understanding this figure is to realize that the lines are coupled together into a jointly resonant structure. You cannot properly analyze just one line without seeing what happens to all the lines. The coupling happens because of the finite output impedance of the driver.

If the driver output impedance R_S were zero (it never is), there would be no cross-coupling between lines, and you could simply use a separate series-terminating resistor of value $R_1 = Z_0$ on each line. Unfortunately, the reality of finite driver impedance forces us to contemplate joint resonance. The paragraphs below show how to jointly analyze the system.

Skipping ahead to the answer, *multiple source termination with a nonzero driver impedance works only if the lines are equally long and the loads at each end are balanced.* The source-termination resistors must equal

$$R_1 = Z_0 - R_S N \qquad [12.6]$$

where R_S = output resistance of driver, Ω,
Z_0 = transmission line characteristic impedance, Ω,
R_1 = value of resistance added to each trace, and
N = number of driven lines.

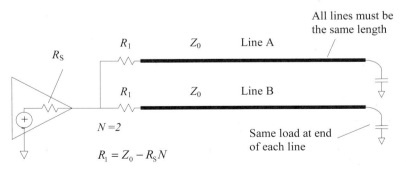

Figure 12.34—A single driver can drive multiple source-terminated loads only under restricted conditions.

When driving one line ($N = 1$), [12.6] matches the total source impedance ($R_S + R_1$) to the characteristic impedance Z_0. This is a normal source termination. When driving multiple lines, [12.6] prescribes smaller source-terminating resistors. With N too large, [12.6] goes negative, implying that no practical solution exists.

Let's analyze the lines in Figure 12.34 one at a time to see what happens. In Figure 12.34, a pulse travels down line A toward the load. This pulse reflects off the far end of line A, returning to the driver. In the usual application of source termination, the source termination matches the characteristic impedance of the line, eliminating the reflection at the driver. In Figure 12.34, however, the effective source impedance is not matched; it is set slightly lower than the characteristic impedance of the line. The returning pulse on line A therefore bounces off the driver, producing a negative reflection. So far, the negative reflection looks like a problem.

Another effect occurs at the same time. As current from the returning pulse on line A surges into the driver chip and through R_S, it generates a voltage at the driver output pin. This voltage couples into line B. The polarity of the crosstalk pulse coupled onto line B is positive.

So far, the consequences of the returning reflected pulse on line A seem to include a negative reflection on line A and a positive crosstalk pulse on line B.

Now imagine what happens if the returning signals reflected off the far ends of lines A and B arrive at the same time. Each signal will induce on its own line a negative reflection and on the other line a positive amount of crosstalk. If you choose the resistor values carefully (according to [12.6]), you can get the negative reflection and positive crosstalk to cancel exactly. The result is a perfectly damped system.

The conditions under which perfect cancellation may be achieved are very restrictive:

> The lines must be equally long (this guarantees the reflected pulses will arrive at the same time).
> The loads must be balanced (this guarantees the reflected pulses will have the same shape).
> The resistors must be calculated according to [12.6].

Equation [12.6] sets the source-terminating resistance so that line A experiences a negative reflection pulse exactly compensated for by the positive crosstalk pulse from line B. Equation [12.6] works with any number of lines, as long as they are equal in length and identically loaded.

Perfect balance rarely occurs in practice. If the lines are not perfectly balanced, the reflections and crosstalk from each line will not cancel. Incomplete cancellation makes the system ring.

POINT TO REMEMBER

> A single driver can service two or more source-terminated lines only under limited conditions.

12.9.1 To Tee or Not To Tee

Article first published in *EDN Magazine*, February 2, 1998

The net topology shown in Figure 12.35 cannot be terminated satisfactorily. What I mean is, you can't simultaneously achieve these four objectives:

- A crisp first incident wave,
- Of full size,
- With no residual reflections,
- That meets the demands of good circuit-design practice.

You can satisfy any combination of three, but not all four, of the above requirements.

[Ed. Note—The following simulations were generated by HyperLynx software. The driver in each case is a simple 3.3-V CMOS model, with 10-Ω output impedances in both HI and LOW states, 6-nH package inductance (BGA), and a 1-ns rise/fall time (10% to 90%). All traces are 50-Ω configurations. Voltages shown are at the driver and receiver locations. Both receivers are the same, with a 3-pF input capacitance. A step-response waveform is shown on the left, and the first three cycles of a 66-MHz clock waveform on the right.]

The ground rules for this discussion are that all three branches of the circuit in Figure 12.35 are long compared to the length of a rising edge. In the simulations, the signal delay on each branch (1 ns) equals the signal rise and fall time (also 1 ns). Such a net, if left unterminated, displays nasty transmission-line characteristics like overshoot, undershoot, and ringing (Figure 12.36).

A slower driver improves the ratio of line delay to risetime, resulting in a better-damped waveform. For example, a 15-ns driver is slow enough to damp out the ringing and reflections, whether you terminate the line or not (Figure 12.37). Unfortunately, this approach gives up on the first criteria—the response here is so slow, it won't work at 66 MHz.

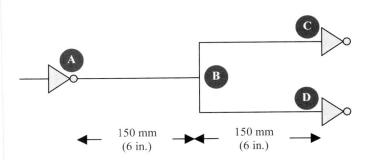

Figure 12.35—This topology, if all three branches are long compared to the signal rise (or fall) time, cannot be terminated satisfactorily.

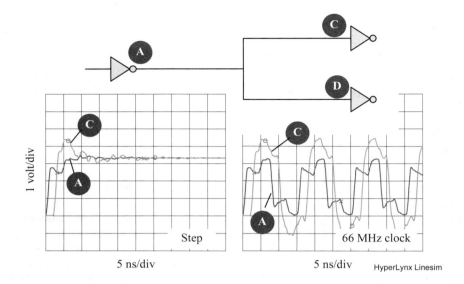

Figure 12.36— If left unterminated, the structure displays overshoot, undershoot, and ringing.

Attenuation can sometimes help. For example, a combination of a 50-Ω series termination at **A** plus 50-Ω end terminations at both the receivers will damp all reflection modes (Figure 12.38). Unfortunately, this approach shrinks the received signal to only 1/3 of normal size. With specialized receivers, this architecture can work wonders. With ordinary single-ended logic receivers, the diminutive received signal is useless.

A weak termination will calm, but not totally cure, the ringing behavior. In Figure 12.39, weak terminations (100-Ω each) placed at each receiver improve the amplitude of the first incident wave, but after a while the reflections trapped between the low-impedance driver at **A** and the mismatch at junction **B** cause the

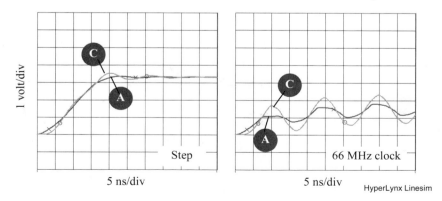

Figure 12.37—A slower driver produces a better-damped waveform, but it's too slow to work at 66MHz.

12.9.1 • To Tee or Not

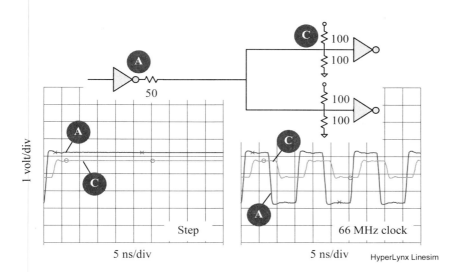

Figure 12.38—Appropriately-placed attenuating networks can damp all the oscillatory modes at the expense of shrinking the received signal to only 1/3 of normal size.

received signal to overshoot, crest, and rattle about. It's not perfect, but at least in the steady-state condition the signal does eventually reach full amplitude.

If you are willing to employ a sneaky trick, you can satisfy the first three conditions (Figure 12.40). This tricky circuit implements segment **AB** as a 50-Ω line while implementing segments **BC** and **BD** as 100-Ω lines (it takes really skinny

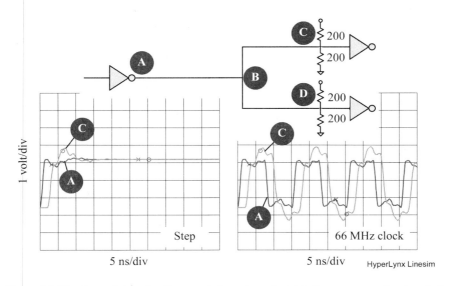

Figure 12.39—A weak termination can help reduce, but totally cure, overshoot and ringing.

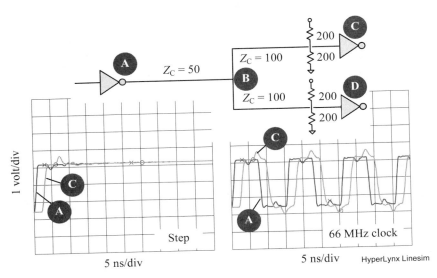

Figure 12.40—A sneaky adjustment of the characteristic impedance on each segment renders this net terminable.

microstrips to get this to work, but it's possible). When the signal from **A** hits junction **B**, it sees two 100-Ω loads in parallel, which is a good match for the 50-Ω segment **AB**. No reflections result. The signal at **B** cleaves perfectly, with half the current flowing down each path. One-hundred-ohm end-terminations at each receiver now perfectly terminate the whole net.

Figure 12.40 looks pretty good, but it's not perfect yet. The little blip 4 ns into the step response at each receiver is caused by the parasitic capacitances of the receivers (set to 3-pF each for this simulation). This parasitic capacitance interferes with the action of the end termination, causing a reflection that eventually returns to haunt the received waveform. If you convert the topology into a source-terminated configuration, and *if the line lengths are identical,* even that tiny effect goes away. To implement this idea, set trace impedance **AB** to 50 Ω and traces **BC** and **BD** to 100 Ω. Apply a single 40-Ω series resistor at point **A**. The value of the resistor is calibrated so that its 40-Ω resistance plus the natural 10-ohm output impedance of driver equals the 50-Ω impedance of line (Figure 12.41). For this topology to work, segments **BC** and **BD** must be the same length.

Figure 12.41 delivers the best-looking waveforms, but let me show you what goes wrong with source-termination if the trace lengths are not the same or if the loads are imbalanced. Using the same setup as in Figure 12.41, stretch the length of segment **BC** to 1.25 ns and shrink segment **BD** to 0.75 ns. That's a difference of only 0.500 ns, or 1/2 of a risetime, but it's enough to totally destroy the signal quality. In Figure 12.42 one of the two resulting received signals is shown. This system is pretty sensitive to delay, isn't it! The same general type of deterioration happens if you implement unbalanced gate capacitances.

Any time you connect up a hairball network like Figure 12.42, you should always check the performance assuming one gate is at its maximum input

12.9.1 • To Tee or Not

capacitance and line length while the other is at its minimum. Simulate with a risetime as fast as you anticipate seeing from any chip over the useful production life of the product.

An end-terminated topology, even using a weak end termination, is not quite so sensitive to delay as the source-terminated topology. This advantage accrues to

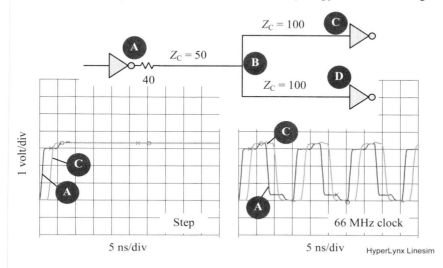

Figure 12.41—The mixed-impedance idea combined with a source termination delivers almost perfect signals to the endpoints.

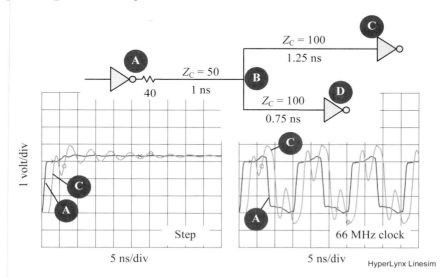

Figure 12.42—The performance of this hairball network is sensitive to the balance between the line lengths leading to C and D.

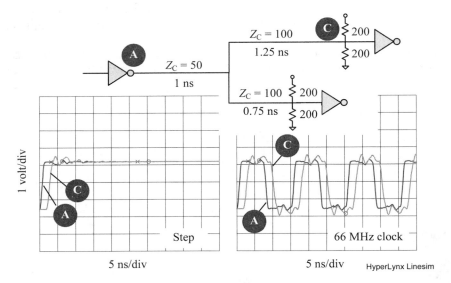

Figure 12.43—The weak-end-terminator approach is not quite as sensitive to line-length imbalance.

the end-terminated topology because the end-termination impedances appear at the receivers in a position where they help prevent the signal from bouncing back and forth between **C** and **D**. In Figure 12.43 I retain the unbalanced line lengths (0.75 and 1.25 ns from **BC** and **BD** respectively), but revert to a 100-Ω termination at each endpoint and a direct hook-up (no resistor) at the driver.

Do any of these techniques constitute good design practice? That depends on your company's internal design procedures. Most companies do not have a good way to document, track, and enforce tricky high-speed design rules. For example, if you write a little thesis on your schematic about a high-speed design trick you've used, it's unlikely that your layout people will ever see it. It's not their job to read your schematic. It's their job to hook up the net list with the part footprints in their database and the standard design rules enforced in their shop at the time of layout. Even if you participate in the layout so it comes out perfectly, the next time your design is revised, it will get screwed up.

Unless you work with an integrated CAD system that automatically keeps track of tricky constraints, you should avoid the tee. Tricks like the tee circuit are too dangerous for practical use. A different approach, like splitting driver **A** into a pair of low-skew drivers with an independent point-to-point link to each load would remove the tricky constraints. A dual-driver topology is the kind of design that will work now and in the future when some kid who inherits your design tries to figure out what you did.

POINTS TO REMEMBER

> ➤ A slow driver can damp ringing, but it may need to be *too* slow for your circuit.
> ➤ Appropriately placed attenuating networks can damp all the oscillatory modes at the expense of shrinking the received signal.
> ➤ A weak termination can help reduce, but totally cure, overshoot and ringing.
> ➤ Test all combinations of maximum and minimum load capacitance and line length.
> ➤ Eventually, someone will inherit your hairball design and try to figure out what you did. Keep it simple.

12.9.2 Driving Two Loads

Article first published in *EDN Magazine*, July 19, 2001

The split-tee configuration (Figure 12.44, omitting R_2 and R_3) conveniently drives two CMOS receivers from one output. If you keep the stub traces connecting receivers IC_2 and IC_3 sufficiently short, the agglomeration of stubs and receivers at the end of the line acts as a single lumped-element capacitive load. Any reflections that bounce off this combination load return to the driver where source-terminating resistor R_1 extinguishes them. The receivers therefore see only one event in response to each step change at the driver. The heavy capacitance at the end of the line may affect the risetime of that event, but the result is monotonic with no lingering residual reflections.

Sounds good, but how short must you keep the stubs? That decision depends on the risetime of the driver and the degree of *balance* between the load impedances. In general, I advocate limiting the stub delay to 1/6 the risetime, which usually works, but I always simulate it to make sure. The circuit also works with longer stubs, but as you stretch the limit, the circuit becomes increasingly

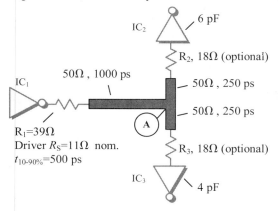

Figure 12.44—The split-tee configuration drives two CMOS receivers from one output.

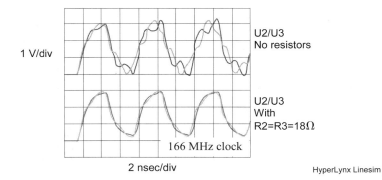

Figure 12.45—Hidden within the split-tee is an unconstrained resonance.

susceptible to imbalance between the loads. Figure 12.45 illustrates the effect of such an imbalance.

Ignoring R_2 and R_3 for the moment, imagine you suddenly inject a positive step at IC_2 and a simultaneous negative step at IC_3. The signal from IC_2 heads south toward IC_3, while the signal at IC_3 heads north toward IC_2. These two signals propagate toward each other, simultaneously crossing at point **A**. At this point, the voltages from the two signals, being opposite, perfectly cancel. No current couples onto the main trace at **A**. The two signals continue on their way, slamming into opposite ends of the stub traces, reflecting, and bouncing back and forth between IC_2 and IC_3. I call this scenario an unconstrained resonance.

The mode is unconstrained because the terminator at IC_1 has no opportunity to damp these oscillations. As long as the system is perfectly balanced, the two opposite signals from IC_2 and IC_3 always perfectly cancel, coupling zero current onto the main trace. Resistor R_1 sees nothing. If the loads at IC_2 and IC_3 are perfectly reactive, the reflections may continue for many cycles. This circuit works like a child's seesaw. Imagine two kids furiously working the plank up and down. Standing at the fulcrum, you can't stop them. You have to stand near one end where the plank is moving to have any significant effect.

> **Hidden within every split-tee network is an unconstrained resonance.**

Given zero coupling between the desired signal-propagation mode (current through R_1) and the IC_2-IC_3 resonance, your signals should theoretically never excite the IC_2-IC_3 resonance, so it should cause no problems. Unfortunately, zero coupling requires perfect balance between the load capacitances at IC_2 and IC_3.

The circuit in Figure 12.44 is not perfectly balanced. Receiver IC_2 contributes 6 pF of loading, whereas receiver IC_3 contributes only 4 pF. The reflections between IC_2 and IC_3 therefore don't perfectly cancel. Coupling does occur between the current in R_1 and the IC_2-IC_3 resonance, and the resulting received signals exhibit the horrible ringing shown in Figure 12.45 (the top traces assume $R_2 = R_3 = 0$). As you can see, the resonant mode occurs at about 500 MHz, corresponding to the third harmonic of the 166-MHz clock. As the clock starts, the heavy third overtone builds to ridiculous levels, causing nonmonotonic behavior and possibly double-clocking in the receivers.

To defeat the resonance, I've added 18-Ω resistors in series with each receiver. That's just enough in this example to knock down the Q of the highly resonant IC_2-IC_3 mode, eliminating the wiggles in the output signal without unnecessarily degrading the received risetime (the bottom trace assume $R_2 = R_3 = 18\,\Omega$).

Any time you build a split-tee, always simulate the circuit with a maximal degree of imbalance. For CMOS loads, that scenario means using the maximum load capacitance at one receiver and the minimum (sometimes zero) at the other. Look at the step response to see whether an observable resonance exists. If it does, simulate the circuit with a clock waveform at that resonant frequency (or 1/3 or 1/5 of that amount). Small series resistors in series with each gate input can sometimes extend the length at which the split-tee safely functions.

H-pattern distributions exhibit similar sensitivity to imbalance. The more symmetrical you make your layout and loading, the further you can stretch the lengths of the H-branches.

POINT TO REMEMBER

➤ Hidden within every split-tee network is an unconstrained resonance.

12.10 DAISY-CHAIN CLOCK DISTRIBUTION

The daisy-chain configuration illustrated in Figure 12.46 distributes one clock signal to multiple receivers. Provided that the receivers do not distort the transmitted signal as it passes by in front of them, the signal should slide cleanly down the structure, delivering an unmolested copy of the original transmitted waveform to each receiver, only with increasing delay as the signal approaches the endpoint.

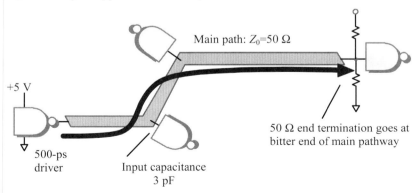

Figure 12.46—A daisy-chain connection works only if the taps do not distort the signal passing along the main path.

Unfortunately, as the signal passes along the main path, each tap generates a small negative reflection. The duration of the reflected pulse is roughly equal to the rise (or fall) time of the incoming signal, and the amplitude of the pulse is approximately given by equation [12.7] (see also Section 5.3.1.2, "Pcb: Lumped-Element Reflections").

The reflected pulses propagate backwards along the line toward the source end of the line. At the source, these pulses bounce off the driver, returning later to the far end of the line where they interfere with reception.

Example Showing Pulses Reflected from Daisy-Chain Load

t_r 10% to 90% risetime of the driver = 500 ps

ΔV amplitude of the incoming step = 2.5V

C lumped-element capacitance = 3 pF

Z_0 characteristic impedance of transmission line = 50 Ω

τ time constant $(1/2)Z_0 C = 75$ ps

The peak amplitude a of the reflected pulse is given approximately by

$$a \approx \Delta V \frac{\tau}{t_r} = 2.5 \frac{75 \text{ ps}}{500 \text{ ps}} = 375 \text{ mV} \quad [12.7]$$

The reflected pulse amplitude amounts to 15% of the signal swing. This is enough to preclude first-incident wave switching with many logic families. To achieve first-incident-wave switching (a requirement for clock signals), each rising edge must immediately proceed to a level above V_{IH} and stay there, and each falling edge must drop below V_{IL} and stay there. Assuming on the falling edge a worst-case (max) V_{OL} from the driver, a late reflection magnitude of 15% could for many single-ended logic families pop the signal back up across the V_{IL} threshold, causing double-clocking.

Five means of reducing the reflected pulse height and thereby improving the system step response, are

Slow the risetime of the driver. According to [12.7], this directly shrinks the reflected pulse. This item points out a prime disadvantage of using logic too fast for your application. The optimum driver for any clock distribution application is just fast enough to meet your clock skew budget, but no faster.

Lower the capacitance of each tap. This reduces the value of the intermediate time constant τ. To the load capacitance, you also need to add the parasitic capacitance of any connectors and the capacitance of any pc trace stub leading to the receiver.

Lower the characteristic impedance of the clock distribution line. This method also reduces the value of the intermediate time constant τ. The lowest valued end termination a driver can safely operate while meeting V_{OH} and V_{OL} on every edge equals the spread between the datasheet values of its V_{OH} and V_{OL} divided by the spread between its I_{OH} and I_{OL}.

12.10.1 Case Study of Daisy-Chained Clock

Isolate each receiver from the bus with a series resistor having a value at least as large as the characteristic impedance of the line. The resistor presents a higher impedance load to the daisy chain, reducing reflections on the main pathway, but degrading the risetime of the signal at each receiver. CMOS circuits, which draw very little DC bias current, work well with this approach. Bipolar circuits that require larger amounts of input current do not.

Compensate for the capacitance at each tap by adjusting the trace width near the tap. This approach is discusses further in the examples below and in Section 5.3.1.3, "Potholes."

POINT TO REMEMBER

➢ Five things reduce the reflection from an isolated, lumped-element capacitive load: slow the risetime, lower the capacitance, lower the characteristic impedance of the trace, isolate the load with a big resistor, or compensate for the capacitance by modulating the trace width.

12.10.1 Case Study of Daisy-Chained Clock

This section examines in minute detail the distributed effect of multiple loads daisy-chained on a clock net. The clock source drives five loads. They are interconnected using an FR-4 stripline daisy chain. Each load is separated by 5 cm from its neighbor (for a raw trace delay between loads of 345 ps). Each load has 3 pF of input capacitance. The driver has a 10-Ω output impedance and a 500-ps rise/fall time.

Figure 12.47 demonstrates the effect of just one of the loads on the step response of the stripline. In the figure the overall length of trace simulated is 25 cm (10 in.). The figure reveals the relationship between voltage and time as observed at six discrete positions along the line, corresponding to distances of 0, 5, 10, 15, 20, and 25 cm from the source. These

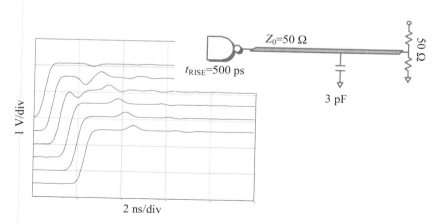

Figure 12.47—A long, uniform transmission line with one capacitive load displays lumps in its step response.

positions correspond to the locations of the source and the five loads to be placed on the structure.

The previous example concluded that the size of the reflections generated by a 3-pF load under similar conditions should be about 15% of the signal swing. That value agrees generally with the size of the worst of the humps and lumps in the waveforms depicted in Figure 12.47. Equation [12.7] is a reasonable approximation for loads widely separated in comparison to the length of your signal's rise or fall time.

In Figure 12.48 the circuit is now burdened with all five loads. Did you expect the signals to look five times worse? They don't. Part of the reason the signals look so good is because in this particular topology the signal risetime of 500 ps exceeds the trace delay between the taps of 345 ps. In systems where the risetime exceeds the tap spacing the signal begins to perceive the loads not as individual potholes, but more as a continuum of capacitance distributed uniformly along the transmission structure. I imagine in this case each rising or falling edge surfing along the loads, hitting the tops but not falling into the troughs.

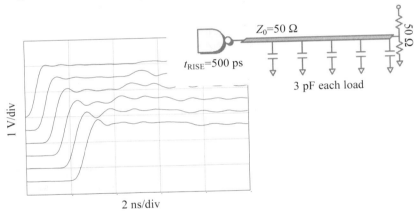

Figure 12.48—In this particular topology adding four more loads doesn't much worsen the signal waveform.

I can exaggerate the surfing effect by switching to a different driver with a slower risetime. Figure 12.49 illustrates the effect of a slower, 1000-ps driver. Its output signal is so slow it averages together the actions of several loads at a time as it surfs down the structure.

Now that the signal is somewhat smoothed out, a new artifact becomes visible. At the far end of the line (bottommost waveform in Figure 12.49) the signal appears to initially overshoot by about 10%, but then after one round-trip delay, the signal falls back to a normal amplitude. Overshoot in an end-terminated configuration is normally considered a sign of a *weak termination* (resistance too high). Yet in this case the characteristic impedance of the raw trace is 50 Ω and the termination is also 50 Ω—the problem is that the effective loaded impedance of the transmission structure formed as a result of adding the extra capacitive loads to the raw trace is no longer 50 Ω. Smearing extra capacitance along the line changes the impedance of the structure according to the formula $Z = \sqrt{L/C}$, where L and C are the total series inductance and shunt capacitance of the entire structure,

12.10.1 • Case Study of Daisy-Chained Clock

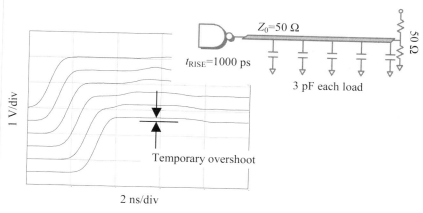

Figure 12.49—A slower risetime averages together the effect of the individual loads, smoothing out the ripples.

including its loads. According to the formula, when you *increase* the C, you *decrease* the Z. By whatever ratio the loads *increase* the natural capacitance of the line, by the square root of that ratio they *decrease* the impedance. This principle is encoded in the following shortcut method for computing the effective impedance of a uniformly loaded structure.

$$Z_{LOADED} = Z_0 \sqrt{\frac{C_{LINE}}{C_{LINE} + C_{LOAD}}} \qquad [12.8]$$

where Z_{LOADED} is the effective characteristic impedance of the loaded structure (Ω),

Z_0 is the characteristic impedance of the raw transmission line out of which the structure is built (Ω),

C_{LINE} is the natural capacitance of the raw transmission line (F). It may be calculated according to $C_{LINE} = t_{PROP}/Z_0$, where t_{PROP} is the one-way propagation delay of the raw unloaded transmission line, and

C_{LOAD} is the total aggregate capacitance of all loads uniformly distributed along the line (F).

NOTE: C_{LINE} and C_{LOAD} may be defined as the capacitances associated with (1) the raw line (full length) and all the loads, or (2) the raw line capacitance per tap and the load capacitance per tap, or (3) the raw line capacitance per unit length and the load capacitance per unit length, using any consistent unit length.

NOTE: This formula applies only to situations where the loads are uniformly spaced, with a spacing whose delay is short compared to the rise and fall time of the driving waveform.

In the present case study the line velocity v is assumed equal to $1.44 \cdot 10^8$ m/s, and the overall length x equals 25 cm, for a total line delay of $x/v = 1.73$ ns. The total line capacitance is (1.73 ns/50 Ω) = 34.6 pF. To that amount the loads have added 15 more pF, bringing the total to 49.6. The square root of 34.6/49.6, multiplied times the original raw trace impedance of 50 Ω, reveals an actual loaded impedance of 41.7 Ω. If that is the true impedance of the structure, why not try a 40-Ω end termination? The results of this experiment appear in Figure 12.50.

The waveforms in Figure 12.50 look clean and usable. They are particularly impressive given that the signal is daisy-chained across five loads. You'll get the same good daisy-chain performance if you follow these three rules:

1. Uniformly space the loads,
2. With a spacing whose delay is small compared to the signal rise and fall time, and
3. Terminate the structure with a resistance that matches the effective impedance of the loaded structure you've built, not just the impedance of the raw trace you started with.

What if your timing budget can't afford a slower driver, or if no such driver is available in your logic (or ASIC) family? In this case you must do something to iron out the bumps in the signal pathway. One obvious improvement would be to reduce the spacing between the taps. The shorter the spacing, the faster a rise or fall time you can use. Unfortunately, this approach has the disadvantage of spreading the same total load capacitance across a transmission line of lesser total length, thereby more severely reducing the loaded impedance of the structure. Also, in many cases you may already have reduced the line to its bare minimum length, so that no further shrinkage may be possible. Have hope, dear reader, as all hope is not lost. Section 5.3.1.3, "Potholes," describes one approach that can partially compensate for the capacitive lumps in your highway. Check out Section 5.3.1.3 for the details of how to do the calculations; here I'll just show a couple of results.

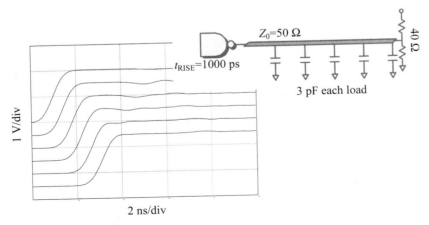

Figure 12.50—Changing the termination to better match the loaded impedance of the structure (not just the impedance of the raw transmission line) eliminates overshoot.

12.10.1 • Case Study of Daisy-Chained Clock

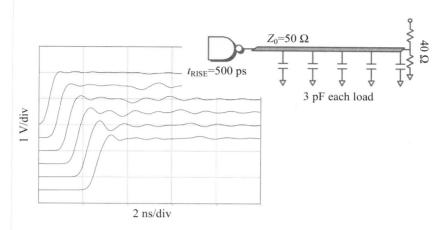

Figure 12.51—Leave the 40-ohm termination in place, but revert back to the 500-ps driver.

First, let's keep the 40-Ω end termination, but revert to a 500-ps driver (Figure 12.51). This signal displays the familiar lumpy pattern from Figure 12.48, but without the 10% overshoot. The last figure of this group (Figure 12.52) modifies the line widths to produce two new impedances of 40 Ω and 80 Ω. The 40-Ω segments form the main body of the structure. Wherever a load appears, it is bracketed by a short section of 80-Ω trace. The length of each 80-Ω section is 11.6 mm. The structure in Figure 12.52 supports rise and fall times twice as fast as the structure in Figure 12.50. The new structure can be reliably clocked at 500 MHz or even faster.

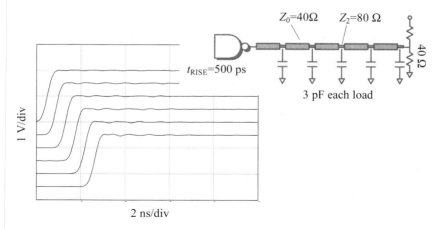

Figure 12.52—Modulate the trace width to partially compensate for the excess capacitance at each load.

POINT TO REMEMBER

> Rules for good daisy-chaining—Uniformly space the loads, with a spacing whose delay is small compared to the signal rise and fall time, and terminate the structure with a resistance that matches the effective impedance of the loaded structure you've built, not just the impedance of the raw trace you started with.

12.11 THE JITTERS

Article first published in *Electronic Design Magazine*, January, 1997

Have you noticed what's happening in the world of clock specifications? It used to be that the only things that mattered in a clock specification were the frequency and the duty cycle. Between these two specifications, vendors were really just setting limits on the minimum time high and the minimum time low each period. A few complications arose with the advent of DRAM technology, like the need for maximum bounds on the clock period, but that was about it.

Recently the whole clock scenario has undergone massive change, due mostly to the widespread use of PLL-based clock recovery schemes used in serial data communications equipment, PLL-based clock multipliers, and PLL-based clock regenerators. The basic premise of a PLL is that it carefully adjusts its own clock, called the local oscillator, to bring it into precise alignment with some external signal, usually called the reference clock.

The PLL concept was originally developed for use in radio and later adapted for use in serial data communications. In the serial data communications application, the reference clock is often embedded, sometimes in very subtle ways, in a stream of data bits. It is the job of the PLL in the clock recovery subsystem to align its local oscillator with the reference clock information embedded in the data stream. Once properly aligned, the local oscillator can be used to clock bits out of the data stream, sampling each data baud right in the center, at the point of maximum noise immunity.

In a clock recovery application, any imperfections in the transmit clock used to construct the data stream may compromise the ability of the PLL to properly align its local oscillator. Improper alignment results in bit errors.

The various possible imperfections in the transmit clock are sometimes classified as frequency offsets, wander, and jitter.

The term *frequency offset* refers to any long-term deviation between the actual transmitted clock frequency and the ideal. For example, crystal-controlled transmission systems can be expected to attain frequency offsets as low as a few hundred parts per million. This sort of specification is measured with a frequency counter, averaging all clock pulses over a period of perhaps many seconds.

A PLL-based clock recovery subsystem is designed to accurately lock in to any reference signal within the permitted frequency offsets. The frequency offset specification often has more to do with whether a PLL will lock in than with the quality of clock recovery, once lock-in has occurred.

Clock wander refers to the tendency of a clock reference to exhibit short-term frequency variations. A PLL is designed to track the short-term wander, provided that it does not slew too fast or wander too far afield. The permitted amount of wander, the rate of which a signal may wander up and down across the permitted frequency rate, and the slew rate of the wander are often key components of a good wander specification.

Jitter refers to the fastest variations in clock frequency—variations too fast to expect a PLL to track. Because a PLL can't track jitter, it always directly affects the accuracy of the timing relation between the reference clock and the local oscillator. In a data communications application, excessive jitter causes bit errors.

Okay, so clock purity is important in data communications applications, we all knew that; but what does clock purity have to do with plain old digital design? Plenty, as we will see, because the same PLL-based clock recovery technology is being widely used to generate multi-hundred megaHertz, very low-skew processor clocks in the latest generation of clock-generator chips from AMCC, Chrontel, PLX, Quality Semiconductor, Triquint, and many others.

> **Jitter refers to variations in clock frequency too fast to expect a PLL to track.**

These new clock generators are flexible, fast, and packed with features. Most incorporate three basic ideas: a reference clock, a PLL clock multiplication circuit, and a means of maintaining very low skew among multiple clock outputs.

In a typical clock multiplier application, the reference clock is often sourced at about 10 MHz from a traditional crystal oscillator. Ten MHz is a very comfortable range for crystals, and it's a good bet you already have one in your system.

To multiply the clock, it is run into a PLL-based clock multiplication circuit. In a multiply-by-ten circuit, for example, the PLL aligns every tenth edge of the local oscillator to the reference clock, thus generating a 100-MHz output. PLL technology can also be used to create zero-delay clock buffers, automatically adaptive skew correction circuits, and other neat features. The combination of PLL, output drivers, and skew correction circuitry is fabricated as a single chip.

What can go wrong? Plenty. Suppose we are feeding rotten power to the crystal source (maybe it has 100-KHz switching noise on it from the power system). If the crystal output violates the offset, wander, or jitter tolerance of the PLL circuit, the 100-MHz output goes nuts. It may fail to lock, drifting to one end or the other of its range; it may flagellate up and down; or, depending on the PLL architecture, it may detect an absence-of-lock condition and just shut off.

What if the clock multiplier is built inside your processor (as with an Intel Pentium processor)? Then the quality of the incoming clock has everything to do with the quality of the resulting system.

If you are using a clock multiplier or a PLL-based clock regenerator, make sure to comply with the specifications for offset, wander, and jitter on the reference clock input. If you have the specifications, test them; if you don't have the specifications, get them; and if your vendor won't fork them over, think carefully about the consequences before you move ahead with your system design.

POINT TO REMEMBER

> ➢ PLL-based clock generators require a stable, low-jitter reference clock.

12.11.1 When Clock Jitter Matters

Clock jitter comes into play whenever you transfer data between synchronous domains that are controlled by independent clocks. At the boundary between the two domains there will inevitably occur at least one synchronizing register that accepts data from one domain yet is clocked by the other. If the relative clock jitter between the two domains is too great, it will violate the timing margins on the synchronizing register.

12.11.1.1 Clock Jitter Rarely Matters within the Boundaries of a Synchronous State Machine

In a simple, synchronous state machine with only one clock, what matters most is the duration of each individual clock period. An adequate measure of jitter in such a system would be a histogram of the clock intervals. A timing interval analyzer is an appropriate instrument for producing such a histogram. Some oscilloscopes can be configured to produce a clock-interval histogram.

Other than the clock interval being too short (or in machines that use poor digital design practices, being too long), no particular pattern of successive long and short intervals is any more damaging to ordinary synchronous logic than any other pattern.

Such is not the case when considering PLL-based architectures.

12.11.1.2 Clock Jitter Propagation

To understand the effect of jitter on a PLL (phase-locked loop), you must first understand three general properties shared by all PLL circuits: the tracking range, the filtering range, and the implications of resonance with the PLL feedback control system. To explain these three concepts I'm going to introduce an analogy to an integrating control system with which you are probably already very familiar—your car (Figure 12.53).

The steering wheel, through a complicated system of linkages and mechanical actions, controls the *angle of travel* of your vehicle. If you steer straight down the roadway, your lateral position doesn't change. If you steer somewhat to the left and keep moving at the same speed, the car moves linearly to the left (up in the picture) towards increasing values of y. Mathematically speaking, your lateral position $y(t)$ along the roadway at any moment is the *integral* of your direction of travel. If this isn't clear to you, don't worry too much about the mathematics—all you need to know is that there is a complicated and time-delayed relation between how you handle the wheel and where your car goes.[121]

---

[121] Those steeped in the art of control system design will recognize that the steering-wheel input determines the rate of change of the angle of travel, so that the entire relation between steering-wheel input and lateral position is that

12.11.1 • When Clock Jitter Matters

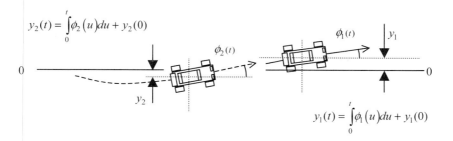

Figure 12.53—The state of each car along the roadway is described by its lateral position $y(t)$ and angle of travel.

Now let's play a high-performance racing game. Imagine you are drafting at 100 mph just inches behind the next driver on a long, straight section of interstate highway. It's your job to follow (track) the movements of the other vehicle as precisely as possible. The other driver is turning his wheel this way and that, trying to throw you off his tail.

If your opponent moves his wheel gradually, you have no difficulty tracking his movements. You see and respond to the graceful movements of his vehicle and have no difficulty following where he's going. This is your *tracking* behavior.

If your opponent grabs his wheel and violently shakes it, without changing the overall average direction of his vehicle, it makes almost no difference to your strategy. His car may vibrate terribly, but as long as you follow his average direction, you'll still probably be close enough to draft effectively. This is your *filtering* behavior. You don't even try to duplicate the shaking motion, you just filter it out.

Figure 12.54 decomposes your opponent's trajectory into its high- and low-frequency components. You track the low-frequency part of his motions. These are the long, slow sweeping turns. You ignore his high-frequency behavior (the rapid shaking).

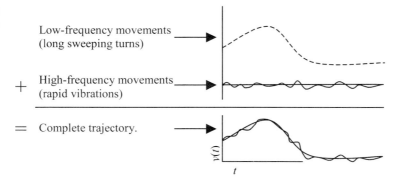

Figure 12.54—A complete trajectory is decomposed into a combination of low-frequency and high-frequency movements.

---

of a double-integral. It is the existence of this double-integration, plus a little bit of delay in your brain, that opens up the possibility of resonance.

Let's chart the frequency response of your steering system. To do this, have your opponent first begin moving his vehicle back and forth across the road in a slow, undulating motion $y_1(t) = a_1 \sin(\omega t)$. Record the frequency ω of his undulations, the amplitude a_1 of his undulations, and the amplitude a_2 of your response. As your opponent slowly increases his rate of undulation from slow to very, very rapid, make a chart showing the system gain a_2/a_1 versus frequency.

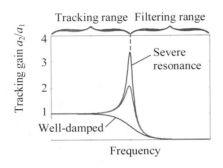

At frequencies within your tracking range, you expect the amplitudes to match perfectly, so the gain is flat (unity gain) in this area. At frequencies within your filtering range, the gain should descend rapidly to zero, because in that area you don't respond. The interesting part happens at the boundary between these two ranges. Most drivers, as the lead car's undulations approach some critical rate, develop acute difficulties. Their response may lag significantly the motions of the lead car, and in their anxious attempts to make up for this delay they will *overshoot the mark* at the apogee of each excursion. As a result, the frequency-response chart exhibits a gain *greater than unity* at some particular frequencies. Severe overshoot appears as a large resonant peak in the frequency-response diagram. A system lacking any resonant peak is said to be *well-damped*.

A mild resonance at the tracking boundary can in some cases help minimize the average tracking error. The practice of causing a mild resonance at the crossover frequency is called PLL *peaking*. A *peaking* feature would be a good thing if yours is the only car in the experiment, but any sort of resonance, even a tiny one, spells disaster for a highly cascaded system.

For example, imagine a long chain of N cars drafting each other on the highway. Suppose the first car commences gyrations having a peak-to-peak amplitude of 1 cm precisely at the resonant frequency. If the overshoot of each car at resonance amounts to 10% (a gain of 1.1 at resonance), the gyrating amplitude of car number 2 will be 1.1, car number three will be 1.21, and so on until at car N the gyrating amplitude will be 1.1^N. Fifty cars down the line the peak-to-peak amplitude works out to 117 cm (if they don't careen off the road).

Chaining PLL circuits exponentially exacerbates the effect of resonance. A PLL designed for a chained application must be well damped (no resonance) at all frequencies.

In this analogy please note that you can measure the system gain either by looking at the ratio of amplitudes of the lateral positions of the cars or alternately by looking at the ratio of amplitudes of the steering-wheel inputs. Both measurements return precisely the same frequency-response graph. This works (for identical vehicles) assuming that for each car i, at each frequency, the relation between steering input s_i and lateral response a_i is the same.

$$\frac{a_2(f)}{s_2(f)} = \frac{a_1(f)}{s_1(f)} \quad \Rightarrow \quad \frac{a_2(f)}{a_1(f)} = \frac{s_2(f)}{s_1(f)} \qquad [12.9]$$

12.11.1 • When Clock Jitter Matters

where a_1 and a_2 are the amplitudes of the lateral position undulations of cars 1 and 2 respectively, and

s_1 and s_2 are the amplitudes of the steering-wheel inputs required in cars 1 and 2 respectively to attain the lateral-position amplitudes a_1 and a_2.

This principle of similarity extends to measurements made of any matching quantities within the steering control system: steering-wheel inputs, hydraulic-fluid pressures, tie-rod displacements, wheel angles, vehicle angles of travel, or vehicle lateral positions on the roadway. In a PLL the chart of tracking gain versus frequency is called the jitter transfer function.

Before I start to sound too much like Click and Clack, the Tappet brothers,[122] I'd better tie this analogy back to PLL design. Figures 12.55 and 12.56 illustrate the analogous relation between steering systems and clock recovery systems. In Figure 12.55 the angle of travel controls the lateral position of the car with an integrating action. Your eyes compare the position of the lead car with your own, and your brain determines how to best steer the vehicle.

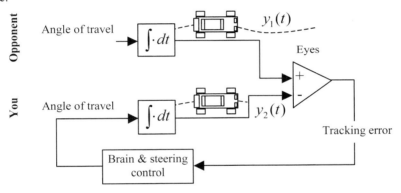

Figure 12.55—The racing game is described as a linear system.

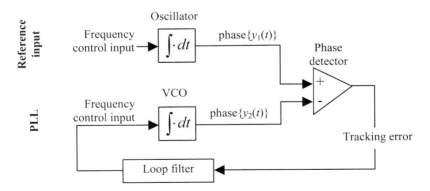

Figure 12.56—A PLL may be described as a linear system.

---

[122] A car-repair radio show popular in the United States.

Figure 12.56 illustrates the analogous control system used in a simple frequency-tracking PLL. In this case it is the relative *phase* of the two input signals that the PLL is designed to control. In the top half of the diagram, the relation between the frequency of oscillation and the phase is shown as an integral. This is derived from the equation for the oscillator, $y(t) = a \sin \omega t$, where if the frequency input ω is held constant, the phase ωt grows linearly without bound. This type of integrating relation holds between the frequency control input and the output phase of any VCO.

A PLL exhibits many characteristics similar to the highway racing game. It has a tracking range and a filtering range. At the boundary between the two ranges, the PLL control loop may resonate.

What's confusing about PLL terminology is that the main variable of interest is itself a frequency (the reference oscillator frequency), so when analyzing the circuit you have to contemplate the frequency of the variations in the reference oscillator frequency. In physical terms, if you imagine the reference input being FM-modulated, any FM-modulation waveform that occurs at a frequency below the tracking bandwidth is tracked, provided you don't exceed the maximum slew rate specification for the PLL.

The *maximum slew rate* is the maximum permitted rate of change of the VCO frequency. It is limited by the physical implementation of the VCO circuit, loop filter, and phase detector.

FM-modulation of the reference input at any frequency above the tracking bandwidth is filtered out. High-frequency modulation, because it occurs in the reference signal but not in the reconstituted VCO output, comprises a source of phase error. In a digital receiver, if the phase error exceeds ±1/2 of a data interval, the receiver cannot properly decode the data.[123]

FM-modulation applied at a frequency in the transition band between the tracking and filtering range may result in control-loop resonance, exacerbating the degree of phase error at that frequency, particularly in chained systems.

Many variations of the basic PLL architecture are possible, including types that compare the internal VCO against multiples or submultiples of the reference clock or against various features extracted from data waveforms (see [94], [95]).

POINT TO REMEMBER

> Any sort of resonance in a PLL, even a tiny one, spells disaster for a highly cascaded system.

12.11.1.3 Variance of the Tracking Error

The tracking behavior of a PLL is equivalent to a linear filtering operation. The PLL acts like a low-pass filter. For example, in the racing game you track the low-frequency part of your opponent's motions. These are the long, slow, sweeping turns. You ignore the high-frequency behavior (the rapid shaking).

---

[123] In a practical system the limit is usually much less than ±1/2 of a bit interval—more like ±10 or ±20 percent.

12.11.1 • When Clock Jitter Matters

In the frequency domain, let the low-pass filter $F(\omega)$ represent your tracking abilities, and let the function $Y(\omega)$ represent the Fourier transform of your opponent's trajectory. The Fourier transform of your trajectory $Z(\omega)$ is therefore a low-pass-filtered version of your opponent's trajectory:

$$Z(\omega) = F(\omega)Y(\omega) \qquad [12.10]$$

where   filter $F(\omega)$ represents your tracking abilities,

function $Y(\omega)$ represents the Fourier transform of your opponent's trajectory, and

function $Z(\omega)$ represents the Fourier transform of your trajectory.

The tracking error $E(\omega)$ is the *difference* between your motion and the motion of your opponent.

$$E(\omega) = Y(\omega) - F(\omega)Y(\omega) \qquad [12.11]$$

where   function $E(\omega)$ represents the Fourier transform of the tracking error.

The tracking error may be expressed differently as a filter $[1 - F(\omega)]$ applied to your opponent's trajectory.

$$E(\omega) = [1 - F(\omega)]Y(\omega) \qquad [12.12]$$

where   filter $[1 - F(\omega)]$ represents the tracking-error filter function,

function $Y(\omega)$ represents the Fourier transform of your opponent's trajectory, and

function $E(\omega)$ represents the Fourier transform of your tracking error.

If the filter $F(\omega)$ is a low-pass filter, then the filter $[1 - F(\omega)]$ must be a high-pass filter, in which case you may recognize that the tracking error is nothing more than the high-frequency part of your opponent's trajectory. It is a theorem of control systems analysis, therefore, that

> *The variance of the tracking error equals the variance of that part of your opponent's signal that falls above the tracking range of your filter.*

Applied to a PLL circuit, this theorem relates the power spectrum $|Y(\omega)|^2$ of the reference phase jitter, the gain of the tracking filter $F(\omega)$, and the variance σ_E^2 of the tracking error:

$$\sigma_E^2 = \begin{cases} \dfrac{1}{2\pi} \displaystyle\int_{-\infty}^{\infty} |1-F(\omega)|^2 |Y(\omega)|^2 \, d\omega\,, & \text{or} \quad \dfrac{1}{\pi} \displaystyle\int_{0}^{\infty} |1-F(\omega)|^2 |Y(\omega)|^2 \, d\omega \\[2ex] \displaystyle\int_{-\infty}^{\infty} |1-F(f)|^2 |Y(f)|^2 \, df\,, & \text{or} \quad 2 \displaystyle\int_{0}^{\infty} |1-F(f)|^2 |Y(f)|^2 \, df \end{cases}$$ [12.13]

where   filter $F(\omega)$ represents the gain of the tracking filter,

function $Y(\omega)$ represents the Fourier transform of the reference phase jitter, and

σ_E^2 represents the variance of the tracking error.

Equation [12.13] appears in four formats. The top two formats integrate the power spectrum of the signal with respect to the frequency variable ω, in rad/s. The bottom two formats integrate with respect to the frequency variable f, in Hertz, where $2\pi f = \omega$. The form of the integration is similar in both cases, but the constant term differs. This difference points out the importance of knowing whether the horizontal axis of a frequency-domain plot is expressed in units of rad/sec or Hertz.

In each row of [12.13], the left-hand expression shows integration over all positive and negative frequencies. This technique is called *two-sided integration*. The right-hand expressions shows integration over only positive frequencies with the results then doubled. The doubling trick works for the evaluation of power associated with real-valued signals, because the power spectrum of a real-valued signal is strictly real and an even function of ω (or f).

The following equations appear in only the top-right format, as *one-sided integrations* with respect to frequency ω in rad/s. You may convert them to any of the four formats shown in [12.13].

Equation [12.13] is often simplified by assuming filter $F(\omega)$ is a perfect low-pass filter with a brick-wall cutoff at some frequency B; in this case the integration need only be carried out from the cutoff frequency B to infinity.[124]

$$\sigma_E^2 \approx \frac{1}{\pi} \int_{B}^{\infty} |Y(\omega)|^2 \, d\omega$$ [12.14]

where   filter $F(\omega)$ is assumed to have unity gain below B and zero gain above B,

the cutoff frequency B is in rad/sec

function $Y(\omega)$ represents the Fourier transform of the reference phase jitter, and

σ_E^2 represents the variance of the tracking error.

In cases where the reference signal is a stochastic signal (as opposed to a deterministic signal) the calculation [12.14] is modified as follows:

---

[124] A two-sided integration would carry from $-\infty$ to $-B$, and then again from B to ∞.

12.11.1 • When Clock Jitter Matters

$$\sigma_E^2 \approx \frac{1}{\pi} \int_B^\infty S(\omega)\, d\omega \qquad [12.15]$$

where function $S(\omega)$ represents the spectral power density of the reference phase jitter, and

σ_E^2 represents the variance of the tracking error.

The power spectrum $S(\omega)$, already being a measure of power, does not need to be squared.

POINT TO REMEMBER

> The variance of the tracking error in a PLL circuit represents all the power in the input reference signal that falls above the tracking range of the PLL.

12.11.1.4 Clock Jitter in FIFO-Based Architectures

Suppose digital state machines A and B each independently use PLL circuits to synchronize their clocks to a common reference (Figure 12.57). Let the common reference frequency be 8 kHz.[125] The clock frequency in each section is 622 MHz, roughly 77,750 times the reference frequency. Data proceeds from section **A**, through the FIFO, into section **B**. Theoretically, once the FIFO gets started, it should stay filled at a constant level because the input and output rates are the same.

In practice, however, the two clocks are hardly the same. The common timing reference signal comes along only once every 77,750 clocks, leaving plenty of time for the two clocks to diverge between reference edges. In the highway racing analogy, this architecture is the equivalent of putting a blindfold over your eyes and permitting you only one quick glimpse of the car in front once every 77,750 car lengths. Obviously, substantial errors may accumulate.

Short-term frequency variations between the two clocks cause the number of words held in the FIFO to gyrate wildly. In general, the greater the ratio of frequencies between the FIFO clock and the reference clock, the greater the gyrations. If the gyrations become too wild, the FIFO either overflows or runs empty.

The maximum deviation in the FIFO corresponds to the maximum *phase* difference between the two clocks, not the maximum frequency difference. Those familiar with the calculus of PLL circuits may recall that phase is the integral of the frequency. In other words, if the frequency difference between the two clocks diverges by x rad/s and holds at that level for t seconds, the accumulated phase difference during interval t would be xt. For example, a frequency offset of just one part in 10^4, averaged over a period of 77,750 cycles, would result in 7.775 clocks of phase offset by the time the next clock arrived.

A good measure of performance in this system would be the *frequency stability* over a period of time T. The frequency stability Δf may be defined as the worst-case difference

---

[125] A common telecommunications reference clock frequency.

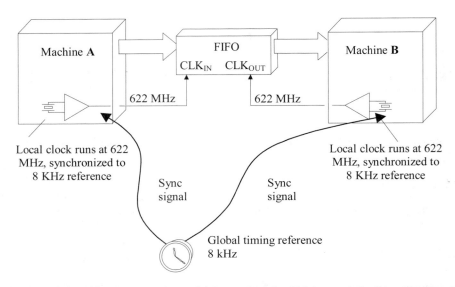

Figure 12.57—Jitter between imperfectly synchronized high-speed clocks causes the number of words held in the FIFO to fluctuate.

between the minimum and maximum number of clock cycles within period T divided by the length of the period T. Appropriate units for Δf are cycles/sec (Hz).

In burst-oriented systems the period of time T usually corresponds to one packet, or one complete data transaction. As long as the clocks don't drift *with respect to each other* more than N complete cycles within a packet, a FIFO of length $2N$ is sufficient to couple the systems (where $N = \Delta f T$).[126]

PLL designers shudder when they see block diagrams like Figure 12.57. Reducing the ratio between the FIFO clock and the reference clock (i.e., distributing an 8 MHz reference instead of an 8 KHz reference) would significantly relax the requirements for PLL stability in this system.

POINT TO REMEMBER

➤ A large ratio between the reference clock frequency and the PLL output frequency requires a very stable VCO.

12.11.1.5   What Causes Jitter

Most oscillators include at least one resonant circuit (or delay element) and one amplifier (or comparator). Jitter in such an oscillator results from at least four superimposed noise sources. First, if you are using a crystal oscillator, noise emanates from the random

---

[126] Preload at least N words into the FIFO before starting your transfer. If the receiver is fast, the FIFO will run completely dry at the end of the transfer. If the transmitter is fast, the FIFO will build to $2N$ words by the end of the transfer.

12.11.1 • When Clock Jitter Matters

movement of electrons within the crystal (*thermal noise*).[127] Second, any mechanical vibrations or perturbations of the crystal cause noise (*microphonic noise*). The third noise source stems from the amplifier or comparator used to construct the oscillator (*self-noise*). The amplifier's contribution is often larger than thermal and mechanical noise from a crystal. The last and potentially most troublesome noise comes from the power supply. Any coupling of an oscillator's power terminals to its sensitive amplifier input sends power supply noise roaring through the amplifier, causing massive amounts of jitter. An oscillator that couples power supply noise into its output is said to have poor *power supply immunity*. Many oscillators do.

These four sources of noise appear together at the output of every oscillator or PLL circuit. Because an oscillator always involves feedback circuits, the same noise is also coupled back into the resonant circuit (or delay element) used to produce the oscillations in such a way that it influences future behavior. In this manner the noise causes both short-term and long-term frequency perturbations. The statistics of such fluctuations are beyond the scope of this book.

In addition to the intrinsic jitter from its internal oscillator, a PLL circuit will propagate any jitter from the reference source that falls within the tracking bandwidth of the PLL.

POINT TO REMEMBER

➢ Jitter in the output of a PLL comes from internal sources plus noise coupled from the power system and noise propagated from the reference input.

12.11.1.6   Random and Deterministic Jitter

Many circuits produce a repetitive, predictable jitter. This effect happens in cheesy clock-multiplier circuits and poorly equalized data recovery units. The predictable component of jitter in these circuits is called deterministic jitter. The remaining components of jitter are called random jitter. The presumption is usually made that the deterministic and random jitter components are not correlated ([94], [96], and [97]).

To measure the deterministic jitter on a clock (or data) waveform, you must trigger your oscilloscope at a rate commensurate with the source of the deterministic jitter. For example, in an 8B10B-coded data waveform transmitting a repetitive 10-bit test pattern, a trigger frequency of 1/10 the data baud rate would be appropriate. For another example, in a clock-multiplier circuit the input reference clock frequency would be appropriate.

The scope must be set to average its measured results, which nulls out all the random jitter, leaving you with a clean picture of a repetitive (though slightly distorted) time-domain waveform. The deterministic jitter is the difference, at each transition in the repetitive sequence, between the *actual* time at which the transition occurred and the ideal time, in a perfect system, at which the transition *should* have occurred.

---

[127] Oscillator circuits using LC tanks, delay lines, or semiconductor delay elements all display similar electrical and mechanical noise effects.

The vector of differences is processed to find the average value, and then the variance, of the measured points. This process puts you in possession of one piece of information: the variance of the deterministic jitter.

By measuring the variance of the overall jitter waveform (including both deterministic and random jitter) you can then infer the variance for the random jitter alone:

$$\sigma^2_{\text{RANDOM}} = \sigma^2_{\text{TOTAL}} - \sigma^2_{\text{DETERMINISTIC}} \qquad [12.16]$$

where σ^2_{RANDOM}, σ^2_{TOTAL}, and $\sigma^2_{\text{DETERMINISTIC}}$ are the variances of the random, deterministic, and total components of jitter.

Deterministic jitter comes from many sources, including duty-cycle distortion (DCD), intersymbol interference (ISI), and word-synchronized distortion due to imperfections within a data serializer (e.g., bit 3 of each data word always appears early).

The point of separating jitter into random and deterministic components is that the deterministic components have a lower ratio of peak value to standard deviation than do the random components. Measured only according to the standard deviation, a certain amount of deterministic jitter doesn't hurt as much as a similar quantity of random jitter.

In a system that combines deterministic and random jitter, therefore, a single specification of the acceptable standard deviation of jitter will always be overly stringent.

Example Showing Calculation of Standard Deviation for Deterministic Jitter

I'll show the evaluation of standard deviation for a very simple case first, and then a more complicated one.

Let r represent a discrete random variable that takes on only two values, x_1 and x_2, remaining at each value on average half the time.

The mean value of r is

$$m_r = (x_1 + x_2)/2$$

The peak excursion of r on either side of the mean is

$$|r - m_r|_{\text{PEAK}} = |x_1 - x_2|/2$$

The variance of r is

$$\sigma_r^2 = \frac{1}{2}(x_1 - m_r)^2 + \frac{1}{2}(x_2 - m_r)^2 = \frac{(x_1 - x_2)^2}{4}$$

The standard deviation of r (square root of variance) is

$$\sigma_r = |x_1 - x_2|/2$$

The ratio of peak magnitude to standard deviation is

12.11.1 · When Clock Jitter Matters

$$\frac{|r - m_r|_{\text{PEAK}}}{\sigma_r} = 1$$

Given a more complicated discrete random variable r that takes on values x_i with probabilities p_i, and has mean value m_r, the variance of r is calculated

$$\sigma_r^2 = \sum_i p_i (x_i - m_r)^2$$

If r were distributed in a Gaussian manner, the peak magnitude could (theoretically) range upward without bound. In the analysis of practical systems the ratio of the *effective* peak magnitude to standard deviation is calculated depending on the BER (bit-error rate) at which the system must operate. Table 12.4 indicates the ratio of the effective peak magnitude to standard deviation for Gaussian waveforms assuming various standard BER values. The table values are computed such that the magnitude of Gaussian noise will not on average exceed the stated peak value more often than once every 1/BER bits.

Table 12.4—Gaussian Waveform Probabilities

BER	Ratio of peak deviation to standard deviation
1E-04	3.891
1E-05	4.417
1E-06	4.892
1E-07	5.327
1E-08	5.731
1E-09	6.109
1E-10	6.467
1E-11	6.807
1E-12	7.131
1E-13	7.441
1E-14	7.739

Example Showing Acceptable Standard Deviation for Jitter

Assume you are willing to accept data errors at a BER of 1E-12. Make the worst-case assumption that all your jitter is Gaussian.

Under these assumptions, if your circuit can tolerate a peak phase error of 0.3 radian before making an error, then to achieve your target BER you must limit the standard deviation of total jitter to a level (according to Table 12.4) not exceeding 0.3/7.131 = 0.042 radians. For a system that combines random (Gaussian) and deterministic jitter, this specification is overly stringent.

To derive a better specification, first subtract from your worst-case total noise budget the known, worst-case amount of deterministic jitter. There is no need to multiply this component of jitter by 7.131 in the budget. It already represents a "worst-case" event.

Divide the remaining noise budget by 7.131 to establish a limit on the standard deviation of random jitter. The overall solution will meet your BER requirement.

For example, a specification calling for no more than 0.1 radian of deterministic noise plus random noise with a standard deviation not to exceed 0.028 radians meets the BER requirement of 1E-12 while providing a reasonable budget for deterministic noise.

POINT TO REMEMBER

> ➤ The point of separating jitter into random and deterministic components is to avoid overly stringent specifications for deterministic jitter.

12.11.2 Measuring Clock Jitter

There are many approaches to measuring clock jitter, including spectral analysis, direct phase measurement, differential phase measurement, BERT scan, and Timing Interval Analysis.

Spectral analysis is performed with a high-quality spectrum analyzer. The spectrum of a perfect clock consists of infinitely thin spectral peaks at harmonics of the fundamental frequency. Close examination of a jittery clock spectrum reveals a tiny amount of spreading around the fundamental frequency and around each harmonic. This spreading relates to clock jitter. Simply put, when a clock spends part of its time at frequency F, we see a peak there corresponding to what percentage of the time it lingered at that frequency. Spectral analysis is very popular with communications engineers.

The problem with spectral analysis is that it does not directly address the issue of phase error. The spectrum tells us what frequencies the clock visited but not how long it stayed. For example, a clock that lingers too long away from its center frequency accumulates a big phase error. A clock that deviates back and forth quickly about its center frequency may visit the same frequency for the same proportion of time, but stay so briefly each visit that it accumulates almost no phase error. From the spectrum alone, you cannot determine the maximum phase deviation from ideal unless you are willing to make the narrowband phase modulation assumption:

Assume the clock never deviates more than one radian from the ideal.

Under this assumption you can model a phase-modulated clock as if it were a perfect sinusoidal clock at frequency f to which you have added a small amount of noise, also at frequency f, like this:

$$y(t) = \sin(\omega t) + a(t)\cos(\omega t) \qquad [12.17]$$

As long as $a(t)$ remains less than one radian, the zero crossings of the signal [12.17] will occur at almost the same locations as in the following phase-modulated signal, where $\theta(t)$ equals $a(t)$.

$$z(t) = \sin(\omega t + \theta(t)) \qquad [12.18]$$

12.11.2 • Measuring Clock Jitter

Using [12.17] you can model any sinusoidal signal $a(t)$ with *small amounts* of phase modulation as a combination of one main sinusoidal carrier plus another amplitude-modulated carrier at the same frequency, but in quadrature with the first signal. The instantaneous magnitude of the modulating signal $a(t)$ is the same as the instantaneous phase jitter $\theta(t)$, where $\theta(t)$ is taken in units of radians.

What this all means is that when you look at the spectrum of a phase-jittery clock, what you see is one big whopping peak near the fundamental and a lower-level spreading of power around the peak. The big peak represents the power in the main sinusoidal carrier. The spreading represents the power present in the noise process $a(t)\cos(\omega t)$. The ratio of the power in the noise process $a(t)\cos(\omega t)$ to the total power in the main carrier precisely equals the variance of the phase jitter.

To obtain from the spectrum a *maximum* phase error (which is what one needs to solve certain FIFO problems), you must combine the power spectrum measurement with some assumption about the nature of the underlying probability distribution of the phase jitter. From the power spectrum measurement, you compute the variance of the distribution, and from knowledge of the properties of the assumed distribution, you may then compute the probability that the phase error $\theta(t)$ will exceed some arbitrary limit (see "Jitter and Phase Noise" article below).

Direct phase measurement requires access to an ideal clock that is compared to your jittery clock with a phase detector. The phase detector output shows just what you want to know: how much the clock jitters. The obvious difficulty with this approach is getting an ideal clock. You might try filtering the jittery clock through a PLL to create a smooth clock having the same average frequency. The phase error output from the PLL will be the jitter signal you seek. This is known as the "golden PLL" method.

If you are measuring jitter from a high-quality frequency source, it may not be easy to build a golden PLL with significantly less intrinsic jitter than your source. This method develops difficulties when measuring phase errors that exceed the bit interval. To solve that problem, try working with a divided-down clock. Measured in units of clock intervals, an error of x in the main clock produces an error in a divided-by-n clock of only x/n.

Differential phase measurement compares a jittery clock not to an ideal clock but to a delayed version of itself. At a large enough delay, the delayed waveform may become uncorrelated with the original, giving you the effect of two similar, but different, jittery clocks. The resulting differential jitter is twice the actual jitter. The advantage of using a delayed version of the original clock is that it naturally has the correct average frequency.

A differential jitter measurement requires an oscilloscope with a delayed time-base sweep feature. First set your oscilloscope to trigger on the clock waveform. Then, using the delayed time-base sweep, take a close look at the clock some hundreds, thousands, or ten-thousands of clock cycles later. Jitter shows up as a blur in the displayed waveform.

Before assuming the blur comes from jitter on the clock, take a look at a stable clock source using the same setup. If it looks clean, you can then assume your scope time base is accurate enough to perform this measurement.

While adjusting the delay interval, you may notice that the jitter gets worse or better. This is normal. Clock jitter normally is worse in some frequency bands, which leads to maxima in the expected differential jitter at certain time delays. Beyond some maximum time delay, the jitter becomes completely uncorrelated and there is no longer any change in jitter with increasing delay.

If through some test procedure, you have intentionally created a large amount of jitter (i.e., FM-modulation of the clock) with a particular period T, the greatest jitter in the output will be observed at time $T/2$ (and successive odd multiples of $T/2$).

If the peak-to-peak amplitude of the phase jitter amounts to more than half a clock period, successive edges will blur together, becoming very difficult to see. In that case, divide the clock by 2, 4, or more using a counter circuit before displaying it. The division doesn't change the worst-case jitter on individual clock edges, but it does lengthen the space between nominal clock transitions so that you can see the jitter.

Jitter measurements on precise crystal clocks require an extremely stable time base and can take a long time to perform. Jitter measurements performed on noncrystal oscillators used in serial data transmission are much easier to do, owing to the much greater intrinsic jitter of those sources.

BERT scan measurements are used to quantify the jitter present on serial data transmission systems. In these methods a serial data stream with a known pseudorandom data pattern is fed into the BERT test instrument. The BERT contains a golden PLL capable of perfectly extracting an (ideally) zero-jitter clock from even the noisiest waveform. The golden PLL clock edge is adjustable within the data window.

The BERT attempts to recover the data, adjusting its ideal clock back and forth across the data window, producing a graph showing the bit-error rate as a function of the clock position. The bit-error rate graph thus produced is called a *BER bathtub curve*. It is so called because at either extreme, as the clock approaches the transition period leading to the next bit, the BER jumps to nearly unity, while in the middle of the curve there is (one would hope) a flat region of zero errors. The shape of the curve resembles a bathtub.

From the slope of the sides of the bathtub curve you may extract information about the statistics of jitter. The ANSI study [97] goes into great detail about the extrapolation of actual BER performance data based on limited measurements of BER bathtub curves.

Timing interval analysis accumulates a histogram of the intervals between successive clock (or data) edges. For example, an accumulation of the histogram of the fine variations in spacing between clock edges separated by a large interval T is equivalent to the information gathered by a differential phase measurement at delay T, but with the advantage that the data is recorded in a form from which the statistics may be easily derived.

Of all the types of measurements mentioned, the manufacturers of time-interval analysis equipment seem most interested in providing tools and software useful for the analysis of jitter.

PLL loop testing is possible if the oscillator under test is controllable with an input voltage. This test uses the oscillator under test as the VCO in an artificially constructed laboratory-grade PLL. An ideal clock is fed into the artificial VCO as a reference. The loop bandwidth of the artificial PLL must be much less than the bandwidth of the VCO phase jitter that you propose to measure.

The PLL structure eliminates low-frequency wander in the oscillator under test, making it easier to see the phase jitter of interest. The output of the artificial PLL phase detector (with a suitable low-pass filter) is your direct phase error measurement. This output can be observed using a low-bandwidth spectrum analyzer or oscilloscope with FFT processing. This approach is very closely related to the golden PLL method described previously.

12.11.2.1 Jitter Measurement

High-Speed Digital Design Online Newsletter, Vol. 3, Issue 22
Ravi writes

I would like to know which is the best way to measure signal jitter using a digital oscilloscope. Here's my situation:

- An Hsync signal is fed to my system from a VGA cable.
- There is a PLL clock generator in my system.
- I want to measure/characterize the jitter coming out of the PLL clock generator.

What type of jitters can be measured using what types of oscilloscopes, and how should one go about it?

Reply

Thanks for your interest in *High-Speed Digital Design*.
  There are (at least) three jitter topics that might interest you:

A. The jitter transfer function (that is, how the PLL amplifies jitter present within the Hsync signal),
B. The power supply sensitivity of the PLL (that is, how the PLL amplifies jitter arriving through its power supply terminals), and
C. The intrinsic residual jitter (noise) from the PLL circuit itself.

A full model of the PLL noise output combines all three effects. Before I describe the measurement techniques in detail, let me make a general point about the relative difficulty of these three measurements.
  Measurements **A** and **B** are made by injecting a known disturbance into your system and observing the result. In the test setup for **A** and **B** you have the freedom to inject a rather large disturbance, which simplifies the measurement task (because you will be looking at a big result). Measurement **C** will be more difficult, because you will be observing very tiny amounts of phase modulation, and it may be difficult to determine the source of the noise.
  Now let's go on to the details. To measure **A**, you need to generate a fake Hsync signal. The fake Hsync signal is phase-modulated with an adjustable sinusoidal source. Call the modulation rate *MR* and the modulation amplitude (in peak-to-peak radians) *MA*. Many high-quality RF signal generators can be FM-modulated (or PM-modulated) in this way and used as a fake Hsync source. While you apply the fake Hsync signal, observe the PLL output with a scope. If the PLL produces a high multiple of the Hsync clock rate, you might want to use a divide-by-*N* counter to reduce the PLL output to a more manageable frequency. Set the scope to trigger on the PLL output, delay by 1/2 the period of the FM modulation (that's 0.5/MR), and then display the PLL output.
  If the modulation frequency is low enough for the PLL to track it, any modulation in the Hsync input will appear directly in the PLL output. Using a horizontal time-base delay of (0.5/MR), you will see the displayed edge switch at a

range of times corresponding to the maximum peak-to-peak phase modulation amplitude (MA) of the fake Hsync source.

If the modulation frequency MR is high enough that the PLL filters it out, the display will appear rock steady. The frequency at which a PLL begins to filter out jitter in the input signal is called its *cutoff frequency*.

At intermediate frequencies, you may find a large peak in the PLL transfer function (a place where the ratio of output phase deviation to input phase deviation exceeds unity). PLL circuits for data communications applications shouldn't have such a peak. The location of the cutoff between the tracking frequency and the filtering frequency, and the magnitude of the intermediate peak (if any), together constitute a good way to characterize the PLL *jitter transfer function*. This function is usually plotted on log-log paper showing the jitter transfer function (ratio of output phase deviation to input phase deviation) as a function of frequency.

If you are planning to chain your PLL circuits, you must ensure that the jitter transfer function does not have any peak or resonance at intermediate frequencies. For example, the on-chip PLL circuits used in the original version of the IBM token-ring LAN were "peaked," creating a small resonance near the cutoff frequency. This is a common technique used in control-circuit design. It tends to improve the lock-on characteristics, reducing the amount of time needed for the circuit to lock onto a fresh input signal.

The disadvantage of peaking, in the token-ring example, is that by the time the standards committee completed its work on the standard, the number of elements allowed in the ring had been enlarged from the original 16 to a new value of 256. Obviously, such a change is quite good for marketing purposes, but very bad for the jitter transfer function. If, for example, each PLL in the original design had on average only about 1/2-dB of peaking near the cutoff frequency, then when you chain together 256 such parts, with each PLL synchronizing on the data signal passed around the ring from the previous station, the total gain at cutoff would be 128 dB. In this type of circuit even the tiniest intrinsic jitter at a frequency near cutoff would be amplified 128 dB as it passed around the ring, causing total system failure. As a consequence of this and other mistakes made in the implementation of the early token-ring circuits, token ring lost the LAN wars and we have an Ethernet-dominated LAN landscape today.

To measure **B**, you will make a test much like **A**, but instead of modulating the Hsync input, this time you will modulate the PLL power supply voltage (Figure 12.58). Do this by injecting sinusoidal noise directly onto the V_{CC} terminal of the PLL. If there is more than one V_{CC} input, then test each input independently. If your circuit incorporates a good power supply filter that prevents you from injecting noise into the V_{CC} terminal of the PLL, remove some of the bypass capacitors on the PLL side of the filter until you are easily able to inject substantial amounts of noise. Always AC-couple your sinusoidal source to the V_{CC} terminal of the PLL using a time constant R_1C_1 sufficiently large to pass the lowest frequency of interest.

While you apply sinusoidal V_{CC} noise at frequency F and amplitude X, observe the PLL output with a scope. As before, set the scope to trigger on the PLL output, delay by 1/2 of the period of F, and then display the PLL output. I like to do this test starting with F below the tracking bandwidth of the PLL and sweeping up to

12.11.2 • Measuring Clock Jitter

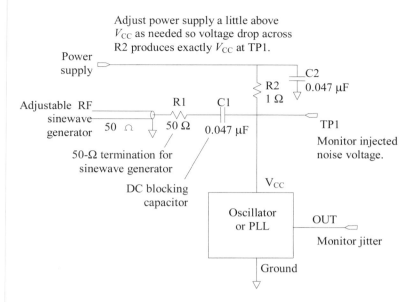

Figure 12.58—This test circuit can be used to measure the power supply noise tolerance of an oscillator or PLL.

somewhat beyond the PLL output rate. At each frequency F, adjust the scope horizontal time-base delay to match $0.5/F$, and then tweak the amplitude X of the sinusoidal V_{cc} noise until you get a standard amount of objectionable phase jitter in the output. You can use any standard objectionable output jitter level, perhaps 0.1 times the output clock period or some other amount that you suspect might begin to cause a problem. The amount you set as your standard objectionable level should (hopefully) be large enough to easily measure.

Make a plot showing, as a function of frequency F, the maximum amplitude X of V_{cc} noise the circuit can tolerate before the output jitter comes just up to the objectionable level. You may combine this basic data with another plot that shows how much noise is already present in your power system as a function of frequency to tell you how much power supply filtering you will need and what must be its frequency response. This is the only rational way I know to *design* a power filter for a PLL.

Sometimes you find a frequency range where the oscillator becomes very sensitive to power supply noise. This effect usually results from insufficient power supply filtering inside the oscillator. The poor tolerance curve shown in Figure 12.59 displays symptoms of ineffective power filtering.

Another, more serious, effect is *squelching*. At some injected noise frequency the power supply filtering components internal to the oscillator may resonate. A low injected noise voltage at this frequency causes extreme amounts of jitter. A high injected noise voltage at this frequency may disrupt the action of the internal amplifier, stopping oscillation altogether. A stopped oscillator is said to be

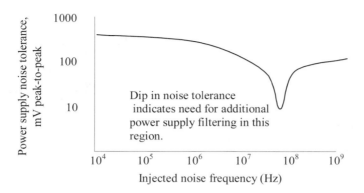

Figure 12.59—A noise tolerance chart shows how much power supply noise your circuit can tolerate at each frequency.

squelched. Sometimes it takes a while after squelching for the oscillator to start working again.

To measure **C**, you can try using the scope to make phase-deviation measurements at various delay intervals, but the results will likely be unsatisfactory. Measurements **A** and **B** are easy to make because you are injecting a *huge* phase deviation (perhaps as much as 0.1 bit interval or more), and the resulting phase jitter is easy to see. For test **C** you need an instrument that can measure tiny amounts of jitter. Use either a timing interval analyzer or a spectrum analyzer. Either can be used to make a measurement of the total variance of the total phase deviation. When you make this test, you will want to carefully filter the PLL power supply and use an ultra-clean, low-jitter Hsync clock, to eliminate noise from those two sources so you can see the remaining intrinsic noise of the PLL circuit.

POINT TO REMEMBER

➢ The noise properties of a PLL are characterized by the intrinsic internal jitter, the power supply sensitivity, and a jitter transfer function.

12.11.2.2   Jitter and Phase Noise

High-Speed Digital Design Online Newsletter, Vol. 4, Issue 7
Bill Stutz writes

I don't know if this falls into your areas of expertise, though your excellent articles and book lead me to believe you might be able to help!

My question has to do with jitter. In many serial digital systems jitter is specified. The specification is usually given in units of absolute time. For example, SMPTE specifies the jitter on the parallel clock of an SDI serializer as 370 picoseconds peak-to-peak for a clock frequency of 27 MHz. When serializing a 10-

12.11.2 • Measuring Clock Jitter

bit data stream at 270 Mb/sec, this amounts to +/− 0.1 UI (unit interval) of jitter. This jitter is specified for offset frequencies between 10 Hz and 1/10 the serial clock rate.

I intend to make my 27-MHz clock from the horizontal sync frequency of my baseband video using a PLL. The PLL will be based on a VCO, for which I have a plot of phase noise in dBc versus frequency.

Can I calculate what the intrinsic jitter of this oscillator will be from its phase noise plot? How would I do that? Any help or light you can shed on this problem would be appreciated.

Reply

Thanks for your interest in *High-Speed Digital Design*.

I've always wanted to know how to do the same calculation, so I researched the math and came up with some good information for you.

Here's what you need to know. For small amounts of jitter (like 0.1 UI or less), you can use what is called the *narrowband phase modulation* assumption to perform your analysis. What this says is that you can model a clock system as if it were receiving a sinusoidal clock at frequency f, to which you have added a small amount of noise, also at frequency f.

The noise has *two* important properties. First, the noise is assumed to be in quadrature (90 degrees out-of-phase) to the main clock sinusoid. Second, the noise is amplitude modulated. If you get out a piece of paper and draw a phasor diagram, you will see that the addition of small amounts of quadrature noise to a sinusoid merely accomplishes a little bit of phase modulation. In other words, you can model any sinusoidal signal with small amounts of phase modulation (which is what you have) as a combination of one main sinusoid and another amplitude-modulated carrier at the same frequency, but in quadrature with the first signal.

What this all means is that when you look at the spectrum of a phase-jittery clock, what you will see is one big whopping peak near the fundamental and a lower-level spreading of energy around the peak. The spreading represents the energy present in the modulating signal. Now your PLL circuit has a certain tracking bandwidth that will filter out all the phase noise within a certain bandwidth B of the first harmonic. This part of the noise is of no concern.

The only phase noise that will escape your PLL is the phase noise that lies further away than B from the main fundamental. In your case, you have told me that the tracking bandwidth of the relevant circuit is 10 Hz, meaning that all the noise further away than 10 Hz from the main peak will add to your jitter.

To find the total power of the modulating signal, you will have to integrate (by hand, with a calculator) the power in the noise surrounding the main signal. If the spectrum analyzer is adjusted to read out in units of decibels per square-root-of-Hertz, you just take samples of the noise level every so often, convert each reading to watts/Hz, multiply each reading by the number of Hertz between readings, and add up the results (in units of watts). That's how you perform the integration.[128]

---

[128] To find the total power you must integrate over both positive and negative frequencies. Alternately, you can just integrate over only positive frequencies (one-sided integration) and then double the result. If all you want are ratios, then you may skip the doubling.

To find the total power in the main signal, you use the same integration method, but this time integrating the power over the big fundamental peak. Use a lot of points for this integration on a spacing that is narrow compared to the bandwidth of the instrument.

The ratio of the noise power to the power in the fundamental equals the variance (standard deviation squared) of the phase modulation in units of radians squared. Take the square root of this ratio to find the standard deviation of the phase modulation in units of radians. This is the RMS value of the noise signal. Now you need to translate this standard deviation into a peak-to-peak value.

To do that, you will need to make an assumption about the statistics of the noise. Assuming the noise is Gaussian (and not the result of some deterministic, predictable phase wander), one normally figures that if the BER of the system is specified at 1E-12, then it's okay to violate the phase jitter spec one time out of every 10^{12}. In numerical terms, what I'm saying is that it's probably okay if the phase jitter occasionally exceeds +/– 0.05 UI (that's 0.1 UI peak-to-peak) as long as it doesn't do so more often than one time in 10^{12}. The peak-to-peak spread between the 1E-12 probability tails on a Gaussian distribution is about 14.3 standard deviations (twice the value in Table 12.4 for a BER of 1E-12).

If you want the peak-to-peak deviation (at 1E-12 BER) to equal 0.1 UI, you require a standard deviation of less than 0.1/14.3 UI, or when translated into radians, a standard deviation of less than $(0.1 \cdot 2\pi)/14.3$. For different BER levels you have to adjust the factor of 14.3 according to Table 12.4. More details are available about this method in [94], [97], and [98].

POINT TO REMEMBER

> You can calculate the variance of jitter using a spectrum analyzer.

12.12 POWER SUPPLY FILTERING FOR CLOCK SOURCES, REPEATERS, AND PLL CIRCUITS

If your oscillator has poor power supply immunity or if it must work inside a noisy system, give it some extra power supply filtering. The amount of filtering required depends on how much a reduction in jitter you must achieve. Determining a precise value for required jitter reduction is almost impossible because all the parameters vary:

> Jitter performance is not specified on many clock sources. When your purchasing department buys a different brand of oscillator, the jitter will change.

> Noise in a system changes when different brands of integrated circuits (perhaps faster switching ones) are assembled.

It should have SMA coaxial-cable connectors attached to both ends of the capacitor with 50-Ω traces. A well-constructed black box should easily produce good performance through about 1 GHz.

Checking your power system's health always returns useful information. If you see too much noise, you know you have some serious work cut out for you. If you see very little noise, you may have the opportunity to save some money, space, and weight on your pcb by stripping out some of the bypass capacitors or reducing their sizes. Either way, measuring the noise on the power system gives you useful information that will help improve your design.

POINT TO REMEMBER

> Observing the noise between V_{CC} and ground always returns useful information.

12.12.2 Clean Power

Article first published in *EDN Magazine*, August 3, 2000

Figure 12.63 illustrates the typical setup used to provide so-called quiet power for a sensitive analog circuit. Good applications for this LC-filter structure include oscillators, PLLs, and fiber-optic receivers. This filter reduces the differential noise that the analog component X between terminals AVCC and DGND perceives.

What about the absolute noise on AVCC? Compared with a true center-of-the-Earth ground-reference point, does the filter reduce the absolute noise on AVCC? Careful consideration of this question may lead you to a better understanding of the purpose of power supply filtering.

First, consider the matter of a 0-V potential reference. First-year electrical-engineering texts normally teach this concept in conjunction with the study of Kirchoff's laws, which form the basis of all modern electrical engineering. According to this method, every circuit contains one 0-V reference node. The equations then

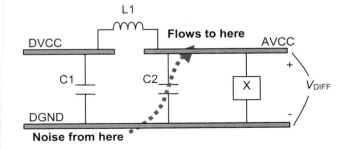

Figure 12.63— Noise from DGND flows through C2, making DGND and AVCC the same at high frequencies, thus eliminating the differential noise V_{DIFF} across circuit X.

define all other voltages in terms of their potential differences from the reference node. Kirchoff's equations are so generally useful and so widely taught that engineers rarely stop to question their applicability.

Unfortunately, the problems of ground noise, electromagnetic radiation, and ESD susceptibility do not succumb to Kirchoff's analysis. These problems violate one of Kirchoff's first and most important assumptions: that all the electromagnetic fields in a circuit must be well-contained within compact, discrete circuit elements. When electromagnetic fields ravage the territory *between* circuit elements, the zero-potential concept evaporates.

For example, when measuring the potential difference between two points on the ground plane of a high-speed digital processor card, you must connect wires (or probes) to these two points and then feed the wires over to the inputs of your measuring equipment. Already, you have a problem. As any EMI professional will tell you, the space surrounding any processor card is filled with intense, high-frequency electromagnetic fields. These fields interact with your wires, inducing noise. The induced noise shows up in your measurement, and there's no way to eliminate it. Even worse, when you move the wires, the noise changes. The measurement and the measurement technique influence each other. Just as in relativistic physics, you, the observer, become part of the circuit.

> **A simple low-pass filter does not eliminate noise on AVCC. It merely makes AVCC and DGND the *same*.**

With electromagnetic noise present, you can talk sensibly about potential differences only between points that are collocated—that is, points so close that the total field strength between those points is negligible. Global 0-V reference potentials do not exist within large, high-speed digital systems.

Lacking a good global 0-V reference, then, does it make sense to talk about reducing the noise on AVCC? Yes, provided that you are interested only in reducing the *differential* noise between AVCC and DGND in the local vicinity of X, a job that the circuit in Figure 12.63 admirably performs.

To see how this circuit works, assume that at operational frequencies, the impedance of L1 is much greater than the impedance of X, and the impedance of C2 is much less than that of X. Component L1 thus operates as an open circuit, and node AVCC is shorted more or less directly through C2 to DGND. Given unavoidable high-frequency noise on DGND, C2 serves to inject that same noise directly onto AVCC, ensuring that AVCC and DGND perfectly track each other at high frequencies.

This filter does not eliminate noise on AVCC. It merely makes AVCC and DGND the *same,* reducing the differential noise between the two. Power supply filters always work that way. They copy junk from one circuit onto another so that the two match. For circuits such as oscillators, PLLs, and fiber-optic receivers, which don't reference other external grounds, noise matching between AVCC and DGND is generally all you need.

Circuits such as A/D converters that reference two ground systems may impose additional constraints. When working with an A/D converter, you need to connect all the relevant grounds together while ensuring that no high-speed currents can flow through the attachment point. The absence of high-speed currents eliminates local magnetic fields, ensuring the applicability of Kirchoff's laws near the attachment point. All the grounded metal near your A/D converter then truly rests at the same potential, and the circuit works.

Engineers often talk about "cleaning up" the AVCC supply. Power supply filters don't do that. If you want to clean up your AVCC plane, use soap. If you want to minimize the differential noise between AVCC and DGND, use a filter.

POINT TO REMEMBER

> ➢ A power-supply filter does not eliminate noise—it merely copies junk from one circuit node to another, eliminating the difference between them.

12.13 INTENTIONAL CLOCK MODULATION

Article first published in *EDN Magazine*, August 3, 1998

An ideal digital clock, from the standpoint of system timing, is an infinite succession of very fast-edged, identical pulses with a perfectly repeating structure. Unfortunately, from an electromagnetic compatibility (EMC) perspective, such a clock is also the worst of all possible signals. It radiates like crazy. This situation is calmly referred to in polite engineering circles as a *fundamental tradeoff*. Late at night, when we engineers let our hair down, I've heard other terms used to describe it.

The problem with a simple, repetitive signal like a clock is that all its power can become concentrated at a relatively small number of discrete frequencies. When these discrete frequencies leak out of your product's packaging into the outside world, all the radiated clock power is concentrated in a small number of radiated modes.

Data signals don't do this. Random data signals spread their power among a much larger number of radiated modes, each with a smaller average power. That's better, because both FCC and EN emissions regulations are written to penalize the worst-case (peak) radiation in any given mode.

> **Clock modulation of any kind complicates the attachment of your product to any form of truly synchronous logic.**

Although the data nets in your product undoubtedly radiate more total power than do the clocks, the data nets usually contribute less to the FCC/EN peak radiation measurements, because the radiation from the data nets is spread evenly at a relatively low level across the vast territory of the electromagnetic spectrum.

Over the years, various techniques have been proposed for modulating, or dithering, the clock frequency in order to break up the accumulated power into a larger number of new modes, each with a reduced power content. If the new modes are separated from each other by more than 100 KHz, which is the effective bandwidth used for FCC/EN spectral power density measurements, the peak power measured within each 100 KHz band will be reduced. Such proposals are backed by solid theoretical reasoning and, for the most part, they are technically sound. Modulating the clock really does reduce the peak measured radiation. Unfortunately, the practical realization of this technique comes at a very high cost. A proper appreciation of the architectural cost of a modulated clock may be gained by considering the many uses to which a clock may be put.

First and foremost, the clock directs the synchronous neural firings of your product's digital brain. When working with a purely digital product architecture, you might conclude that you need merely to guarantee a *minimum* clock period. Any modulation or dither above and beyond the minimum period should, theoretically, have no impact on the correctness of the computed results. Dither may perturb the timing of the final result (it will always be slower than if you had run the machine continuously at full speed), but it should not affect the correctness—at least that's the theory.

The practical side of the matter is that intentional clock modulation of any kind complicates the attachment of your product to any form of truly synchronous logic. For example, a modulated clock can never be used as the reference clock input to any advanced data communication transceiver (Ethernet, Fibre Channel, FDDI, ATM, SONET, or ADSL). These parts require a pristine, jitter-free reference clock. When connected to a jittery clock, these transceivers may fail to lock or may lock poorly, leading to data errors or other flaky behavior. Never use an intentionally modulated clock as the reference input for any kind of data communication transceiver.

For similar reasons, you'll find a modulated clock unsuitable as a main system clock for any modern high-performance CPU. These parts all contain internal clock multiplier circuits (PLLs) which are very sensitive to jitter in the reference clock. The Semiconductor Industry Association roadmap[129] implies that more and more components will incorporate clock multipliers in future years. This is a trend you won't want to forgo.

Finally, in case you are not yet convinced, I'd like to point out that wireless communication is becoming progressively more important in many applications. Modulated clocks should not be used as a reference source for RF-communication systems. Especially for direct-sequence spread spectrum links, where the data rate and the communications modulation rate (the chip rate) are related in a fixed manner, it is important to have a stable, jitter-free system clock for the transfer of data to and from the RF subsystem.

In each of the three cases I've outlined here, it is of course still possible to connect a jittery clock domain to a purely synchronous subsystem. The connection requires a clean reference clock for the synchronous side of your product (in addition to the jittery modulated clock you already have), plus a dual-ported asynchronous FIFO to connect the jittery clock domain to the purely synchronous domain. Why bother with this sort of architecture?

---

[129] www.sematech.org/public/roadmap/index.htm

12.16 • Reducing Emissions

You can reduce the level of stray currents and thus emissions in your designs by reducing the level of intentional signal current. This process is easy if your simulator shows you the current waveforms.

Figures 12.65 and 12.66 illustrate the simulated voltage and current waveforms respectively for a 30.5-cm (12-in.), 133-MHz clock net. The net is source-terminated. The voltage waveform at the receiver is shown for several source-termination-resistor values starting at 10 Ω. As the source-termination-resistor value increases in steps to 39 Ω, some pulse-amplitude risetimes lengthen, but the signal is still acceptable in all cases.

Looking at the current waveforms, you can see that the 10-Ω resistor allows much more current to flow than the other values allow. Further examination reveals that the 22-Ω and 25-Ω waveforms contain extra current glitches that are missing for larger resistors. At high harmonics of the clock frequency, the larger resistors

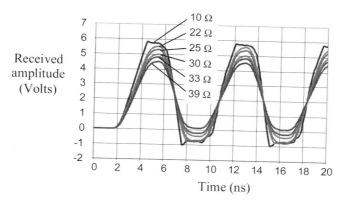

Figure 12.65—Changing the source resistance does not much affect the recevied waveform.

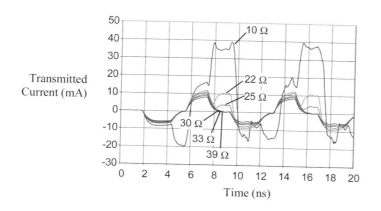

Figure 12.66—Changing the source resistance dramatically affects the transmitted current.

reduce the current amplitude by 10 dB to 20 dB with a concomitant reduction in EMI by that same amount.

If you take the time to look at current waveforms in your simulated data, you may find similar ways to improve your emissions.

Bruce Archambeault, PhD, is an EMI specialist at IBM and the author of the EMI/EMC Computational Handbook, *Kluwer Academic Publishers, 1998.*

POINT TO REMEMBER

- On a short line, if a range of series termination values will work, the biggest value minimizes the transmitted current and therefore the emissions.

For further study see: www.sigcon.com

CHAPTER **13**

TIME-DOMAIN SIMULATION TOOLS AND METHODS

13.1 RINGING IN A NEW ERA

Article first published in *EDN Magazine*, **October 9, 1997**

The time has come. After more than 50 years of progress in the field of digital electronics, we have reached a breakthrough moment. I am ready now, today, right here, to declare the start of a new era. Just to make things official, here's my statement:

> *Be it known in all corners of the globe that from this day forward, there is absolutely, completely, totally no longer any excuse whatsoever for system problems, glitches, data errors, or other artifacts related to ringing in digital signals.*

Before we all throw our pocket protectors into the air and shout "hooray!" let's take a good close look at the wording of my pronouncement. After all, you may be wondering, has there been any change in the laws of physics? Has ringing just gone away? Can we henceforth just ignore it?

The answer to these questions is, of course, an emphatic "No, no, and no."

I didn't say ringing would no longer exist. Nor did I imply that it would not occasionally raise its ugly head to munch on a tender signal or two. Far from it. All I said was that there would no longer be any *excuse* for problems caused by ringing. I feel confident making this statement because ringing is a totally preventable system problem.

The most important thing to realize about ringing is that it is a deterministic, predictable, system artifact that can be simulated with incredible accuracy. All you have to do is run the ringing simulations. If the simulator says your circuit won't work, don't build it.

> **Computer-automated simulation of ringing, even with all its warts, is far better than simple rules of thumb.**

Please don't misinterpret this as a blanket endorsement of simulation for all digital design problems. In some applications, today's simulation technology just doesn't work. For example, try simulating the crosstalk on traces that pass over a split-plane boundary. In that sort of complex, three-dimensional electromagnetic field application, most simulators don't have enough muscle to do the job. However, the simulation of ringing on pcb traces, in the presence of a solid ground plane, with known source and load impedances, and with a known risetime, is a well-known, easily calculable problem. If the simulation says a circuit is okay, it will probably work.

Computer simulation of ringing, even with all its warts, is far better than the kind of simple rules of thumb I see used in design shops all around the world. Old rules like "3-inch trace stubs are okay," which may have served well in the past, don't even come close to cutting it with today's super-fast digital logic. If you really want to know how far you can push a trace, simulate it.

I can't tell you how many engineers have contacted me with some weird bus configuration, wanting to know if it's going to work. The answer: If you have to ask, simulate it.

What if you don't have a simulator? That brings me to my last point, the "no excuse" part.

Simulation technology is widespread, easy to use, and cheap. If by using a simulator you can save one design spin on one circuit board, the simulator pays for itself. If you don't have a simulator, get one. If you already have one but aren't satisfied with it, check out the latest new products. The technology is rapidly improving.

POINT TO REMEMBER

> ➤ If by using a simulator you can save one design spin on one circuit board, the simulator pays for itself.

13.2 SIGNAL INTEGRITY SIMULATION PROCESS

A good simulator is a *predictive* tool, and good predictions don't come easily. The process involves a lot of steps, each one of which controls the efficacy of the overall result.

Signal integrity simulation at the pcb level begins with electrical descriptions of the IC die involved, the IC packages, and the traces on the pcb (Figure 13.1). The objective at this stage is to capture a reasonable description of the components involved, a description

13.2 • Signal Integrity Simulation Process

sufficiently accurate to permit good-quality simulations, and yet not so complicated that it becomes difficult to manage. The process of distilling from the plethora of available data those pieces of information most relevant to the modeling task at hand is called *parameter extraction*.

For an individual IC die, one extracts parameters relevant to the operation of the I/O circuits. These parameters may be rendered in the form of a SPICE circuit description file or (more appropriate for large-scale simulation) an IBIS specification file.

For a chip package, one begins with a physical description and extracts from it information about the mutual inductance and capacitance between every pair of pins. For small packages, this information may be encoded in the form of two matrices of mutual coupling terms (one for inductance and one for capacitance) plus a resistance vector (one value per pin). For larger packages (or at extremes of speed), the coupling information may be represented as a collection of coupled transmission line models interconnected by mutual-coupling coefficients.

Prediction is very difficult, especially about the future.
—Neils Bohr

At the board level, one extracts a collection of trace impedances, trace lengths, trace topologies, and coupling functions representing all traces and connectors.

The set of all electrical models for the chips, packages, and traces on a pcb constitutes, for signal integrity purposes, a complete electrical model of the board. The most common errors arising at this stage are:

1. Choosing an inappropriate degree of modeling for the problem at hand, or
2. Goofing up the data entry, assigning incorrect or out-of-date values to parameters.

Figure 13.1—Signal integrity simulation at the pcb level begins with electrical descriptions of the IC die involved, the IC packages, and the traces on the pcb.

In the first case a model that is too simplistic will gloss over the fine details, often missing important aspects of system performance. On the other hand, a model that is overly complex will take so long to put together that you may never finish it. Finding the right balance is a matter of experience.

The problem of errors is dealt with by having a second individual double-check all the sources of model parameters. Does this sound like a lot of work? It is.

In a mature signal-integrity department, where full ringing and crosstalk analyses are run on each pcb, expect to find about one signal-integrity specialist for every five digital-circuit designers. A large, well-organized department has individuals who specialize in model-building, chip-level packaging, connectors, and so forth.

13.2.1 How Much Modeling Do You Need?

The extent of modeling required has to do with the risetime of your components, the distances your signals must traverse, and the accuracy required of the model.

For ordinary digital logic on fine-pitch pc-boards no more than 25 cm (10 inches) across, here are some generic guidelines for the required simulation complexity as a function of risetime.

> ➤ **3 ns**—Lossless transmission lines suffice for most pcb problems. IC package modeling is not necessary.
> ➤ **1 ns**—You may notice some side effects (resonance) due to the packaging and want to include IC package models for your fast signals.
> ➤ **300 ps**—Extensive modeling of the packaging, vias, all interconnection discontinuities, skin effect, and dielectric losses becomes essential.

When you add a new level of complexity, always compare the new simulation with your old one to see if it makes any difference. This is one way to determine when new levels of complexity are required.

13.2.2 What Happens After Parameter Extraction?

Once the electrical parameter extraction is complete, it's time to get down to simulation. When I write about signal-integrity simulations, I'm thinking about time-domain waveforms showing ringing, crosstalk, or ground bounce waveforms. These simulations may be produced using specialized signal-integrity analysis software from the major CAD vendors.

Signal-integrity simulations may be performed in either of two distinct modes. There's the *what-if* mode, intended for use by digital designers at the early stages of product architecture, and there's the *post-processing* mode, normally run after the conclusion of trace routing to validate the final design.

Tools developed for what-if analysis should include a comfortable schematic capture interface, a broad library of standard parts, and a good help system. These are the tools to which every digital designer should have unimpeded access. These tools typically produce individual time-domain waveforms or overlays of multiple simulation runs, which the designer evaluates by hand.

Tools developed for post-processing analysis should include extensive support for library management, flexible reporting, and good integration with your pcb layout system.

These tools compute time-domain results for the thousands of nets on your board in a batch-processing mode. Taking data directly from the finished layout, a post-processing tool will simulate every net, computing the complete received waveform at every node, using every possible combination of drivers. On a big board, the volume of output from a post-processing tool can be overwhelming.

The post-processing output typically feeds into a software analysis module, which flags nets in violation of specified criteria like percentage overshoot, percentage ringback, nonmonotonic behavior, and peak crosstalk. This is where the good tools really distinguish themselves in terms of evaluating, prioritizing, and reporting the results.

After post-analysis, all nets in violation of the specified criteria must be reworked by the designer. Once the problem is cleared, the affected net may be resubmitted for verification. That's the post-analysis approach to problem-solving. Post-analysis routines typically provide this sort of information:

> List of nets in violation of ringing criteria
> List of nets in violation of crosstalk criteria
> List of nets in violation of settling time criteria
> List of recommendations for termination
> List of places where terminations are not necessary

Tool sets are highly differentiated according to their degree of *software integration*. Highly integrated tools tightly link all the modules so that, for instance, a post-routing change in termination strategy is automatically back-annotated to the schematic and bill of materials. Organizations that grind out lots of designs each year commonly pay big bucks to obtain such integration.

13.2.3 A Word of Caution

Pcb routing software is getting much smarter. Given sufficient computing resources, crosstalk can be calculated on the fly during routing. Traces in violation of crosstalk constraints can be ripped up and moved. Ringing can be handled the same way. As traces stretch beyond the unterminated-line limit, terminations can be inserted automatically, back-annotated to the schematic, and added to the bill of materials. Vendors of pcb layout software already offer such features. In the not-too-distant future *all* routers will identify and react to signal-integrity problems during the routing process.

Nifty stuff, but keep in mind that as with any automated process, the quality of your input determines the quality of the final result. When the input suffers, the program spews out a mass of garbage without hinting that anything has gone wrong. Automated tools can be as dangerous as they are powerful and easy to use. Go slow, and double-check your results frequently.

POINTS TO REMEMBER

> Signal-integrity simulations may be performed in what-if mode or post-processing mode.

➢ Tool sets are highly differentiated according to their degree of *software integration*.

➢ Automated tools can be as dangerous as they are powerful and easy to use.

13.3 THE UNDERLYING SIMULATION ENGINE

All signal-integrity time-domain tools use iterative circuit-solution techniques pioneered by SPICE. An understanding of how SPICE operates therefore tells you a lot about how signal-integrity time-domain analysis algorithms behave (even if they are based on different software).

At its heart, SPICE is a guess-and-iterate algorithm. When simulating the performance of the simple circuit in Figure 13.2, SPICE performs the following steps.

1. Assume initial values for voltages v_{FET} and v_{R1}.
2. Compute what currents flow as a consequence of the assumed voltages.
3. If the currents i_1 and i_2 are not equal, adjust the assumed voltages.
4. Repeat from step 2 until voltages converge on the correct values.

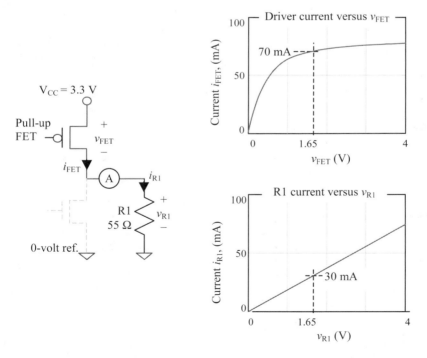

Figure 13.2—According to the I-V tables, with $v_{FET} = v_{R1} = 1.65$, more current would have to be flowing through the pull-up FET than through R1.

13.3 • The Underlying Simulation Engine

Figure 13.2 shows the I-V characteristics of the driver (which is pulling UP at the moment) and the resistive load. The I-V characteristics are crucial to the simulation process and must be supplied for each component.

Because the driver and load are wired in series, the simulator knows that at any given moment the voltage across the driver (v_{FET}) plus the voltage across the load (v_{R1}) must equal the power supply voltage (V_{CC}). Imagine the algorithm starts by assuming v_{FET} and v_{R1} each equal to half of V_{CC}.

The driver output current associated with $v_{FET} = 1.65$ V is given in the figure as 70 mA. The resistor current associated with $v_{R1} = 1.65$ V is given in the figure as 30 mA.

With this selection of voltages the driver current exceeds the resistor current by 40 mA, a clear violation of Kirchoff's current law, "The sum of currents into any node must equal zero", at node **A**. Therefore, the voltages must be wrong. You need to adjust the voltages in the direction of causing less driver current and more resistor current. This is done by lowering v_{FET} and raising v_{R1} (the two are constrained to sum to V_{CC}, so lowering one always raises the other).

How, you might ask, does the SPICE algorithm know in which direction to move v_{FET}? It doesn't; it merely tries both directions and then picks whichever direction seems to best improve the imbalance in current between the driver and the resistive load.

Figure 13.3 illustrates the result of lowering v_{FET} and raising v_{R1}. The driver current is diminished, while the resistor current increased, bringing them closer to balance. If the iteration routine is smart (and SPICE is), it will keep adjusting the node voltages until it arrives at a point where the currents are acceptably equal. That's the gist of SPICE.

Note that the iteration idea does not require that the I-V curves be linear, only that they be reasonably smooth. Also note that as long as the curves are monotonic, just about any searching strategy should lead to the correct answer after enough iterations. It might take a lot of iterations, but one of the great things about computers is that they have a lot of patience, so they can keep going around the try-and-evaluate loop over and over until they converge to an acceptable answer.

SPICE-like algorithms use mathematical techniques related to the Newton-Raphson algorithm to decide in which direction to move each guess value and how far to move it to obtain convergence. The Newton-Raphson algorithm is much faster than a simple binary

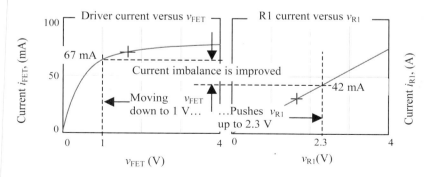

Figure 13.3—Decreasing v_{FET} (and simultaneously increasing voltage v_{R1}) decreases the current in the FET while increasing current in R1, bringing the circuit closer to balance.

search procedure, yielding very accurate answers within a small number of iterations for functions that are *well behaved*. For a circuit to be well behaved means that the I-V functions within it are monotonic and differentiable, plus a few other topological conditions. Most digital circuits, in the absence of feedback, have these properties.[130]

13.3.1 Evolving Forward

When you first engage a SPICE simulator, it attempts to derive steady-state initial circuit conditions. This is done by assuming all sources to be fixed at their initial values, all capacitances to be open circuits, and all inductances to be short circuits, and deriving according to the algorithm described previously the steady-state node voltages and currents everywhere. The algorithm is then ready to *evolve forward*, step-by-step, computing the state of the circuit at all future points in time. The evolution process, like everything else in SPICE, is iterative.

At each point after the algorithm has converged to a satisfactory set of node voltages for time t_n, it evolves forward by a small time-step Δt to a new time t_{n+1}. The evolution process involves four steps:

- Time-varying input waveforms are advanced by one time-step.
- The voltage $v_{C,k}(n)$ stored on every capacitor C_k at time t_n is advanced[131] according to the current $i_{C,k}(n)$ then flowing through it:

$$v_{C,k}(n+1) = v_{C,k}(n) + \Delta t \left(\frac{i_{C,k}(n)}{C_k} \right).$$

- The current $i_{L,m}(n)$ flowing through every inductor L_m at time t_n is advanced according to the voltage $v_{L,m}(n)$ then impressed upon it:

$$i_{L,m}(n+1) = i_{L,m}(n) + \Delta t \left(\frac{v_{L,m}(n)}{L_m} \right).$$

- SPICE then accepts the new capacitor voltages (and new inductor currents) as fixed quantities and iterates the other node voltages and currents to find a complete solution for circuit behavior at time t_{n+1}.

13.3.2 Pitfalls of SPICE-Like Algorithms

One significant complicating factor for SPICE is circuit complexity. Our basic example had only two components, with one unknown value and one constraint. When modeling a real circuit, the SPICE algorithm must take into account hundreds or thousands of nodes and I-V relationships. The software to handle these features can become very complex, but it's not any different in philosophy from the simple two-element example in Figure 13.2. Regardless of the number of nodes, SPICE still operates by guesses and iteration.

---

[130] An example of a very non-smooth I-V relationship is $i(v) = \sin(1/v)$. This function drives SPICE absolutely wild.
[131] The evolution equations for capacitors and inductors are discrete-time, integral forms of the familiar relationships $i_C = C(dv_C/dt)$ and $v_L = L(di_L/dt)$.

POINT TO REMEMBER

➤ When you first start working with any simulator, begin by setting up some simple, low-frequency test circuits for which you can predict the response by hand calculations.

13.3.5 Using SPICE Intelligently

SPICE does a superb job simulating the behavior of integrated circuits. It is quite capable of simulating the behavior of every transistor inside a logic circuit, even including logic functionality deep inside the chip. Unfortunately, if you try to do a complete circuit simulation in this manner, taking into account every transistor in every chip on a big board, the simulation may not complete within a reasonable amount of time.

If you *need* to simulate every transistor, there are no alternatives; you just have to turn on your computer and let it grind (sometimes for weeks). However, if your mission is merely to simulate ringing and crosstalk on connections between integrated circuits, consider using the modeling technique popularized by the I/O Buffer Information Specification (IBIS) standard. It is simpler and faster than a full transistor-level model. For ringing and crosstalk problems, it delivers suitably accurate answers.

The IBIS technique assumes foreknowledge of I-V tables and certain sampled waveforms representing your integrated circuits. It also assumes the calculations for each net are *separable*. That is, it assumes no knowledge of coincidental switching or of functionality within each chip. The technique merely steps through each net, trying each driver, calculating the received waveforms for both high and low transitions at all the receiver locations. This set of received waveforms is then analyzed to extract worst-case values for overshoot, ringback, settling time, and crosstalk on each net.

13.4 IBIS (I/O BUFFER INFORMATION SPECIFICATION)

Article first printed in *PC Design*, April 1997

Dealing with ringing and crosstalk in fast digital systems has never been an easy task. Especially today, with 150-MHz processors, new chips at 300-ps edge rates, and digital designers that want to carpet your board with 128-bit buses flying every which way, sometimes it's a wonder anything *ever* works.

On top of that, every year, relentless progress in the density of high-speed integrated circuits makes the situation worse. You can literally watch it happening. To see the effect, compare a new PC motherboard today with one from just a few years ago. The new one will have a lot more terminators. That is the direct, incontrovertible evidence that signal-integrity problems are growing more prevalent with every new product generation.

Whenever I speak to a group of digital designers, I ask "How many of you have ever had to add terminators to a board during debug to get it to work?" Without hesitation, everybody's hand goes up.

It's an endemic problem and an indication of the rather crude state of the art of signal-integrity design at many companies. Terminations are one of the key tools available to help fix problems with ringing, yet many digital designers don't know how to tell when ringing will occur, what kind of termination will be required to fix the problem, and where it must be placed.

Too often have I seen a board laid out with termination mounting pads provided on every net, with the assumption that a debug technician will test every signal by hand, apply terminators as needed, and then update the net list. Can't we do better than this? Isn't there some way to automate the process? *IBIS to the rescue!*

13.4.1 What Is IBIS?

The I/O Buffer Information Specification (IBIS) is an international standard for the electrical specification of chip drivers and receivers. It provides a standard file format for recording parameters like driver output impedances and waveforms, input loading, package parasitics, and pcb descriptions, all of which may then be used by any software application.

The parameters provided by an IBIS data file are ideally suited for automatic calculation of ringing and crosstalk.

The IBIS file structure makes it easy to specify the behavior of large chips with lots of I/O. IBIS I/O specifications are like macros that may be easily assigned to individual package pins. This hierarchical structure keeps the I/O specifications somewhat distinct from the package specifications, which helps when specifying very large devices or devices that may be packaged in different ways. The proponents of IBIS call it *component-centric,* which means that the root level of specification is at the component level (or packaged-chip level), which is the same level at which pcb layout tools operate. By way of contrast, a SPICE model usually focuses more on the detailed operation of individual I/O circuits rather than on the specification of a whole component.

Another difference between IBIS and SPICE is the form in which information appears. You can look in an IBIS file format and directly read out the worst-case V_{OH} value at a specified output current. In that sense, IBIS provides *specifications*. With a SPICE model, you must run a SPICE simulation with various combinations of circuit parameters to discover the same information. SPICE provides *models*. With SPICE, there's no way to represent a circuit that operates at one level of performance today, while reserving some headroom in the specification for future changes in the chip production process. An IBIS specification can do that.

13.4.2 Who Created IBIS?

The IBIS file format was originally created by an industry group called the IBIS Open Forum and later adopted by the American National Standards Institute (ANSI) and also the Electronic Industries Alliance (EIA) (see Section 13.5.1, "IBIS Historical Overview"). Information about the latest state of the standard is maintained by the EIA [105].

Keep in mind that IBIS by itself is nothing but a file format. It specifies *how* to record the various parameters of a chip driver or receiver in a standard IBIS file, but it does not specify *what* to do with them once they have been recorded. That's up to the simulation tools that use IBIS models.

To effect practical simulations using IBIS, you need four things:

1. A source of raw information about your chip drivers and receivers,
2. A way to translate that raw data into IBIS format,
3. A machine-readable version of the trace layout you wish to simulate, and
4. A software tool that understands IBIS and your trace layout format, and that can do the calculations you want.

13.4.3 What Is Good About IBIS?

IBIS is a fairly simple, straightforward file format. It is well suited for use by SPICE-like circuit simulation tools, but it is not SPICE-compliant, because the file format is not directly readable by all versions of SPICE, although this is changing rapidly with the introduction from several SPICE vendors of new **B** models that directly accept IBIS parameters.

IBIS provides a behavioral description of a driver or receiver without revealing proprietary details of how the circuit is internally fabricated. In other words, vendors can use IBIS models to specify how their great new gate designs work without giving away too much information to their competitors. Also, because it is a simplified model, it is reported to require on the order of 10 to 15 times less computation time than an equivalent SPICE transistor-level model when simulating typical digital configurations.

IBIS provides for specification of a complete I-V table representing a driver in it's high state, another I-V table to represent the driver in it's low state, plus some other information that tells it how to morph from one to the other at a defined rate of transition. The use of I-V tables is what gives IBIS the ability to easily model nonlinear effects like protection diodes, TTL totem-pole drivers, and emitter-follower outputs.

IBIS can be used to produce accurate, detailed simulations of high-speed ringing and crosstalk behavior. It can be used to examine signal behavior under worst-case risetime conditions, something impossible to manage with physical testing.

Lastly, because IBIS is a file format, not a procedural specification, you can use it for lots of stuff. Right now, it's being built into many of the tools you already use on a daily basis. Don't be surprised if all layout tools of the future calculate ringing and crosstalk on the fly as they route your traces, identifying and fixing signal integrity violations during auto-routing.

13.4.4 What's Wrong with IBIS?

Of course, IBIS is not perfect. There are some problems, but in my opinion, none significant enough to imperil the status of IBIS as the best, most comprehensive, and genuinely useful piece of signal-integrity technology to come along in a great while. With that said, here's my list of flaws:

- First and foremost, there is a distinct lack of support for IBIS models among many chip vendors. And IBIS tools won't work without IBIS model files. It's true that IBIS files may be constructed by hand or automatically converted from SPICE circuits, but all the translation tools in the world won't help if you can't pry a minimum risetime number out of your chip vendor.
- IBIS doesn't gracefully handle some forms of controlled risetime drivers, especially those that incorporate sophisticated feedback circuits.
- You will hear people say that IBIS is lacking in its ability to model ground bounce. What IBIS contains is a way to specify the mutual inductance of various pin-pair combinations from which can be extracted some very useful ground bounce information. What it doesn't do is model the way that large ground-bounce voltages can modify the behavior of an output driver as it moves from the high state to the low state.

I don't view any of these issues as major impediments to eventual acceptance of the IBIS technology. Most engineers today get almost no support when it comes to ringing, crosstalk, and ground bounce, and are suffering because of it. If IBIS helps, I say more power to it.

13.4.5 What You Can Do to Help

IBIS is coming. IBIS is going to solve a lot of common, everyday, high-speed design problems, but first we have to get our chip vendors to provide IBIS model files for every part they make.

When you talk to chip vendors about library files, please indicate your interest in IBIS. Let them know you think it's important. Let them know you need it. And, if you are planning to buy a lot of high-speed parts, let them know that you value working with a vendor that understands the importance of signal integrity in high-speed digital design.

POINTS TO REMEMBER

- ➤ IBIS is an international standard for the electrical specification of chip drivers and receivers.
- ➤ IBIS specifies *how* to record the various parameters of a chip driver or receiver, but it does not specify *what* to do with them.
- ➤ IBIS is the best, most comprehensive, and genuinely useful piece of signal-integrity technology to come along in a great while.
- ➤ We need our chip vendors to provide IBIS model files for every part they make.
- ➤ At the time of publication, the IBIS committee maintained work-in-progress copies of its latest draft standards at the Electronic Design Automation (EDA)

and Electronic Computer-Aided Design (ECAD) one-stop standards resource: *http://www.eda.org/pub/ibis*.

13.5 IBIS: HISTORY AND FUTURE DIRECTION

Bob Ross, Past Chair of the EIA IBIS Open Forum, June 3, 2002

The IBIS format is now a widely supported modeling format used by semiconductor manufacturers and EDA products, including several SPICE simulators. Many concerns expressed in Dr. Johnson's earlier article on IBIS have been addressed. Solutions to others are being considered.

IBIS is thriving not because it is always the best format, but because it provides the right balance between comprehensive detail, sufficient accuracy, and model availability. Users support IBIS because it gives fast, reliable, and accurate board-level simulation results. Semiconductor manufacturers support IBIS because it specifies device behavior without revealing proprietary device process and architectural information. The industrial infrastructure, driven by common needs, has enabled IBIS acceptance with freeware and commercial utilities. It has opened the pipeline for models developed by simulation and measurement.

IBIS models are available from at least three sources:

1. Many semiconductor manufacturers provide free, downloadable IBIS models, and others provide them through direct contact.
2. Commercial vendors, including prominent EDA vendors, are providing IBIS libraries and modeling services.
3. Users are also developing IBIS models and utilities, and managing internal IBIS libraries.

Since its beginnings in 1993, IBIS has evolved, keeping up with advances in digital circuitry and simulation needs. IBIS has been applied in higher speed applications than originally thought possible. Work continues on further enhancements and new directions.

13.5.1 IBIS Historical Overview

In the early 1990s, Intel Corporation initiated a spreadsheet, table-based model to convey the requirements of PCI bus drivers and provide a format among different divisions for external communication. Several vendor-specific formats did exist at that time, making data preparation and model availability primary issues. Intel invited EDA vendors to help define a common model format for industry. As a result of this collaboration, IBIS Version 1.0 was issued in June 1993 in an ASCII text–based model format, and a clarification update Version 1.1 was introduced in August 1993. The group formed the IBIS Open Forum to keep up with technical needs and promote the availability of IBIS models. IBIS Version 1.1 provided the initial framework with the understanding that it would be expanded in an upward compatible manner. The IBIS Open Forum issued an official syntax checker and electronic parser known as ibis_chk. North Carolina State University (NCSU) started work on a SPICE-to-IBIS model generation utility designated as s2ibis. These utilities along with a cookbook and public IBIS Open Forum support helped seed the industry to produce IBIS models.

More EDA vendors, semiconductor manufacturers, and users joined the Forum, and through cooperative efforts IBIS Version 2.0 was presented in June 1994 with many practical extensions. In February 1995 the Forum became affiliated with the (renamed to) Electronic Industries Alliance (EIA) as the EIA IBIS Open Forum. With formal review and letter ballot processes, IBIS Version 2.1 was issued and approved in December 1995 by the American National Standards Institute (ANSI) as ANSI/EIA-656. IBIS Version 2.1 provided a rich baseline set of features suitable for accurately modeling most digital buffers for many years. The EIA IBIS Open Forum upgraded the syntax checker and parser to ibischk2, and NCSU produced a corresponding s2ibis2 utility.

The EIA IBIS Open Forum continued making additions and improvements that led to the adoption of IBIS Version 3.2 nationally as ANSI/EIA-656-A by ANSI in September 1999 and internationally as IEC 62014-1 in April 2001 by the International Electrotechnical Commission (IEC). IBIS Version 3.2 included formats for advanced electronic features in some popular devices. It also provided extended package model features and electrical board descriptions. A corresponding ibischk3 was issued. Private and commercial utilities and services emerged for IBIS model development.

13.5.2 Comparison to SPICE

Like SPICE models, IBIS models are formatted in human-readable ASCII text. However, IBIS models do not require manufacturers to disclose proprietary information about how their circuits are designed. The critical electrical characteristics are described in IBIS by a purely behavioral model. Some manufacturers view this as a distinct advantage. IBIS models leave room for a chip manufacturer to create new implementations and process variations that still fit within the parameters of the published IBIS model (and are still therefore guaranteed to work in a user's circuit). That's a significant point in favor of IBIS. Lastly, SPICE models tend to be buffer-centric—the model often represents only one of several inputs or outputs on a component. IBIS models are component-centric—the model describes all the pins of a physical component. EDA vendors like IBIS because it integrates complete electrical component descriptions with the preexisting physical data already used to describe the thousands of connections in a large design. As a result, models of current devices are now becoming more readily available in IBIS than in SPICE formats.

13.5.3 Future Directions

Many people have contributed to the evolution of IBIS. From the very beginning, certain format inconsistencies and even a few mistakes have been introduced. Shorter-term practical commercial reality drove the evolution more than did technical purity. However, the existing format is capable of serving industry for many years.

New IBIS versions may appear, focusing less on technical advances and more on informational and datasheet content. In fact, the future directions are to link IBIS with other simulator formats, pursue related technical areas, and promote better models. The EIA IBIS Open Forum has been directly and indirectly involved with groups doing the following:

- Developing a multilingual extension format to support technical extensions and other features within a co-simulation environment with other languages.
- Advancing an interconnect format for both package and connector modeling improvements.
- Using IBIS within the Integrated Circuits Electromagnetic Model (ICEM) pending IEC standard for core noise generation.
- Tracking frequency-domain modeling details to provide better resonance detection.
- Improving IBIS models by forming a Quality Working Group to work on checklists and other support material.

The EIA IBIS Open Forum has investigated how to quickly add new technical features. It originally considered developing a new macro language for configuring buffer models. However, it is now considering leveraging off of established languages such as SPICE, VHDL-AMS, and Verilog-AMS in a co-simulation setting where needed. SPICE features can be used for advanced buffer detail and internal die interconnections. The analog and mixed-signal (AMS) extensions open the door to new capability with equation-based models and integration with logic design. Complex IBIS elements can be created using smaller IBIS building blocks and logic controls. IBIS can be the model container for analog circuits such as regulators, timers, operational amplifiers, et cetera, and for digital logic blocks. The multilingual approach will allow IBIS to enable solutions for SSO analysis, buffer interaction on internal die-level timing, and EMI core noise modeling. New buffer features can be implemented quickly within existing executable code.

No one can really predict how IBIS will evolve. IBIS might even be discarded and replaced with a new way of modeling. However, IBIS more likely will continue to advance in the broadest sense based on commercial interests and driven by individuals facing increasingly difficult challenges.

13.6 IBIS: Issues with Interpolation

An IBIS model specifies two I-V tables for each driver, one representing the static behavior of the driver when switched to the low state and another I-V curve representing the driver in the high state. In addition, the file provides other information that says how to *morph* from one I-V table to the other (see Figure 13.5). The purpose of morphing is to produce, at each point in time, a complete I-V table that a SPICE-like simulation engine can use to compute circuit results.

The form of the morphing information you will see inside an IBIS file depends on which version of the IBIS format you are using. The oldest (and simplest) IBIS specification included nothing more than a transition rate. The transition rate provides some gross

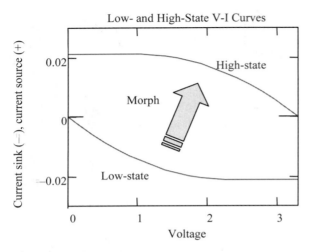

Figure 13.5—IBIS specifies a rate at which the low-state I-V curve morphs into the high-state I-V curve.

information about how to morph from the low table to the high table, but as with any interpolation process, the exact form of interpolation used affects the accuracy of the final result. For example, software vendors that implement the transition as a strictly linear ramp create waveforms significantly different from those that use quadratic smoothing at the corners of each rising or falling edge.

More recent IBIS specifications include rising and falling waveform specifications. These specifications include samples of the (rising or falling) waveform as observed under particular measurement conditions. The measurement conditions used are called out as part of each waveform specification. There may be more than one waveform specification.

For CMOS totem-pole drivers, a *good* IBIS model provides two rising edge waveforms—one measured with a 50-Ω load to ground and one measured with a 50-Ω load to V_{CC}. The model also provides two falling waveforms, for a total of four combinations of behavior showing rising and falling waveforms under both load conditions.

I rate such a model as good because I believe that such measurement conditions span the range of circumstances under which critical evaluation of signal shape is most important for CMOS drivers. For emitter-follower type drivers (ECL, PECL, and some GaAs), one waveform of each type measured with a 50-Ω load to the appropriate termination voltage will suffice. For pull-down only drivers like RAMBUS, GTL, and BTL one waveform of each type measured with a 25-Ω load (or whatever is appropriate in your system) to the appropriate termination voltage will suffice.

If only one waveform of each type is to be measured for a CMOS driver it should be measured using a symmetrically split end-termination.

An IBIS simulator uses the measured waveforms to improve the accuracy of its calculations. To understand how, you need a little information about the IBIS chip model.

Your IBIS simulator assumes each chip incorporates a totem-pole driver, some protection diodes, and a lumped-element capacitance representing the aggregate capacitance of the FETs and diodes (Figure 13.6). The totem-pole stage and the protection diodes are

13.6 • IBIS: Issues with Interpolation

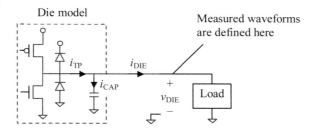

Figure 13.6—The IBIS die model includes a capacitor whose action must be de-embedded from the I-V table.

assumed to be memory-less, meaning that their actions can be completely described by an I-V curve. The curve may vary with time, but at any instant it has a particular shape. The capacitance is assumed to be both linear and time-invariant.

Your IBIS simulator is faced with the task of discovering a time-varying I-V function that satisfies four constraints:

1. In the steady-state high condition it mimics the high I-V table provided in the IBIS specification.
2. In the steady-state low condition it mimics the low I-V table provided in the IBIS specification.
3. During a rising edge, with the load called out in the specification, it mimics the rising waveform samples provided in the IBIS specification.
4. During a falling edge, with the load called out in the specification, it mimics the falling waveform samples provided in the IBIS specification.

One possible way to solve this problem is to concoct a hypothetical I-V function that is a linear combination of the low and high I-V tables provided.[134]

$$i_{TP}(v,t) = \alpha(t) i_{LOW}(v) + \beta(t) i_{HIGH}(v) \qquad [13.1]$$

where $i_{TP}(v,t)$ represents the time-varying I-V function associated with the totempole and diode components in Figure 13.6,

i_{LOW} represents the low I-V table provided in the IBIS specification,

i_{HIGH} represents the high I-V table provided in the IBIS specification, and

$\alpha(t)$ and $\beta(t)$ are scalar functions of time.

The functions $\alpha(t)$ and $\beta(t)$ are discovered by first de-embedding the die capacitance from the measurements and then solving for the functions $\alpha(t)$ and $\beta(t)$ that satisfy constraints numbered 3 and 4. The complete die model is composed of the time-varying I-V relationship [13.1] placed in parallel with the shunt die capacitance. This method was described by

---

[134] Not all simulator vendors use this method.

Bernhard Unger and Manfred Maurer of Siemens AG at the February 1998 IBIS summit meeting, SI-Analysis with HSPICE based on IBIS Models (see IBIS archives at [104]).

The de-embedding of the capacitance has a direct bearing on the accuracy of the results, because the rising and falling waveforms are defined at the die, *including* the capacitor (v_{DIE} in Figure 13.6), but the I-V curves you seek are defined at the output of the totem-pole and diode stage *prior* to the capacitor.

Using the rising waveform specified as part of the IBIS model, and given knowledge of the load conditions, your IBIS simulator can compute the current i_{DIE} flowing out of the die and into the load at every point in time during one rising edge. Your simulator can also compute the current flowing through the capacitor, i_{CAP}, assuming it knows accurately the dv/dt of the measured waveform and the value of the die capacitance. The simulator then adds i_{DIE} and i_{CAP} to find the current i_{TP} coming out of the totem-pole stage. This calculation is called the *de-embedding process*. After de-embedding the die capacitance the simulator knows, for each point in time during the rising edge, a pair of current and voltage values (i_{TP} and v_{DIE}) which together represent one point on the I-V table for the totem-pole stage at that particular time with that particular load.

If only one rising waveform is available the simulator uses the de-embedded information to solve for $\alpha(t)$ and $\beta(t)$ under the additional assumption that the two functions sum to unity.

If a second rising waveform is available measured under different conditions of loading, then your simulator can deduce two points on the totem-pole I-V curve for every moment in time, which is just enough information to solve independently for the coefficients $\alpha(t)$ and $\beta(t)$. This dual-waveform solution produces a rising edge model better suited for extrapolation to other conditions of loading.

The procedure is repeated to compute values of $\alpha(t)$ and $\beta(t)$ for falling edges.

If the *assumed* die capacitance doesn't match the *actual* die capacitance present during laboratory measurement (or SPICE simulation), then the coefficients $\alpha(t)$ and $\beta(t)$ will not come out quite right. The result of that inaccuracy is that waveforms predicted under conditions similar to the measurement conditions will come out just fine, but as you move away from the measurement conditions to other types of loads, the predicted waveforms become increasingly erroneous. This effect highlights a very important concept relevant to all interpolation algorithms:

> *Always specify circuit behavior under conditions similar to the actual conditions present in your application.*

For example, high-speed drivers rarely encounter 50-pF lumped capacitances in real-world applications; therefore, the old 50-pF timing specification makes little or no sense. For high-speed parts, it is much better to specify the performance of your driver into a 50-Ω load and then extrapolate from there to determine what would happen with the occasional 50-pF load.

If your waveform measurements are taken on a *packaged* component (at point v_{IO} in Figure 13.7), then the entire package model must first be de-embedded to determine the die waveforms. This operation is generally beyond the scope of an IBIS simulator, although it is certainly within the realm of reason that someone would take packaged-chip measurements and from that attempt to intuit the die waveforms.

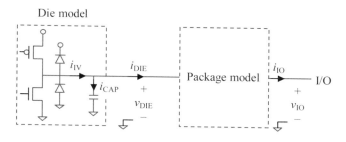

Figure 13.7—IBIS models a chip die and its associated package separately.

POINT TO REMEMBER

> ➤ Specify circuit behavior under conditions similar to the actual conditions present in your application.

13.7 IBIS: Issues with SSO Noise

A powerful logic driver, when it switches from one state to another, draws a huge surge of current through its power pins. Some of this current is due to overlap in the ON conditions of the two transistors in a totem-pole driver circuit. This overlap current flows from V_{CC} straight through to GND.

In addition to the overlap current, any current flowing to or from the load must also traverse the power pins (see Figure 13.8). When pulling low, current surges from the load, through the I/O pin, through the low-side FET, and from there flushes out the ground pin of the package back to the ground plane on the pcb and back to the load. When sourcing logic high, current surges from the power rail, through the power pin of the package, through the high-side FET, and from there through the I/O pin to the load. The point is, whatever current traverses an I/O pin must also somehow pass through the power or ground pins of the package.

As current surges through the finite inductance of the chip's power and ground connections (the *lead inductance* of the package), it perturbs the voltage between the true ground plane on the pcb and the chip's internal ground substrate (or between the true power plane on the pcb and the chip's internal power net). These voltages are called *simultaneous switching output noise* (SSO noise) [101],[102]. On the ground rail inside the chip, the magnitude of the SSO noise equals

$$v_{SSO}(t) = L_{GROUND} \frac{d}{dt}\left(i_{GROUND}(t)\right) \qquad [13.2]$$

where v_{SSO} is the noise voltage on the ground substrate inside the chip package,

i_{GROUND} is the current flowing through the ground pins of the chip package, and

L_{GROUND} is the effective inductance of the array of ground pins.

Equation [13.2] indicates that an SSO simulator needs to know (a) the effective cumulative inductance of all the ground pins in the package, and (b) the rate of change of current flowing through the ground pins. So far, so good. Here's the problem:

Item (a) can be measured in a variety of ways, but often the measured results are not available. Without this crucial piece of information, you can't make ground-bounce calculations. In this author's opinion, information about ground-pin inductance should be provided as a matter of routine by all package vendors. What I specifically want is a complete matrix of pin inductances and mutual pin inductances for the chip package. From this information, anyone can compute the effective inductance of any combination of ground pins (or power pins). There are optional provisions made in IBIS for incorporating such packaging information, and *I wish everyone would provide the data.*

Item (b) may be approximated if one knows the core currents required by the chip and also the sum of all I/O currents being drawn low by the output drivers. The conventional approach to computing this quantity is to first neglect the core currents on the assumption that the core currents are small compared to the massive I/O currents surging through the power and ground pins (a reasonable assumption for any but the largest VLSI circuits). Then

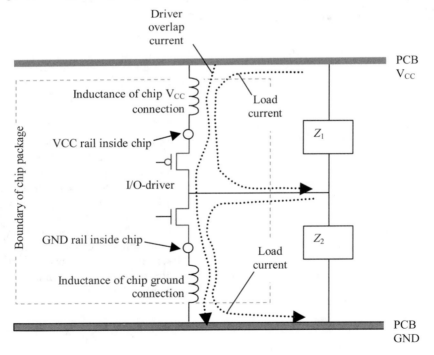

Figure 13.8—Current surging through the I/O pin of the device must also traverse the power and ground connections, creating noise on the internal power and ground rails as it passes.

use a ringing simulator to calculate the output current waveform for each output. From the output current waveforms, the peak di/dt due to each individual output may be estimated. The worst-case sum of all peak di/dt values may then be used for item (b), *provided that* the output currents sum linearly. The problem is, they don't. The drivers interact.

The driver interaction is caused by the shared power rails, which are bussed among I/O cells. When one output switches, it reduces the on-chip supply voltage available to the other drivers. When a driver is faced with a reduced power voltage, it switches more slowly than one with a full power voltage available. Therefore, when the second driver switches, the net maximum di/dt from the pair is less than twice the di/dt from a single driver. In other words, the assumption of independence used by IBIS simulators overestimates ground bounce, sometimes dramatically.

You can get a feeling for the sensitivity of a driver to ground bounce by simply operating the chip at a reduced DC power supply voltage. As you reduce the voltage, the output switching time usually slows down.

The IBIS committee has investigated the possibility of improving its specification file to incorporate the driver-to-driver interactions necessary to properly compute SSO noise, but to date has chosen not to do so.

POINT TO REMEMBER

> IBIS simulators don't yet properly compute SSO noise.

13.8 NATURE OF EMC WORK

The problem with electromagnetic compatibility (EMC) is not the difficulty of calculating the emissions from any particular radiating structure. The problem is much more one of recognizing the various types of radiating modes, *deciding which is the worst case,* and then knowing how to fix it. Identifying the worst-case radiating mode is quite important, because any problem you fix that isn't the worst-case mode makes zero difference to your compliance.

Don't expect anybody's software to tell you what is going to be the worst-case radiating mode in your design. Experts can't even do that.[135]

The ability to discover which is the worst offender is 90% of the battle. Once the worst source is identified and fixed, the second-worst source looms as the next problem. For a product that is 20 dB out of spec, there may be 10 different fixes, each worth 2 to 3 dB, required to fully address all the problems. It's a long, serial process.

---

[135] Put any three EMC specialists in a room, have them review your design, and ask them to agree on what is the most significant flaw. Don't interpret the resulting conflagration as a reflection of their skill; merely accept it as an indication of the difficulty of the problem.

13.8.1 EMC Simulation

Article first published in *EDN Magazine*, March 2, 1998

Let me give you some advice: Real live EMI problems are much too complex for even the best software tools. As much as I wish the situation weren't true, at this point the best tool is still experience. Many aspects of the EMI problem make prediction a difficult task, especially when working at the system level.

First, the process involves simulation of 3-D wave patterns over a rather large area. That is, your computer is going to spend a lot of time grinding numbers to get even the most rudimentary results.

Second, every bit of metal in the product matters: every via, trace, pad, bonding wire, connector, and cable. Many times, the system parts you choose not to include in the model turn out to be the very ones that create the worst EMI headaches. That's the nature of the problem. You seldom know beforehand what parts of a system will turn out to be the worst EMI offenders.

Third, EMI is a strong function of switching speed, data patterns, and precise data timing. That's right, data patterns and timing. If you don't believe me, find out what kind of software people ship to the EMI test range—some companies send several versions to see which works best.

> **EMI problems are much too complex for even the best software tools.**

The EMI problem is so complex, it's naïve to think you can just type in a few parameters and get meaningful results. For example, how should you model the split-plane zone on a mixed 3.3V/2.5V processor board? Obviously, the board stack-up, the shape of the 3.3V and 2.5V regions, and the trace layout matter. Did you realize that the placement and layout of bypass capacitors will markedly influence the result? How about power supply noise? You'd have to model the power supply noise, including phase, at all frequencies from 30 MHz through several GHz to get meaningful answers. Oops, that would be a function of the software running on the board, wouldn't it? Guess you'll have to model that too. Now add little features such as power supply wiring. It is well known that enclosing your power supply wiring in a good metal conduit can reduce radiation from the wiring. How would you like to spend your Easter break modeling the contact resistance and inductance of the screws used to hold the conduit in place?

Get the picture? You can't model everything. It's too complex. You can't leave anything out, either, because you never fully know what is going to matter. So, what can you do? Apply these time-tested, common-sense rules:

1. Limit your risetimes where practicable.
2. Use your experience. If last year's design ran at speed x, and today you are planning to do $2x$, you are likely to need another 6 dB of protection.
3. Run a quick EMI scan as early as possible.

The most promising new EMI tools will be more like expert systems than simulators. They will give you advice and provide reference information, but they won't make rash predictions. If these tools are worth their salt, the first three bits of advice they will provide are listed in 1, 2, and 3 above.

POINT TO REMEMBER

> ➢ Real live EMI problems are much too complex for even the best software tools.

13.9 POWER AND GROUND RESONANCE

Article first published in EDN Magazine, September 1, 1998

Many engineers believe that solid power and ground planes in a digital pcb should act as a large, perfect, lumped-element capacitor. Ideally, a perfect lumped-element capacitor of this size should provide a very low impedance between V_{CC} and ground at very high frequencies (several hundred MHz and higher). When a gate demands current from the power system, such a capacitor should supply that current with very little short-term voltage sag, or droop, according to the rule $i = C\,dv/dt$.

I wish the world were that simple. The problem is, the planes do not form a perfect lumped-element capacitor. Instead, they comprise a distributed system of surprising complexity.

The distinction between a distributed system and a nondistributed (or lumped-element) system involves the relationship between the time delay of the system and the risetime of your signals. As long as the time delay is sufficiently short, you might reason that it hardly affects the results and may be safely ignored. For example, if your pcb dimensions are 6 in. by 6 in., the time delay for signals trapped between the V_{CC} and ground planes as they travel from one side of the board to the other is approximately 1 nsec (for an FR-4 substrate). If you are using logic with a rise/fall time in the vicinity of 5 nsec, the lumped-element condition is mostly satisfied, and the planes will act to your benefit as one, large, parallel-plate capacitor.

What happens with much faster logic? With 200 psec rise/fall times, the drivers perceive the power-and-ground structure as a distributed object with a significant delay. This delay causes a number of artifacts. First, during an individual rising or falling edge, only the

> **Fast drivers perceive the power-and-ground structure as a distributed object with a significant delay.**

portion of the planes located within a small radius of the driver can react before the rising (or falling) edge is over and done. As a result, the initial noise spike may be much larger than you had anticipated. Next, the residual power-system noise signals from that first event travel across the board, bump into the edges of the board, reflect, and then finally return, a couple of nanoseconds later, to the driver location. If at that precise moment, the driver switches a second time, you will see both the old reflected response from the first signal and a response to the second signal superimposed together. If the phases add, you get more noise at the second edge than at the first. If the driver continues to act in a repetitive manner (like a clock), the reflected noise will build to a significant degree. This behavior exemplifies a *resonant mode* in the power system. In severe cases, resonance in the power system can cause your product to fail to function.

If your next system design depends on the natural power-plane capacitance to limit power supply noise, check to see how the round-trip delay across your board compares to your clock period. If it's close, consider building a test board mock-up (just use a plain, double-sided FR-4 core). With this setup, you can directly measure the impedance between the planes using a network analyzer to check for resonance. Another possibility is to take advantage of the latest power-system simulation tools. Several CAD vendors have developed packages that can simulate the distributed nature of a power-plane pair. These packages can help you evaluate the effectiveness of bypass capacitor placement, and also detect any resonance due to the shape and configuration of the planes.

Figure 13.9 illustrates the output from one such package. The structure it depicts contains 6 metal layers, (Signal, Power, Signal, Signal, Ground, Signal) as shown in the bottom part of the figure. The graph above shows the spatial distribution of voltage between the power and ground planes during transient simulation. The ripples near the center of the board emanate from an active driver. The de-coupling capacitors mounted on the board are placed to minimize the

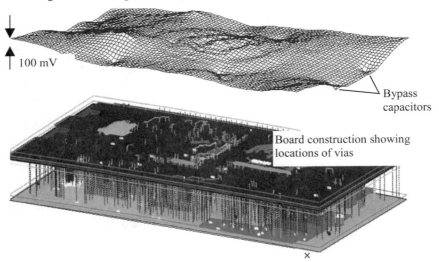

Figure 13.9—The magnitude of noise varies in a complicated way across the surface of a power plane. (Picture courtesy of Sigrity, provider of software tools for the detailed analysis and design of signal and power delivery systems in high-speed digital products.)

fluctuations. You can identify the locations of the decoupling capacitors from the dents in the graph.

If you haven't seen this category of tools, ask your CAD vendor for a demo. It's time well spent (and the 3-D graphics look really cool).

POINT TO REMEMBER

> ➤ If your system design depends on the natural power-plane capacitance, compare the roundtrip delay across your board to your clock period—you could be headed for resonance problems.

For further study see: www.sigcon.com

COLLECTED REFERENCES

Chapter 1: Fundamentals

[1] Donald Christiansen, ed., *Electronics Engineers' Handbook*, 4th ed., McGraw-Hill, 1997, ISBN 0-07-021077-2.

[2] Yusuf Billah and Bob Scanlan, "Resonance, Tacoma Narrows Bridge Failure, and Undergraduate Physics Textbooks," *Am. J. Phys. 59* (2), February 1991.

Chapter 2: Transmission Line Parameters

[3] Henry Ott, *Noise Reduction Techniques in Electronic Systems,* John Wiley & Sons, 1988, 2nd ed., ISBN 0-471-85068-3.

[4] Clayton R. Paul, *Introduction to Electromagnetic Compatibility,* John Wiley & Sons, 1992, ISBN 0-471-54927-4.

[5] Constantine A. Balanis, *Advanced Engineering Electromagnetics,* John Wiley & Sons, 1989, ISBN 0-471-62194-3.

[6] N. W. McLachlan, *Bessel Functions for Engineers*, Oxford Press, London, 1934 (see also 2nd ed. with corrections, 1961).

[7] Herbert B. Dwight, *Tables of Integrals,* The Macmillan Company, New York, 1961.

[8] Stephen H. Hall, Garrett W. Hall, and James A. McCall, *High-Speed Digital System Design,* John Wiley & Sons, Inc., 2000, ISBN 0-471-36090-2.

[9] Samuel P. Morgan, Jr., "Effect of Surface Roughness on Eddy Current Losses," *Journal of Applied Physics,* Vol. 20, April, 1949, pp. 352–362.

[10] Patrick J Zabinski, "Surface roughness of pcb tracks…," correspondence to the SI-LIST, 11 Jun 2001.

[11] Alina Deutsch, *Broadband Characterization of Low Dielectric Constant and Low Dielectric Loss CYTUF™ Cyanate Ester Printed Circuit Board Material,* Trans. CPMT-part B, Vol. 19, No. 2, May 1996.

[12] Hendrik W. Bode, *Network Analysis and Feedback Amplifier Design,* D. Van Nostrand Co., 1945.

[13]   Athanasios Papoulis, *The Fourier Integral and Its Applications*, McGraw-Hill, 1962.

[14]   Christer Svensson and Greg Dermer, "Time Domain Modeling of Lossy Interconnects," *IEEE Trans. Advanced Packaging*, Vol. 24, No. 2, May 2001.

[15]   R. S. Ramo, J. R. Whinnery, and T. Van Duzer, *Fields and Waves in Communication*, 2nd ed., Wiley, New York, 1984.

[16]   Ashok K. Goel, *High-Speed VLSI Interconnections*, John Wiley & Sons, Inc., 1994.

[17]   Y. R. Kwon, V. M. Hietala, and K. S. Champlin, "Quasi-TEM analysis of 'Slow-Wave' Mode Propagation on Coplanar Microstructure MIS Transmission Lines," *IEEE Trans. Microwave Theory Tech.*, MTT-35(6), pp. 545–551, June 1987.

[18]   H. Hasegawa and S. Seki, "Analysis of Interconnection Delay on Very High-Speed LSI/VLSI Chips Using an MIS Microstrip Line Model," *IEEE Trans. Electron. Devices*, ED-31, pp. 1954–1960, Dec. 1984.

[19]   Edward C. Jordan and Keith G. Balmain, *Electromagnetic Waves and Radiating Systems*, 2nd ed., Prentice-Hall, 1968.

[20]   T. C. Edwards, *Foundations of Interconnect and Microstrip Design*, 3rd ed., John Wiley and Sons, 2000, ISBN 0-471-60701-0.

[21]   Victor Brzozowski, chapter 8 of Charles A. Harper, *Electronic Packaging and Interconnection Handbook*, 2nd ed., McGraw-Hill, 1997, ISBN 0-07-026694-8.

[22]   Clyde F. Coombs, Jr., *Printed Circuits Handbook*, 4th ed., McGraw-Hill, 1996, ISBN 0-07-012754-9.

[23]   Bruce Archambeault, *EMI/EMC Computational Modeling Handbook*, Kluwer Academic Publishers, 1997, ISBN: 0-412-12541-2.

[24]   Alain Bossavit, *Computational Electromagnetism*, Academic Press, 1998, ISBN 0-12-118710-1.

[25]   Daniel Zwillinger, ed., *Standard Mathematical Tables and Formulae*, CRC Press, 1996, ISBN 0-8493-2479-3.

Chapter 3: Performance Regions

[26]   Clayton R. Paul, *Introduction to Electromagnetic Compatibility*, John Wiley & Sons, 1992, p. 143.

[27]   C. R. Paul and S. A. Nasar, *Introduction to Electromagnetic Fields*, 2nd ed., McGraw-Hill, 1987.

[28]   Frederick Terman, *Radio Engineers Handbook*, McGraw-Hill, 1943; out of print, but great if you can find it used.

[29]   Howard Johnson and Martin Graham, *High-Speed Digital Design*, Prentice-Hall, 1993, ISBN 0-13-395724-1.

[30]   W. C. Elmore, "The transient response of damped linear networks," *J. Appl. Phys.*, vol *19*, pp 55–63, January 1948.

[31] Yehea I. Ismail, Eby G. Friedman, and Jose L. Neves, "Equivalent Elmore Delay for RLC Trees," *IEEE Trans. Computer-Aided Design of Integrated Circuits and Systems 19* (1), January 2000.

Chapter 4: Frequency-Domain Modeling

[32] C. S. Burrus and T. W. Parks, *DFT/FFT and Convolution Algorithms,* John Wiley & Sons, 1985, ISBN 0-471-81932-8.

[33] Ronald N. Bracewell, *The Fourier Transform and Its Applications*, McGraw-Hill, 1978, ISBN 0-07-007013-X.

[34] Douglas F. Elliott and K. Ramamohan Rao, *Fast Transforms: Algorithms, Analysis, Applications,* Academic Press, 1982, ISBN 0-12-237080-5.

[35] Alan V. Oppenheim & Ronald W. Schafer, *Digital Signal Processing,* Prentice-Hall, 1975, ISBN 0-13-214635-5.

Chapter 5: Pcb (printed-circuit board) Traces

[36] Victor Brzozowski, chapter 8 of Charles A. Harper, *Electronic Packaging and Interconnection Handbook,* 2nd ed., McGraw-Hill, 1997, ISBN 0-07-026694-8.

[37] Alina Deutsch, "Broadband Characterization of Low Dielectric Constant and Low Dielectric Loss CYTUF Printed Circuit Board Material," *IEEE Trans. Components, Packaging, and Manufacturing Technology,* Part B, Vol. 19, No. 2., May 1996.

[38] T. C. Edwards, Foundations of Interconnect and Microstrip Design, 3rd ed., John Wiley & Sons, 2000, ISBN 0-471-60701-0.

[39] K. C. Gupta, Ramesh Garg, et al., Microstrip Lines and Slotlines, 2nd ed., Artech House, 1996, ISBN 0-89006-766-X.

[40] Constantine A. Balanis, *Advanced Engineering Electromagnetics,* John Wiley & Sons, 1989, ISBN 0-471-62194-3.

[41] Brian Wadell, *Transmission Line Design Handbook,* Artech House, 1991, ISBN 0-89006-436-9.

[42] Clyde F. Coombs, Jr., *Printed Circuits Handbook,* 4th ed., McGraw-Hill, 1996, ISBN 0-07-012754-9.

[43] Teradyne, "Design Considerations for Gigabit Backplane Systems," IEC Web Proforum Tutorials, *http://www.iec.org/online/tutorials/design_backplane/topic05.html.*

[44] Clyde F. Coombs, *Printed Circuits Handbook,4th ed.* McGraw-Hill, 1996, ISBN 0-07-012754-9.

[45] R. A. Pucel, D. J. Masse, and C. P. Hartwig, "Losses in Microstrip," *IEEE Trans. Microwave Theory Tech.,* MTT-16, 342, 1968.

[46] John Lau, *Flip-Chip Technology,* McGraw-Hill 1996, ISBN 0-07-036609-8.

[47] John Lau, *Ball Grid Array Technology,* McGraw-Hill, 1995, ISBN 0-07-036608-X.

[48] Brock J. LaMeres, *Characterization of a Printed Circuit Board Via*, B.S.E.E., Montana State University Technical Report EAS_ECE_2000_09, 1998, reprinted in part in *Microwave Journal,* November 2000, and later in *Conformity,* Vol. 6, No. 7, July, 2001.

[49] Henry Ott, *Noise Reduction Techniques in Electronic Systems,* 2nd ed., John Wiley & Sons, 1988, ISBN 0-471-85068-3.

[50] I. J. Bahl and R. Garg, "Simple and Accurate Formulas for Microstrip with Finite Strip Thickness," *Proc. IEEE,* Vol. 65, 1977, pp. 1611–1612.

[51] Howard Johnson and Martin Graham, *High-Speed Digital Design,* Prentice-Hall, 1993, ISBN 0-13-395724-1.

[52] Schneider, M. V., et al., "Microwave and Millimeter Wave Hybrid Integrated Circuits for Radio Systems," *The Bell System Technical Journal,* July–August 1969, pp. 1703–1726.

[53] Gautam Patel and Katie Rothstein, "Signal Integrity Characterization of Printed Circuit Boards", DesignCon 1999, San Jose, CA

[54] Israel, Paul, *Edison: A Life of Invention,* John Wiley & Sons, 1998, ISBN 0-471-52942-7.

Chapter 6: Differential Signaling

[55] Robert A. Pease, *Troubleshooting Analog Circuits,* Butterworth-Heinemann, 1991, chapter 9, ISBN 0-7506-9184-0.

[56] Howard Johnson, "Rainy Day Fun," *EDN,* March 4, 1999.

[57] Clayton Paul, *Introduction to Electromagnetic Compatibility,* John Wiley & Sons, 1992, ISBN 0-471-54927-4.

[58] Brian C. Wadell, *Transmission Line Design Handbook,* Artech House, 1991, ISBN 0-89006-436-9.

[59] Ramesh Senthinathan and John L. Prince, *Simultaneous Switching Noise of CMOS Devices and Systems,* Kluwer Academic Publishers, 1994, ISBN 0-7923-9400-3.

[60] Henry Ott, *Noise Reduction Techniques in Electronic Systems,* John Wiley & Sons, 1988, ISBN 0-471-85068-3.

[61] Howard Johnson and Martin Graham, *High-Speed Digital Design,* Prentice-Hall 1993, ISBN 0-13-395724-1.

[62] ANSI/IEEE 1596.3-1995, "Draft Standard for Low-Voltage Differential Signals (LVDS) for Scalable Coherent Interface (SCI)," Draft 1.3, November 27, 1995.

Chapter 7: Generic Building-Cabling Standards

[63] Howard Johnson, *Fast Ethernet: Dawn of a New Network,* Prentice-Hall, 1996, ISBN: 0-13-352643-7.

[64] TIA/EIA-568-B.1-2001 Commercial Building Telecommunications Cabling Standard, Part 1: General Requirements.

[65] ISO/IEC 11801:2002 Information Technology—Generic Cabling for Customer Premises.

Chapter 8: 100-Ohm Balanced Twisted-Pair Cabling

[66] Maurice G. Bellanger, *Adaptive Digital Filters and Signal Analysis,* Marcel Dekker, Inc., 1987, ISBN 0-8247-7784-0.

[67] John A. C. Bingham, *The Theory and Practice of Modem Design,* Wiley-Interscience, 1988, ISBN 0-471-85108-6.

[68] Edward A. Lee and David G. Messerschmitt, *Digital Communication,* 2nd ed., Kluwer Academic Publishers, 1994, ISBN 0-7923-9391-0.

[69] Constantine A. Balanis, *Advanced Engineering Electromagnetics,* John Wiley & Sons, 1989, ISBN 0-471-62194-3.

[70] TIA/EIA-568-B.1-2001, Commercial Building Telecommunications Cabling Standard Part 1: General Requirements.

[71] TIA/EIA-568-B.2, April 23, 2001, Commercial Building Telecommunications Cabling Standard Part 2: Balanced Twisted-Pair Cabling Components (Categories 3, and 5e).

[72] TIA/EIA-568-B.2-1, June 20, 2002, Commercial Building Telecommunications Cabling Standard Part 2: Balanced Twisted-Pair Cabling Components, Addendum 1—Transmission Performance Specifications for 4-Pair 100 Ω Category 6 Cabling.

[73] ISO/IEC 11801:2002 Information technology—Generic Cabling for Customer Premises.

[74] IEC61156-1 (2001-07) Multicore and symmetrical pair/quad cables for digital communications, Part 1: Generic specification.

[75] IEC61156-2 (2001-09) Multicore and symmetrical pair/quad cables for digital communications, Part 2: Horizontal floor wiring—Sectional specification (Category 3).

[76] IEC61156-3 (2001-09) Multicore and symmetrical pair/quad cables for digital communications, Part 3: Work area wiring—Sectional specification (Category 3).

[77] IEC61156-4 (2001-09) Multicore and symmetrical pair/quad cables for digital communications, Part 4: Riser cables—Sectional specification (Category 3).

[78] IEC61156-5 (2002-03) Multicore and symmetrical pair/quad cables for digital communications, Part 5: Symmetrical pair/quad cables with transmission characteristics up to 600 MHz—Horizontal floor wiring—Sectional specification (Categories 5e, 6 and 7).

[79] IEC61156-6 (2002-03) Multicore and symmetrical pair/quad cables for digital communications, Part 6: Symmetrical pair/quad cables with transmission characteristics up to 600 MHz—Work area wiring—Sectional specification (Categories 5e, 6 and 7).

Chapter 9: 150-Ohm STP-A Cabling

[80] Robert C. Pritchard, Douglas C. Smith, "A Comparison of the Susceptibility Performance of Shielded and Unshielded Twisted-Pair Cable for Data Transmission," IEEE 1992.

Chapter 10: Coaxial Cabling

[81] Constantine A. Balanis, *Advanced Engineering Electromagnetics,* John Wiley & Sons, 1989, ISBN 0-471-62194-3.

[82] Trompeter Electronics, Inc., *Coaxial Connector Catalog* (818) 707-2020.

[83] Charles A. Harper, *Passive Electronic Component Handbook*, 2nd ed., McGraw-Hill, 1997, ISBN 0-07-026698-0.

[84] *Electronic Warfare and Radar Systems Engineering Handbook*, originally published by the Avionics Department of the Naval Air Warfare Center Weapons Division in 1992 under report number TS 92-78. Since that time five editions were published before changing the document number to TP 8347. See *http://ewhdbks.mugu.navy.mil/coax_con.htm*.

Chapter 11: Fiber-Optic Cabling

[85] J. Wilson, J. F. B. Hawkes, *Optoelectronics, An Introduction*, 2nd ed., Prentice-Hall, 1989, ISBN 0-13-638461-7.

[86] IEC 793-2 (1992) *Optical Fibres—Part 2: Product Specifications*.

[87] IEC 60825-1 *Safety of Laser Products—Part I: Equipment Classification, Requirements and User's Guide*, 1993.

[88] Howard Johnson and Martin Graham, *High-Speed Digital Design*, Appendix B Prentice-Hall, 1993, ISBN 0-13-395724-1.

[89] Joseph C. Palais, *Fiber Optic Communications*, 4th ed., Prentice-Hall, 1998, ISBN 0-13-895442-9.

[90] Govind P. Agrawal, *Fiber-Optic Communication Systems*, 2nd ed., John Wiley & Sons, 1997, ISBN 0-471-17540-4.

[91] Paul A. Tipler, *Physics*, Worth Publishers, Inc., 1976, ISBN 0-87901-041-X.

[92] G. Mahlke and P. Gössing, *Fiber Optic Cables*, 3rd ed., Publicis MCD Verlag, Erlangen (for Siemans Aktiengesellschaft), 1997, ISBN 3-89578-068-5.

[93] David Cunningham and William G. Lane, *Gigabit Ethernet Networking*, Macmillan Technical Publishing, 1999, ISBN 1-57870-062-0.

Chapter 12: Clock Distribution

[94] Floyd Gardner, *Phaselock Techniques*, John Wiley & Sons, 1988, 0-471-04294-3.

[95] John Bingham, *The Theory and Practice of Modem Design*, John Wiley & Sons, 1998, ISBN 0-471-851108-6.

[96] Patrick R. Trischitta, *Jitter in Digital Transmission Systems*, Artech House, 1989, ISBN 0-89006-248-X.

[97] Fibre Channel—Methodologies for Jitter Specification, ANSI draft proposed NCITS Technical Report, Feb. 8, 1998.

[98] Hewlett-Packard Associates (recently Agilent), "Phase Noise Characterization of Microwave Oscillators," Product Note 11729C-2, Palo Alto, CA September, 1985.

[99] Charles Babbage Institute, Center for the History of Information Technology, University of Minnesota, *www.cbi.umn.edu*.

Chapter 13: Time-Domain Simulation Tools and Methods

[100] Ron Kielkowski, *Inside SPICE*, 2nd ed., McGraw-Hill, 1998, ISBN 0-07-913712-1.

Collected References

[101] Howard Johnson and Martin Graham, *High-Speed Digital Design,* Prentice-Hall, 1993, ISBN 0-13-395724-1.

[102] Ramesh Senthinathan and John L. Prince, *Simultaneous Switching Noise of CMOS Devices and Systems,* Kluwer Academic Press, 1994, ISBN 0-7923-9400-3.

[103] Donald Christiansen, *Electronics Engineers' Handbook,* 4th ed., McGraw-Hill, 1996, ISBN 0-07-021077-2.

[104] IBIS summit-meeting presentation archives: *http://www.eda.org/pub/ibis/summits/feb98.*

[105] Official IBIS standard web site: http://www.eigroup.org/ibis/ibis.htm.

For further study see: www.sigcon.com

POINTS TO REMEMBER

- 1.1 The *knee frequency*, $f_{\text{knee}} \triangleq 0.5/t_{10\text{-}90\%}$, is a crude estimate of the highest frequency content within a particular digital signal.
- 1.2 One neper equals 8.68588963806 dB.
- 1.3.1 Large objects have more inductance and capacitance than small ones.
- 1.3.1 High-frequency connectors must be very small to push the parasitic resonances up to frequencies above the bandwidth of the signal.
- 1.3.1.2 Simultanesouly enlarging the height and width of a transmission line has no effect on the characteristic impedance or per-unit-length delay.
- 1.3.2 Lower-voltage logic is remarkably power-efficient.
- 1.3.3 Shrinking *every parameter* of an unterminated structure speeds its settling time in direct proportion to the scale factor.
- 1.3.3 Terminated structures circumvent the link between physical size and signal quality.
- 1.3.4 A lower-impedance source coupled to a lower-impedance transmission line can drive larger capacitive loads.
- 1.3.5 Reducing the dielectric constant of your transmission-line substrate increases the characteristic impedance and decreases the delay.
- 1.3.6 Adjustments to magnetic permeability are rarely made in digital circuits.
- 1.4 Low-Q, dissipative circuits can't resonate. This is a desirable feature for a digital transmission path.
- 1.5 Resonance affects all physical structures, including bridges.
- 2.1 Two long conductors insulated from each other with a uniform cross section make a good transmission line.
- 2.1 The telegrapher's equations represent only the TEM mode of signal propagation.
- 2.1 You can model almost any transmission line with the telegrapher's equations.
- 2.2.1 The telegrapher's equations are derived from a cascaded lumped-element equivalent circuit model.
- 2.2.2 In typical digital transmission situations on pcbs the characteristic impedance changes fairly slowly over the relevant frequency range.
- 2.1.3 At frequencies above the LC and skin-effect mode onset, but below the onset of multiple waveguide modes of operation, the characteristic impedance is relatively flat and $Z_0 \approx \sqrt{L/C}$
- 2.2.4 Signals propagating on a transmission line decay exponentially with distance.
- 2.2.4 The per-unit-length attenuation factor $H(\omega)$ is called the *propagation function* of a transmission line.

2.2.4	The propagation coefficient $\gamma(\omega)$ is defined as the (negative of the) natural logarithm of $H(\omega)$.
2.2.4	The propagation coefficient $\gamma(\omega)$ may be broken down into its real and imaginary parts (α and β).
2.2.4	The real part of $\gamma(\omega)$ defines the attenuation per unit length of a transmission structure in nepers per unit length.
2.2.4	The imaginary part $\gamma(\omega)$ defines the phase shift per unit length of a transmission structure in radians per unit length.
2.3	A lossless transmission line requires $R = G = 0$.
2.3	For the special case of a lossless line $Z_C = \sqrt{L/C}$ and $\gamma(\omega) = j\omega\sqrt{LC}$.
2.4	The nonzero resistance of practical transmission lines dissipates a portion of the signal power, causing both attenuation (loss) and distortion in propagating signals.
2.5	Shunt conductance G is practically zero at DC for the types of insulators used in most modern digital transmission applications.
2.5	Dielectric loss models for high-frequency applications incorporate AC dielectric losses into the definition of complex permittivity, creating a capacitance term C with both real and imaginary parts.
2.6	Magnetic fields within a conductor adjust the distribution of high frequency current, forcing it to flow only in a shallow band just underneath the surface of the conductor.
2.6	The effective depth of penetration of current is called the *skin depth*.
2.6	The increase in the apparent resistance of the conductor caused by this redistribution of current is called the *skin effect*.
2.6	At frequencies above the skin-effect onset frequency ω_δ the effective series resistance of a conductor rises with the square root of frequency.
2.7	The distribution of current at high frequencies minimizes inductance.
2.7	At DC, the path of least DC resistance creates a slightly higher inductance.
2.7	Good models for skin effect take into account changes in both resistance and inductance with frequency.
2.8	Merely adding the DC and AC models of resistance produces substantial errors at frequencies near the onset of the skin effect, and predicts the wrong value of internal inductance.
2.8	The second-order approximation [2.52] better matches both the real and imaginary parts of the skin effect at frequencies near the transition region.
2.10	The proximity effect distributes AC current unevenly around the perimeter of a conductor.
2.10	The proximity factor increases the apparent AC resistance of a conductor above and beyond what you would expect from the action of the skin effect alone.
2.10	Above that frequency where the proximity effect takes hold, the distribution of current around the perimeter of the conductor attains a minimum-inductance configuration and does not vary further with frequency.

2.10 The skin effect and the proximity effect are two manifestations of the same principle: that magnetic lines of flux cannot penetrate a good conductor.

2.10 Field simulators base their calculations on many assumptions, and don't always produce the right answers.

2.11 At a microscopic level, all materials exhibit surface irregularities and bumps.

2.11 Rough toothing profiles are purposefully etched into copper layers prior to lamination to facilitate adhesion between layers.

2.11 Roughness on a scale comparable to the skin depth increases the mean length of the path of current, increasing the resistance.

2.12 All insulators exhibit some degree of dielectric loss.

2.12 Make sure you know the frequency at which a value of dielectric constant is specified.

2.12.1 Dielectric losses in a transmission line scale in proportion to both frequency and length.

2.12.1 The dielectric loss tangent is the tangent of the phase angle formed by the real and imaginary components of complex permittivity.

2.12.1 For small loss tangents, l is approximately the same as the ratio of the imaginary part to the real part of complex permittivity.

2.12.2 A mixed dielectric carries a permittivity ϵ equal to the weighted average, on a volumetric basis, of the permittivities of the constituent materials:

2.12.2 Any air or water present in a dielectric mixture will change the dielectric constant of the resulting mixture.

2.12.3 The loss tangent for a mixed dielectric can be calculated from the loss tangents and filling factors of the constituent materials.

2.12.4 The filling factor for a dielectric-air mixture may be inferred from the velocity of propagation.

2.12.5 The real and imaginary portions of any realizable network function bear certain subtle yet incontrovertible relations to each other. Specifying one without the other leads to non-realizeable circuit results.

2.12.6 You can calculate a variation in dielectric constant to match any specified loss tangent.

2.12.5 The Kramers-Kronig relations constrain the behavior of the real and imaginary parts of complex permittivity.

2.13 An impedance in series with the return path affects the signal just as much as an impedance in series with the signal conductor.

2.14 In an on-chip MIS configuration, if the electric fields penetrate to a depth of h_1, while the magnetic fields penetrate to a futher depth h_2, the resulting combination of large capacitance and large inductance creates an absurdly slow velocity of signal propagation.

3.1 The signal propagation model computes the transfer function and impedance of cables made in multiwire, ribbon, UTP, STP, or coaxial format. It also works for pcb traces, both striplines and microstrips, up to a frequency of approximately 10 GHz.

3.1 The parameter R_{DC} provides an amount of loss that is flat with frequency.

3.1 The parameter R_0 provides an amount of loss that grows (in dB) in proportion to the square root of frequency

Points to Remember

3.1   The parameter θ_0 provides an amount of loss that grows (in dB) in direct proportion to frequency

3.1   At all frequencies the magnitude and phase responses match to produce a causal, minimum-phase response

3.2   Sweeping from low frequencies to high, the loss curve for a transmission line changes in a predictable way as you pass the onset of various regions of operation.

3.2   The distinguishing features of each region are determined by the propagation coefficient, propagation function, and characteristic impedance.

3.2   The regions usually appear in this order: lumped-element, RC, LC, skin-effect, dielectric, and waveguide.

3.2.1   A pcb trace of any length always remains a transmission line, supportig two modes of propagation (out and back).

3.2.1   When a transmission line is short, two modes of propagation still exist, only their temporal superposition creates the illusion of a direct connection between source and load.

3.3   The undistorted conveyance of a signal from source to load requires a propagation function that remains flat over a band of frequencies covering the bulk of the spectral content of the data signal.

3.4   The classification of a transmission line in the lumped-element region does not determine how the line is going to act. It determines merely how the line may be analyzed.

3.4.1   A transmission line can always be shortened to a length below which it operates in the lumped-element region.

3.4.1   Transmission lines short enough to operate in the lumped-element region rarely require termination.

3.4.2   The pi model applies to any transmission line electrically short compared to the signal wavelength, and where the time constant $l^2 R_{DC} C$ remains small compared to the signal period.

3.4.3   Within the lumped-element region you may use Taylor-series expansions for H and H^{-1}.

3.4.4   The input impedance of a short, unloaded transmission line looks entirely capacitive.

3.4.4   The input impedance of a short, grounded pcb-trace looks entirely inductive.

3.4.5   Any transmission line can be shortened to the point where it acts as a perfect connection.

3.4.5   If the source can't drive the load in the first place, then hooking the source and load together with a transmission line isn't likely to make things better.

3.4.5   Conditions necessary such that a short, lumped-element transmission line not affect signal quality are given by [3.39] through [3.41].

3.4.6   Even a short transmission line may resonate horribly if used to interconnect a ferociously reactive combination of source and load.

3.5   The terms RC transmission line, dispersive transmission line, and diffusion line all mean the same thing.

3.5.1   The RC region extends in frequency from DC up to that point ω_{LC} where the magnitude of the line inductance ($\omega_{LC} L$) equals the DC resistance (R_{DC}).

3.5.1	For any transmission line there exists a critical length below which you need *never* concern yourself with the distributed RC mode of operation.
3.5.2	The input impedance of a line *without reflections* is predictable and independent of line length.
3.5.2	An RC transmission line may be equalized using a suitable reactive source or load impedance network.
3.5.3	Within the RC region, characteristic impedance is a complex function of frequency with a phase angle of –45° and a magnitude slope of -10 dB per decade.
3.5.4	A perfectly-matched end termination applied to an RC transmission line renders the input impedance of the structure indepedant of line length. This advantage comes at the expense of a terrible degradation of the transfer response.
3.5.4	A fixed resistance at the end of an RC transmission line flattens the gain curve, providing more usable bandwidth at the expense of a reduction in the received signal amplitude, and a greater variation with line length in the input impedance of the structure.
3.5.4	A purely resistive end termination equal to $Z_0 = \sqrt{L/C}$ eliminates reflections within the LC band while also providing a relatively flat propagation function within the RC band.
3.5.6	A resistive termination at both ends of an RC-LC mixed-mode transmission line provides flatter gain than termination at only one end or the other.
3.5.7	A fixed resistance at the end of an RC transmission line improves the settling time at the expense of a reduction in the received signal amplitude.
3.5.8	The speed of operation achievable within the RC region scales *inversely* with the square of transmission-line length.
3.5.10	The Elmore delay approximation takes into account only the resistance and capacitance of a transmission configuration. It applies to well-damped circuits composed of any number of series resistances, shunt capacitances, and distributed RC transmission lines.
3.5.10	The Elmore delay approximation does not apply circuits involving inductance, resonance, overshoot, or any form of poorly damped or nonmonotonic behavior.
3.5.10	The Elmore delay for a lumped resistance R feeding a total downstream capacitance of C is RC.
3.5.10	The Elmore delay for a distributed RC transmission line having total resistance R and distributed capacitance C is $(1/2)RC$.
3.6.1	The LC region begins where the magnitude of the line inductance (ωL) exceeds the DC resistance (R_{DC}).
3.6.1	Within the LC region the line attenuation does not much vary with frequency.
3.6.1	Unfortunately, in the practical transmission lines used in most digital designs the LC region is relatively narrow (or sometimes nonexistent).
3.6.2	At frequencies far above ω_{LC} the characteristic impedance of a pure LC-mode transmission line asymptotically approaches $Z_0 \triangleq \sqrt{L/C}$.
3.6.2	At a frequency ten times greater than ω_{LC} the complex value of characterisitic impedance differs from Z_0 by only about 4%.

Points to Remember

3.6.3   A fixed series resistance induces an upward tilt in the TDR measurement, indicating a gradually increasing impedance at lower frequencies.

3.6.4   The propagation function for an LC-mode transmission line is a simple delay with a fixed attenuation.

3.6.4   Doubling the length of an LC-mode transmission line doubles the delay, and doubles the attenuation (in dB).

3.6.5   In the LC region a signal can accumulate a substantial amount of phase delay without suffering much attenuation. This property indicates that a transmission line can act as an extremely high-Q resonant circuit.

3.6.5   LC-mode resonance affects only signals whose bandwidth extends into the resonant region

3.6.6   The source, end, and both-ends terminations can all be used to eliminate LC-mode resonance.

3.6.6   The end-termination is least sensitive to the series resistance of the transmission line.

3.6.7   The speed of operation achievable within the LC region is not directly limited by transmission-line length.

3.6.8   Transmission lies operated at a length greater than l_{RC} may display characteristics of both LC and RC behavior.

3.7.1   The skin effect region starts when the internal inductance of the signal conductor becomes significant compared to its DC resistance.

3.7.1   Within the skin effect region the characteristic impedance remains fairly flat, while the line attenuation in dB varies in proportion to the square root of frequency.

3.7.2   At frequencies far above ω_δ the characteristic impedance of a skin-effect limited transmission line asymptotically approaches $Z_0 \triangleq \sqrt{L/C}$.

3.7.2   The asymptotic convergence is not quite as fast as for an LC transmission structure with fixed (non-frequency-varying) resistance.

3.7.3   The skin effect induces an upward tilt in the TDR measurement with a steep initial slope, gradually tapering to a more gentle rise.

3.7.4   The attenuation (in dB) within the skin-effect region grows in proportion to the square root of frequency.

3.7.4   Doubling the length of a skin-effect-limited transmission line doubles the attenuation.

3.7.5   Transmission lines in the skin-effect region fall prey to the same resonance difficulties that afflict the LC region and respond to the same means of termination.

3.7.6   The step response associated with the skin effect has a quick rise and a long, sloping tail.

3.7.7   The risetime of a skin-effect-limited channel scales with the square of its length.

3.7.7   The speed of operation within the skin-effect region scales inversely with the square of transmission-line length.

3.7.7   A conductor twice the diameter (or width) has 1/2 the AC resistance and thus 1/4 the skin-effect risetime.

3.7.7   It's rare in pcb problems that you see skin-effect losses without also having to take into account dielectric dispersion.

3.8.1   Skin-effect loss grows only in proportion to the square root of frequency, while the dielectric loss grows in direct proportion to frequency. Above some frequency ω_θ the dielectric loss equals, and then exceeds, the skin-effect loss.

3.8.2   In the vicinity of the skin-effect onset ω_δ the skin effect *increases* characteristic impedance while dielectric loss *decreases* it.

3.8.2   At frequencies above the onset of the dielectric-loss-limited mode ω_θ, dielectric losses ultimately force the characteristic impedance back up above Z_0.

3.8.3   Dielectric losses cause an upward tilt to a plot of characteristic impedance versus frequency. Resistive losses create a neagtive slope. Working together, the two effects can sometimes almost cancel, creating a TDR slope less steep than when either effect is present alone.

3.8.4   The attenuation (in dB) within the dielectric-loss-limited region grows in direct proportion to frequency.

3.8.4   Doubling the length of a dielectric-loss-limited transmission line doubles the attenuation.

3.8.5   The dielectric effect induces a gradual rise in the characteristic impedance Z_C in proportion to the log of freqeuncy.

3.8.5   Transmission lines in the dielectric-loss-limited region fall prey to the same resonance difficulties that afflict the LC region and respond to the same means of termination.

3.8.6   Given two systems with the same –3dB loss at frequency f_1, one system having only dielectric losses and the other having only skin-effect losses, the dielectric step response begins more slowly than the skin-effect response, but finishes sooner.

3.8.7   The risetime of a dielectric-loss-limited channel scales directly with its length.

3.8.7   The speed of operation within the dielectric-loss region scales inversely with transmission-line length.

3.8.7   A dielectric medium with twice the loss tangent incurs twice the loss (in dB) and induces a settling time twice as long.

3.8.7   It's rare in pcb problems that you see skin-effect losses without also having to take into account dielectric dispersion.

3.9.1   If the wavelengths of the signals conveyed approach the dimensions of your conductors, strange modes of propagation begin to appear.

3.9.1   For ordinary digital signaling on FR-4 printed circuit boards at 10 Gbps you may use microstrip trace heights up to 20 mils without encountering significant microstrip dispersion.

3.11   Over the range of frequencies dominated by the skin effect, you can scale the length of one coaxial cable type to cause it to mimic the performance of any other type.

3.12   Five ways to improve the performance of a copper transmission channel: use more copper, don't go as far, use a higher characteristic impedance, add equalization, or use a better dielectric material.

3.13   The performance of a metallic interconnection is heavily affected by its physical construction, which is comparatively well controlled in the manufacturing process. Metallic transmission systems have a relatively hard, fixed upper limit on distance that should never be exceeded.

Points to Remember

3.15 Intersymbol interference may be characterized by a dispersion penalty.
3.15 The dispersion penalty may be circumvented by equalization.
4.1   Frequency-domain simulation gives you incredible control over the exact form of frequency-dependant losses, like the skin effect and dielectric-loss effect.
4.1   Frequency-domain simulators may be easily programmed in any software spreadsheet application (like MatLab, Mathematica, or MathCad), giving you control over every aspect of the simulation, including searching for optimum and worst-case parameter values.
4.1   Frequency-domain simulation applies only to linear systems.
4.2   The DFT is a discrete-time approximation to the Fourier transform.
4.2   The popular Cooley-Tukey FFT algorithm is a clever, highly efficient implemetation of the DFT that works only for N equal to a power of two.
4.3   The FFT requires two parameters: a sample interval ΔT and a sample vector length N.
4.3   The spacing ΔT must be small enough to fairly represent the complete signal waveform without loss of significant information.
4.3   Always provide an N large enough to allow your simulated system to come to a steady-state condition before the end of the FFT time window.
4.4   The FFT requires that your signal waveform have the same value at start and finish.
4.5   Most FFT routines require external scale factors that depend on the sample interval ΔT and sample vector length N.
4.6   Table 4.1 shows how to form FFT frequency vectors for test signals, data patterns, pulses, and feathered edges.
4.7   In simulations of high-speed digital systems an inadequate sampling rate causes a waveform to "wiggle around" as a function of precisely where it is sampled.
4.9   Frequency-domain simulation handles some pretty complicated situations.
4.10 Before using an unfamiliar FFT routine, check the transform of a simple impulse at time 0 and also the transform of the same impulse delayed by one sample.
5   Analysis of pcb performance generally assumes well-defined uniform paths for both signal current and returning signal current, conductors long compared to the spacing between the signal and return paths, and conductors shorter than the critical RC length l_{RC}.
5.1.1 If you don't already have a 2-D field solver, get one.
5.1.2.1 A 1/2-oz copper pcb trace with 100-μm (3.9 mil) width has a DC resistance of 9.6 Ω/m. The DC resistance scales inversely with the width and inversely with the copper plating weight.
5.1.2.2 Low-frequency current in a pcb trace therefore follows the *path of least resistance*, filling the cross-sectional area of the trace,
5.1.2.2 The skin effect confines high-frequency current to a shallow band of depth δ around the perimeter of a conductor.
5.1.2.2 The proximity effect draws signal current towards the side of a microstrip facing the reference plane, or that side of a stripline that faces the nearest reference plane.
5.1.2.2 The increase in resistance of a typical high-speed digital signal conductor due to the proximity effect (above and beyond simple consideration of the skin depth

	and trace circumference assuming a uniform current distribution) typically ranges from 25% to 50%.
5.1.2.2	Another similar-sized increase in resistive dissipation occurs due to the nonuniform distribution of current on the reference plane.
5.1.2.2	Traces with similar ratios of w/h inherit similar values of k_p regardless of the dielectric constant.
5.1.2.6	You can simulate the magnetic field surrounding a pc-board stripline using a rubber sheet and a Popsicle stick.
5.1.2.7	At frequencies on the order of 1 GHz, nickel-plating the top surface of a microstrip cuts in third the effective useful length the trace.
5.1.3	For FR-4 digital circuit board applications with risetimes of 500 ps or slower, at distances up to 10 inches, you may ignore dielectric losses.
5.1.3	At longer distances or at higher speeds, dielectric losses can become quite significant.
5.1.3.1	A microstrip has dielectric properties intermediate between the properties of the dielectric substrate and air.
5.1.3.4	Core and prepreg laminate materials are now available in a staggering array of types and variations.
5.1.3.4	The core or prepreg laminate comprises a fabric of fine threads embedded in a solidified resin.
5.1.3.4	Inhomogeneities in a fabric-resin laminate ultimately limit the size of the thinnest dielectric that can be produced from that combination of materials.
5.1.3.5	The dielectric loss of a backplane may change substantially with temperature.
5.1.3.6	Copper traces on outer layers may be protected from corrosion by passivation or by coating them with an inert material.
5.1.4	The skin-effect step produces a sharper initial rise, but a longer, more lingering tail, than does the dielectric effect.
5.1.5.1	For normal digital signaling on FR-4 pc boards at 10 Gbps, you may use any trace height up to 0.5 mm (0.020 in.) without encountering significant microstrip dispersion.
5.2	As you stretch the channel length to extreme distances, sensitivity-limited systems fail due to insufficient signal amplitude at the receiver.
5.2	Dispersion-limited systems fail due to signal distortion, also called intersymbol interference (ISI).
5.2	Amplifying the received signal does not change the performance of a dispersion-limited system. Equalization is what helps.
5.2	Systems limited by dispersion may sometimes be improved by a change in data coding.
5.2.1	A non-linear DC restoration system can un-do the effects of AC coupling.
5.3.1.1	Pcb traces terminated at *both ends* enjoy a great advantage in immunity to reflections as compared to their singly terminated cousins.
5.3.1.2	A small lumped-element capacitance shunting a transmission line creates a backwards-propagating reflection.
5.3.1.2	A small lumped-element inductance in series with a transmission line does the same, but with the opposite polarity.

Points to Remember

- 5.3.1.3 Adjustments to transmission-line width can partially compensate for one small, isolated capacitive load.
- 5.3.1.4 Adjustments to transmission-line width can partially compensate for one small, isolated series inductance.
- 5.3.1.5 Right-angle bends in pc-board traces perform perfectly well in digital designs in speeds as fast as 2 Gbps.
- 5.3.1.6 Blind or buried vias are smaller and have less effect than full-sized vias.
- 5.3.1.7 Densely packed component pads greatly reduce trace impedance.
- 5.3.1.8 Place a series terminator no more than a small fraction of one risetime away from the driver.
- 5.3.1.9 Place an end terminator no more than a small fraction of one risetime from the end of the line.
- 5.3.1.10 A low-impedance driver combined with a tight-tolerance resistor in series makes an accurate series termination.
- 5.3.1.10 A high-impedance current-source driver combined with a tight-tolerance resistor in shunt across the driver also makes an accurate series termination.
- 5.3.1.11 Impedance translation over any band that includes DC is accomplished using a resistive pad.
- 5.3.2.1 Solid reference planes exist to control crosstalk.
- 5.3.2.2 Crosstalk varies strongly with trace separation and with the trace height above the reference planes.
- 5.3.2.2 A field solver is the best way to estimate crosstalk for general digital purposes, provided that no holes, slots, or gaps in the planes cross the path of either the victim or aggressor trace.
- 5.3.2.3 Crosstalk is highly directional.
- 5.3.2.3 Whether initially headed forward or backward, crosstalk reflects and bounces off any imperfections in the transmission structure, often ending up at both ends of the line.
- 5.3.2.4 For parallel traces shorter than half the signal risetime, near-end crosstalk varies in proportion to the length of parallelism.
- 5.3.2.4 For parallel traces longer than half the signal risetime, near-end crosstalk saturates at a maximum level. The ratio of crosstalk to aggressive step-size at saturation is the NEXT coefficient.
- 5.3.2.4 Saturated NEXT looks like a long, low rectangle with a flat top and a duration equal to twice the trace delay plus one source risetime.
- 5.3.2.5 Far-end crosstalk varies in proportion to the trace length.
- 5.3.2.5 FEXT looks like a short pulse with a duration equal to the source risetime.
- 5.3.2.6 The both-ends terminated stripline architecture greatly reduces, but does not completely eliminate, both FEXT and NEXT.
- 5.3.2.7 Both voltage and current affect crosstalk.
- 5.4.1 Connector crosstalk in open-pin-field connectors acts through a transformer-like principle.
- 5.4.1 Separating the loops of signal current within a connector by providing private power or ground pins for each signal reduces crosstalk.
- 5.4.1 Reducing the current in the aggressive circuit reduces crosstalk.

5.4.2	Never assume your fabrication shop will build the board the way you ask. Always check.
5.4.3	Three primary measures of connector performance are signal fidelity, crosstalk, and EMI.
5.4.4	A tapered transition maintains constant impedance when interconnecting transmission lines having widely different physical scales.
5.4.5	A straddle-mount connector locates all its pins close to the plane of the pcb.
5.4.6	A high-frequency shield needs direct metallic contact with the product chassis, completely surrounding the signal conductors.
5.5.1	The properties of a via are modified by the trace to which it is attached.
5.5.1	Inductance is a property of an entire current pathway (a loop of current). Don't use partical inductance values by themselves.
5.5.1	A via contributes incremental shunt capacitance and incremental series inductance to a trace.
5.5.2	If the incremental capacitance or inductance of your via is not sufficient to cause an objectionable reflection, then no model is required.
5.5.2	A first-order model reduces the via to either a single value of excess shunt capacitance, or a single value of excess series inductance, according to which effect creates the greatest reflection.
5.5.2	If your via is so large compared to the signal risetime that you require anything more than a simple pi-model for the via, then it probably isn't going to work very well for a digital application. Use a smaller via.
5.5.2	Narrowband applications sometimes use large vias at frequencies well beyond the useful band for digital applications.
5.5.3	A long, dangling via can develop a resonance, exacerbating the effects of its capacitance.
5.5.4	The incremental capacitance of a via is affected by the geometry of the via, the surrounding reference planes, the trace width used to connect to the via, and the dielectric constant of the substrate material.
5.5.4	Via capacitance varies in proportion to the overall size of the via.
5.5.5.1	The inductance of a signal via depends on the location of the return path associated with that signal via.
5.5.5.1	A signal via that traverses only one plane keeps the returning signal current close at hand all along the signal pathway.
5.5.5.1	A signal via that traverses two reference planes forces returning signal current through the nearest available interplane connection.
5.5.5.1	If a signal changes reference planes from a ground plane to a power plane (or vice versa), the interplane return path must traverse bypass capacitors.
5.5.5.1	Pcb vendors, often without telling you, make last-minute changes to hole, pad, and clearance sizes in an attempt to improve their finished board yield.
5.5.5.2	Vias that traverse a common stripline cavity (i.e., the space between two reference planes) create crosstalk.
5.5.5.2	The crosstalk voltage induced in a victim circuit equals the rate of change of current in the aggressor times the mutual inductance, L_M, shared between the two circuits.

Points to Remember

6.1 The big advantage of single-ended signaling is that it requires only one wire per signal.
6.1 Single-ended signaling falls prey to disturbances in the reference voltage.
6.1 Single-ended signaling is susceptible to ground bounce.
6.1 Single-ended signaling requires a low-impedance common reference connection.
6.2 Two-wire signaling renders a system immune to disturbances in distribution of global reference voltages.
6.2 Two-wire signaling counteracts any type of interfering noise that affects both wires equally.
6.2 Two-wire signaling counteracts ground bounce (also called simultaneous switching noise) within a receiver.
6.2 Two-wire signaling counteracts ground shifts in connectors.
6.2 Two-wire signaling works when there is no significant stray returning signal current.
6.3 Differential signaling delivers equal but opposite AC voltages *and currents* on two wires.
6.3 Assuming the layout is symmetrical, any AC currents induced in the reference system by one wire are counteracted by equal and opposite signals induced by the complementary wire.
6.3 Differential pcb traces need not be tightly coupled to be effective.
6.3 Differential signaling markedly reduces radiated emissions.
6.4 Differential and common-mode signals are used to describe the voltages and currents on a two-wire transmission system.
6.4 Odd-mode and even-mode signals are yet another way to describe the voltages and currents on a two-wire transmission system.
6.4 Differential receivers cancel common-mode noise.
6.5 Microstrips support slightly different propagation velocities for the differential and common modes. The impact of this difference is not very great.
6.6 Common-mode balance is the ratio of common-mode to differential-mode signal amplitudes.
6.7 Don't violate the common-mode input range specification for a receiver (not even briefly).
6.8 An imbalanced circuit can translate part of a perfectly good differential signal into a common-mode signal, or vice versa.
6.9 Differential impedance is the impedance measured *between* two conductors when they are driven in the differential mode.
6.9 Odd-mode impedance is the impedance measured *on either of* two conductors when they are driven with opposite signals in the differential mode.
6.9 The value of differential-mode impedance is twice the value of odd-mode impedance.
6.9 The differential impedance of two matched, uncoupled transmission lines is double the impedance of either line alone.
6.9 The odd-mode impedance of two matched, uncoupled transmission lines equals the impedance of either line alone.
6.9 Coupling between two parallel pcb traces decreases both differential and odd-mode impedances.

6.9	Common-mode impedance is the impedance measured on two wires in parallel when they are driven together.
6.9	Even-mode impedance is the impedance measured *on either of* two wires when they are driven with identical signals in the common mode.
6.9	The value of common-mode impedance is half the value of even-mode impedance.
6.9.3	Aside from the complications introduced by unbalanced modes, differential transmission lines behave pretty much like single-ended ones.
6.10.2	Differential traces *can* be pushed really, really close together. If you do so, compute a new trace width to compensate for the fact that the differential impedance goes down for closely spaced pairs.
6.10.2	Widely spaced (i.e., loosely-coupled) pairs are not subject to picky, difficult-to-implement spacing and width requirements.
6.10.2	The most important determiner of skin-effect loss is the trace width.
6.10.2	An interpair trace separation of four times h yields about a 6% effect on impedance, a small enough value in many cases to simply ignore.
6.10.2	Matching the elements of each pair to within 1/20 of a risetime limits the common-mode signal contributed by trace skew to less than 2.5% of the single-ended signal amplitude.
6.10.3	If you separate elements of a tightly-coupled pair the differential impedance reverts to twice the uncoupled value of Z_0.
6.10.4	Broadside differential trace impedance is maximized by a trace height equal to 25% of the interplane separation.
6.10.4	The bottom trace of a broadside-coupled differential pair has some extra delay built in at the endpoints.
6.10.4	Avoid broadside-coupled traces unless they are made necessary by routing considerations.
6.11.1	Match the differential characteristic impedance of two pcb traces to the differential characteristic impedance of a balanced cable.
6.11.1	Make the two pcb traces as symmetrical as possible, with equal impedances to ground.
6.11.2	Differential signaling defeats ground bounce.
6.11.3	You need not struggle to place ordinary differential digital traces any closer than 0.5 mm (0.020 in.) for any EMI purpose.
6.11.4	Subject to the limits of common-mode rejection, ground shifts generated within a connector are totally cancelled within a differential receiver.
6.11.5	Differential receivers often have more accurately specified switching thresholds than single-ended receivers.
6.11.5	Uncoupled differential traces need not follow the same path; they just need to have the same delay.
6.11.6	Tightly coupling a differential pair delivers only a modest improvement in crosstalk.
6.11.8	The benefits of differential signaling apply to multidrop configurations.
6.11.9	Every long, differential link needs at least one good differential termination *and also* a reasonable common-mode termination to prevent severe common-mode resonance.

Points to Remember

6.11.10 Visualize the propagation of a differential signal as a quad of four currents.
6.11.11 Chamfering or rounding of differential corners does not eliminate skew.
6.11.12 A pair that starts and ends going north has by definition equal numbers of right-hand and left-hand turns.
6.12 The twisted-pair cable guarantees low crosstalk by virtue of having a different rate of twist on all the pairs within the same jacket.
6.12 Quad cable guarantees low crosstalk by virtue of its unique geometrical alignment.
6.12.1 Ribbon cables can use the same twist pitch on every pair because the wires are held in a rigid geometry.
6.12.2 Never introduce a metallic connection between any two frames powered by different AC power sources.
6.12.2 If you must electrically connect two boxes, make sure that both boxes are served by green-wire grounds connected to the same Earth potential.
6.12.2 Differential signaling with unshielded cables does not require a direct ground connection between the two ends of the link.
6.12.3 To get the best RF-rejection performance from your cabling,
6.12.3 Use a tightly twisted, well-balanced cable. Twisted cables work better than quad cables in this respect.
6.12.3 Don't scrimp on connectors. Buy and use connectors designed to go with the cable.
6.12.3 Use well-balanced circuitry for both transmitter and receiver.
6.12.4 Differential receivers have more accurate switching thresholds than ordinary single-ended logic.
6.13.1 Normal operating voltages for LVDS logic are 1.2 ± 0.2 V on each wire.
6.13.2 LVDS, like most digital transceivers, is not extraordinarily well balanced.
6.13.3 The common-mode noise tolerance for general-purpose LVDS logic is ± 925 mV.
6.13.4 The high noise margin gives LVDS a built-in natural advantage in combating ringing, overshoot, and crosstalk from like devices.
6.13.5 Always provide fast-edged inputs to LVDS logic.
6.13.6 LVDS works best with 100-Ω transmission lines.
6.13.7 You need not struggle to place ordinary differential digital traces any closer than 0.5 mm (0.020 in.) for any EMI purpose.
6.13.8 Always double-check your final artwork to make sure you've met the specifications for skew.
6.13.9 Fail-safe features are permitted by the LVDS standard, *but not required.*
7 Any system that connects from room to room, or from building to building, should use generic building cabling.
7 Building-cabling standards are evolving rapidly. If you want the latest information, order the latest standards.
7.1 Horizontal cabling is the most widely deployed, highest-volume element of the building-cabling architecture.
7.1 New buildings in North America provide two outlets in every work area, with four-pair, 100-Ω UTP, category 5 or better cabling to both outlets.

7.1	Backbone cables are mostly a mix of category 5 cables, multimode fiber (62.5-μm or 50-μm), and some single-mode fiber.
7.1	A weird backbone cabling requirement is a sales obstacle to be overcome. A weird horizontal cabling requirement is a wooden stake in the heart of your project.
7.2	Don't underestimate the complexity of proper SNR budgeting.
7.6	Multi-pair building cables should be installed straight-through with no crossing of the pairs.
7.6	When necessary, an external crossover should be implemented in a short, clearly visible section of cabling and boldly labeled.
7.7	The materials used to make fire-resistance plenum-rated cables are heavy, stiff, and somewhat more expensive than PVC.
7.8	Cable performance must be de-rated to account for operation at the elevated temperatures commonly found in building attics.
8	Cabling standards proliferate faster than bunnies.
8.1	Compared to category 3 cabling, categories 5e and 6 higher have progressively tighter twists and better plastic insulation with less dielectric loss at high frequencies. The resulting cables pick up less noise and have a superior frequency response.
8.1.2	The many possible combinations of surface plating, types of shielding, and dielectric make it difficult to accurately predict the performance of all twisted-pair cables from the basic information provided on a datasheet.
8.1.2	The cable model you use for system simulation should either add another 2 dB of fixed, flat loss to the datasheet attenuation or extend the simulated maximum cable length by another 10% to 20%.
8.2	Timing jitter is improved when all received amplitudes are independent of past history.
8.2	Simple fixed pre-emphasis boosts the maximum operational cable length of a Manchester-coded link by at least 50%.
8.2	A more sophisticated adaptive equalizer can extend operation to even greater distances.
8.3.1	A complete noise budget takes into account all reflections within a cabling system.
8.3.1	Connectors generate reflections that superimpose on the reflections generated by changes in cable impedance.
8.3.2	Bidirectional links must tolerate near-end reflections.
8.3.2.1	A specification of *structural return loss* combined with a specification of the mean value of characteristic impedance is used for old category 3 cables.
8.3.2.1	A single specification of cable *return loss* (as measured with the cable terminated in a 100-Ω load) simultanesouly limits both the mean value and local perturbations in cable impedance.
8.3.2.2	Structural return noise is modeled as a summation of many noise sources with random amplitudes.
8.3.2.2	Structural return noise grows at a rate of 15 dB per decade.
8.3.3	A hybrid circuit makes possible bidirectional full-duplex transmission through a single channel.

Points to Remember

8.3.3	A sufficiently complex adaptive digital filter can compensate for cable roughness and also cable-transition reflections simultaneously. Such a filter is called an adaptive echo cancellation circuit.
8.3.4	NEXT is modeled as a summation of many noise sources with random amplitudes.
8.3.4	NEXT grows at a rate of 15 dB per decade.
8.3.5	Alien crosstalk comes from devices occupying unused pairs within your cable jacket.
8.3.6	FEXT is modeled as a single noise source with a random amplitude.
8.3.6	FEXT grows at a rate of 20 dB per decade.
8.3.7	Within a single jacket there may be one combination of pairs that press up against the limit for pair-to-pair NEXT or ELFEXT, but not all combinations of pairs may do so.
8.3.8	The best antidote for RFI is good signal balance.
8.3.8	A 27-MHz low-pass filter applied to category-3 horizontal cabling should cut RFI to less than 40 mV in most commercial situations.
8.3.8	Categories 5e and 6 cabling pick up less RFI.
8.3.9	The key to obtaining good radiated performance is good common-mode balance.
8.3.9	Scrambling spreads the spectral power density of the transmitted signal, reducing the peak radiation.
8.4	UTP connectors are cheap, and the performance is outstanding.
8.4	Systems that tolerate polarity reversal greatly simplify installation.
8.5	Even though screened cables are heavily favored in Europe, this author does not recommend their use.
8.6	Never use PVC-insulated category 3 cables in an uncooled attic space.
9	Think about 150-Ω STP-A when you need a quick and dirty transceiver for a first product release (or beta-trial).
9.2	When 150-Ω STP-A is used in a unidirectional mode it is not subject to near-end reflections, alien crosstalk, or far-end crosstalk.
9.2.1	Inside a 150-Ω STP-A cable, the signal on one wire of a pair might arrive ahead of the signal on the other wire.
9.3	Pigtail and AC-coupled shields work at audio frequencies, but not at a gigahertz.
9.4	Customers will not maintain the shields on an STP system.
9.5	The equipment-end connector used with FDDI and Ethernet 150-Ω STP-A installations is the shielded DB-9.
10	The electrical performance of coaxial cable is as good as anything else, but physically, coax is difficult to handle.
10	Coaxial cable suffers from an overabundance of standards.
10.1	A good coaxial cable presents a nearly uniform impedance at all frequencies above the onset of the skin effect.
10.1	Coaxial cables formed from foamed, cellular, or helically-wrapped dielectrics exhibit a faster propagation velocity and less high-frequency loss than their solid-dielectric counterparts.
10.1	The step response duration for a coaxial cable scales roughly in proportion to the square of cable length.

10.1.2	A characteristic impedance of approximately 50 Ω minimizes the skin-effect losses in a solid-polyethylene coaxial cable.
10.1.3	Wimpy drivers appreciate higher-impedance transmission lines.
10.1.3	I consider IBM's selection of 150-Ω for STP-A a goof.
10.1.3	50-Ω coax is less sensitive than 75-Ω coax to reflections caused by transceiver taps.
10.2.1	Coaxial cables are generally manufactured to much tighter impedance standards than UTP cables.
10.2.3	RF susceptibility and radiation in coaxial cables result from imperfections in the shield.
10.2.4	If you block the direct path of signal current with an isolating device, such as a transformer, optical isolator, or differential receiver, then you are free, as far as signal integrity is concerned, to disconnect the coax ground from your equipment ground.
10.2.4	A common-mode choke can also block the flow of intercabinet ground current.
10.2.4	DC-balanced signals are perfectly suited for connection through transformers.
10.3	Above 100 MHz, you should always match the characteristic impedance of the connector to the cable.
10.3	Contact plating serves to stave off corrosion and eventual failure of the contacts.
10.3	If it goes on a boat, a car, a plane, or anything that moves, use threaded connectors.
10.3	Crimp-style connectors generally superior to the other types for high-frequency work.
10.3	Always specify heat-treated beryllium-copper for critical contact springs.
11	The bandwidth-carrying capacity of modern fiber-optic cabling greatly exceeds that of any form of copper cabling, an advantage counterbalanced by the high costs and practical difficulties associated with fiber.
11.1	Glass optical fiber is drawn as one continuous thread from a single cylinder of purified glass called a *preform*.
11.2	The key parameter that differentiates fiber in the marketplace is core diameter.
11.3	The optical properties of the fiber are determined almost entirely by the coated glass core.
11.3	The mechanical properties of the cable are determined almost entirely by the buffer and jacket construction.
11.4	The three most popular wavelength windows for glass fiber are (1) 770 nm to 860 nm, (2) 1270 nm to 1355 nm, and (3) 1500 nm to 1600 nm.
11.5	The two most popular standard core diameters for multimode glass fiber are 50 μm and 62.5 μm.
11.5	A graded-index multimode fiber higher bandwidth than a step-index multimode fiber of the same core diameter and quality.
11.5.1	Within a multimode fiber, there exist hundreds of different pathways, or modes of propagation
11.5.1	The multiple modes cause a step input to gradually disperse in time as it travels down the fiber.
11.5.1	Dispersion in a multimode fiber is divided into modal dispersion and chromatic dispersion.

Points to Remember

11.5.1 Modal bandwidth is a function of the refractive index profile of the fiber.

11.5.1 Chromatic dispersion is a function of the material properties of the glass and also the refractive index profile of the fiber.

11.5.2 Carefully grading the profile of the index of refraction greatly improves modal bandwidth.

11.5.3 Internationally recognized specifications for 50 and 62.5 µm multimode optical fibers are provided by IEC 793-2.

11.5.4 Fifty-micron multimode fiber has a generally higher bandwidth and less attenuation than 62.5-µm multimode fiber. These advantages are counterbalanced by the fact that some common LED sources can't couple efficiently into 50-µm core.

11.5.5.1 Dispersion calculations determine the extent of risetime degradation and estimate the impact that degradation will have on signal reception.

11.5.5.2 An attenuation budget allocates attenuation among the long continuous runs of fiber cabling, the short fiber jumpers, and the connectors in a typical installation.

11.5.6 Fiber-optic transmission systems commonly divide the jitter budget into *deterministic jitter* and *random jitter*.

11.5.7 Fiber cabling may be immune to crosstalk and RFI, but your fiber-optic receiver is not.

11.5.8 Never look into the end of a fiber.

11.5.9 The use of laser-diodes on multimode fiber depends on subtle, undocumented, and unspecified features of the multimode fiber.

11.5.10 A VCSEL shines perpendicular to its top surface, just like a surface-emitting LED.

11.5.11 No one has yet designed a satisfactory, easy to install, inexpensive fiber-optic connector.

11.5.11 Optics work well for intersystem connections, but I've not yet seen a cost-effective optical backplane.

11.6.1 Single-mode fiber does not suffer from modal dispersion, differential mode delay, modal noise, or mode partition noise.

11.6.1 The most important optical parameters for a single-mode fiber are the operating wavelength, the attenuation in dB/km, and the chromatic dispersion.

12 Clock signals, because they are so fast, so heavily loaded, and so important for system timing, are subject to special requirements.

12.1 DLL or PLL technology can produce arbitrary, precise, intentional clock skew where and when you need it.

12.2 Timing margin measures the slack, or excess time, remaining in each clock cycle.

12.2 Lowering the clock frequency fixes setup problems, but not hold problems.

12.2 *Clock skew* affects operating speed as much as any other propagation delay.

12.3 The performance of a clock tree structure depends heavily on the input-to-output uncertainty of the clock repeaters.

12.3 Keeping the clock repeater isolated in its own package is a good idea.

12.3.1 A skew-compensated clock repeater architecture does nothing to combat uncertainty in the overall input-to-output delay.

12.3.1 Actively compensated clock repeaters are highly susceptible to power supply noise.
12.3.2 A zero-delay clock buffer directly controls the input-to-output uncertainty.
12.3.3 What you really want is low skew as defined *at the points of usage*.
12.4 Given similar dielectrics, signals propagate faster on a microstrip layer than on a stripline layer. For best speed matching, don't mix the two types.
12.5 For low skew, use the same clock drivers everywhere, source-terminate every driver, and use the same length line with the same impedance and the same loading on every trace.
12.6 The spread between V_{IL} and V_{IH} creates an uncertainty in the exact moment at which a clock receiver will switch.
12.7 Resistive loading attenuates the output of a digital driver, but does not change its rise (or fall) time.
12.8.3 Delay elements are built from three basic building blocks: transmission lines, logic gates, and passive lumped circuits.
12.8.3 A fixed delay cannot cancel variations in board fabrication or active component delay.
12.8.3 An adjustable delay compensates for actual delays, not just nominal delays, elsewhere in the circuit.
12.8.3 Whatever form of delay you choose, incorporate its *uncertainty* in delay into your timing margin calculations.
12.8.5 Avoid long, coupled switchbacks.
12.9 A single driver can service two or more source-terminated lines only under limited conditions.
12.9.1 A slow driver can damp the ringing on a hairball network, but it may need to be *too* slow for your circuit.
12.9.1 Appropriately placed attenuating networks can damp all the oscillatory modes at the expense of shrinking the received signal.
12.9.1 A weak termination can help reduce, but totally cure, overshoot and ringing.
12.9.1 Test all combinations of maximum and minimum load capacitance and line length.
12.9.1 Eventually, someone will inherit your hairball design and try to figure out what you did. Keep it simple.
12.9.2 Hidden within every split-tee network is an unconstrained resonance.
12.10 Five things reduce the reflection from an isolated, lumped-element capacitive load: slow the risetime, lower the capacitance, lower the characteristic impedance of the trace, isolate the load with a big resistor, or compensate for the capacitance by modulating the trace width.
12.10.1 Rules for good daisy-chaining—Uniformly space the loads, with a spacing whose delay is small compared to the signal rise and fall time, and terminate the structure with a resistance that matches the effective impedance of the loaded structure you've built, not just the impedance of the raw trace you started with.
12.11 PLL-based clock generators require a stable, low-jitter reference clock.
12.11.1.2 Any sort of resonance in a PLL, even a tiny one, spells disaster for a highly cascaded system.

Points to Remember

12.11.1.3 The variance of the tracking error in a PLL circuit represents all the power in the input reference signal that falls above the tracking range of the PLL.

12.11.1.4 A large ratio between the reference clock frequency and the PLL output frequency requires a very stable VCO.

12.11.1.5 Jitter in the output of a PLL comes from internal sources plus noise coupled from the power system and noise propagated from the reference input.

12.11.1.6 The point of separating jitter into random and deterministic components is to avoid overly stringent specifications for deterministic jitter.

12.11.2.1 The noise properties of a PLL are characterized by the intrinsic internal jitter, the power supply sensitivity, and a jitter transfer function.

12.11.2.2 You can calculate the variance of jitter using a spectrum analyzer.

12.12 Filters designed for wideband operation are built from a cascade of multiple sections, each section scaled to provide coverage in successively higher frequency bands.

12.12.1 Observing the noise between V_{CC} and ground always returns useful information.

12.12.2 A power-supply filter does not eliminate noise—it merely copies junk from one circuit node to another, eliminating the difference between them.

12.13.1 A modulated clock can never be used as the reference clock input to any advanced data communication transceiver.

12.13.2. A scrambled clock spreads the clock emissions without modulating the mean clock frequency.

12.14 Reduced-voltage clock signaling saves power and cuts EMI at the expense of noise susceptibility.

12.15 The physical means of providing extra crosstalk protection are simple; the logistical means are complex.

12.16 On a short line, if a range of series termination values will work, the biggest value minimizes the transmitted current and therefore the emissions.

13.1 If by using a simulator you can save one design spin on one circuit board, the simulator pays for itself.

13.2 Signal-integrity simulations may be performed in what-if mode or post-processing mode.

13.2 Tool sets are highly differentiated according to their degree of *software integration*.

13.2 Automated tools can be as dangerous as they are powerful and easy to use.

13.3.2 All signal-integrity time-domain analysis tools use simulation techniques pioneered by SPICE.

13.3.2 Especially on circuits containing inductive spikes or hard corners in the I-V curves, SPICE may fail to converge.

13.3.2 Some versions of SPICE have a lower limit on the smallest permissible step size.

13.3.2 Check your documentation to make sure TOL and REFTOL are set properly for your application.

13.3.2 If your parameter extraction efforts fail to properly account for all the significant parasitic elements in a circuit, SPICE results will be incorrect.

13.3.3 The SPICE lossless transmission-line model is computationally efficient.

13.3.3	For typical pcb traces up to 25 cm (10 in.) long, at risetimes of 1 ns or slower, a lossless transmission-line model serves adequately well. Higher speeds and greater distances require the use of a transmission-line model that accounts for the *skin effect* and *dielectric-loss*.
13.3.3	Lossy transmission-line models take a *lot* longer to run.
13.3.4	When you first start working with any simulator, begin by setting up some simple, low-frequency test circuits for which you can predict the response by hand calculations.
13.4.5	IBIS is an international standard for the electrical specification of chip drivers and receivers.
13.4.5	IBIS specifies *how* to record the various parameters of a chip driver or receiver, but it does not specify *what* to do with them.
13.4.5	IBIS is the best, most comprehensive, and genuinely useful piece of signal-integrity technology to come along in a great while.
13.4.5	We need our chip vendors to provide IBIS model files for every part they make.
13.4.5	At the time of publication, the IBIS committee maintained work-in-progress copies of its latest draft standards at the Electronic Design Automation (EDA) and Electronic Computer-Aided Design (ECAD) one-stop standards resource: *http://www.eda.org/pub/ibis*.
13.6	Specify circuit behavior under conditions similar to the actual conditions present in your application.
13.7	IBIS simulators don't yet properly compute SSO noise.
13.8.1	Real live EMI problems are much too complex for even the best software tools.
13.9	If your system design depends on the natural power-plane capacitance, compare the roundtrip delay across your board to your clock period—you could be headed for resonance problems.

APPENDIX A

BUILDING A SIGNAL INTEGRITY DEPARTMENT

Article first published in *EDN Magazine*, June 4, 1998

Signal integrity is no longer a "nice-to-know" subject. It has become essential to the proper operation of every high-speed digital product. Without due consideration of basic signal-integrity issues (ringing, crosstalk, ground bounce, and power supply noise), typical high-speed products can fail to operate on the bench or, worse yet, become flaky or unreliable in the field. To address these concerns, many leading-edge companies are now doing one or more of the following:

- Training their employees to better understand signal-integrity issues,
- Hiring people who are already signal-integrity experts, or
- Acquiring new CAD technology aimed at alleviating signal-integrity problems.

If you believe your company should bring in some signal-integrity experts and tools, you might wonder, Where should I put them? How should I best integrate signal-integrity functions within my company?

There are three basic choices. Some companies assign responsibility for signal-integrity issues to their pcb-layout department. Others assign this responsibility to their digital-circuit designers. The most aggressive engineering managers are choosing to constitute an independent department of signal integrity. The primary advantages of a standalone department are that it can be given a clear mission, and its performance can be tracked.

> In a mature signal-integrity department expect to find about one signal-integrity specialist for every five digital-circuit designers.

What sort of a mission do you give to a department of signal integrity? Try this: *Maximize the performance and minimize the cost of interconnection technology used in high-speed digital designs.* This mission is easy to state and very valuable if implemented correctly.

What about the strategy of a signal-integrity department? That is a matter of considerable debate. I believe that the signal-integrity department should initially act in the role of a consultant to the rest of the design organization—offering alternatives, evaluating solutions, but never mandating compliance. Once the signal-integrity department has established basic credibility by showing that it can deliver useful results on a pilot-program basis, it is ready to insert itself into the mainstream development process. The key to success is to ensure that the individuals working on signal-integrity issues are viewed as helpful, not harmful, and that they are truly well-informed as opposed to merely pedantic.

A signal-integrity specialist should know that most signal-integrity problems are easily observed. A good simulation or a good laboratory demonstration can usually put to rest any question about the efficacy of a particular solution. In this area signal-integrity specialists enjoy a natural advantage over their EMC counterparts.

How should a signal-integrity department be organized? The group should be led by a strong manager who can sell the program to other parts of the company. In its early stages, the signal-integrity department must influence the company's design, layout, and manufacturing processes without any form of direct control over them. This feat is accomplished through the development of personal relationships within the company and by selling the advantages of new approaches to the affected departments. Good managers know how to do that.

In a mature signal-integrity department, where full ringing and crosstalk analyses are run on each pcb, expect to find about one signal-integrity specialist for every five digital-circuit designers. A large department might have individuals who specialize in model-building, chip-level packaging, connectors, and so forth.

Signal integrity is a rapidly growing field. There is no one right way to build your department. The most important thing is to hire intelligent, self-motivated people with a healthy interest in properly balancing your signal-integrity, EMC, and manufacturing-cost objectives. Train them well, constantly keep on the lookout for new tools, and tear apart plenty of other people's products to see what the competition is doing. The payoff is easy to understand: better system-level performance, a more reliable product, and an overall reduction in cost. Who could ask for more?

APPENDIX B

CALCULATION OF LOSS SLOPE

The loss slope, for the purposes of this discussion, is defined as the slope, using a log-log plot, of attenuation a, in dB, versus frequency f, in Hz.

$$\text{Loss slope}(f) \triangleq \frac{d \log a}{d \log f} \qquad [B.1]$$

Notice that since a, in dB, is already the logarithm of signal amplitude v in Volts, the loss-slope definition is making use of a double-logarithm. The next equation shows the relation between loss slope and signal amplitude.

$$\text{Loss slope}(f) \triangleq \frac{d \log(20 \cdot \log(v))}{d \log f} \qquad [B.2]$$

As an example, let's determine the loss slope for a signal amplitude v that varies exponentially with some power η of frequency.

$$v = e^{-kf^\eta} \qquad [B.3]$$

The attenuation a in dB is base-10-logarithmically related to the signal amplitude.

$$a = -20 \cdot \log(v) \qquad [B.4]$$

Substitute [B.3] for v, and simplify using the relation $\log(e^x) = x / \ln(10)$.

$$\begin{aligned} a &= -20 \cdot \log\left(e^{-kf^\eta}\right) \\ &= -20\left(-kf^\eta / \ln(10)\right) \end{aligned} \qquad [B.5]$$

And the log of attenuation is what we display in the loss-slope graph.

$$\begin{aligned}\log a &= \log\left(-20\left(-kf^{\eta}/\ln(10)\right)\right) \\ &= \log\left(f^{\eta}\right) + \log\left(20k/\ln(10)\right) \\ &= \eta\log(f) + \log\left(20k/\ln(10)\right)\end{aligned} \qquad [B.6]$$

The loss slope is defined as $d(\log a)/d(\log f)$. Let $u = \log f$, and evaluate the loss slope as $d(\log a)/du$ using [B.6] for the definition of $\log a$.

$$Loss\ slope = \frac{d}{du}\left[\eta\log(f) + \log(20k/\ln(10))\right] \qquad [B.7]$$

Inside the square brackets the right-hand term remains constant and so contributes nothing to the derivative. In the left-hand term the expression log(f) may be changed to *u*, producing this:

$$Loss\ slope = \frac{d}{du}\eta u = \eta \qquad [B.8]$$

In conclusion, given a linear system whose attenuation in dB varies with f^{η}, the loss slope is η.

In the skin-effect limited region the transmission-line attenuation in dB varies with the square root of *f*, producing a loss slope of 1/2. In the dielectric-loss region the attenuation varies directly with *f*, producing a loss slope of 1.

In a dispersion-limited transmission system the loss slope indicates the severity of the tradeoff between transmission distance and operating speed.

APPENDIX C

TWO-PORT ANALYSIS

Figure C.1 depicts a generalized two-port circuit. The conventions for voltage and current preferred by this author are indicated. The figure assumes that both input and output ports share a common reference terminal, which is not strictly required for two-port analysis to work, but since the assumption conforms to the general situation in transmission-line analysis, the figure has been drawn that way.

Two-port circuits are much studied in the field of electrical engineering, and you will find many references dealing with them. Beware that there is a great variation in the conventions for polarity of input and output currents and voltages, and in the definition of the elements that comprise the four-element matrix used to describe the frequency-response properties of the circuit.

Underlying all of two-port analysis is the assumption that the circuit under study is both linear and time-invariant. This assumption renders possible the use of a frequency-based description of the circuit. Nonlinear circuits do not succumb to linear two-port analysis.

In the figure, the four elements of the *transmission matrix* **A** prescribe the actions of the circuit. Each of the four elements of **A** *is itself a function of frequency*. The matrix specifies, for any particular frequency, the allowed relationships between the input and output currents and voltages *at that frequency*.

Of all the possible two-port definitions, your author has selected the *transmission matrix* version of two-port analysis for this book because it simplifies certain calculations often performed for *cascaded systems*. To explain, suppose two systems represented by two-port matrices **A** and **B** are cascaded as shown in Figure C.2. Matrix **B** converts the output

Transmission matrix representation of circuit

$$\begin{bmatrix} v_1 \\ i_1 \end{bmatrix} = \begin{bmatrix} a_{0,0} & a_{0,1} \\ a_{1,0} & a_{1,1} \end{bmatrix} \begin{bmatrix} v_2 \\ i_2 \end{bmatrix}$$

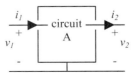

Figure C.1—Any linear, time-invariant, two-port circuit can be represented by a four-element transmission matrix.

parameters (v_3,i_3) into (v_2,i_2), and then matrix **A** converts (v_2,i_2) into (v_1,i_1). The overall two-port description of the system spanning from (v_1,i_1) to (v_3,i_3) is therefore given by the matrix product **AB**. This simple method of chaining together the two-port matrices accounts for the popularity of the transmission-matrix formulation.

Let us derive from the four elements of a transmission matrix **A** the input impedance and gain of the circuit. The input impedance is calculated first assuming the output circuit (v_2,i_2) is open-circuited, meaning that $i_2 = 0$. Under that condition the matrix equations for (v_1,i_1) simplify to

$$v_1 = a_{0,0} v_2$$
$$i_1 = a_{1,0} v_2 \qquad [C.1]$$
$$Z_{\text{in,open-circuit}} \triangleq \left.\frac{v_1}{i_1}\right|_{(i_2=0)} = \frac{a_{0,0}}{a_{1,0}}$$

What if the output of the circuit is shorted to ground? In that case $v_2 = 0$, and the input impedance is again calculated using simplified matrix equations:

$$v_1 = a_{0,1} i_2$$
$$i_1 = a_{1,1} i_2 \qquad [C.2]$$
$$Z_{\text{in,short-circuit}} \triangleq \left.\frac{v_1}{i_1}\right|_{(v_2=0)} = \frac{a_{0,1}}{a_{1,1}}$$

In the case of an open-circuited output, the voltage transfer function is computed from the first line of [C.1].

$$\text{Voltage gain} \triangleq \left.\frac{v_2}{v_1}\right|_{(i_2=0)} = \frac{1}{a_{0,0}} \qquad [C.3]$$

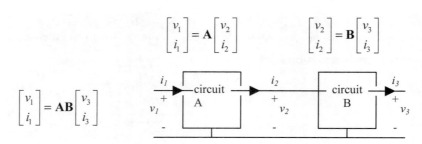

Figure C.2—A cascade of circuits may be represented by multiplying their transmission matrices.

Simple Cases Involving Transmission Lines

Figure C.3 summarizes three classic forms of transmission matrices used to define the overall system gain of a typical transmission-line configuration. In the figure Z_C represents the characteristic impedance and H the one-way transfer function of a transmission line.

The derivations of the top two forms in Figure C.3 are self-evident from the definition of the transmission matrix (Figure C.1). The third form corresponding to a transmission line is developed in the following way.

The general form of solution for the signals on a transmission line is composed of two traveling waves, one propagating to the right and one moving to the left (see Section 2.2.5, Figure 2.6). Suppose that at the right-hand end of the line the signal amplitudes of the two waves are denoted a and b. The currents associated with these two waveforms at that point must then be $+a/Z_C$ (a current moving to the right) and $-b/Z_C$ (representing a current moving to the left). At the right-hand end of the line, the superposition of these waves must generate the voltage and current extant at that end.

$$v_2 = a + b$$
$$i_2 = \frac{a}{Z_C} - \frac{b}{Z_C}$$
[C.4]

The preceding conditions may be inverted to determine the amplitudes a and b.

$$a = \frac{v_2 + Z_C i_2}{2}$$
$$b = \frac{v_2 - Z_C i_2}{2}$$
[C.5]

At the left end of the line the same conditions prevail, except that the amplitudes of the right- and left-traveling waveforms must be adjusted to account for their propagation through the transmission medium. The amplitude of the left-going waveform is diminished by H, the one-way transfer function of the transmission line, while the amplitude of the right-traveling waveform must be *increased* by H^{-1}, so that after traveling to the right end of the line, it will appear at the correct amplitude a. Summing the voltages and currents at the left end of the line produces a relationship between a, b and v_1, i_1.

$$v_1 = aH^{-1} + bH$$
$$i_1 = \frac{a}{Z_C}H^{-1} - \frac{b}{Z_C}H$$
[C.6]

Now substitute for a and b the relations to v_2 and i_2.

A series impedance

$$\begin{bmatrix} v_1 \\ i_1 \end{bmatrix} = \begin{bmatrix} 1 & z \\ 0 & 1 \end{bmatrix} \begin{bmatrix} v_2 \\ i_2 \end{bmatrix}$$

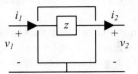

A shunt impedance

$$\begin{bmatrix} v_1 \\ i_1 \end{bmatrix} = \begin{bmatrix} 1 & 0 \\ z^{-1} & 1 \end{bmatrix} \begin{bmatrix} v_2 \\ i_2 \end{bmatrix}$$

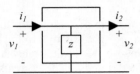

A transmission line

$$\begin{bmatrix} v_1 \\ i_1 \end{bmatrix} = \begin{bmatrix} \dfrac{H^{-1}+H}{2} & Z_C \dfrac{H^{-1}-H}{2} \\ \dfrac{1}{Z_C}\dfrac{H^{-1}-H}{2} & \dfrac{H^{-1}+H}{2} \end{bmatrix} \begin{bmatrix} v_2 \\ i_2 \end{bmatrix}$$

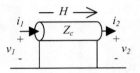

Figure C.3—These three forms of transmission matrix are often used to describe digital transmission circuits.

$$v_1 = \frac{v_2 + Z_C i_2}{2} H^{-1} + \frac{v_2 - Z_C i_2}{2} H$$

$$i_1 = \frac{\frac{v_2 + Z_C i_2}{2}}{Z_C} H^{-1} - \frac{\frac{v_2 - Z_C i_2}{2}}{Z_C} H \qquad \text{[C.7]}$$

Collecting together the terms associated with v_2 and i_2 respectively reveals the form of the transmission matrix.

$$v_1 = \frac{H^{-1}+H}{2} v_2 + Z_C \frac{H^{-1}-H}{2} i_2$$

$$i_1 = \frac{1}{Z_C} \frac{H^{-1}-H}{2} v_2 + \frac{H^{-1}+H}{2} i_2 \qquad \text{[C.8]}$$

Fully Configured Transmission Line

Figure C.4 illustrates a three-way combination of source impedance, transmission line, and load impedance.

The input impedance of the loaded transmission line may be determined from inspection of the cascaded combination of **BC**. This part of the system represents the transmission line and its load.

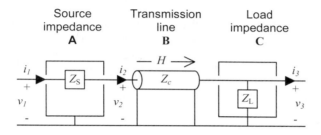

Figure C.4—A transmission line complete with source and load impedances may be modeled as a cascade of three two-port circuits.

$$\mathbf{BC} = \begin{bmatrix} \dfrac{H^{-1}+H}{2} & Z_C \dfrac{H^{-1}-H}{2} \\ \dfrac{1}{Z_C} \dfrac{H^{-1}-H}{2} & \dfrac{H^{-1}+H}{2} \end{bmatrix} \begin{bmatrix} 1 & 0 \\ \dfrac{1}{Z_L} & 1 \end{bmatrix}$$

$$= \begin{bmatrix} \dfrac{H^{-1}+H}{2} + \dfrac{Z_C}{Z_L} \dfrac{H^{-1}-H}{2} & Z_C \dfrac{H^{-1}-H}{2} \\ \dfrac{1}{Z_C} \dfrac{H^{-1}-H}{2} + \dfrac{1}{Z_L} \dfrac{H^{-1}+H}{2} & \dfrac{H^{-1}+H}{2} \end{bmatrix} \quad [\text{C.9}]$$

The input impedance v_2/i_2 equals the ratio $\mathbf{BC}_{0,0}/\mathbf{BC}_{1,0}$.

$$Z_{\text{in,loaded}} = \left\{ \dfrac{\left(\dfrac{H^{-1}+H}{2}\right) + \dfrac{Z_C}{Z_L}\left(\dfrac{H^{-1}-H}{2}\right)}{\dfrac{1}{Z_C}\left(\dfrac{H^{-1}-H}{2}\right) + \dfrac{1}{Z_L}\left(\dfrac{H^{-1}+H}{2}\right)} \right\} \quad [\text{C.10}]$$

Multiplying both numerator and denominator by the factor Z_C simplifies the structure of the fraction somewhat. For now, leave the sum-and-differences of the H terms unmolested, as you will have an opportunity to develop some interesting approximations for these terms later.

$$Z_{in,loaded} = Z_C \left\{ \frac{\left(\frac{H^{-1}+H}{2}\right) + \frac{Z_C}{Z_L}\left(\frac{H^{-1}-H}{2}\right)}{\left(\frac{H^{-1}-H}{2}\right) + \frac{Z_C}{Z_L}\left(\frac{H^{-1}+H}{2}\right)} \right\} \quad [C.11]$$

Some interesting simplifications can be teased out of [C.11] under special conditions. When Z_L is very large, the left-hand terms in the numerator and denominator of [C.11] dominate. When $Z_L/Z_C = 1$, the numerator and denominator exactly cancel. When Z_L is very small, only the right-hand terms matter.

Case: $Z_L \gg Z_C$ $\qquad\qquad Z_{in,open\text{-}circuit} = Z_C \dfrac{H^{-1}+H}{H^{-1}-H} \qquad$ [C.12]

Case: $Z_L = Z_C$ $\qquad\qquad Z_{in,end\text{-}terminated} = Z_C \qquad$ [C.13]

Case: $Z_L \ll Z_C$ $\qquad\qquad Z_{in,short\text{-}circuit} = Z_C \dfrac{H^{-1}-H}{H^{-1}+H} \qquad$ [C.14]

The gain (voltage transfer function v_3/v_1) of the loaded transmission line may be determined from inspection of the cascaded combination of all three parts **ABC**. This matrix represents the combination of source, transmission line, and load.

$$\mathbf{ABC} = \begin{bmatrix} 1 & Z_S \\ 0 & 1 \end{bmatrix} \begin{bmatrix} \dfrac{H^{-1}+H}{2} & Z_C \dfrac{H^{-1}-H}{2} \\ \dfrac{1}{Z_C}\dfrac{H^{-1}-H}{2} & \dfrac{H^{-1}+H}{2} \end{bmatrix} \begin{bmatrix} 1 & 0 \\ \dfrac{1}{Z_L} & 1 \end{bmatrix} \quad [C.15]$$

In phasor notation, the voltage gain $G_{FWD} = v_3/v_1$ equals the inverse of the first element of **ABC**.

$$\frac{v_3}{v_1} = \left[\mathbf{ABC}_{0,0}\right]^{-1}$$
$$= \left[\frac{H^{-1}+H}{2} + \frac{Z_C}{Z_L}\frac{H^{-1}-H}{2} + \frac{Z_S}{Z_C}\frac{H^{-1}-H}{2} + \frac{Z_S}{Z_L}\frac{H^{-1}+H}{2}\right]^{-1} \quad [C.16]$$

The voltage gain expression may be simplified somewhat by factoring related terms.

Appendix C - Two-Port Analysis

$$G_{FWD} = \frac{v_3}{v_1} = \left[\left(\frac{H^{-1}+H}{2} \right)\left(1+\frac{Z_S}{Z_L}\right) + \left(\frac{H^{-1}-H}{2}\right)\left(\frac{Z_S}{Z_C}+\frac{Z_C}{Z_L}\right) \right]^{-1} \quad [\text{C.17}]$$

The response measured by a time-domain reflectometer (TDR) would be the gain from v_1 to v_2. You can compute $G_{TDR} = v_2/v_1$ as the product of v_3/v_1, which is given by [C.17], times the upper-left member of matrix **BC**, which represents the ratio v_2/v_3 under the condition $i_3 = 0$.

$$G_{TDR} = \frac{v_2}{v_1} = \frac{v_3}{v_1} \cdot \left[\frac{H^{-1}+H}{2} + \frac{Z_C}{Z_L}\frac{H^{-1}-H}{2} \right] \quad [\text{C.18}]$$

In expanded form,

$$G_{TDR} = \frac{v_2}{v_1} = \frac{\dfrac{H^{-1}+H}{2} + \dfrac{H^{-1}-H}{2}\left(\dfrac{Z_C}{Z_L}\right)}{\left(\dfrac{H^{-1}+H}{2}\right)\left(1+\dfrac{Z_S}{Z_L}\right) + \left(\dfrac{H^{-1}-H}{2}\right)\left(\dfrac{Z_S}{Z_C}+\dfrac{Z_C}{Z_L}\right)} \quad [\text{C.19}]$$

COMPLICATED CONFIGURATIONS

Two-port matrices may be cascaded ad infinitum to create structures of arbitrary complexity. Presuming a number of two-port sections $\mathbf{A}_1, \mathbf{A}_2 \ldots \mathbf{A}_N$ have been cascaded, you may calculate the overall circuit gain.

$$G_{FWD} = \left[\left(\mathbf{A}_1 \mathbf{A}_2 .. \mathbf{A}_N \right)_{0,0} \right]^{-1} \quad [\text{C.20}]$$

The various sections could represent, for example,

- A series-connected two-port representing the output impedance of the driver on an integrated circuit die: $z_1(\omega) = j\omega L_{DIE} + R_{DIE}$
- A series-connected two-port representing the series impedance of a wire-bond connection to the chip package: $z_2(\omega) = j\omega L_{WIREBOND} + R_{WIREBOND}$
- A shunt-connected two-port representing the wire-bond landing pad capacitance, $z_3^{-1}(\omega) = j\omega C_{\text{WIREBOND PAD ON BGA SUBSTRATE}}$
- A transmission-line two-port representing the characteristic impedance, delay, and loss of the BGA routing track
- A series-connected two-port representing the BGA ball inductance

- ➤ A shunt-connected two-port representing the BGA ball capacitance
- ➤ A series of three two-port models: shunt, series, and shunt, representing a pi-model of the pcb via
- ➤ A transmission-line two-port representing a skinny breakout track as it winds its way out of the BGA ball field
- ➤ A transmission-line two-port representing a regular track proceeding a long distance towards a receiver
- ➤ A series of three two-port models: shunt, series, and shunt, representing a pi-model of an intermediate pcb via,

…and so on until the model is sufficiently rich to satisfy your desire for accuracy.
The TDR response of a complicated two-port model involving N stages, defined as the gain from the input voltage source on the left to a point just to the right of (after) stage M, is

$$G_{TDR} = \frac{\left(\mathbf{A}_{M+1}\mathbf{A}_{M+2}..\mathbf{A}_N\right)_{0,0}}{\left(\mathbf{A}_1\mathbf{A}_2..\mathbf{A}_N\right)_{0,0}} \qquad [C.21]$$

APPENDIX D

ACCURACY OF PI MODEL

The transfer function H for a transmission line may be expressed in terms of the Taylor-series approximation for the exponential function.

$$H = e^{-l\gamma} \approx 1 - (l\gamma) + \frac{(l\gamma)^2}{2!} - \frac{(l\gamma)^3}{3!} + \frac{(l\gamma)^4}{4!} \ldots \qquad [D.1]$$

where γ is the propagation coefficient for the transmission line (complex nepers/m), and

l is the length of the line (m).

The inverse of this function is similarly defined for negative exponents:

$$H^{-1} = e^{-l\gamma} \approx 1 + (l\gamma) + \frac{(l\gamma)^2}{2!} + \frac{(l\gamma)^3}{3!} + \frac{(l\gamma)^4}{4!} \ldots \qquad [D.2]$$

Taking only the first four terms of each expression, you can derive approximate formulas for the forms of H used in the calculation of system gain (Appendix C).

$$\frac{H^{-1} + H}{2} \approx 1 + \frac{(l\gamma)^2}{2} \qquad [D.3]$$

$$\frac{H^{-1} - H}{2} \approx (l\gamma) + \frac{(l\gamma)^3}{6} \qquad [D.4]$$

Now substitute approximations [D.3] and [D.4] into the expression for the gain of a fully configured transmission system [C.17].

$$\frac{v_3}{v_1} = \left[\left(1+\frac{Z_S}{Z_L}\right) + l\gamma\left(\frac{Z_S}{Z_C}+\frac{Z_C}{Z_L}\right) + \frac{(l\gamma)^2}{2}\left(1+\frac{Z_S}{Z_L}\right) + \frac{(l\gamma)^3}{6}\left(\frac{Z_S}{Z_C}+\frac{Z_C}{Z_L}\right)\right]^{-1} \quad [D.5]$$

The next calculations compare [D.5] with the expression for the gain of a pi-model circuit in Section 3.4.2, Figure 3.4. The two-port transmission matrix for the pi-model circuit, fully configured with source and load impedances Z_S and Z_L respectively, is a composite of five two-port mini-models. The first (leftmost) matrix represents the series-connected source impedance. The next three represent the components in the pi-model. The last matrix represents the shunt-connected load impedance.

$$B = \begin{bmatrix} 1 & Z_S \\ 0 & 1 \end{bmatrix}\begin{bmatrix} 1 & 0 \\ \frac{l}{2}(j\omega C) & 1 \end{bmatrix}\begin{bmatrix} 1 & l(j\omega L+R) \\ 0 & 1 \end{bmatrix}\begin{bmatrix} 1 & 0 \\ \frac{l}{2}(j\omega C) & 1 \end{bmatrix}\begin{bmatrix} 1 & 0 \\ \frac{1}{Z_L} & 1 \end{bmatrix} \quad [D.6]$$

The expression for the circuit gain follows:

$$B_{0,0}^{-1} = \left[\begin{array}{c}\left(1+\frac{Z_S}{Z_L}\right) + l\cdot j\omega C Z_S + l(j\omega L+R)\frac{1}{Z_L}+\ldots \\ \ldots l^2\frac{j\omega C}{2}(j\omega L+R)\left(1+\frac{Z_S}{Z_L}\right) + l^3 j\omega C \frac{j\omega C}{4}(j\omega L+R)Z_S\end{array}\right] \quad [D.7]$$

Next use the definitions of $\gamma = \sqrt{(j\omega L+R)j\omega C}$ and $Z_C = \sqrt{(j\omega L+R)/j\omega C}$ to derive the following substitutions:

$$(l\gamma)Z_C = l(j\omega L+R) \quad [D.8]$$

$$(l\gamma)/Z_C = l\cdot j\omega C \quad [D.9]$$

Plugging these new substitutions back into [D.7] leads to the following expression for the gain of the pi-model circuit.

$$B_{0,0}^{-1} = \left[\left(1+\frac{Z_S}{Z_L}\right) + (l\gamma)\left(\frac{Z_S}{Z_C}+\frac{Z_C}{Z_L}\right) + \frac{(l\gamma)^2}{2}\left(1+\frac{Z_S}{Z_L}\right) + \frac{(l\gamma)^3}{4}\frac{Z_S}{Z_C}\right]^{-1} \quad [D.10]$$

Comparing [D.10] with [D.5] reveals a perfect match of all zero-, first-, and second-order terms, with the lowest-order difference showing up in the third-order part of the expression. Assuming a small value of the propagation coefficient $l\gamma < 1/4$, and thereby an overall gain somewhere near unity, you might reasonably conclude that the modeling error should not exceed the difference between [D.5] and [D.10]

$$PI\text{-}model\ Error < \frac{(1/4)^3}{6}\left(-\frac{1}{2}\frac{Z_S}{Z_C} + \frac{Z_C}{Z_L}\right) \quad [\text{D}.11]$$

Further assuming the ratios $Z_S/(2Z_C)$ and Z_C/Z_L to each be less than 3.8, the resulting pi-model error (measured as a function of frequency) should remain less than 1%.

As the coefficient $l\gamma$ grows, so grows the error in proportion to (at least) the third power of $l\gamma$. At a value of $l\gamma < 1/2$, you should expect a modeling accuracy no better than 1 part in 10.

PI-MODEL OPERATED IN THE LC REGION

Suppose you intend to model a transmission line operating at a frequency at or above the LC region boundary. Constrain the line to a length l sufficiently short that the total line delay t_d remains less than 1/6 of the signal risetime t_r.

$$\frac{t_d}{t_r} < \frac{1}{6} \quad [\text{D}.12]$$

The next equations determine the effect of constraint [D.12] on the magnitude of the length-adjusted propagation coefficient $|l\gamma|$. In the LC region the inductive reactance ωL by definition exceeds R, so you may safely approximate the magnitude of the coefficient $l\gamma$ by omitting the R.

$$\left|l\gamma_{LC\,region}\right| \approx \left|l\sqrt{(j\omega L)(j\omega C)}\right| = l\omega\sqrt{LC} \quad [\text{D}.13]$$

The midpoint of the spectrum associated with the rising and falling edges of the signal driving the line is related to the rise and fall time t_r.

$$\omega_{edge} \approx 2\pi\frac{.35}{t_r}\ \text{rad/s} \quad [\text{D}.14]$$

Evaluating [D.13] at ω_{edge}, and recognizing that for an LC-mode transmission line the effective line delay t_d equals $l\sqrt{LC}$,

$$\left|l\lambda_{\text{LC region}}\right| = 2\pi \left(0.35\right)\frac{t_d}{t_r} \qquad [\text{D}.15]$$

At a ratio of $t_d/t_r = 1/6$ the coefficient $\left|l\lambda_{\text{LC region}}\right|$ equals .366, at which value the pi-model error given by [D.11] (assuming the ratios Z_S/Z_C and Z_C/Z_L to each be less than 3.8) works out to about 3%. At a ratio $t_d/t_r = 1/3$ the error soars to 25%. Above $t_d/t_r = 1/2$ the pi model loses all useful predictive power.

APPENDIX E

ERF( )

NOTE: In the mathematical literature you will see many tabulations of the function $erf(\ )$, sometimes with definitions different from what is presented here. Although the definitions may be transformed from one to another fairly easily, it is sometimes confusing when you need one function and have only a table for another. Table E.1 shows how to convert tabulated data (or software functions) provided under an alternate definition into my format. The functions defined here as $erf(\ )$ and $erfc(\ )$ may be computed from any of the other forms:

$$erf(a) = 2\left(erf_2\left(a\sqrt{2}\right) - \tfrac{1}{2}\right) \qquad [\text{E.1}]$$

$$erfc(a) = 2\cdot\left(1 - erf_2\left(a\sqrt{2}\right)\right) \qquad [\text{E.2}]$$

$$erf(a) = 2 \cdot erf_3\left(a\sqrt{2}\right) \qquad [\text{E.3}]$$

$$erfc(a) = 1 - 2 \cdot erf_3\left(a\sqrt{2}\right) \qquad [\text{E.4}]$$

$$erf(a) = 2\left(\tfrac{1}{2} - Q\left(a\sqrt{2}\right)\right) \qquad [\text{E.5}]$$

$$erfc(a) = Q\left(a\sqrt{2}\right) \qquad [\text{E.6}]$$

$$erf(a) = 2\left(pnorm\left(a\sqrt{2}, 0, 1\right) - \tfrac{1}{2}\right) \qquad [\text{E.7}]$$

$$erfc(a) = 2\cdot\left(1 - pnorm\left(a\sqrt{2}, 0, 1\right)\right) \qquad [\text{E.8}]$$

TABLE E.1—Alternate Definitions for the Error Function

Function name	Definition	Range for $a > 0$	Comment
My error function[136,137] $erf(a)$	$\dfrac{2}{\sqrt{\pi}} \displaystyle\int_0^a e^{-x^2} dx$	[0,1]	I like the minimal simplicity of this definition.
My complementary error function $erfc(a)$	$1 - erf(a) = \dfrac{2}{\sqrt{\pi}} \displaystyle\int_a^\infty e^{-x^2} dx$	[0,1]	Used in probability analysis and communication theory to emphasize connection with Gaussian probability density function.
Error function[138] (alt. definition) $erf_2(a)$	$\dfrac{1}{\sqrt{2\pi}} \displaystyle\int_{-\infty}^a e^{-u^2/2} du$	[½,1]	Used in probability analysis and communication theory to emphasize connection with Gaussian probability density function.
Yet another error function[139] $erf_3(a)$	$\dfrac{1}{\sqrt{2\pi}} \displaystyle\int_0^a e^{-u^2/2} du$	[0,½]	Variant; same as $(erf_2(a) - ½)$.
Complementary error function[140,141] $Q(a)$	$\dfrac{1}{\sqrt{2\pi}} \displaystyle\int_a^\infty e^{-u^2/2} du$	[½,0]	Variant; same as $(1 - erf_2(a))$.
MathCad built-in function $Pnorm(a,\mu,\sigma)$	$\dfrac{1}{\sigma\sqrt{2\pi}} \displaystyle\int_{-\infty}^a e^{-(u-\mu)^2/2\sigma^2} du$	[½,1]	For $\mu = 0$ and $\sigma = 1$, same as $erf_2(x)$; NOTE: In versions of MathCad earlier than 2001 do not use the built-in function $erf(\ )$, as it is a totally unrelated function. Use *Pnorm* instead.

---

[136] John M. Wozencraft, Irwin Mark Jacobs, *Principles of Communication Engineering*, John Wiley & Sons, 1965, ISBN 0 471 96240 6

[137] John A. Aseltine, *Transform Method in Linear System Analysis*, McGraw-Hill, 1958, U.S. Lib. Congress cat. no. 58-8038

[138] Harry Van Trees, *Detection, Estimation, and Modulation Theory: Part I*, John Wiley and Sons, 1968 ISBN 471 89955 0

[139] Athanasios Papoulis, *Probability, Random Variables, and Stochastic Processes*, McGraw-Hill, 1965, ISBN 07-048448-1

[140] John M. Wozencraft, Irwin Mark Jacobs, *Principles of Communication Engineering*, John Wiley & Sons, 1965, ISBN 0 471 96240 6

[141] Bernard Sklar, *Digital Communications*, Prentice Hall, 1988, ISBN 0-13-211939-0

Index

Clock jitter
    causes of, 644–45
    digital oscilloscope, using to measure, 651–54
    FIFO-based architectures, 643–44
    jitter-free clocks, 667–68
    measuring, 648–50
    phase noise, 654–56
    PPL technology, 634–36
    propagation, 636–40
    random and deterministic, 645–48
    tracking error, variance of, 640–43
    when it matters, 636
Clock phase adjustments, 605
Clock repeaters
    active skew correction, 593
    chaining, 590, 592
    common characteristics, 589
    line length, compensating for, 595–96
    low-skew outputs, expanding, 590, 591 fig
    multiple outputs and termination, 589–90
    timing performance, selected chips, 590, 591 tbl
    zero delay, 594–95
Clock skew
    automated maximum delay analysis, 587–88
    calculating, 586–87
    intentional adjustments, 582–83, 588. *see also* Intentional delay adjustments
    overall operating speed, impact on, 588
    setup margin, 585–86
    setup-time and hold-time violations, 584–85
    synchronous arrival of transmissions, 580, 582
Clock window penalty, 564
Coaxial cable
    air-dielectric, 526
    connectors, 532–35
    inner conductor, problems accessing, 513
    physical construction, 513
    standards, overabundance of, 514
    as transmission line media, 30

Coaxial cable, signal propagation
    AC resistance, signal and shield, 517
    AC skin-effect resistance, 516–17
    air bubbles, injecting during extrusion, 519
    characteristic impedances, 515
    conductors, surface treatment on, 521–22
    dielectric losses, 518–19
    electrical specifications (Belden types), 515, 516 tbl
    helical wrapping, 518–19
    impedance, above skin effect onset, 515
    metallic-transmission model, 515–22
    optimality of 50-Ω value, 524–25, 527–28
    raw size, 520–21
    stranded and/or plated wires, 517
    stranded-center conductors, 522–23
    sweep-tested cable, 522
    transmission line parameters, worst-case (Belden types), 519–22
Coaxial cable, noise and interference
    common-mode chokes, 530–31
    DC balance, 531–32
    far-end reflected noise, 528
    grounds, treatment of, 530–32
    radiation, 529–30
    radio-frequency interference (RFI), 529
    UTP, compared with, 528
Common impedance coupling, 366–67
Common-mode
    balance, 377–78
    currents, 372, 493
    impedance, 382
    range, 378
    voltage, 374
Common-mode chokes, 530–31
Common-mode rejection ratio (CMRR), 378
Complementary signals, 370
Complex electric permittivity, 96
Complex magnetic permeability, 115
Complex relative permittivity, 97
Component pads, parasitic capacitance of, 306–09
Concentric-ring skin-effect model, 75–76

Conduction current, 95
Conductors, physical scaling of, 7–8
Connection stubs, 309–11
Constant dielectric loss, 215
Constant-impedance plateau, 43
Constant loss configuration, 43
Constant-loss region. *see* LC region
Constant loss tangent, 106–10
Constant voltage scaling law, 13
Cooley-Tukey FFT, 240
Coplanar waveguides (CPW), 332–33
Copper transmission media, scaling
    attenuation, ways of reducing, 227–29
    length, attenuation, and bandwidth, 224–27
Core laminate, 271
Critical-path failures, 584
Crossover wiring, 451–52
Crosstalk. *see also* Pcb connectors, crosstalk; Pcb traces, crosstalk
    adjacent-trace, 282
    alien, 490
    balanced cabling, 422–23
    on clock lines, 669–70
    equal-level far-end (ELFEXT), 490–92
    far-end (FEXT), 321, 490, 492
    near-end (NEXT), 320, 487–89
    predicting worst-case, 27
    and vias, 354–59
Current-loop principle, 360–61
Current ratios, in decibel equivalents, 4–5

D

Dangling vias, 343–44
DB-9 connector, 510–11
DC-balanced data code, 223
DC balanced data code, 531–32
DC conductance, 57–58
DC resistance, 55–57
Decibel notation, 2–5
Decomposition, 374–76
Decrete Fourier Transform (DFT), 240
Delay-locked-loop (DLL) technology, 593
Dermer, G. H., 109

Deterministic jitter, 288, 568
Deutsch, Alina, 102
Dielectric constant, 97–98
Dielectric-constant scaling, 14–15
Dielectric effects, 142
    complex magnetic permeability, 115
    conductors, 95–96
    constant loss tangent, 106–10
    critical frequency, conducting versus insulating mode, 96
    currents, magnitude and phase of, 94–95
    dielectric constant, 97–98
    dielectric loss, 94, 462
    dielectric loss tangent, 97, 98–99, 101–05
    Hilbert-transform pairs, FFT technique for computing, 110–14
    insulators, 96–98
    Kramers-Kronig relations, 114–15
    low-permittivity materials, 97
    mixtures, rule of, 99–101
    network functions relations, 105–06
    residual air, 100–101
    TDR measurements, influence on, 205–06
Dielectric-loss coefficient, 208, 280
Dielectric-loss-limited region
    boundaries, 200–201
    characteristic impedance, 202–04
    distance and speed tradeoffs, 216
    distinguishing feature, 201
    inductance, internal and external, 202–03
    lossy lines, 203–04
    phase and amplitude response, predicting, 207–09
    phase and attenuation, decoupling of, 207
    propagation coefficient, 206–10
    step response. *see as main heading*
    terminations, 207
    termination styles, 211
    and transmission line properties, 209–10
Dielectric loss tangent, 97, 98–99, 101–05
Differential Clocks, 413–14
Differential impedance

Index

common- and differential-modes, 382
coupling, 381
defined, 380
even- and odd-modes, 382
explained, 380–81
reflections, 384–85
uncoupled impedance, 383
Differential (Microstrip) Trace Impedance, 386–89
Differential-mode impedance, 382
Differential-mode modeling, 252–53
Differential phase measurement, 649–50
Differential Reflections, 384–85
Differential signaling. *see also* LVDS signaling
    capacitive imbalance, 380
    common-mode balance, 377–78
    common-mode current, reducing, 372–73
    common-mode range, 378
    common-mode voltage, defined, 374
    decomposition, differential-and-common-mode, 375
    decomposition, even-and-odd mode, 375
    defined, 370
    differential pcb traces, coupling of, 373–74
    differential to common-mode, converting, 378–80
    differential voltage, defined, 374
    explained, 370–71
    precise balance approach, 372–73
    UTP, common-mode rejection of, 373
    velocity, differential and common-mode, 376–77
    weak-coupling approach, 372
Differential signaling, intercabinet applications
    balanced cabling, 422
    crosstalk, 422–23
    Earth potential, 425
    external radio-frequency interference (RFI), rejection of, 426–27
    flipping rule, 422–23
    large ground shifts, immunity to, 424–26
    quad cable, 422
    ribbon-style twisted-pair cables, 423–24
    skin effect, tolerance to, 427–28
    twisted pair cable, 422
Differential signaling, pcb applications
    balanced cabling, connecting to, 404–05
    chamfering corners, effect of, 419–20
    clocks and parallel bus signals, 414
    clock skew, reducing, 409–11
    connector crosstalk, reducing, 407–09
    EMI, reducing, 405–07
    ground bounce, defeating, 405, 413
    intrapair skew, minimizing, 420–22
    local crosstalk, reducing, 411–12, 413–14
    pair-turning skew, 419–20
    termination, 414–16
    U-turn zone, counteracting, 417–18
Differential signaling, pcb configurations
    2-D field solvers, need for, 385–86
    broadside-coupled striplines, 399–403
    edge-coupled striplines, 389–96
    trace impedance, 386–89
    trace pairs, breaking up, 397–99
    trace requirements, 385
Differential Signaling (Through Connectors), 408–09
Differential Termination, 414–16
Differential U-Turn, 417–18
Differential voltage, 374
Diffusion equations, 149
Diffusion line, 149
Directionality of Crosstalk, 323–25
Direct phase measurement, 649
Discrete assumption, 89
Discrete time mapping, 241–42
Dispersion-flattened fiber, 550
Dispersion-limited link, 563
Dispersion penalty, 231–33
Dispersion-shifted fiber, 549
Dispersive region. *see* RC region
Dispersive transmission line, 149
Displacement current, 36–37, 95, 203
Dissipation factor, 98
Double-clocking, 579

Double-treat process, 93
Driving Two Loads, 625–27
Dual-window fibers, 544, 552
Duplex-SC connector, 575
Duty-cycle distortion, 564

E

Earth potential, 425
Echo cancellation, 486–87
Eddy currents
    within conductors, 61–63
    explained, 58–60
    and the proximity effect, 80
    and solid planes, 318
Edge-coupled striplines, 389–96
Edge-current concentration, 264
Edge-emitting LEDs, 554
Edwards, Terry, 304
Eisenhart SMA connectors, 333–34
Electric flux, 36
Electric susceptibility, 98
Electromagnetic compatibility (EMC) simulation, 697–99
ELFEXT (equal-level far-end crosstalk), 490–92
Elmore, W. C., 160
Elmore delay estimation
    accuracy and effectiveness, 161–62
    applications, 160
    continuous transmission lines, 163–64
    multiple RC sections, 162–63
    procedure, 160–61
    step response waveforms, 162
    tree-structured circuits, 164–65
Embedded microstrip, 597
EMC Simulation, 698–99
Equal-level far-end crosstalk (ELFEXT), 490–92
Equivalence principle, coaxial cable, 221–24
ERF(), 747–48
Even-mode impedance, 382
External inductance, 186–87, 202–03
External radio-frequency interference (RFI), 426–27
Extra Fries, Please, 582–83

F

Faraday cages, 337
Far-end reflections, 294
Far-end reflections (UTP), 471–75
Fast-Fourier Transform (FFT), 107, 110–14, 240–41. *see also* Frequency-domain simulation
FDDI MIC connector, 576
Feathering the phase to zero, 108–09
FET gates, 314
FEXT coefficient, 321
FEXT (far-end crosstalk), 321–22
FFT time wrapping, 242, 251
Fiber-optic cabling
    bandwidth-carrying capacity, 537
    construction of, 538–39
    core diameter, MMF versus SMF, 539–41
    cost and practical drawbacks, 537
    dispersion-flattened fiber, 550
    dispersion-shifted fiber, 549
    dispersion-shifting, 544
    dual-window fibers, 544, 552
    impurities, absorption by, 543
    index of refraction, 538–39
    infrared absorption, 543
    mechanical properties, getting information about, 541
    numerical aperture, 553
    optical properties, getting information about, 541
    Rayleigh scattering, 543
    tight and loose buffer methods, 541–43
    wavelengths of operation, 544
Fiber-optic cabling, multimode
    50-micron fiber, use of, 554
    attenuation budget, 566–68
    connectors, 575–76
    dispersion budget. *see* Multimode dispersion budget
    fiber safety, 571
    graded-index versus step-index fiber, 551
    index of refraction, 545

jitter, 568–69
laser-diode transmitters, 571–73
LED transmitters, 554
modal noise, 572–73
noise and interference, 570–71
optical differences, 545
optical performance budget, 555
signal propagation. *see* Multimode signal propagation
standards and specifications, 552–53
VCSEL laser diodes, 571, 573–74
Fiber-optic cabling, single-mode
connectors, 578
defined, 576–77
fiber safety, 578
noise and interference, 578
signal propagation, 577–78
Fiber-resin laminates, 271–72, 273–74
First-incident-wave switching, 133
Fixed delay, 605–07
Fly-by termination, 313
Foamed dielectrics, 519
Forward crosstalk, 321–22
The Fourier Integral and Its Applications (Papoulis), 105
Fourier transform, 239–40
Fourier transform pairs, 245–46
Frequency-domain simulation, 234–35
aliasing, 245, 247
arbitrary data sequences, transforming, 249–50
checking FFT routine outputs, 253–54
circular-shifting effect, 242, 251
decrete Fourier transform (DFT), 240
DFT normalization factors, deriving, 244–45
differential-mode modeling, 252–53
discrete time mapping, 241–42
fast-Fourier transform (FFT), 240–41
FFT method, 238–39
Fourier transform, 239–40
implementing, 249–50
inadequate sampling rate, effect of, 247–48
limitations of the FFT, 243
linear analysis versus SPICE, 237
network functions relations, 107–08
normalization of FFT routines, 243–44
reasons for using, 237
sampling sufficiency, testing, 242
time-domain waveforms, shifting, 252
Fringing-field assumption, 87

G

Gain-bandwidth tradeoff, 154
Galvanizing process, 278
Garg, R., 269
Gaussian filter, 245
Gaussian waveforms, 558
Generic building-cabling standards
building-cabling practices, FAQs, 449–51
cable combinations, preferred, 449
cabling terms, glossary of, 446–49
compliance, benefits of, 439
crossover wiring, 451–52
defined, 440
older cable types, FAQs, 453–55
plenum-rated cables, 452–53
purchasing, 440
selecting the appropriate, 441–42
SNR budgeting, 446
standards organizations, 441
TIA/EIA and ISO/IEC, major points of agreement, 440
uncooled attic space, 453
Generic cabling architecture
backbone cables, 442
horizontal cables, 442, 443, 445
star-wiring topology, 442, 443 fig
testable specifications (cable, permanent link, and channel), 444–45
Gibb's phenomena, 110, 245, 248
Going Nonlinear, 237–39
Graded index, 545
Ground-bounce, 142
Ground bounce, 367–68

Ground shift, 368, 369
Ground-transfer impedance, 331
Gupta, K. C., 269

H

Hanning window, 110
Healthy Power, 659–61
Helical wrapping, 518–19
Hermaphroditic design, 509–10
High-dielectric-constant materials, 14
High-Q resonators, 173
Hilbert-transform pairs, FFT technique for computing, 110–14
Hold-time violations, 584–85
Horizontal cables, 442, 443
How Close Is Close Enough?, 309
Hudale, Matt, 347
Hybrid circuits. *see* UTP hybrid circuits
Hybrid return loss, 483
Hysteresis, 431–32

I

IBIS: History and Future Direction, 689–91
IBIS (I/O Buffer Information Specification), 685–88
IBIS (I/O Buffer Information Specification)
    creators of, 686
    defined, 686
    future directions, 690–91
    historical overview, 689–90
    interpolation, issues with, 691–95
    positive and negative aspects, 687–88
    promoting use of, 688
    ringing and crosstalk, 685–86
    simulations, requirements for performing, 687
    SPICE, comparison to, 690
    SSO noise, issues with, 695–97
IBIS model, 245
IBM Type-I cable. *see* 150-Ω STP-A cabling
IEEE 802.3z Gigabit Ethernet committee, 552
Impedance matching pad, 330–31
Impedance scaling, 12–14

Impedance slope, calculation of, 171
Index of refraction, 538–39
Inductive potholes, 303
Inductors, physical scaling of, 8
Infrared absorption, 543
In-phase current, 95
Input impedance, lumped-element region
    reflected-wave effect, 140
    short, unloaded transmission lines, 140–41
    traces leading to ground, 141–42
Inside vapor deposition, 538
Insulation displacement connectors, 498
Intentional Clock Modulation, 663–66
Intentional delay adjustments
    adjustable delays, 607–09
    automatically programmable delays, 607–09
    fixed delay, 605–07
    serpentine delays, 610–12
    switchback coupling, 612–16
Internal impedance, modeling
    above skin-effect onset, 69
    approximations, simple versus better, 70–73
    below skin-effect onset, 68
    internal inductance and resistance, 67–69
    rectangular conductors, 73–75
    round-wire values, 69–70
Internal inductance, 187, 203
Intersymbol interference (ISI), 288, 556–57
Intrapair skew, 420–22

J

Jitter and Phase Noise, 654–56
Jitter-Free Clocks, 667–68
Jitter Measurement, 651–54
The Jitters, 634–36
Junction-matching circuitry, 315–16

K

King, W. Michael, 493–94
Kirchoff, Robert Gustav, 36, 37, 364
Kirchoff's current law (KCL), 36, 364
Knee frequency, 2, 134

Kramers-Kronig relations, 114–15

L

LaMeres, Brock J., 345
Laminate materials, properties of, 271–75
LC region
    attenuation versus frequency, 166
    boundaries, 166–67
    characteristic impedance, 167–68
    distance and speed tradeoffs, 183
    high-Q resonance, 173, 176–79
    mixed mode operation (LC and RC regions), 183
    Pi model, 745–46
    propagation coefficient, 173–76
    termination. *see* Termination, LC transmission lines
    transition impedance, RC to LC mode, 168
Leapfrogging, 32, 37
LED transmitters, 554
Linear, time-invariant, lumped-element circuits
    capacitors, impedance magnitude of, 1–2
    inductors, impedance magnitude of, 1
    parasitic impedances, 2
Linear-analysis method. *see* Frequency-domain simulation
Linear equalization, 230–34
Linear-ramp filter, 245
Linear systems, maximizing response (digital output)
    crosstalk, predicting worst-case, 27
    impulse response, negative at intervals, 24–25
    impulse response, purely positive, 24
    mathematical model, 23–24
    non-linear totem-pole drivers, 27
    power-supply excursions, predicting worst-case, 28
    practical application, 26–28
Load impedance, 151
Long-path failures, 584
Loose buffer method, 541, 542–43
Lossless circuits, 6

Loss slope, calculation of, 733–34
Loss tangent, 97, 98–99, 101–05, 270–71
Low-loss structures, 133
Low-Voltage Differential Signaling. *see* LVDS signaling
Luminance signals, 16
Lumped-element circuits, 35–36, 38
Lumped-element reflections, pcb traces, 297–300
Lumped-element region
    boundaries, approximating, 136
    input impedance. *see* Input impedance, lumped-element region
    mathematical extent of, 135–36
    maximum frequency, determining, 137
    pi model, 137–38
    step response, 10-pF load capacitance, 147–48
    step response, no load capacitance, 145–47
    Taylor-series approximation of H, 139–40
    termination of transmission lines, 135, 137
    transfer function, 143–45
LVDS signaling
    common-mode noise tolerance, 430–31
    common-mode output, 430
    differential-mode noise tolerance, 431
    fail-safe, 436–38
    hysteresis, 431–32
    impedance control, 432–35
    input capacitance, 435
    output levels, 429–30
    risetime, 435
    skew, 435–36
    trace radiation, 435

M

Magnetic permeability
    of nickel, 266
    scaling, 15–16
Making an Accurate Series Termination, 314–15
Manchester coding, 465–67, 470
Matching Pads, 315–17
MathCad, 70, 123, 249–50
Maxwell's equations, 36

Measuring Connectors, 330–31
Medium Interface Connector (MIC), 509–10
Metallic-transmission model, 462–65
Microphonic noise, 645
Microstrip dispersion, 217, 284–86
Micro-via process, 346
Min-delay failures, 584
Minimum attenuation, 193–94
Mixing zone, skin- and dielectric-effect regions, 201, 202 tbl, 215–16, 281–82
Mixture dielectric, 269
Modal dispersion, 547–48
Modal noise, 572–73
Modeling Skin Effect, 76–78
Multimode dispersion budget
    assumptions, 557
    attenuation-limited link, 563
    clock window penalty, 564
    cumulative effect of components, 561
    dispersion formula, 560
    dispersion-limited link, 563
    dispersion penalty, 563
    duty-cycle distortion, 564
    FDDI dispersion penalties, example calculation of, 565–66
    Gaussian assumption, use of, 558
    generally accepted shortcut, assumptions for, 557
    parameters required, 555–56
    risetime at point TP3b, approximating, 558–59
    risetimes of components, 559–61
    test points, 556
    waveform, expected under worst-case-ISI condition, 562
    worst-case data pattern for ISI, 561
    worst-case intersymbol interference, limiting, 556–57
Multimode dispersion (fiber optics), 284
Multimode fiber (MMF), 539–41
Multimode fiber-optic cables, scaling, 229–30
Multimode signal propagation
    chromatic dispersion, 548–50

    dispersion effect, 546–47
    independent modal propagation, 546
    modal dispersion, 547–48
    overall fiber dispersion, 550
Mutual Understanding, 326–27

N

Nahman, R. S., 196n
Napier, Sir Charles James (1782-1853), 5
Napierian logarithm, 5
Napiers. see Nepers
National Semiconductor LVDS logic family, 436–37
Near-end reflections (UTP), 475–77
Nearly-planar waveguides (NPW), 333, 334
Neper, John (1550-1617), 5
Nepers, 5
Network Analysis and Feedback Amplifier Design (Bode), 105
Newton, Sir Isaac, 549
NEXT coefficient, 320–21
NEXT (near-end crosstalk), 320–21
Nickel-plated traces, 266–68
Noise margin, 366
Noisy ground, 368
Non-linear DC restorer, 292–94
Non-linear totem-pole drivers, 27
Non-TEM modes
    3-D electromagnetic field simulators, 286
    big-fat-trace approach, 282
    microstrip dispersion, 284–86
    non-TEM behavior, simulating, 286–88
    and stripline traces, 285–86
    waveguide-dispersion region, 216–17
Non-TEM structures, 37
No-storage principle, 35–36
Notational conventions, x
Nyquist rate, 248

O

Odd-mode impedance, 382
Offset stripline, 596
On-chip interconnections, future of, 359–61

Optical power budget, 547
Outside vapor deposition (OVD), 538
Overall bulk transport delay, 191
Overfilled launch, 547, 571
Oversampling ratio, 251
Overshoot, in resonant circuits driven by step input, 20–21
Oxide treatments (black, brown, and red), 93

P

Pair-turning skew, 419–20
PAM-4 data coding, 234–35
Parameter extraction, 675
Parasitic capacitance, 8
Parasitic impedances, 2
Parasitic Pads, 306–09
Passivation
 and AC resistance, 277–78
 and DC resistance, 277
 defined, 277
 explained, 278
Passive circuits, 6
Path of least inductance, 258
Path of least resistance, 258
Pcb connectors
 cable shield grounding, 336–38
 EMI, measuring, 331
 NPW tapers, 333, 334
 signal-fidelity test, 330–31
 straddle-mount, 332–37
 tapered transitions, 332–34
Pcb connectors, crosstalk
 coupling, 326–27
 layout specifications, checking, 328–29
 measuring, 331
 mutual inductance versus parasitic capacitance, 327
 pinpointing, 326
 through-hole clearance, 328–29
Pcb traces. *see also* Vias, modeling
 AC resistance, 258–61
 characteristic impedance and delay, 257
 DC resistance, 258
 geometry, assumptions about, 255–56
 impedances less than 50 Ω, 523
 nickel-plated, 266–68
 non-TEM modes. *see as main heading*
 on-chip interconnections, future of, 359–61
 perimeter, calculation of, 261
 Popsicle-stick analysis, 262–65
 RC region, operation in, 256
 skin effect and dielectric loss, mixtures of, 281–82
 skin-effect loss coefficient, calculating, 262
 solid reference planes, 256
 trace height, limiting, 285, 288, 523
 as transmission line media, 30
 transmission line structures, 255
 very low impedance, 262
Pcb traces, crosstalk
 2-D field solvers, use of, 319
 coupled crosstalk, 320
 directionality, 319–20, 323–25
 far-end crosstalk (FEXT), 321–22
 FEXT and NEXT, reducing, 322–23
 near-end crosstalk (NEXT), 320–21
 solid plane layers, purpose of, 318
 trace separation and height, effects of, 318–19, 322
Pcb traces, dielectric effects
 ceramics, 272–73
 dielectric-loss coefficient, calculating, 280
 effective dielectric constant, estimating, 269–70
 effective loss tangent, calculating, 270–71
 effective permittivity calculations, 269
 fiber-resin laminates, 271–72, 273–74
 microwave versus digital design, 268
 passivation. *see as main heading*
 practical laminates, 274–75
 soldermask overlays, 279–80
 temperature, variations with, 275–77
Pcb traces, distance limitations

760 Index

dispersion-limited systems, 288, 291
poor sensitivity, causes of, 288
sensitivity-limited systems, 288, 290–91
signal distortion, causes of, 288–89
signal loss, estimating, 289–90
SONET data coding, 291–94
transceivers, single-ended CMOS and bipolar TTL, 290

Pcb traces, reflections
both-ends termination, 294–97
component pads, parasitic capacitance of, 306–09
connection stubs, 309–11
end termination, placement of, 312–14
explained, 294
inductive potholes, 303
junction-matching circuitry, 315–16
lumped-element, 297–300
resistive matching pads, 316–17
right-angle bends, 304–05
series termination, accurately constructing, 314–15
skinny-trace compensation technique, 300–302
stubs and vias, 305–06

Performance regions. *see also individual regions*
adaptive equalization, 234–36
breakpoints between, summary of, 218–21
for copper media, listed, 121
copper transmission media, scaling, 224–29
equivalence principle, coaxial cable, 221–24
hierarchy and distinguishing features, 128–30
input impedance, 132–33
linear equalization, 230–34
multimode fiber-optic cables, scaling, 229–30
signal propagation model. *see as main heading*
transfer function, 133–35

Permanent link specifications, 444–45
Phase-locked-loop (PLL) technology, 593, 634–36
Physical size, scaling of
conductors, 7–8

examples, 8
inductors, 8
passive and lossless circuits, 6
resistive scaling, 6–8
transmission-line dimensions, 8

Pi model
accuracy of, 743–45
dangling stubs, 344
defined, 137–38
LC region, operation in, 745–46
step response, lumped-element region, 147–48
vias, 342

Placement of End Termination, 312–13
Plated-through-hole (PTH) vias, 346
Plenum-rated cables, 452–53, 502
Plenum-return system, 453
PLL loop testing, 650
Polarity reversal (RJ-45 connectors), 499
Popsicle-Stick Analysis, 262–65
Potholes, 300–302
Power and Ground Resonance, 699–701
Power factor, 98–99
Power ratios, in decibel equivalents
equal impedances, in two circuits, 3–4
general formula, 2
nepers, 5
unequal impedances, in two circuits, 4
voltages and currents, expression of, 4–5

Power scaling, 9
PPL peaking, 638
Precise balance approach, 372–73
Pre-emphasis circuits, 467–69
Preforms, 538
Prepreg laminate, 271
Propagation coefficient
LC region, 173–76
RC region, 155
skin-effect region, 189–92
transmission lines, 44–48
Propagation delay, 50
Propagation function

informIT

www.informit.com

YOUR GUIDE TO IT REFERENCE

Articles

Keep your edge with thousands of free articles, in-depth features, interviews, and IT reference recommendations – all written by experts you know and trust.

Online Books

Answers in an instant from **InformIT Online Book's** 600+ fully searchable on line books. Sign up now and get your first 14 days **free**.

POWERED BY

Catalog

Review online sample chapters, author biographies and customer rankings and choose exactly the right book from a selection of over 5,000 titles.

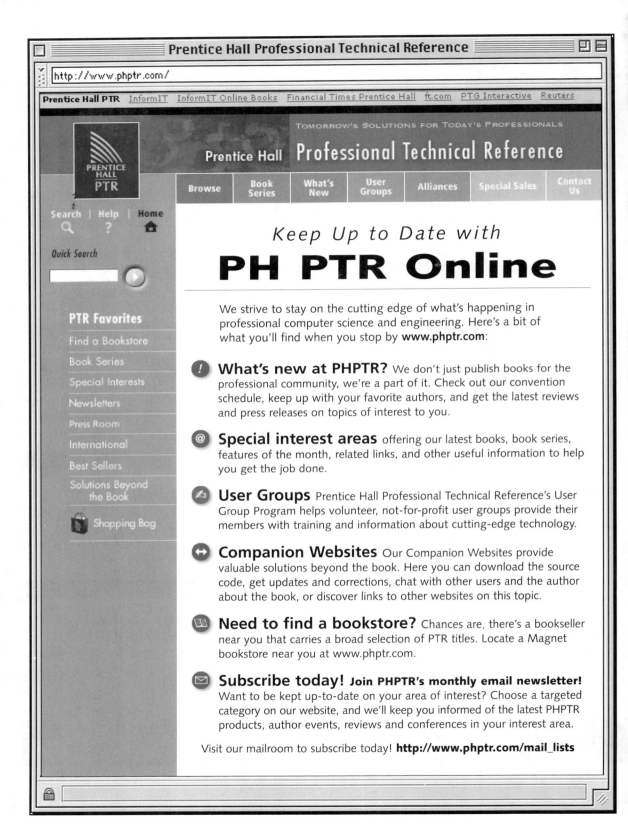